T0189177

CRC Handbook of Basic Tables for Chemical Analysis

CRC Handbook of Basic Tables for Chemical Analysis
Data-Driven Methods and Interpretation

Fourth Edition

Thomas J. Bruno and Paris D.N. Svoronos

CRC Press
Taylor & Francis Group
Boca Raton London New York

CRC Press is an imprint of the
Taylor & Francis Group, an **informa** business

Fourth edition published 2020
by CRC Press
6000 Broken Sound Parkway NW, Suite 300, Boca Raton, FL 33487-2742

and by CRC Press
2 Park Square, Milton Park, Abingdon, Oxon, OX14 4RN

© 2021 Taylor & Francis Group, LLC

First edition published by CRC Press 1989
Second edition published by CRC Press 2003
Third edition published by CRC Press 2010

CRC Press is an imprint of Taylor & Francis Group, LLC

Library of Congress Cataloging-in-Publication Data
Names: Bruno, Thomas J., author. | Svoronos, Paris D. N., author.
Title: CRC handbook of basic tables for chemical analysis / Thomas J. Bruno, Paris D.N. Svoronos.
Other titles: Handbook of basic tables for chemical analysis
Description: Fourth edition. | Boca Raton, FL : CRC Press, 2020. | Includes
bibliographical references and index.
Identifiers: LCCN 2020012170 (print) | LCCN 2020012171 (ebook) |
ISBN 9781138089044 (hardback) | ISBN 9781315109466 (ebook)
Subjects: LCSH: Chemistry, Analytic—Tables.
Classification: LCC QD78 .B78 2020 (print) | LCC QD78 (ebook) |
DDC 543.02/1—dc23
LC record available at https://lccn.loc.gov/2020012170
LC ebook record available at https://lccn.loc.gov/2020012171

ISBN: 978-1-138-08904-4 (hbk)
ISBN: 978-0-367-51719-9 (pbk)
ISBN: 978-1-315-10946-6 (ebk)

Typeset in Times
by codeMantra

*We dedicate this work to our wives, Clare and Soraya,
and our children, Kelly Anne, Alexandra, and Theodore,
sons-in-law Jesse and Dean, and grandchild Noah.*

Contents

Chapter 13

Chapter 14

Notice

Certain commercial equipment, instruments, or materials are identified in this handbook in order to provide an adequate description. Such identification does not imply recommendation or endorsement by the National Institute of Standards and Technology, the City University of New York, or Georgetown University, nor does it imply that the materials or equipment identified are necessarily the best available for the purpose. The authors, publishers, and their respective institutions are not responsible for the use in which this handbook is made. Occasional use is made of non-SI units, in order to conform to the standard and accepted practice in modern analytical chemistry.

Preface to the First Edition

This work began as a slim booklet prepared by one of the authors (TJB) to accompany a course on Chemical Instrumentation presented at the National Institute of Standards and Technology, Boulder Laboratories. The booklet contained tables on chromatography, spectroscopy, and chemical (wet) methods, and was intended to provide the students with enough basic data to design their own analytical methods and procedures. Shortly thereafter, with the co-authorship of Professor Paris D.N. Svoronos, it was expanded into a more extensive compilation entitled *Basic Tables for Chemical Analysis*, published as a National Institute of Standards and Technology Technical Note (number 1096). That work has now been expanded and updated into the present body of tables. Although there have been considerable changes since the first version of these tables, the aim has remained essentially the same. We have tried to provide a single source of information for those practicing scientists and research students who must use various aspects of chemical analysis in their work. In this respect, it is geared less toward the researcher in analytical chemistry than to those practitioners in other chemical disciplines who must make routine use of chemical analysis. We have given special emphasis to those "instrumental techniques" which are most useful in solving common analytical problems. In many cases, the tables contain information gleaned from the most current research papers and provide data not easily obtainable elsewhere. In some cases, data are presented which are not available at all in other sources. An example is the section covering supercritical fluid chromatography, in which a tabular P-Δ-T surface for carbon dioxide has been calculated (specifically for this work) using an accurate equation of state.

While the authors have endeavored to include data, which they perceive to be most useful, there will undoubtedly be areas which have been slighted. We therefore ask you, the user, to assist us in this regard by informing the corresponding author (TJB) of any topics or tables which should be included in future editions.

The authors would like to acknowledge some individuals who have been of great help during the preparation of this work. Stephanie Outcalt and Juli Schroeder, chemical engineers at the National Institute of Standards and Technology, provided invaluable assistance in searching the literature and compiling a good deal of the data included in this book. Teresa Yenser, manager of the NIST word processing facility, provided excellent copy despite occasional disorganization on the part of the authors. We owe a great debt to our board of reviewers, who provided insightful comments on the manuscript: Professors D.W. Armstrong, S. Chandrasegaran, G.D. Christian, D. Crist, C.F. Hammer, K. Nakanishi, C.F. Poole, E. Sarlo, Drs. R. Barkley, W. Egan, D.G. Friend, S. Ghayourmanesh, J.W. King, M.L. Loftus, J.E. Mayrath, G.W.A. Milne, R. Reinhardt, R. Tatken, and D. Wingeleth. The authors would like to acknowledge the financial support of the Gas Research Institute and the United States Department of Energy, Office of Basic Energy Sciences (TJB) and the National Science Foundation, and the City University of New York (PDNS). Finally, we must thank our wives, Clare and Soraya, for their patience throughout the period of hard work and late nights.

Preface to the Fourth Edition

It has been 34 years since the first edition of *CRC Handbook of Basic Tables for Chemical Analysis* was completed and 9 years since the third edition. All of these editions have enjoyed the positive acceptance and use of the scientific community, but major changes to this edition will, in our opinion, be even more valuable to the user. First, we have undertaken the reconciliation of all numerical data on thermophysical properties with the *CRC Handbook of Chemistry and Physics, 100th Ed.* and with available standard reference databases. While an occasional oversight may be found, the user can be confident that such numerical data are the best available. Second, as a departure from our philosophy of the first three editions, we have now included substantially more explanatory and descriptive text which will be especially welcome to students. Our prior editions emphasized presentation of information for scientists at decision points in chemical analyses, for example: what stationary phase to choose, what temperature, what pressure, etc. This is especially important for researchers who must make use of chemical analysis techniques but whose research areas lie elsewhere. While that is still the main purpose of this work, much more explanation is now provided. Third, we have significantly updated all sections, with the major changes to be found in chromatographic methods, UV-visible, and laboratory safety chapters. Moreover, all these sections are supported by an update of the Solutions chapter, which serves to support many of the other chapters. Fourth, we have purposely withdrawn materials from the prior editions that are no longer relevant or which have been superseded by updated or better information sources. An example is the density table for carbon dioxide, which while present in the third edition, can now be easily calculated by accessible and reliable equations of state.

The user might notice that we often use the most familiar nomenclature when listing specific chemical compounds, which may not always correspond to the IUPAC name. We have been gradually moving toward IUPAC names through the course of these four editions. When, in our opinion, the common or familiar name has become increasingly archaic, we have provided the IUPAC name instead, or parenthetically. This has been a choice we made from the first edition, in hopefully assisting the user to find information.

We have endeavored to bring the user a reference work that we hope will "go to work with them" every day, and we repeat the request that we made in 1986 with our first offering. We ask that you, the user, let one of us know of any useful additions and changes that we should consider for the future.

Thomas J. Bruno
Paris D.N. Svoronos

Acknowledgments

The authors would like to thank some individuals who have been of great help during the preparation of this work, especially this fourth edition. The input of Thomas W. Grove, Sonja G. Ringen, and Beverly L. Smith in developing aspects of the Safety chapter is gratefully acknowledged. The assistance of Marilyn Yetzbacher and Megan Harries with figure preparation in this and prior editions is gratefully acknowledged. The assistance of Clare F. Bruno with formatting, proofreading, printing, and scanning the final copy is gratefully acknowledged. We also thank our editors at CRC, Barbara Knott, and Assunta Petrone for their careful work. In prior editions, we have acknowledged the many scientists who have reviewed sections of this work as it has developed over the past 34 years. While space does not permit a complete listing, we are in your debt, especially our professors and colleagues from Georgetown University. Finally, we thank our wives Clare and Soraya who had to put up with our late nights and weekends, and the occasional squabble between TJB and PDNS.

Authors

Thomas J. Bruno was group leader in the Applied Chemicals and Materials Division at NIST, Boulder, Colorado, before retiring in 2019. He received his BS in Chemistry from Polytechnic Institute of Brooklyn, and his MS and PhD in Physical Chemistry from Georgetown University. Dr. Bruno has done research on fuels, explosives, forensics, and environmental pollutants. He was also involved in research on chemical separations, development of novel analytical methods, and novel detection devices for chromatography. Among his inventions are the Advanced Distillation Curve Method (for fuel characterization) and PLOT-cryoadsorption (for vapor sampling). He has published approximately 270 research papers, 7 books, and has been awarded 10 patents. He serves as associate editor of the *CRC Handbook of Chemistry and Physics*, and as associate editor for *Fuel Processing Technology*. Dr. Bruno was awarded the Department of Commerce Bronze Medal in 1986 for his work on "the thermophysics of reacting fluids" and the Department of Commerce Silver Medal in 2010 for "a new method for analyzing complex fluid mixtures for of new fuels into the U.S. energy infrastructure." He was named a Distinguished Finalist for the 2011 CO-Labs Governor's Award for High Impact Research and received the American Chemical Society Colorado Section Research Award in 2015. He has served as a forensic consultant and/or an expert witness for the US Department of Justice (notably during the federal trial of Terry Nichols for the Oklahoma City bombing), various US Attorney's offices, and various offices of the US Inspectors General. He received a letter of commendation from Department of Justice for these efforts in 2002. He is currently working to develop science and technology education programs for the American judiciary. In this capacity, he serves on the boards of the National Courts and Science Institute and the Bryson Institute for Judicial Education.

Paris D. N. Svoronos is a professor of chemistry at CUNY-Queensborough Community College. He earned a BS in Chemistry and Physics from the American University of Cairo and a PhD in Organic Chemistry from Georgetown University. Among his research interests are the synthesis of organosulfur chemistry, determination of antioxidants in beverages, electrochemistry, organic structure determination, and trace analysis. He is particularly interested in undergraduate education and has authored and co-authored several widely used laboratory manuals and workbooks in addition to several dozen publications. He was awarded several grants, both as a principal investigator including a $2,000,000 NSF grant, as well as a co-principal investigator on $760,000 Department of Education and a $800,000 NSF-ATE grant. He was selected as the 2003 Outstanding Professor of the Year by the CASE (Council for the Advancement and Support of Education) and Carnegie Foundation. He has been in the governing board of the American Chemical Society—New York section serving, among others, as its chair (2015), the chair of the Long Island subsection (2002), the co-chair of the Microwave topical group (2015–present), and chair of the Chemistry and History topical group (2018–present). He was in the board of the MARM organizing committee held in New York twice (2008 and 2016). He was awarded several ACS awards including the Ann Nalley award (2016), the Stanley Israel award for promoting diversity in the chemical sciences (2018), the National ACS fellowship (2018), and the inaugural ACS-NY section Community College Professor of the Year award (2019) and the James Flack Norris Award for Outstanding Achievement in the Teaching of Chemistry (2020). He is particularly proud of his students' academic progress and research accomplishments that include professional conference presentations and publications.

CHAPTER 1

Chemical Analysis Basics

The purpose of this chapter is to provide some fundamental knowledge about chemical analysis, primarily for practitioners in fields outside of research in analytical chemistry. As we have stated in the preface, this book is intended for researchers who must make use of chemical analysis techniques, but whose research areas lie elsewhere. Herein is provided a discussion of some of the major instrumental techniques and also a listing of the most common abbreviations used in analytical chemistry.

SOME BASIC TECHNIQUES IN CHEMICAL ANALYSIS

The following section provides a very brief description of the major instrumental methods of chemical analysis. Please note that these paragraphs are general and are not meant to convey a comprehensive knowledge on these topics. The reader is referred to one of many excellent texts on instrumental methods of chemical analysis for additional details.

SUGGESTED READING

Skoog, D.A., Holler, F.J., and Crouch, S.R., *Principles of Instrumental Analysis*, 7th ed., Thomson Brooks/ Cole, Belmont, 2018.

Robinson, J.R., Skelly-Frame, E.M., and Frame, G.M., *Undergraduate Instrumental Analysis*, 7th ed., CRC Taylor & Francis Group, Boca Raton, FL, 2014.

Pungor, P. *A Practical Guide to Instrumental Analysis*, CRC Press, Boca Raton, FL, 1994.

Bruno, T.J. and Svoronos, P.D.N., in *CRC Handbook of Chemistry and Physics*, 99th ed., J. Rumble., ed., CRC Press, Boca Raton, FL, 2018.

Separation Methods

Gas Chromatography (GC)

A separation method in which the sample or solute is vaporized (usually in a solvent, but sometimes neat or free of solvent) and passed through a medium under the influence of a carrier gas. The medium is called the stationary phase, in contrast to the carrier gas, which is mobile. The most common modern stationary phases are based on open tubular or capillary columns, in which the separation medium coats the inside periphery of a tube (typically tenths of millimeters in inside diameter) that is between 25 and 60 m long. Older media are packed columns, consisting of packed beds, which are still used for gas analysis. In these applications, a solid sorbent is very common, and this is called gas–solid chromatography. Some open tubular columns are available with solid sorbents as well. Interactions of the solute with the separation medium affect the separation of the components of the mixture. A wide variety of detectors are available for general or specific applications. One of the most useful combinations is gas chromatography coupled with mass spectrometry (GC-MS). Solutes amenable to analysis by GC are usually of moderate volatility and relative molecular mass, usually not exceeding a relative molecular mass of 400. The most common stationary phases are cross-linked polymers based on dimethyl polysiloxane, the backbone of which can be derivatized with ligands to provide specific interactions. It is also possible to incorporate stereogenic (chiral) stationary phases as well.

Liquid Chromatography (LC, HPLC)

A separation method in which a sample or solute (usually in a solvent, but sometimes neat or free of solvent) is passed through a medium under the influence of a carrier liquid. The medium is called the stationary phase, in contrast to the carrier, which is the mobile phase. Unlike GC, where the carrier gas plays a little role other than mass transfer, the mobile phase in LC is a controllable variable whose polarity and other properties are varied, in addition to the interactions with the stationary phase, to affect separation. The stationary phase in LC is usually a micrometer size particle packed bed that requires a high-pressure solvent system to cause mass flow. Liquid chromatographic systems have therefore been called high-pressure liquid chromatography (HPLC), although the acronym is usually taken to mean high-performance liquid chromatography. Separations in HPLC are designed by considerations of sample polarity, molecular charge, and molecular size. Normal-phase and reversed-phase HPLC separations are based on polarity. Ion exchange chromatography is based on molecular charge and several techniques take advantage of molecular size. Many variations have been developed for specific analysis. For example, gel permeation chromatography is an adaptation used for the separation of polymers. Affinity chromatography is similar in concept to gel permeation chromatography, but uses the specific interaction between an antibody and an antigen. The use of stereogenic (or chiral) stationary phases is also an important development, especially in the analysis of pharmaceuticals.

Thin-Layer Chromatography (TLC)

A separation method in which a stationary phase (typically a polar adsorbent such as alumina or silica gel) coated on a sheet of plastic, aluminum, or glass is used with a mobile phase usually consisting of a solvent or mixture of solvents in a beaker. The solute is applied as a blotted spot just above the end of the adsorbent-coated plate, and then, the end is immersed into the solvent (but not so far as to immerse the solute spot). Commercial plates are robust, plastic sheets that can be cut to the size desired. In earlier applications, filter paper has been used in TLC, giving rise to the term "paper chromatography." This is rarely used today. Solvent is then drawn up through the adsorbent coating by capillary action. Separation results from a combination of interactions with the adsorbent and solvent. Commonly, the separated components are rendered visible by spraying stains or reactants on the plate after separation has been completed. It is also common to view the "developed" plate under an ultraviolet (UV) lamp, to visualize spots that may be fluorescent.

Supercritical Fluid Chromatography and Extraction

A separation technique similar to other extraction and chromatographic methods, but in which the mobile phase is actually a fluid in its supercritical fluid state. A supercritical fluid is a fluid that is held above its critical temperature and pressure, for which no application of additional pressure can result in the development of a liquid phase. Supercritical fluids are unique in that while they are chosen to possess liquid-like densities, the mass transfer properties are very much like those of liquids. Supercritical fluid chromatography remains a niche method that is applicable to pharmaceuticals and other high relative molecular mass solutes. Supercritical extraction, on the other hand, is more widely used as a sample preparation method, especially in pharmaceutical analyses, polymers, and environmental analyses.

Electrophoresis

A family of separation methods based on the motion induced in particles by an applied, uniform electric field. The most common application of electrophoresis is gel electrophoresis, in which sample (typically proteins, amino acids, nucleic acids, etc.) is applied to a channel that is formed in a cross-linked polymer, usually polyacrylamide or agarose (the gel). The speed at which the individual species move through the gel under the influence of the field is determined largely by the size of the specie, as expressed by the mass-to-charge ratio. After separation, the individual species usually appear as discrete bands that may be better visualized by staining with ethidium bromide, silver, or Coomassie Brilliant Blue dye. Other related and more specific techniques include isoelectric focusing, pulsed field gel electrophoresis, immunoelectrophoresis, and isotachophoresis.

Purge and Trap Sampling

Purge and trap sampling is a family of methods that are used to capture the headspace above a condensed phase for subsequent analysis, most often for complex mixtures, environmental samples, etc. The headspace is the vapor space that develops above any condensed (solid or liquid) phase. Thermodynamics assures us that the concentration of a particular analyte found in the headspace will be different than that found in the condensed phase, but often the relationship is predictable. The value in the method comes from the simplicity; sample preparation is usually far simpler than the cleanup that is typically required for many complex mixtures. Purge and trap methods fall into two general categories: static and dynamic. The dynamic method typically uses a sweep gas to continuously purge vapor analytes into a cold (cryogenic) trap or an adsorbent. Modern dynamic purge and trap methods include porous layer open tubular (PLOT) column cryoadsorption, which uses a combined adsorbent and cryotrapping approach on a high-efficiency platform. Static methods typically employ a syringe to pressurize the headspace above a condensed phase, followed by uptake of the pressurized headspace into a trap. A modern static method (that usually is done without pressurization) is solid-phase microextraction (SPME). This method utilizes a fiber coated with a stationary phase (similar to stationary phases used for gas chromatography and liquid chromatography) at the end of a wire mounted in a syringe needle.

Spectroscopic Methods

Mass Spectrometry (MS)

An analytical technique in which charged particles or radical ions are produced from a sample by either electron impact (bombardment with a stream of electrons) or chemical ionization (interaction with a small charged ion). Analysis on the basis of mass-to-charge ratio is performed on fragments of the molecule that develop after the initial ionization. The method is very useful for mass determination, and structure determination on the basis of the induced fragmentation pattern. The charged fragments can be separated or analyzed by a magnetic sector, a quadrupole, an ion trap, time of flight, or cyclotroning. When coupled to separation techniques such as gas or liquid chromatography, a nearly universal qualitative detection capability is provided. Structure determination can be performed by comparison with well-known fragmentation patterns or characteristics.

When a mass spectrometer is capable of high mass resolution, a nearly unequivocal identification of a compound formula is possible. While direct interfaces and gas chromatographs are the most common sample introduction techniques, many others are available for specific applications. Matrix assisted laser desorption and ionization MALDI is often used in time-of-flight instruments and is especially useful for the analysis of biopolymers. Inductively coupled plasma ionization is capable of producing mass spectra at high sensitivity for many metals and some nonmetals. When coupled to LC instrumentation, thermospray and electrospray methods have been used with HPLC for analytes that are not amenable to GC separation.

Ultraviolet Spectrophotometry (UV, UV-Vis)

A spectroscopic technique that focuses on electronic transitions with the visible and UV regions of the electromagnetic spectrum for excitation and detection. The practical UV region extends from 190 to 400 nm in wavelength. The UV region can be divided into subranges: near UV: 300–400 nm, mid UV: 300–200 nm, far UV: 200–122 nm, vacuum UV: 200–100 nm. Other divisions are possible, but these are less important for analytical chemistry. The visible region, so called because of the response of human vision, extends from about 390 to 750 nm in wavelength. Although it is possible to use UV visible spectrophotometry for structure determination, it is most often used as a quantitative tool. The wavelength–structure correlations are not as detailed, nor the spectra as sharp, as with other spectroscopic methods such as IR spectrophotometry. The utility of UV-Vis absorptions for many organic compounds has led to this instrument being adapted as a detector for liquid chromatography. Related to UV spectrophotometry are fluorescence spectroscopic methods. In these methods, the energy emitted is at a different wavelength (usually longer, of lower energy), and this is typically detected perpendicularly to the incident excitation beam. This makes fluorescence spectroscopy more sensitive than UV-Vis spectroscopy. This type of instrument is also incorporated as a detector for liquid chromatography.

Infrared Spectrophotometry (IR, FTIR)

A spectroscopic technique that focuses on molecular vibrations (with a concurrent change in dipole moment) with the IR region of the electromagnetic spectrum for excitation and detection. This region is further divided into three separate but overlapping ranges. The near IR (high-energy IR, approximately $14,000–4,000 \, cm^{-1}$, 0.8–2.5 µm wavelength) is used to study overtone or harmonic vibrations. The mid-range IR (mid-range-energy IR, approximately $4,000–400 \, cm^{-1}$, 2.5–25 µm wavelength) is used to study the fundamental vibrations and associated rotational-vibrational combinations. The far-IR (low-energy IR, approximately $400–10 \, cm^{-1}$, 25–1,000 µm wavelength) is close to the microwave region and is used to study rotational transitions. Most modern instruments use the Fourier transform technique to record the spectrum over all wavelengths, rather than by scanning through the wavelengths. The absorbances of the IR radiation are associated with specific chemical moieties, and a study of the spectra can be used to aid in structure determination. One often uses structure correlation charts to aid in assignment of absorbance bands. An analysis of the intensity of the absorptions can also be used for quantitative analysis.

Nuclear Magnetic Resonance Spectrometry (NMR)

A spectroscopic method that takes advantage of magnetic nuclei (nuclei with an odd number of protons and/or neutrons, having an intrinsic magnetic moment and angular momentum), placed in a magnetic field, will absorb pulses of electromagnetic radiation, and then radiate this energy back out. For these nuclei, the energy and signal intensity are proportional to the applied magnetic field. The power of NMR results from the ability to probe the molecular environment around a particular nucleus, thus making it long a tool of the organic chemist. This is done by measurement of the chemical shift (or frequency) of an absorption and by the analysis of splitting patterns, which are caused by the influence of adjacent nuclei. New high-field, high-sensitivity instruments have given this technique more applications in analytical chemistry, however. The most commonly studied nuclei are ^{1}H (proton, the most NMR-sensitive isotope after the radioactive ^{3}H) and ^{13}C. With high-field instruments, additional nuclei are accessible: ^{2}H, ^{10}B, ^{11}B, ^{14}N, ^{15}N, ^{17}O, ^{19}F, ^{23}Na, ^{29}Si, ^{31}P, ^{35}Cl, ^{113}Cd, ^{129}Xe, and ^{195}Pt.

Raman Spectroscopy

A vibrational spectroscopic method that arises from the inelastic scattering of monochromatic radiation by molecules that undergo a change in polarizability during the vibration. This is in contrast to IR spectrophotometry, in which a change in the dipole moment occurs during the vibration. When radiation (typically light from a laser in the visible, near-IR, or near-UV range) is scattered, a small fraction of the scattered radiation is observed to have a different frequency (the Raman effect). The variations of Raman spectroscopy are used to locate functional groups or chemical bonds in molecules. There are several variations in the approach to Raman spectroscopy. In resonance Raman spectroscopy, the excitation wavelength is matched to an electronic transition of the molecule, enhancing the vibrational modes. In coherent anti-Stokes Raman spectroscopy (CARS), two laser beams are used to generate a coherent anti-Stokes frequency beam. In surface-enhanced Raman spectroscopy (SERS), surface plasmons (a quantum of plasma oscillation) on a silver or gold colloid on a surface (such as a mirror) are excited by the laser, resulting in an increase in the electric fields surrounding the metal.

Atomic Absorption and Emission Spectroscopy (AA, AES)

Two related spectroscopic methods applied primarily to the analysis of inorganic compounds. Atomic absorption procedures use the absorption of optical radiation (light) by free atoms in the gaseous state. The light can be produced by a hollow cathode lamp, an electrodeless discharge lamp, or a deuterium lamp. The light is absorbed by the analyte during an electronic transition, the wavelength of which corresponds to only one element in the analyte, and the width of an absorption line is only of the order of a few picometers. This method can be used for the quantitative determination (on the basis of a calibration curve) of approximately 70 different elements in solution or directly in solid samples. Atomic emission spectroscopy (AES) uses the light emitted by a vaporized sample in a flame, plasma, arc, spark, or laser, at a particular wavelength, to determine the atomic spectrum (for determination of the elemental composition) quantity of an element in a sample. The wavelength of the atomic spectral line gives the identity of the element, while the intensity of the emitted light is proportional to the number of atoms of the element. No single source, as described above, is optimal for a given sample, and it is the choice of source that distinguishes the various techniques.

Miscellaneous Methods

Colorimetry, Spot Tests, and Presumptive Tests

Rapid, simple tests based on color change are frequently used as the basis of preliminary conclusions or approximate concentration measurement. Colorimetric methods use simple comparative instruments to determine the concentration of colored compounds in solution. These devices, called colorimeters, can be manual or automatic, and use filtered light in the visible region (between 400 and 700 nm). In both cases, the operation depends upon having multiple solutions, including a blank, for comparison with a solution of unknown concentration. Manual colorimeters function by measuring the variable light path through the unknown solution as compared to a known solution until a match is achieved visually. The product of concentration and path length matches when the concentrations are the same, so an unknown concentration may be obtained by a proportion. Automated devices function similarly but with photocells. Colorimetric tests are usually done by following four basic protocols: (1) an unknown can be treated/reacted with a reagent to form a new compound which is colored; (2) a chelate complex is formed having a different color than the starting compound; (3) a colored compound is oxidized or "bleached" by another compound, resulting is less color; or (4) an intermediate is formed that can be oxidized or reduced later to a colored compound. Related to colorimetric methods are spot tests, often called presumptive tests in forensic science (since they are used to establish probable cause for ordering and performing more sophisticated tests), which are the observation of color changes upon the addition of one or more reagents to an unknown. Spot tests are inherently simple and are done with a drop or two of reagent(s), on a filter paper or other suitable medium, and generate minimal waste. Usually instrumentation is not used, one merely notes a color change, although simple devices such as handheld color space analyzers (used in the paint industry) can be helpful. Other approaches include the recording of a digital photograph of the spot and the determination of the L*, a*, and b* axes in LAB color space. An example of spot testing for presumptive purposes includes the addition of cobalt thiocyanate ($CoC_2N_2S_2$) to cocaine, producing a blue color (the Scott test). These tests are presumptive rather than confirmatory because of the potential of false positives. Thus, the Scott test will also produce a blue coloration in the presence of lidocaine and diphenylhydramine.

Refractometry

Refractometry is the measurement of the degree to which the path of electromagnetic radiation, specifically light, is bent upon traversing from one medium to another. Indirectly, a refractometer measures the speed of light since the refractive index, the measurand provided by a refractometer, is defined as the speed of light in vacuo divided by the speed of light in a particular medium (which can be a gas, liquid, or solid). Refractive index typically varies between 1.3 and 1.7 for most compounds, and being a ratio, it is dimensionless. The refractive index of vacuum is by definition unity. Refraction is the result of differing densities of media; light passes through dense media slower than it passes through less dense media. Refractive index is dependent on temperature and the wavelength of light used for the measurement, and when both are known for a media, it can be used as an indication of composition or concentration. Thus, the refractive index is typically reported measured with white light from the sodium D-line (589 nm), and often at a temperature that is near ambient (20 °C). The measurand is thus indicated as n_D^{20}. As an analytical tool, refractometry is used in four primary ways. First, it is used to help identify compounds by comparison with known values. Second, it is used to assess purity by comparison with the refractive index of a pure material. Third, it is used as a measure of solute concentration of a solution, by reference to a calibration curve. Finally, it is used as the basis of universal detection in HPLC. Refractometers are commercially available as small handheld instruments, both manual and digital (that can be carried in a pocket), Abbe refractometers (the typical lab bench device usually equipped with a thermostat) and in-line refractometers such as those used for HPLC detectors.

Thermal Analysis (TA) Methods

A family of analytical techniques in which various properties of a sample are examined as a function of changing temperature at a particular rate of change. For chemical analysis, the most common thermal analysis techniques include differential thermal analysis (DTA) and thermogravimetric analysis (TGA). There are many other thermal analysis methods available, perhaps the most common of which is DTA, used for the determination of phase transitions. In chemical analysis, TGA is used to determine a mass change as a function of temperature, and to determine decomposition or degradation temperatures, moisture content of materials (although Karl Fisher coulombic titrimetry is also used for this), the level of inorganic and organic components in materials, decomposition points of explosives, and solvent residues in materials. DTA, on the other hand, monitors temperature change rather than mass change. In this respect, it is useful as a complementary technique to probe the energetic of decomposition, moisture loss, etc. Other thermal analysis methods, such as differential scanning calorimetry (DSC), are not primarily analytical tools but rather thermophysical property measurement tools.

Electrochemical Methods

A family of techniques that analyze the effect and role of electricity in either creating or serving as an outcome of a chemical change. These techniques include electrolysis, electrogravimetry, cyclic voltammetry, linear sweep voltammetry, electrochemical titrations, and the newer area of nanoelectrochemistry. Electrolysis leads to the separation and isolation of metals originally in a molten or a solution (ionic) mixture on an electrode using a direct current (DC) and a voltage called the decomposition potential. Electrogravimetry includes electroplating, electrophoretic deposition, and underpotential deposition. These methods are closely related to coulometry, which quantitatively measures the amount of matter transformed during electrolysis by using Faraday's laws. Cyclic voltammetry is a potentiodynamic electrochemical measurement where the electrode potential is ramped linearly vs. time. Once the set potential is reached, the working electrode's potential is ramped toward the opposite direction to return to the initial potential, thus creating a cycle which can be repeated at will. Linear sweep voltammetry involves the measurement of the current at a working electrode, while the potential between the working electrode and a reference electrode is plotted linearly against time. Electrometric titration refers to any technique that uses an electrometer or an instrument that determines, or even detects, the magnitude of a potential difference or charge by the different electrostatic forces between charged species. Nanoelectrochemistry is a recent branch of electrochemistry that studies the electrochemical properties of nanometer-sized materials. It uses nanoelectrodes whose size is in the order of 1–100 nm and which are made of metals or semiconducting materials. It has created a significant impact in the development of many sensors and the study of reactions that involve extremely low concentrations.

X-Ray Methods

A family of techniques that utilize radiation in the X-ray region, with wavelengths between 0.01 and 10 nm (corresponding to frequencies in the range of 3×10^{16} to 3×10^{19} Hz) to analyze for the presence of elements in a sample. In X-ray fluorescence spectroscopy (XRS), short-wavelength X-rays are used to excite secondary or fluorescent X-rays from a sample. The wavelength of X-rays used to excite the sample must be shorter than the expected fluorescence wavelength. The sample is typically presented as a powdered solid on a glass plate. Spectra are generated by changing the incident angle between the source and the sample with a device called a goniometer. The fluorescent signals obtained are very precise and specific, and can rival wet chemical methods for the identification of elements. Auger electron spectroscopy AES is a surface analysis method that measures the electrons emitted from a surface by electron bombardment of the surface. This method is considered to be in the same family as X-ray methods because during electron bombardment, the surface can lose energy by either electron emission (the Auger effect via Auger electron emission) or X-ray emission. These methods are in contrast to X-ray diffraction methods, in which a crystal structure measurement is desired.

Flow Injection Analysis (FIA)

An analytical protocol that seeks to replace the manual "test tube and beaker" aspects of wet chemical analysis by injecting an analyte in a flowing stream of carrier reactant. As the analyte flows with the reactant stream, it diffuses into the reactant and product forms. Ultimately, the product zone, under the influence of the moving reactant, is passed into a detector section. The detection devices can consist of the same wide variety as is used in HPLC. The major advantage of FIA is the automation and decreased uncertainty associated with sampling and reagent addition. Strict control of reagent concentration, flow rate, and analyte volume is possible. Modern applications of FIA include the sequential addition of analyte and reactant in a stream so that the two are "stacked" in an inert carrier. They then mix by the parabolic flow profile of a laminar flowing stream in a tube. This arrangement can be miniaturized within a sampling valve, forming the so-called lab-in-a-valve approach.

ABBREVIATIONS USED IN ANALYTICAL CHEMISTRY

One frequently encounters acronyms and abbreviations in the literature of analytical chemistry, standards documents, and company procedures. While these acronyms and abbreviations should be defined upon their first use in a document, this is not always done and even when done, it is convenient to have a source of the most common at one's fingertips [1].

1. Bruno, T.J. *CRC Handbook of Chemistry and Physics*, 99th ed., Rumble, J., ed., CRC Press, Boca Raton, FL, 2018.

A	chromatographic peak area; surface area of solid granular adsorbent
A	eddy diffusion term in the van Deemter equation
AAA	absolute activation analysis
AAD	atomic absorption detector
AAS	atomic absorption spectroscopy
AC	alternating current, affinity chromatography
ACP	alternating current plasma
ADXPS	angular-dependent X-ray photoelectron spectroscopy
AED	atomic emission detector
AEM	analytical electron microscope (microscopy)
AES	Auger electron spectroscopy, atomic emission spectroscopy
AFID	alkali flame ionization detector
AFM	atomic force microscopy
AFS	atomic force spectroscopy
AIS	average of individual samples
AL	action level
AM	amplitude modulation
AMS	accelerator mass spectrometry
AN	area normalization
ANRF	area normalization with response factors
AOTF	acousto-optical tunable filter
AP	analytical pyrolysis
APCI	atmospheric pressure chemical ionization
APD	azimuthal photoelectron diffraction
API	atmospheric pressure ionization

APSTM	analytical photon scanning tunneling microscope
ARAES	angle resolved Auger electron spectroscopy
ARF	absolute response factor
ARM	atomic resolution microscopy
ARPES	angle resolved photoelectron spectroscopy
ARUPS	angle-resolved ultraviolet photoelectron spectroscopy
ASE	accelerated solvent extraction
AsFIFFF	asymmetrical flow field flow fractionation (AF4)
ATD	above-threshold dissociation
ATI	above-threshold ionization
ATR	attenuated total reflection
BB	band broadening
BE	magnetic sector–electric sector tandem mass spectrometer (note: also called a MIKE spectrometer)
BEE	magnetic sector–electric sector–electric sector mass spectrometer
BET	Brunauer–Emmett–Teller (adsorption isotherm)
BIFL	burst integrated fluorescence lifetime
BIS	Bremsstrahlung isochromat spectroscopy
BJH	Barrett–Joyner–Halenda (method)
BL	bioluminescence
BLRF	bispectral luminescence radiance factor
BQQ	magnetic sector-double quadrupole mass spectrometer
BTOF	magnetic sector–time-of-flight tandem mass spectrometer
CAD	collision-activated dissociation
CAR	continuous addition of reagent
CARS	coherent anti-Stokes Raman spectroscopy
CCC	counter-current chromatography
CCD	charge-coupled device
CCT	constant current topography
CD	circular dichroism
CE	capillary electrophoresis, counter electrode
CEC	capillary electrokinetic chromatography, capillary electrochromatography
CED	cohesive energy density
CFA	continuous flow analysis
CF-FAB	continuous flow-fast atom bombardment
CFM	chemical force microscopy
CGE	capillary gel electrophoresis
CHEMFET	chemical-sensing field effect transistor CI chemical ionization
CID	collision-induced dissociation
CIEF	capillary isoelectric focusing
CITP	capillary isotachophoresis
CL	chemiluminescence
CLLE	continuous liquid–liquid extraction
CMA	cylindrical mirror analyzer
COSY	correlation spectroscopy
CPAA	charged particle activation analysis
CP/MAS	cross-polarization/magic angle spinning
CRDS	cavity ring-down spectroscopy
CRF	chromatographic response function
CRM	certified reference material

CS	carbon strip (adsorbent, as used for headspace analysis)
CT	cryogenic trapping
CTD	charge transfer device
CV	cyclic voltammetry
CV-ASS	cold vapor atomic absorption spectrometry CVD chemical vapor deposition
CW	continuous wave
CZE	capillary zone electrophoresis
DA	diode array
DAD	diode array detector (UV-Vis)
DADI	direct analysis of daughter ions
dB	de Boar t-plot
DBE	double bond equivalent
DC	direct current
DCI	desorption chemical ionization
DCP	direct-current plasma
DEP	differential electrolytic potentiometry
DEPT	distortionless enhancement by polarization transfer
DETA	dielectric thermal analysis
DIN	direct injection nebulizer
DLI	direct liquid introduction
DLS	dynamic light scattering
DMA	dynamic mechanical analysis
DME	dropping mercury electrode
DNMR	dynamic nuclear magnetic resonance
DPP	differential pulse polarography
DRIFT	diffuse-reflectance infrared Fourier transform
DSC	differential scanning calorimetry
DTA	differential thermal analysis
DTC	differential thermal calorimetry
EAES	electron-excited Auger electron spectroscopy
EB	electric sector–magnetic sector tandem mass spectrometer
EBE	electric sector–magnetic sector-electric sector tandem mass spectrometer
EC	electrochemical
ECD	electron capture detector
ECMS	electron capture mass spectrometry
ECNIMS	electron capture negative ionization mass spectrometry
EDL	electrodeless discharge lamp
EDS	energy-dispersive spectrometer
EDXRF	energy-dispersive X-ray fluorescence
EELS	electron energy-loss spectroscopy
EFFF	electric field flow fractionation
EG	electrogravimetry
EGA	evolved gas analysis
EIA	enzyme-linked immunoassay
EI(I)	electron impact (ionization)
EIMS	electron impact mass spectrometry
ELCD	electrolytic conductivity detector
ELISA	enzyme-linked immunosorbent assay
ELSD	evaporative light scattering detector
EM	electron microscopy

EMIRS	electrochemically modulated IR spectroscopy
EOF	electro-osmotic flow
EPL	enhanced photoactivated luminescence
EPMA	electron-probe microanalysis
EPR	electron paramagnetic resonance
EPXMA	electron-probe X-ray microanalysis
EQL	estimated quantitation limit
ERD	elastic recoil detection
ESA	electrostatic analyzer
ESCA	electron spectroscopy for chemical analysis
ESEM	environmental scanning electron microscope
ESI	electrospray ionization
ESP	electrospray
ESR	electron spin resonance
ET	electrometric titration
ETA	electrothermal analyzer, emanation thermal analysis
EXAFS	extended X-ray absorption fine structure
FAA	flame atomic absorption
FAAS	flame atomic absorption spectroscopy
FABMS	fast-atom bombardment mass spectrometry
FAES	flame atomic emission spectroscopy
FAFS	flame atomic fluorescence spectroscopy
FAM	field analytical method
FAS	flame absorption spectroscopy
FD	field desorption
FD/FI	field desorption/field ionization
FES	flame emission spectroscopy
FFEM	freeze-fracture electron microscopy
FFF	field flow fractionation
FFFF	flow field flow fractionation
FFM	friction force microscopy
FFS	flame fluorescence spectroscopy
FFT	fast Fourier transform
FGC	fast gas chromatography
FI	flow injection, field ionization
FIA	flow injection analysis
FIB	focused ion beam
FID	flame ionization detector, free-induction decay
FIM	field ion microscopy
FNAA	fast neutron activation analysis
FOCS	fiber optic chemical sensor
FPD	flame photometric detector
FSOT	fused silica open tubular (column)
FT	Fourier transform
FT-ICR	Fourier transform ion cyclotron resonance
FT-IR	Fourier transform infrared (often "FT/IR," "FTIR," "FT IR")
FT-IRRAS	FT-IR reflection-absorption spectroscopy
FT-MS	Fourier transform mass spectrometry
FWHM	full-width half-maximum
GC	gas chromatography

GC-IR	gas chromatography–infrared spectrometry
GCMS	gas chromatography–mass spectrometry
GDL	glow discharge lamp
GDMS	glow discharge mass spectrometry
GE	gel electrophoresis, gradient elution
GEMBE	gradient elution moving boundary electrophoresis
GFAAS	graphite furnace atomic absorption spectroscopy
GLC	gas–liquid chromatography
GPC	gel permeation chromatography
GS	Gram–Schmidt (algorithm)
GSC	gas–solid chromatography
GSED	gaseous secondary electron detector
HCL	hollow cathode lamp
HCOT	helically coiled open tubular (column)
HDC	hydrodynamic chromatography
HETCOR	heteronuclear correlation
HETP	height equivalent of (a) theoretical plate(s)
HG	hydride generation
HIC	hydrophoric interaction chromatography
HMBC	heteronuclear multiple-bond correlation
HPAC	high-performance affinity chromatography
HPIAC	high-performance immunoaffinity chromatography
HPLC	high-performance liquid chromatography and/or high-pressure liquid chromatography
HPTLC	high-performance thin-layer chromatography
HRCGC	high-resolution capillary gas chromatography
HRGC	high-resolution gas (or liquid) chromatography
HS	headspace
HSA	hemispherical analyzer
HSC	heteronuclear shift correlation
HSQC	heteronuclear single quantum coherence
HTC	high-temperature combustion
IA	isocratic analysis
IAC	immunoaffinity chromatography
IAES	ion-excited Auger electron spectroscopy
IC	ion chromatography
ICMS	ion chromatography mass spectrometry
ICP	inductively coupled plasma
ICP-OES	ICP optical emission spectrometry
ICR	ion cyclotron resonance
IDMS	isotope dilution mass spectrometry
IEC	ion-exchange chromatography
IEF	isoelectric focusing
IF	intermediate frequency
IGC	inverse gas chromatography
IGF	inert gas fusion
ILDA	intensified linear diode array
IMAC	immobilized metal-ion affinity chromatography
INADEQUATE	incredible natural abundance double-quantum transfer experiment INEPT insensitive nuclei enhancement by polarization transfer

INAA	instrumental neutron activation analysis
IP	ion pairing
IPC	ion-pair chromatography
IPG	immobilized pH gradient
IPMA	ion probe microanalysis
IR	infrared (spectrophotometry)
IRN	indicator radionuclide(s)
IRS	internal reflection spectroscopy
ISCA	ionization spectroscopy for chemical analysis
ISE	ion selective electrode
ISP	ion spray
ISS	ion scattering spectrometry
LAMMS	laser micromass spectrometry
LARIMS	laser atomization resonance ionization mass spectrometry
LARIS	laser atomization resonance ionization spectroscopy
LASER	light amplification by stimulated emission of radiation
LBB	Lambert–Beer–Bouguer law
LC	liquid chromatography
LC-LS	multidimensional liquid chromatography
LDMS	laser desorption mass spectrometry
LDR	linear dynamic range
LEAFS	laser-excited atomic fluorescence spectrometry
LED	light-emitting diode
LEED	low-energy electron diffraction
LEEM	low-energy electron microscopy
LEI	laser-enhanced ionization
LEISS	low-energy ion scattering spectrometry
LESS	laser-excited Shpol'skii spectroscopy
LFM	lateral force microscopy
LIDAR	light detection and ranging
LIFD	laser-induced fluorescence detection
LIMS	laboratory information management system
LLC	liquid–liquid chromatography
LLD	lower-limit detection
LLE	liquid–liquid extraction
LNRI	laser non-resonant ionization
LO	local oscillator
LOC	lab on a chip
LOD	limit of detection
LPDA	linear photodiode array
LPSIRS	linear potential-sweep IR reflectance spectroscopy
LRI	laser resonance ionization
LRMA	laser Raman microanalysis
LSC	liquid–solid chromatography
LSE	liquid–solid extraction
LTP	low-temperature phosphorescence
MAE	microwave-assisted extraction
MALDI	matrix-assisted laser desorption/ionization
MAS	magic angle spinning
MCD	magnetic circular dichroism

MCP	microchannel plate
MDGC	multidimensional gas chromatography
MDL	method detection limit
MDM	minimum detectable mass
MDQ	minimum detectable quantity
MEIS	medium-energy ion scattering
MEKC	micellar electrokinetic chromatography
MFM	magnetic force microscopy
MID	multiple ion detection
MIKE	mass analyzed ion kinetic energy mass spectrometry
MIP	microwave-induced plasma, mercury intrusion porosimetry
MIRS	multiple internal reflection spectroscopy
MLC	micellar liquid chromatography
MLLSQ	multiple linear least squares
MMF	minimum mass fraction
MMLLE	microporous membrane liquid–liquid extraction
MPD	microwave plasma detector
MPI	multiphoton ionization
MRDL	maximum residual disinfectant level (in water analysis)
MRI	magnetic resonance imaging
MS	mass spectrometry
MS-MS	tandem mass spectrometry
MSPD	matric solid-phase dispersion
MSRTP	micelle-stabilized room-temperature phosphorescence
MWD	microwave (assisted) digestion
NAA	neutron activation analysis
NCIMS	negative chemical ionization mass spectrometry
NDP	neutron depth profiling
NEXAFS	near edge X-ray absorption fine structure
NHE	normal-hydrogen electrode
NICI	negative ion chemical ionization
NIR	near-infrared or near-IR
NIRA	near-infrared reflectance analysis
nm	nanometer
NMR	nuclear magnetic resonance
NOE (nOe)	nuclear Overhauser effect
NOESY	nuclear Overhauser effect spectroscopy
NPD	nitrogen-phosphorus detector, normal photoelectron diffraction
NPLC	normal-phase liquid chromatography
ODMR	optically detected magnetic resonance
ODS	octadecylsilane
OES	optical emission spectrometry, optical emission spectroscopy
OID	optoelectronic imaging device
OMA	optical multichannel analyzer
OPO	optical parametric oscillator
OPTLC	over-pressured thin-layer chromatography
ORD	optical rotary dispersion
OTE	optically transparent electrodes
PA	proton affinity
PAA	photon activation analysis

PAGE	polyacrylamide gel electrophoresis
PAH	polycyclic aromatic hydrocarbon
PAS	photoacoustic spectroscopy
PB	particle beam
PC	paper chromatography
PCA	principal component analysis
PCR	polymerase chain reaction
PCS	photon correlation spectroscopy
PCSE	partially coherent solvent evaporation
PD	plasma desorption
PDA	photodiode array
PDHID	pulsed discharge helium ionization detector
PDMS	plasma desorption mass spectrometry, polydimethyl siloxane
PED	pulsed electrochemical detection, plasma emission detector, photoelectron diffraction
PES	photoelectron spectroscopy
PET	positron emission tomography
PFIA	process flow injection analysis
PGC	packed-column gas chromatography
pH	negative logarithm of hydrogen ion concentration
PICI	positive ion chemical ionization
PID	photoionization detector
PIXE	particle-induced X-ray emission
pK	negative logarithm of an equilibrium constant
PLE	pressurized liquid extraction
PLOT	porous-layer open tubular
PLOT-cryo	porous-layer open tubular (column) cryoadsorption
PMT	photomultiplier tube
ppb	parts per billion
ppm	parts per million
ppt	parts per thousand, parts per trillion
PSD	position-sensitive detector
PTFE	polytetrafluoroethylene
PTR	proton transfer reaction (in mass spectrometry)
PTV	programmable temperature vaporizer
PVD	pulsed voltammetric detection, physical vapor deposition
QCL	quantum cascade laser
QCM	quartz-crystal microbalance
QFAA	quartz furnace atomic adsorption
QIT	quadrupole ion trap
qNMR	quantitative nuclear magnetic resonance
QQQ	triple quadrupole mass spectrometer
QqQ	triple quadrupole mass spectrometer
Q1qQ3	triple quadrupole mass spectrometer
QTH	quartz tungsten halogen
QTOF	tandem quadrupole time-of-flight mass spectrometer
RAA	running annual average
RBS	Rutherford backscattering spectrometry
REELS	reflection electron energy loss spectrometry
RES	reflection electron spectrometry
RF	radio frequency

RHEED	reflection high-energy electron diffraction
RIC	reconstructed ion chromatogram
RI	refractive index, retention index
RID	refractive-index detector
RIMS	resonance ionization mass spectrometry
RIS	resonance ionization spectroscopy, range of individual samples
RM	reference material
RNAA	radiochemical neutron activation analysis
ROA	Raman optical activity
ROESY	rotating frame nuclear Overhauser effect spectroscopy
RPLC	reversed-phase liquid chromatography
RRDE	rotating ring-disk electrode
RS	Raman spectroscopy
RSF	relative sensitivity factor
RTIL	room-temperature ionic liquid
RTP	room-temperature phosphorescence
S/N	signal-to-noise ratio
SAE	sonication assisted extraction
SAM	scanning Auger microscopy, self-assembly monolayers
SANS	small-angle neutron scattering
SAW	surface acoustic wave
SAXS	small-angle X-ray scattering
SBSE	stir bar sorptive extraction
SCE	standard calomel electrode, saturated calomel electrode
SCF	supercritical fluid
SCOT	support-coated open tubular
SDD	silicon drift detector
SdFFF	sedimentation field flow fractionation
SEC	size-exclusion chromatography
SEM	scanning electron microscope
SERS	surface-enhanced Raman spectroscopy
SFC	supercritical-fluid chromatography
SFE	supercritical-fluid extraction
SFFF	sedimentation field flow fractionation
SF-MS	sector field mass spectrometry
SFS	synchronous fluorescence spectroscopy
SHE	standard hydrogen electrode
SIA	sequential injection analysis
SIDA	stable isotope dilution assay
SIMS	secondary ion mass spectrometry
SIRIS	sputter-initiated resonance ionization spectroscopy
SMCL	secondary maximum contaminant level (in water analysis)
SMDE	static mercury drop electrode
SMSS	spark source mass spectrometry
SNIFTIRS	subtractively normalized interfacial FT-IR spectroscopy
SNMS	sputtered neutral mass spectrometry
SPAES	spin-polarized Auger electron spectroscopy
SPE	solid-phase extraction
SPME	solid-phase microextraction
SPR	surface plasmon resonance

SRE	stray radiant energy
SRM	standard reference material
SSMS	spark source mass spectrometry
SSRTF	solid-surface room-temperature fluorescence
SSRTP	solid-surface room-temperature phosphorescence
STEM	scanning transmission electron microscope
STM	scanning tunneling microscope
SVE	solvent vapor exit
SWE	supercritical-water extraction
TCA	thermochemical analysis
TCD	thermal-conductivity detector
TCT-GC-MS	thermal cold trap gas chromatography mass spectrometry thermodilatometry
TD	thermodilatometry
TDL	tunable diode laser
TEA	thermal energy analyzer
TED	thermionic emission detector
TEELS	transmission electron energy loss spectrometry
TEM	transmission electron microscope
TET	thermometric enthalpimetric titration
TFFF	thermal field flow fractionation
TGA	thermogravimetric analysis
TGA-IR	thermogravimetric analysis-infrared
THEED	transmission high-energy electron diffraction
ThFFF	thermal field flow fractionation
TIC	total ion current chromatogram, tentatively identified compound
TIMS	thermoionization mass spectrometry
TLC	thin-layer chromatography
TLE	thin-layer electrode
TLM	thermal lens microscopy
TLV	threshold limit value
TMA	thermomechanical analysis
TMS	tetramethylsilane
TOCSY	total correlation spectroscopy
TOF	time-of-flight
TOF-MS	time-of-flight mass spectrometry
TSP	thermospray
UHV	ultrahigh vacuum
USE	ultrasonic extraction
UV	ultraviolet
UVPES, UPS	ultraviolet photoelectron spectroscopy
UV-Vis	ultraviolet-visible
VAR	variable angle reflectance
Vis	visible (radiation)
VOC	volatile organic compound(s)
VOX	volatile organic halogens
VUV	vacuum ultraviolet
W	Wien filter (used in mass spectrometry)
WCOT	wall-coated open tubular
WDS	wavelength dispersive spectrometer
WWCOT	whisker-wall-coated open tubular (column)

WWPLOT	whisker-wall-coated porous layer open tubular (column)
WWSCOT	whisker-wall support-coated open tubular (column)
XAES	X-ray excited Auger electron spectroscopy, X-ray adsorption edge spectrometry
XANES	X-ray absorption near-edge spectroscopy
XPS	X-ray photoelectron spectroscopy
XRD	X-ray diffraction
XRF	X-ray fluorescence
XRFS	X-ray fluorescence spectroscopy
XRS	X-ray spectroscopy
ZAF	Z (element number) absorption fluorescence

Symbols used in analytical chemistry:

α	Auger yield, fine structure constant
a_0	Bohr's radius
A	absorbance
A	peak asymmetry factor
B	magnetic field strength
$[c]$	concentration of component c
d_p	particle diameter (HPLC stationary phase)
D_{ab}	diffusion coefficient (of component a into component b)
e	electron elementary charge
ε	extinction coefficient
E	energy
E	electrode potential
E_b	binding energy
E_{ea}	electron affinity
E_i	ionization energy
ν	frequency
γ	gyromagnetic ratio
n	refractive index
h	Planck's constant
H	enthalpy, plate height
I_0	incident intensity
J	coupling constant
k	coverage factor
k'	capacity factor
λ	wavelength, thermal conductivity
m/z	mass-to-elementary charge ratio (mass spectrometry)
Q_{crit}	Q value (outlier test)
q	quadrupole parameter (mass spectrometry)
R	resolution, correlation coefficient
R_∞	Rydberg's constant
ρ	density
s	standard deviation
s^2	variance
δ	chemical shift
δ^*	solubility parameter
τ	true value of a measured quantity
τ_{crit}	Chauvenet's criterion (outlier test)
$t_{1/2}$	half-life

t_M	mobile-phase hold up
t_R	retention time
t_R^0	specific retention time
T	transmittance
T1	spin–lattice relaxation time
T2	spin–spin relaxation time
$\bar{\mu}$	carrier phase velocity
V_M	carrier hold-up volume
V_R	retention volume
V_R^0	specific retention

ANALYTICAL STANDARDIZATION, CALIBRATION, AND UNCERTAINTY

Most modern instrumental techniques used in analytical chemistry produce an output or signal that is not absolute; the signal or peak is not a direct quantitative measure of concentration or target analyte quantity. Thus, to perform quantitative analysis, one must convert the raw output from an instrument (information) into a quantity (knowledge). This is done by standardizing or calibrating the raw response from an instrument [1–10]. Here, we briefly summarize the most common methods applied in analytical chemistry, recognizing that this is a very large field. We note that the common use of the term "standardization" is not to be confused with the application of standard methods as specified by regulatory or consensus standard organizations.

REFERENCES

1. Chalmers, R.A., *Chapter 2: Standards and standardization in chemical analysis*, Vol. 3, Elsevier, Amsterdam, 1975.
2. Danzer, K. and Currie, L.A., Guidelines for calibration in analytical chemistry. Part I. Fundamentals and single component calibration (IUPAC Recommendations 1998), *Pure Appl. Chem.*, 70, 993, 1998.
3. Danzer, K., Otto, M., and Currie, L.A. Guidelines for calibration in analytical chemistry. Part 2. Multispecies calibration. (IUPAC Technical Report). *Pure Appl. Chem.*, 76, 1215, 2004.
4. Woodget, B.W. and Cooper, D., *Samples and Standards, Analytical Chemistry by Open Learning*, John Wiley and Sons, Chichester, 1987.
5. Gy, P. *Sampling for Analytical Purposes*, John Wiley and Sons, Chichester, 1998.
6. Vitt, J.E. and Engstrom, R.C., Effect of sample size on sampling error: an experiment for introductory analytical chemistry. *J. Chem. Edu.* 76, 99, 1999.
7. Horowitz, W. Nomenclature for sampling in analytical chemistry (Recommendations 1990), *Pure Appl. Chem.*, 62, 1193, 1990.
8. Grob, R.L. *Modern Practice of Gas Chromatography*, Wiley Interscience, New York, 1995.
9. Inczedy, J., Lengyel, T., Ure, A.M., Compendium of analytical nomenclature, *IUPAC*, 3rd. ed., 1997.
10. Bruno, T.J. *Calibration and Experimental Uncertainty, ASM Handbook, Vol 10, Materials Characterization*, ASM Handbook Committee, ASM International, Materials Park, 2019.

Samples

In all of the discussion to follow, we assume that the sample has been properly drawn from the parent population material and properly prepared. Clearly, the most precise analytical methods and the most painstaking calibration methods are useless if applied to a sample that does not represent reality. Nevertheless, the term "sampling," which describes the process of obtaining the sample (from the population material), implies the existence of a sampling uncertainty (arising mainly from population material heterogeneity) [5,6]. Thus, the analytical result is an estimate of what would be obtained from the parent population material. The theory, concepts, and nomenclature regarding samples and sampling constitute a complex, statistically based subspecialty of analytical chemistry well beyond the scope presented here. We begin with some simplified definitions [7]:

Amount of Substance: The amount of substance is the fundamental quantity of material measured in the number of moles.

Analyte: The analyte is the target component or compound in the sample for which one desires a measurement.

Aliquot: An aliquot is a known fraction of a homogeneous mass or volume.

Bias: As applied to sampling, bias refers to a systematic displacement, error, uncertainty, or mistake caused by a flaw in the sampling procedure. Determinate error and bias are related terms that describe uncertainty that arises from a fixed cause and that can, in principle, be eliminated if recognized. Determinate error (or systematic error) is most often associated with a measurement, whereas bias can be associated with either a measurement or the sampling procedure.

Convenience sample: A sample chosen on the basis of accessibility, expediency, cost, efficiency, or other reasons is not directly concerned with sampling parameters.

Determination: The determination is the entire analytical procedure or method performed on a test portion.

Matrix: The matrix is the background or carrier material of the sample that includes all components except the analyte(s) of interest.

Phase: The phase describes the physical state of a substance, primarily solid, liquid, and gas, but this term might include more detailed descriptions to include supercritical fluid and plasma. Note that the term "vapor" typically refers to a gas phase above and in equilibrium with a condensed phase (often called the headspace in chemical analysis).

Population Material: The population material is the entirety of the bulk material from which the sample is drawn. This might be a plot of earth, a warehouse full of sugar, or a tank of jet fuel.

Quantities: The quantity refers to the specific mass or volume of a substance.

Random Sample: A sample selected so that any portion of the sample has an equal or known chance of being chosen for measurement.

Representative Sample: A sample resulting from a sampling process that can be expected to adequately reflect the properties of interest of the parent population. A representative sample may be a random sample. The degree of representativeness of the sample is limited by cost or convenience.

Selective sample: A sample that is deliberately chosen by using a sampling plan that eliminates materials with certain characteristics or selects only material with other relevant characteristics desired for an analysis.

Test Portion: The test portion is the actual material removed from a sample for analysis.

Umpire sample: A sample taken, prepared, and stored in an agreed-upon manner for the purpose of settling a dispute, arrived at by agreement that will include the test method and procedure, serving as the basis for acceptance, rejection, or economic adjustment. This is sometimes called a referee sample.

Unknown: The unknown is a term that describes the target measurement or unknown quantity that is desired for the analyte or the analyte itself.

Sampling uncertainty is that part of the total uncertainty in an analytical procedure or determination that results from using only a fraction of the population material. In this respect, sampling by any method is an extrapolation process. Since the sampling uncertainty is usually ignored for an individual analysis on an individual test portion, the sampling uncertainty is considered as being due entirely to the variability of the test portion. It is therefore assessed, when necessary, by replication of the sampling from the parent population material and statistically isolating the uncertainty, thus introduced by analysis of variance. Typically, the problems associated with liquid population material are less complex but must not be ignored. Sample stratification, concentration and thermal gradients, poor mixing, and gradients associated with flow are all real effects that must be considered. Sampling uncertainty is often minimized by field and laboratory processing, with procedures that can include mixing, reduction, coning and quartering, riffling, milling, and grinding.

Another aspect that must be considered subsequent to sampling is sample preservation and handling. The integrity of the sample must be preserved during the inevitable delay between sampling and analysis. Sample preservation may include the addition of preservatives or buffer solutions, pH adjustment, use of an inert gas "blanket," and cold storage or freezing.

Calibration and Standardization

External Standard Methods

The external standard method can be applied to nearly all instrumental techniques, within the general limits discussed here, and the specific limitations that may be applicable with individual techniques. This method is conceptually simple; the user constructs a calibration curve of instrumental response with prepared mixtures containing the analyte(s) at a range of concentrations, an example of which is shown in Figure 1.1a. Thus, the curve represents the raw instrumental response of the analyte as a function of analyte concentration or amount. Each point on this plot must be measured several times so that the repeatability can be assessed. Repeatability in this context describes the precision under constant operating conditions over a short period of time. Only random uncertainty should be observed in the replicates; trends of increasing or decreasing response (hysteresis) must be remedied by identifying the source and adjusting the method accordingly. The calibration solutions should be randomized (that is, measured in random order). Despite being called a calibration "curve," ideally the signal vs. concentration plot is linear or substantially linear (that is, areas of nonlinearity are unimportant; otherwise, they are localized and minor, and properly treated by the measurement technique). In some cases, the response may be linearizable (for example, by calculating the logarithm of the raw response). If a curve shows nonlinearity in an area that is important for the analysis, one must measure more concentrations (data points) in the region of curvature.

In practice, the line that results from the calibration is fit with an appropriate program, and the desired value for the unknown concentration is calculated. The curve can be used graphically

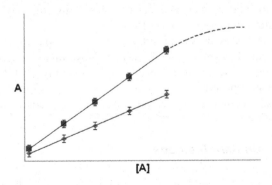

Figure 1.1a   An example calibration curve prepared by use of the external standard method. The instrument response is represented by A, and the concentration resulting in that response is [A]. While curves for two analytes are shown, in principle one can plot as many analytes as desired. While five points per analyte have been shown, one can measure as many as required. Note that a region of nonlinearity is shown in the latter part of the curve for one of the components. One would require a larger number of points to adequately represent and fit any nonlinear areas.

if approximation suffices. Mixtures prepared for external standard calibration can contain one or many analytes. Once a calibration curve is prepared, it can often be used for some time period, provided such a procedure has been previously validated (that is, the stability of the standards and the instrument over the time of use has been assessed). Otherwise, it is best to measure the unknown and the standards within a short period of time. Moreover, if any major change is made to the instrumentation (changing a detector or detector parameters, changing a chromatographic column, etc.), the standards must be re-measured.

To successfully use the method, the standard mixtures must be in a concentration range that is comparable to that of the unknown analyte, and ideally should bracket the unknown. Multiple measurements of each standard mixture should be made to establish repeatability of points on the curve. Many instrumental methods have operation ranges (frequency, temperature, etc.) in which the uncertainty is minimized, so components and concentrations for standard mixtures must respect this. The standard mixtures should be in the same matrix as the unknown, and the matrix must not interfere with the unknown or other standard mixture components. Any pretreatment of the unknown must be reflected in the standard mixtures. As with any calibration method, components in the standard mixtures must be available at a high (or at least known) purity, they must be stable during preparation and must be soluble in the required matrix. Unless the physical phenomenon of a measurement is well understood, extrapolation beyond the curve is not recommended (and indeed is usually strongly discouraged); nevertheless, extrapolation is occasionally done in practice. In those cases, one must be cautious, report exactly how the extrapolation was done, and assess any increase in uncertainty that may result. Note that the curve might not extrapolate through the origin. This is usually the result of adsorption (of components on container walls), carryover hysteresis, absorption (of components in seals or septa), or component degradation or evaporation.

A major consideration with external standardization is that typically, the sample size (for example, the injection volume in chromatography) must be maintained constant for standard mixtures and the unknowns. If the sample size varies slightly, it is often possible to apply a correction to the raw signal. One should not attempt to generate a calibration curve by varying the sample size (that is, for example, injecting increasing volumes into a gas chromatograph). This caution does not preclude serial dilution methods (see below), in which multiple solutions are generated for separate measurement. Other issues that can hinder successful application of the external standards method include instrumental aspects that might not be readily apparent. In chromatographic methods, for example, one can overload the column or detector. In older instruments, settings of signal attenuation were typically made manually, whereas in newer instruments, this may occur through software, sometimes without operator interaction or knowledge.

Note, *inter alia,* that in Figure 1.1a (and indeed all the examples presented here), uncertainty is only indicated for the variable on the y-axis. In reality, we must recognize that there is uncertainty for the values plotted on the x-axis as well, but we often only treat the largest uncertainty or the uncertainty that is most important for our application. Note, also, that it is critical to maintain the integrity of standards; decomposition, degradation, moisture uptake, etc. will adversely affect the validity of the calibration.

Abbreviated External Standard Methods

In many situations in chemical analysis, a full calibration curve is not prepared because of the complexity, time, or cost. In such situations, abbreviated external standard methods are often used. Under no circumstances can an abbreviated method be used if the raw signal response is nonlinear. Moreover, these methods are not generally appropriate for analyses in regulatory, forensic, or health care environments where the consequences can be far-reaching.

Single Standard

This method uses a simple proportion approach to standardize an instrument response. It can be used only when the system has no constant, determinate error, or bias,[1] and when the reagents used give a zero blank response (that is, the instrument response from the matrix and measurement system only, without the analyte). A standard should be prepared such that the concentration is close to that of the unknown. One then calculates the concentration of the unknown, [X], as

$$[X] = (A_x/A_s)[S], \tag{1}$$

where A_x is the instrument response of the unknown, A_s is the instrument response of the standard, and [S] is the concentration of the standard.

Single Standard Plus Assumed Zero

This method, illustrated schematically in Figure 1.1b, assumes that the blank reading will be zero. One uses a two-point calibration in which the origin is included as the first point. It is important to ensure, by experiment or experience, that such a method is adequate to the task.

Single Standard plus Blank

If the analytical method has no determinate error or bias, but does produce a finite blank value, then one must also perform a blank measurement, which is subtracted from the instrument response of the standard and the unknown. Then, the same procedure (Equation 1) is used for the single standard. If multiple samples are to be measured, it is important to measure the blank between each measurement.

Two Standards plus Blank

When the analytical method has both a determinate error (or bias) and a finite blank value, at least three calibrations must be made: two standards and one blank. The standard concentrations are

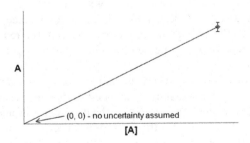

Figure 1.1b An example of a single point calibration curve. The instrument response is represented by A, and the concentration resulting in that response is [A]. The origin (0,0) is assigned as part of the curve and is assumed to have no uncertainty.

---

[1] Determinate error and bias are related terms that describe uncertainty that arises from a fixed cause, and that can, in principle, be eliminated if recognized. Determinate error (or systematic error) is most often associated with a measurement, while bias can be associated with either a measurement or with the sampling procedure.

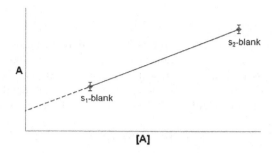

Figure 1.1c   An example of two standards plus a blank calibration curve. The blank is subtracted from each
of the standards. The instrument response is represented by A, and the concentration resulting
in that response is [A].

typically prepared widely spaced in concentration, and the higher concentration should be chosen to represent the limit of linearity of the instrument or method. If this is not practical, the higher concentration should simply be the highest expected concentration of the analyte (unknown). This method is illustrated schematically in Figure 1.1c. If multiple samples are to be measured, it is important to measure the blank between each measurement.

Internal Normalization Method

As mentioned above, the raw signal from an analytical instrument is typically not an absolute measure of concentration of the analyte(s), because the instrument may respond differently to each component. In some cases, such as with chromatographic methods, it is possible to apply response factors, determined from a standard mixture containing all constituents of the unknown sample, for standardization [8]. The standard mixture is gravimetrically prepared (with known mass percents for each component), and the instrument response is measured, for example, as chromatographic areas. The total mass percent and the total area percent each sum to 100. One calculates the ratio of each mass percent to each area, choosing *one* component as the reference, which is assigned a response factor of unity. To obtain the response factors of all the other components, one divides its (mass % to area ratio) with that of the reference. This is done for all components, producing a response factor for all components, except of course for the reference, defined as unity. When the unknown sample is measured, the response factor is multiplied by each raw area, and the resulting area percent provides the normalized mass percents of each component in the unknown.

This method corrects for minor variations in sample size (earlier defined as the test portion), although large differences in sample size must be avoided so that one is assured of consistent instrument performance. Although the method corrects for the different response of samples, large differences must be avoided. This also means that the detector must respond linearly to the concentrations of each component, even if the concentrations are very different. This may require dilution or concentration of the sample in some situations. In chromatographic applications, all components of a mixture must be analyzed and standardized, since normalization must be performed on the entire sample.

Some techniques, such as GC with flame ionization detection and thermal conductivity detection, have well-defined physical phenomena associated with output signals. With these techniques, there are some limited, published response factor data that can be used in an approximate way to standardize the response from these devices.

In Situ Standardization

While it is rare that an analytical method can be calibrated by use of a single solution, some instances of spectrophotometry and electroanalytical methods can qualify. To use this method, one sequentially and incrementally adds known masses of standard analytes to a solution, with an instrument response being measured after each addition. This procedure can only be used if the analytical method itself does not change the analyte concentration (nondestructive) and does not lead to a loss of solution volume. A solid crystalline analyte is an example. One must also minimize changes in solution volume over the course of the standardization.

Standard Addition Methods

Samples presented for analysis often are contained in complex matrices with many impurities that may interact with the analyte, potentially enhancing or diminishing a signal from an instrumental technique. In such cases, the preparation of an external standard calibration curve will be impossible, because it might be very difficult to reproduce the matrix. In these cases, the standard addition method may be used. A standard solution containing the target analyte is prepared and added to the sample, thus accounting for the unknown impurities and their effects. While the quantity of target analyte in the target sample is unknown, the added quantity is known, and its incremental additive effect on the instrument signal can be measured. Then, the quantity of the unknown analyte is determined by what is effectively an extrapolation. In practice, the volume of standard solution added is kept small to avoid dilution of the unknown impurities by no more than 1 % of the total signal. This method can only be used if there is a verified linear relationship between the signal and quantity of analyte. If a determinate error is present, then the slope of the line must be known. Moreover, the sample cannot contain any components that can respond as the analyte (that is, masquerade).

Single Standard Addition

In the simplest case, one addition of analyte is made after first measuring the response of the analyte in the unknown sample. Thus, two measurements are required:

$$A_{xo} = m[X_0] \tag{2}$$

$$A_{xi} = m([X_o] + [S]), \tag{3}$$

where A_{xo} is the instrument response of the analyte in the unknown sample, $[X_0]$ is the concentration in the unknown sample, and A_{xi} is the instrument response upon the addition of the standard, $[S]$ (additive in equation because X and S are the same compound). The assumed slope is the proportionality constant, m. The two equations are solved simultaneously for $[X_0]$. This technique is very rapid and economical, but there are serious drawbacks. There is no built-in check for mistakes on the part of the analyst, there is no means to average random uncertainties, and there is no way to detect interference (mentioned above as masquerade).

Multiple Standard Addition

This standard addition method alleviates some of the problems inherent in single standard addition. Here, the unknown sample is first measured in the instrument. Then, that sample is "spiked" with incrementally increasing concentrations of the analyte, generating a curve such as that shown in Figure 1.1d. The curve should extrapolate to zero signal at zero concentration. The concentration of the analyte in the unknown is read or calculated from the abscissa (x-axis).

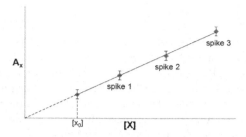

Figure 1.1d   An example of calibration by multiple standard addition. Three additions (spikes) of the analyte X are shown, as is the extrapolation to the unknown concentration, X_0.

Internal Standard Methods

An internal standard is a compound added to a sample at a known concentration, the purpose of which is to exhibit a similar signal when measured in an instrument, but be distinguishable from the signal of the desired analyte. It provides the highest level of reliability in quantitation by chromatographic methods and is not affected by large differences in sample size [8]. Unlike the internal normalization method, it is not necessary to elute or measure all the components of the sample, one needs focus only on the component(s) of interest. In atomic spectrometry, this method is not affected by changes in gas flow rates, sample aspirations rates, and flame suppression or enhancement. Another situation in which this method is valuable is when the sample matrix is either unknown or very complex, precluding the preparation of external standards.

Multiple Internal Standards

A set of calibration solutions is prepared by mass, containing the target analyte, X, and a standard that is not present in the unknown sample, A. The instrument response (for example, a chromatographic area) is measured for each calibration solution, and a plot is made to establish linearity as shown in Figure 1.1e. The ordinate axis is the ratio of the response of the unknown analyte component, A_x, to the response of the chosen standard, A_s. The abscissa is the ratio of mass of X to the mass of S for that standard mixture. Once the linearity is confirmed in the concentration range of interest, the unknown is spiked with a known mass of S, the instrument response is measured, and the area ratio A_x/A_s is calculated. Either the graph or a fit of the data on Figure 1.1e is then used to determine the corresponding mass fraction, from which the mass of X may be determined.

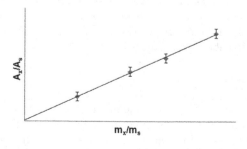

Figure 1.1e   An example of the multiple internal standard method. The ordinate (y) axis is the ratio of the response of the unknown analyte component, A_x, to the response of the chosen standard, A_s. The abscissa (x) axis is the ratio of mass of X to the mass of S for that standard mixture.

Note that the calculations could be simplified if the same mass per volume of internal standard is added to both the unknown samples and the calibration standards.

Single Internal Standard

In practice, once the linearity is established for a given mixture, it is no longer necessary to use multiple standards, although this is the most precise method. Subsequent to the verification of linearity, one standard solution can be used to fix the slope, provided it is close in concentration to that of the target analyte. In this case, the mass of the unknown can be found from:

$$X / S = (A_x / A_s)(1/R),$$ (4)

where X is the mass of the unknown analyte in the sample, S is the mass of the added internal standard in the sample, A_x and A_s are the instrument responses (areas) of the unknown and internal standard, respectively. R is a ratio determined from the standard solution prepared with both X and S: (mass, unknown analyte/mass, internal standard)/(signal, unknown analyte/signal, internal standard) = R.

$$\left(\frac{\left(\frac{\text{mass, unknown analyte}}{\text{mass, internal standard}} \right)}{\left(\frac{\text{signal, unknown analyte}}{\text{signal, internal standard}} \right)} \right) = R$$ (5)

Since R is the slope of the calibration curve discussed above, once linearity is established, one solution suffices. There are many conditions that must be fulfilled in order to use the internal standard method, and it is rare that all of them can actually be met. Indeed, in practice, one tries to meet as many as possible, but those that are mandatory are underlined. The compound chosen *must not be present* already in the unknown. The compound chosen *must be separable from the analyte* present in the unknown. An exception occurs when an isotopically labeled standard is used, in conjunction with mass discrimination or radioactive counting detection. In a chromatographic measurement, this is typically at least baseline resolution, although this would be a minimally acceptable degree of separation. On the other hand, the unknown analyte peak and the internal standard peak should be close to each other (temporally) on the chromatogram. The compound chosen *must be miscible* with the solvent at the temperature of reagent preparation and measurement. The compound chosen must not react chemically with the sample or solvent, or interfere in any way with the analysis. It is critical to maintain the integrity of standards; decomposition, degradation, moisture uptake, etc. will adversely affect the validity of the calibration. In the case of a chromatographic measurement, the same applies to interactions with the stationary phase. The compound chosen must be chemically similar (for example, in functionality, thermophysical properties) to the analyte. If such a compound is not available (for example, in a chromatographic measurement), an appropriate hydrocarbon should be chosen as a surrogate. The standard solution should be prepared at a similar concentration as in the unknown matrix; ratio correction of large differences is no substitute for an appropriate concentration. In a chromatographic measurement, the compound chosen must elute as closely as possible to the analyte and should not be the last peak to elute (the final peak often shows different geometry such as tailing). The compound chosen must be sufficiently nonvolatile to allow for storage as needed. When there is the potential for the unknown analyte to be lost by adsorption, absorption, or some other interaction with the matrix or container, a compound called a carrier is sometimes added in large excess. The carrier is similar, chemically and physically, to the unknown analyte, but easily separated from it. Its purpose is to saturate or season the matrix and prevent analyte loss.

Serial Dilution

Serial dilution is less a standardization method as it is a method of generating solutions to be used for standardizations. Nevertheless, its importance and utility, as well as the popularity of its application, warrants mention in this section. A serial dilution is the stepwise dilution of a substance, observant of a specified, constant progression, usually geometric (or logarithmic). One first prepares a known volume of stock solution of a known concentration, followed by withdrawing some small fraction of it to another container or vial. This subsequent container is then filled to the same volume as the stock solution with the same solvent or buffer. The process is then repeated for as many standard solutions as are desired. A ten-fold serial dilution could be 1 M, 0.1 M, 0.01 M, 0.001 M, etc. A ten-fold dilution for each step is called a logarithmic dilution or log-dilution, a 3.16-fold ($10^{0.5}$-fold) dilution is called a half-logarithmic dilution or half-log dilution, and a 1.78-fold ($10^{0.25}$-fold) dilution is called a quarter-logarithmic dilution or quarter-log dilution. In practice, the ten-fold dilution is the most common. The serial dilution procedure is not only used in chemical analysis but also in serological preparations in which cellular materials such as bacteria are diluted. A critical aspect of serial dilution is that the initial solution concentration must be prepared and determined with great care, since any mistake here will be propagated into all resulting solutions.

Traceability

Analytical measurements and certifications often contain a statement of traceability. Traceability describes the "result or measurement whereby it can be related to appropriate standards, generally international or national standards, through an unbroken chain of comparisons" [9]. It typically includes the application of reference materials (RM) or standard reference materials (SRMs) for instrument calibration before standardization for the analytes of interest. The **true value** of a measured quantity (τ) cannot typically be determined. The true value is defined as characterizing a quantity that is perfectly defined. It is an ideal value which could be arrived at only if all causes of measurement uncertainty were eliminated, and the entire population was sampled.

Uncertainty

As stated in Section 2, the result of a measurement is only an approximation or estimate of the true value of the measurand or quantity subject to measurement. In the determination of the combined standard uncertainty and ultimately, the expanded uncertainty, it is critical to include the uncertainty of calibration in the process, as discussed above. The process of arriving at the uncertainty U_y of a quantity y that is based upon measured quantities $x_i, \ldots, x_z$ is called the propagation of uncertainty. A full discussion of propagation of uncertainty is beyond the scope of this section; a simplified prescription, in the form of general and specific formulae, is provided here. In general, the propagated random uncertainty in y can be determined from the following:

$$U_y = \sqrt{\left(\frac{\partial y}{\partial x_i} U_{x_i} \right)^2 + \cdots \left(\frac{\partial y}{\partial x_z} U_{x_z} \right)^2} \tag{6}$$

This approach can be used when the uncertainties are random (not systematic), are relatively small, and are independent or uncorrelated (that is, in the absence of covariance). Relatively large uncertainties (such as those approaching the magnitude of the measurand itself) cannot be treated with this approach, especially if the measurand is a nonlinear function of the measured quantity. Note that by convention, U_y denotes the expanded uncertainty, which is the uncertainty multiplied

Table 1.1   Specific Formulae for the Propagation of Random, Independent Uncertainty

Measurand Argument	Arithmetic Uncertainty Formula
Y (where y is a counted random event over a time interval)	$U_y = \sqrt{y}$
$y = A \times x$ (where A is a constant with no uncertainty)	$U_y = |A| \times U_x$
$y = x_1 + x_2$	$U_y = \sqrt{U_{x_1}^2 + U_{x_2}^2}$
$y = x_1/x_2$ $y = x_1 \times x_2$	$\dfrac{U_y}{|y|} = \sqrt{\left(\dfrac{U_{x_1}}{x_1}\right)^2 + \left(\dfrac{U_{x_2}}{x_2}\right)^2}$
$y = (x_1 \times x_2)/x_3$	$\dfrac{U_y}{|y|} = \sqrt{\left(\dfrac{U_{x_1}}{x_1}\right)^2 + \left(\dfrac{U_{x_2}}{x_2}\right)^2 + \left(\dfrac{U_{x_3}}{x_3}\right)^2}$
$y = \log(x)$	$U_y = \left(\dfrac{1}{2.303}\right)\left(\dfrac{U_x}{x}\right)$
$y = \ln(x)$	$U_y = \dfrac{U_x}{x}$
$y = e^x$	$U_y = y \times U_x$
$y = x^a$	$\dfrac{U_y}{|y|} = |a| \times \left(\dfrac{U_x}{x}\right)$
$y = 10^x$	$\dfrac{U_y}{|y|} = 2.303 \times U_x$

by a coverage factor k in excess of unity (for the 95 % confidence level, the coverage factor is 2). A coverage factor k = 1 represents the 68 % confidence level. In scientific and technical reports and publications, the goal is to report measurements and the standard uncertainty (k = 1) or the expanded uncertainty (k > 1).

It is possible to reduce this general formulation to more specific formulae in the cases of common mathematical operations. These are provided in Table 1.1.

COMPLIANCE AGAINST LIMITS

It is often necessary or desirable to compare an analytical result against some established regulatory standard or limit. For example, it might be important to determine if the concentration of a toxic substance falls above or below a legal limit. It is imperative to consider the uncertainty when determining compliance against limits; indeed, most established limits are set with some consideration or allowance for uncertainty. A decision rule is often built into tests of compliance limits. A common decision rule is that a measured result indicates non-compliance if the measurand exceeds the limit by the expanded uncertainty. A similar approach is applied for measurands that fall below an established limit.

UNCERTAINTY AND ERROR RATES

A disconnect often occurs when scientific personnel attempt to convey their measured results in legal or regulatory venues. For example, the courts in the United States routinely must deal with scientific testimony based on measurements, and scientists are often asked to characterize their measurements in terms of error rates.

Statistically, the error rate is the frequency of type I and type II errors in null hypothesis significance testing. This has importance in forensic chemistry, for example, when the blood alcohol content (BAC) of a sample is measured. Here, a null hypothesis might be: the BAC of sample X is not below 0.08 % (mass/mass). A measurement above that level, and therefore a failure to reject the null hypothesis, can result in a legal finding of intoxication. A type I error occurs when a rejected null hypothesis is correct (false positive); a type II error occurs when the accepted null hypothesis is false (false negative). Independent of the frequency of type I and type II errors (the statistical error rate), each measurement of BAC has an uncertainty. The uncertainty of each measurement is determined by the propagation of the contributions to uncertainty that is represented by the uncertainty budget, multiplied by the appropriate coverage factor. The error rate of a particular laboratory or technique is not so easily determined. In some large state forensic laboratories, error rates can be approximated by inserting known standard samples anonymously into the normal workflow, but even this approach has limitations.

It is important to understand that the concept of error rate is distinct from the frequency at which an analytical instrument "throws an error." For example, in the headspace gas chromatographic analysis of BAC, the instrument might report a sampling error, and the operator might notice a damaged needle, which is then replaced. The frequency of this type of error is different from the error rate mentioned above.

FIGURES OF MERIT

In the interpretation of results from a particular technique, one often must evaluate the performance of the measurement on the basis of objective measures that are called figures of merit [1,2]. The following table provides a description of the more common figures of merit used in analytical chemistry. Even more important than the evaluation of performance is the validation of metrology and the assessment of validation. The same terms used as figures of merit are critical in the validation process. Since many standardization bodies will issue their own definitions regarding these terms, it is not practical to be inclusive; the reader must be aware that other terms may sometimes be substituted for those used here.

REFERENCES

1. General requirements for the competence of testing and calibration laboratories, ISO/IEC 17025, International Organization for Standardization, 2005.
2. General Chapter 1058 Analytical Instrument Qualification, USP 32 – NF 27, U.S. Pharmacopeia Convention, 2009.

Parameter or Metric	Definition	Comments
Accuracy	The deviation of a measured value from the true value (τ), which cannot typically be determined, but which is often approached by use of a standard method under standard conditions.	Can be expressed as accuracy $= (x_{av} - \tau)/\tau$, where x_{av} is the mean of a series of measurements.
Precision	The reproducibility of a series of replicate measurements, typically under the same operating conditions in a relatively short period of time.	High precision does not imply high accuracy; reproducibility and repeatability often are used interchangeably, but the former typically refers to precision between two instruments or laboratories.
Sensitivity	Typically, the ability to distinguish between small differences in concentration between samples at a desired confidence level.	Often specified by response to a de facto or consensus standard; a simple measure is the slope of a calibration curve adjusted for recovery in the sampling process.
Limit of detection (LOD)	The lowest measurable concentration of an analyte by use of a particular metrology at a desired confidence level.	Typically defined as a response that is some multiplicative factor of the noise level; specified SNRs of 2–4 are typical; as with sensitivity, must adjust for recovery in sample preparation processes.
Linear dynamic range (LDR)	The linear range of a calibration curve obtainable with a particular metrology and analyte.	Within the LDR, the difference in response for two concentrations of a given compound is proportional to the difference in concentration of the two samples.
Selectivity (α)	The ability to distinguish the signal of an analyte from the signal of interferences, the matrix, or degradants.	Typically expressed as a ratio, such as the ratio of capacity factors in chromatography, sometimes called specificity.
Analysis speed	The time required for sample preparation and measurement.	Sometimes necessary to include data processing time.
Throughput	The number of samples that can be run in a given time.	Related to ease of automation, below.
Ease of automation	The ability of a measurement to be performed without operator attention.	Related to throughput, above, since high throughput typically requires automation and minimal operator attention.
Ruggedness	The response of an instrument or technique to adverse conditions.	Adverse conditions include extremes in temperature, humidity, and dust; and rough handling, closely related to reproducibility, since it carries similar information content; distinguished from robustness, which is the ability of an instrument to remain unaffected by a small but intentional change in operating parameter, and still perform within specification.
Robustness	The capacity of an analytical device or method to be unaffected by small, deliberate changes in method parameters.	Robustness describes the reliability of the technique under normal usage; parameters that may be varied can include pH, flow rates, and temperatures, etc.
Portability	The ability to use an instrument in other than a fixed location or installation.	Can include the ability to use a device in the field in addition to the laboratory.
Environmental acceptability	The efficiency in terms of low waste generation and low power consumption.	Also described as sustainability and often focused on the minimization of use of hazardous substances.
Economics	The sum of costs required to operate the sample preparation, analysis, and data processing steps of a measurement.	Typically includes the costs of equipment, supplies, labor, utilities, and insurance; labor cost must consider the skill level required of personnel.
Stability	The property or attribute of a material to resist change or decomposition due to internal reaction or the action of air, heat, light, pressure, etc.	Includes consideration of samples, standards, stock solutions; short-term and long-term effects, and freeze–thaw cycle effects.

DETECTION OF OUTLIERS IN MEASUREMENTS

The field of outlier detection and treatment is considerable and a rigorous mathematical discussion is well beyond any treatment that is possible here. Moreover, the practice in the treatment of analytical results is usually simplified, since the number of observations is often not very large. The two most common methods used by analysts to detect outliers in measured data are versions of the Q-test [1–3] and Chauvenet's criterion [4], both of which assume that the data are sampled from a population that is normally distributed.

REFERENCES

1. Dean, R.B. and Dixon, W.J., Simplified statistics for small numbers of observations. *Anal. Chem.*, 23, 636–639, 1951.
2. Day, R.A. and Underwood, A.L., *Quantitative Analysis*, 6th ed., Prentice Hall, Englewood Cliffs, 1991.
3. Taylor, J.R., *An Introduction to Error Analysis*, 2nd ed., University Science Books, Sausolito, 1997.
4. Bruno, T.J. and Svoronos, P.D.N., *CRC Handbook of Basic Tables for Chemical Analysis*, 3rd ed., CRC Press, Boca Raton, FL, 2011.

The Q-Test

To perform the Q-test, one calculates the Q value given by

$$Q = Q_{gap}/R,$$

where Q_{gap} is the difference between the suspected outlier and the measured value closest to it, and R is the range of all the measured values in the data set. One then compares the calculated Q value with the critical Q values (Q_{crit}) in the following table:

Number of Observations	Q_{crit}, 90 % Confidence Level	Q_{crit}, 95 % Confidence Level	Q_{crit}, 99 % Confidence Level
3	0.941	0.970	0.994
4	0.765	0.829	0.926
5	0.642	0.710	0.821
6	0.560	0.625	0.740
7	0.507	0.568	0.680
8	0.468	0.526	0.634
9	0.437	0.493	0.598
10	0.412	0.466	0.568

If the calculated value of Q is greater than the appropriate value of Q_{crit}, then the value is a suspected outlier.

Chauvenet's Criterion

To perform Chauvenet's test on a set of measurements, one first must calculate the mean and standard deviation of the data. Then, one calculates

$$\tau = (x_i - x_{ave})/\sigma,$$

where x_i is the suspected outlier, x_{ave} is the average of all the measurements, and σ is the standard deviation. One then compares the calculated value of τ with τ_{crit} in the following table:

Number of Observations N	τ_{crit}
5	1.65
6	1.73
7	1.81
8	1.86
9	1.91
10	1.96
15	2.12
20	2.24
25	2.33
50	2.57
100	2.81
150	2.93
200	3.02
500	3.29
1,000	3.48

If the calculated value of τ is greater than the value of τ_{crit}, then the value is a suspected outlier. For numbers of observations between those given in the table, especially the larger numbers of observations, one may use the following chart to approximate the value of Chauvenet's τ_{crit}:

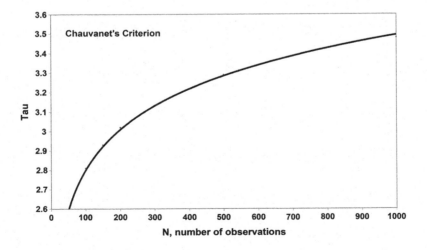

CHAPTER 2

Gas Chromatography

CARRIER GAS PROPERTIES

The following table gives the properties of common (and less common) gas chromatographic carrier and make-up gases [1]. These properties are those used most often in designing separation and optimizing detector performance. With few exceptions (indicated by an asterisk), all the properties provided in this table have been derived from the Helmholtz free energy equation of state and appropriate transport property models that are based upon the most reliable experimental measurements. Unless indicated, the pressure in each case was 101.325 kPa, or standard atmospheric pressure. For gases in which the heat capacity showed a very slight temperature dependence, only selected values are provided.

In addition to the pure gases, two mixtures are included. The hydrogen + helium mixture is sometimes used to obtain positive peaks when using the thermal conductivity detector (TCD) in the analysis of mixtures containing hydrogen. The argon + methane mixture is sometimes used with electron capture detection.

REFERENCE

1. Lemmon, E.W., Huber, M.L., McLinden, M.O., *REFPROP, Reference Fluid Thermodynamic and Transport Properties*, NIST Standard Reference Database 23, Version 9.1, National Institute of Standards and Technology, Gaithersburg, MD, 2013.

Carrier Gas	Relative Molecular Mass (RMM)	Density (kg/m³)	Thermal Conductivity (mW/m K)	Viscosity (µPa s)	Sound Speed (m/s)	Heat Capacity, C_p (kJ/kg K)	Heat Capacity Ratio, C_p/C_v
Hydrogen	2.016						
25°C		0.08235	184.88	8.9154	1,315.4	14.306	1.4054
100°C		0.0658	221.42	10.389	1,468.2		
200°C		0.0519	267.83	12.206	1,652		
Helium	4.003						
25°C		0.16353	155.31	19.846	1,016.4	5.193	1.6665
100°C		0.13068	181.41	23.154	1,137		
200°C		0.10307	213.93	27.294	1,280.2		
Nitrogen	28.016						
25°C		1.1452	25.835	17.805	352.07	1.0413	1.4013
100°C		0.91469	31.038	21.101	393.73		
200°C		0.72124	37.418	25.066	442.56		
Argon	39.94						
25°C		1.6339	17.746	22.624	321.67	0.52156	1.6696
100°C		1.3048	21.311	27.167	359.92		
200°C		1.0288	25.635	32.684	405.31		
Methane	16.04						
25°C		0.65688	33.931	11.067	448.47	2.2317	1.3062
100°C		0.52429	45.282	13.395	495.46	2.4433	1.2711
200°C		0.41324	63.134	16.214	548.7	2.7999	1.228
Ethane	30.07						
25°C		1.2385	20.984	9.3541	311.31	1.7572	1.1939
100°C		0.9857	31.643	11.485	344.31	2.0668	1.1576
200°C		0.7757	48.771	14.111	383.29	2.4866	1.1264

(Continued)

Carrier Gas	Relative Molecular Mass (RMM)	Density (kg/m³)	Thermal Conductivity (mW/m K)	Viscosity (µPa s)	Sound Speed (m/s)	Heat Capacity, C_p (kJ/kg K)	Heat Capacity Ratio, C_p/C_v
Ethene	28.05						
25°C		1.1533	20.326	10.318	329.92	1.5373	1.2461
100°C		0.9187	30.73	12.769	363.37	1.7981	1.2006
200°C		0.7234	45.734	15.758	403.21	2.1414	1.162
Propane	44.09						
25°C		1.832	18.31	8.1463	248.63	1.6847	1.1364
100°C		1.4516	27.362	10.129	276.92	2.0121	1.1074
200°C		1.1398	41.954	12.643	310.06	2.4404	1.0085
n-Butane	58.12						
25°C		2.4493	16.564	7.4054	210.49	1.7317	1.1053
100°C		1.9252	24.869	9.2534	236.8	2.0436	1.0807
200°C		1.5064	38.722	11.625	266.67	2.4596	1.0638
Oxygen	32						
25°C		1.3088	26.34	20.55	328.72	0.91963	1.3967
100°C		1.0452	32.061	24.507	366.63		
200°C		0.8241	39.206	29.283	410.47	0.96365	1.3701
Carbon dioxide	44.01						
25°C		1.808	16.643	14.932	268.62	0.85085	1.2941
100°C		1.4407	22.875	18.475	297.63		
200°C		1.1346	31.274	22.891	332.02	0.99708	1.2356
Carbon monoxide	28.01						
25°C		1.1453	26.478	17.649	352.03	1.0421	1.4013
100°C		0.9147	31.411	20.813	393.58		
200°C		0.7212	37.631	24.592	442.07	1.059	1.3904

(Continued)

Carrier Gas	Relative Molecular Mass (RMM)	Density (kg/m³)	Thermal Conductivity (mW/m K)	Viscosity (μPa s)	Sound Speed (m/s)	Heat Capacity, C_p (kJ/kg K)	Heat Capacity Ratio, C_p/C_v
Sulfur hexafluoride	146.05						
25 °C		6.0383	12.06 (0 °C)*	1.42 (0 °C)*	134.98	0.669	1.0984
100 °C		4.7954	14.06 (25 °C)*	1.61 (25 °C)*	150.81	0.7702	
200 °C		3.7698			169.55	1.8617	1.0718
Hydrogen in helium (8.5 % mol/mol**)							
25 °C		0.1566	162.65	19.159	1027.7	5.6002	1.6319
100 °C		0.1252	194.87	22.18	1149.2		
200 °C		0.0987	237.32	25.88	1293.8		
Methane in argon (5 % mass/mass)							
25 °C		1.5215	20.053	22.049	325.12	0.6055	1.5883
100 °C		1.2149	24.777	26.49	361.97		
200 °C		0.9582	31.121	31.882	404.4		

* Note for this mixture, the composition specification is provided on a molar basis, which is how the mixture is typically sold. The reader should be aware that the composition specification is also commonly expressed on a volume basis, making the ideal gas assumption. If there is ambiguity about a particular mixture, the user should discuss the specific composition specification with the supplier.

CARRIER GAS VISCOSITY

The following table provides the viscosity of common carrier gases, in $\mu Pa \cong s$, used in gas chromatography (GC) [1,2]. The values were obtained with a corresponding states approach with high-accuracy equations of state for each fluid. Carrier gas viscosity is an important consideration in efficiency and in the interpretation of flow rate data as a function of temperature. In these tables, the temperature, T, is presented in °C, and the pressure, P, is given in kilopascals and in pounds per square inch (absolute). To obtain the gauge pressure (that is, the pressure displayed on the instrument panel of a gas chromatograph), one must subtract the atmospheric pressure. Following the table, the data are presented graphically.

REFERENCES

1. Lemmon, E.W, Peskin, A. P., McLinden, M.O., and Friend, D.G., *Thermodynamic and Transport Properties of Pure Fluids* (NIST Standard Reference Database 12); Version 5.0; National Institute of Standards and Technology, Gaithersburg, 2000.
2. Lemmon, E.W., Huber, M.L, and McLinden, M.O., REFPROP, *Reference Fluid Thermodynamic and Transport Properties*, NIST Standard Reference Database 23, Version 9.1. National Institute of Standards and Technology, Gaithersburg, 2013.

P = 204.8 kPa, 29.7 psia

T (°C)	He	H_2	Ar	N_2	Air	Ar/CH$_4$ (90/10)	Ar/CH$_4$ (95/5)
0	18.699	8.3996	20.979	16.655	17.277	20.013	20.505
10	19.163	8.6088	21.625	17.129	17.775	20.625	21.134
20	19.621	8.8154	22.264	17.597	18.266	21.229	21.755
30	20.076	9.0197	22.894	18.058	18.75	21.826	22.369
40	20.527	9.2218	23.517	18.513	19.228	22.415	22.975
50	20.974	9.4216	24.133	18.962	19.699	22.998	23.574
60	21.418	9.6194	24.742	19.404	20.165	23.573	24.166
70	21.858	9.8152	25.344	19.842	20.624	24.142	24.751
80	22.294	10.009	25.939	20.273	21.078	24.705	25.329
90	22.727	10.201	26.527	20.7	21.526	25.261	25.901
100	23.157	10.391	27.109	21.121	21.969	25.811	26.467
110	23.583	10.58	27.685	21.538	22.407	26.355	27.027
120	24.007	10.767	28.255	21.949	22.84	26.893	27.581
130	24.427	10.952	28.819	22.357	23.268	27.426	28.129
140	24.845	11.136	29.378	22.759	23.691	27.953	28.671
150	25.26	11.318	29.931	23.157	24.11	28.474	29.209
160	25.672	11.498	30.479	23.552	24.524	28.991	29.74
170	26.082	11.678	31.021	23.942	24.934	29.502	30.267
180	26.489	11.856	31.558	24.328	25.34	30.008	30.788
190	26.894	12.033	32.09	24.71	25.742	30.51	31.305
200	27.296	12.208	32.618	25.089	26.14	31.006	31.817

(Continued)

P = 204.8 kPa, 29.7 psia (*Continued*)

T (°C)	He	H$_2$	Ar	N$_2$	Air	Ar/CH$_4$ (90/10)	Ar/CH$_4$ (95/5)
210	27.696	12.382	33.14	25.464	26.534	31.499	32.324
220	28.094	12.555	33.658	25.835	26.924	31.986	32.826
230	28.49	12.727	34.172	26.203	27.311	32.47	33.325
240	28.883	12.898	34.681	26.568	27.695	32.949	33.818
250	29.274	13.068	35.186	26.93	28.075	33.424	34.308
260	29.664	13.236	35.687	27.288	28.451	33.894	34.793
270	30.051	13.404	36.183	27.644	28.825	34.361	35.275
280	30.436	13.571	36.676	27.996	29.195	34.824	35.752
290	30.82	13.736	37.164	28.346	29.562	35.284	36.226
300	31.201	13.901	37.649	28.692	29.927	35.739	36.696

P = 308.2 kPa, 44.7 psia

T (°C)	He	H$_2$	Ar	N$_2$	Air	Ar/CH$_4$ (90/10)	Ar/CH$_4$ (95/5)
0	18.704	8.4024	21.001	16.672	17.296	20.033	20.527
10	19.167	8.6114	21.647	17.146	17.794	20.644	21.155
20	19.625	8.8179	22.285	17.613	18.284	21.248	21.775
30	20.08	9.0222	22.915	18.074	18.767	21.844	22.388
40	20.531	9.2241	23.537	18.528	19.244	22.433	22.993
50	20.978	9.4239	24.152	18.977	19.715	23.015	23.592
60	21.421	9.6217	24.76	19.419	20.18	23.59	24.183
70	21.861	9.8174	25.361	19.856	20.639	24.158	24.768
80	22.297	10.011	25.956	20.287	21.092	24.72	25.346
90	22.73	10.203	26.544	20.713	21.54	25.276	25.917
100	23.159	10.393	27.126	21.134	21.982	25.825	26.483
110	23.586	10.582	27.701	21.55	22.42	26.369	27.042
120	24.009	10.769	28.271	21.962	22.852	26.907	27.596
130	24.43	10.954	28.835	22.369	23.28	27.439	28.143
140	24.847	11.137	29.393	22.771	23.703	27.966	28.685
150	25.262	11.319	29.945	23.169	24.121	28.487	29.222
160	25.675	11.5	30.493	23.563	24.535	29.003	29.754
170	26.084	11.68	31.035	23.953	24.945	29.514	30.28
180	26.491	11.857	31.572	24.338	25.351	30.02	30.801
190	26.896	12.034	32.103	24.72	25.752	30.521	31.317
200	27.298	12.21	32.631	25.099	26.15	31.018	31.829
210	27.698	12.384	33.153	25.474	26.544	31.51	32.336
220	28.096	12.557	33.671	25.845	26.934	31.997	32.838
230	28.492	12.729	34.184	26.213	27.321	32.48	33.336
240	28.885	12.899	34.693	26.577	27.704	32.959	33.829
250	29.276	13.069	35.198	26.939	28.084	33.434	34.319
260	29.666	13.238	35.698	27.297	28.46	33.904	34.804
270	30.053	13.405	36.194	27.652	28.834	34.371	35.285
280	30.438	13.572	36.687	28.005	29.204	34.834	35.763
290	30.822	13.738	37.175	28.354	29.571	35.293	36.236
300	31.203	13.903	37.66	28.701	29.935	35.749	36.706

P = 446.1 kPa, 64.7 psia

T (°C)	He	H₂	Ar	N₂	Air	Ar/CH₄ (90/10)	Ar/CH₄ (95/5)
0	18.71	8.406	21.032	16.696	17.322	20.061	20.556
10	19.172	8.6149	21.676	17.169	17.818	20.671	21.183
20	19.63	8.8213	22.313	17.636	18.307	21.274	21.802
30	20.085	9.0254	22.942	18.096	18.79	21.869	22.414
40	20.535	9.2273	23.563	18.549	19.266	22.457	23.019
50	20.982	9.427	24.178	18.997	19.736	23.038	23.616
60	21.425	9.6246	24.785	19.439	20.2	23.612	24.207
70	21.865	9.8203	25.385	19.875	20.658	24.18	24.79
80	22.301	10.014	25.979	20.306	21.111	24.741	25.368
90	22.734	10.206	26.567	20.731	21.558	25.296	25.939
100	23.163	10.396	27.148	21.152	22	25.845	26.504
110	23.59	10.584	27.723	21.567	22.437	26.388	27.062
120	24.013	10.771	28.292	21.978	22.869	26.925	27.615
130	24.433	10.956	28.855	22.385	23.296	27.457	28.163
140	24.851	11.14	29.413	22.786	23.719	27.983	28.704
150	25.266	11.322	29.965	23.184	24.137	28.504	29.24
160	25.678	11.502	30.512	23.578	24.55	29.02	29.771
170	26.088	11.682	31.053	23.967	24.96	29.53	30.297
180	26.495	11.86	31.59	24.353	25.365	30.036	30.818
190	26.899	12.036	32.121	24.734	25.766	30.537	31.334
200	27.302	12.212	32.648	25.113	26.164	31.033	31.845
210	27.701	12.386	33.17	25.487	26.557	31.524	32.352
220	28.099	12.559	33.687	25.858	26.947	32.012	32.854
230	28.495	12.731	34.2	26.226	27.334	32.494	33.351
240	28.888	12.901	34.709	26.59	27.717	32.973	33.844
250	29.279	13.071	35.213	26.951	28.096	33.447	34.333
260	29.668	13.24	35.713	27.309	28.472	33.918	34.818
270	30.056	13.407	36.209	27.664	28.845	34.384	35.299
280	30.441	13.574	36.702	28.016	29.215	34.847	35.776
290	30.824	13.74	37.19	28.366	29.582	35.306	36.25
300	31.206	13.904	37.674	28.712	29.946	35.761	36.719

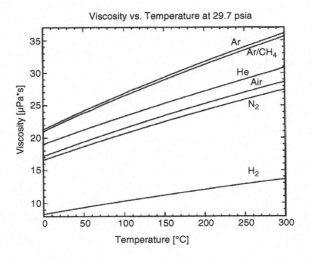

Viscosity vs. Temperature at 29.7 psia

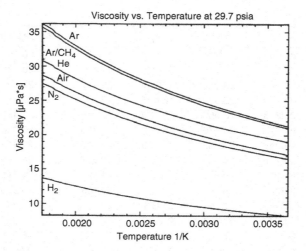

Viscosity vs. Temperature at 29.7 psia

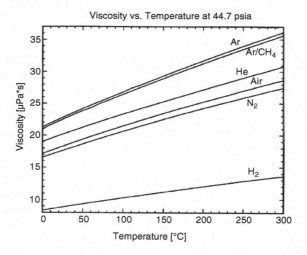

Viscosity vs. Temperature at 44.7 psia

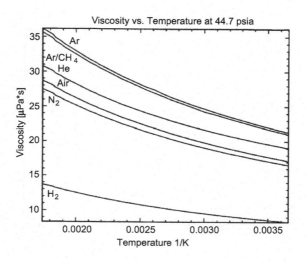

Viscosity vs. Temperature at 44.7 psia

GAS-CHROMATOGRAPHIC SUPPORT MATERIALS FOR PACKED COLUMNS

The following table lists the more common solid supports used in packed-column gas chromatography and preparative-scale gas chromatography, along with relevant properties [1–5]. The materials are also used in packed capillary columns. The performance of several of these materials can be improved significantly by acid washing and treatment with DMCS (dimethyldichlorosilane) to further deactivate the surface. The nonacid-washed materials can be treated with hexamethyl-disilane to deactivate the surface; however, the deactivation is not as great as that obtained by an acid wash followed by DMCS treatment. Most of the materials are available in several particle size ranges. The use of standard sieves will help insure reproducible sized packings from one column to the next. Data are provided for the Chromosorb family of supports since they are among the most well characterized. It should be noted that other supports are available to the chromatographer, with a similar range of properties provided by the Chromosorb series.

REFERENCES

1. Poole, C.F. and Schuette, S.A., *Contemporary Practice of Chromatography*, Elsevier, Amsterdam, 1984.
2. Gordon, A.J. and Ford, R.A., *The Chemist's Companion*, John Wiley & Sons, New York, 1972.
3. Heftmann, E., ed., *Chromatography: A Laboratory Handbook of Chromatographic and Electrophoretic Methods*, 3rd ed., Van Nostrand Reinhold, New York, 1975.
4. Poole, C.F., *The Essence of Chromatography*, Elsevier, Amsterdam, 2003.
5. Grant, D.W., *Gas-Liquid Chromatography*, Van Nostrand Reinhold, London, 1971.

Support Name	Support Type	Density (Free Fall) g/mL	Density (Packed) g/mL	pH	Surface Area (m²/g)	Maximum Liquid Loading	Color	Notes
Chromosorb A	Diatomite	0.40	0.48	7.1	2.7	25 %	Pink	Most useful for preparative GC; high strength; high liquid-phase capacity; low surface activity.
Chromosorb G	Diatomite	0.47	0.58	8.5	0.5	5 %	Oyster white	High mechanical strength; low surface activity; high density.
Chromosorb P	Diatomite firebrick	0.38	0.47	6.5	4.0	30 %	pink	High mechanical strength; high liquid capacity; moderate surface activity; for separations of moderately polar compounds.
Chromosorb W	Diatomite	0.18	0.24	8.5	1.0	15 %	White	Lower mechanical strength than pink supports; very low surface activity; for polar compound separation.
Chromosorb 750	Diatomite	0.33	0.49		0.75	7 %	White	Highly inert surface; useful for biomedical and pesticide analysis; mechanical strength similar to Chromosorb G.
Chromosorb R-4670-1	Diatomite				5–6	Low	White	Ultra-fine particle size used to coat inside walls of capillary columns; typical particle size is 1–4 μm.
Chromosorb T*	Polytetra-fluoro-ethylene	0.42	0.49		7.5	5 %	White	Maximum temperature of 240 °C; handling is difficult due to static charge; tends to deform when compressed; useful for analysis of high-polarity compounds.
Kel-F*	Chloro-fluoro-carbon				2.2	20 %	White	Hard, granular chlorofluorocarbon; mechanically similar to Chromosorbs; generally gives poor efficiency; use below 160 °C, very rarely used.
Fluoropak-80*	Fluoro-carbon resin				1.3	5 %	White	Granular fluorocarbon with sponge-like structure; low liquid phase capacity; use below 275 °C.
Teflon-6*	Polytetra-fluoro-ethylene				10.5	20 %	White	Usually 40–60 (US) mesh size; for relatively nonpolar liquid phases; low mechanical strength; high inert surface; difficult to handle due to static charge; difficult to obtain good coating of polar phases due to highly inert surface.
T-Port-F*	Polytetra-fluoro-ethylene	0.5					White	Use below 150 °C.
Porasil (types A through F)	Silica				2–500, type dependent	40 %	White	Rigid, porous silica bead; controlled pore size varies from 10–150 mm; highly inert; also used as a solid adsorbent.

* The fluorocarbon supports can be difficult to handle since they develop electrostatic charge easily. This is especially problematic in dry climates. It is generally advisable to work with them below 19 °C (solid transition point), using polyethylene laboratory ware.

MESH SIZES AND PARTICLE DIAMETERS

The following tables give the relationship between particle size diameter (in μm) and several standard sieve sizes. The standards are as follows:

United States Standard Sieve Series, ASTM E-11-61
Canadian Standard Sieve Series, 8-GP-16
British Standards Institution, London, BS-410-62
Japanese Standard Specification, JI S-Z-8801
French Standard, AFNOR X-11-501
German Standard, DIN-4188

Particle Size (μm)	US Sieve Size	Tyler Mesh Size	British Sieve Size	Japanese Sieve Size	Canadian Sieve Size
4,000	5	–	–	–	–
2,000	10	9	8	9.2	8
1,680	12	10	–	–	–
1,420	14	12	–	–	–
1,190	16	14	–	–	–
1,000	18	16	–	–	–
841	20	20	18	20	18
707	25	24	–	–	–
595	30	28	25	28	25
500	35	32	–	–	–
420	40	35	36	36	36
354	45	42	–	–	–
297	50	48	52	52	52
250	60	60	60	55	60
210	70	65	72	65	72
177	80	80	85	80	85
149	100	100	100	100	100
125	120	115	120	120	120
105	140	150	150	145	150
88	170	170	170	170	170
74	200	200	200	200	200
63	230	250	240	250	240
53	270	–	300	280	300
44	325	–	350	325	350
37	400	–	–	–	–

French and German Sieve Sizes

Particle Size (µm)	Sieve Size
2,000	34
800	30
500	28
400	27
315	26
250	25
200	24
160	23
125	22
100	21
80	20
63	19
50	18
40	17

Mesh-Size Relationships

Mesh Range	Top Screen Opening (µm)	Bottom Screen Opening (µm)	Micron Screen (µm)	Range Ratio
10/20	2,000	841	1,159	2.38
10/30	2,000	595	1,405	3.36
20/30	841	595	246	1.41
30/40	595	420	175	1.41
35/80	500	177	323	2.82
45/60	354	250	104	1.41
60/70	250	210	40	1.19
60/80	250	177	73	1.41
60/100	250	149	101	1.68
70/80	210	177	33	1.19
80/100	177	149	28	1.19
100/120	149	125	24	1.19
100/140	149	105	44	1.42
120/140	125	105	20	1.19
140/170	105	88	17	1.19
170/200	88	74	14	1.19
200/230	74	63	11	1.17
230/270	63	53	10	1.19
270/325	53	44	9	1.20
325/400	44	37	7	1.19

PACKED-COLUMN SUPPORT MODIFIERS

During the analysis of strongly acidic or basic compounds, peak tailing is almost always a problem, especially when using packed columns. Pretreatment of support materials, such as acid washing and treatment with DMCS, will usually result in only modest improvement in performance. A number of modifiers can be added to the stationary phase (in small amounts, 1 %–3 %) in certain situations to achieve a reduction in peak tailing. The following table provides several such reagents [1,2]. It must be remembered that the principal liquid phase must be compatible with any modifier being considered. Thus, the use of potassium hydroxide with polyester or polysiloxane phases would be inadvisable since this reagent can catalyze the depolymerization of the stationary phase. It should also be noted that the use of a tail-reducing modifier may lower the maximum working temperature of a particular stationary phase.

REFERENCES

1. Poole, C.F. and Schuette, S.A., *Contemporary Practice of Chromatography*, Elsevier, Amsterdam, 1984.
2. Poole, C.F., *The Essence of Chromatography*, Elsevier, Amsterdam, 2003.

Compound Class	Modifier Reagents	Notes
Acids	Phosphoric acid, FFAP (Carbowax-20M-terephthalic acid ester), trimer acid	These modifiers will act as subtractive agents for basic components in the sample; FFAP will selectively abstract aldehydes; phosphoric acid may convert amides to the nitrile (of the same carbon number), desulfonate sulfur compounds, and may esterify or dehydrate alcohols.
Bases	Potassium hydroxide, polyethyleneimine, polypropyleneimine, N,N'-bis-l-methylheptyl-p-phenylenediamine, sodium metanilate, THEED (tetrahydroxyethylenediamine)	These modifiers will act as subtractive agents for acidic components in the sample; polypropyleneimine will selectively abstract aldehydes, polyethyleneimine will abstract ketones.

PROPERTIES OF CHROMATOGRAPHIC COLUMN MATERIALS

The following table provides physical, mechanical, electrical, and (where appropriate) optical properties of materials commonly used as chromatographic column tubing [1–6]. The data will aid the user in choosing the appropriate tubing material for a given application. The mechanical properties are measured at ambient temperature unless otherwise specified. The chemical incompatibilities cited are usually only important when dealing with high concentrations which are normally not encountered in GC. Caution is urged nevertheless.

REFERENCES

1. Materials Engineering - Materials Selector, Penton/IPC, Cleveland, 1986.
2. Khol, R., ed., *Machine Design (Materials Reference Issue)*, 58(8), 1986.
3. Polar, J.P., *A Guide to Corrosion Resistance*, Climax Molybdenum Co., Greenwich, 1981.
4. Fontana, M.G. and Green, N.D., *Corrosion Engineering*, McGraw-Hill Book Co., New York, 1967.
5. Shand, E.B., *Glass Engineering Handbook*, McGraw-Hill Book Co., New York, 1958.
6. Fuller, A., *Corning Glass Works*, Science Products Division, Corning, NY, Private Communication, 1988.

Aluminum (Alloy 3003)

Density	2.74 g/mL
Hardness (Brinell)	28–55
Melting range	643.3°C –654.4°C
Coefficient of expansion (20°C –100°C)	2.32×10^{-5} °C^{-1}
Thermal conductivity (20°C, annealed)	193.14 W/(m·K)
Specific heat (100°C)	921.1 J/(kg·K)
Tensile strength (hard)	152 MPa
Tensile strength (annealed)	110 MPa

Notes: Soft and easily formed into coils; high thermal conduction; and incompatible with strong bases, nitrates, nitrites, carbon disulfide, and diborane.
Actual alloy composition: Mn = 1.5 %; Cu = 0.05 %–0.20 %; balance is Al.

Copper (Alloy C12200)[a]

Density	8.94 g/mL
Hardness (Rockwell-f)	40–45
Melting point	1,082.8°C
Coefficient of expansion (20°C –300°C)	1.76×10^{-5} °C^{-1}
Thermal conductivity (20°C)	339.22 W/(m·K)
Specific heat (20°C)	385.11 J/(kg·K)
Tensile strength (hard)	379 MPa
Tensile Strength (annealed)	228 MPa
Elongation (in 0.0508 m annealed)%	45

Notes: Copper columns often cause adsorption problems and incompatible with amines, anilines, acetylenes, terpenes, steroids, and strong bases.
[a] High-purity phosphorus deoxidized copper.

Borosilicate Glass

Density	2.24 g/rnL
Hardness (Knh 100)	418
Young's modulus (25 °C)	62 GPA
Poisson's ratio (25 °C)	0.20
Softening point	806.9 °C
Annealing point	565 °C
Melting point	1,600 °C
Strain point	520 °C
Coefficient of expansion (av.)	$3 \times 10^{-6} °C^{-1}$
Thermal conductivity	1.26 W/(m·K)
Specific heat	710 J/(kg·K)
Refractive index*	1.473
Normal service temperature (annealed)	215 °C
Extreme service temperature (annealed)	476 °C
Critical surface tension	750 mN/m

Notes: Has been used for both packed columns and capillary columns and incompatible with
fluorine, oxygen difluoride, and chlorine trifluoride.
* Clear grade, at 588 mm.

Fused Silica (SiO$_2$)

Density	2.15 g/mL
Hardness (Moh)	6
Young's Modulus (25 °C)	72 GPa
Poisson's ratio (25 °C)	0.14
Softening point	1,590 °C
Annealing point	1,105 °C
Melting point	1,704 °C
Strain point	1,000 °C
Coefficient of expansion (av.)	$5 \times 10^{-7} °C^{-1}$
Thermal conductivity	1.5 W/(m·K)
Specific heat	1,000 J/(kg·K)
Refractive index (588 nm)	1.458
Normal service temperature (annealed)	886 °C
Extreme service temperature (annealed)	1,086 °C
Critical surface tension	760 mN/m

Notes: Used for capillary columns; typical inside diameters range from 5 to 530 µm; coated on
outside surface by polyimide or aluminum to prevent surface damage; and incompatible with fluorine, oxygen difluoride, chlorine trifluoride, and hydrogen fluoride.

Nickel (Monel R-405)

Density	8.83 g/mL
Hardness (Brinell, 21 °C)	110–245
Melting range	1,298 °C –1,348 °C
Coefficient of expansion (21 °C –537 °C)	$1.64 \times 10^{-5} °C^{-1}$
Thermal conductivity (21 °C)	21.81 W/(m·K)
Specific heat (21 °C)	427.05 J/(kg·K)
Tensile strength (hard)	483 MPa
Tensile strength (annealed)	793 MPa
Elongation (in 2 inches, 21.1 °C)	15 %–50 %

Notes: Provides excellent corrosion resistance and no major chemical incompatibilities.
Actual alloy composition: Ni = 66 %; Cu = 31.5 %; Fe = 1.35 %, C = 0.12 %; Mn = 0.9 %;
S = 0.005 %; and Si = 0.15 %.

Polytetrafluorethylene (Teflon)

Specific gravity	2.13–2.24
Hardness (Rockwell-d)	52–65
Melting range	Decomposes
Coefficient of expansion	1.43×10^{-4} °C^{-1}
Thermal conductivity (21 °C)	2.91 W/m·K
Specific heat (21 °C)	1,046.7 J/kg·K
Tensile strength	17–45 MPa
Refractive index[a]	1.35

Notes: Flexible and easy to use; cannot be used above 230 °C; thermal decomposition products are toxic; and tends to adsorb many compounds which may increase tailing. No major chemical incompatibilities.

[a] Using sodium-D line, as per ASTM standard test D542-50.

Stainless Steel (304)

Density	7.71 g/mL
Hardness (Rockwell B)	149
Melting range	1,398.9 °C –1,421.1 °C
Coefficient of expansion (0 °C –100 °C)	1.73×10^{-5} °C^{-1}
Thermal conductivity (0 °C)	16.27 W/(m·K)
Specific heat (0 °C –100 °C)	502.42 J/(kg·K)
Tensile strength (hard)	758 MPa
Tensile strength (annealed)	586 MPa
Elongation (in 2 in)	60 %

Notes: Good corrosion resistance; easily brazed using silver-bearing alloys; and high nickel content may catalyze some reactions at elevated temperatures. No major chemical incompatibilities.
Actual alloy composition: C = 0.08 %; Mn = 2 % (max); Si = 1 % (max); P = 0.045 % (max); S = 0.030 (max); Cr = 18 %–20 %; Ni = 8 %–12 %; and balance is Fe. The low carbon alloy, 304L, is similar except for C = 0.03 % max and is more suitable for applications involving welding operations and where high concentrations of hydrogen are used.

Stainless Steel (316)

Density	7.71 g/mL
Hardness (Rockwell B)	149
Melting range	1,371 °C –1,398 °C
Coefficient of expansion (0 °C –100 °C)	7.17×10^{-5} °C^{-1}
Thermal conductivity (0 °C)	16.27 W/(m·K)
Specific heat (0 °C –100 °C)	502.42 J/(kg·K)
Tensile strength (annealed)	552 MPa
Elongation (in 2 inches)	60 %

Notes: Best corrosion resistance of any standard stainless steel, including the 304 varieties, especially in reducing and high-temperature environments.
Actual alloy composition: C = 0.08 % (max), Mn = 2 % (max); Si = 1 % (max); P = 0.045 % (max); S = 0.030 (max); Cr = 16 %–18 %; Ni = 10 %–14 %; Mo = 2 %–3 %; and balance is Fe. The low carbon alloy, 316L, is similar except for C = 0.03 % max and is more suitable for applications involving welding operations and where high concentrations of hydrogen are used.

PROPERTIES OF SOME LIQUID PHASES FOR PACKED COLUMNS

The following table lists some of the more common gas-chromatographic liquid phases that have been used historically, along with some relevant data and notes [1–3]. Most of these phases have been superseded by silicone phases used in capillary columns, but these liquid phases still find application in some instances. This is especially true with work involving established protocols, such as ASTM[1] or Association of Official Analytical Chemists (AOAC) methods. Moreover, the data are still useful in interpreting analytical results in the literature. The minimum temperatures, where reported, indicate the point at which some of the phases approach solidification or when the viscosity increases to the extent that performance is adversely affected. The maximum working temperatures are determined by vapor pressure (liquid-phase bleeding) and chemical stability considerations. The liquid phases are listed by their most commonly used names. Where appropriate, chemical names or common generic names are provided in the notes.

The McReynolds constants (a modification of the Rohrschneider constant) tabulated here are based on the retention characteristics of the following test probe samples:

Constant	Test Probe
X	Benzene
Y	1-Butanol
Z	3-Pentanone
U	1-Nitropropane
S	Pyridine

Compounds which are chemically similar to these probe solutes will show similar retention characteristics. Thus, benzene can be thought of as representing lower aromatic or olefinic compounds. Higher values of the McReynolds constant usually indicate a longer retention time (higher retention volume) for a compound represented by that constant, for a given liquid (stationary) phase.

Solvents
Ace—acetone
Chlor—chloroform
Pent—n-pentane
DMP—dimethylpentane
EAC—ethyl acetate

MeCl— methylene chloride
Tol—toluene
MeOH—methanol
H_2O—water

Polarity
N—nonpolar
P—polar
I—intermediate polarity
HB—hydrogen bonding
S—specific interaction

REFERENCES

1. McReynolds, W.O., Characterization of some liquid-phases, *J. Chromatogr. Sci.*, 8, 685, 1970.
2. McNair, H.M. and Bonelli, E.J., *Basic Gas Chromatography*, Varian Aerograph, Palo Alto, 1968.
3. Heftmann, E., *Chromatography: A Laboratory Handbook of Chromatographic and Electrophoretic Methods*, 3rd. ed., Van Nostrand Reinhold Co., New York, 1975.

[1] ASTM standards and protocols are consensus standards that are extensively reviewed and balloted among committee members. Historically, this was under the purview of the American Society for Testing and Materials (ASTM), but since 2001 the organization is called ASTM International.

Liquid Phase	T_{min} (°C)	T_{max} (°C)	Polarity	Solvents	McReynolds Constant					Notes
					X	Y	Z	U	S	
Acetonyl acetone (2,5-hexanedione)	−4	25	I	Ace						
Acetyl tributyl citrate	25	180	I	Ace	135	268	202	314	233	
Adiponitrile	5	50	I	Chlor MeCl						1,4-Dicyanobutane
Alka terge-T, amine surfactant	59	75	I	Chlor MeCl MeOH						60 % oxazoline, weakly cationic
Amine 220	0	180	P	Chlor MeCl	117	380	181	293	133	2-(8-Heptadecenyl)-2-imidazoline-ethanol
Ansul ether		80	P	MeOH						Tetraethylene glycol dimethyl ether, used for hydrocarbons
Apiezon H	50	275	N	Chlor	59	56	81	151	129	Low vapor pressure hydrocarbon oil
Apiezon J	50	300	N	Chlor MeCl	38	36	27	49	57	Low vapor pressure hydrocarbon oil
Apiezon L	50	300	N	Chlor MeCl	32	22	15	32	42	Low vapor pressure hydrocarbon oil
Apiezon M	50	275	N	Chlor MeCl	31	22	15	30	40	Low vapor pressure hydrocarbon oil
Apiezon N	50	300	N	Chlor MeCl	38	40	28	52	58	Low vapor pressure hydrocarbon oil
Apiezon K	50	300+	N	Chlor						Low vapor pressure hydrocarbon oil
Apiezon W	50	275	N	Chlor	82	135	99	155	154	Low vapor pressure hydrocarbon oil
Apolane-87	30	280	N	Tol	21	10	3	12	35	24,24-Diethyl-19,29-dioctadecyl heptatetracontane, C-87 hydrocarbon
Armeen SD		100	P, S, HB	Chlor MeCl						Primary aliphatic amine
Armeen 12D		100	P, HB	Chlor MeCl						
Armeen		125	P, HB	Tol						Secondary aliphatic amine
Armeen 2HT		100	P, HB	Chlor						
Arneel DD		100	P	MeOH						Aliphatic nitrile
Arochlor 1242		125		Ace						Chlorinated polyphenyl, used for gases, may be carcinogenic
Asphalt		300	N	Chlor MeCl						Complex mixture of aliphatic, aromatic, and heterocyclic compounds
Atpet 80			I	Chlor						Sorbitan partial fatty acid esters
p,p-Azoxydiphenetol	130	140	I	Chlor						
Baymal		300		Tol						Colloidal alumina

(Continued)

Liquid Phase	T_{min} (°C)	T_{max} (°C)	Polarity	Solvents	McReynolds Constant					Notes
					X	Y	Z	U	S	
Beeswax	20	200		Chlor	43	110	61	88	122	For essential oils
Bentone-34		200	S	Tol						Dimethyl dioctadecylammonium bentonite
7,8-Benzoquinoline		150	I	Chlor						For hydrocarbons, aromatics, heterocycles, and sulfur compounds
Benzylamine adipate		125		Chlor						
Benzyl cellosolve		50	I	Ace						2-(benzyloxy ethanol), for hydrocarbons
Benzyl cyanide		50	I	MeOH						Phenyl acetonitrile
Benzyl cyanide-AgNO$_3$		25	S	MeCl						
Benzyl diphenyl		100	I	Ace						
Benzyl ether		50	I	Chlor MeCl						Dibenzyl ether
Bis (2-butoxyethyl) phthalate		175	I	MeOH	151	282	227	338	267	
Bis (2-ethoxyethyl) phthalate					214	375	305	446	364	
Bis (2-ethoxyethyl) sebacate					151	306	211	320	274	
N,N-bis (2-cyano-ethyl formamide)	0	125	I	MeOH	690	991	853	110	000	
Bis (2-ethoxyethyl) adipate	0	150	I	Ace						
Bis (2-methoxyethyl) adipate	20	150	I	Ace Chlor						
Bis (2-ethylhexyl) tetrachlorophthalate	0	150	I	Chlor MeCl	112	150	123	108	181	
Butanediol adipate	60	225	I,P	Chlor MeCl						
Butanediol 1,4-succinate		225	I,P	Chlor	370	571	488	651	611	(BDS) Craig polyester, for alcohols, aromatics, heterocycles, fatty acids and esters, and hydrocarbons
Bis[2-(2-methoxy-ethoxy) ethyl] ether		50	I	Chlor						Tetraethylene glycol dimethyl ether
Carbitol		100	P	Ace						Glycol ether (mol. mass 134) for aldehydes and ketones

(Continued)

Liquid Phase	T$_{min}$ (°C)	T$_{max}$ (°C)	Polarity	Solvents	McReynolds Constant					Notes
					X	Y	Z	U	S	
Carbowax 300	10	100	P	MeCl						Polyethylene glycol, av. mol. mass <380
Carbowax 400	10	125	P	MeCl	333	653	405			Polyethylene glycol, av. mol. mass 380–420
Carbowax 400 monooleate	10	125	P	MeCl						
Carbowax 550	20	125	P	MeCl						
Carbowax 600	30	125	P	MeCl	323	583	382			Polyethylene glycol, av. mol. mass 570–630
Carbowax 600 monostearate		125	P	MeCl						
Carbowax 750	25	150	P	MeCl						Methoxy polyethylene glycol, av. mol. mass 715–785
Carbowax 1000	40	175	P	MeCl	347	607	418	626	589	Polyethylene glycol, av. mol. mass 950–1,050
Carbowax 1500 (or Carbowax 540)	40	200	P	MeCl						Polyethylene glycol, av. mol. mass 500–600
Carbowax 1540	40	200	P	MeCl	371	639	453	666	641	Polyethylene glycol, av. mol. mass 1,300–1,600
Carbowax 4000 (or 3350)	60	200	P	MeCl	317	545	378	578	521	Polyethylene glycol, av. mol. mass 3,000–3,700
Carbowax 4000 TPA		175	P	MeCl MeOH						Terminated with terephthalic acid
Carbowax 4000 monostearate	60	220	P	MeCl	282	496	331	517	467	
Carbowax 6000	60	200	P	MeCl	322	540	369	577	512	Polyethylene glycol, av. mol. mass 6,000–7,500
Carbowax 8000	60	120	P	Chlor	322	540	369	577	512	Polyethylene glycol, av. mol. mass 7,000–8,500
Carbowax 20M	60	250	P	MeCl	322	536	368	572	510	Polyethylene glycol, av. mol. mass 15,000–20,000
Carbowax 20M-TPA	60	250	P	MeCl	321	537	367	573	520	Terminated with terephthalic acid
Castorwax	90	200	P	MeCl	108	265	175	229	246	Triglyceride of 12-hydroxysteric acid (hydrogenated castor oil)
Citroflex A-4		150	I	MeOH	135	286	213	324	262	Tributyl citrate
Chlorowax 70		130	P	MeCl						Chlorinated paraffin, 70 % (wt/wt) Cl; for hydrocarbons
1-Chloronaphthalene		75	I	Tol						
Cyanoethyl sucrose	20	175	P	Ace	647	919	043	976		Vitrifies at –10 °C
Cyclodextrin acetate		250	I	Ace						For fatty acids and esters
Cyclohexane dimethanol succinate	100	210	I	Chlor	269	446	328	498	481	
n-Decane		30	N	MeCl						For inorganic and organometallic compounds

(Continued)

Liquid Phase	T_min (°C)	T_max (°C)	Polarity	Solvents	McReynolds Constant					Notes
					X	Y	Z	U	S	
Di(ethoxyethoxy-ethyl) phthalate					233	408	317	470	389	
Di(butoxyethyl) adipate	-10	150	P	Ace	137	278	198	300	235	
Di(butoxyethyl) phthalate	-30	200	P	Tol	157	292	233	348	272	
Di-n-butyl cyanamide		50		MeOH						For gases
Di-n-butyl maleate	0	50	P,I	Tol						For halogenated compounds
Di-n-butyl phthalate	-20	100	I	Tol						For aldehydes, ketones, halogenated compounds, hydrocarbons, and phosphorus compounds
Dibutyltetrachlorophthalate	0	150	I	Tol						
Didecyl phthalate	20	150	I	Tol	136	255	213	320	235	
Dicyclohexyl phthalate					146	257	206	316	245	
Diethylene glycol adipate	0	200	I	MeCl	378	603	460	665	658	DEGA; for aldehydes, ketones, esters, fatty acids, and pesticides
Diethylene glycol glutarate		225	I	MeCl						
Diethylene glycol sebacate	80	190	I	MeCl						DEGSB
Diethylene glycol succinate	20	190	P	MeCl	496	746	590	837	835	DEGS; for alcohols, aldehydes, ketones, amino acids, essential oils, steroids, esters, phosphorus compounds, and sulfur compounds
Diethylene glycol stearate					64	193	106	143	191	
Di(2-ethylhexyl) phthalate	20	150	P	Tol	135	254	213	320	235	
Di(2-ethylhexyl) adipate	-30	250	P	Ace	76	181	121	197	134	Dioctyl adipate
Di(2-ethylhexyl) sebacate	-20	125	I	Tol	72	168	108	180	125	For alcohols, drugs, alkaloids, esters, fatty acids, halogenated compounds, and blood gases
Diethyl-D-tartarate		125	P,S	MeCl						For alcohols
Diglycerol	20	120	HB	MeCl MeOH	371	826	560	676	854	For alcohols, aldehydes, ketones, aromatics, heterocycles, and hydrocarbons
Dilauryl phthalate		150	I	Tol	79	158	120	192	158	
Diisodecyl adipate	-10	175	P	Ace	71	171	113	185	128	
Diisooctyl adipate	90	150	P	Ace	78	187	126	204	140	
Diisodecyl phthalate	0	150	I	Tol Ace	84	173	137	218	155	For alcohols, aromatics, heterocycles, essential oils, esters, halogen and sulfur compounds, and hydrocarbons

(Continued)

Liquid Phase	T_{min} (°C)	T_{max} (°C)	Polarity	Solvents	McReynolds Constant					Notes
					X	Y	Z	U	S	
Diisooctyl sebacate		175	I	Ace						For aldehydes, ketones, and hydrocarbons
2,4-Dimethyl sulfolane	0	50	P	Chlor						For hydrocarbons, inorganic, and organometallic compounds
Dimer acid		100	I	MeCl						C_{36} dicarboxylic acid
Diisooctyl phthalate	0	175	I	Tol	94	193	154	243	202	
Dimethyl formamide	−20	20	P	Ace						DMF
Dimethyl sulfoxide	20	30	P	MeCl						DMSO; for gases
Dinonyl phthalate	20	150	I	Tol	83	183	147	231	159	For aromatics, heterocycles, and halogen compounds
Dioctyl phthalate	−20	150	I	Tol	92	186	150	230	167	For aromatics, heterocycles, and halogen compounds
Dioctyl sebacate		100	I	MeCl	72	168	108	180	123	
Diphenyl formamide	75	100	I	Tol						
Di-n-propyl tetrachloro phthalate	10	75	I	Tol						
Ditridecyl phthalate	−10	225	P	Tol	75	156	122	195	140	
Emulphor ON-870	0	200	I	Chlor	202	395	251	395	344	Aryloxy polyethylene oxyethanol; for aromatics, heterocycles, essential oils, and halogen compounds
EPON 1001	60	225	P	MeCl hot	284	489	406	539	601	Epichlorohydrin-bisphenol A resin, av. mol. mass 900; for steroids and pesticides
Ethofat 60/25	50	125	I	MeCl hot	191	382	244	380	333	Polyethylene oxyglycol stearate; for aldehydes and ketones
Ethomeen S/25		75	P	MeCl	186	395	242	370	339	Polyethoxylated aliphatic amine
Ethyl benzoate		150	I	MeOH						For hydrocarbons
Ethylene glycol adipate	100	225	I,P	MeCl	372	576	453	655	617	For alcohols, aromatics, heterocycles, bile/urinary compounds, drugs, alkaloids, essential oils, nitrogen, and sulfur compounds
Ethylene glycol phthalate	100	200	I,P	Tol	453	697	602	816	872	For nitrogen compounds and steroids
Ethylene glycol succinate	100	200	I,P	Ace	537	787	643	903	889	
Ethylene glycol glutarate		225	I,P	MeCl						
Ethylene glycol sebacate		200	I,P	MeCl hot						
Ethylene glycol tetra-chlorophthalate	120	200	P	Tol	307	345	318	428	466	
Ethylene glycol		30	HB	MeOH						

(Continued)

Liquid Phase	T_{min} (°C)	T_{max} (°C)	Polarity	Solvents	McReynolds Constant X	Y	Z	U	S	Notes
Ethylene glycol silver nitrate		30	S	Ace						
Eutectic (LiNO$_3$-NaNO$_3$-KNO$_3$/27.3-18.2-54.5)		400	—	H$_2$O						For aromatic hydrocarbons and heterocycles
Eutectic (KCl-CdCl$_2$/33-67)		400	—	H$_2$O						For aromatic hydrocarbons and heterocycles
Eutectic (NaCl-AgCl/41-59)		400	—	H$_2$O						For aromatic hydrocarbons and heterocycles
Eutectic (BiCl$_3$-PbCl$_3$/89-11)		400	—	H$_2$O						For aromatic hydrocarbons and heterocycles
Flexol 8N8		180	P	Ace	96	254	164	260	179	2,2'-(2-Ethyl hexynamido)-diethyl-di-2-ethylhexanoate; for alcohols and nitrogen compounds
Fluorolube HG-1200		100	I	Ace	51	68	114	144	116	Polymers of trifluorovinyl chloride; for halogenated compounds
Formamide	20	50	I	MeOH						For alcohols
Glycerol	20	100	HB	MeOH						
Fluorad FC-431	40	200	I	EAC	281	423	297	509	360	Fluorocarbon surfactant
Hallcomid M-18	40	150	I	MeCl	79	580	397	602	627	Dimethylsteramide; for alcohols, ketones, aldehydes, and esters
Hallcomid M-18-OL	8	150	I	MeCl	89	280	143	239	165	Dimethyloleamide; for alcohols, ketones, aldehydes, and fatty acids
Halocarbon 10-25	20	100	I	Chlor	47	70	108	113	111	
Halocarbon K 352	0	250	I		47	70	73	238	146	
Halocarbon W9X (600)	50	150		Ace	55	71	116	143	123	
Halocarbon-1321	0	100		Ace						
Halocarbon-11-14	0	100		Ace						
HMPA	20	35	P	Chlor						Hexamethylphosphoramide
Hi-Eff-1 AP	20	210	I,P	Chlor	378	603	460	665	658	Diethylene glycol adipate
Hi-Eff-2 AP	100	210	I,P	Chlor	372	576	453	655	617	
Hi-Eff-8 BB	100	250	I,P	Chlor	271	444	333	498	463	Cyclohexane dimethanol succinate
Hi-Eff-1 BP	20	200	I,P	Chlor	499	751	593	840		Diethylene glycol succinate
Hi-Eff-2 BP	100	200	I,P	Chlor	537	787	643	903	889	Ethylene glycol succinate

(Continued)

Liquid Phase	T_{min} (°C)	T_{max} (°C)	Polarity	Solvents	McReynolds Constant					Notes
					X	Y	Z	U	S	
Hi-Eff-3 AP	50	230	I,P	Chlor						Neopentyl glycol adipate
Hi-Eff-8 AP	100	250	I,P	Chlor						Cyclohexane dimethanol adipate
Hi-Eff-9 AP	100	250	I,P	Chlor						Tetramethyl cyclobutanediol adipate
Hi-Eff-3 BP			I,P							Neopentyl glycol succinate
Hi-Eff-4 BP	50	230	I,P	Chlor						Butane-1,4-diol succinate
Hi-Eff-10 BP	20	230	I,P	Chlor						Phenyl diethanolamine succinate
Hi-Eff-2 CP	100	200	I,P	Chlor						Ethylene glycol sebacate
Hi-Eff-3 CP	50	230	I,P	Chlor						Neopentyl glycol sebacate
Hi-Eff-2 EP	100	210	I,P	Chlor						Ethylene glycol isophthalate
Hi-Eff-26 P	100	210	I,P	Chlor						Ethylene glycol phthalate
Hyprose-SP-80		225	P	MeOH	336	742	492	639	727	Octakis (2-hydroxy propyl) sucrose
1,2,3,4,5,6 Hexakis-(2-cyanoethoxy-cyclohexane)	125	150	I,P	Tol	567	825	713	978	901	
Hercoflex 600		150	P	MeCl	112	234	168	261	194	High boiling ester of pentaerythritol and a saturated aliphatic acid
n-Hexadecane	20	50	N	Pent						
Hexadecane	20	50	N	Pent						Isomeric mixture
1-Hexadeconal		35	I	MeOH						Cetyl alcohol; for halogenated compounds and hydrocarbons
Hexatricontane	80	150	N	MeCl	12	2	−3	1	11	$C_{36}H_{74}$
IGEPAL CO-880	100	200	I	MeCl hot	259	461	311	482	426	Nonyl phenoxy poly(ethyleneoxyethanol) n=30; for alcohols
IGEPAL CO-990	100	200	I	MeCl hot	298	508	345	540	475	Nonyl phenoxy poly(ethyleneoxyethanol) n=100; for alcohols
IGEPAL CO-630	100	200	I	MeCl hot	192	381	253	382	344	Nonyl phenoxy poly(ethyleneoxyethanol) n=9; for alcohols
IGEPAL CO-730		200	I		224	418	279	428	379	
IGEPAL CO-710	100				205	397	266	401	361	
β,β-Iminodipropionitrile		110	I	MeOH						For halogenated compounds
Isoquinoline		50	I,P	MeCl						For hydrocarbons

(Continued)

Liquid Phase	T_{min} (°C)	T_{max} (°C)	Polarity	Solvents	McReynolds Constant					Notes
					X	Y	Z	U	S	
Lexan	220	270	P	DMP hot						Polycarbonate resin
Mannitol	170	200	HB	H$_2$O						For sugars
Montan wax		175	I	Chlor						For halogenated compounds
Naphthylamine		150	I	Chlor						For aromatics and heterocycles
Neopentyl glycol adipate	50	240	I	MeCl	234	425	312	402	438	NPGA; for amino acids, drugs, alkaloids, pesticides, and steroids
Neopentyl glycol isophthalate	50	240	I	MeCl						
Neopentyl glycol sebacate	50	225	I	MeCl	172	327	225	344	326	NPGSB; for amino acids and steroids
Neopentylglycol succinate	50	225	I	MeCl	272	469	366	539	474	NPGS; for amino acids, bile and urinary compounds, esters, and inorganics
Nitrobenzene		150	I	MeOH						For hydrocarbons, inorganic, and organometallic compounds
Nujol		100	N	Pent	9	5	2	6	11	Paraffin oil, mineral oil; for hydrocarbons
n-Octadecane	30	55	N	Pent						For inorganic and organometallic compounds
Octyl decyl adipate		175	P	Ace	79	179	119	193	134	
Oronite NIW		170	P		180	370	242	370	327	Complex mixture of petroleum liquids
β,β′-Oxydipropionitrile		100	P	Ace						For halogenated compounds
Phenyl diethanolamine succinate		225	P	Ace	386	555	472	674	654	For drugs, alkaloids, and hydrocarbons
Polyethylene imine	0	250	P	MeOH	322	800	—	573	524	
Poly-m-phenylxylene	125	375	I	Tol	257	355	348	433	—	PPE-20
Poly-m-phenyl ether		250	I	Tol	176	227	224	306	283	Five rings; for aromatics and heterocycles
Poly-m-phenyl ether	0	300	I	Ace, Tol	182	233	228	313	293	Six rings; for alcohols, essential oils, and esters
Poly-m-phenyl ether	50	400	I	Tol						High polymer
Poly-m-phenyl ether with squalane	50	100	I	MeCl						Six rings
Polypropylene glycol	0	150	HB	MeOH	128	294	173	264	226	Av. mol. mass = 2,000; for drugs, alcohols, and alkaloids
Polypropylene glycol sebacate	20	225	I	Chlor	196	345	251	381	328	

(Continued)

Liquid Phase	T_{min} (°C)	T_{max} (°C)	Polarity	Solvents	McReynolds Constant					Notes
					X	Y	Z	U	S	
Polypropylene glycol silver nitrate	20	75	S	MeCl						PEG/AgNO$_3$-3/1; for unsaturated hydrocarbons
Polypropylene imine	0	200	I,P	Chlor	122	425	168	263	224	
Propylene carbonate	0	60	P	MeCl						1,2 Propanediol cyclic carbonate; for gases and hydrocarbons
Polysulfone	0	315	I	Ace						
Polyvinyl pyrrolidone	80	225	HB	MeOH						
Quadrol	0	150	HB	Chlor	214	571	357	472	489	Consists of N,N,N',N'-tetrakis (2-hydroxy-propyl) ethylenediamine; for alcohols, aldehydes, ketones, amino acids, and essential oils
Reoplex 400	0	200	I	MeCl	364	619	449	647	671	Poly(propylene glycol adipate); for aromatics, heterocycles, vitamins, sulfur, and phosphorus compounds
Reoplex 100	0	200	I	MeCl						Poly(propylene glycol sebacate)
Renex-678	0	200	HB	MeOH	223	417	278	427	381	Ethylene oxide-nonylphenol surfactant; for alcohols
Sebaconitrile		150	P							
Squalane	20	100	N	Pent	0	0	0	0	0	For hydrocarbons, organic vapors, nitrogen, sulfur, and phosphorus compounds
Squalene	0	100	N,I	Pent	152	341	238	329	344	For hydrocarbons, gases, nitrogen, sulfur, and phosphorus compounds
Sorbitol	15	150	P	Chlor	232	582	313			Hexahydric alcohol, C$_6$H$_6$(OH)$_6$
STAP	100	255	P	Chlor	345	586	400	610	627	Steroid analysis phase
Siponate-DS-10	20	210	I,P	MeOH						Sodium dodecylbenzene sulfonate
Sorbitan monooleate	20	150	P	Chlor	97	266	170	216	268	SPAN-80
Sorbitol hexaacetate					335	553	449	652	543	
Sucrose acetate isobutyrate	0	200	I,P	MeCl	172	330	251	378	295	SAIB; for alcohols, essential oils
Sucrose octaacetate	90	250	I,P	Ace	344	570	461	671	569	
Tergitol Nonionic NP-35	10	175	P	Chlor	197	380	258	389	351	Surfactant mixture

(Continued)

Liquid Phase	T_min (°C)	T_max (°C)	Polarity	Solvents	McReynolds Constant					Notes
					X	Y	Z	U	S	
TCEPE	30	175	P,S	MeCl	526	782	677	920	837	Tetracyanoethylatedpentaerythritol; for fatty acids and esters
Terephthalic acid	100	250	P,I	Tol						For hydrocarbons
Tetraethylene glycol		70	P	MeCl						For nitrogen compounds
Tetraethylene-pentamine		150	HB	MeOH						
1,2,3,4-Tetrakis-(2-cyanoethyl)butane	110	200	I,P	Chlor	617	860	773	48	941	
THEED	0	125	HB	Chlor	463	942	626	801	893	Tetrahydroxyethylenediamine; for alcohols, hydrocarbons, and nitrogen compounds
β,β'-Thiodipropionitrile		100	P	MeOH						For hydrocarbons
Triacetin		60	P	MeOH						For gases
Tributyl phosphate	20	125	I	Ace						For gases
Tricresyl phosphate	20	215	I	MeOH	176	321	250	374	299	Tritolyl phosphate
Triethanolamine		100	HB	MeOH						For alcohols and gases
Trimer acid	20	200	HB	MeOH	94	271	163	182	378	C_{54} tricarboxylic acid; for alcohols
1,2,3-tris(2-cyano-ethoxy)propane	30	150	P	MeOH	594	857	759	31	917	For alcohols, aldehydes, ketones, halogen compounds, and inorganic and organometallic compounds
Tris(tetrahydrofur-furyl) phosphate	20	125	I	Ace						
Tris(2-cyanoethyl) nitromethane	20	140	I,P	Chlor						
Triton X-100	20	190	P	MeCl	203	399	268	402	362	Octylphenoxypolyethyl ethanol, for aromatics and heterocycles
Triton X-305	20	250	P	Ace	262	467	314	488	430	Octylphenoxypolyethyl ethanol, for alcohols
Trixylol phosphate	20	250	I,P	Ace						
TWEEN 20	20	150	P	MeOH						Polyethoxysorbitan monolaurate; for essential oils
TWEEN 80	20	160	P	MeOH	227	430	283	438	396	Polyethoxysorbitan monooleate; for fatty acids, esters, pesticides

(Continued)

Liquid Phase	T_{min} (°C)	T_{max} (°C)	Polarity	Solvents	McReynolds Constant					Notes
					X	Y	Z	U	S	
UCON LB-550-X	0	200	P	Chlor	118	271	158	243	206	10 % polyethylene glycol, 90 % propylene glycol
UCON 50-HB-280-X	0	200	P	Chlor	177	362	227	351	302	30 % polyethylene glycol, 70 % propylene glycol; for alcohols, fatty acids, and esters
UCON 50-HB-2000	0	200	P	Chlor	202	394	253	392	341	40 % polyethylene glycol, 60 % propylene glycol; for alcohols, aldehydes, and ketones
UCON 50-HB-5100	20	200	P	MeCl	214	418	278	421	375	50 % polyethylene glycol and 50 % propylene glycol
UCON LB-1715	20	200	I	MeCl	132	297	180	275	235	For alcohols, ketones, and nitrogen compounds
UCON 75-H-90,000	20	200	P	MeCl	255	452	299	470	406	80 % polyethylene glycol, 10 % propylene glycol
Versamide 900	190	250	P	MeCl						Polyamide resin; for alcohols
Versamide 940	115	200	P	MeCl	109	314	145	212	209	Polyamide resin; for alcohols
Versamide 930	115	150	P	MeCl	109	313	144	211	209	Polyamide resin
Versamide 940		200	P	See Notes	109	314	145	212	209	Soluble in hot chloroform butanol, 50/50 v/v, for aromatics, heterocycles, pesticides, and nitrogen compounds
Xylenyl phosphate		175	I	MeCl						
Zonyl E7		200	I	MeCl	223	359	468	549	465	Fluoroalkyl ester
Zonyl E91		200	I	MeCl	130	250	320	377	293	Fluoroalcohol camphorate
Zinc stearate	135	175	I	Ace (warm)	61	231	59	98	544	

SELECTION OF STATIONARY PHASES FOR
PACKED-COLUMN GAS CHROMATOGRAPHY

The following stationary phases have been shown to be of value in the separation of these major classes of compounds, using packed columns of typical dimensions (4–10 m in length, 0.32 cm diameter) [1–10]. They have also been useful in packed capillary columns. The resolution will undoubtedly be lower than that obtainable with capillary columns, which have superseded packed columns in many applications. The two main exceptions are in the analysis of permanent gases and preparative-scale gas chromatography. Data on the packed column stationary phases are included since they still find use in many laboratories. This table is meant to provide only a rough guide. The additional data which can be found in the preceding stationary phase data table will aid in determining the final choice. Note that in some instances the use of a surfactant is specified. More information on surfactants can be found in the table entitled Surface Active Chemicals (Surfactactants) in the solutions Chapter 14 of this book.

REFERENCES

1. Heftmann, E., ed., *Chromatography—A Laboratory Handbook of Chromatographic and Electrophoretic Methods*, 3rd. ed., Van Nostrand Reinhold Co., New York, 1975.
2. Grant, D.W., *Gas-Liquid Chromatography*, Van Nostrand Reinhold Co., London, 1971.
3. McNair, H.M. and Bonelli, E.J., *Basic Gas Chromatography*, Varian Aerography, Palo Alto, 1969.
4. Grob, R.L., ed., *Modern Practice of Gas Chromatography*, 2nd ed., Wiley Interscience, New York, 1985.
5. Poole, C.F. and Schuette, S.A., *Contemporary Practice of Chromatography*, Elsevier, Amsterdam, 1984.
6. Mann, J.R. and Preston, S.T., Selection of preferred liquid phases, *J. Chromatogr. Sci.*, 11, 216, 1973.
7. Coleman, A.E., Chemistry of liquid phases, other silicones, *J. Chromatogr. Sci.*, 11, 198, 1973.
8. Yancey, J.A., Liquid phases used in packed gas chromatographic columns—part 1—polysiloxane phases, *J. Chromatogr. Sci.*, 23, 161, 1985.
9. Yancey, J.A., Liquid phases used in packed gas chromatographic columns—part 2—use of liquid phases which are not polysiloxanes, *J. Chromatogr. Sci.*, 23, 370, 1985.
10. McReynolds, W.O., Characterization of some liquid phases, *J. Chromatogr. Sci.*, 8, 685, 1970.

Compound	Suggested Stationary Phases
Alcohols C_1–C_5	Apiezon L; Apiezon M; benzyldiphenyl; butanediol succinate (Craig polyester); Carbowax 400, 600, 750, 1000, 1000 (monostearate); diethylene glycol succinate; di(2-ethylhexyl) sebacate; diethyl-d-tartrate; di-n-decyl-phthalate; diglycerol; diisodecyl phthalate; dinonyl phthalate; ethylene glycol succinate; Flexol 8N8; Hallcomid M-18-OL; Quadrol; Renex 678; Sorbitol; tricresyl phosphate; triethanolamine.
C_5–C_{18}	Butanediol succinate (Craig polyester); Carbowax 1500, 1540, 4000, 4000 (dioleate), 4000 (monostearate), 6000, 20M, 20M-TPA; ethylene glycol adipate; Igepal series; Ucon series; Versamid series.
Aldehydes (and ketones)	Apiezon L, M; Carbowax 400, 750, 1000, 1500, 1540; di-n-butyl phthalate; diethylene glycol succinate; ethylene glycol succinate; Hallcomid M-18; squalene; tricresyl phosphate; 1,2,3-tris (2-cyanoethoxy) propane; Ucon series.
Alkaloids (includes drugs and vitamins)	Apiezon L; Carbowax 20M; di(2-ethylhexyl) sebacate; ethylene glycol adipate; ethylene glycol succinate, neopentyl glycol adipate; phenyl diethanolamine succinate; SE-30 (methyl silicone phases).

(Continued)

Compound	Suggested Stationary Phases
Amides	Carbowax 600 (on Chromosorb T); diethylene glycol succinate; ethylene glycol succinate; neopentyl glycol sebacate; Versamid 900; SE-30 (methyl silicone phases).
Amino acids (and derivatives)	Carbowax 600; diethylene glycol succinate (stabilized); Ethofat (on Chromosorb T); ethylene glycol succinate; neopentyl glycol adipate; SE-30; XE-60 (methyl silicone phases).
Amines	Penwalt 213; Chromosorb 103. (See support modifiers.)
Boranes	Apiezon L; beeswax; Carbowax 400, 1540, 4000, 20M; Castorwax; diethylene glycol succinate; di-n-decyl phthalate; diisodecyl phthalate; Emulphor-ON-870; ethylene glycol adipate; free fatty acid phase (FFAP), polyphenyl ether (4 or 5 ring); Quadrol; Reoplex 400; SE-30; XE-60; sucrose acetate isobutyrate; tricresyl phosphate; Ucon series.
Esters	Apiezon L; benzyldiphenyl; Carbowax 20M; cyclodextrin acetate; diethylene glycol adipate; di(2-ethylhexyl) sebacate; diisodecyl phthalate; dimer acid/OV-1 (50/50, V/V); Hallcomid M-18; neopentyl glycol succinate; propylene glycol; SE-30; SE-52; XE-60; Friton X-100; Tween 80.
Ethers	Apiezon L; Carbowax 1500, 1540, 4000, 20M; diethylene glycol sebacate; ethylene glycol adipate.
Glycols	Porapak-Q, Porapak-1S; QF-1.
Halogenated compounds	Bentone 34; benzyldiphenyl; butanediol succinate (Craig polyester); Carbowax 400, 1000, 4000, 20M; dibutyl phthalate; diethylene glycol succinate; di (2-ethylhexyl) sebacate; di-n-decyl phthalate; dinonyl phthalate; dioctyl phthalate; β,β'-iminodipropionitrile; β,β'-oxydipropionitrile; SE-30; squalane; Tween 80.
Inorganic compounds (includes organometallic compounds)	n-Decane; di-n-decyl phthalate; dimethyl sulfolane; neopentyl glycol succinate; 1,2,3-Trix (2-cyanoethoxy) propane; SE-30 (methyl silicone phases).
Hydrocarbons C_1–C_5 (aliphatic)	Carbowax 400–1500; most branched and substituted phthalate, sebacate, succinate and adipate phases; octadecane; squalane (boiling point separations); methyl silicones.
Above C_5 (aliphatic)	Apiezon phases; Carbowax 1500, 1540, 4000, 6000, 20M; most of the high temperature substituted adipates, phthalates, succinates and sebacates (boiling point separations); methyl silicones.
(Aromatic)	Apiezon phases; bentone-34; Carbowax phases; substituted adipates, phthalates, succinates and sebacates; tetracyanoethylated pentaerythritol; liquid crystalline phases; phenyl methyl silicone phases.
Nitrogen compounds	Apiezon L; Armeen SD; butanediol succinate (Craig polyester); Carbowax 400, 1500, 20M; ethylene glycol adipate; propylene glycol; tetraethylene glycol dimethyl ether; THEED; Ucon phases.
Pesticides	Carbowax-20M; diethylene glycol adipate; Epon 1001; neopentyl glycol adipate; methyl silicone phases, including gum viscosities.
Phosphorus compounds	Apiezon L; Carbowax 20M; di-n-butyl phthalate; diethylene glycol succinate; Emulphor-ON-870; ethylene glycol succinate; Reoplex-400; methyl silicone phases, including gum viscosities; squalane; STAP.
Silanes	Methyl silicone phases; STAP.
Sugars	Apiezon L; butanediol succinate; Carbowax 4000, Hyprose SP80; mannitol; methyl silicone phases.
Sulfur compounds	Apiezon L; 7,8-benzoquinoline; Carbowax 1500, 20 M; diethylene glycol succinate; diisodecyl phthalate; methyl silicone phases; Reoplex-400; tricresyl phosphate.
Urinary and bile compounds	Ethylene glycol adipate; methyl silicone-nitrile phases.

ADSORBENTS FOR GAS–SOLID CHROMATOGRAPHY

The following table lists the more common adsorbents used in gas–solid chromatography, along with relevant information on separation and technique [1–3]. The adsorbents are used chiefly for the analysis of gaseous mixtures. The maximum temperatures listed represent the point of severe resolution loss. The materials are often chemically stable to much higher temperatures. The 60–100 mesh sizes (US) are most useful for chromatographic applications. All of these materials must be activated before being used, and the degree of activation will influence the retention behavior. The user should also be aware that the adsorption of water during use will often change retention characteristics dramatically, sometimes resulting in a reversal of positions of adjacent peaks. Due to surface adsorption of solutes, some experimentation with temperature may be necessary to prevent tailing or to avoid statistical correlation (or a propagating error) among replicate analyses [4].

REFERENCES

1. Jeffery, P.G. and Kipping, P.J., *Gas Analysis by Gas Chromatography*, Pergamon Press, Oxford, 1972.
2. Cowper, C.J. and DeRose, A.J., *The Analysis of Gases by Chromatography*, Pergamon Press, Oxford, 1983.
3. Breck, D.W., *Zeolite Molecular Sieves*, John Wiley & Sons, New York, 1973.
4. Bruno, T.J., An apparatus for direct fugacity measurements on mixtures containing hydrogen, *J. Res. Natl. Bur. Stand.*, 90(2), 1127, 1985.

Packing Name	Max. Temp. (°C)	Separation Affected	Notes
Silica gel	300	H_2, air, CO, C_1–C_4, normal hydrocarbons, alkenes, and alkynes	Used often as a second column (with a molecular sieve); very hydrophilic; requires activation; can be unpredictable; largely replaced by porous polymers
Porous silica	300	Same as silica gel	Higher surface area than silica gel; often used with a humidified carrier gas; can be coated with a conventional liquid phase; Spherosil and Porasil are examples
Alumina	300	Light hydrocarbons at ambient temperature (C_1–C_5), H_2 and light hydrocarbons at subambient temperature	Often useful with controlled water preadsorption after activation; can be coated with a conventional liquid phase; a variety of polarities are available depending on the washing technique used
Activated carbon	300	H_2, CO, CO_2, C_1–C_3 alkanes, alkenes, and alkynes	Requires oxygen-free carrier gas; largely replaced by porous polymers
Cyclo-dextrin	260	Light hydrocarbons (C_1–C_{10}) and halocarbons	α and β cyclodextrins have been used; care should be taken with halocarbon analysis, due to the potential of HF contamination of the sample
Graphite	300	Light hydrocarbons, H_2S, SO_2, CH_3SH, sour gas	Often modified with small quantities (1.5 %–5 %) of conventional liquid phases; requires oxygen-free carrier
Carbon molecular sieve	300	H_2 (O_2, N_2 co-elute), CO, CH_4, H_2O, CO_2, C_1–C_3 alkanes, alkenes, and alkynes	High affinity for hydrocarbons; requires oxygen-free carrier
Molecular sieve, 5A	225	Air and light gas analysis; H_2, O_2, N_2, (CH_4, CO, NO, SF_6 co-elute)	Synthetic calcium alumino-silicate (zeolite) having an effective pore diameter of 0.5 nm CO_2 is adsorbed strongly; 5A usually gives the best results of all synthetic zeolites; should be activated before use and used above critical adsorption temperature; 21.6 % (mass/mass) water capacity
Molecular sieve, 13X	200	Same as 5A, but with C_1–C_4, alkanes, alkenes, and alkynes being separated as well	Sodium alumino-silicate (zeolite), having a larger pore size than 0.5 nm, thus producing lower retention times and less resolution; 28.6 % (mass/mass) water capacity
Molecular sieve, 3A	200	Light permanent gases	Potassium alumino-silicate (zeolite) 20 % (mass/mass) water capacity, smaller pore size than 0.5 nm, thus different retention characteristics
Molecular sieve, 4A	200	Light permanent gases	Sodium alumino-silicate (zeolite); 22 % (mass/mass) water capacity; retention characteristics differ from 5A due to smaller pore size

POROUS POLYMER PHASES

Porous polymer phases, first reported by Hollis [1–3], are of great value for a wide variety of separations. They are usually white in color but may darken during use especially at higher temperatures. This darkening does not affect their performance. High temperature conditioning is required to drive off solvent and residual monomer. The polymers may either swell or shrink with heating; thus, flow rate changes must be anticipated. The retention indices reported here are from the work of Dave [2]. The use of these indices is the same as for the packed-column liquid phases, provided in an earlier table. The Porapak and HayeSep serve similar roles; the Porapak materials have been reported to deliver a lower back pressure through the column per unit length than the similar HayeSep material.

Index	Test Probe
W	Benzene
X	t-Butanol
Y	2-Butanone
Z	Acetonitrile

The physical property data were taken from the work of Poole and Schuette [3].

REFERENCES

1. Hollis, O.L., Separation of gaseous mixtures using porous polyaromatic polymer beads, *Anal. Chem.*, 38, 309, 1966.
2. Dave, S., A comparison of the chromatographic properties of porous polymers, *J. Chromatogr. Sci.*, 7, 389, 1969.
3. Poole, C.F. and Schuette, S.A., *Contemporary Practice of Chromatography*, Elsevier, Amsterdam, 1984.

Phase Name	Maximum Temperature (°C)	Material Type	Free Fall Density (g/cm³)	Surface Area (m²/g)	Pore Diameter Av. (μm)
Chromosorb 101	275	Styrene–divinylbenzene copolymer	0.30	<50	0.3–0.4

Retention Indices

W	X	Y	Z	Separation Affected	Notes
745	565	645	580	Free fatty acids, glycols, alcohols, alkanes, esters, aldehydes, ketones, ethers	Hydrophobic; condition at 250°C; not recommended for amines or anilines, lower retention times than obtained with Chromosorb 102

Phase Name	Maximum Temperature (°C)	Material Type	Free Fall Density (g/cm³)	Surface Area (m²/g)	Pore Diameter Av. (μm)
Chromosorb 102	250	Styrene–divinylbenzene copolymer	0.29	300–500	0.0085

Retention Indices

W	X	Y	Z	Separation Affected	Notes
650	525	570	460	Subambient temperature: H_2, O_2, N_2, Ar, NO, CO; ambient temperature: H_2, (Air + Ar + NO + CO), CH_4, CO_2, H_2O, N_2O, C_2H_6; above ambient temperature: C_1–C_4 hydrocarbons, H_2S, COS, SO_2, esters, ethers, alcohols, ketones, aldehydes, glycols	May entrain some species; hydrophobic; condition at 225 °C; not recommended for amines or nitriles; little tailing of water or oxygenated hydrocarbons

Phase Name	Maximum Temperature (°C)	Material Type	Free Fall Density (g/cm³)	Surface Area (m²/g)	Pore Diameter Av. (μm)
Chromosorb 103	275	Polystyrene cross-linked	0.32	15–25	0.3–0.4

Retention Indices

W	X	Y	Z	Separation Affected	Notes
720	575	640	565	Ammonia, light amines, light amides, alcohols, aldehydes, hydrazines	Hydrophobic; high affinity for basic species; not recommended for acidic species, glycols, nitriles, nitroalkanes

Phase Name	Maximum Temperature (°C)	Material Type	Free Fall Density (g/cm³)	Surface Area (m²/g)	Pore Diameter Av. (μm)
Chromosorb 104	250	Acrylonitrile divinylbenzene copolymer	0.32	100–200	0.06–0.08

Retention Indices

W	X	Y	Z	Separation Affected	Notes
845	735	860	885	Sulfur gases, ammonia, nitrogen oxides, nitriles, nitroalkanes, xylenols, water in benzene	Hydrophobic; condition at 225 °C; not recommended for glycols and amines; moderately polar

Phase Name	Maximum Temperature (°C)	Material Type	Free Fall Density (g/cm³)	Surface Area (m²/g)	Pore Diameter Av. (μm)
Chromosorb 105	250	Acrylic ester (polyaromatic)	0.34	600–700	0.04–0.06

Retention Indices

W	X	Y	Z	Separation Affected	Notes
635	545	580	480	Permanent and light hydrocarbon gases; aqueous solutions of light organics such as formalin	Hydrophobic; less polar than Chromosorb 104; condition at 225 °C; not recommended for acidic species, glycols, amines, and amides

Phase Name	Maximum Temperature (°C)	Material Type	Free Fall Density (g/cm³)	Surface Area (m²/g)	Pore Diameter Av. (μm)
Chromosorb 106	250	Polystyrene cross-linked	0.28	700–800	0.05

Retention Indices

W	X	Y	Z	Separation Affected	Notes
605	505	540	405	Fatty acids from fatty alcohols, up to C_5; benzene from nonpolar organic compounds	Hydrophobic; not recommended for glycols and amines

Phase Name	Maximum Temperature (°C)	Material Type	Free Fall Density (g/cm³)	Surface Area (m²/g)	Pore Diameter Av. (μm)
Chromosorb 107	250	Acrylic ester cross-linked	0.30	400–500	0.8

Retention Indices

W	X	Y	Z	Separation Affected	Notes
660	620	650	550	Aqueous solutions of formaldehyde; alkynes from alkanes	Hydrophobic; moderately polar; not recommended for glycols and amines

Phase Name	Maximum Temperature (°C)	Material Type	Free Fall Density (g/cm³)	Surface Area (m²/g)	Pore Diameter Av. (μm)
Chromosorb 108	250	Acrylic ester cross-linked	0.30	100–200	0.25

Retention Indices

W	X	Y	Z	Separation Affected	Notes
710	645	675	605	Polar materials such as water, alcohols, aldehydes, glycols	Hydrophobic; condition at 250°C

Phase Name	Maximum Temperature (°C)	Material Type	Free Fall Density (g/cm³)	Surface Area (m²/g)	Pore Diameter Av. (μm)
Hayesep A	165	Divinylbenzene/ethylene glycol-dimethacrylate (high purity)	0.356	526	—

Retention Indices

W	X	Y	Z	Separation Affected	Notes
				Separates permanent gases at ambient temperatures and is useful for hydrocarbons to C_2, H_2S, H_2O at elevated temperatures	Relatively high polarity

Phase Name	Maximum Temperature (°C)	Material Type	Free Fall Density (g/cm³)	Surface Area (m²/g)	Pore Diameter Av. (μm)
HayeSep B	190	Divinylbenzene/polyethyleneimine	0.33	608	—

Retention Indices[a]

W	X	Y	Z	Separation Affected	Notes
				Separates C_1 and C_2 amines and trace levels of NH_3 and H_2O	High polarity

[a] Retention indices are not available for these porous polymers, but a table of relative retentions on some representative solutes is included at the end of this section.

Phase Name	Maximum Temperature (°C)	Material Type	Free Fall Density (g/cm³)	Surface Area (m²/g)	Pore Diameter Av. (μm)
HayeSep C	250	Divinylbenzene/acrylonitrile	0.322	442	—

Retention Indices[a]

W	X	Y	Z	Separation Affected	Notes
				Separates polar hydrocarbons, also HCN, NH_3, H_2S, H_2	Moderate polarity, with separation characteristics similar to Chromosorb 104

Phase Name	Maximum Temperature (°C)	Material Type	Free Fall Density (g/cm³)	Surface Area (m²/g)	Pore Diameter Av. (μm)
HayeSep D	290	Divinylbenzene (high purity)	0.3311 (av.)	795 (av.)	0.0308–0351

Retention Indices[a]

W	X	Y	Z	Separation Affected	Notes
				Separates light gases; CO, CO_2, C_2H_2, C_2 hydrocarbons, H_2S, H_2O	Low-polarity polymer available in four formulations of different surface area (771–802 m²/g), density (0.3283–0.3834 g/mL), and porosity (64.2–70.4 %)

[a] Retention indices are not available for these porous polymers, but a table of relative retentions on some representative solutes is included at the end of this section.

Phase Name	Maximum Temperature (°C)	Material Type	Free Fall Density (g/cm³)	Surface Area (m²/g)	Pore Diameter Av. (μm)
HayeSep N	165	Divinylbenzene/ethylene glycol-dimethacrylate (high purity)	0.355	405	—

Retention Indices[a]

W	X	Y	Z	Separation Affected	Notes
				Separation similar to Porapak materials; moderately high H_2O retention; see retention table	Low-polarity polymer

Phase Name	Maximum Temperature (°C)	Material Type	Free Fall Density (g/cm³)	Surface Area (m²/g)	Pore Diameter Av. (μm)
HayeSep P	250	Divinylbenzene/styrene	0.420	165	—

Retention Indices[a]

W	X	Y	Z	Separation Affected	Notes
				Separation of low molecular mass materials containing halogens, sulfur, water, aldehydes, ketones, alcohols, esters, and fatty acids	Moderate to low polarity

[a] Retention indices are not available for these porous polymers, but a table of relative retentions on some representative solutes in included at the end of this section.

Phase Name	Maximum Temperature (°C)	Material Type	Free Fall Density (g/cm³)	Surface Area (m²/g)	Pore Diameter Av. (μm)
HayeSep Q	275	Divinylbenzene	0.351	582	—

Retention Indices[a]

W	X	Y	Z	Separation Affected	Notes
				Separation similar to Hayesep P; see retention table	Low polarity

Phase Name	Maximum Temperature (°C)	Material Type	Free Fall Density (g/cm³)	Surface Area (m²/g)	Pore Diameter Av. (μm)
HayeSep R	250	Divinylbenzene/ N-vinyl-2-pyrollidinone	0.324	344	—

Retention Indices[a]

W	X	Y	Z	Separation Affected	Notes
				Separation similar to Hayesep P; see retention table	Moderate polarity

[a] Retention indices are not available for these porous polymers, but a table of relative retentions on some representative solutes is included at the end of this section.

Phase Name	Maximum Temperature (°C)	Material Type	Free Fall Density (g/cm³)	Surface Area (m²/g)	Pore Diameter Av. (μm)
HayeSep S	250	Divinylbenzene/4-vinyl-pyridine	0.334	583	—

Retention Indices[a]

W	X	Y	Z	Separation Affected	Notes
				Separation similar to Hayesep P; see retention table	Moderate polarity

Phase Name	Maximum Temperature (°C)	Material Type	Free Fall Density (g/cm³)	Surface Area (m²/g)	Pore Diameter Av. (μm)
HayeSep T	165	Ethylene glycol-dimethacrylate (high purity)	0.381	250	—

Retention Indices[a]

W	X	Y	Z	Separation Affected	Notes
				See retention table	High polarity

[a] Retention indices are not available for these porous polymers, but a table of relative retentions on some representative solutes is included at the end of this section.

Phase Name	Maximum Temperature (°C)	Material Type	Free Fall Density (g/cm³)	Surface Area (m²/g)	Pore Diameter Av. (μm)
Porapak-Q	250	Ethylvinylbenzene–divinylbenzene copolymer	0.35	500–700	0.0075

Retention Indices

W	X	Y	Z	Separation Affected	Notes
630	538	580	450	Similar to Chromosorb 102	Similar to Chromosorb 102; condition at 250 °C; most popular of all porous polymer phases

Phase Name	Maximum Temperature (°C)	Material Type	Free Fall Density (g/cm³)	Surface Area (m²/g)	Pore Diameter Av. (μm)
Porapak-P	250	Styrene–divinylbenzene copolymer	0.28	100–200	—

Retention Indices

W	X	Y	Z	Separation Affected	Notes
765	560	650	590	Similar to Porapak-Q	Hydrophobic; low polarity; larger pore size than Porapak-Q, thus lower retention times are observed; not recommended for amines or anilines; condition at 250 °C

Phase Name	Maximum Temperature (°C)	Material Type	Free Fall Density (g/cm³)	Surface Area (m²/g)	Pore Diameter Av. (μm)
Porapak-N	200	Vinylpyrolidone	0.39	225–350	—

Retention Indices

W	X	Y	Z	Separation Affected	Notes
735	605	705	595	Similar to Chromosorb 105; high water retention; CO_2, NH_3, H_2O, C_2H_2, from light hydrocarbons	Condition at 175 °C; not recommended for glycols, amines, or acidic species

Phase Name	Maximum Temperature (°C)	Material Type	Free Fall Density (g/cm³)	Surface Area (m²/g)	Pore Diameter Av. (µm)
Porapak-R	250	Vinylpyrolidone	0.33	300–450	0.0076

Retention Indices

W	X	Y	Z	Separation Affected	Notes
645	545	580	455	Ethers, esters, H$_2$O from chlorine gases (HCl, Cl$_2$), nitriles, and nitroalkanes	Moderately polar; condition at 250 °C; not recommended for glycols and amines

Phase Name	Maximum Temperature (°C)	Material Type	Free Fall Density (g/cm³)	Surface Area (m²/g)	Pore Diameter Av. (µm)
Porapak-S	250	Vinyl pyridine	0.35	300–450	0.0076

Retention Indices

W	X	Y	Z	Separation Affected	Notes
645	550	575	465	Normal and branched alcohols, aldehydes, ketones, halocarbons	High polarity; not recommended for acidic species and amines; condition at 250 °C

Phase Name	Maximum Temperature (°C)	Material Type	Free Fall Density (g/cm³)	Surface Area (m²/g)	Pore Diameter Av. (µm)
Porapak-T	200	Ethylene glycol-dimethacrylate	0.44	250–300	0.009

Retention Indices

W	X	Y	Z	Separation Affected	Notes
–	675	700	635	Water in formalin (and other aqueous organic mixtures), retention characteristics similar to Chromosorb 107	Condition at 180 °C; highest polarity of Porapak series; not recommended for glycols and amines

Phase Name	Maximum Temperature (°C)	Material Type	Free Fall Density (g/cm³)	Surface Area (m²/g)	Pore Diameter Av. (µm)
Porapak-QS	250	Ethylvinylbenzene–divinylbenzene copolymer	—	—	—

Retention Indices

W	X	Y	Z	Separation Affected	Notes
625	525	565	445	Similar to Porapak-Q at lower operating temperatures but useful for higher molecular weight solutes	Silanized Porapak-Q, reduces tailing of high-polarity compounds; condition at 250 °C

Phase Name	Maximum Temperature (°C)	Material Type	Free Fall Density (g/cm³)	Surface Area (m²/g)	Pore Diameter Av. (µm)
Porapak-PS	250	Styrene–divinylbenzene copolymer	—	—	—

Retention Indices				Separation Affected	Notes
W	X	Y	Z		
—	—	—	—	Similar to Porapak-P	Silanized Porapak-P; condition at 250 °C

Phase Name	Maximum Temperature (°C)	Material Type	Free Fall Density (g/cm³)	Surface Area (m²/g)	Pore Diameter Av. (µm)
Tenax-GC	375	p-2,6 diphenyl-phenylene oxide polymer	0.37	18.6	—

Retention Indices				Separation Affected	Notes
W	X	Y	Z		
—	—	—	—	Similar to Porapak-Q	Highest thermal stability of all porous polymers

RELATIVE RETENTION ON SOME HAYESEP POROUS POLYMERS

The following table provides relative retention values for Hayesep polymers N, Q, R, S, and T. These data were obtained using a 2 m long, 0.32 cm OD stainless steel column, using helium as the carrier gas.

HayeSep Polymer

Compound	N	Q	R	S	T
Hydrogen	0.19	0.143	0.17	0.19	0.21
Air	0.23	0.186	0.2	0.21	0.25
Nitric oxide	0.25	0.217	0.21	0.23	0.33
Methane	0.3	0.256	0.28	0.3	0.35
Carbon dioxide	0.71	1.15	0.5	0.52	0.85
Nitrous oxide	0.8	1.43	0.59	0.59	—
Ethylene	0.83	0.74	0.78	0.78	0.9
Acetylene	1.41	0.74	1	0.87	2.11
Ethane	1	1	1	1	1
Water	10.1	1.45	0.68	4.12	19.1
Hydrogen sulfide	2.1	1.4	1.73	1.87	2.88
Hydrogen cyanide	1.93	2.31	15.6	8.26	28.8
Carbonyl sulfide	2.82	2.33	2.46	2.63	3.4
Sulfur dioxide	12	3.05	9.78	17.8	19
Propylene	4.66	3.2	3.45	3.65	4.91
Propane	4.66	3.67	3.88	4.1	4.63
Propadiene	6.5	4.12	4.39	4.7	7.55
Methylacetylene	9.5	4.12	4.84	5.14	11.3
Methyl chloride	7.43	3.93	4.67	4.92	9.2
Vinyl chloride	14.9	6.04	9.04	9.7	17.3
Ethylene oxide	17.7	6.06	8.78	9.7	23.3
Ethyl chloride	35	12.25	19.3	20.7	43.2
Carbon disulfide	—	32.4	—	—	40.7

STATIONARY PHASES FOR POROUS-LAYER OPEN TUBULAR COLUMNS

The practical application of solid adsorbents is commonly in porous-layer open tubular (PLOT) columns. In this table, several of the more common PLOT column stationary phases are listed, along with the separations that may be affected and some additional information [1–4]. The maximum temperatures listed represent the point of severe resolution loss. The materials are often chemically stable to much higher temperatures. The user should also be aware that the adsorption of water during use will often change retention characteristics dramatically, sometimes resulting in a reversal of positions of adjacent peaks. Due to surface adsorption of solutes, some experimentation with temperature may be necessary to prevent tailing or to avoid statistical correlation (or a propagating error) among replicate analyses [5].

REFERENCES

1. Jeffery, P.G. and Kipping, P.J., *Gas Analysis by Gas Chromatography*, Pergamon Press, Oxford, 1972.
2. Cowper, C.J. and DeRose, A.J., *The Analysis of Gases by Chromatography*, Pergamon Press, Oxford, 1983.
3. Breck, D.W., *Zeolite Molecular Sieves*, John Wiley & Sons, New York, 1973.
4. Poole, C.F., *The Essence of Chromatography*, Elsevier, Amsterdam, 2003.
5. Bruno, T.J., An apparatus for direct fugacity measurements on mixtures containing hydrogen, *J. Res. Natl. Bur. Stand.*, 90(2), 1127, 1985.

Phase	Max. Temp. (°C)	Separation Affected	Notes
Silica gel	250	H_2, Air, CO, C_1–C_4, normal hydrocarbons, alkenes and alkynes, inorganic gases, volatile ethers	Very hydrophilic; requires activation; can be unpredictable; largely replaced by porous polymers; bonded versions are suitable for use with GC-mass spectrometry (MS), because of the absence of particles.
Alumina, deactivated with KCl	300	C_1–C_8 hydrocarbons, especially useful for resolution of propadiene and butadiene from ethylene and propylene.	Least polar of the alumina phases; lowest retention of olefins relative to the corresponding paraffin; specified in many standard methods.
Alumina, deactivated with Na_2SO_4	300	C_1–C_8 hydrocarbons, resolves acetylene from n-butane and propylene from isobutane	Medium- and high-polarity phases are available among the alumina phases; specified in many standard methods.
Cyclo dextrin	260	Light hydrocarbons (C_1–C_{10}) and halocarbons	α and β cyclodextrins have been used; care should be taken with halocarbon analysis, due to the potential of HF contamination of the sample.
Styrene–divinylbenzene	250	C_1–C_3 hydrocarbons; paraffins up to C_{12}; CO from air, ethers, sulfur gases, water	See the information on porous polymers.
Divinylbenzene ethylene glycol-dimethacrylate		C_1–C_7 hydrocarbon isomers; CO_2, CH_4, amines, common solvents, alcohols, aldehydes, ketones	More polar than styrene–divinylbenzene phases.
Molecular sieve, 5A	350	Air and light gas analysis; H_2, O_2, N_2, (CH_4, CO, NO, SF_6 co-elute); thick film phase can resolve Ar from O_2 at 35 °C	Synthetic calcium alumino-silicate (zeolite) having an effective pore diameter of 0.5 nm CO_2 is adsorbed strongly; 5A usually gives the best results of all synthetic zeolites; thick and thin film columns are available.
Molecular sieve, 13X		Same separations as those performed on 5A but with C_1–C_4, alkanes, alkenes, and alkynes being separated as well	Sodium alumino-silicate (zeolite), having a larger pore size than 0.5 nm, thus producing lower retention times and less resolution; 28.6 % (mass/mass) water capacity
Monolithic carbon	350	C_1–C_5 hydrocarbon isomers; acetylene in ethylene; methane	Phase consists of a bonded carbon monolith; suitable for use with GC-MS, because of the absence of particles.

SILICONE LIQUID PHASES

The following table lists the chromatographic properties of some of the more popular polysiloxane-based liquid phases [1–8]. The polysiloxanes are the most widely used stationary phases in GC and are especially applicable to capillary columns. The listing provided here is far from exhaustive. Since it is impractical to present the structures of all polysiloxane-based phases, the OV phases (originally from the Ohio Valley Chemical Co) have been chosen as representative since their properties are among the most well characterized. The phases which are listed in the notes as "similar phases" have thermal and chromatographic properties which are similar to the phase described. In modern applications of capillary-column gas chromatography, silicone phases are cross-linked to provide stability. Cross-linking can change the properties of a phase to some extent, but often this is relatively minor.

The reader should note that there are many commercial variations of silicone liquid phases available. In compiling properties such as those listed in this table, one must strike a balance between general usefulness and simply providing information that is contained in vendor catalogs, promotional brochures, and web sites. In that context, this table is meant to serve as a starting point for the design of an analysis.

The McReynolds constants are indices with respect to the following test probe compounds:

McReynolds Constant	Test Probe
1	benzene
2	1-butanol
3	2-pentanone
4	1-nitropropane
5	pyridine
6	2-methyl-2-pentanol
7	1-iodobutane
8	2-octyne
9	1,4-dioxane
10	cis-hydrindane

The use of these constants is described in the table entitled "Properties of Some Liquid Phases for Packed Columns." The viscosity data, where available, are presented in cSt, which is $10^{-6} m^2/s$. Cross-linked silicone phases based on the silicones are especially valuable for capillary GC. They are not specifically treated in this table since the differences in many properties are quite often subtle. The cross-linked phases have much longer lifetimes due to the effective immobilization.

Abbreviations:
  Solvents:

    Ace—acetone   Chlor—chloroform
    Tol—toluene
    (When a silicone fluid is cross-linked, it will be insoluble.)

Note: N denotes a phenyl group in a structure.
        N, nonpolar; I, intermediate polarity; P, polar

REFERENCES

1. Yancey, J.A., Liquid phases used in packed gas chromatographic columns. Part I. Polysiloxane liquid phases, *J. Chromatogr. Sci.*, 23, 161, 1985.
2. McReynolds, W.O., *J. Chromatogr. Sci.*, 8, 685, 1970.
3. Mann, J.R. and Preston, S.T., *J. Chromatogr. Sci.*, 11, 216, 1973.
4. Trash, C.R., *J. Chromatogr. Sci.*, 11, 196, 1973.
5. McNair, H.M. and Bonelli, E.J., *Basic Gas Chromatography*, Varian Aerograph, Palo Alto, 1969.
6. Heftmann, E., *Chromatography: A Laboratory Handbook of Chromatographic and Electrophoretic Methods*, 3rd ed., Van Nostrand Reinhold, New York, 1975.
7. Grant, D.W., *Gas Liquid Chromatography*, Van Nostrand Reinhold, London, 1971.
8. Coleman, A.E., *J. Chromatogr. Sci.*, 11, 198, 1973.

Liquid Phase	Solvent	Av. Mol. Mass	Viscosity	T_{min} (°C)	T_{max} (°C)	Polarity	McReynolds Constants										Structure	Notes
							1	2	3	4	5	6	7	8	9	10		
OV-1, dimethyl silicone (gum)	Tol	$>10^6$	Gum	100	350	N	16	55	44	65	42	32	4	23	45	−1		100 % methyl, low selectivity, boiling point separations; similar phases: UCC-L45, UCC-W-98, SE-30, DB-1, HP-1
OV-101, dimethyl silicone fluid	Tol	3×10^4	1,500	20	350	N	17	57	45	67	43	33	4	23	46	−2		100 % methyl, low selectivity, boiling point separations; similar phases: DC-11, DC-200, DC-550, SF-96, SP-2100, STAP
Phenylmethyl-dimethyl silicone	Tol	2×10^4		20	350	I	33	72	66	99	67	46	24	36	68	10		5 % phenyl methyl, boiling point separations; similar phases: DB-5, HP-5. This does not have a corresponding OV identification number because it was formulated later than the other fluids. This phase is probably the most common starting phase for most analyses and one that is specified in many standard protocols; similar to SE-54

(Continued)

Liquid Phase	Solvent	Av. Mol. Mass	Viscosity	T_{min} (°C)	T_{max} (°C)	Polarity	McReynolds Constants										Structure	Notes
							1	2	3	4	5	6	7	8	9	10		
OV-3, phenylmethyl-dimethyl silicone	Ace	2×10^4	500	20	350	I	44	86	81	124	88	55	39	46	84	17		10 % phenylmethyl; similar to SE-52
OV-7, phenylmethyl-dimethyl silicone	Ace	1×10^4	500	20	350	I	69	113	111	171	128	77	68	66	120	35		20 % phenylmethyl
OV-11, phenylmethyl-dimethyl silicone	Ace	7×10^3	500	0	350	I	102	142	145	219	178	100	103	92	164	59		35 % phenylmethyl; similar phases: DC-710
OV-17, phenylmethyl silicone	Ace	4×10^3	1,300	20	350	I	119	158	162	243	202	112	119	105	184	69		50 % methyl, similar phases: SP-2250
OV-22, phenylmethyl-diphenyl silicone	Ace	8×10^3	>50,000	20	350	I	160	188	191	283	253	133	152	132	228	99		65 % phenyl
OV-25, phenylmethyl-diphenyl silicone	Ace	1×10^4	100,000	20	350	I	178	204	208	305	280	144	169	147	215	113		75 % phenyl
OV-61, diphenyldimethyl silicone	Tol	4×10^4	>50,000	20	350	I	101	143	142	213	174	99	—	86	—	—		33 % phenyl

(Continued)

Liquid Phase	Solvent	Av. Mol. Mass	Viscosity	T_{min} (°C)	T_{max} (°C)	Polarity	McReynolds Constants										Structure	Notes
							1	2	3	4	5	6	7	8	9	10		
OV-73, diphenyldimethyl silicone gum	Tol	8×10^5	Gum	20	350	I	40	6	76	114	85	57	—	39	—	—		5.5 % phenyl, similar phases: SE-52, SE-54
OV-105, cyano propylmethyl-diemthyl silicone	Ace		1,500	20	250	N,I	36	108	93	139	86	74	—	29	—	—		
OV-202, trifluoropropyl-methyl silicone	Chlor	1×10^4	500	0	275	I,P	146	238	358	468	310	202	139	56	283	60		50 % trifluoropropyl fluid, similar phases: SP-2401; phases can be prone to oxidation at the Si-C bond
OV-210, trifluoropropyl-methyl silicone	Chlor	2×10^5	10,000	20	275	I,P	146	238	358	468	310	206	139	56	283	60		50 % trifluoropropyl, similar phases: QF-1, FS-1265, SD-2401; phases can be prone to oxidation at the Si-C bond
OV-215, trifluoropropyl-methyl silicone gum			Gum			I,P	149	240	363	478	315	208	—	56	—	—		50 % trifluoropropyl; phases can be prone to oxidation at the Si-C bond
OV-225, cyanopropyl-methylphenyl methylsilicone	Ace	8×10^3	9,000	20	275	I,P	228	369	338	492	386	282	226	150	342	117		25 % phenyl, 25 % cyanopropylmethyl; similar phases: EX-60, AN-600

(Continued)

Liquid Phase	Solvent	Av. Mol. Mass	Viscosity	T_{min} (°C)	T_{max} (°C)	Polarity	McReynolds Constants										Structure	Notes
							1	2	3	4	5	6	7	8	9	10		
OV-275, dicyanoallyl silicone	Ace	5×10^3	20,000	20	275	P	781	1,006	885	1,177	1,089	—	—	—	—	—		
Dexsil 300 copolymer	Chlor	16,000–20,000	Waxy solid	50	450	I	47	80	103	148	96	—	—	—	—	—		Carborane-methyl silicone; siloxane-to-carborane ratio, 4:1; used for methyl esters, aromatic amines, halogenated alcohols, pesticides, polyphenyl ethers, silicone oils
Dexsil 400	Chlor	12,000–16,000	—	20	375	I	60	115	140	188	174	—	—	—	—	—		Carborane-methyl phenyl silicone copolymer; siloxane-to-carborane ratio, 5:1
Dexsil 410	Chlor	9,000–12,000		20	375	I	85	165	170	240	180	—	—	—	—	—		Carborane-methyl-β-silicone cyanoethyl copolymer; siloxane-to-carborane ratio, 5:1

PROPERTIES OF COMMON CROSS-LINKED SILICONE STATIONARY PHASES

The preceding table on the silicone stationary phases provides a useful means of comparing the various silicone stationary phases that are the most widely used in GC. As was noted, the fluids that were used for those measurements were not cross-linked but rather were coated on a packing. Cross-linking or application to an open tubular (or capillary) column will not change the chromatographic behavior (that is reflected in retention indices such as those of Kovats and McReynolds) to any significant extent; however, in the following table, we provide chromatographic data on the two most common cross-linked phases [1]. We note that the efficiency (in terms of the number of theoretical plates) of a typical commercial open tubular column is much higher than that of a column prepared with a coated packing. The retention indices presented in these tables were measured at 120 °C isothermally. Retention indices are temperature dependent, the temperature dependence of the Kovats indices has been studied for many compounds [2].

REFERENCES

1. Vickers, A. *Life Sciences and Chemical Analysis*, Agilent Technologies, Folsom, CA, personal communication, 2009.
2. Bruno, T.J., Wertz, K. H., and Caciari, M., Kovats retention indices of halocarbons on a hexafluoro-propylene epoxide modified graphitized carbon black, *Anal. Chem.*, 68(8), 1347–1359, 1996.

Phase: 5 % Phenyl Dimethylpolysiloxane

Temperature ranges:
−60 °C to 325 °C isothermally, −60 °C to 350 °C programmed for <0.32 mm I.D. columns.
−60 °C to 300 °C isothermally, −60 °C to 320 °C programmed for 0.53 mm I.D. columns.
−60 °C to 260/280 °C for >2.0 μm films.
Note: I.D. refers to the internal diameter of the column.

Similar Phases

DB-5, Ultra-2, SPB-5, CP-Sil 8 CB, Rtx-5, BP-5, OV-5, 007-2 (MPS-5), SE-52, SE-54, XTI-5, PTE-5, HP-5MS, ZB-5, AT-5, and MDN-5

Notes: This phase is probably the most commonly used stationary phase in GC, since it combines boiling point separation with a minor contribution of a specific interaction; typically used as the first phase in any method development; versatile for hydrocarbons and more polar compounds; and other varieties of this phase.

Probe Compound	McReynolds Constant	McReynolds Code	Kovats' Retention Index
n-Hexane			600
1-Butanol	66	y′	656
Benzene	31	x′	684
2-Pentanone	61	z′	688
n-Heptane			700
1,4-Dioxane	64	L	718
2-Methyl-2-pentanol	41	H	731
1-Nitropropane	93	u′	745
Pyridine	62	s′	761
n-Octane			800
Iodobutane	22	J	840
2-Octyne	35	K	876
n-Nonane			900
n-Decane			1,000

Phase: Dimethylpolysiloxane

Temperature range:
−60 °C to 325 °C for normal operations, periodic operation to 350 °C can be used to facilitate column clean-up.
−60 °C to 260/280 °C for >2.0 μm films.

Similar Phases

DB-1, OV-1, HP-1, DB-1ms, HP-1ms, Rtx-1, Rtx-1ms, CP-Sil 5 CB Low Bleed/MS, MDN-1, AT-1.

Notes: Useful for the separation of hydrocarbons, pesticides, PCBs, phenols, sulfur compounds, flavors and fragrances, and some amines; columns are typically stable and low bleed; a good all-purpose column used to begin method development protocols.

Probe Compound	McReynolds Constant	McReynolds Code	Kovats' Retention Index
n-Hexane			600
1-Butanol	54	y′	644
Benzene	16	x′	669
2-Pentanone	44	z′	671
n-Heptane			700
1,4-Dioxane	46	L	700
2-Methyl-2-pentanol	31	H	721
1-Nitropropane	62	u′	714
Pyridine	44	s′	743
n-Octane			800
Iodobutane	3	J	821
2-Octyne	23	K	864
n-Nonane			900
n-Decane			1,000

MESOGENIC STATIONARY PHASES

The following table lists the liquid crystalline materials that have found usefulness as gas-chromatographic stationary phases in both packed and open tubular column applications. In each case, the name, structure, and transition temperatures are provided (where available), along with a description of the separations which have been done using these materials. The table has been divided into two sections. The first section contains information on phases that have either smectic or nematic phases or both, while the second section contains mesogens that have a cholesteric phase. It should be noted that each material may be used for separations other than those listed, but the listing contains the applications reported in the literature.

It should be noted that some of the mesogens listed in this table are not commercially available and must be prepared synthetically for laboratory use. The reader is referred to the appropriate citation for details.

REFERENCES

1. Panse, D.G., Naikwadi, K.P., Bapat, B.V., and Ghatge, B.B., Applications of laterally mono and disubstituted liquid crystals as stationary phases in gas liquid chromatography, *Ind. J. Tech.*, 19, 518–521, Dec 1981.
2. Grushka, E. and Solsky, J.F., p-Azoxyanisole liquid crystal as a stationary phase for capillary column gas chromatography, *Anal. Chem.*, 45(11), 1836, 1973.
3. Witkiewicz, Z. and Stanislaw, P., Separation of close-boiling compounds on liquid-crystalline stationary phases, *J. Chromatogr.*, 154, 60, 1978.
4. Naikwadi, K.P., Panse, D.G., Bapat, B.V., and Ghatge, B.B., I. Synthesis and application of stationary phases in gas-liquid chromatography, *J. Chromatogr.*, 195, 309, 1980.
5. Dewar, M. and Schroeder, J.P., Liquid crystals as solvents. I. The use of nematic and smectic phases in gas-liquid chromatography, *J. Am. Chem. Soc.*, 86, 5235, 1964.
6. Dewar, M., Schroeder, J.P., and Schroeder, D., Molecular order in the nematic mesophase of 4,4'-di-n-hexyloxyazoxybenzene and its mixture with 4,4'-dimethoxyazoxybenzene, *J. Org. Chem.*, 32, 1692, 1967.
7. Naikwadi, K.P., Rokushika, S., and Hatano, H., New liquid crystalline stationary phases for gas chromatography of positional and geometrical isomers having similar volatilities, *J. Chromatogr.*, 331, 69, 1985.
8. Richmond, A.B., Use of liquid crystals for the separation of position isomers of disubstituted benzenes, *J. Chrom. Sci.*, 9, 571, 1971.
9. Witkiewicz, Z., Pietrzyk, M., and Dabrowski, R., Structure of liquid crystal molecules and properties of liquid-crystalline stationary phases in gas chromatography, *J. Chromatogr.*, 177, 189, 1979.
10. Ciosek, M., Witkiewicz, Z., and Dabrowski, R., Direct gas—chromatographic determination of 2-napthylamine in 1-napthylamine on liquid-crystalline stationary phases, *Chemia Analityczna*, 25, 567, 1980.
11. Jones, B.A., Bradshaw, J.S., Nishioka, M., and Lee, M.L., Synthesis of smectic liquid-crystalline polysiloxanes from biphenylcarboxylate esters and their use as stationary phases for high-resolution gas chromatography, *J. Org. Chem.*, 49, 4947, 1984.
12. Porcaro, P.J. and Shubiak, P., Liquid crystals as substrates in the GLC of aroma chemicals, *J. Chromatogr. Sci.*, 9, 689, 1971.
13. Witkiewicz, Z., Suprynowicz, Z., Wojcik, J., and Dabrowski, R., Separation of the isomers of some disubstituted benzenes on liquid crystalline stationary phases in small-bore packed micro-columns, *J. Chromatogr.*, 152, 323, 1978.
14. Dewar, M. and Schroeder, J.P., Liquid-crystals as solvents. II. Further studies of liquid crystals as stationary phases in gas-liquid chromatography, *J. Org. Chem.*, 30, 3485, 1965.

15. Witkiewicz, Z., Szule, J., Dabrowski, R., and Sadowski, J., Properties of liquid crystalline cyanoazoxybenzene alkyl carbonates as stationary phases in gas chromatography, *J. Chromatogr.*, 200, 65, 1980.
16. Witkiewicz, Z., Suprynowicz, Z., and Dabrowski, R., Liquid crystallinecyanoazoxybenzene alkyl carbonates as stationary phases in small-bore packed micro-columns, *J. Chromatogr.*, 175, 37, 1979.
17. Lochmüller, C.H. and Souter, R.W., Direct gas chromatographic resolution of enantiomers on optically active mesophases, *J. Chromatogr.*, 88, 41, 1974.
18. Markides, K.E., Nishioka, M., Tarbet, B.J., Bradshaw, J.S., and Lee, M.L., Smectic biphenylcarboxylate ester liquid crystalline polysiloxane stationary phase for capillary gas chromatography, *Anal. Chem.*, 57, 1296, 1985.
19. Vetrova, Z.P., Karabanov, N.T., Shuvalova, T.N., Ivanova, L.A., and Yashin, Ya.I., The use of p-n-butyl oxybenzoic acid as liquid crystalline sorbent in gas chromatography, *Chromatographia*, 20, 41, 1985.
20. Cook, L.E. and Spangelo, R.C., Separation of monosubstituted phenol isomers using liquid crystals, *Anal. Chem.*, 46(1), 122, 1974.
21. Kong, R.C. and Milton, L.L., Mesogenic polysiloxane stationary phase for high resolution gas chromatography of isomeric polycyclic aromatic compounds, *Anal. Chem.*, 54, 1802, 1982.
22. Bartle, K.D., El-Nasri, A.I., and Frere, B., *Identification and Analysis of Organic Pollutants in Air*, Ann Arbor Science, Ann Arbor, 183, 1984.
23. Finklemann, H., Laub, R.J., Roberts, W.E., and Smith, C.A., Use of mixed phases for enhanced gas chromatographic separation of polycyclic aromatic hydrocarbons. II. Phase transition behavior, mass-transfer non-equilibrium, and analytical properties of a mesogen polymer solvent with silicone diluents, *Polynucl. Aromat. Hydrocarbons: Phys. Biol. Chem. 6th Int. Symp.*, M. Cooke, ed., 275–285, 1982.
24. Janini, G.M., Recent usage of liquid crystal stationary phases in gas chromatography, *Adv. Chromatogr.*, 17, 231, 1979.
25. Witkiewicz, Z. and Waclawczyk, A., Some properties of high-temperature liquid crystalline stationary phases, *J. Chromatogr.*, 173, 43, 1979.
26. Zielinski, W.L., Johnston, R., and Muschik, G.M., Nematic liquid crystal for gas-liquid chromatographic separation of steroid epimers, *Anal. Chem.*, 48, 907, 1976.
27. Smith, and Wozny, M.E., Gas chromatographic separation of underivatized steroids using BPhBT liquid crystal stationary phase, HRC CC, *J. High Resolut. Chromatogr.*, 3, 333, 1980.
28. Barrall, E.M., Porter, R.S., and Johnson, J.F., Gas chromatography using cholesteryl ester liquid phase, *J. Chromatogr.*, 21, 392, 1966.
29. Heath, R.R. and Dolittle, R.E., Derivatives of cholesterol cinnamate. A comparison of the separations of geometrical isomers when used as gas chromatographic stationary phases, HRC CC, *J. High Resolut. Chromatogr.*, 6, 16, 1983.
30. Sonnet, P.E. and Heath, R.R., Aryl substituted diastereomeric alkenes: gas chromatographic behavior on a non-polar versus a liquid crystal phase, *J. Chromatogr.*, 321, 127, 1985.

Name: 2-chloro-4'-n-butyl-4-(4-n-butoxybenzoyloxy) azobenzene
Structure:

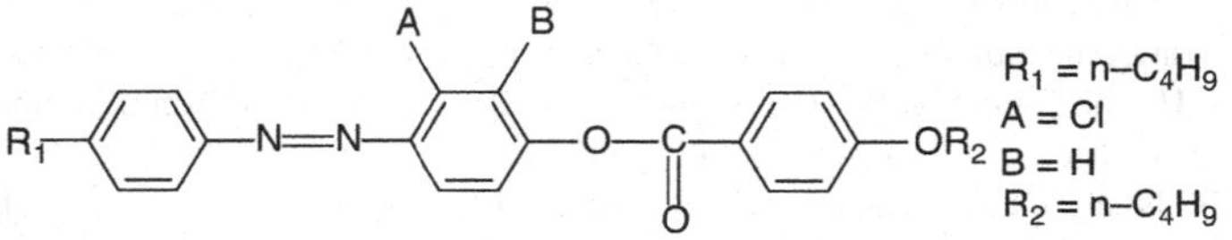

$R_1 = n\text{--}C_4H_9$
$A = Cl$
$B = H$
$R_2 = n\text{--}C_4H_9$

Thermophysical Properties:
solid   → nematic 87.2 °C
nematic → isotropic 168 °C
Analytical Properties: separation of close-boiling disubstituted benzenes
Reference: [1]

Name: p-azoxyanisole (4,4'-dimethoxyazoxybenzene)
Structure:

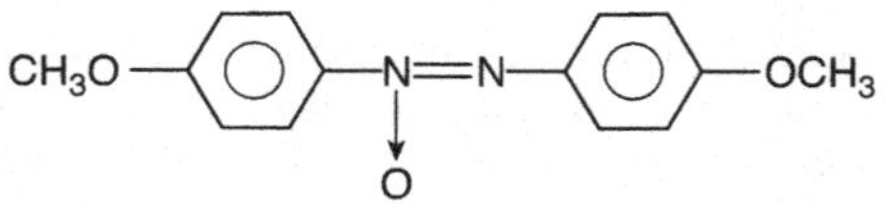

Thermophysical Properties:
solid   → nematic 118 °C
nematic → isotropic 135 °C

Note: Supercooling has been noted at 110 °C by observing nematic-like properties.
Liquid crystalline behavior can sometimes persist to 102 °C

Analytical Properties: separation of xylenes, separation of lower molecular weight aromatic hydrocarbon isomers, especially at the lower area of the nematic region
Reference: [2]

Name: 2-chloro-4'-n-butyl-4-(4-methylbenzoyloxy) azobenzene
Structure:

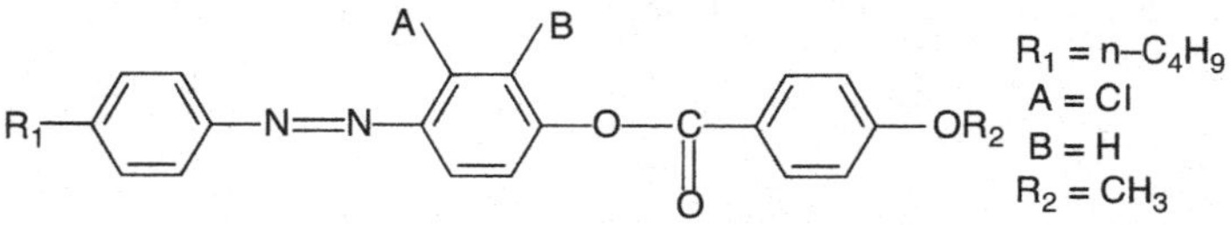

$R_1 = n\text{--}C_4H_9$
$A = Cl$
$B = H$
$R_2 = CH_3$

Thermophysical Properties:
solid   → nematic 92.5 °C
nematic → isotropic 176 °C

Note: Supercooling has been noted at 110 °C by observing nematic-like properties.
Liquid crystalline behavior can sometimes persist to 102 °C

Analytical Properties: separation of close-boiling disubstituted benzenes
Reference: [1]

Name: 2-chloro-4′-ethyl-4-(4-n-butoxybenzoyloxy) azobenzene
Structure:

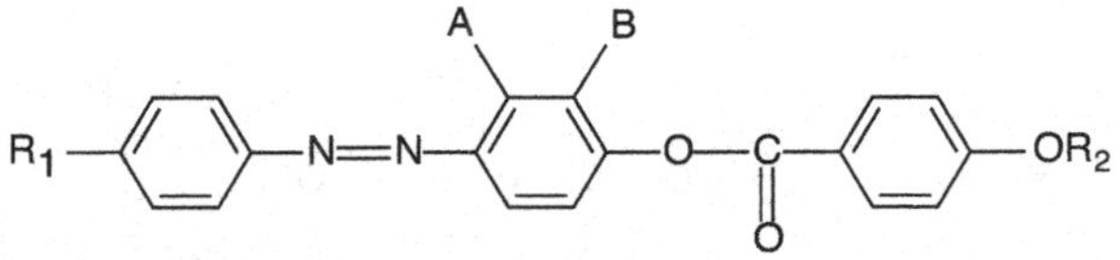

$R_1 = C_2H_5$
$A = Cl$
$B = H$
$R_2 = n–C_4H_9$

Thermophysical Properties:
solid   → nematic 117 °C
nematic → isotropic 172 °C

Note: Supercooling has been noted at 110 °C by observing nematic-like properties.
Liquid crystalline behavior can sometimes persist to 102 °C

Analytical Properties: separation of close-boiling disubstituted benzenes
Reference: [1]

Name: 2-chloro-4′-n-butyl-4-(4-ethoxybenzoyloxy) azobenzene
Structure:

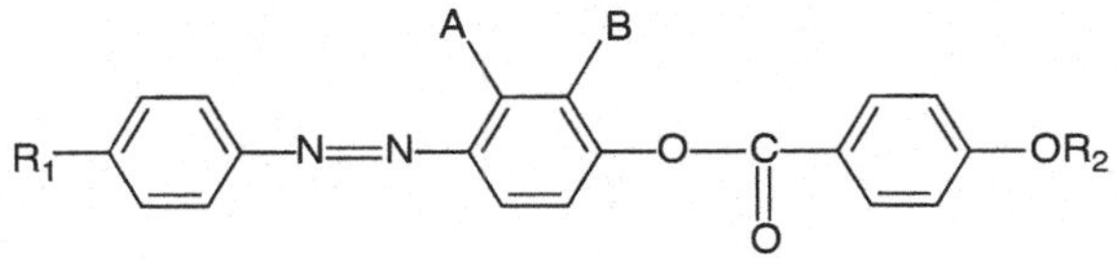

$R_1 = n–C_4H_9$
$A = Cl$
$B = H$
$R_2 = C_2H_5$

Thermophysical Properties:
solid   → nematic 89.7 °C
nematic → isotropic 170 °C
Analytical Properties: separation of close-boiling disubstituted benzenes
Reference: [1]

Name: 2-chloro-4′-methyl-4(4-n-butoxybenzoyloxy) azobenzene
Structure:

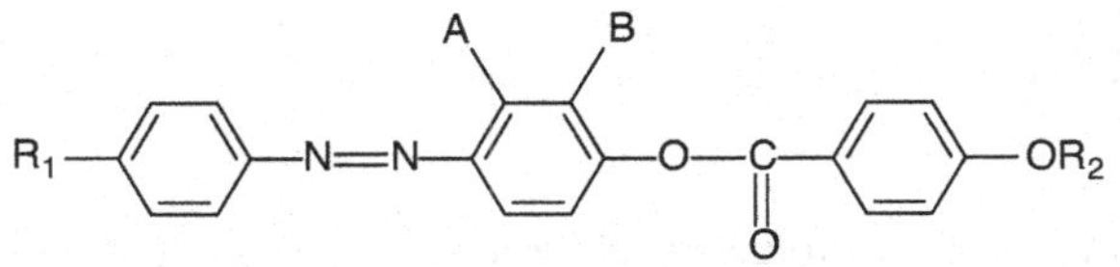

$R_1 = CH_3$
$A = Cl$
$B = H$
$R_2 = n–C_4H_9$

Thermophysical Properties:
solid   → nematic 112 °C
nematic → isotropic 165 °C
Analytical Properties: separation of close-boiling disubstituted benzenes
Reference: [1]

Name: 2-chloro-4′-n-methyl-4-(4-ethoxybenzoyloxy) azobenzene
Structure:

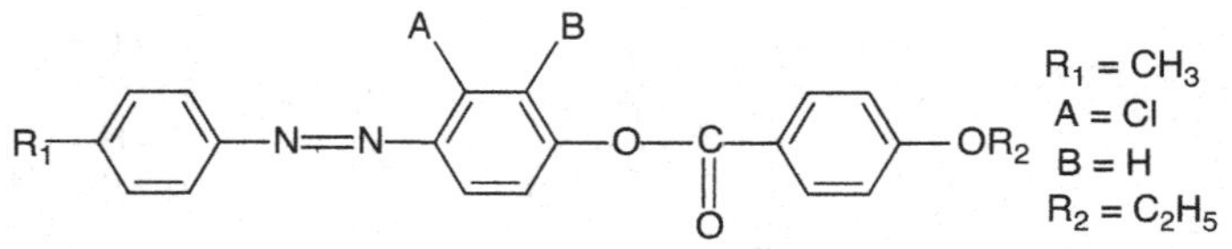

$R_1 = CH_3$
$A = Cl$
$B = H$
$R_2 = C_2H_5$

Thermophysical Properties:
solid → nematic 128.3 °C
nematic → isotropic 185 °C
Analytical Properties: separation of close-boiling disubstituted benzenes
Reference: [1]

Name: p-cyano-p′-pentoxyazoxybenzene
Structure:

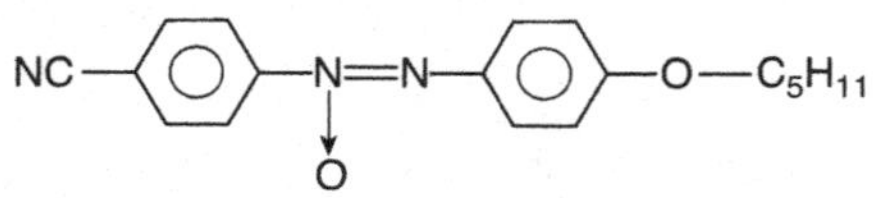

Thermophysical Properties:
solid → nematic 124 °C
nematic → isotropic 153 °C
Analytical Properties: complete separation of ethyltoluenes, chlorotoluenes, bromotoluenes, and dichlorobenzenes. Also, ethylbenzene from xylenes and propylbenzene from ethyltoluenes
Reference: [3]

Name: p-cyano-p′-pentoxyazobenzene
Structure:

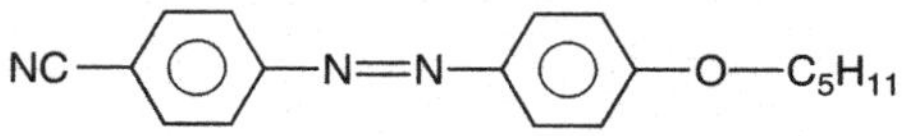

Thermophysical Properties:
solid → nematic 106 °C
nematic → isotropic 116.5 °C
Analytical Properties: separation of ethyltoluenes, chlorotoluenes, bromotoluenes, and dichloro-benzenes. Also, ethylbenzenes from xylenes and propylbenzenes from ethylbenzenes
Reference: [3]

Name: p-cyano-p′-pentoxyazoxybenzene (mixed isomers)
Structure:

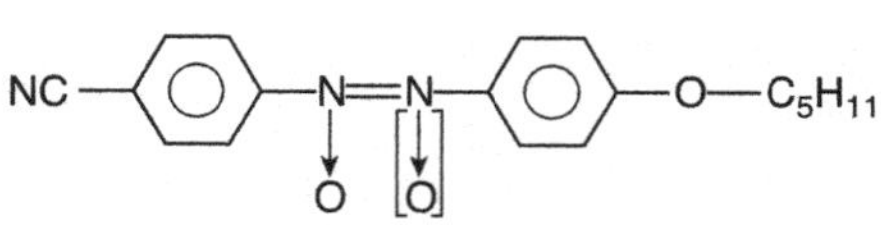

Thermophysical Properties:
solid → nematic 93.5 °C
nematic → isotropic 146.5 °C
Analytical Properties: complete separation of ethyltoluenes, chlorotoluenes, bromotoluenes, and dichlorobenzenes. Also, ethylbenzene from xylenes and propylbenzene from ethyltoluenes
Reference: [3]

Name: p-cyano-p′-octoxyazoxybenzene (mixed isomers)
Structure:

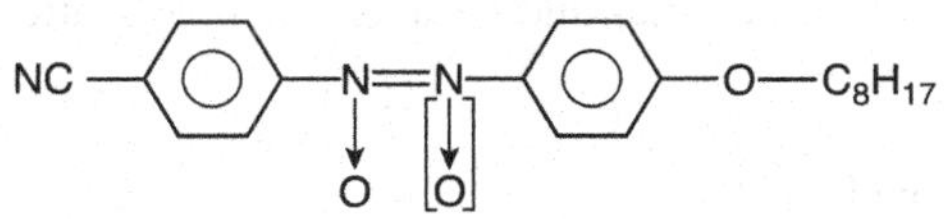

NC —⟨○⟩— N=N —⟨○⟩— O—C_8H_{17}
         ↓   ↓
         O [O]

Thermophysical Properties:
solid → smectic 71 °C
smectic → nematic 117 °C
nematic → isotropic 135 °C
Analytical Properties: separation of ethyltoluenes, chlorotoluenes, bromotoluenes, and dichlorobenzenes. Also, ethylbenzene from xylenes and propylbenzene from ethylbenzenes
Reference: [3]

Name: p-cyano-p′-octoxyazoxybenzene
Structure:

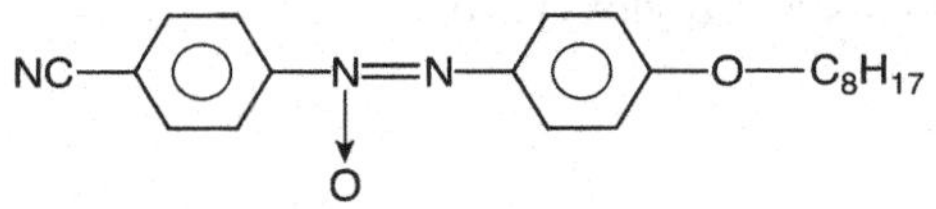

NC—⟨○⟩—N=N—⟨○⟩—O—C_8H_{17}
           ↓
           O

Thermophysical Properties:
solid → smectic 100.5 °C
smectic → nematic 138.5 °C
nematic → isotropic 148.5 °C
Analytical Properties: separation of ethyltoluenes, chlorotoluenes, bromotoluenes, and dichlorobenzenes. Also, ethylbenzene from xylenes and propylbenzene from ethylbenzenes
Reference: [3]

Name: 4′-n-butyl-4(4-n-butoxybenzoyloxy) azobenzene
Structure:

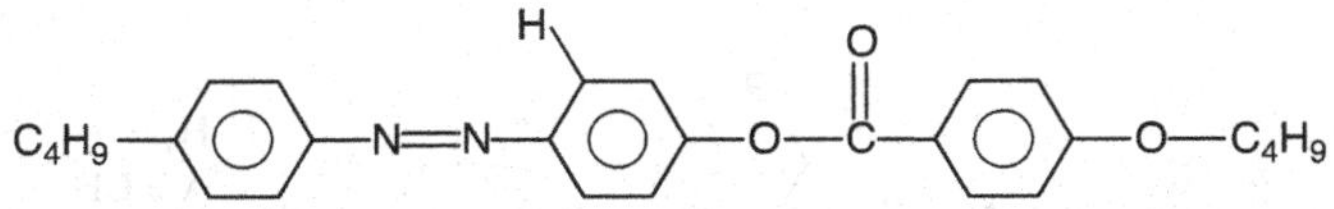

C_4H_9—⟨○⟩—N=N—⟨○⟩—O—C(=O)—⟨○⟩—O—C_4H_9

Thermophysical Properties:
solid → nematic 94 °C
nematic → isotropic 234 °C
Analytical Properties: separation of chlorinated biphenyls
Reference: [4]

Name: 4-4′-di-n-heptyloxyazoxybenzene
Structure:

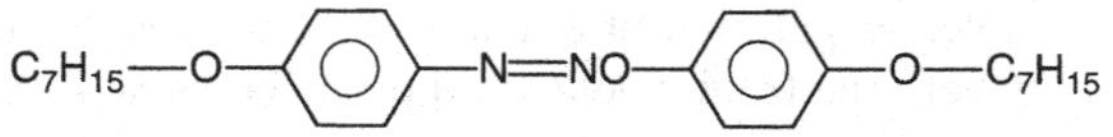

Thermophysical Properties:
solid → nematic 95 °C
nematic → isotropic 127 °C
Analytical Properties: separation of meta- and para-xylene in nematic region
Reference: [5]

Name: 4,4′-di-n-hexyloxyazoxybenzene
Structure:

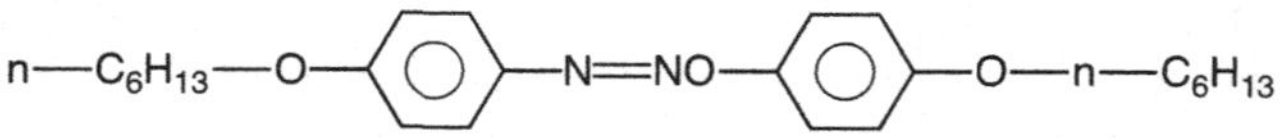

Thermophysical Properties:
solid → nematic 81 °C
nematic → isotropic 129 °C
Analytical Properties: separation of meta- and para-xylene using GC
Reference: [5,6]

Name: 4′-methoxy-4-(-4-n-butoxybenzoyloxy) azobenzene
Structure:

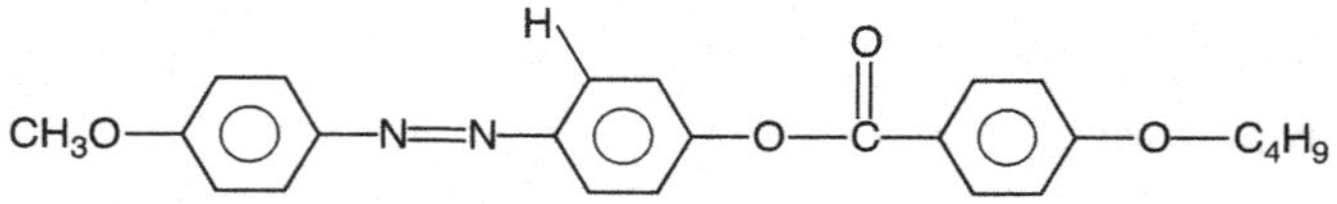

Thermophysical Properties:
solid → nematic 116 °C
nematic → isotropic 280 °C
Analytical Properties: separation of chlorinated biphenyls
Reference: [4]

Name: 2-methyl-4′-n-butyl-4-(4-n-butoxybenzoyloxy) azobenzene
Structure:

$R_1 = n–C_4H_9$
$A = CH_3$
$B = H$
$R_2 = n–C_4H_9$

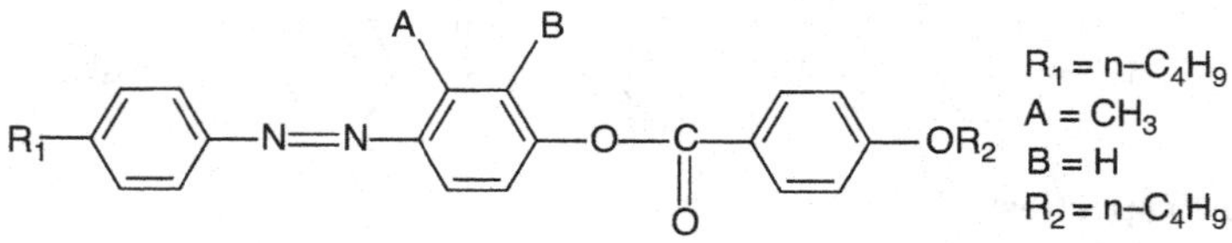

Thermophysical Properties:
solid   → nematic 90 °C
nematic → isotropic 175 °C
Analytical Properties: separation of close-boiling disubstituted benzenes
Reference: [1]

Name: 2-methyl-4′-n-butyl-4-(p-methoxycinnamoyloxy) azobenzene
Structure:

Thermophysical Properties:
solid   → nematic 109 °C
nematic → isotropic 253 °C
Analytical Properties: separation of positional isomers of aromatic hydrocarbons
Reference: [7]

Name: 2-methyl-4′-methoxy-4-(4-ethoxybenzoyloxy) azobenzene
Structure:

Thermophysical Properties:
solid   → nematic 125 °C
nematic → isotropic 244 °C
Analytical Properties: separation of chlorinated biphenyls
Reference: [4]

Name: 2-methyl-4′-methoxy-4-(4-methoxybenzoyloxy) azobenzene
Structure:

Thermophysical Properties
solid   → nematic 160 °C
nematic → isotropic 253 °C
Analytical Properties: separation of chlorinated biphenyls
Reference: [4]

Name: 2-methyl-4′-ethyl-4-(4′-methoxycinnamyloxy) azobenzene
Structure:

Thermophysical Properties:
solid   → nematic 126 °C
nematic → isotropic 262 °C
Analytical Properties: separation of polyaromatic hydrocarbons and insect sex pheromones
Reference: [5]

Name: 2-methyl-4′-methoxy-4-(p-methoxycinnamoyloxy) azobenzene
Structure:

Thermophysical Properties:
solid   → nematic 149 °C
nematic → isotropic 298 °C
Analytical Properties: separation of positional isomers of aromatic compounds and geometrical isomers of sex pheromones
Reference: [7]

Name: 2-methyl-4′-methyl-4-(4-ethoxybenzoyloxy) azobenzene
Structure:

Thermophysical Properties:
solid   → nematic 125 °C
nematic → isotropic 220 °C
Analytical Properties: separation of chlorinated biphenyls
Reference: [4]

Name: 4,4′-azoxyphenetole
Structure

Thermophysical Properties:
solid → nematic 138 °C
nematic → isotropic 168 °C
Analytical Properties: separation of meta and para isomers of disubstituted benzenes
Reference: [8]

Name: 4,4-biphenylylene-bis-[p-(heptyloxy) benzoate]
Structure:

Thermophysical Properties:
solid → smectic 150 °C
smectic → nematic 211 °C
nematic → isotropic 316 °C
Analytical Properties: separation of meta and para isomers of disubstituted benzenes
Reference: [8]

Name: p′-ethylazoxybenzene p-cyanobenzoate (mixed isomers)
Structure:

Thermophysical Properties:
melting range → 114 °C –136 °C
nematic → isotropic >306 °C
Analytical Properties: separation of substituted xylenes
Reference: [9]

Name: p′-ethylazoxybenzene p-cyanobenzoate (pure isomer)
Structure:

Thermophysical Properties:
solid → nematic 115 °C
nematic → isotropic 294 °C
Analytical Properties: separation of nitronaphthalenes
Reference: [10]

Name: p′-ethylazobenzene p′-cyanobenzoate
Structure:

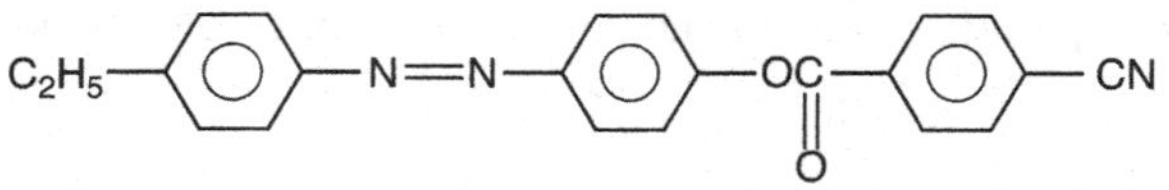

Thermophysical Properties:
solid   → nematic 138 °C –140 °C
nematic → isotropic 292 °C
Analytical Properties: separation of substituted xylenes
Reference: [9]

Name: p′-ethylazobenzene p-methylbenzoate
Structure:

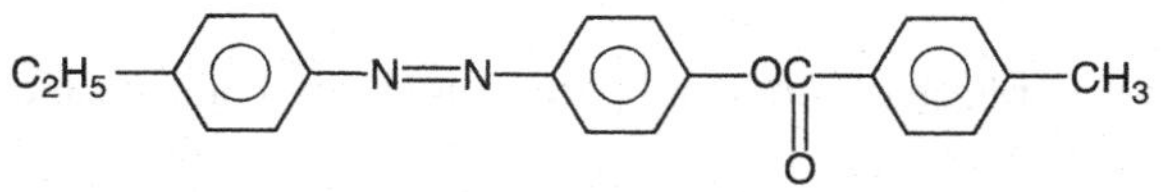

Thermophysical Properties:
solid   → nematic 108 °C
nematic → isotropic 230 °C
Analytical Properties: separation of nitronaphthalenes
Reference: [10]

Name: p-ethylazoxybenzene p′-methylbenzoate (mixed isomers)
Structure:

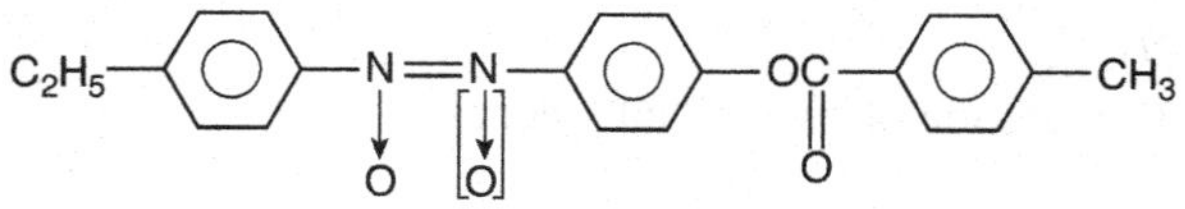

Thermophysical Properties:

(directly after crystallization)   (After melting and cooling)
crystal   → nematic 97.5 °C   Crystal   → nematic 87.5–97.5 °C
nematic → isotropic 250.5 °C   Nematic → isotropic 250.5 °C

Analytical Properties: separation of substituted xylenes
Reference: [9]

Name: 4′-methoxybiphenyl-4,4-[-(allyloxy)phenyl] benzoate
Structure:

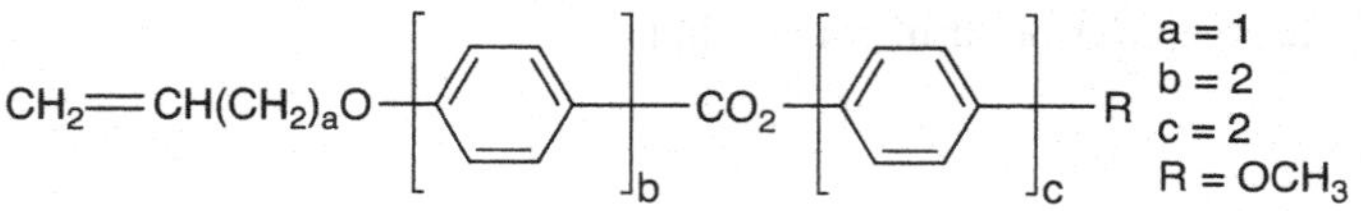

Thermophysical Properties:
solid   → nematic 214 °C
nematic → isotropic 290 °C
Analytical Properties: suggested for separation of polycyclic aromatic compounds
Reference: [11]

Name: (S)-4-[(2-methyl-1-butoxy)carbonyl]phenyl 4-[4-(4-pentenyloxy)phenyl] benzoate
Structure:

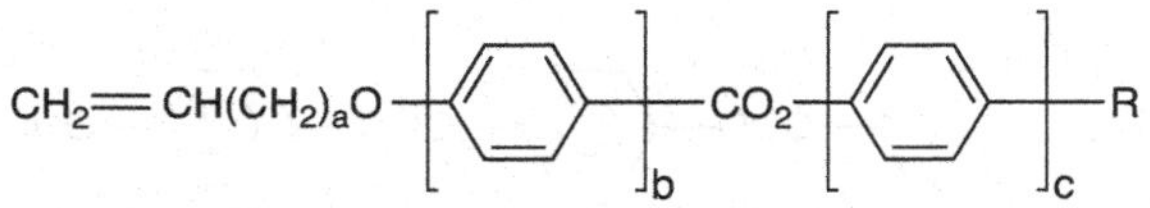

a = 3
b = 2
c = 1
R = COOCH$_2$C*H(CH$_3$)CH$_2$CH$_3$

Thermophysical Properties:
solid   → smectic 105 °C
smectic → isotropic 198 °C
Analytical Properties: suggested for separation of polycyclic aromatic compounds
Reference: [11]

Name: 4-methoxyphenyl (4-[4-(allyloxy) phenyl] benzoate
Structure:

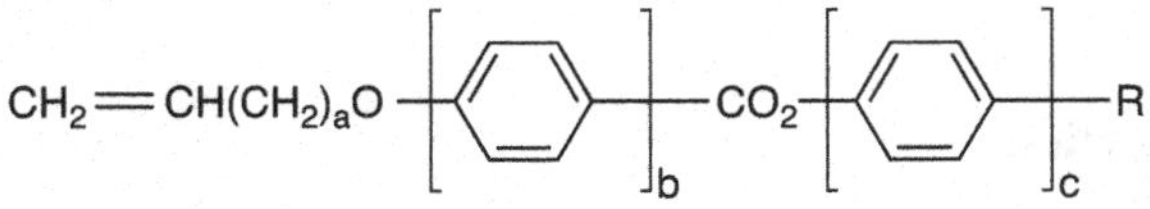

a = 1
b = 2
c = 1
R = OCH$_3$

Thermophysical Properties:
solid   → nematic 137 °C
nematic → isotropic 243 °C
Analytical Properties: suggested for the separation of polycyclic aromatic compounds
Reference: [11]

Name: 4-methoxyphenyl 4-[4-(4-pentenyloxy) phenyl] benzoate
Structure:

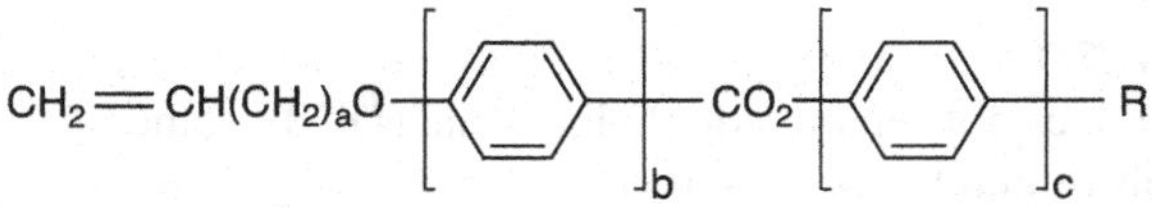

a = 3
b = 2
c = 1
R = OCH$_3$

Thermophysical Properties:
solid   → smectic 133 °C
smectic → nematic 172 °C
nematic → isotropic 253 °C
Analytical Properties: suggested for separation of polycyclic aromatic compounds
Reference: [11]

Name: p-phenylene-bis-4-n-heptyloxybenzoate
Structure:

$C_7H_{15}O$—⬡—COO—⬡—OOC—⬡—$OH_{15}C_7$

Thermophysical Properties:
solid   → smectic 83 °C
smectic → nematic 125 °C
nematic → isotropic 204 °C
Analytical Properties: separation of 1- and 2-ethylnapthalene; baseline separation of pyrazines
Reference: [12]

Name: 4-[(4-dodecyloxyphenyl)azoxy]-benzonitrile
Structure:

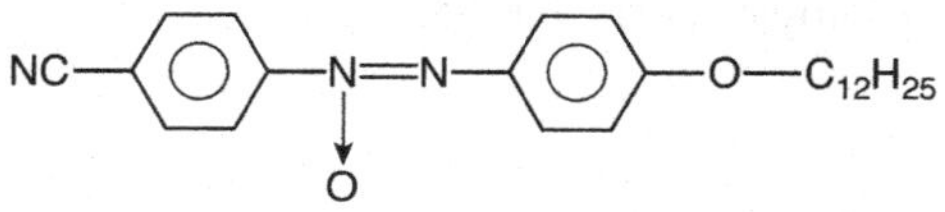

NC—⬡—N=N—⬡—O—$C_{12}H_{25}$
           ↓
           O

Thermophysical Properties:
solid   → smectic 106 °C
smectic → isotropic 147 °C
Analytical Properties: marginal effectiveness in separating disubstituted benzene isomers
Reference: [13]

Name: 4-[(4-pentyloxyphenyl)azoxy]-benzonitrile (mixed isomers)
Structure:

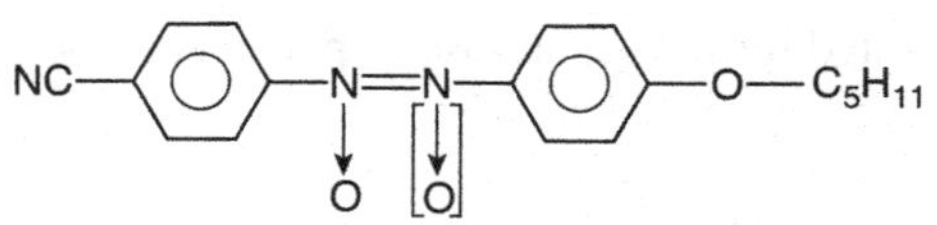

NC—⬡—N=N—⬡—O—C_5H_{11}
        ↓ ↓
        O [O]

Thermophysical Properties:
solid   → nematic 94 °C
nematic → isotropic 141.5 °C
Analytical Properties: does not separate diethylbenzene (DEB) isomers
good separation of disubstituted benzene isomers
Reference: [13]

Name: 4-[(4-octyloxyphenyl)azoxy]-benzonitrile
Structure:

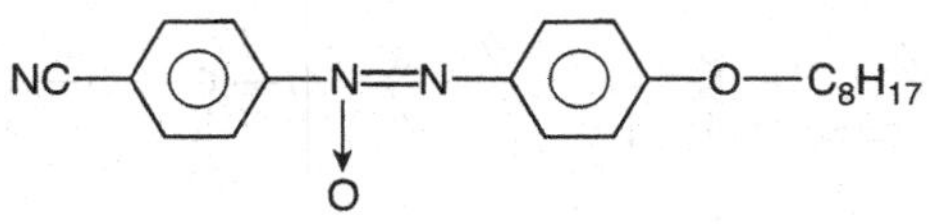

NC—⬡—N=N—⬡—O—C_8H_{17}
          ↓
          O

Thermophysical Properties:
solid → smectic 101.5 °C
smectic → nematic 137 °C
nematic → isotropic 151.5 °C
Analytical Properties: separates DEB isomers
Reference: [13]

Name: 4-[(4-pentyloxyphenyl) azoxy]-benzonitrile (pure isomers)
Structure:

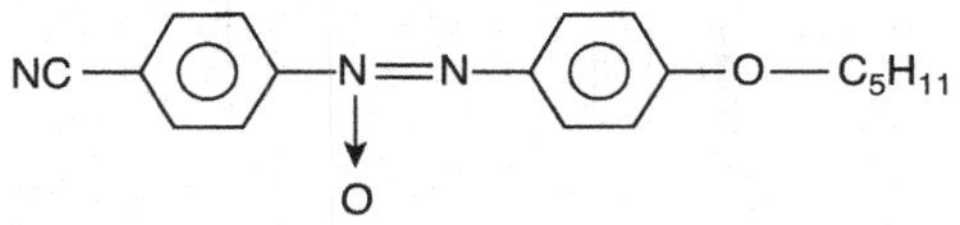

NC—⬡—N=N—⬡—O—C_5H_{11}
          ↓
          O

Thermophysical Properties:
solid → nematic 124 °C
nematic → isotropic 153 °C
Analytical Properties: complete separation of dichlorobenzene or bromotoluene isomers at 126 °C. Complete separation of chlorotoluene isomers at 87 °C; partial separation of m- and p-xylenes at 87 °C
Reference: [13]

Name: 4,4'-bis(p-methoxybenzylidene amino)-3,3'-dichloro biphenyl
Structure:

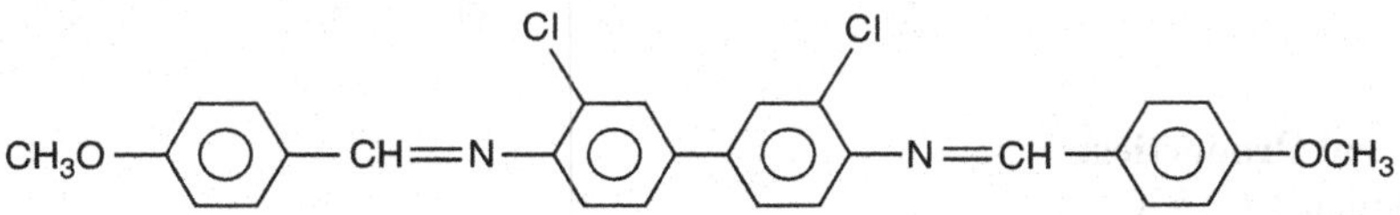

                        Cl   Cl
CH_3O—⬡—CH=N—⬡—⬡—N=CH—⬡—OCH_3

Thermophysical Properties:
solid → nematic 154 °C
nematic → isotropic 344 °C
Analytical Properties: separation of dimethyl benzene isomers, dihalo benzene isomers (Cl, Br), halo-ketone benzene isomers, and dimethoxy benzene isomers
Reference: [14]

Name: azoxybenzene p-cyano-p′-heptyl carbonate
Structure:

NC—⟨C₆H₄⟩—N=N—⟨C₆H₄⟩—O—C(=O)—O—C₇H₁₅
(with N→O azoxy oxygens)

Thermophysical Properties:
solid → nematic 66 °C
Analytical Properties: separation of disubstituted benzene isomers
Reference: [15]

Name: azoxybenzene p-cyano-p′-octyl carbonate (mixed isomers)
Structure:

NC—⟨C₆H₄⟩—N=N—⟨C₆H₄⟩—O—C(=O)—O—C₈H₁₇
(with N→O azoxy oxygens)

Thermophysical Properties:
solid → smectic 60.5 °C
smectic → nematic 119.5 °C
Analytical Properties: separation of ethyltoluenes, chlorotoluenes, bromotoluenes, and dichlorobenzenes. Also, ethylbenzenes from xylenes and propylbenzene from ethylbenzenes
Reference: [15]

Name: azoxybenzene p-cyano-p′-pentyl carbonate (pure isomer)
Structure:

NC—⟨C₆H₄⟩—N=N—⟨C₆H₄⟩—O—C(=O)—O—C₅H₁₁
(with N→O azoxy oxygen)

Thermophysical Properties:
solid → nematic 60.5 °C
nematic → isotropic 132 °C
Analytical Properties: separation of ethyltoluenes, chlorotoluenes, bromotoluenes, and dichlorobenzenes. Also, ethylbenzenes from xylenes and propylbenzene from ethylbenzenes
Reference: [3]

Name: azoxybenzene p-cyano-p′-pentyl carbonate (mixed isomers)
Structure:

NC—⟨C₆H₄⟩—N=N—⟨C₆H₄⟩—O—C₅H₁₁
(with N→O azoxy oxygens)

Thermophysical Properties:
solid → nematic 96 °C –100 °C
Analytical Properties: separation of disubstituted benzene isomers
Reference: [15]

Name: cyanoazoxybenzene decyl carbonate
Structure:

$$CN-\text{C}_6\text{H}_4-N=N-\text{C}_6\text{H}_4-O-\underset{\underset{O}{\|}}{C}-O-C_{10}H_{21}$$

(with O substituents on the N=N azoxy group)

Thermophysical Properties:
solid → smectic 74 °C
smectic → isotropic 125.5 °C
Analytical Properties: separation of polycyclic hydrocarbons
Reference: [16]

Name: cyanoazoxybenzene hexyl carbonate (mixed isomers)
Structure:

$$CN-\text{C}_6\text{H}_4-N=N-\text{C}_6\text{H}_4-O-\underset{\underset{O}{\|}}{C}-O-C_6H_{13}$$

(with O substituents on the N=N azoxy group)

Thermophysical Properties:
solid → nematic 73 °C –76 °C
nematic → isotropic 137 °C
Analytical Properties: separation of xylene and ethyltoluene isomers
Reference: [16]

Name: cyanoazoxybenzene nonyl carbonate (mixed isomers)
Structure:

$$CN-\text{C}_6\text{H}_4-N=N-\text{C}_6\text{H}_4-O-\underset{\underset{O}{\|}}{C}-O-C_9H_{19}$$

(with O substituents on the N=N azoxy group)

Thermophysical Properties:
solid → smectic 61 °C
smectic → nematic 124 °C
nematic → isotropic 127 °C
Analytical Properties: separation of polycyclic hydrocarbons
Reference: [16]

Name: p-(p-ethoxyphenylazo) phenyl crotonate
Structure:

$$CH_3CH_2O-\langle\bigcirc\rangle-N=N-\langle\bigcirc\rangle-O-\overset{\overset{O}{\|}}{C}-CH=CH-CH_3$$

Thermophysical Properties:
solid   → nematic 110 °C
nematic → isotropic 197 °C
Analytical Properties: separation of aromatic isomers
Reference: [12]

Name: carbonyl-bis-(D-leucine isopropyl ester)
Structure:

$$\begin{array}{c} CH_3\,CH_3 \qquad\qquad CH_3\,CH_3 \\ CH \qquad\qquad\quad CH \\ | \qquad\qquad\qquad | \\ CH_3 \qquad\quad CH_2 \quad O \quad\quad CH_2 \qquad CH_3 \\ CH-O-C-{}^*C-N-C-N-{}^*C-C-O-CH \\ CH_3 \qquad O \;\; H \;\; H \quad\; H \;\; H \;\; O \qquad CH_3 \end{array}$$

Thermophysical Properties:
solid   → smectic 55 °C
smectic → isotropic 110 °C
Analytical Properties: baseline and near-baseline separations of racemic mixtures of N-perfluoroacyl-2-aminoethyl benzenes, trifluoroacetyl (TFA), pentafluoropropionyl (PFP), and heptafluorobutyl (HFB)
Reference: [17]

Name: carbonyl-bis-(L-valine isopropyl ester)
Structure:

$$\begin{array}{c} CH_3\,CH_3 \qquad\qquad CH_3\,CH_3 \\ CH \quad O \qquad\quad CH \\ | \qquad\qquad\qquad | \\ CH_3 \qquad\quad\quad\quad\quad\quad\quad\quad CH_3 \\ CH-O-C-{}^*C-N-C-N-{}^*C-C-O-CH \\ CH_3 \qquad O \;\; H \;\; H \quad\; H \;\; H \;\; O \qquad CH_3 \end{array}$$

Thermophysical Properties:
solid   → smectic[1] 91 °C   This compound exhibits two stable smectic
smectic[1] → smectic[2] 99 °C   states prior to melting
smectic[2] → isotropic 109 °C

Analytical Properties: separation of enantiomers
Reference: [17]

Name: carbonyl-bis-(L-valine t-butyl ester)
Structure:

Thermophysical Properties:
solid → smectic 98 °C
smectic → isotropic 402 °C
Analytical Properties: separation of enantiomers
Reference: [17]

Name: carbonyl-bis-(L-valine ethyl ester)
Structure:

Thermophysical Properties:
solid → smectic 88 °C
smectic → isotropic 388 °C
Analytical Properties: separation of enantiomers
Reference: [17]

Name: carbonyl-bis-(L-valine methylester)
Structure:

Thermophysical Properties:
solid → smectic 382 °C
smectic → isotropic 415 °C
Analytical Properties: separation of enantiomers
Reference: [17]

Name: phenylcarboxylate ester (systematic name not available)
Structure:

Thermophysical Properties:
solid   → smectic 118 °C
smectic → sotropic 300 °C
Analytical Properties: separation of three- and four-member methylated polycyclic aromatic hydrocarbons (PAHs) on the basis of length-to-breadth ratio (l/b); as l/b increases, retention time decreases, cross-linking increases retention times and separation of methylcrypene isomers
Reference: [18]

Name: p-cyano-p′-octoxyazobenzene
Structure:

Thermophysical Properties:
solid   → nematic 101 °C
nematic → isotropic 111 °C
Analytical Properties: separation of ethyltoluenes, chlorotoluenes, bromotoluenes, and dichlorobenzenes. Also, ethylbenzenes from xylenes and propylbenzene from ethylbenzenes
Reference: [3]

Name: p-n-butoxy benzoic acid
Structure:

Thermophysical Properties:
solid 100 °C
mesomorphous 150 °C (not well characterized)
isotropic 160 °C
Analytical Properties: separation of methyl and monoalkyl substituted benzenes as well as orga-
noelemental compounds (for example, dimethyl mercury)
Reference: [19]

Name: p-[(p-methoxybenzylidene)-amino]phenylacetate
Structure:

Thermophysical Properties:
solid   → nematic 80 °C
nematic → isotropic 108 °C
Analytical Properties: separation of substituted phenols, selectivity is best at the lower end of the
nematic range
Reference: [20]

Name: poly (mesogen/methyl) siloxane (PMMS)—compound has not been named
Structure:

Thermophysical Properties:
solid   → nematic 70 °C
nematic → isotropic 300 °C
high thermal stability
Analytical Properties: separation of methylchrysene isomers
Reference: [21]

Name: N,N′-bis-(p-butoxybenzylidene)-bis-p-toluidine (BBBT)
Structure:

Thermophysical Properties:
solid   → smectic 159 °C
smectic → nematic 188 °C
nematic → isotropic 303 °C
Analytical Properties: separation of polycyclic aromatic hydrocarbons on the basis of length-to-breadth ratio
Reference: [23]

Name: N,N′-bis(ethoxy-benzylidene)- α,α″-bi-p-toluidine (BEBT)
Structure:

Thermophysical Properties:
solid   → nematic 173 °C
nematic → isotropic 341 °C
Analytical Properties: separation of polynuclear aromatic hydrocarbons
Reference: [24]

Name: N,N′-bis(n-heptoxy-benzylidene)- α,α″-bi-p-toluidine (BHpBT)
Structure:

Thermophysical Properties:
solid   → smectic 119 °C
smectic → nematic 238 °C
nematic → isotropic 262 °C
Analytical Properties: separation of polynuclear aromatic hydrocarbons
Reference: [24]

Name: N,N′-bis(n-hexoxy-benzylidene) -α,α′- bi-p-toluidine (BHxBT)
Structure:

Thermophysical Properties:
solid   → smectic 127 °C
smectic → nematic 229 °C
nematic → isotropic 276 °C
Analytical Properties: separation of methyl and nitro derivatives of naphthalene; separation of higher hydrocarbons
Reference: [25]

Name: N,N′-bis (p-methoxybenzylidene)- α,α′-bi-p-toluidine (BMBT)
Structure:

$$CH_3O — \phi — CH=N — \phi — CH_2 — CH_2 — \phi — N=CH — \phi — OCH_3$$

$$\phi = \langle \bigcirc \rangle$$

Thermophysical Properties:
solid → nematic 181 °C
nematic → isotropic 320 °C
Analytical Properties: separation of androstane and cholestane alcohols and ketones, good separation of azaheterocyclic compounds, and column bleed of BMBT can occur during prolonged periods of operation at elevated temperatures
Reference: [26]

Name: N,N-bis(n-octoxy-benzylidene)-α,α′-bi-p-toluidine (BoBT)
Structure:

$$C_8H_{17}O — \phi — CH=N — \phi — CH_2 — CH_2 — \phi — N=CH — \phi — OC_8H_{17}$$

$$\phi = \langle \bigcirc \rangle$$

Thermophysical Properties:
solid → smectic 118 °C
smectic → nematic 244 °C
nematic → isotropic 255 °C
Analytical Properties: separation of polynuclear aromatic hydrocarbons
Reference: [24]

Name: N,N-bis(n-pentoxy-benzylidene)-α,α′-bi-p-toluidine (BPeBT)
Structure:

$$C_5H_{11}O — \phi — CH=N — \phi — CH_2 — CH_2 — \phi — N=CH — \phi — OC_5H_{11}$$

$$\phi = \langle \bigcirc \rangle$$

Thermophysical Properties:
solid → smectic 139 °C
smectic → nematic 208 °C
nematic → isotropic 283 °C
Analytical Properties: separation of polynuclear aromatic hydrocarbons
Reference: [24]

Name: N,N′-bis (p-phenylbenzylidene)-α,α′-bi-p-toluidine (BphBT)
Structure:

Thermophysical Properties:
solid → nematic 257 °C
nematic → isotropic 403 °C
Analytical Properties: separation of unadulterated steroids; used chromatographically in the temperature range of 260 °C –270 °C
Reference: [27]

Name: N,N′-bis(n-propoxy-benzylidene)-α,α′-bi-p-toluidine (BPrBT)
Structure:

$$C_3H_7O - \phi - CH = N - \phi - CH_2 - CH_2 - \phi - N = CH - \phi - OC_3H_7$$

$$\phi = \langle\bigcirc\rangle$$

Thermophysical Properties:
solid → smectic 169 °C
smectic → nematic 176 °C
nematic → isotropic 311 °C
Analytical Properties: separation of polynuclear aromatic hydrocarbons
Reference: [24]

Cholesteric Phases

Name: cholesteryl acetate
Structure:

Thermophysical Properties:
solid   → cholesteric 94.5 °C
cholesteric → isotropic 116.5 °C
Analytical Properties: separation of aromatics and paraffins
Reference: [28]

Name: (S)-4′-[(2-methyl-1-butoxy)carbonyl] biphenyl-4-yl 4(allyloxy) benzoate
Structure:

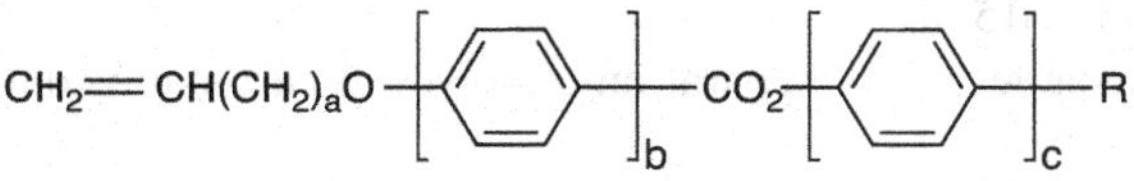

$$CH_2 = CH(CH_2)_aO - \left[\text{C}_6\text{H}_4 \right]_b - CO_2 - \left[\text{C}_6\text{H}_4 \right]_c - R$$

a = 1
b = 1
c = 2
R = COOCH_2C*H(CH_3)CH_2CH_3

Thermophysical Properties:
solid → smectic 100 °C
smectic → cholesteric 150 °C
cholesteric → isotropic 188 °C
Analytical Properties: suggested for separation of polycyclic aromatic compounds
Reference: [11]

Name: (S)-4′-[(2-methyl-1-butoxy)carbonyl] biphenyl-4-yl 4-[4-(allyloxy) phenyl] benzoate
Structure:

$$CH_2 = CH(CH_2)_aO - \left[\text{C}_6\text{H}_4 \right]_b - CO_2 - \left[\text{C}_6\text{H}_4 \right]_c - R$$

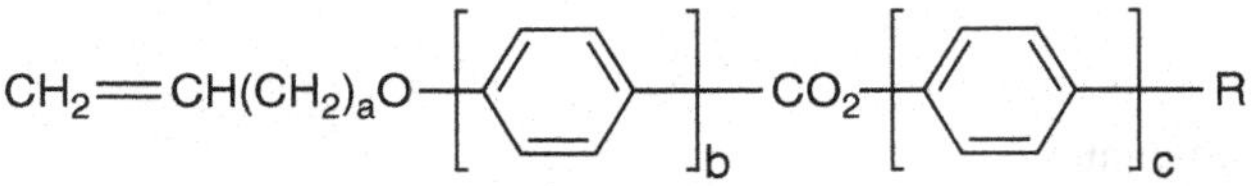

a = 1
b = 2
c = 2
R = COOCH_2C*H(CH_3)CH_2CH_3

Thermophysical properties:
Solid → smectic 152 °C
Smectic → cholesteric 240 °C
Cholesteric → isotropic 278 °C
Analytical Properties: suggested for separation of polycyclic aromatic compounds
Reference: [11]

Name: (S)-4′-[(2-methyl-1-butoxy)biphenyl-4-yl 4-[4-4-pentenyloxy) phenyl] benzoate
Structure:

$$CH_2 = CH(CH_2)_aO - \left[\text{C}_6\text{H}_4 \right]_b - CO_2 - \left[\text{C}_6\text{H}_4 \right]_c - R$$

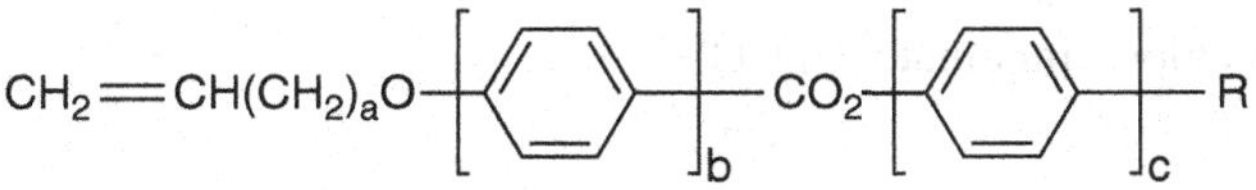

a = 3
b = 2
c = 2
R = COOCH_2C*H(CH_3)CH_2CH_3

Thermophysical Properties:
solid   → smectic 135 °C
smectic   → cholesteric 295 °C
cholesteric   → isotropic 315 °C
Analytical Properties: suggested for separation of polycyclic aromatic compounds
Reference: [11]

Name: (S)-4-[(2-methyl-1-butoxy)carbonyl]phenyl 4-[4-(allyloxy)phenyl] benzoate
Structure:

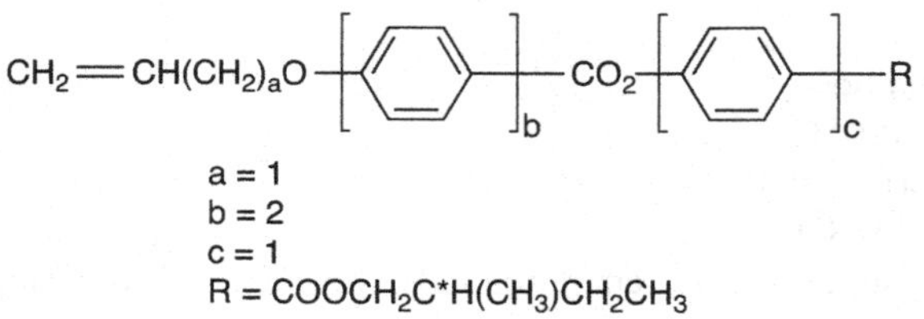

a = 1
b = 2
c = 1
R = COOCH$_2$C*H(CH$_3$)CH$_2$CH$_3$

Thermophysical Properties:
solid   → smectic 118 °C
smectic   → cholesteric 198 °C
cholesteric   → isotropic 213 °C
Analytical Properties: suggested for separation of polycyclic aromatic compounds
Reference: [11]

Name: cholesterol cinnamate
Structure:

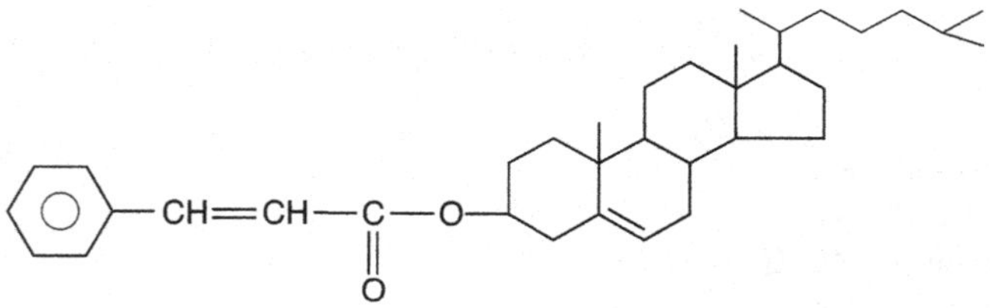

Thermophysical Properties:
solid   → cholesteric 160 °C
cholesteric   → isotropic 210 °C
Analytical Properties: separation of olefinic positional isomers
Reference: [12,29]

Name: cholesterol-p-chlorocinnamate (CpCC)
Structure:

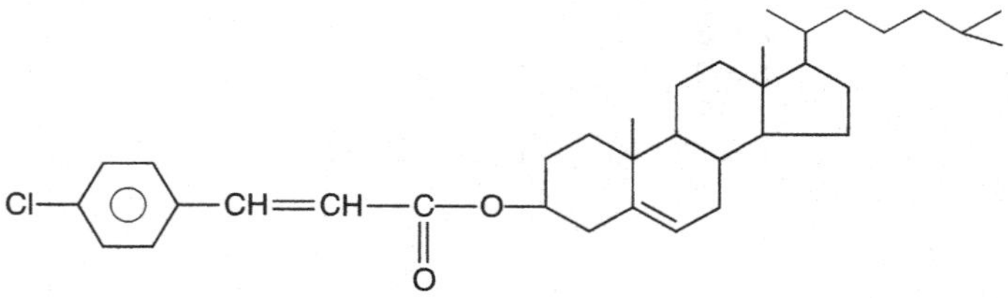

Thermophysical Properties:
solid   → cholesteric 144 °C
cholesteric   → isotropic 268 °C
Analytical Properties: separation of diastereomeric amides and carbamates; the separation of olefinic geometrical isomers is dependent upon the position of the double bond
Reference: [29,30]

Name: cholesterol-p-methylcinnamate
Structure:

Thermophysical Properties:
solid   → cholesteric 157 °C
cholesteric   → isotropic 254 °C
Analytical Properties: separation of olefinic positional isomers
Reference: [29]

Name: cholesterol-p-methoxycinnamate
Structure:

Thermophysical Properties:
solid   → cholesteric 165 °C
cholesteric   → isotropic 255 °C
Analytical Properties: separation of olefinic positional isomers
Reference: [29]

Name: cholesterol p-nitro cinnamate
Structure:

Thermophysical Properties:

solid   → cholesteric 167 °C
cholesteric   → isotropic 265 °C

Analytical Properties: separation of geometrical isomers (2 and 3 octadecene) using p-substituted cholesterols. (Best separation) p-NO$_2$>p-MeO>cholesterol cinnamate>p-Me>p-Cl (worst separation) for unsaturation occurring within 4 carbon atoms from the terminal methyl, the above order holds for separations of tetradecen-1-ol acetates; for unsaturation on carbons 5–12 from the terminal methyl of the tetradecen-1-ol of acetates, the best separation is the reverse of the above

Reference: [29]

Name: cholesteryl nonanoate
Structure:

$$CH_3(CH_2)_7 - \overset{\overset{\displaystyle O}{\|}}{C} - O-$$

Thermophysical Properties:

solid   → smectic 77.5 °C
smectic   → cholesteric 80.5 °C
cholesteric   → isotropic 92 °C

Analytical Properties: separation of aromatics and paraffins
Reference: [28]

Name: cholesteryl valerate
Structure:

$$CH_3(CH_2)_3 - \overset{\overset{\displaystyle O}{\|}}{C} - O-$$

Thermophysical Properties:

solid   → cholesteric 93 °C
cholesteric   → isotropic 101.5 °C

Analytical Properties: separation of aromatics and paraffins
Reference: [28]

TRAPPING SORBENTS

The following table provides a listing of the major types of sorbents used in sampling, concentrating, odor profiling, and air and water pollution research [1–6]. These materials are useful in a wide variety of research and control applications. Many can be obtained commercially in different sizes, depending upon the application involved. The purpose of this table is to aid in the choice of a sorbent for a given analysis. Information that is specific for solid-phase mircroextraction (SPME) is provided elsewhere in this chapter.

REFERENCES

1. Borgstedt, H.U., Emmel, H.W., Koglin, E., Melcher, R.G., Peters, and Sequaris, J.M.L., *Analytical Problems*, Springer-Verlag, Berlin, 1986.
2. Averill, W. and Purcell, J.E., Concentration and gc determination of organic compounds from air and water, *Chrom. Newsletter*, 6(2), 30, 1978.
3. Gallant, R.F., King, J.W., Levins, P.L., and Piecewicz, J.F., Characterization of sorbent resins for use in environmental sampling, Report EPA-600/7-78-054, March 1978.
4. Chladek, E. and Marano, R.S., Use of bonded phase silica sorbents for the sampling of priority pollutants in waste waters, *J. Chromatogr. Sci.*, 22, 313, 1984.
5. Good, T.J., Applications of bonded-phase materials, *Am. Lab.*, 36, July 1981.
6. Beyermann, K., *Organic Trace Analysis*, Halsted Press (of John Wiley & Sons), New York, 1984.

Sorbent	Desorption Solvents	Applications
Activated carbon	Carbon disulfide, methylene chloride, diethyl ether, diethyl ether with 1 % methanol, diethyl ether with 5 % 2-propanol (caution: CS_2 and CH_3OH can react in the presence of charcoal)	Used for common volatile organics; examples include methylene chloride, vinyl chloride, chlorinated aliphatics, aromatics, acetates; more data is provided in the table entitled "Adsorbents for Gas Chromatography"

Notes: Metallic or salt impurities in the sorbent can sometimes cause the irreversible adsorption of electron-rich oxygen functionalities; examples include 1-butanol, 2-butanone, and 2-ethoxyacetate; recovery rate is often poor for polar compounds.

Sorbent	Desorption Solvents	Applications
Graphitized carbon black	Carbon disulfide, methylene chloride, diethyl ether (or thermal desorption can be used)	Used for common volatile aliphatic and aromatic compounds, organic acids and alcohols, and chlorinated aliphatics; more data is provided in the table entitled "Adsorbents for Gas Chromatography"

Notes: These sorbents are hydrophobic and are not very sensitive to moisture; the possibility of thermal desorption makes them of value for "trace-level" analyses.

Sorbent	Desorption Solvents	Applications
Silica gel	Methanol, ethanol, water, diethyl ether	Used for polar compound collection and concentration; examples include alcohols, phenols, chlorophenols, chlorinated aromatics, aliphatic and aromatic amines, nitrogen dioxide; more data is provided in the table entitled "Adsorbents for Gas Chromatography"

Notes: Useful for compounds which can't be recovered from the charcoal sorbents; the most serious problem with silica is the effect of water, which can cause desorption of the analytes of interest, and the heating effect involved can sometimes initiate reactions such as polymerization of the analyte.

Sorbent	Desorption Solvents	Applications
Activated alumina	Water, diethyl ether, methanol	Used for polar compounds such as alcohols, glycols, ketones, aldehydes; has also been used for polychlorinated biphenyls and phthalates; more data is provided in the table entitled "Adsorbents for Gas Chromatography"

Notes: Similar in application to silica gel.

Sorbent	Desorption Solvents	Applications
Porous polymers	Hexane, diethyl ether, alcohols (thermal desorption also possible in some cases)	Used for a wide range of compounds which include phenols, acidic and basic organics, pesticides, priority pollutants; more data is provided in the table entitled "Porous Polymer Phases"

Notes: The most commonly used porous polymer sorbent is Tenax-GC, although the Porapak and Chromosorb Century series have also been used; Tenax-GC has been used with thermal desorption methods but can release toluene, benzene, and trichloroethylene residues at the higher temperatures; in addition to Tenax-GC, XAD 2–8, Porapak-N, and Chromosorbs 101, 102, 103, and 106 have found applications, sometimes in "stacked" sampling devices (for example, a sorbent column of Tenax-GC and Chromosorb 106 in tandem); Chromosorb 106, a very low-polarity polymer, has the lowest retention of water with respect to organic materials and is well suited for use as a back-up sorbent.

Sorbent	Desorption Solvents	Applications
Bonded phases	Methanol, hexane, diethyl ether	Used for specialized applications in pesticides, herbicides, and polynuclear aromatic hydrocarbons

Notes: Most expensive of the common sorbents; useful for the collection of organic samples from water.

Sorbent	Desorption Solvents	Applications
Molecular sieves	Carbon disulfide, hexane diethyl ether	Have been used for the collection of aldehydes, alcohols, and for acrolein

Notes: Molecular sieve 13-X is the main molecular sieve to be used as a trapping adsorbent; the sorbents will also retain water.

COOLANTS FOR CRYOTRAPPING

The following table provides fluids (in some cases mixtures) that can be used to chill trapping sorbents (or any cold trap) for the purge and trap sampling of solutes [1–3]. In each case, the ratio is mass/mass.

REFERENCES

1. Gordon, A.J. and Ford, R.A., *The Chemists Companion—a Handbook of Practical Data, Techniques and References*, Wiley Interscience, New York, 1972.
2. Bruno, T.J., Chromatographic cryofocusing and cryotrapping with the vortex tube, *J. Chromatogr. Sci.*, 32(3), 112, 1994.
3. Bruno, T.J., Simple, quantitative headspace analysis by cryoadsorption on a short alumina PLOT column, *J. Chromatogr. Sci*, 47(8), 1, 2009.

Coolant	Temperature (°C)
Crushed ice + sodium chloride (3:1)	−21
Crushed ice + calcium chloride (1.2:2)	−39
Vortex tube	−40
Liquid nitrogen + n-butyl amine (slush)	−50
Crushed ice + calcium chloride (1.4:2)	−55
Liquid nitrogen + chloroform (1:1)	−63
Liquid nitrogen + t-butyl amine (slush)	−68
Dry ice	−78
Liquid nitrogen + acetone (slush)	−95
Liquid nitrogen + ethanol (slush)	−120
Liquid nitrogen + methyl cyclohexane (slush)	−126
Liquid nitrogen + n-pentane (slush)	−131
Liquid nitrogen + 1,5-hexadiene (slush)	−141
Liquid nitrogen + isopentane (slush)	−160
Liquid argon	−186
Liquid nitrogen	−196

SORBENTS FOR THE SEPARATION OF VOLATILE INORGANIC SPECIES

The following sorbents have proven useful for the adsorptive separation of volatile inorganic species [1]. This material is used with permission of John Wiley & Sons.

REFERENCE

1. MacDonald, J.C., *Inorganic Chromatographic Analysis—Chemical Analysis Series, Vol 78*, John Wiley & Sons, New York, 1985.

Separation Material	Typical Separations
Alumina	O_2, N_2, CO_2
Beryllium oxide	H_2S, H_2O, NH_3
Silica gel	O_2/N_2, CO_2, O_3, H_2S, SO_2
Chromium(III) oxide	O_2, N_2, Ar, He
Clay minerals (attapulgite, sepiolite)	O_2, N_2, CO, CO_2
Kaolin	He, O_2, N_2, CO, CO_2
Sodium fluoride, lithium fluoride, alumina	MoF_6, SbF_5, UF_6, F
Quartz granules	Ta, Re, Ru, Os, Ir: oxides, hydroxides
Chromosorb 102	Element hydrates
Graphite	NH_3, N_2, H_2
Synthetic diamond	CF_2O, CO_2
Molecular sieve	Hydrogen isotopes
Carbon molecular sieve	O_2, N_2, CO, CO_2, N_2O, SO_2, H_2S
XAD resins	NH_3, SO_2, H_2S, CO, CO_2, H_2O
Porapak Q	GeH_4, SnH_4, AsH_3, SbH_3, $Sn(CH_3)_4$
Porapak QS polymers	H_2S, CH_3SH, $(CH_3)_2S$, $(CH_3)_2S_x$, SO_2
Porapak P	Chlorides of Si, Sn, Ge, P, As, Ti, V, Sb
Teflon	F, MoF_6, SbF_6, SbF_3

ACTIVATED CARBON AS A TRAPPING SORBENT FOR TRACE METALS

Activated carbon, which is a common trapping sorbent for organic species, can also be used for trace metals [1]. This material is typically used by passing the samples through a thin layer (50–150 mg) of the activated carbon that is supported on a filter disk. It can also be used by shaking 50–150 mg of activated carbon in the solution containing the heavy metal and then filtering the sorbent out of the solution.

REFERENCE

1. Alfasi, Z.B. and Wai, C.M., *Preconcentration Techniques for Trace Elements*, CRC Press, Boca Raton, 1992.

Matrices	Trace Metals	Complexing Agents
Water	Ag, Bi, Cd, Co, Cu, Fe, In, Mg, Mn, Ni, Pb, Zn	(NaOH; pH 7–8)
Water	Ag, As, Ca, Cd, Ce, Co, Cu, Dy, Fe, La, Mg, Mn, Nb, Nd, Ni, Pb, Pr, Sb, Sc, Sn, U, V, Y, Zn	8-Quinolinol
Water	Ba, Co, Cs, Eu, Mn, Zn	APDC, DDTC, PAN, 8-quinolinol
Water	Hg, Methyl mercury	—
Water	Hg (halide)	—
Water	Hg (halide)	—
Water	U	L-ascorbic acid
HNO_3, water, Al, KCl	Ag, Bi, Cd, Cu, Hg, Pb, Zn	Dithizone
Mn, MnO_3, Mn salts	Bi, Cd, Co, Cu, Fe, In, Ni, Pb, Tl, Zn	Ethyl xanthate
Co, $Co(NO_3)_2$	Ag, Bi	APDC
Ni, $Ni(NO_3)_2$	Ag, Bi	APDC
Mg, $Mg(NO_3)_2$	Ag, Cu, Fe, Hg, In, Mn, Pb, Zn	(pH 8.1–9)
Al	Cd, Co, Cu, Ni, Pb	Thioacetamide
Ag, $TlNO_3$	Bi, Co, CU, Fe, In, Pb	Xenol orange
Cr salts	Ag, Bi, Cd, Co, Cu, In, Ni, Pb, Tl, Zn	HAHDTC
Co, In, Pb, Ni, Zn	Ag, Bi, Cu, Tl	DDTC
Se	Cd, Co, Cu, Fe, Ni, Pb, Zn	DDTC
$NaClO_4$	Ag, Bi, Cd, Co, Cu, Fe, Hg, In, Mn, Ni, Pb	(pH 6)

APDC, ammonium pyrrolidinecarbodithioate; DDTC, diethyldithiocarbamate; HAHDTC, hexamethyleneammonium hexaethylenedithiocarbamate; PAN, 1-(2-pyridylazo)-2-naphthol

REAGENT-IMPREGNATED RESINS AS TRAPPING
SORBENTS FOR HEAVY METALS

Reagent-impregnated resins can be used as trapping sorbents for the preconcentration of heavy metals [1]. These materials can be used in the same way as activated carbons.

REFERENCE

1. Alfasi, Z.B. and Wai, C.M., *Preconcentration Techniques for Trace Elements*, CRC Press, Boca Raton, 1992.

Reagents	Adsorbents	Metals
TBP	Porous polystyrene DVB resins	U
TBP	Levextrel (polystyrene DVB resins)	U
DEHPA	Levextrel	Zn
DEHPA	XAD-2	Zn
Alamine 336	XAD-2	U
LIX-63	XAD-2	Co, Cu, Fe, Ni, etc.
LIX-64N, -65N	XAD-2	Cu
Hydroxyoximes	XAD-2	Cu
Kelex 100	XAD-2	Co, Cu, Fe, Ni
Kelex 100	XAD-2,4,7,8,11	Cu
Dithizone, STTA	Polystyrene DVB resins	Hg
Dithizone (acetone)	XAD-1,2,4,7,8	Hg, methyl mercury
DMABR	XAD-4	Au
Pyrocatechol violet	XAD-2	In, Pb
TPTZ	XAD-2	Co, Cu, Fe, Ni, Zn

TBP, tributyl phosphate; DEHPA, di-ethylhexyl phosphoric acid; STTA, monothiothenolytrifluoroacetone; DMABR, 5-(4-dimethylaminobenzylidene)-rhodanine; TPTZ, 2,4,6-tri(2-pyridyl)-1,3,5-triazine; LIX 63, aliphatic α-hydroxyoxime; LIX 65N, 2-hydroxy-5-nonylbensophenoneoxime; LIX 64N, a mixture of LIX 65N with approximately 1 % (vol/vol) of LIX-63.

REAGENT IMPREGNATED FOAMS AS TRAPPING SORBENTS FOR INORGANIC SPECIES

Reagent impregnated foams can be used as trapping sorbents for the preconcentration of heavy metals [1]. These materials can be used in the same way as activated carbons.

REFERENCE

1. Alfasi, Z.B. and Wai, C.M., *Preconcentration Techniques for Trace Elements*, CRC Press, Boca Raton, 1992.

Matrices	Elements	Conc.	Foam Type	Reagents
Water	^{131}I, ^{203}Hg	Traces	Polyether	Alamine 336
Natural water				
Water	Bi, Cd, Co, Cu, Fe, Hg, Ni, Pb, Sn, Zn	Traces	Polyether	Amberlite LA-2
Water	Co, Fe, Mn	Traces to μg/1	Polyether	PAN
Natural water	Cd	μg/1	Polyether	PAN
Water	Au, Hg	μg/1	Polyether	PAN
Water	Ni	Traces to μg/1	–	DMG, α-benzyldioxime
Water	Cr	μg/1	Polyether	DPC
Water	Hg, methyl-Hg, phenyl-Hg	μg/1	Polyether	DADTC
Natural water	Sn	Traces	Polyether	Toluene-3,4-dithiol
Water	Cd, Co, Fe, Ni	Traces	Polyether	Aliquot
Water	Th	Traces	Polyether	PMBP HDEHP-TBP
Water	PO_4^{3-}	Traces		Amine-molybdate-TBP

PAN, 1-(2-pyridylazo)-2-naphthol; DMG, dimethylglyoxime; DPC, 1,5-diphenylcarbazide; DADTC, diethylammonium diethyldithiocarbamate; PMBP, 1-phenyl-3-methyl-4-bensoyl-pyrazolone-5; HDEHP, bis-[2-ethylhexyl]phosphate; TBP, tributyl phosphate.

CHELATING AGENTS FOR THE ANALYSIS OF
INORGANICS BY GAS CHROMATOGRAPHY

The following table provides guidance in choosing a chelating agent for the analysis of inorganic species by GC [1–3]. The key to the abbreviation list is provided below.

REFERENCES

1. Guiochon, G. and Pommier, C., *Gas Chromatography of Inorganics and Organometallics*, Ann Arbor Science Publishers, Ann Arbor, 1973.
2. Robards, K., Patsalides, E., and Dilli, S., Review—Gas chromatography of metal beta-diketonates and their analogues, *J. Chromatogr.* 41, 1–41, 1987.
3. Robards, K. and Patsalides, E., Comparison of the liquid and gas chromatography of five classes of metal complexes, *J. Chromatogr. A.*, 844, 181–190, 1999.

acac = acetylacetonate
dibm = 2,6-dimethyl-3,5-heptanedionate
fod = 1,1,1,2,2,3,3,-heptafluoro-7,7dimethyl-4,6-octanedionate
hfa = hexafluoroacetylacetonate
tacac = monothioacetylacetonate
tfa = trifluoroacetylacetonate
thd = 2,2,6,6-tetramethyl-3,5-heptadionate
tpm = 1,1,1-trifluoro-5,5dimethyl-2,4-hexanedionate

Aluminum

In Mixture with	Complex
Be, Sc	acac
Be	acac
Cr	acac
Be, Cr	acac
Be, Sc	tfa
Be, Rh	tfa
Cr, Rh	tfa
Cr, Rh	tfa
Cu, Fe	tfa
Ga, In	tfa
Fe	tfa
Cr, Rh, Zr	tfa
Be, Ga, In, Tl	tfa
Be, Cr	hfa
Be, Cr, Cu	hfa
Be, Cr, Fe	hfa
Be, Cu, Cr, Fe, Pd, Y	fod
Cr, Fe	tpm
Cr, Fe, Cu	tpm
Be, Cr, Fe, Ni	dibm
Traces on U	tfa
Traces	tfa, hfa

Beryllium

In Mixture with	Complex
Al, Sc	acac
Cu	acac
Al, Cr	acac
Al, Sc	tfa
Al, Ga, Tl, In	tfa
Al, Cr	hfa
Al, Cr, Cu	hfa
Al. Cr, Fe	hfa
Al, Cu, Cr, Fe, Pd, Y	fod
Al, Cr, Fe, Ni	dibm
Traces	tfa

Chromium

In Mixture with	Complex
Al, Be	acac
Al	acac
Al, Rh	tfa
Al, Rh, Zr	tfa
Al, Be	hfa
Al, Be, Cu	hfa
Al, Be, Fe	hfa
Fe, Rh	hfa
Ru	hfa, tpm
Al, Fe	tpm
Al, Fe, Cu	tpm
Al, Be, Cu, Fe, Pd, Y	fod
Al, Be, Fe, Ni	dibm
Traces in Fe	tfa
Traces	tfa, hfa

Cobalt

In Mixture with	Complex
Ru	tfa, hfa
Ni, Pd	tacac
Traces	fod

Copper

In Mixture with	Complex
Be	acac
Al, Fe	tfa
Fe	tfa
Al, Be, Cr	hfa
Fe	hfa
Al, Cr, Fe	tpm
Al, Be, Cr, Fe, Pd, Y	fod

Gallium

In Mixture with	Complex
Al, In	tfa
Al, Be, In, Tl	tfa

Indium

In Mixture with	Complex
Al, Ga	tfa
Al, Be, Ga, Tl	tfa

Iron

In Mixture with	Complex
Al, Cu	tfa
Al	tfa
Cr	tfa
Cu	tfa
Al, Be, Cr	hfa
Cu	hfa
Cr, Rh	hfa
Al, Cr	tpm
Al, Cr, Cu	tpm
Al, Be, Cr, Cu, Pd, Y	fod
Al, Be, Cr, Ni	dibm

Nickel

In Mixture with	Complex
Co, Pd	tacac
Al, Be, Cr, Fe	dibm

Palladium

In Mixture with	Complex
Al, Be, Cr, Cu, Fe, Y	fod
Co, Ni	tacac

Rare Earths

In Mixture with	Complex
Sc	thd
Sc, Y	tpm
Sc, Y	fod

Rhodium

In Mixture with	Complex
Al, Cr	tfa
Al, Cr	tfa
Al, Cr, Zr	tfa
Cr, Fe	hfa
Traces	tfa

Ruthenium

In Mixture with	Complex
Co	tfa, hfa
Cr	hfa

Scandium

In Mixture with	Complex
Al, Be	acac
Al, Be	tfa
Rare earths	thd
Rare earths, Y	tpm
Rare earths, Y	fod

Thallium

In Mixture with	Complex
Al, Be, Ga, In	tfa

Thorium

In Mixture with	Complex
U	fod

Uranium

In Mixture with	Complex
Th	fod

Yttrium

In Mixture with	Complex
Sc, rare earths	tpm
Sc, rare earths	fod
Al, Be, Cr, Cu, Fe, Pd	fod

Zirconium

In Mixture with	Complex
Al, Cr, Rh	tfa

BONDED PHASE-MODIFIED SILICA SUBSTRATES
FOR SOLID-PHASE EXTRACTION

The following table provides the most commonly used bonded phase-modified silica substrates used in solid-phase extraction, reproduced with permission from Ref. 1. Additional information on many of these materials can be found in the table entitled "More Common HPLC Stationary Phases" in Chapter 3 in this book.

REFERENCE

1. Fritz, J.S., *Solid Phase Extraction*, Wiley-VCH, New York, 1999.

Phase	Polarity of Phase	Designation
Octadecyl, endcapped	Strongly apolar	C18ec
Octadecyl	Strongly apolar	C18
Octyl	Apolar	C8
Ethyl	Slightly polar	C2
Cyclohexyl	Slightly polar	CH
Phenyl	Slightly polar	PH
Cyanopropyl	Polar	CN
Diol	Polar	2OH
Silica gel	Polar	SiOH
Carboxymethyl	Weak cation exchanger	CBA
Aminopropyl	Weak anion exchanger	NH_2
Propylbenzene sulfonic acid	Strong cation exchanger	SCX
Trimethylaminopropyl	Strong anion exchanger	SAX

SOLID-PHASE MICROEXTRACTION SORBENTS

The following tables provide information on the selection and optimization of solid-phase microextraction fibers [1]. The reader is also advised to consult the tables for headspace analysis in this chapter.

REFERENCE

1. Shirey, R., Supelco Corp., Bellefonte, Private Communication, 2009.

Fiber Selection Criteria

The main fiber selection parameters are polarity and relative molecular mass. This table provides general guidelines on the applicability of available fibers relative to these two parameters. The fibers are characterized by the extraction mechanism, either adsorption or absorption. Adsorbent fibers contain particles suspended in PDMS or Carbowax.

Fiber	Type of Fiber	Polarity	RMM Range
7 µm PDMS	Absorbent	Nonpolar	150–700
30 µm PDMS	Absorbent	Nonpolar	80–600
85 µm polyacrylate	Absorbent	Moderately polar	60–450
100 µm PDMS	Absorbent	Nonpolar	55–400
50 µm Carbowax (PEG)	Adsorbent	Polar	50–400
PDMS-DVB	Adsorbent	Bipolar	50–350
Carbowax-DVB	Adsorbent	Polar	50–350
PDMS-DVB-Carboxen	Adsorbent	Bipolar	40–270
PDMS-Carboxen	Adsorbent	Bipolar	35–180
Carbopak Z-PDMS	Adsorbent	Nonpolar	50–500

PDMS, polydimethylsiloxane; DVB, divinylbenzene (3–5 µm particles); PEG, polyethylene glycol; Carboxen, Carboxen 1006 (contains micro-, meso-, and macro- tapered pores) (3–5 µm particles); RMM Range, relative molecular mass range that is the ideal range for optimum extraction. Ranges can be extended by varying extraction times, but results will not be optimized.

Phase Material Characteristics

Polydimethylsiloxane (PDMS)

Similar in properties to the OV-1 or SE-30 phases discussed in the tables on silicone liquid phases; nonpolar fluid suitable for nonpolar or slightly polar analytes; thicker coatings extract more analyte but require longer extraction times.

Polyacrylate

Rigid solid material; moderate polarity; diffusion of analytes through bulk is relatively slow because of rigidity of material; relatively higher desorption temperatures required because of rigidity of material; can be oxidized easily at higher temperatures; must use oxygen-free carrier gas and ensure GC system is leak-free; fibers are very solvent resistant; darkens to a brown color upon exposure to temperatures in excess of 280 °C, but fiber is generally still usable until color becomes black.

Carbowax (Polyethylene Glycol, PEG)

Similar in properties to the PEG coatings used extensively in chromatography and descrbed elsewhere in this book; moderately polar; highly cross-linked to counteract water solubility.

Sensitive to attack by oxygen at temperatures in excess of 220 °C, at which point the fiber will darken and become powdery; it requires use of high-purity carrier gas (typically He at 99.999 % purity) treated for oxygen contamination.

Divinlybenzene (DVB)

Similar to the properties of divinylbenzene porous polymer phases described elsewhere in this book; higher polarity than Carbowax, and when combined with Carbowax results in a more polar phase; like polyacrylate, it is a solid particle which must be carried in a liquid to coat on a fiber.

Carboxen

Similar to the material used in Carboxen PLOT columns.

Structure has an approximately even distribution of macro-, meso- ,and micropores, making it valuable for smaller analytes; larger analytes can show hysteresis which must be addressed by desorption at 280 °C.

EXTRACTION CAPABILITY OF SOLID-PHASE MICROEXTRACTION SORBENTS

This table shows the extraction capability of the fibers for acetone, a small moderately polar analyte, for 4-nitrophenol, a medium-size polar analyte, and benzo(GHI)perylene, a large nonpolar analyte. This provides a general guideline for fiber selection.

Fiber	Approx. Linear Conc. Range Acetone 10-min ext[a] (FID)	Approx. Linear Conc. Range 4-Nitrophenol 20-min ext[b] (GC/MS)	Approx. Linear Conc. Range Benzo(GHI)perylene 20-min Ext
7 μm PDMS	100 ppm and up	Not extracted	100 ppt to 500 ppb
30 μm PDMS	10 ppm and up	10 ppm and up	100 ppt to 10 ppm
85 μm polyacrylate	1 ppm to 1,000 ppm	5 ppb to 100 ppm	500 ppt to 10 ppm
100 μm PDMS	500 ppb to 1,000 ppm	500 ppb to 500 ppm	500 ppt to 10 ppm
50 μm Carbowax (PEG)	1–1,000 ppm	5 ppb to 50 ppm	25 ppb to 10 ppm
PDMS-DVB	50 ppb to 100 ppm	25 ppb to 10 ppm	10 ppb to 1 ppm
Carbowax-DVB	100 ppb to 100 ppm	5 ppb to 10 ppm	50 ppb to 5 ppm
PDMS-DVB-Carboxen	25 ppb to 10 ppm	50 ppb to 10 ppm	100 ppb to 1 ppm poorly desorbed
PDMS-Carboxen	5 ppb to 5 ppm	100 ppb to 10 ppm	Not desorbed
Carbopak Z-PDMS	10–500 ppm	5–100 ppm	500 ppt to 100 ppb

Note: In each case, the concentration is expressed on a mass basis (for example, ppm mass/mass)
[a] Water sample contains 25 % NaCl (mass/mass)
[b] Water sample contains 25 % NaCl (mass/mass) acidified to pH = 2 with 0.05 M phosphoric acid
1 ppm = 1 part in 1×10^6
1 ppb = 1 part in 1×10^9
1 ppt = 1 part in 1×10^{12}

Typical Phase Volumes of SPME Fiber Coatings

Fiber Coating Thickness/Type	Type of Fiber Core	Fiber Core Diameter (mm)	Phase Volume (m³ or μl)
100 μm PDMS	Fused silica	0.110	0.612
100 μm PDMS	Metal	0.130	0.598
30 μm PDMS	Fused silica	0.110	0.132
30 μm PDMS	Metal	0.130	0.136
7 μm PDMS	Fused silica	0.110	0.028
7 μm PDMS	Metal	0.130	0.030
85 μm PA	Fused silica	0.110	0.543
60 μm PEG	Metal	0.130	0.358
15 μm Carbopack Z/PDMS	Metal	0.130	0.068
65 μm PDMS/DVB	Fused silica	0.120	0.418
65 μm PDMS/DVB	Proprietary	0.130	0.440
65 μm PDMS/DVB	Metal	0.130	0.440
75 μm Carboxen–PDMS	Fused silica	0.120	0.502
85 μm Carboxen–PDMS	Proprietary	0.130	0.528
85 μm Carboxen–PDMS	Metal	0.130	0.528
50/30 μm DVB/Carboxen	Metal		
Carboxen layer		0.130	0.151
DVB layer		0.190	0.377
50/30 μm DVB/Carboxen	Metal		
Carboxen layer		0.130	0.151
DVB layer		0.190	0.377
60 μm PDMS–DVB HPLC	Proprietary	0.160	0.459

SALTING OUT REAGENTS FOR HEADSPACE ANALYSIS

The following table provides data on the common salts used for salting out in chromatographic headspace analysis, as applied to direct injection methods and to solid-phase microextraction [1,2]. Data are provided for the most commonly available salts, although others are possible. Sodium citrate, for example, occurs as the dihydrate and the pentahydrate. The pentahydrate is not as stable as the dihydrate, however, and dries out on exposure to air, forming cakes. Potassium carbonate occurs as the dihydrate, trihydrate, and sesquihydrate; however, data is provided only for the anhydrous material. The solubility is provided as the number of grams that can dissolve in 100 mL of water at the indicated temperature. The vapor enhancement cited is the degree of increase of the concentration of vapor over the solution of a 2 % (mass/mass) ethanol solution in water at 60 °C [3,4].

REFERENCES

1. Rumble, J, ed., *CRC Handbook for Chemistry and Physics*, 100th. ed., CRC Press, Boca Raton, 2019.
2. NIST Chemistry Web Book, NIST Standard Reference Database Number 69 (www.webbook.nist.gov/chemistry/), accessed November, 2019.
3. Machata, G., The advantages of automated blood alcohol determination by head space analysis. *Clin. Chem. Newsletter*, 4(2), 29, 1972.
4. Ioffe, B.V. and Vitenberg, A.G., *Head Space Analysis and Related Methods in Gas Chromatography*, Wiley Interscience, New York, 1983.

Salt	Formula	Rel. Mol. Mass	Density	Solubility Cold Water	Solubility Hot Water	Vapor Enhancement
Potassium carbonate	K_2CO_3	138.21	2.428 at 14 °C	112[a]	156[b]	8
Ammonium sulfate	$(NH_4)_2SO_4$	132.13	1.769 at 50 °C	70.6[c]	103.8[b]	5
Sodium citrate (dihydrate)	$Na_3C_6H_5O_7 \cdot 2H_2O$	294.10		72[d]	167[b]	5
Sodium chloride	N_aCl	58.44	2.165[e]	37.5[a]	39.12[b]	3
Ammonium chloride	NH_4Cl	53.49	1.527	29.7[c]	75.8[b]	2

[a] 20 °C
[b] 100 °C
[c] 0 °C
[d] 25 °C
[e] Specific gravity, 25/4 °C

PARTITION COEFFICIENTS OF COMMON FLUIDS IN AIR–WATER SYSTEMS

The following table provides the partition coefficients (or distribution coefficients), $K = C_s/C_v$, (solid/vapor) at various temperatures, for application in gas chromatographic headspace analysis [1,2]. The values marked with an asterisk were determined from a linear regression of experimental data.

1. Ioffe, B.V. and Vitenberg, A.G., *Head Space Analysis and Related Methods in Gas Chromatography*, Wiley Interscience, New York, 1983.
2. Kolb, B. and Ettre, L.S., *Static Headspace Gas Chromatography- Theory and Practice*, Wiley-VCH, New York, 1996.

Partition Coefficient, K

Fluid	20 °C	25 °C	30 °C	40 °C	50 °C	60 °C
Cyclohexane				0.077	0.055*	0.040
n-Hexane				0.14	0.068*	0.043
Tetrachloroethylene				1.48	1.28*	1.27
1,1,1-Trichloromethane				1.65	1.53*	1.47
o-Xylene				2.44	1.79*	1.31
Toluene	4.6	3.6	2.9	2.82	2.23*	1.77
Benzene	4.8	4.0	3.4	2.90	3.18*	2.27
Dichloromethane				5.65	4.29*	3.31
n-Butylacetate	126	87	59	31.4	20.6*	13.6
Ethylacetate	210	150	108	62.4	42.7*	29.3
Methyl ethylketone	600	380	283	139.5	109*	68.8
n-Butanol	4,660	3,600	2,710	647	384*	238
Ethanol	7,020	5,260	4,440	1,355	820*	511
1,4-Dioxane	8,000	5,750	4,330	1,618	1,002*	624
m-Xylene	5.9	4.0	3.9			
1-Propanol	5,480	4,090	3,210		479*	
Acetone	752	551	484			

VAPOR PRESSURE AND DENSITY OF SATURATED WATER VAPOR

The following table provides the temperature dependence of the saturated vapor pressure and vapor density of water. This information is useful in gas-chromatographic headspace analysis, and for SPME sampling [1,2].

REFERENCES

1. Kolb, B. and Ettre, L.S., *Static Headspace Gas Chromatography: Theory and Practice*, Wiley-VCH, New York, 1997.
2. Rumble, J, ed, *CRC Handbook for Chemistry and Physics*, 100th. ed., CRC Press, Boca Raton, 2019.

°C	$p°$ (kPa)	$p°$ (torr)	d (µg/mL)
10	1.2	9.2	9.4
20	2.3	17.5	17.3
30	4.2	31.8	30.3
40	7.4	55.3	51.1
50	12.3	92.5	83.2
60	19.9	149.4	130.5
70	31.1	233.7	198.4
80	47.2	355.1	293.8
90	69.9	525.8	424.1
100	101.1	760.0	598.0
110	142.9	1,074.5	826.5
120	198.1	1,489.1	1,122.0

SOLVENTS FOR SAMPLE PREPARATION FOR
GAS CHROMATOGRAPHIC ANALYSIS

Many different solvents are used to prepare samples for analysis by GC, and it would be impossible to list all of them in one place. In this table, the most common solvents are provided, along with relevant properties. The solubility parameter and the dielectric constant for each solvent are used to choose the best match for the solutes present in the sample, based on polarity considerations. Unless otherwise indicated, the dielectric constant is provided at 20 °C. The conventional units for the solubility parameter are $cal^{1/2}cm^{-3/2}$. Conversion to SI units is

$$1\,cal^{1/2}cm^{-3/2} = \left(4.184\ J\right)^{1/2}\left(0.01\ m\right)^{-3/2} = 2.045\times10^{3}\ J^{1/2}m^{-3/2} = 2.045\ MPa^{1/2}.$$

The solvent viscosity is most commonly used to optimize the operation of automatic samplers. A delay must be programmed in the filling sequence for highly viscous solvents. Unless otherwise indicated, the viscosity is provided at 20 °C. The normal boiling temperature is provided to guide the selection of injector temperatures and pressure programs (along with sample decomposition considerations). It is also used in optimization of automatic sampler programs, since highly volatile solvents can form bubbles in the syringe barrel if repeat pumps are not programmed. The recommended starting oven temperatures are provided to guide the development of temperature programs for splitless injector operations to take advantage of solvent focusing at the head of the column. Temperature ranges marked with an asterisk indicate that subambient starting temperature will be needed, most easily obtained with a vortex tube or a cryoblast valve installed on the gas chromatograph.

REFERENCES

1. Willard, H.H., Merritt, L.L., Dean, J.A., and Settle, F.A., *Instrumental Methods of Analysis*, 6th ed., Wadsworth Publishing Co., Belmont, 1981.
2. Dreisbach, R.R., Physical properties of chemical compounds, Number 22 of the Advances in Chemistry Series, American Chemical Society, Washington DC, 1959.
3. Krstulovic, A.M. and Brown, P.R., *Reverse Phase High Performance Liquid Chromatography*, John Wiley & Sons (Interscience), New York, 1982.
4. Rumble, J, ed, *CRC Handbook for Chemistry and Physics*, 100th. ed., CRC Press, Boca Raton, 2019.
5. Poole, C.F. *The Essence of Chromatography*, Elsevier, Amsterdam, 2003.
6. Hoy, K.L., New values of the solubility parameters from vapor pressure data, *J. Paint Technol.*, 42, 541, 1970.
7. Barton, A.F.M., *Handbook of Solubility Parameters and Other Cohesion Parameters*, 2nd ed., CRC Press, Boca Raton, 1991.

Solvent	Solubility Parameter, δ, cal$^{1/2}$cm$^{-3/2}$	Viscosity mPa•s (20 °C)	Dielectric Constant (20 °C)	Normal Boiling Temperature (°C)	Recommended Starting Oven Temperature (°C)
Acetone	9.62	0.30(25 °C)	20.7(25 °C)	56.3	35–45*
Acetonitrile	12.11	0.34(25 °C)	37.5	81.6	55–65
Benzene	9.16	0.65	2.284	80.1	55–65
1-Butanol	11.60	2.95	17.8	117.7	85–100
2-Butanol	11.08	4.21	15.8(25 °C)	99.6	75–85
n-Butyl acetate	8.69	0.73		126.1	100–115
n-Butylchloride	8.37	0.47(15 °C)		78.4	55–65
Carbon tetrachloride	8.55	0.97	2.238	76.8	55–65
Chlorobenzene	9.67	0.80	2.708	131.7	100–120
Chloroform	9.16	0.58	4.806	61.2	30–50*
Cyclohexane	8.19	0.98	2.023	80.7	55–65
Cyclopentane	8.10	0.44	1.965	49.3	20–35*
o-Dichlorobenzene	10.04	1.32(25 °C)	9.93(25 °C)	180.5	155–165
N,N-Dimethylacetamide		2.14	37.8	166.1	135–145
Dimethylformamide	11.79	0.92	36.7	153.0	125–140
dimethyl sulfoxide	12.8	2.20	4.7	189.0	165–175
1,4-Dioxane	10.13	1.44(15 °C)	2.209(25 °C)	101.3	75–85
2-Ethoxyethanol		2.05		135.6	100–120
Ethyl acetate	8.91	0.46	6.02(25 °C)	77.1	50–65
Diethyl ether	7.53	0.24	4.335	34.6	10–25
Glyme (ethylene glycol dimethyl ether)		0.46(25 °C)		93.0	65–75
n-Heptane	7.50	0.42	1.92	98.4	60–80
n-Hexadecane		3.34		287.0	250–270
n-Hexane	7.27	0.31	1.890	68.7	40–60
Isobutyl alcohol	11.24	4.70(15 °C)	15.8(25 °C)	107.7	70–90
Methanol	14.50	0.55	32.63(25 °C)	64.7	35–55
2-Methoxyethanol	11.68	1.72	16.9	124.6	95–110
2-Methoxyethyl acetate				144.5	120–135
Methylene chloride	9.88	0.45(15 °C)	9.08	39.8	10–35
Methylethylketone	9.45	0.42(15 °C)	18.5	79.6	50–70
n-Nonane	7.64	0.72	1.972	150.8	125–140
n-Pentane	7.02	0.24	1.84	36.1	10–25*
Petroleum ether		0.30		30–60	10–30*
1-Propanol	12.18	2.26	20.1(25 °C)	97.2	70–85
2-Propanol	11.44	2.86(15 °C)	18.3(25 °C)	82.3	55–70
Pyridine	10.62	0.95	12.3(25 °C)	115.3	95–110
Tetrachloroethylene	9.3	0.93(15 °C)		121.2	90–110
Tetrahydrofuran	9.1	0.55	7.6	66.0	35–55*
Toluene	8.93	0.59	2.379(25 °C)	110.6	80–100
Trichloroethylene	9.16	0.57	3.4(16 °C)	87.2	60–75
1,2,2-Trichloro-1,2,2-trifluoroethane		0.71		47.6	25–35*
2,2,4-Trimethylpentane	6.86	0.50	1.94	99.2	70–85
o-Xylene	9.06	0.81	2.568	144.4	120–135
p-Xylene			2.270	138.5	120–135

DERIVATIZING REAGENTS FOR GAS CHROMATOGRAPHY

The following table lists some of the more common derivatizing reagents used in GC for the purposes of (l) increasing sample volatility, (2) increasing sample thermal stability, (3) reducing sample-support interactions, and (4) increasing sensitivity toward a particular detector. The table is divided into reagents for acylation, alkylation, esterification, pentafluorophenylation, and silylation. The conditions and concentrations used in derivatization must be carefully considered, since one can often cause more problems than one cures using these methods. Such problems include poor peak resolution, incomplete reactions and side products, and less than stoichiometric yields of products. The reader is referred to the citation list for more detail on the reagents, conditions, and difficulties.

REFERENCES

General References

1. Blau, K. and King, G.S., (eds.), *Handbook of Derivatives for Chromatography*, Heyden, London, 1978.
2. Knapp, D.R., *Handbook of Analytical Derivatization Reactions*, John Wiley & Sons, New York, 1979.
3. Drozd, J., *Chemical Derivatization in Gas Chromatography*, Elsevier, Amsterdam, 1981.
4. Poole, C.F. and Schutte, S.A., *Contemporary Practice of Chromatography*, Elsevier, Amsterdam, 1984.
5. Grob, R.L., *Modern Practice of Gas Chromatography*, John Wiley & Sons, New York, 1985.
6. Braithwaite, A. and Smith, F.J., *Chromatographic Methods*, Chapman and Hall, London, 1985.
7. Merritt, C., in *Ancillary Techniques of Gas Chromatography*, L.S. Ettre, and W.H. McFadder (eds.), Wiley Interscience, New York, 1969.
8. Hammarstrand, K. and Bonelli, E.J., *Derivative Formation in Gas Chromatography*, Varian Aerograph, Walnut Creek, 1968.
9. Vanden Heuvel, W.J.A., *Gas Chromatography of Steroids in Biological Fluids*, Plenum Press, New York, 1965.

Acylating Reagents

1. Brooks, C.J.W. and Horning, E.C., Gas chromatographic studies of catecholamines, tryptamines, and other biological amines, *Anal. Chem.*, 36(8), 1540, 1964.
2. Imai, K., Sugiura, M., and Tamura, Z., Catecholamines in rat tissues and serum determined by gas chromatographic method, *Chem. Pharm. Bull.*, 19, 409–411, 1971.
3. Scoggins, M.W., Skurcenski, L., and Weinberg, D.S., Gas chromatographic analysis of geometric diamine isomers as tetramethyl derivatives, *J. Chromatogr. Sci.*, 10, 678, 1972.

Esterification Reagents

1. Shulgin, A.T., Separation and analysis of methylated phenols as their trifluoroacetate ester derivatives, *Anal. Chem.*, 36(4), 920, 1964.
2. Argauer, R.J., Rapid procedure for the chloroacetylation of microgram quantities of phenols and detection by electron—capture gas chromatography, *Anal. Chem.*, 40(1), 122, 1968.
3. Vanden Heuvel, W.J.A., Gardiner, W.L., and Horning, E.C., Characterization and separation of amines by gas chromatography, *Anal. Chem.*, 36(8), 1550, 1964.
4. Änggård, E. and Göran, S., Gas chromatography of catecholamine metabolites using electron capture detection and mass spectrometry, *Anal. Chem.*, 41(10), 1250, 1969.

5. Alley, C.C., Brooks, J.B., and Choudhary, G., Electron capture gas-liquid chromatography of short chain acids as their 2,2,2-trichloroethyl esters, *Anal. Chem.*, 48(2), 387, 1976.

6. Godse, D.D., Warsh, J.J., and Stancer, H.C., Analysis of acidic monoamine metabolites by gas chromatography-mass spectrometry, *Anal. Chem.*, 49(7), 915, 1977.

7. Matin, S.B. and Rowland, M., Electron-capture sensitivity comparison of various derivatives of primary and secondary amines, *J. Pharm. Sci.*, 61(8), 1972.

8. Bertani, L.M., Dziedzic, S.W., Clarke, D.D., and Gitlow, S.E., A gas-liquid chromatographic method for the separation and quantitation of nomethanephrine and methanephrine in human urine, *Clin. Chem. Acta*, 30, 227–233, 1970.

9. Kawai and Tamura, Z., Gas chromatography of catecholamines as their trifluoroacetates, *Chem. Pharm. Bull.*, 16(4), 699, 1968.

10. Moffat, A.C. and Horning, E.C., A new derivative for the gas-liquid chromatography of picogram quantities of primary amines of the catecholamine series, *Biochem. Biophys. Acta.*, 222, 248, 1970.

11. Lamparski, L.I. and Nestrick, T.J., Determination of trace phenols in water by gas chromatographic analysis of heptafluorobutyl derivatives, *J. Chromatogr.*, 156, 143, 1978.

12. Mierzwa, S. and Witek, S., Gas-liquid chromatographic method with electron-capture detection for the determination of residues of some phenoxyacetic acid herbicides in water as their 2,2-trichloroethyl esters, *J. Chromatogr.*, 136, 105, 1977.

13. Hoshika, Y., Gas chromatographic separation of lower aliphatic primary amines as their sulphur-containing schiff bases using a glass capillary column, *J. Chromatogr.*, 136, 253, 1977.

14. Brooks, J.B., Alley, C.C., and Liddle, J.A., Simultaneous esterification of carboxyl and hydroxyl groups with alcohols and heptafluorobutyric anhydride for analysis by gas chromatography, *Anal. Chem.*, 46(13), 1930, 1974.

15. Deyrup, C.L., Chang, S.M., Weintraub, R.A., and Moye, H.A., Simultaneous esterification and acylation of pesticides for analysis by gas chromatography. 1. Derivatization of glyphosate and (aminomethyl) phasphonic acid with fluorinated alcohols-perfluoronated anhydrides, *J. Agric. Food Chem.*, 33(5), 944, 1985.

16. Samar, A.M., Andrieu, J.L., Bacconin, A., Fugier, J.C., Herilier, H., and Faucon, G., Assay of lipids in dog myocardium using capillary gas chromatography and derivatization with boron trifluoride and methanols, *J. Chromatogr.*, 339(1), 25–34, 1985.

Pentaflouro Benzoyl Reagents

1. Mosier, A.R., Andre, C.E., and Viets, F.G., Jr., Identification of aliphatic amines volatilized from cattle feedyard, *Envir. Sci. & Tech.*, 7(7), 642, 1973.

2. DeBeer, J., Van Peteghem, C., and Heyndridex, Al., Electron capture-gas-liquid chromatography (EC-GLC) determination of the herbicidal monohalogenated phenoxyalkyl acid mecoprop in tissues, urine and plasma after derivatization with pentafluorobenzylbromide, *Vet. Human Toxicol.*, 21, 172, 1979.

3. Davis, B., Crown ether catalyzed derivatization of carboxylic acids and phenols with pentafluorobenzyl bromide for electron capture gas chromatography, *Anal. Chem.*, 49(6), 832, 1977.

4. Avery, M.J. and Junk, G.A., Gas chromatography/mass spectrometry determination of water-soluble primary amines as their pentafluorobenzaldehyde imines, *Anal. Chem.*, 57(4), 790, 1985.

Silyating Reagents

1. Metcalfe, L.D. and Martin, R.J., Gas chromatography of positional isomers of long chain amines and related compounds, *Anal. Chem.*, 44(2), 403, 1972.

2. Sen, H.P. and McGeer, P.L., Gas chromatography of phenolic and catecholic amines as the trimethylsilyl ethers, *Biochem. Biophys. Res. Commun.*, 13(5), 390, 1963.

3. Fogelgvist, Josefsson, B. and Roos, C., Determination of carboxylic acids and phenols in water by extractive alkylation using pentafluorobenzylation, glass capillary g.c. and electron capture detection, HRC CC, *J. High Resolut. Chromatogr., Chromatogr. Commun.*, 3, 568, 1980.

4. Poole, C.F.C, Sye, W.F., Singhawangcha, S., Hsu, F., Zlatkis, A., Arfwidsson, A., and Vessman, J., New electron-capturing pentafluorophenyldialkylchlorosilanes as versatile derivatizing reagents for gas chromatography, *J. Chromatogr.*, 199, 123, 1980.
5. Quilliam, M.A., Ogilvie, K.K., Sadana, K.L., and Westmore, J.B., Study of rearrangement reactions occurring during gas chromatography of tert-butyl-dimethylsilyl ether derivatives of uridine, *J. Chromatogr.*, 194, 379, 1980.
6. Poole, C.F. and Zlatkis, A., Trialkylsilyl ether derivatives (other than TMS) for gas chromatography and mass spectrometry, *J. Chrom. Sci.*, 17 (3), 115, 1979.
7. Francis, A.J., Morgan, E.D., and Poole, C.F., Flophemesyl derivatives of alcohols, phenols, amines and carboxylic acids and their use in gas chromatography with electron-capture detection, *J. Chromatogr.*, 161, 111, 1978.
8. Harvey, D.J., Comparison of fourteen substituted silyl derivatives for the characterization of alcohols, steroids and cannabinoids by combined gas-liquid chromatography and mass spectrometry, *J. Chromatogr.*, 147, 291, 1978.
9. Quilliam, M.A. and Yaraskavitch, J.M., Tertbutyldiphenylsilyl derivatization for liquid chromatography and mass spectrometry, *J. Liq. Chromatogr.*, 8(3), 449, 1985.

Derivatizing Reagent	Structure/Formula	Notes
	Acylating Reagents	
Acetic anhydride	$(CH_3CO)_2O$	Used for amino acids, steroids, urinary sugars, pesticides and herbicides, and narcotics.
Chloracetic anhydride	$(CH_2ClCO)_2O$	Useful for electron capture detection of lower aliphatic primary amines.
2,4'-Dibromoacetophenone	$BrCH_2-C=O$ (with Br-substituted benzene ring)	Used for short- and medium-chain aliphatic carboxylic acids.
Heptafluorobutyric anhydride	$(CF_3CF_2CF_2CO)_2O$	Used in basic solution for alcohols, amines, nitrosamines, amino acids, and steroids; heptafluorobutylimidazole is used in a similar fashion in the analysis of phenols.
Pentafluorobenzaldehyde	(pentafluorophenyl–CHO structure)	Useful for electron capture detection of several primary amines.
Pentafluorobenzoyl chloride	(pentafluorophenyl–CCl=O structure)	Useful for electron capture detection of several primary amines.
Pentafluoropropionic anhydride	$(CF_3CF_2CO)_2O$	Used for aromatic monoamines and their metabolites.
Propionic anhydride	$(CH_3CH_2CO)_2O$	Used for amines, amino acids, narcotics.
Pivalic anhydride	$[(CH_3)_3CCO]_2O$	Used for hormone analysis.

(Continued)

Derivatizing Reagent	Structure/Formula	Notes
2-Thiophene aldehyde	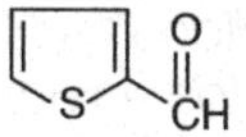	Used for electron capture detection of lower aliphatic primary amines.
Trifluoroacetic anhydride	$(CF_3CO)_2O$	Used for phenols, amines, amino acids, amino phosphoric acids, saccharides, and vitamins.
N-trifluoroacetyl-imidazole	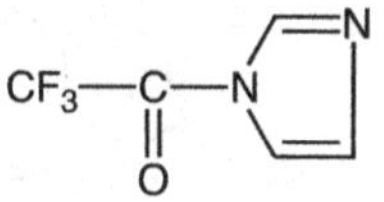	Useful for the relatively straightforward acylation of hydroxyl groups, and secondary or tertiary amines.

Acylating Reagents

Derivatizing Reagent	Structure/Formula	Notes
Diazomethane	$CH_2=N=N$ $+$ $-$	Used as a common alkylating agent; acts on acidic and enolic groups rapidly and more slowly on other groups with replaceable hydrogens (the use of a Lewis acid catalyst such as BF_3 is sometimes helpful). All diazoalkanes are toxic and sometimes explosive and are used in microscale operations only.
Trimethylanilinium hydroxide (TMAH) (in methanol)	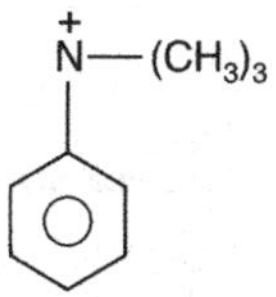	Useful for methylation of amines.
Pentafluorobenzyl bromide	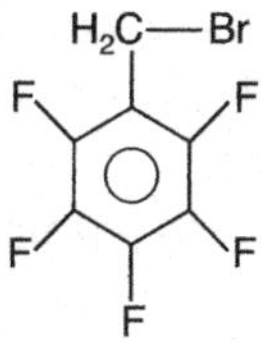	Useful for the derivatization of acids, amides, and phenols, providing great increase in sensitivity toward electron capture detection.

Esterification Reagents

Derivatizing Reagent	Structure/Formula	Notes
Boron trifluoride + methanol	$BF_3 + CH_3OH$	Useful for carboxylic acids (aromatic and aliphatic), fatty acids, fatty acid esters, Krebs cycle acids.
Boron trifluoride + n-propanol	$BF_3 + CH_3(CH_2)_2OH$	Useful for fatty acid, lactic acid, and succinic acid.

(Continued)

Derivatizing Reagent	Structure/Formula	Notes
N,N-Dimethyl-formamide dimethyl acetal	$\begin{array}{cc} OCH_3 & CH_3 \\ | \quad & | \\ C-CH-N \\ | \quad & | \\ OCH_3 & CH_3 \end{array}$	Useful in the formation of fatty acid esters and for N-protected amino acids, sulfonamides, barbiturates.
2-Bromopropane	$(CH_3)_2CHBr$	Used for amino acids and amides.
n-Butanol	$CH_3(CH_2)_3OH$	Used for carboxylic acids and amino acids.
Hydrogen chloride+methanol	$HCl+CH_3OH$	Useful for carboxylic acids, branched-chain fatty acids, oxalic acid, amino acids, and lipids; HCl serves as a catalytic agent.
Sodium methoxide	CH_3ONa in CH_3OH	Used for the transesterification of lipids.
Sulfuric acid+methanol	$H_2SO_4+CH_3OH$	Useful for carboxylic and fatty acids.
Tetramethyl ammonium hydroxide	$(CH_3)_4NOH$ in CH_3OH	Useful for carboxylic acids, fatty acids, and alkyd and polyester resins.
Thionyl chloride	$SOCl_2$	Useful in the formation of esters of carboxylic acids and other acidic functional groups.
2,2,2-Trichloroethanol	$\begin{array}{c} H \\ | \\ CCl_3-C-OH \\ | \\ H \end{array}$	Useful in the esterification of short chain acids followed by electron capture detection; sometimes used with trifluoroacetic anhydride in the presence of H_2SO_4.
Triethyl orthoformate	$HC(OC_2H_5)_3$	Used for aminophosphoric acids.
Trimethylphenyl-ammonium hydroxide	$(CH_3)_3N^+$ ⬡ OH^- in CH_3OH	Used for fatty acids, aromatic acids, herbicides, and pesticides.

Pentafluorophenyl Reagents

α-Bromopentafluorotoluene	CH_2Br on pentafluorophenyl ring (F at all ring positions)	Used to etherify sterols and phenols, in diethyl ether with the presence of potassium t-butanolate.

(Continued)

Derivatizing Reagent	Structure/Formula	Notes
Pentafluorobenzaldehyde		Used in derivatizing primary amines; greatly enhances electron capture detector (ECD) response (to the picogram level).
Pentafluorobenzyl alcohol		Used in derivatizing carboxylic acids.
Pentafluorobenzyl bromide		Used in the derivatization of carboxylic acids, phenols, mercaptans, and sulfamides; lachrymator; potentially unstable; high sensitivity for electron capture detection; not usable for formic acid.
Pentafluorobenzyl chloride		Used in the derivatization of amines, phenols, and alcohols; used in a solution of NaOH.

(Continued)

Derivatizing Reagent	Structure/Formula	Notes
Pentafluorobenzyl chloroformate		Used in derivatization of tertiary amines.
Pentafluorobenzyl hydroxylamine		Used in derivitization of ketones; can form both syn- and anti-isomers (two peaks).
Pentafluorophenacetyl chloride		Used in derivatization of alcohols, phenols, and amines.
Pentafluorophenylhydrazine		Used in derivatization of ketones; can form both the syn- and anti-isomers, resulting in two peaks.

(Continued)

Derivatizing Reagent	Structure/Formula	Notes
Pentafluorophenoxyacetyl chloride		Used in derivatization of alcohols, phenols, and amines.
Silylating Reagents		
Bis(dimethylsilyl) acetamide (BSDA)	$CH_3-C=N-Si-(CH_3)_2$ with O, H, and $H-Si-(CH_3)_2$	Similar in use and application to DMCS (see below).
N,N-Bis (trimethyl-silyl)-acetamide (BSA)	$Si(CH_3)_3$, $CH_3-C=N-Si(CH_3)_3$ with O	More reactive than HMDS (see below) or TMCS but forming essentially similar derivatives; useful for alcohols, amines, amino acids, carboxylic acids, penicillic acid, purine, and pyrimidene bases.
Bis(trimethylsilyl) trifluoroacetamide (BSTFA)	$Si-(CH_3)_3$, $CF_3-C=N-Si(CH_3)_3$ with O	Similar in use and application to BSA, but the derivatives are more volatile; by-products often elute with the solvent front; reacts more strongly than HMDS or TMCS; may promote enol-TMS formation unless ketone groups are protected.
Dimethylchlorosilane (DMCS)	$(CH_3)_2SiHCl$	Similar in use and application to TMCS and HMDS but usually forming more volatile and less thermally stable derivatives; also finds use in surface deactivation of chromatographic columns and injectors.
1,1,1,3,3,3-hexamethyl disilizane (HMDS)	$(CH_3)_3-Si-NH-Si(CH_3)_3$	Useful for such compounds as sugars, phenols, alcohols, amines, thiols, and steroids; especially recommended for citric acid cycle compounds and amino acids; reaction is often carried out in pyridine or dimethyl formamide (the latter being preferred for 17-keto steroids); care must be taken to eliminate moisture; lowest silyl donating strength of all common silating reagents.

(Continued)

Derivatizing Reagent	Structure/Formula	Notes
1,1,3,3,3-Hexamethyl disiloxane (HMDSO)	$(CH_3)_3Si-O-Si-(CH_3)_3$	Similar in use and application to HMDS (see above).
N-Methyl-N-(trimethylsilyl)-acetamide (MSTA)	$CF_3-\overset{\overset{CH_3}{\|}}{\underset{\underset{O}{\|\|}}{C}}-N-Si(CH_3)_3$	Similar in use and application to HMDS but somewhat higher "silyl donating" strength.
Silyating Reagents		
N-Methyl-N-(trimethylsilyl)tri-fluoroacetamide (MSTFA)	$CF_3-\overset{\overset{CH_3}{\|}}{\underset{\underset{O}{\|\|}}{C}}-N-Si(CH_3)_3$	Similar to MSTA but produces the most volatile derivatives of all common silylating agents; particularly useful with low molecular mass derivatives.
Tetramethyldisilazane (TMDS)	$(CH_3)_2-\overset{\overset{H}{\|}}{Si}-NH-\overset{\overset{H}{\|}}{Si}-(CH_3)_2$	Similar in use and application to DMCS.
N-Trimethylsilyl diethylamine (TMSDEA)	$(CH_3)_3-Si-N-(C_2H_5)_2$	Similar in use and application to DMCS.
N-Trimethylsilyl imidazole (TMSIM)	$(CH_3)_3-Si-$ [imidazole]	Generally useful reagent with a high silyl donor ability; will not react with amino groups; will not cause formation of enol-ether on unprotected ketone groups; especially useful for ecdysones, norepinephrine, dopamine, steroids, sugars, sugar phosphates, and ketose isomers.
Trimethylchlorosilane (TMCS)	$(CH_3)_3SiCl$	Similar properties and applications as for HMDS; useful for amino acid analyses; provides good response for electron capture detection; has relatively low silyl donating ability and is usually used in the presence of a base such as pyridine; may cause enol-ether formation with unprotected ketone groups; often used as a catalyst with other silylating reagents.
Halomethylflophemesyl reagents	$C_6F_5-\overset{\overset{CH_3}{\|}}{\underset{\underset{R}{\|}}{Si}}-Y$ $R = CH_2Cl$ $Y = Cl$	Similar in use and applications to the flophemesyl and alkylflophemesyl reagents.

(Continued)

Derivatizing Reagent	Structure/Formula	Notes
Silylating Reagents		
Halomethyldimethyl silyl reagents	$CH_2X-Si-Y$ with CH_3 groups; $X = Cl, Br, I$; $Y = Cl, N(C_2H_5)_2, NHSi(CH_3)_2CH_2X$	Family of derivatizing agents which improve sensitivity of analyte to the ECD; the response enhancement is in the order expected: $I > Br > Cl \gg F$, reverse order of the volatility of these compounds. The iodomethyldimethylsilyl reagents are unstable, and these derivatives are usually prepared *in situ*.
Flophemesyl reagents	C_6F_5-Si-Y with CH_3 and R; $R = CH_3$; $Y = Cl, NH_2, N(C_2H_5)_2$	Family of reagents forming derivatives which have stabilities similar to those produced by TMSIM, BSA, MSTFA, BSTFA, with additional electron capture detection sensitivity enhancement; usually used in pyridine as a solvent; reactions subject to steric considerations.
Alkylflophemesyl reagents	C_6F_5-Si-Y with CH_3 and R; $R = CH(CH_3)_2, C(CH_3)_3$; $Y = Cl$	Family of reagents forming derivatives of somewhat higher stability than the flophemesyl reagents; reactions subject to steric considerations.
Miscellaneous Reagents		
Boronation reagents	$(OH)_2B-R$ $R=CH_3, -C(CH_3)_3$	Used to block two vicinal hydroxy groups, derivatives have very distinctive mass spectra which are easily identified.
Carbon disulfide	CS_2	Used to derivatize primary amines to yield isothiocyanates.
Dansyl chloride	SO_2Cl / naphthalene / $N(CH_3)_2$	Used for derivatization of tripeptides; provides high sensitivity toward spectrofluorimetric detection.
Dimethyldiacetoxysilane	$(Cl)_2Si(CH_3)_2$	Used in similar applications as the boronation reagents in pyridene or trimethylamine solvent.

(Continued)

Derivatizing Reagent	Structure/Formula	Notes
2,4-Dinitrophenylhydrazone	$H-N=NH_2$ (2,4-dinitrophenyl, NO_2, NO_2)	Useful in derivatizing carbonyl compounds, and also provides a "spot test" for these compounds.
l-Fluoro-2,4-dinitro-fluorobenzene	F (2,4-dinitrophenyl, NO_2, NO_2)	Useful for derivatizing C_1–C_4 primary and secondary amines, providing high ECD response; this reagent is also useful for primary alicyclic amines.
Girard reagent T	$(CH_3)_3N\text{-}Cl\text{-}CH_2CONHNH_2$	Useful for derivatization of saturated aldehydes.
Hydrazine	NH_2NH_2	Used for the analysis of C-terminal peptide residue species.
Methyl iodide + silver oxide	$CH_3I + Ag_2O$ (in di-methylformamide)	Used to convert polyhydroxy compounds to the methyl ethers.
Methyloxamine hydrochloride	$CH_3\text{-}O\text{-}NH \cdot HCl$	Used in derivatization of steroids and carbohydrates.
2-Methylthioaniline	NH_2, SCH_3 (on benzene ring)	Used to form sulfur-bearing derivatives of benzaldehydes.
Phenyl isocyanate	(phenyl) $N=C=O$	Used for derivatization of N-terminal peptide residue
2,4,6-Trichlorophenylhydrazine	$HN-NH_2$ (2,4,6-trichlorophenyl, Cl, Cl, Cl)	Used for derivatization of carbonyl compounds.

DETECTORS FOR GAS CHROMATOGRAPHY

The following table provides some comparative data to aid in interpreting results from the more common detectors applied to capillary and packed-column gas chromatography [1–8]. For more detailed information regarding operation and interpretation of results, see Ref. [8].

1. Hill, H.H. and McMinn, D., eds., *Detectors for Capillary Chromatography*, Wiley-Interscience, John Wiley & Sons, Inc., 1992.
2. Buffington, R. and Wilson, M.K., *Detectors for Gas Chromatography—A Practical Primer*, Hewlett Packard Corp, Avondale, 1987.
3. Buffington, R., *GC-Atomic Emission Spectroscopy using Microwave Plasmas*, Hewlett Packard Corp, Avondale, 1988.
4. Liebrand, R.J., ed., *Basics of GC/IRD and GC/IRD/MS*, Hewlett Packard Corp, Avondale, 1993.
5. Bruno, T.J., A review of hyphenated chromatographic instrumentation, *Sep. Purif. Methods*, 29(1), 63–89, 2000.
6. Bruno, T.J., A review of capillary and packed column chromatographs, *Sep. Purif. Methods*, 29(1), 27–61, 2000.
7. Sevcik, J., Detectors in Gas Chromatography, *J. Chromatogr. Lib.*, Vol. 4, Elsevier, Amsterdam, 1976.
8. Bruno, T.J. and Svoronos, P.D.N., *CRC Handbook of Chemistry and Physics*, 100th ed., Rumble, J., ed., CRC Press, Boca Raton, 2019.

Detector	Limit of Detection	Linearity	Selectivity	Comments
Thermal conductivity detector (TCD, katharometer)	1×10^{-10} g propane (in helium carrier gas)	1×10^6	Universal response, concentration detector	• Ultimate sensitivity depends on analyte thermal conductivity difference from carrier gas. • Since thermal conductivity is temperature dependent, response depends on cell temperature. • Wire selection depends on chemical nature of analyte. • Helium is recommended as carrier and make-up gas. When analyzing mixtures containing hydrogen, one can use a mixture of 8.5 % (mass/mass) hydrogen in helium.
Gas density balance detector (GADE)	1×10^{-9} g, H_2 with SF_6 as carrier gas	1×10^6	Universal response, concentration detector	• Response and sensitivity are based on difference in relative molecular mass of analyte from that of the carrier gas; approximate calibration can be done on the basis of relative density. • The sensing elements (hot wires) never touch sample, thus making GADE suitable for the analysis of corrosive analytes such as acid gases; gold-sheathed tungsten wires are most common. • Best used with SF_6 as a carrier gas, switched between nitrogen when analyses are required. • Detector can be sensitive to vibrations and should be isolated on a cushioned base.
Flame ionization detector (FID)	1×10^{-11} to 1×10^{-10} g	1×10^7	Organic compounds with C-H bonds	• Ultimate sensitivity depends on the number of C-H bonds on analyte. • Nitrogen is recommended as carrier gas and make-up gas to enhance sensitivity. • Sensitivity depends on carrier, make-up, and jet gas flow rates. • Column must be positioned 1–2 mm below the base of the flame tip. • Jet gases must be of high purity.
Nitrogen–phosphorus detector (NPD, thermionic detector, alkali FID)	4×10^{-13} to 1×10^{-11} g of nitrogen compounds 1×10^{-13} to 1×10^{-12} g of phosphorus compounds	1×10^4	10^5–10^6 by mass selectivity of N or P over carbon	• Does not respond to inorganic nitrogen such as N_2 or NH_3. • Jet gas flow rates are critical to optimization. • Response is temperature dependent. • Used for trace analysis only and is very sensitive to contamination. • Avoid use of phosphate detergents or leak detectors. • Avoid tobacco use nearby. • Solvent quenching is often a problem.
ECD	5×10^{-14} to 1×10^{-12} g	1×10^4	Selective for compounds with high electron affinity, such as chlorinated organics; concentration detector	• Sensitivity depends on number of halogen atoms on analyte. • Used with nitrogen or argon/methane (95/5, mass/mass) carrier and make-up gases. • Carrier and make-up gases must be pure and dry. • The radioactive ^{63}Ni source is subject to regulation and periodic inspection in many jurisdictions.

(Continued)

Detector	Limit of Detection	Linearity	Selectivity	Comments
Flame photometric detector (FPD)	2×10^{-11} g of sulfur compounds, 9×10^{-13} g of phosphorus compounds	1×10^3 for sulfur compounds 1×10^4 for phosphorus compounds	10^5–1 by mass selectivity of S or P over carbon	• Hydrocarbon quenching can result from high levels of CO_2 in the flame. • Self-quenching of S and P analytes can occur with large samples. • Gas flows are critical to optimization. • Response is temperature dependent. • Condensed water can be a source of window fogging and corrosion.
Photoionization detector (PID)	1×10^{-12} to 1×10^{-11} g	1×10^7	Depends on ionization potentials of analytes	• Used with lamps with energies of 10.0–10.2 eV • Detector will have response to ionizable compounds such as aromatics and unsaturated organics, some carboxylic acids, aldehydes, esters, ketones, silanes, iodo- and bromoalkanes, alkylamines and amides, and some thiocyanates
Sulfur chemilumenescence detector (SCD)	1×10^{-12} g of sulfur in sulfur compounds	1×10^4	10^7 by mass selectivity of S over carbon	• Equimolar response to all sulfur compounds to within 10 %. • Requires pure hydrogen and oxygen combustion gases. • Instrument generates ozone *in situ*, which must be catalytically destroyed at detector outlet • Catalyst operates at 950 °C–975 °C. • Detector operates at reduced pressure (10^{-3} Pa).
Electrolytic conductivity detector (ECD, Hall detector)	10×10^{-13} to 1×10^{-12} g of chlorinated compounds, 2×10^{-12} g of sulfur compounds, 4×10^{-12} g of nitrogen compounds	1×10^6 for chlorinated compounds; 10^4 for sulfur and nitrogen compounds	10^6 by mass selectivity of Cl over carbon. 10^5–10^6 by mass selectivity of S and N over carbon	• Only high-purity solvents should be used. • Carbon particles in conductivity chamber can be problematic. • Frequent cleaning and maintenance is required. • Often used in conjunction with a PID. • For chlorine, use hydrogen as the reactant gas and n-propanol as the electrolyte. • For nitrogen or sulfur, hydrogen or oxygen can be used as reactant gas, and water or methanol as the electrolyte. • Ultrahigh purity reactant gases are required.
Ion mobility detector (IMD)	1×10^{-12} g	1×10^3 to 1×10^4	10^3	• Amenable to use in handheld instruments. • Linear dynamic range of 10^3 for radioactive sources and 10^5 for photoionization sources. • Selectivity depends on mobility differences of ions. • Has been used for a wide variety of compounds including amino acids, halogenated organics, and explosives. • The radioactive ^{63}Ni source is subject to regulation and periodic inspection.

(Continued)

Detector	Limit of Detection	Linearity	Selectivity	Comments
Mass selective detector (MSD, mass spectrometer, MS)	1×10^{-11} g (single ion monitoring) 1×10^{-8} g (scan mode)	1×10^5	Universal	• Single quadrupole (SQ), multiple quadrupole, ion trap (IT), time of flight, and magnetic sector instruments available (see separate entry entitled "Varieties of Hyphenated Gas Chromatography with Mass Spectrometry"). • Must operate under moderate vacuum (1×10^{-4} Pa). • Requires a molecular jet separator to operate with packed columns. • Amenable to library searching for qualitative identification. • Requires tuning of electronic optics over the entire m/e range of interest.
Infrared detector (IRD)	1×10^{-9} g of a strong infrared absorber	1×10^3	Universal for compounds with mid-infrared active functionality	• A costly and temperamental instrument that requires high-purity carrier gas, a nitrogen purge of optical components (purified air will, in general, not be adequate). • Must be isolated from vibrations. • Presence of carbon dioxide causes a typical impurity band at 2,200–2,300 cm^{-1}. • Requires frequent cleaning and optics maintenance. • Amenable to library searching for qualitative identification.
Atomic emission detector (AED)	1×10^{-13} to 2×10^{-11} g of each element	1×10^3 to 1×10^4	10^3–10^5, element to element	• Requires the use of ultrahigh purity carrier and plasma gases. • Plasma produced in a microwave cavity operated at 2,450 MHz. • Scavenger gases (H_2, O_2) are used as dopants. • Photodiode array is used to detect emitted radiation.
Vacuum ultraviolet (VUV) absorption detector	1×10^{-11} to 1×10^{-9} g	1×10^3 to 1×10^4	Universal except for He, H_2, Ar, and N_2	• Wavelength range of 120–430 nm, filter selectable. • Operable to 430°C to prevent condensation of low-volatility compounds. • Amenable to library search, though current libraries are limited. • Software can deconvolute multiple overlapping peaks. • Requires a ≈2 mL/min make-up gas of Ar, He, H_2 or N_2, which maintains constant pressure in flow cell.

VARIETIES OF HYPHENATED GAS CHROMATOGRAPHY
WITH MASS SPECTROMETRY

Because of the very powerful and common application of MS with GC, the various technologies warrant more detailed consideration [1–4]. The following table provides basic information on the capabilities and applicability of the most common approaches. Clearly, the use of the SQ is by far the most prevalent and economical method, but the other techniques are important and advantageous. We exclude methods that are typically not interfaced with gas chromatographic separations and that are highly specific in research settings (such as ion cyclotron mass spectrometry, and combined quadrupoles beyond the triple quad, and the various hybrid sector-quad-time of flight-ion trap combinations) that are used in fundamental ion chemistry research. We also exclude magnetic sector instruments that are seldom used with chromatography.

REFERENCES

1. March, R.E. and Todd, J.F., *Quadrupole Ion Trap Mass Spectrometry*. Wiley-Intersceince: NewYork, 2005.
2. Portoles, T., Sancho, J.V., Hernandez, A., Newton, A., and Hancock, P., Potential of atmospheric chemical ionizationsource in GC-QTOF for pesticide residue analysis. *J. Mass Spectrom.*, 45, 926–936, 2010.
3. Busch, K.L., Glish, G.L., and McLuckey, S.A., *Mass Spectrometry Mass Spectrometry: Techniques and Applications of Tandem Mass Spectrometry*. VCH Publishers, New York, 1988.
4. Wong, P.S. and Cooks, R.G., Ion trap mass spectrometry, *Curr. Sep.*, 16, 85–92, 1997.

Method (with Accepted Acronyms and Abbreviations)	Modes	Advantages	Limitations
Gas chromatography–mass spectrometry GC-MS (single quadrupole, SQ)	Scan, selected ion monitoring (SIM) and scan/SIM	Relatively simple and relatively inexpensive; compound identification by library search; dynamic range = 10^5, but typically limited to >10^4; mass/charge range = 10^3–10^4, resolution (at m/z = 1,000) 10^3–10^4 for most ions; well-developed hardware and software; used in standard protocols.	Co-elution of compounds compromise library identifications; often user must interpret fragmentation patterns; SIM mode provides higher sensitivity but for target ions only, sensitivity decreases with increasing number of SIM ions; scan/SIM methods must balance scan and SIM sensitivity requirements.
Gas chromatography–mass spectrometry GC-MS (ion trap, IT)	SIM; tandem MS-MS; gas or liquid chemical ionization	High sensitivity, compact design; tandem mass spectrometry is possible; dynamic range = 10^4, but typically limited to <10^4; mass/charge range = 10^4–10^5; resolution (at m/z = 1,000) 10^4; well-developed hardware and software; used in standard protocols.	Space charge effects can lead to relatively poor dynamic range; however, for "clean" samples, the dynamic range can be as high as single quadrupole units; specific libraries are limited; SIM is for target ions only with a sensitivity lower than that obtainable by an SIM analysis via single quadrupole, above; MS-MS analyses are limited to approximately 120–150 compounds per analysis; MS-MS analysis is typically slower than that done with a tandem MS-MS (process is done in time rather than space).

(Continued)

Method (with Accepted Acronyms and Abbreviations)	Modes	Advantages	Limitations
Gas chromatography (time of flight) mass spectrometry, low-resolution GC-TOF	Scan or selected ion monitoring	Identification possible with good chromatographic separations, dynamic range = 10^4, mass/charge range = 10^5, resolution (at m/z = 1,000) 10^3–10^4	Unit mass resolution limits identification capability; libraries are limited; if chromatographic separation is poor, comprehensive (GCXGC) might be needed; cannot distinguish neutral losses.
Comprehensive gas chromatography (time of flight) mass spectrometry, low-resolution GCXGC-TOF	Scan or selected ion monitoring; normal column configuration (nonpolar–polar) or reversed column configuration (polar–nonpolar)	Identification possible with good chromatographic separations in two dimensions on a nonpolar (long) and a polar (short) column; identification of families with the help of principal component analysis tools; dynamic range = 10^4, mass/charge range = 10^5, resolution (at m/z = 1,000) 10^3–10^4, well-developed hardware and software; used in standard protocols.	Unit mass resolution limits identification capability; libraries are limited; GCXGC may not remedy all aspects of component co-elution.
Gas chromatography mass Spectrometry (triple quadrupole) GC-MS (QQQ, or QqQ)	Multiple and selected reaction monitoring	Provides product and precursor ion scans; very sensitive for target compounds or functional groups; developed to provide enhanced daughter ion resolution; relatively simple construction with straightforward scanning procedures; no high-voltage arcing; dynamic range = 10^5, mass/charge range = 10^3–10^4, resolution (at m/z = 1,000) 10^3–10^4.	Unit mass, making identification ambiguous due to multiple structures as source of breakdown mass; empirical formula determination can be difficult, spectra and fragmentations must often be interpreted manually; full scan data can be acquired (similar to the single quad procedure) by turning off the collision cell.
Gas chromatography (time of flight) mass spectrometry, high-resolution GC-TOF	Scan or selected ion monitoring, tandem MS-MS	Might obviate the need for GCXGC separations with accurate mass determinations; deconvolution software can aid in identification of multiple components under peaks; sensitivity intermediate between multiple reaction monitoring and product ion scan of a QQQ; dynamic range = 10^4, mass/charge range = 10^5, resolution (at m/z = 1,000) 10^3–10^4.	Very large data files (currently approaching 2 Gb), software is currently developmental.
Gas chromatography tandem mass spectrometry GC-QTOF	Scan or selected ion monitoring, tandem MS-MS	Might obviate the need for GCXGC separations with accurate mass determinations; isolation of parent ions and subsequent fragmentation provides identification; will detect any daughter ion passed into the TOF; sensitivity intermediate between multiple reaction monitoring and product ion scan of a QQQ dynamic range = 10^4, mass/charge range = 10^5, resolution (at m/z = 1,000) 10^3–10^4.	Large data files are produced; sophisticated software is needed for processing and deconvolution; requires accurate mass and high resolution.

RECOMMENDED OPERATING RANGES FOR HOT WIRE
THERMAL CONDUCTIVITY DETECTORS

The following table provides guidance in the operation of hot wire TCDs. The operating trances are provided in mA dc for detector cells operated between 25 °C and 200 °C [1]. The current ranges and the cold resistances provided are for typical wire lengths and configurations. In many modern gas chromatographs, there is limited control over TCD operation.

1. Gow-Mac Instrument Company Manual SB-13, Thermal Conductivity Detector Elements for Gas Analysis, 1995.

Substance	Carrier Gas				
	H_2mA-dc	HemA-dc	N_2mA-dc	CO_2, ArmA-dc	Cold Resistance Ohms, 25 °C
Tungsten, W	250–500	250–400	100–175	90–130	18
Tungsten-rhenium, WX (97 %–3 %)	250–400	230–375	100–150	90–130	26–32
Nickel, Ni 99.8 %	300–500	300–450	125–150	100–130	12.5
Gold sheathed tungsten AuW	250–400	250–375	100–150	75–120	24

CHEMICAL COMPATIBILITY OF THERMAL CONDUCTIVITY DETECTOR WIRES

The following table provides guidance in the selection of hot wires for use in TCDs [1–3]. This information is applicable to the operation of packed and open tubular columns. Some of the entries in this table deal with analytes, and others deal with solutions that might be used to clean the TCD cell.

REFERENCES

1. Gow-Mac Instrument Company Manual SB-13, *Thermal Conductivity Detector Elements for Gas Analysis*, Gow-Mac, Bethlehem, PA, 1995.
2. Seveik, J., *Detectors in Gas Chromatography*, Elsevier Scientific Publishing Co., Amsterdam, 1976.
3. Lawson, A.E. and Miller, J.M., Thermal conductivity detectors in gas chromatography, *J. Gas Chromatogr.*, 4(8), 273–284, 1966.

Substance	Tungsten (W)	Rhenium–Tungsten (WX)	Nickel (Ni)	Gold-sheathed tungsten (AuW)
Air/oxygen	Good	Good	Good	Very good
Water	Good	Good	Good	Good
Steam	Good below 700 °C	Good below 700 °C	Good	Good
Ammonia/amines	Good	Good	Poor in presence of water	Poor[a]
Carbon monoxide/ carbon dioxide	Good	Good	Good	Good
Hydrogen	Good	Good	Good	Good
Nitrogen	Good	Good	Good	Fair
Fluorine	Poor (fluoride forms at 20 °C)	Poor (fluoride forms at 20 °C)	Good	Poor
Chlorine	Fair	Fair	Good	Fair
Bromine	Fair	Fair	Good	Fair
Iodine	Fair	Fair	Good	Fair
Sulfur	Fair	Good	Poor	Good
Hydrogen sulfide/sulfur dioxide (sulfuric acid)	Fair	Fair	Poor	Good
Hydrogen chloride	Fair	Fair	Good	Fair
Aqua regia	Fair	Fair	Poor	Poor
Hydrogen fluoride	Fair	Fair	Good	Fair
Hydrogen fluoride/ nitric acid	Poor	Poor	Good	Poor

[a] Gold-sheathed tungsten filaments are attacked by amines, but the process is somewhat reversible. The baseline departure will recover, but the peak will develop a significant tail.

DATA FOR THE OPERATION OF GAS DENSITY DETECTORS

The following data provide useful guidance in the operation and optimization of procedures with the gas density balance detector in GC [1]. The property values were calculated with the REFPROP database [2].

1. Nerheim, A.G., A gas density detector for gas chromatography, *Anal. Chem.*, 35, 1640, 1963.
2. Lemmon, E. W., Huber, M.L., and McLinden, M.O., REFPROP, Reference fluid thermodynamic and transport properties, NIST Standard Reference Database 23, Version 9.2. National Institute of Standards and Technology, Gaithersburg, 2017.

Argon, Ar, 24 psia

Temp. (°C)	Density (g/L)	Cp/Cv	Viscosity (µPa s)
30	2.6251	1.6712	22.887
60	2.3877	1.6703	24.735
90	2.1899	1.6697	26.521
120	2.0224	1.6692	28.249
150	1.8787	1.6688	29.925

Carbon Dioxide, CO_2, 24 psia

Temp. (°C)	Density (g/L)	Cp/Cv	Viscosity (µPa s)
30	2.912	1.295	15.179
60	2.6441	1.2802	16.614
90	2.4221	1.2679	18.018
120	2.235	1.2576	19.391
150	2.0749	1.2487	20.731

Helium, He, 24 psia

Temp. (°C)	Density (g/L)	Cp/Cv	Viscosity (µPa s)
30	0.26258	1.6665	20.075
60	0.23895	1.6665	21.417
90	0.21923	1.6665	22.726
120	0.20251	1.6665	24.006
150	0.18816	1.6665	25.259

Hydrogen, H_2, 24 psia

Temp. (°C)	Density (g/L)	Cp/Cv	Viscosity (µPa s)
30	0.13222	1.4047	9.0188
60	0.12032	1.4015	9.6186
90	0.11039	1.3997	10.2
120	0.10197	1.3987	10.766
150	0.09475	1.3982	11.317

Nitrogen, N$_2$, 24 psia

Temp. (°C)	Density (g/L)	Cp/Cv	Viscosity (µPa s)
30	1.8396	1.4022	18.052
60	1.6734	1.4013	19.399
90	1.5348	1.4002	20.695
120	1.4175	1.3989	21.945
150	1.3169	1.3973	23.153

Sulfur Hexafluoride, SF$_6$, 24 psia

Temp. (°C)	Density (g/L)	Cp/Cv	Viscosity (µPa s)
30	9.7615	1.0997	15.646
60	8.8401	1.0913	17.105
90	8.0832	1.0851	18.514
120	7.4489	1.0804	9.869
150	6.909	1.0768	21.177

1,1,1,2-Tetrafluoroethane, R134a, CF$_3$CFH$_2$, 24 psia

Temp. (°C)	Density (g/L)	Cp/Cv	Viscosity (µPa s)
30	6.9186	1.1247	12.013
60	6.2347	1.1116	13.209
90	5.6842	1.1021	14.365
120	5.2285	1.0949	15.486
150	4.8436	1.089	16.575

COMMON SPURIOUS SIGNALS OBSERVED IN GC-MS

The following table provides guidance in the recognition of spurious signals (m/z peaks) that will sometimes be observed in measured mass spectra [1]. Often, the occurrence of these signals can be predicted by the recent history of the instrument or the method being used. This is especially true if the MS is interfaced to a gas chromatograph.

1. Maintaining your GC-MS system Agilent Technologies, Applications manual, 2001, available online at www.agilent.com/chem, accessed December, 2019.

Ions Observed, m/z	Possible Compound	Possible Source
13, 14, 15, 16	Methane[a]	Chlorine reagent gas
18	Water[a]	Residual impurity, outgasing of ferrules, septa, and seals.
14, 28	Nitrogen[a]	Residual impurity, outgasing of ferrules, septa, and seals; leaking seal.
16, 32	Oxygen[a]	Residual impurity, outgasing of ferrules, septa, and seals; leaking seal.
44	Carbon dioxide[a]	Residual impurity, outgasing of ferrules, septa, and seals; leaking seal; note it may be mistaken for propane in a sample.
31, 51, 69, 100, 119, 131, 169, 181, 214, 219, 264, 376, 414, 426, 464, 502, 576, 614	Perfluorotributyl amine (PFTBA) and related ions	This is a common tuning compound; may indicate a leaking valve.
31	Methanol	Solvent; can be used as a leak detector.
43, 58	Acetone	Solvent; can be used as a leak detector.
78	Benzene	Solvent; can be used as a leak detector.
91, 92	Toluene	Solvent; can be used as a leak detector.
105, 106	Xylenes	Solvent; can be used as a leak detector.
151, 153	Trichloroethane	Solvent; can be used as a leak detector.
69	Fore pump fluid, PFTBA	Back diffusion of fore pump fluid, possible leaking valve of tuning compound vial.
73, 147, 207, 221, 281, 295, 355, 429	Dimethylpolysiloxane	Bleed from a column or septum, often during high-temperature program methods in GC-MS.
77, 94, 115, 141, 168, 170, 262, 354, 446	Diffusion pump fluid	Back diffusion from diffusion pump, if present.
149	Phthalates	Plasticizer in vacuum seals, gloves.
X–14 peaks	Hydrocarbons	Loss of a methylene group indicates a hydrocarbon sample.

[a] It is possible to operate the analyzer to ignore these common background impurities. They will be present to contribute to poor vacuum if these impurities result from a significant leak.

PHASE RATIO FOR CAPILLARY COLUMNS

The phase ratio is an important parameter used in the design of capillary (open tubular) column separations [1]. This quantity relates the partition coefficient (K) to the partition ratio (k):

$$K = k\beta$$

where β is the phase ratio, defined as the ratio of the volume occupied by the gas or mobile phase (V_m) relative to that occupied by the liquid or stationary phase (V_s). For wall-coated open tubular columns, the phase ratio can be found from

$$\beta = r/2d_f$$

where r is the internal radius of the column and d_f is the thickness of the stationary phase film. The following table provides the phase ratio for common combinations of column internal diameter and stationary-phase film thickness. These values are given to the nearest whole number, since only an approximate value is needed for most analytical applications.

REFERENCE

1. Sandra, P., *High Resolution Gas Chromatography*, 3rd ed., Hewlett Packard Corporation, Avondale, 1989.

Phase Ratio for Capillary Columns

Film Thickness (μm)	Column Inside Diameter (mm)							
	0.05	0.10	0.20	0.30	0.32	0.40	0.50	0.53
0.03	417	833	1,667	2,500	2,667	3,333	4,167	4,417
0.06	208	417	833	1,250	1,333	1,667	2,083	2,208
0.1	125	250	500	750	800	1,000	1,250	1,325
0.2	63	125	250	375	400	500	625	663
0.3	42	83	167	250	267	333	417	442
0.4	31	63	125	188	200	250	313	331
0.5	25	50	100	150	160	200	250	265
0.6	21	42	83	125	133	167	208	221
0.7	18	36	71	107	114	143	179	189
0.8	16	31	63	94	100	125	156	166
0.9	14	28	56	83	89	111	139	147
1.0	13	25	50	75	80	100	125	133
1.5	8	17	34	50	53	67	83	88
2.0	6.3	13	25	38	40	50	63	66
2.5	5	10	20	30	34	40	50	53
3.0	4	8	17	25	27	33	42	44
3.5	4	7	14	21	23	29	18	38
4.0	3	6	13	19	20	25	32	33
4.5	3	6	11	17	18	22	29	29
5.0	2.5	5	10	15	16	20	25	27
5.5	2	5	9	14	15	18	23	24
6.0	2	4	8	13	13	17	21	22
6.5	2	4	8	12	12	15	19	20
7.0	2	4	7	11	11	14	18	19
7.5	2	3	7	10	11	13	17	18
8.0	2	3	6	9	10	13	16	17
8.5	1	3	6	9	9	12	15	16
9.0	1	3	6	8	9	11	14	15

PRESSURE DROP IN OPEN TUBULAR COLUMNS

The pressure drop across an open tubular or capillary column is often important for optimization of chromatographic analyses. Column prformance is typically assesed by the height equavalent to a theoretocal plate (HETP), which is based on the average linear carrier gas velocity. As the average linear velocity increases, the head pressure and carrier gas flow rate increases as well. One may express the pressure drop across the colunn as

$$\Delta p = p_i - p_o,$$

where Δp is the pressure drop, p_i is the inlet or head pressure, and p_o is the outlet pressure. The head pressure is typically a gauge pressure measured electronically, while the outlet pressure is the barometric pressure, which can be measured electronically, with a mercury barometer, or with an aneroid barometer. Concern for the spillage of mercury has caused an increase in the number of laboratories employing an electronic measure for this. In relation to the average carrier gas velocity

$$\Delta p = 8\eta Lu/r_c^2,$$

where η is the carrier gas viscosity, L is the column length, and r_c is the column internal radius. For helium carrier gas at $100\,^{\circ}\mathrm{C}$, the following tables provide the pressure drop in units of psig and kPa.

REFERENCE

1. Hinshaw, J.V., Open tubular column pressures and flows, GC troubleshooting, *LC-GC*, 7(3), 237–239, 1989.

For 10 m Columns

Diameter, $2r_c$ (mm)	0.750	0.530	0.320	0.200	0.100
Carrier Gas Velocity, u (cm/sec)	Pressure Drop (psig)				
10	0.19	0.38	1.0	2.7	10.0
20	0.38	0.75	2.1	5.3	21.2
30	0.56	1.1	3.1	7.9	31.8
40	0.75	1.5	4.1	10.6	42.3
60	1.1	2.3	6.2	15.9	63.5
80	1.5	3.0	8.3	21.3	84.7

For 25 m Columns

Diameter, $2r_c$ (mm)	0.750	0.530	0.320	0.200	0.100
Carrier Gas Velocity, u (cm/sec)	Pressure Drop (psig)				
10	0.47	0.94	2.6	6.6	26.5
20	0.94	1.9	5.2	13.3	52.9
30	1.4	2.8	7.8	19.8	79.4
40	1.9	3.8	10.3	26.5	
60	2.8	5.7	15.5	39.7	
80	3.8	7.5	20.7	52.9	

For 50 m Columns

Diameter, $2r_c$ (mm)	0.750	0.530	0.320	0.200	0.100
Carrier Gas Velocity, u (cm/sec)	Pressure Drop (psig)				
10	0.94	1.9	5.2	13.2	52.9
20	1.9	3.8	10.3	26.5	
30	2.8	5.7	15.5	39.7	
40	3.8	7.5	20.7	52.9	
60	5.6	11.3	31.0	79.4	
80	7.5	15.1	41.3		

For 10 m Columns

Diameter, $2r_c$ (mm)	0.750	0.530	0.320	0.200	0.100
Carrier Gas Velocity, u (cm/sec)	Pressure Drop, kPa, Gauge				
10	1.3	2.6	6.9	18.6	69.0
20	2.6	5.2	14.5	36.5	146.2
30	3.9	7.6	21.4	54.5	219.3
40	5.2	10.3	28.3	73.1	291.7
60	7.6	15.9	42.7	109.6	437.8
80	10.3	20.7	57.2	146.9	584.0

For 25 m Columns

Diameter, $2r_c$ (mm)	0.750	0.530	0.320	0.200	0.100
Carrier Gas Velocity, u (cm/sec)	Pressure Drop, kPa, Gauge				
10	3.2	6.5	17.9	45.5	182.7
20	6.5	13.1	35.9	91.7	364.7
30	9.7	19.3	53.8	136.5	547.5
40	13.1	26.2	71.0	182.7	
60	19.3	39.3	106.9	273.7	
80	26.2	51.7	142.7	364.7	

For 50 m Columns

Diameter, $2r_c$ (mm)	0.750	0.530	0.320	0.200	0.100
Carrier Gas Velocity, u (cm/sec)	Pressure Drop, kPa, Gauge				
10	6.5	13.1	35.9	91.0	364.7
20	13.1	26.2	71.0	182.7	
30	19.3	39.3	106.9	273.7	
40	26.2	51.7	142.7	364.7	
60	38.6	77.9	213.7	547.5	
80	51.7	104.1	284.8	0.0	

MINIMUM RECOMMENDED INJECTOR SPLIT
RATIOS FOR CAPILLARY COLUMNS

In order to avoid overloading high-efficiency open tubular or capillary columns (with theoretical plate counts between 400,000 and 600,000), it is necessary to split the flow in the injector. Split ratios that are too low will result in distorted peak shapes and poor analyses. As a first approximation, the lowest split ratio that can be used is dependent upon the column internal diameter. Secondary factors then include the solute properties (polarity, etc.), column temperature (or temperature program), liner volume, injector volume, and stationary phase properties. The following table provides the minimum split ratios that should be considered for typical capillary columns [1].

REFERENCE

1. Rood, D., Gas chromatography problem solving and troubleshooting, *J. Chromatogr. Sci.*, 36(9), 476, 1998.

Column Diameter (mm)	Minimum Split Ratio
0.18	1:25
0.20	1:20
0.25	1:15–1.20
0.32	1:10–1:12
0.53	1:3–1:5

MARTIN-JAMES COMPRESSIBILITY FACTOR AND
GIDDINGS PLATE HEIGHT CORRECTION FACTOR

The following table provides the Martin-James compressibility factor, j [1], and the Giddings plate height correction factor, f [2], for chromatographically useful pressures. These quantities are defined as

$$j = 3/2 \left[\frac{\left[\left(P_i^{abs}/P_o \right)^2 - 1 \right]}{\left[\left(P_i^{abs}/P_o \right)^3 - 1 \right]} \right]$$

$$f = 9/8 \left[\frac{\left[\left(P_i^{abs}/P_o \right)^4 - 1 \right] \left[\left(P_i^{abs}/P_o \right)^2 - 1 \right]}{\left[\left(P_i^{abs}/P_o \right)^3 - 1 \right]^2} \right]$$

where P_i is the absolute inlet pressure, and P_o is the outlet pressure.

The inlet pressures listed in the table are gauge pressures; the pressures used in the calculations of j and f are absolute pressures. Thus, atmospheric pressure had already been accounted for in the inlet pressure. The outlet pressure is taken as standard atmospheric pressure. As an example, for a measured gauge pressure of 137.9 kPa (20 psig), the ratio P_i^{abs}/P_o is 2.361. The actual value of the atmospheric pressure will vary day to day and with altitude; thus, if an exact value for j or f is desired, local pressure measurements must be made.

REFERENCES

1. Grob, R.L., *Modern Practice of Gas Chromatography*, 2nd ed., John Wiley & Sons (Wiley Interscience), New York, 1985.
2. Lee, M.L., Yang, F.J., and Bartle, K.D., *Open Tubular Column Gas Chromatography*, John Wiley & Sons (Wiley Interscience), New York, 1984.

Pressure	j	f
15.0	0.638	1.034
16.0	0.622	1.037
17.0	0.606	1.039
18.0	0.592	1.042
19.0	0.578	1.044
20.0	0.564	1.046
25.0	0.505	1.057
30.0	0.456	1.066
35.0	0.416	1.074
40.0	0.381	1.080
45.0	0.352	1.085
50.0	0.327	1.090
55.0	0.305	1.093
60.0	0.286	1.096

Gas Hold-Up Volume

There are a few instances in which it is important to determine the gas hold-up volume of the chromatographic system consisting of the injector, column, and detector swept volumes. Noxious volumes are by definition unswept and are generally minimized by design. The most common application of the gas hold-up volume or measurement is in the determination of the average column flow rate with the following equation:

$$F_{ave} = \pi r^2 L / t_m,$$

where F_{ave} is the average flow rate, L is the length of the column in cm, and t_m is the average retention time of a marker compound that is minimally retained. The following table provides potential minimally retained markers for various detectors:

Detector	Minimally Retained Marker Compound
FID	Methane, n-butane
TCD	Methane, n-butane, air
MSD	Methane, n-butane, air
ECD	Sulfur hexafluoride, methylene chloride
NPD	Acetonitrile

Methane is usually easily obtainable from a natural gas line, and n-butane is easily obtained from a disposable lighter. For the liquids, it is important to only use an aliquot of the headspace or to use a permeation vial [1].

Another application in which the gas hold-up volume is needed is in the use of chromatographic retention parameters for solute identification. Chromatographic parameters include net retention volumes, relative retentions, specific retention volumes, and retention indices. Here, it is important to evaluate the applicability of a minimally retained marker in each case, since even a very light solute such as methane can show retentive behavior. It is usually best to use an extrapolative method to estimate the hold-up, although the chromatographic behavior of methane is often used in these procedures as well [2]. A convenient way to dispense the methane is with a permeation tube methanizer [3].

REFERENCES

1. Bruno, T.J., Permeation tube approach to long-term use of automatic sampler retention index standards. *J. Chromatogr. A.*, 704(1), 157–62, 1995.
2. Miller, K.E. and Bruno, T.J., Isothermal Kovats retention indices of sulfur compounds on a poly(5 % phenyl-95 % dimethyl siloxane) stationary phase. *J. Chromatogr. A.*, 1007, 117–125, 2003.
3. Bruno, T.J., Simple and efficient methane-marker devices for chromatographic samples. *J. Chromatogr. A.*, 721(1), 157–164, 1996.

CRYOGENS FOR SUBAMBIENT TEMPERATURE GAS CHROMATOGRAPHY

The following table lists properties of common cryogenic fluids used to produce subambient temperatures for gas chromatographic columns [1–5]. These properties are of value in designing low-temperature chromatographic experiments efficiently and safely. Due to the potential dangers in handling extremely low temperatures and high pressures, appropriate precautions must be observed. These precautions must include protective clothing and shielding to prevent frostbite. Most cryogenic fluids can create a health hazard if they are vaporized in an inhabited area. Even small quantities can contaminate and displace air in a relatively short period of time. It may be advisable to locate a self-contained breathing apparatus immediately outside the laboratory in which the cryogens are being used. The effect of low temperatures on construction materials (of G.C. ovens and columns, for example) should also be considered. In this respect, differential expansion and tensile strength changes are pertinent issues. A dew point vs. moisture content table is also provided to allow the user to estimate the effects of ambient and impurity water. The viscosity data are provided in cP, which is equivalent to mPa·s, the appropriate SI unit. The freezing points are reported at 0.101325 MPa (1 atm), and the expansion ratios are reported at standard temperature and pressure (STP).

It is recommended that one consult the tables on cryogen and cryogen container safety in Chapter 15 before using these cyrogens.

If temperatures no lower than approximately −40 °C are required, the use of the Ranque–Hilsch vortex tube should be considered [6–8]. This device requires a source of clean, dry compressed air at a pressure of approximately 0.70 MPa (100 psi) for proper operation. The flow rate of air which is required depends on the volume of space to be cooled.

REFERENCES

1. Zabetakis, M.G., *Safety with Cryogenic Fluids*, Plenum Press, New York, 1967.
2. Cook, G.A., ed., *Argon, Helium and the Rare Gases*, John Wiley & Sons (Interscience) New York, 1961.
3. Brettell, T.A. and Grob, R.L., Cryogenic techniques in gas chromatography. Part two: cryofocusing and cryogenic trapping, *Am. Lab.*, 17(10), 19, 1985.
4. Cowper, C.J. and DeRose, A.J., *The Analysis of Gases by Chromatography*, Pergamon Press, Oxford, 1983.
5. Braker, W., Mossman, A.L., *Matheson Gas Data Book*, 4th ed., The Matheson Company, East Rutherford, 1966.
6. Bruno, T.J., Vortex cooling for subambient temperature gas chromatography, *Anal. Chem.*, 58(7), 1596, 1986.
7. Bruno, T.J., Vortex refrigeration of HPLC components, LC, *Liq. Chromatogr. HPLC Mag.*, 4(2), 134, 1986.
8. Bruno, T.J., Laboratory applications of the vortex tube, *J. Chem. Educ.*, 64(11), 987, 1987.

Cryogen Name	Relative Molecular Mass	Freezing Point °C, (K)	Heat of Fusion, J/g	Normal Boiling Point °C, (K)	Heat of Vaporization J/g
Argon Ar	39.948	−189.4 (83.8)	27.6	−185.84 (87.302)	163.2

Critical Temperature °C, (K)	Critical Pressure MPa	Critical Density g/L	Vapor Pressure MPa	Gas Density g/L	Liquid/Gas Expansion Ratio
−122.3 (150.9)	4.89	530.5	a	1.63	860

Heat Capacity C_p J/(kg·K)	Heat Capacity C_v (J/(kg·K)	Thermal Conductivity × 10^{-2} w/(m·K)	Viscosity Pa·s × 10^5 (cP)	Solubility in Water 0 °C, V/V
523.8 (21 °C)	313.8 (15.6 °C)	1.44 (233 K)	2.21 (21 °C)	0.056

Cryogen Name	Relative Molecular Mass	Freezing Point °C, (K)	Heat of Fusion, J/g	Normal Boiling Point °C, (K)	Heat of Vaporization J/g
Carbon Dioxide CO_2	44.01	−78.5[b] (194.7)	198.7	−56.6 (216.6)	151.5

Critical Temperature °C, (K)	Critical Pressure MPa	Critical Density g/L	Vapor Pressure MPa	Gas Density g/L	Liquid/Gas Expansion Ratio
31.1 (304.2)	7.38	468	5.72 (21 °C)	1.98	790

Heat Capacity C_p J/(kg·K)	Heat Capacity C_v (J/(kg·K)	Thermal Conductivity × 10^{-2} w/(m·K)	Viscosity Pa·s × 10^5 (cP)	Solubility in Water 0 °C, V/V
831.8 (15.6 °C)	638.8 (15.6 °C)	1.17 (233 K)	1.48 (21 °C)	0.90

Cryogen Name	Relative Molecular Mass	Freezing Point °C, (K)	Heat of Fusion, J/g	Normal Boiling Point °C, (K)	Heat of Vaporization J/g
Helium ^{4}He	4.003	−272[b] (1)	[b]	−268.920 (4.2238)	23.0 (15 °C)

Critical Temperature °C, (K)	Critical Pressure MPa	Critical Density g/L	Vapor Pressure MPa	Gas Density g/L	Liquid/Gas Expansion Ratio
−268.0 (5.2)	0.23	69.3	[a]	0.16	780

Heat Capacity C_p J/(kg·K)	Heat Capacity C_v (J/(kg·K)	Thermal Conductivity × 10^{-2} w/(m·K)	Viscosity Pa·s × 10^5 (cP)	Solubility in Water 0 °C, V/V
5221.6 (21 °C)	3146.4 (15.6 °C)	12.76 (233 K)	1.96 (21 °C)	0.0086

Cryogen Name	Relative Molecular Mass	Freezing Point °C, (K)	Heat of Fusion, J/g	Normal Boiling Point °C, (K)	Heat of Vaporization J/g
Methane CH_4	16.04	−182.6 (90.6)	58.6	−161.5 (87.3)	510.0

Critical Temperature °C, (K)	Critical Pressure MPa	Critical Density g/L	Vapor Pressure MPa	Gas Density g/L	Liquid/Gas Expansion Ratio
−82.1 (190.1)	4.64	162.5	[a]	0.7174	650

Heat Capacity C_p J/(kg·K)	Heat Capacity C_v J/(kg·K)	Thermal Conductivity × 10^{-2} w/(m·K)	Viscosity Pa·s × 10^5 (cP)	Solubility in Water 0 °C, V/V
2,205.4 (15.6 °C)	1,687.0 (15.6 °C)	2.57 (233 K)	1.20 (21 °C)	

Cryogen Name	Relative Molecular Mass	Freezing Point °C, (K)	Heat of Fusion, J/g	Normal Boiling Point °C, (K)	Heat of Vaporization J/g
Nitrogen N_2	28.013	−210.1 (63.1)	25.5	−195.79 (77.355)	199.6

Critical Temperature °C, (K)	Critical Pressure MPa	Critical Density g/L	Vapor Pressure MPa	Gas Density g/L	Liquid/Gas Expansion Ratio
−146.9 (150.9)	3.4	311	a	1.14	710

Heat Capacity C_p J/(kg·K)	Heat Capacity C_v (J/(kg·K)	Thermal Conductivity × 10^{-2} w/(m·K)	Viscosity Pa·s × 10^5 (cP)	Solubility in Water 0 °C, V/V
1,030.6 (21 °C)	738.6 (21 °C)	2.11 (233 K)	1.744 (15 °C)	0.023

Cryogen Name	Relative Molecular Mass	Freezing Point °C, (K)	Heat of Fusion, J/g	Normal Boiling Point °C, (K)	Heat of Vaporization J/g
Oxygen O_2	31.999	−218.8 (54.4)	13.8	−182.96 (90.188)	213.0

Critical Temperature °C, (K)	Critical Pressure MPa	Critical Density g/L	Vapor Pressure MPa	Gas Density g/L	Liquid/Gas Expansion Ratio
−118.4 (154.8)	5.04	410	a	1.3	875

Heat Capacity C_p J/(kg·K)	Heat Capacity C_v J/(kg·K)	Thermal Conductivity × 10^{-2} w/(m·K)	Viscosity Pa·s × 10^5 (cP)	Solubility in Water 0 °C, V/V
910.9 (15 °C)	650.2 (15 °C)	2.11 (233 K)	2.06 (20 °C)	0.0489

[a] Fluid is supercritical at ambient temperature.
[b] Helium will not solidify at 1 atmosphere pressure (0.101325 MPa). The approximate pressure at which solidification can occur is calculated to be 2,535 kPa.

Dew Point–Moisture Content

Dew Point (°F)	Dew Point (°C)	Moisture ppm (vol/vol)	Dew Point (°F)	Dew Point (°C)	Moisture ppm (vol/vol)
−130	−90.0	0.1	−83	−63.9	6.20
−120	−84.4	0.25	−82	−63.3	6.60
−110	−78.9	0.63	−81	−62.8	7.20
−105	−76.1	1.00	−80	−62.2	7.80
−104	−75.6	1.08	−79	−61.7	8.40
−103	−75.0	1.18	−78	−61.1	9.10
−102	−74.4	1.29	−77	−60.6	9.80
−101	−73.9	1.40	−76	−60	10.50
−100	−73.3	1.53	−75	−59.4	11.40
−99	−72.8	1.66	−74	−58.9	12.30
−98	−72.2	1.81	−73	−58.3	13.30
−97	−71.7	1.96	−72	−57.8	14.30
−96	−71.7	2.15	−71	−57.2	15.40
−95	−70.6	2.35	−70	−56.7	16.60
−94	−70.0	2.54	−69	−56.1	17.90
−93	−69.4	2.76	−68	−55.6	19.20
−92	−68.9	3.00	−67	−55.0	20.60
−91	−68.3	3.28	−66	−54.4	22.10
−90	−67.8	3.53	−65	−53.9	23.60
−89	−67.2	3.84	−64	−53.3	25.60
−88	−66.7	4.15	−63	−52.8	27.50
−87	−66.1	4.50	−62	−52.2	29.40
−86	−65.6	4.78	−61	−51.7	31.70
−85	−65.0	5.30	−60	−51.1	34.00
−84	−64.4	5.70			

COMMON SYMBOLS USED IN GAS (AND LIQUID) CHROMATOGRAPHIC SCHEMATIC DIAGRAMS

The literature of gas and liquid chromatography frequently contains schematic diagrams that depict analytical apparatus and peripherals. The interpretation of such diagrams can be facilitated by the following graphics that show the most common symbols used to represent chromatographic components.

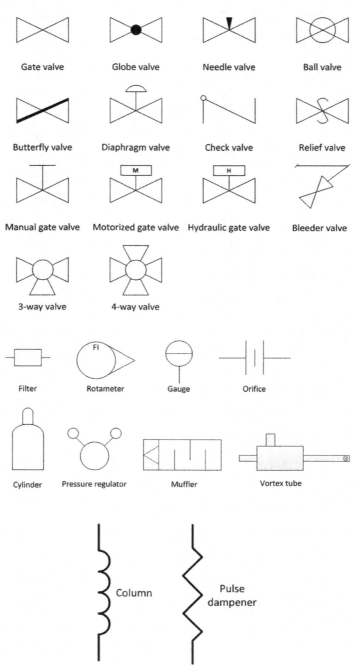

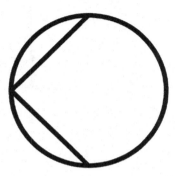

Pump, general

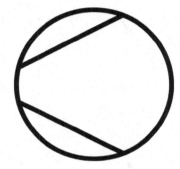

Vacuum pump or compressor

Pump, peristaltic

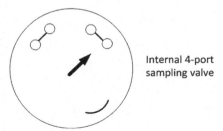

Internal 4-port
sampling valve

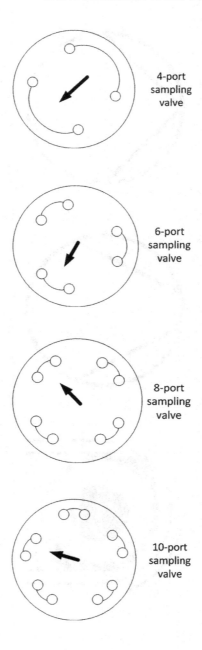

4-port
sampling
valve

6-port
sampling
valve

8-port
sampling
valve

10-port
sampling
valve

CHAPTER 3

High-Performance Liquid Chromatography

MODES OF LIQUID CHROMATOGRAPHY

The following flow chart provides a rough guide among the various liquid chromatographic techniques based on sample properties [1,2].

REFERENCES

1. Courtesy of Millipore Corporation, Waters Chromatography Division.
2. Jeerage, K. and Dooley, G. *Liquid Chromatography*, ASM Handbook, Vol. 10, Materials Characterization, ASM Handbook Committee, ASM International, Materials Park, OH, 2019.

These chromatographic modes can be used as a starting point in the design of a separation by high-performance liquid chromatography (HPLC). In general, separations in HPLC are designed by consideration of polarity, charge, and molecular size. Separation is affected by the design of mobile and stationary phase to suit a solute mixture.

Polarity, High to Low	Molecular Structure
	Salt
	Acid
	Alcohol
	Ketone
	Ether
	Halogenated hydrocarbons
	Aromatic hydrocarbons
	Aliphatic hydrocarbons
	Perfluorinated hydrocarbons

The same basic principles used to guide separations in gas chromatography can be applied to HPLC, with the complication of the stationary phase playing a critical role in addition to mass transfer. A high-polarity solute will be attracted to and retained on a high-polarity stationary phase and vice versa. Separation is obtained by balancing the attractiveness of the stationary phase with that of the mobile phase to provide favorable retention in a tractable time.

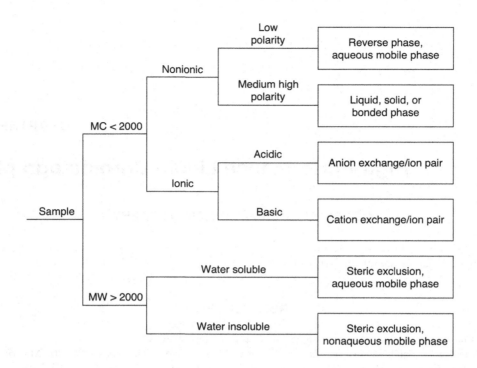

SOLVENTS FOR LIQUID CHROMATOGRAPHY

The following table provides the important physical properties for the selection of solvent systems for HPLC [1–8]. These properties are required for proper detector selection and the prediction of expected column pressure gradients. The values of the dielectric constant aid in estimating the relative solubilities of solutes and other solvents. Data on adsorption energies of useful HPLC solvents on silica and alumina (the eluotropic series) can be found in Chapter 4. Here we present the values for alumina, $\varepsilon°$, not because this is a common surface encountered in HPLC, but because there is more data on this surface than for silica. These numbers should be used for trend analysis. The data presented were measured at 20 °C, unless otherwise indicated (in parentheses). The solubility parameters, δ, defined fundamentally as the cohesive energy per unit volume, were calculated from vapor pressure data [8] or estimated from group contribution methods [9]. Those values obtained by group contribution are indicated by an asterisk.

REFERENCES

1. Willard, H.H., Merritt, L.L., Dean, J.A., and Settle, F.A., *Instrumental Methods of Analysis*, 6th ed., Wadsworth Publishing Co., Belmont, 1981.
2. Snyder, L.R. and Kirkland, J.J., *Introduction to Modern Liquid Chromatography*, 2nd ed., John Wiley & Sons (Interscience), New York, 1979.
3. Dreisbach, R.R., *Physical Properties of Chemical Compounds*, Number 22 of the Advances in Chemistry Series, American Chemical Society, Washington DC, 1959.
4. Krstulovic, A.M. and Brown, P.R., *Reverse Phase High Performance Liquid Chromatography*, John Wiley & Sons (Interscience), New York, 1982.
5. Rumble, J., ed., *CRC Handbook for Chemistry and Physics*, 100th ed., CRC Press, Boca Raton, FL, 2019.
6. Poole, C.F. and Shuttle, S.A., *Contemporary Practice of Chromatography*, Elsevier, Amsterdam, 1984.
7. Braithwaite, A. and Smith, F.J., *Chromatographic Methods*, 4th ed., Chapman and Hale, London, 1985.
8. Hoy, K.L, New values of the solubility parameters from vapor pressure data, *J. Paint Technol.*, 42, 541, 1970.
9. Barton, A.F.M., *Handbook of Solubility Parameters and Other Cohesion Parameters*, 2nd ed., CRC Press, Boca Raton, FL, 1991.

Solvent	$\varepsilon°$	δ	Viscosity (mPa·s) (20 °C)	UV Cutoff (nm)	Refractive Index (20 °C)	Normal Boiling Temperature (°C)	Dielectric Constant (20 °C)
Acetic acid	1.0	13.01	1.31(15 °C)		1.3720	117.9	6.15
Acetone	0.56	9.62	0.30(25 °C)	330	1.3588	56.08	20.7(25 °C)
Acetonitrile	0.65	12.11	0.34(25 °C)	190	1.34423	81.6	37.5
Benzene	0.32	9.16	0.65	278	1.5011	80.08	2.284
1-Butanol		11.60	2.95	215	1.3988	117.6	17.8
2-Butanol		11.08	4.21	260	1.3978	99.4	15.8(25 °C)
n-Butyl acetate		8.69	0.73	254	1.3941	126.0	
n-Butyl chloride		8.37	0.47(15 °C)	220	1.402	78.4	7.39 (25 °C)
Carbon tetrachloride	0.18	8.55	0.97	263	1.4601	76.7	2.238
Chlorobenzene	0.30	9.67	0.80	287	1.5241	131.6	2.708
Chloroform	0.40	9.16	0.58	245	1.4459	61.2	4.806
Cyclohexane	0.04	8.19	0.98	200	1.4235	80.7	2.023
Cyclopentane	0.05	8.10	0.44	200	1.4065	49.2	1.965
o-Dichlorobenzene		10.04	1.32(25 °C)	295	1.551	180.19	9.93(25 °C)
N,N-Dimethylacetamide			2.14	268	1.4375	165.1	37.8
Dimethylformamide		11.79	0.92	268	1.4305	152.8	36.7
Dimethyl sulfoxide	0.62	12.8	2.20	286	1.4793	191.9	4.7
1,4-Dioxane	0.56	10.13	1.44(15 °C)	215	1.4224	101.3	2.209(25 °C)
2-Ethoxyethanol			2.05	210	1.408	135.6	
Ethyl acetate	0.58	8.91	0.46	256	1.3723	77.1	6.02(25 °C)
Ethyl ether (diethyl ether)	0.38	7.53	0.24	218	1.3526	34.4	4.335
Glyme (ethylene glycol dimethyl ether)			0.46(25 °C)	220	1.380	85	
n-Heptane	0.01	7.50	0.42	200	1.3855	98.38	1.92
n-Hexadecane			3.34	200	1.4329	286.9	2.06 (25 °C)
n-Hexane	0.01	7.27	0.31	200	1.3727	68.72	1.890
Isobutanol (2-methylpropan-1-ol, isobutyl alcohol)		11.24	4.70(15 °C)	220	1.396	107.89	15.8(25 °C)
Methanol	0.95	14.50	0.55	205	1.3288	64.5	32.63(25 °C)
2-Methoxyethanol		11.68	1.72	210	1.4024	124.3	16.9
2-Methoxyethyl acetate				254	1.402	144.5	
Methylene chloride (dichloromethane)	0.42	9.88	0.45(15 °C)	233	1.4244	39.6	9.08
Methylethylketone (butanone)	0.51	9.45	0.42(15 °C)	329	1.3788	79.6	18.5
Methylisoamylketone		8.65		330	1.406	−144.0	
Methylisobutylketone	0.43	8.58	0.54(25 °C)	334	1.3958	116.5	
N-Methyl-2-pyrrolidone			1.67(25 °C)	285	1.488	202.0	32.0
n-Nonane		7.64	0.72	200	1.4058	150.8	1.972
n-Pentane	0.00	7.02	0.24	200	1.3575	36.06	1.84
Petroleum ether	0.01		0.30	226		30–60	
β-Phenethylamine				285	1.529(25 °C)	197–198	
1-Propanol	0.82	12.18	2.26	210	1.3850	97.04	20.1(25 °C)
2-Propanol	0.82	11.44	2.86(15 °C)	205	1.3776	82.21	18.3(25 °C)
Propylene carbonate		13.3			1.4189	241.6	

(Continued)

Solvent	$\varepsilon°$	δ	Viscosity (mPa·s) (20°C)	UV Cutoff (nm)	Refractive Index (20°C)	Normal Boiling Temperature (°C)	Dielectric Constant (20°C)
Pyridine	0.71	10.62	0.95	330	1.5095	115.2	12.3(25°C)
Tetrachloroethylene		9.3	0.93(15°C)	295	1.5059	121.2	
Tetrahydrofuran	0.45	9.1	0.55	212	1.4050	66.0	7.6
Tetramethyl urea				265	1.449(25°C)	175.2	23.0
Toluene	0.29	8.93	0.59	284	1.4941	110.60	2.379(25°C)
Trichloroethylene		9.16	0.57	273	1.4773	86.8	3.4(16°C)
1,1,2-Trichloro-1,2,2-trifluoroethane			0.71	231	1.356(25°C)	47.6	
2,2,4-Trimethylpentane	0.01	6.86	0.50	215	1.391	99.3	1.94
Water	large	23.53	1.00	<190	1.33336	100.0	80.0
o-Xylene	0.26	9.06	0.81	288	1.5018	144.4	2.568
p-Xylene				290	1.5004	138.3	2.270

INSTABILITY OF HPLC SOLVENTS

Solvents that are commonly used in HPLC frequently have inherent chemical instabilities that must be considered when designing an analysis or in the interpretation of results [1,2]. In many cases, such solvents are obtainable with stabilizers added to control the instability or to slow the reaction. Reactive solvents that do not have stabilizers (or solvents that have had the stabilizers intentionally removed by the user) must be used quickly or be given proper treatment. In either case, it is important to understand that the solvents (as they may be used in an analysis) are not necessarily pure materials.

REFERENCES

1. Sadek, P.C., *The HPLC Solvent Guide*, 2nd ed., Wiley Interscience, New York, 2002.
2. Bruno, T.J. and Straty, G.C., Thermophysical property measurement on chemically reactive systems: A case study, *J. Res. Nat. Bur. Stds. (U.S.)*, 91(3), 135–138, 1986.

Solvent	Contaminants, Reaction Products	Stabilizers
Ethers		
Diethyl ether	Peroxides[a]	2 %–3 % (vol/vol) ethanol[b] 1–10 ppm (mass/mass) BHT (1.5 %–3.5 % ethanol)+(0.2 %–0.5 % water)+(5–10 ppm (mass/mass) BHT)
Isopropyl ether	Peroxides[a]	0.01 % (mass/mass) hydroquinone 5–100 ppm (mass/mass) BHT
1,4-Dioxane	Peroxides[a]	25–1,500 ppm (mass/mass) BHT
Tetrahydrofuran	Peroxides[a]	25–250 ppm (mass/mass) BHT
Chlorinated Alkanes		
Chloroform	Hydrochloric acid, chlorine, phosgene (ccl_2o)	0.5 %–1 % (vol/vol) ethanol 50–150 ppm (mass/mass) amylene[c] Various ethanol amylene blends
Dichloromethane	Hydrochloric acid, chlorine, phosgene (ccl_2o)	25 ppm (mass/mass) amylene 25 ppm (mass/mass) cyclohexene 400–600 ppm (mass/mass) methanol Various amylene methanol blends
Alcohols		
Ethanol	Water, numerous denaturants are commonly added	
Methanol	Water; formal dehydrate (at elevated temperature)	
Acetone	Diacetone alcohol and higher oligomers	

[a] The peroxide concentration that is usually considered hazardous is 250 ppm (mass/mass).
[b] Ethanol does not actually stabilize diethyl ether nor is it a peroxide scavenger, although it was thought to be so in the past. It is still available in chromatographic solvents to preserve the utility of retention relationships and analytical methods.
[c] Amylene is a generic name for 2-methyl-2-butene.
BHT, 2,6-di-t-butyl-p-cresol.

ULTRAVIOLET ABSORBANCE OF REVERSE-PHASE MOBILE PHASES

The following table provides guidance in the selection of mobile phases that are to be used in conjunction with ultraviolet spectrophotometric detection [1,2]. The data in this table differ from the other solvent tables in this volume in that the wavelength dependence of absorbance is provided here. Moreover, common mixed mobile phases are considered here. The percentages that are given are on the basis of (vol/vol). This material is used by permission of John Wiley & Sons Inc.

REFERENCES

1. Snyder, L.R., Kirkland, J.J., and Glajch, J., *Practical HPLC Method Development*, John Wiley & Sons, New York, 1997.
2. Li, J.B., Signal to noise optimization in HPLC UV detection, LC/GC, 10, 856, 1992.

	Absorbance (AU) at Wavelength (nm) Specified									
	200	205	210	215	220	230	240	250	260	280
Solvents										
Acetonitrile	0.05	0.03	0.02	0.01	0.01	<0.01				
Methanol	2.06	1.00	0.53	0.37	0.24	0.11	0.05	0.02	<0.01	
Degassed	1.91	0.76	0.35	0.21	0.15	0.06	0.02	<0.01		
Isopropanol	1.80	0.68	0.34	0.24	0.19	0.08	0.04	0.03	0.02	0.02
Tetrahydrofuran										
Fresh	2.44	2.57	2.31	1.80	1.54	0.94	0.42	0.21	0.09	0.05
Old[a]	>2.5	>2.5	>2.5	>2.5	>2.5	>2.5	>2.5	>2.5	2.5	1.45
Acids and Bases										
Acetic acid, 1 %	2.61	2.63	2.61	2.43	2.17	0.87	0.14	0.01	<0.01	
Hydrochloric acid, 6 mM (0.02 %)	0.11	0.02	<0.01							
Phosphoric acid, 0.1 %	<0.01									
Trifluoroacetic acid (TFA)										
0.1 % in water	1.20	0.78	0.54	0.34	0.20	0.06	0.02	<0.01		
0.1 % in acetonitrile	0.29	0.33	0.37	0.38	0.37	0.25	0.12	0.04	0.01	<0.01
Ammonium phosphate, dibasic, 50 mM	1.85	0.67	0.15	0.02	<0.01					
Triethylamine, 1 %	2.33	2.42	2.50	2.45	2.37	1.96	0.50	0.12	0.04	<0.01
Buffers and Salts										
Ammonium acetate, 10 mM	1.88	0.94	0.53	0.29	0.15	0.02	<0.01			
Ammonium bicarbonate, 10 mM	0.41	0.10	0.01	<0.01						
EDTA (ethylenediaminetetraacetic acid), disodium, 1 mM	0.11	0.07	0.06	0.04	0.03	0.03	0.02	0.02	0.02	0.02

(Continued)

	Absorbance (AU) at Wavelength (nm) Specified									
	200	205	210	215	220	230	240	250	260	280
HEPES [N-(2-hydroxyethyl) piperazine-N'-2-ethanesulfonic acid], 10mM pH 7.6	2.45	2.50	2.37	2.08	1.50	0.29	0.03	<0.01		
MES [2-(N-morpholino) ethanesulfonic acid], 10mM, pH 6.0	2.42	2.38	1.89	0.90	0.45	0.06	<0.01			
Potassium phosphate										
Monobasic, 10mM	0.03	<0.01								
Dibasic, 10mM	0.53	0.16	0.05	0.01	<0.01					
Sodium acetate, 10mM	1.85	0.96	0.52	0.30	0.15	0.03	<0.01			
Sodium chloride, 1mM	2.00	1.67	0.40	0.10	<0.01					
Sodium citrate, 10mM	2.48	2.84	2.31	2.02	1.49	0.54	0.12	0.03	0.02	0.01
Sodium formate, 10mM	1.00	0.73	0.53	0.33	0.20	0.03	<0.01			
Sodium phosphate, 100mM, pH 6.8	1.99	0.75	0.19	0.06	0.02	0.01	0.01	0.01	0.01	<0.01
Tris-HCl, 20mM										
pH 7.0	1.40	0.77	0.28	0.10	0.04	<0.01				
pH 8.0	1.80	1.90	1.11	0.43	0.13	<0.01				

[a] For additional information, see the table entitled "Instability of HPLC Solvents" in this book.

ULTRAVIOLET ABSORBANCE OF NORMAL-PHASE MOBILE PHASES

The following table provides guidance in the selection of mobile phases that are to be used in conjunction with ultraviolet spectrophotometric detection [1].

REFERENCE

1. Snyder, L.R., Kirkland, J.J., and Glajch, J., *Practical HPLC Method Development,* John Wiley & Sons, New York, 1997.

Solvent	Absorbance (A) at Wavelength (nm) Indicated						
	200	210	220	230	240	250	260
Ethyl acetate	>1.0	>1.0	>1.0	>1.0	>1.0	>1.0	0.10
Ethyl ether	>1.0	>1.0	0.46	0.27	0.18	0.10	0.05
Hexane	0.54	0.20	0.07	0.03	0.02	0.01	0.00
Methylene chloride	>1.0	>1.0	>1.0	1.4	0.09	0.00	0.00
Methyl-t-butyl ether	>1.0	0.69	0.54	0.45	0.26	0.11	0.05
n-Propanol (1-Propanol)	>1.0	0.65	0.35	0.15	0.07	0.03	0.01
i-Propanol (2-propanol)	>1.0	0.44	0.20	0.11	0.05	0.03	0.02
Tetrahydrofuran	>1.0	>1.0	0.70	0.50	0.30	0.16	0.09

SOME USEFUL ION-PAIRING AGENTS

The following table provides a short list of ion pair chromatographic modifiers, for use in the separation ionic or ionizable species [1–4]. The use of these modifiers can often greatly improve the chromatographic performance of both normal and reverse phase systems. In many cases, new column technology has superseded the use of ion-pairing agents, especially when mass spectrometry is used with HPLC. Ion-pairing agents can cause numerous difficulties when the column is interfaced with a mass spectrometer. Modern stationary phases with embedded polar groups can eliminate the need for ion-pairing agents, and these stationary phases should be considered as an alternative.

REFERENCES

1. Poole, C.F. and Schuette, S.A., *Contemporary Practice of Chromatography*, Elsevier Science Publishers, Amsterdam, 1984.
2. Snyder, L.R. and Kirkland, J.J., *Introduction to Modern Liquid Chromatography*, John Wiley & Sons, New York, 1979.
3. Krstulovic, A.M. and Brown, P.R., *Reversed-Phase High Performance Liquid Chromatography*, John Wiley & Sons, 1982.
4. Basic Principles in Reversed Phase Chromatography, Amersham Biosciences Online Education Centre, 2002.

Ion Type/Examples	Applications/Notes
Perchloric acid	Used for a wide range of basic analytes; typically used at 0.1 M concentration in reverse-phase solvent system and at approximately the same concentration in a water buffer system on the stationary phase in normal mode. See the compatibilities information presented in the next table for cautions.
TFA	One of the most common ion-pairing agents used in HPLC; used for solutes that form positive ions; it is volatile and is therefore often easily removed; it has low absorption within detection wavelengths.
Heptafluorobutyric acid (HFBA)	Used with analytes that form positive ions.
Pentafluoropropionicacid (PFPA)	Used with analytes that form positive ions.
Bis-(2-ethylhexyl) phosphate	Used for cationic species of intermediate polarity, such as phenols; typically used in reverse phase, on bis-(2-ethylhexyl) phosphoric acid/chloroform stationary phase at a pH ≈ 3.8.
N,N-Dimethylprotriptyline	Used for carboxylic acids; typically used in normal phase, with a basic (pH ≈ 9) buffered stationary phase and an organic mobile phase.
Quarternary amines: tetramethyl, tetrabutyl, palmityltrimethyl-ammonium salts, usually in chloride or phosphate forms.	Used for strong and weak acids, sulfonated dyes, carboxylic acids, in normal-phase applications, typical buffer pH values are between 6 and 8.5, with an organic mobile phase; in reverse phase, the mobile phase is typically aqueous plus a polar organic modifier at nearly neutral pH values.
Tertiary amines: tri-n-octyl amine	Used for carboxylic acids and sulfonates; used in reverse-phase mode with a water+buffer+approximately 0.05 M perchloric acid mobile phase.
Sulfonates, alkyl and aromatic: methane or heptanesulfonate, camphorsulfonic acid	Used for strong and weak bases, benzalkonium salts, and catecholamines.
Alkyl sulfates: octyl, decyl, dodecyl, and lauryl sulfates	Used in similar applications as the sulfonates but provide a different selectivity; typically used in reverse-phase mode, often using a water+methanol+sulfuric acid mobile phase.

MATERIALS COMPATIBLE WITH AND RESISTANT TO 72 % PERCHLORIC ACID

The perchloric acid mentioned in the previous table on ion-pairing agents must be handled with great care since it can be a very powerful oxidizing agent. Cold perchloric acid at a concentration of 70 % (mass/mass) or less is not considered a very strong oxidizing agent. At concentrations of 73 % or higher, or at lower concentrations but at higher temperatures, perchloric acid is a powerful oxidant. The following table provides some guidance in handling this material in the laboratory [1].

REFERENCE

1. Furr, A.K., ed., *CRC Handbook of Laboratory Safety*, 5th ed., CRC Press, Boca Raton, FL, 2000.

Material	Comments
Elastomers	
Gum Rubber	Each batch must be tested to determine compatibility
Vitons	Slight swelling only
Metals and Alloys	
Tantalum	Excellent
Titanium (chemically pure grade)	Excellent
Zirconium	Excellent
Columbium (Niobium)	Excellent
Hastelloy	Slight corrosion rate
Plastics	
Polyvinyl chloride	
Teflon	
Polyethylene	
Polypropylene	
Kel-F	
Vinylidine fluoride	
Saran	
Epoxies	
Others	
Glass	
Glass-lined steel	
Alumina	
Fluorolube	

Incompatible

Plastics

Polyamide (nylon)

Modacrylic ester, Dynel (35 %–85 %) acrylonitrile

Polyester (dacron)

Bakelite

Lucite

Micarta

Cellulose-based lacquers, metals

Copper

Copper alloys (brass, bronze, etc.) for very shock-sensitive perchlorate salts

Aluminum (dissolves at room temperature)

High nickel alloys (dissolves), others

Cotton

Wood

Glycerin-lead oxide (letharge)

MORE COMMON HPLC STATIONARY PHASES

The following table provides a summary of the general characteristics of the most popular stationary phases used in modern HPLC [1–8]. The most commonly used phases are the bonded reverse-phase materials, in which separation control is a function of the mobile (liquid) phase. The selection of a particular phase and solvent system is an empirical procedure involving survey analyses. The references provided below will assist the reader in this procedure.

REFERENCES

1. Snyder, L.R. and Kirkland, J.J., *Introduction to Modern Liquid Chromatography*, 2nd ed., John Wiley & Sons, New York, 1979.
2. Poole, C.F. and Schuette, S.A., *Contemporary Practice of Chromatography*, Elsevier, Amsterdam, 1984.
3. Krstulovic, A.M. and Brown, P.R., *Reverse-Phase High Performance Liquid Chromatography*, John Wiley & Sons (Interscience), New York, 1982.
4. Berridge, J.C., *Techniques for the Automated Optimization of HPLC Separations*, John Wiley & Sons, Chichester, 1985.
5. Braithwaite, A. and Smith, F.J., *Chromatographic Methods*, 4th ed., Chapman and Hall, London, 1985.
6. Sander, L.C., Sharpless, K.E., and Pursch, M., C30 stationary phases for the analysis of food by liquid chromatography, *J. Chromatogr. A*, 880, 189–202, 2000.
7. Snyder, L.R., Kirkland, J.J., and Glajch, J., *Practical HPLC Method Development*, John Wiley & Sons, New York, 1997.
8. Taylor, T., Choosing the right HPLC stationary phase, *LCGC North America*, 33(3), 218, 2015.

Phase Type	Bond Type	Functional Group	Separation Mode	Notes and Applications
			Solid Sorbents	
Silica (pure)	SiO_2	Si–OH	Adsorption	Usually used with nonpolar mobile phase, since it is the most polar sorbent; selectivity is based on differences in number and location of polar groups; results can be unpredictable due to changes in the surface due to adsorption; water or acetic acid is often added (in low concentrations) to the mobile phase to better control surface characteristics; usually the best choice for normal phase and preparative scale separations.
Controlled pore glass	Deglassed borosilicate		Adsorption	Made by deglassing borosilicate glass and subsequent removal of B_2O_3; pressure stable; can be used in acid and alkali but not strong alkali; can be sterilized; can be derivatized; narrow pore sizes, rigidity, high pore volumes are advantages, used for macromolecular samples, especially biologicals, used mainly in preparatory or industrial scales, rather than analytical scales.
Alumina, acidic	Al_2O_3–A	—	Adsorption-normal phase	Similar in characteristics and application to silica; a classic Lewis acid, lacking two electrons in the Al center, having an approximate pH of 4.5; this phase can be treated to make it more retentive to electron-rich species.
Alumina, neutral	Al_2O_3		Adsorption-normal phase	Prepared as a neutral surface, approximate pH=7.5, used for separation of aromatics and moieties that contain electronegative groups such as oxygen; should not be heated above 150°C; more prone to chemisorption than silica; somewhat loser efficiency (plate height) than silica.
Alumina, base treated	Al_2O_3–B		Adsorption-normal phase, weak cation exchange	Base treatment makes the phase suitable for separation of hydrogen bonding or cationic species; approximate pH=10; weak cation exchanger, should not be heated above 150°C.
Zirconia (pure)	ZrO_2	—	Normal phase, ion exchange	Can be used over the entire pH range, 1–14; can be heated to 200°C; has a stable particle size that will not shrink; surface is free of silanol groups; can function in ion exchange mode since it is a Lewis acid.
Titania	TiO_2		Normal and reverse phase	Has basic –OH groups on the surface; stable at high pH; has been used for the separation of phosphopeptides; has been modified to sol–gel phases modified with poly(dimethylsiloxane).

(Continued)

Phase Type	Bond Type	Functional Group	Separation Mode	Notes and Applications
Porous graphitic carbon	Intertwined ribbons of porous graphite	Carbon network	Primarily reverse phase, can be used in normal phase	Spherical macrostructure with a crystalline graphitic surface; stable over wide pH range; thermally stable to high temperatures; lower capacity and efficiency than silica phases; unique selectivity; can be used to separate very polar compounds; can be modified.
Magnesium silicate (florisil)	MgO_3Si	—	Polar adsorbent	Florisil is a magnesium substituted silica with a highly polar surface; typically has a relatively large particle size (approximately 200 μm), and therefore, high flow rates are possible even with viscous samples; used with many official methods, and in cases where the Lewis acidity of alumina would be problematic; must be used with caution since aromatics, amines esters, and other compounds can be chemisorbed.
Hydroxyapatite	$Ca_{10}(PO_4)_6(OH)_2$	—	Polar adsorbent; specific interactions	Hexagonally crystallized calcium phosphate; pressure stable to 15 MPa; typically used with a linear gradient of a potassium or sodium phosphate buffer at a pH of approximately 6.8; useful for the separation of proteins and other biopolymers, nucleic acids, viruses; see the entry under specialized HPLC phases.

Polymeric Phases

Styrene-divinyl benzene			Reverse phase	Polymer must have at least 8 % divinyl benzene to be suitable for high pressure; bed volume will change with solvent or ionic strength of mobile phase, once a solvent is chosen, it generally cannot be changed; structure can be microporous and macroporous, allowing larger molecules to enter structure; stable at pH 1–13; chromatographic behavior similar to ODS but with specific interactions (π–π) for aromatics; can be modified for ion exchange.
Agarose				Cross-linked polysaccharide stable over pH=1–14; can be derivatized for affinity chromatography.

(Continued)

Phase Type	Bond Type	Functional Group	Separation Mode	Notes and Applications
Bonded Phases, Straight Chain				
ODS	Si–O–Si–C	Octadecyl, n–C_{18} $(CH_2)_{17}CH_3$, hydrocarbon chain	Bonded, reverse phase	Octadecylsilane; most common material used in HPLC; high resolution possible; pH must be maintained between 2 and 7.
C2	Si–O–Si–C	–CH_2CH_3	Moderately polar bonded, reverse phase	A moderately polar phase that is used for aqueous samples, blood and urine samples; moderate polarity derives from the polar substrate, silica; has a polarity similar to a cyclohexyl bonded phase.
OS	Si–O–Si–C	Octyl, n–C_8 hydrocarbon chain	Bonded, reverse phase	Octylsilane; lower resolution and retention than the octadecyl bonded phase; useful when separations involve species of greatly different polarity.
C30	Si–O–Si–C	Triacontyl, C30 hydrocarbon chain	Bonded, reverse phase	A polymeric phase useful for the separation of caratenoid compounds, fullerenes.
TMS	Si–O–Si–C	Methyl, CH_3	Bonded, reverse phase	Tetramethylsilane; lowest resolution of reverse-phase packings; useful for "survey" separations and for large molecules.
ODA	Al–O–Si–R	Octadecyl, n–C_{18} $(CH_2)_{17}CH_3$, hydrocarbon chain	Bonded, reverse phase	Far less used than the silica bonded phases, although alumina chemistry can be fore facile than silica chemistry; has been used for separation of small and larger peptide molecules.
OA	Al–O–Si–R	Octyl, n–C_8 hydrocarbon chain	Bonded, reverse phase	Far less used than the silica bonded phases, although alumina chemistry can be fore facile than silica chemistry; has been used for separation of small and larger peptide molecules.
Bonded Phases, Functionalized[a]				
Bonded diol	Si–O–Si–C	OH OH | | –C—C– |	Polar bonded phase	A polar phase that has a hydrogen bonding capability similar to that of unbonded silica; useful in size-exclusion chromatography and in the analysis of glycols and glycerol, oils, lipids, and related compounds.
Carboxyl acid, CBA	Si–O–Si–C	–CH_2CH_2COOH	Polar bonded phase	Medium-polarity phase that has a weak cation exchange capability useful for strong cations; above pH=4.8, most of the functional groups are negatively charged, and therefore, the phase can be used for cationic compounds; lowering pH to 2.8 elutes retained analytes.

(Continued)

Phase Type	Bond Type	Functional Group	Separation Mode	Notes and Applications
Cyclyhexyl, CH	Si–O–Si–C	$-C_6H_9$	Moderately polar bonded phase	A moderately polar phase that is used for aqueous samples; moderate polarity derives from the polar substrate, silica; has a polarity similar to a C2 bonded phase.
Bonded nitrile	Si–O–Si–C	$-CH_2CH_2CH_2-C\equiv N$	Moderately polar bonded phase	Moderate polar phase but with selectivity modified with respect to silica; less sensitive to mobile-phase impurities than silica; less retentive than OS; many nitrile phases are less stable than OS; also called cyanopropyl phase.
Bonded nitro	Si–O–Si–C	$-NO_2$	Polar bonded phase	
Bonded amine	Si–O–Si–C	$CH_2CH_2CH_2-NH_2$	Polar bonded phase	Selectivity is modified with respect to silica through the aminopropyl functionality; the propyl linkage can interact with nonpolar interactions; highly polar phase overall; phase is less stable than cyano or diol phases; can utilize hydrogen bonding and ion exchange mechanisms; protonates below pH=9.8; useful for sugar and carbohydrate separations; not recommended for samples which contain aldehydes and ketones.
Phenyl	Si–O–Si–C	Often represented as ϕ	Normal or reverse phase	Lower efficiency than other bonded phases; more polar than ODS, OS, and tetramethylsilane (TMS) phases; used with both normal- and reverse-phase solvent systems.
Polybutadiene	Zr–C	$[CH_2=C-C=C\overset{CH_{2\}n}{\underset{H}{}}]$	Reverse phase	Similar to ODS in separation characteristics; can be used up to 150 °C.
Carbon on zirconia	Zr–C		Reverse phase	This elemental carbon on zirconia is useful in the separation of diastereomers.
Polystyrene on zirconia	Zr–C		Reverse phase	Separations are similar to those obtained with phenyl-bonded silica; can be used up to 150 °C.

a In this context, functionalized refers to functionalization beyond straight-chain hydrocarbons.

(Continued)

Bonded Phases, Ion Exchange[a]

Phase Type	Bond Type	Functional Group	Separation Mode	Notes and Applications
Bonded amine	Si–O–Si–C	$CH_2CH_2CH_2–NH_2$	Polar bonded phase	Selectivity is modified with respect to silica through the aminopropyl functionality; the propyl linkage can interact via nonpolar interactions; highly polar phase overall; phase is less stable than cyano or diol phases; can utilize hydrogen bonding and ion exchange mechanisms, and as such is a weak anion exchanger; protonates below pH = 9.8; useful for sugar and carbohydrate separations; not recommended for samples which contain aldehydes and ketones.
Benzene sulfonic acid	Si–O–Si–C	$CH_2CH_2N–SO_3^-H^+$	Ion exchange	Separates cations, with divalent ions more strongly retained than monovalent ions; phosphate buffer systems are often used, sometimes with low concentrations of polar nonaqueous modifiers added; the presence of the benzene group on the benzenesulfonic acid moiety gives this phase a dual nature and the ability to separate based upon nonpolar interactions.
Propyl, ethylene diamine	Si–O–Si–C	$–CH_2CH_2CH_2–$ $NHCH_2CH_2NH_2$	Ion exchange	Weak anion exchange phase for aqueous and biological samples; incorporates a bidentate ligand to form chelate complexes useful for metal separations; less polar than the propylamine bonded phase.
Propyl sulfonic acid	Si–O–Si–C	$–CH_2CH_2CH_2–$ $SO_3^-Na^+$	Ion exchange	Strong cation exchange substrate for aqueous and biological samples; effective for the separation of weaker cations such as pyridinium compounds.
Propyl, trimethylamino	Si–O–Si–C	$–CH_2CH_2CH_2–$ $N^+Cl^-(CH_3)_3$	Ion exchange	Strong anion exchange phase for aqueous and biological samples suitable for weaker anions such as carboxylic acids; properties may be modified or conditioned by proper formulation of buffer mobile phases (see the appropriate table in Chapter 14).

[a] Note that while the principal separation mechanism is ion exchange, the organic moieties on many of these phases can interact through nonpolar interactions as well. Thus, many phases are mixed mode. Buffer solutions are used to achieve column starting conditions, separations, and regeneration. Some ion exchange buffer solutions are provided in the appropriate table in Chapter 14.

ELUOTROPIC VALUES OF SOLVENTS ON OCTADECYLSILANE

The following table provides, for comparative purposes, eluotropic values on octadecyl silane (ODS) and octyl silane (OS) for common solvents [1,2].

REFERENCES

1. Krieger, P.A., *High Purity Solvent Guide*, Burdick and Jackson Laboratories, McGaw Park, WI, 1984.
2. Ahuja, S., *Trace and Ultratrace Analysis by HPLC*, John Wiley & Sons, New York, 1992.

Solvent	Eluotropic Value, ODS	Eluotropic Value, OS
Acetic acid	–	2.7
Acetone	8.8	9.3
Acetonitrile	3.1	3.3
1,4-Dioxane	11.7	13.5
Dimethyl-formamide	7.6	9.4
Methanol	1.0	1.0
Ethanol	3.1	3.2
n-Propanol	10.1	10.8
2-Propanol	8.3	8.4
Tetrahydrofuran	3.7	–

MESH-SIZE RELATIONSHIPS

The following table provides the relationship between particle sizes and standard sieve mesh sizes. It should be noted, however, that the trend in HPLC has been toward shorter columns containing much finer particles than the standard sieves will separate. These values will be of use when packing relatively large diameter columns for bench-top elutions.

Mesh Range	Top Screen Opening (μm)	Bottom Screen Opening (μm)	Micron Screen (μm)	Range Ratio
80/100	177	149	28	1.19
100/120	149	125	24	1.19
100/140	149	105	44	1.42
120/140	125	105	20	1.19
140/170	105	88	17	1.19
170/200	88	74	14	1.19
200/230	74	63	11	1.19
230/270	63	53	10	1.19
270/325	53	44	9	1.20
325/400	44	37	7	1.19

EFFICIENCY OF HPLC COLUMNS

The efficiency of a column used for HPLC describes the ability of the column to produce sharp narrow peaks. Typically, the efficiency is represented at the plate number, N. The plate number can be estimated by

$$N = 3500L/d_p,$$

where L is the column length in cm, and d_p is the particle diameter in (μm). The following table provides the plate number for optimized test conditions for various combinations of column length and particle diameter. It therefore represents the upper limit of efficiency and can be used as a column diagnostic measurement [1].

REFERENCE

1. Snyder, L.R., Kirkland, J.J., and Glajch, J.L., *Practical HPLC Method Development*, 2nd. Ed. John Wiley & Sons, New York, 1997, reproduced with permission.

Particle Diameter (μm)	Column Length (cm)	Plate Number
10	15	6,000–7,000
10	25	8,000–10,000
5	10	7,000–9,000
5	15	10,000–12,000
5	25	17,000–20,000
3	5	6,000–7,000
3	7.5	9,000–11,000
3	10	12,000–14,000
3	15	17,000–20,000

COLUMN FAILURE PARAMETERS

The point at which a column used for HPLC will fail depends largely upon how the operator uses it. Eventually, however, all HPLC columns will fail. The onset of column failure can be monitored by two common failure parameters, the peak asymmetry factor, A_s, and the peak tailing factor. These parameters are defined according to the figure below [1]:

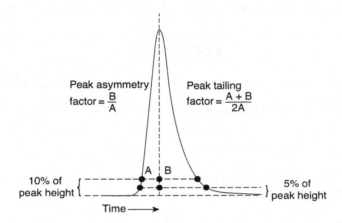

The two parameters are related, and the following table provides the interconversion [1].

REFERENCE

1. Snyder, L.R., Kirkland, J.J., and Glajch, J.L., *Practical HPLC Method Development*, 2nd ed. John Wiley & Sons, New York, 1997, reproduced with permission.

Peak Asymmetry Factor (A_s, 10 %)	Peak Tailing Factor (5 %)
1.0	1.0
1.3	1.2
1.6	1.4
1.9	1.6
2.2	1.8
2.5	2.0

To put these factors into context, column performance can be described as in the following table:

Peak Asymmetry Factor (A_s, 10 %)	Column Performance
1.0–1.05	Excellent, new
1.2	Acceptable
2.0	Degraded, approaching poor
4.0	Not usable

COLUMN REGENERATION SOLVENT SCHEMES

When HPLC columns become fouled or inefficient, it is sometimes possible to regenerate performance to some degree. The following solvent schemes have been found helpful [1,2]. Some workers recommend daily flushing of columns to maintain performance and avoid problems. Another cause of efficiency loss can be settling of the packing, especially if the column has been in service for several months. In some cases, it is possible to top off the columns with packing material. If packing material is not available, glass beads of the appropriate size can be used.

REFERENCES

1. Meyer, V.R., *Practical High-Performance Liquid Chromatography*, 4th ed., John Wiley & Sons, Chichester, 2004.
2. Majors, R.E., The cleaning and regeneration of reverse phase HPLC columns, LC-GC Europe, July 2003.

Silica Adsorbent Columns

Pump the following solvents sequentially at the rate of 1–3 mL/min.
75 mL of tetrahydrofuran
75 mL of methanol

If acidic impurities are suspected:
75 mL of 1 %–5 % (vol/vol) of pyridine in water

If basic impurities are suspected:
75 mL of 1 %–5 % (vol/vol) of acetic acid in water
75 mL of tetrahydrofuran
75 mL of t-butylmethyl ether
75 mL of n-hexane or hexanes.

Bonded Phase Columns (OS, ODS, Phenyl, and Nitrile Phases)

Pump the following solvents sequentially at the rate of 0.5–2 mL/min.
75 mL of water, while injecting 100 µL of dimethyl sulfoxide four times
75 mL of methanol
75 mL of chloroform
75 mL of methanol.

Another sequence that is possible, usually offline from the HPLC instrument, is the following:

Water
M sulfuric acid
Water.

It is recommended that this be done with a spare pump used only for this solvent sequence.

When proteins have been separated on a reverse phase column, the following solvent systems can be used for regeneration and cleaning:

1 % (vol/vol) acetic acid in water
1 % TFA in water
0.1 % TFA + propanol, 40:60 (vol/vol).

Note: this mixture is relatively viscous, so a low flow rate should be used.

Triethylamine + propanol 40:60 (vol/vol)
(Adjust triethylamine to pH = 2.5 with 0.25 N phosphoric acid before mixing)
Aqueous urea or guanidine, 5–8 M (adjusted to pH 6–8)
Aqueous sodium chloride, sodium phosphate or sodium sulfate at 0.5–1.0 M
Dimethyl sulfoxide + water, 50:50 (vol/vol)
Dimethylformamide + water, 50:50 (vol/vol).

When metal ions have been introduced into a reverse-phase column, the organic solvents listed above are often ineffective. In those cases, the following mixture may be useful:

0.05 M EDTA
Water flush.

Anion Exchange Columns

Pump the following solvents sequentially at the rate of 0.5–2 mL/min.
    75 mL of water
    75 mL of methanol
    75 mL of chloroform, then, apply a methanol to water gradient.

Cation Exchange Columns

Pump the following solvents sequentially at the rate of 0.5–2 mL/min.
    75 mL of water with 100 μL dimethyl sulfoxide injected four times
    75 mL of tetrahydrofuran
    Water flush.

Styrene-Divinylbenzene Polymer Columns

Pump the following solvents sequentially at the rate of 0.5–2 mL/min.
    40 mL toluene, or 40 mL tetrahydrofuran (peroxide free).

SPECIALIZED STATIONARY PHASES FOR LIQUID CHROMATOGRAPHY

The following table provides information on the properties and application of some of the more specialized bonded, adsorbed, and polymeric phases used in modern HPLC. In many cases, the phases are not commercially available, and the reader is referred to the appropriate literature citation for details on the synthesis.

REFERENCES

1. Pietrzyk, D.J. and Cahill, W.J., Amberlite XAD-4 as a stationary phase for preparative liquid chromatography in a radially compressed column, *J. Liq. Chromatogr.*, 5(4), 781–795, 1982.
2. Nikolov, Z., Meagher, M., and Reilly, P., High-performance liquid chromatography of trisaccharides on amine-bonded silica columns, *J. Chromatogr.*, 321, 393–399, 1985.
3. Ascalone, V. and Dal Bo, L., Determination of Ceftriaxone, a novel cephalosporin, in plasma, urine and saliva by high-performance liquid chromatography on an NH_2 bonded-phase column, *J. Chromatogr.*, 273, 357, 1983.
4. Pharr, D.Y., Uden, P.C., and Siggia, S., A 3-(p-acetylphenoxy) propylsilane bonded phase for liquid chromatography of basic amines and other nitrogen compounds, *J. Chromatogr. Sci.*, 23, 391, 1985.
5. Felix G. and Bertrand, C., Separation of polyaromatic hydrocarbons on caffeine-bonded silica gel, *J. Chromatogr.*, 319, 432, 1985.
6. Felix, G., Bertrand, C., and Van Gastel, F., A new caffeine bonded phase for separation of polyaromatic hydrocarbons and petroleum asphaltenes by high-performance liquid chromatography, *Chromatographia*, 20(3), 155, 1985.
7. Bruner, F., Bertoni, G., and Ciccioli, P., Comparison of physical and gas chromatographic properties of Sterling FT and Carbopack-C graphitized carbon blacks, *J. Chromatogr.*, 120, 307–319, 1976.
8. Bruner, F., Ciccioli, P., Crescentini, G., and Pistolesi, M.T., Role of the liquid phase in gas-liquid-solid chromatography and its influence on column performance - an experimental approach, *Anal. Chem.*, 45(11), 1851, 1973.
9. Ciccioli, P. and Liberti, A., Microbore columns packed with graphitized carbon black for high-performance liquid chromatography, *J. Chromatogr.*, 290, 173, 1984.
10. DiCorcia, A., Liberti, A., and Samperi, R., Gas-liquid-solid chromatography-theoretical aspects and analysis of polar compounds, *Anal. Chem.*, 45(7), 1228, 1973.
11. Hatada, K., Kitayama, T., Shimizu, S-I., Yuki, H., Harris, W., and Vogl, O., High-performance liquid chromatography of aromatic compounds on polychloral, *J. Chromatogr.*, 248, 63–68, 1982.
12. Abe, A., Tasaki, K., Inomata, K., and Vogl, O., Conformational rigidity of polychloral: Effect of bulky substituents on the polymerization mechanism, *Macromolecules*, 19, 2707, 1986).
13. Kubisa, P., Corley, L.S., Kondo, T., Jacovic, M., and Vogl, O., Haloaldehyde polymers. XXIII: Thermal and mechanical properties of chloral polymers, *Pol. Eng. Sci.*, 21(13), 829, 1981.
14. Veuthey, J.-L., Bagnoud, M.-A., and Haerdi, W., Enrichment of amino and carboxylic acids using copper-loaded silica pre-columns coupled on-line with HPLC, *Intern. J. Envirn. Anal. Chem.*, 26, 157–166, 1986.
15. Guyon, F., Foucault, A., Caude, M., and Rosset, R., Separation of sugars by HPLC on copper silicate gel, *Carbohyd. Res.*, 140, 135–138, 1985.
16. Leonard, J.L., Guyon, F., and Fabiani, P., High-performance liquid chromatography of sugars on copper (11) modified silica gel, *Chromatographia*, 18, 600, 1984.
17. Miller, N.T. and Shieh, C.H., Preparative hydrophobic interaction chromatography of proteins using ether based chemically bonded phases, *J. Liq. Chromatogr.*, 9(15), 3269, 1986.
18. Williams, R.C., Vasta-Russell, J.F., Glajch, J.L., and Golebiowski, K., Separation of proteins on a polymeric fluorocarbon high-performance liquid chromatography column packing, *J. Chromatogr.*, 371, 63–70, 1986.

19. Hirayama, C., Ihara, H., Yoshinga, T., Hirayama, H., and Motozato, Y., Novel packing for high pressure liquid chromatography. Partially alkylated and cross-linked PMLG spherical particles, *J. Liq. Chromatogr.*, 9(5), 945–954, 1986.

20. Kawasaki, T., Kobayashi, W., Ikeda, K., Takahashi, S., and Monma, H., High-performance liquid chromatography using spherical aggregates of hydroxyapatite micro-crystals as adsorbent, *Eur. J. Biochem.*, 157, 291–295, 1986.

21. Kawasaki, T. and Kobayashi, W., High-performance liquid chromatography using novel square tile-shaped hydroxyapatite crystals as adsorbent, *Biochem. Int.*, 14(1), 55–62, 1987.

22. Funae, Y., Wada, S., Imaoka, S., Hirotsune, S., Tominaga, M., Tanaka, S., Kishimoto, T., and Maekawa, M., Chromatographic separation of α,-acid glyco-protein from α-antitrypsin by high-performance liquid chromatography using a hydroxyapatite column, *J. Chromatogr.*, 381, 149–152, 1986.

23. Kadoya, T., Isobe, T., Ebihara, M., Ogawa, T., Sumita, M., Kuwahara, H., Kobayashi, A., Ishikawa, T., and Okuyama, T., A new spherical hydroxyapatite for high performance liquid chromatography of proteins, *J. Liq. Chromatogr.*, 9(16), 3543–3557, 1986.

24. Kawasaki, T., Niikura, M., Takahashi, S., and Kobayashi, W., High-performance liquid chromatography using improved spherical hydroxyapatite particles as adsorbent: Efficiency and durability of the column, *Biochem. Int.*, 13(6), 969–982, 1986.

25. Bernardi, G., Chromatography of nucleic acids on hydroxyapatite columns, *Methods Enzymol.*, 21D, 95–139, 1971.

26. Bernardi, G., Chromatography of proteins on hydroxyapatite, *Methods Enzymol.*, 27, 471–479, 1973.

27. Bernardi, G., Chromatography of proteins on hydroxyapatite, *Methods Enzymol.*, 22, 325–339, 1971.

28. Figueroa, A., Corradini, C., Feibush, B., and Karger, B., High-performance immobilized-metal affinity chromatography of proteins on iminodiacetic acid silica-based bonded phases, *J. Chromatogr.*, 371, 335–352, 1986.

29. Danielson, N.D., Ahmed, S., Huth, J.A., and Targrove, M.A., Characterization of organomagnesium modified Kel-f polymers as column packings, *J. Liq. Chromatogr.*, 9(4), 727–743, 1986.

30. Taylor, P.J. and Sherman, P.L., Liquid crystals as stationary phases for high performance liquid chromatography, *J. Liq. Chromatogr.*, 3(1), 21–40, 1980.

31. Taylor, P.J. and Sherman, P.L., Liquid crystals as stationary phases for high performance liquid chromatography, *J. Liq. Chromatogr.*, 2(9), 1271–1290, 1979.

32. Felix, G. and Bertrand, C., HPLC on pentafluorobenzamidopropyl silica gel, HRC&CC, *J. High Resolut. Chromatogr., Chromatogr. Commun.*, 8, 362, 1985.

33. Kurosu, Y., Kawasaki, H., Chen, X-C., Amano, Y., Fang, Y-I., Isobe, T., and Okuyama, T., Comparison of retention times of polypeptides in reversed phase high performance liquid chromatography on polystyrene resin and on alkyl bonded silica, *Bunseki Kagaku*, 33, E301–E308, 1984.

34. Yang, Y-B. and Verzele, M., New water-compatible modified polystyrene as a stationary phase for high-performance liquid chromatography, *J. Chromatogr.*, 387, 197, 1987.

35. Nieminen, N. and Heikkila, P., Simultaneous determination of phenol, cresols and xylenols in workplace air, using a polystyrene-divinylbenzene column and electrochemical detection, *J. Chromatogr.*, 360, 271, 1986.

36. Yang, Y.B., Nevejans, F., and Verzele, M., Reversed-phase and cation-exchange chromatography on a new poly(styrenedivinylbenzene) high capacity, weak cation-exchanger, *Chromatographia*, 20(12), 735, 1985.

37. Tweeten, K.A. and Tweeten, T.N., Reversed-phase chromatography of proteins on resin-based wide-pore packings, *J. Chromatogr.*, 359, 111, 1986.

38. Lee, D.P. and Lord, A.D., A high performance phase for the organic acids, *LC-GC*, 5(3), 261, 1987.

39. Miyake, K., Kitaura, F., Mizuno, N., and Terada, H., Determination of partition coefficient and acid dissociation constant by high-performance liquid chromatography on porous polymer gel as a stationary phase, *Chem. Pharm. Bull.*, 35(1), 377, 1987.

40. Joseph, J.M., Selectivity of poly(styrene-divinylbenzene) columns, *ACS Symp. Ser.*, 297, 83–100, 1986.

41. Cope, M.J. and Davidson, I.E., Use of macroporous polymeric high-performance liquid chromatographic columns in pharmaceutical analysis, *Analyst*, 112, 417, 1987.

42. Werkhoven-Goewie, C.E., Boon, W.M., Praat, A.J.J., Frei, R.W., Brinkman U.A. Th., and Little, C.J., Preconcentration and LC analysis of chlorophenols, using a styrene-divinyl-benzene copolymeric sorbent and photochemical reaction detection, *Chromatographia*, 16, 53, 1982.

43. Smith, R.M., Selectivity comparisons of polystyrenedivinylbenzene columns, *J. Chromatogr.*, 291, 372–376, 1984.

44. Köhler, J., Poly (vinylpyrrolidone)-coated silicia: A versatile, polar stationary phase for H.P.L.C., *Chromatographia*, 21, 573, 1986.

45. Murphy, L.J., Siggia, S., and Uden, P.C., High-performance liquid chromatography of nitroaromatic compounds on an N-propylaniline bonded stationary phase, *J. Chromatogr.*, 366, 161, 1986.

46. Felix, G. and Bertrand, C., HPLC on n-propyl picryl ether silica gel, HRC&CC, *J. High Resolut. Chromatogr., Chromatogr. Commun.*, 7, 714, 1984.

47. Risner, C.H. and Jezorek, J.R., The chromatographic interaction and separation of metal ions with 8-quinolinol stationary phases in several aqueous eluents, *Anal. Chem. Acta*, 186, 233, 1986.

48. Shahwan, G.J. and Jezorek, J.R., Liquid chromatography of phenols on an 8-quinolinol silica gel-iron (III) stationary phase, *J. Chromatogr.*, 256, 39–48, 1983.

49. Krauss, G.-J., Ligand-exchange H.P.L.C. of uracil derivatives on 8-hydroxyquinoline-silica-polyol, HRC&CC, *J. High Resolut. Chromatogr., Chromatogr. Commun.*, 9, 419–420, 1986.

50. Hansen, S.A., Helboe, P., and Thomsen, M., High-performance liquid chromatography on dynamically modified silica, *J. Chromatogr.*, 360, 53–62, 1986.

51. Helboe, P., Separation of corticosteroids by high-performance liquid chromatography on dynamically modified silica, *J. Chromatogr.*, 366, 191–196, 1986.

52. Hansen, D.H., Helboe, P., and Thomsen, M., Dynamically modified silica-the use of bare silica in reverse-phase high-performance liquid chromatography, *Trends in Anal. Chem. TrAC*, 4(9), 233, 1985.

53. Flanagan, R.J., High-performance liquid chromatographic analysis of basic drugs on silica columns using non-aqueous ionic eluents, *J. Chromatogr.*, 323, 173–189, 1985.

54. Vespalec, R., Ciganková, M., and Viska, J., Effect of hydrothermal treatment in the presence of salts on the chromatographic properties of silica gel, *J. Chromatogr.*, 354, 129, 1986.

55. Unger, K.K., Jilge, G., Kinkel, J.N., and Hearn, M.T.W., Evaluation of advanced silica packings for the separation on biopolymers by high-performance liquid chromatography, *J. Chromatogr.*, 359, 61–72, 1986.

56. Lullmann, C., Genieser, H-G., and Jastorff, B., Structural investigations on reversed-phase silicas, *J. Chromatogr.*, 354, 434–437, 1986.

57. Schou, O. and Larsen, P., Preparation of 6,9,12-trioxatridecylmethylsilyl substituted silica, a new stationary phase for liquid chromatography, *Acta Chem. Scand.*, B35, 337, 1981.

58. Desideri, P.G., Lepri, L., Merlini, L., and Checchini, L., High-performance liquid chromatography of amino acids and peptides on silica coated with ammonium tungstophosphate, *J. Chromatogr.*, 370, 75, 1986.

Name: amberlite XAD-4

Structure: macroporous polystyrene-divinylbenzene nonpolar adsorbent, 62–177 μm particle size

Analytical Properties: used mainly in preparative-scale HPLC; stable over entire pH range; (1–13) sometimes difficult to achieve column to column reproducibility due to packing the irregular particles. Relatively lower efficiency than alkyl bonded phases; particles tend to swell as the organic content of the mobile phase increases

Reference: [1]

Name: amine bonded phase

Structure: NH_2 functionality with a Si–O–Si–C or Si–C linkage

Analytical Properties: polar phase useful for sugar and carbohydrate separation; not recommended for samples which contain aldehydes and ketones

Reference: [2,3]

Name: 3-(p-acetophenoxy) propyl bonded phase

Structure:

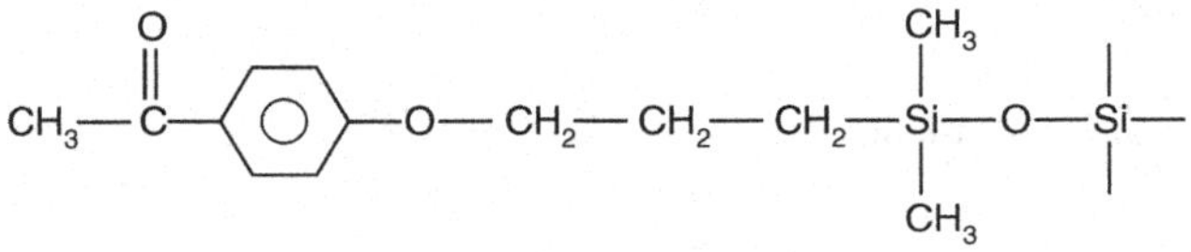

Analytical Properties: selective for aromatic amines, with the selectivity being determined by the interactions with the carbonyl group

Reference: [4]

Name: caffeine bonded phase

Structure:

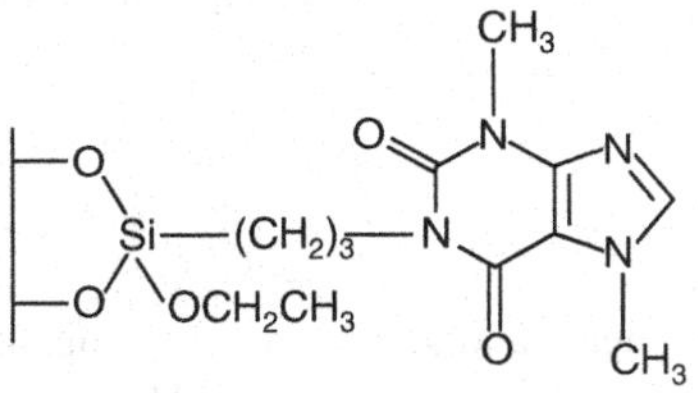

Analytical Properties: separation of polynuclear aromatic hydrocarbons (of the type often encountered in petroleum residue work) by donor–acceptor complex formation.

References: [5,6]

Name: graphitized carbon black

Structure: carbon subjected to +1,300 °C in helium atmosphere, resulting in a graphite-like structure in the form of polyhedra, with virtually no unsaturated bonds, ions, lone electron pairs, or free radicals

Analytical Properties: especially for use in microbore columns; suggested for lower aromatics but with some potential for higher molecular mass compound separations

References: [7–10]

Name: polychloral (polytrichloroacetaldehyde)

Structure:

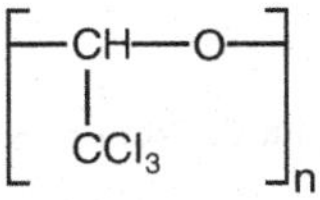

Analytical Properties: separation of lower aromatic hydrocarbons and small fused ring systems using toluene and hexane methanol as the stationary phases; the relatively low-pressure rating on the polymeric phase limits solvent flow rate
References: [11–13]

Name: cyclam-copper-silica
Structure:

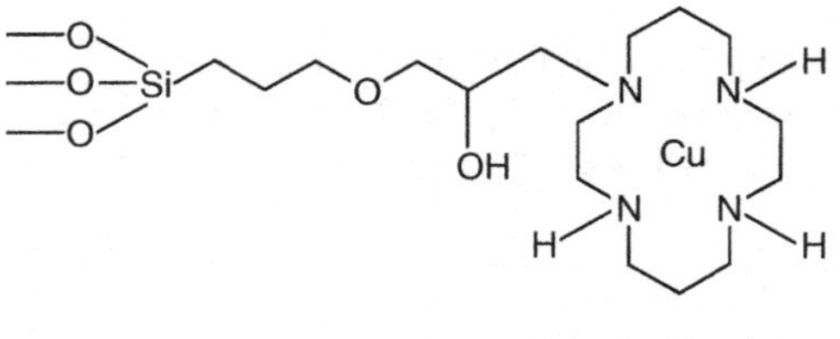

Analytical Properties: this phase has found use in preconcentrating carboxylic acids on pre-columns
Reference: [14]

Name: bis-dithiocarbamate-copper-silica
Structure:

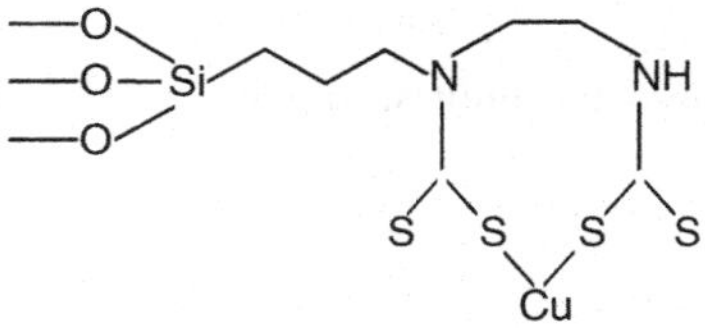

Analytical Properties: this phase has found use in preconcentrating amino acids on pre-columns
Reference: [14]

Name: copper (II)-coated silica gel
Structure: $(\equiv Si-O)_2 Cu(NH_3)_x (H_2O)_y$
  x = 1 or 2
Analytical Properties: separation of sugars and amino sugars by ligand exchange or partitioning interactions using water + acetonitrile + ammonia liquid phases. The phase is usually prepared by treating silica gel with ammoniacal copper sulfate solution prior to packing
Reference: [15,16]

Name: ether bonded phase
Structure:

$$CH_3(O{-}CH_2{-}CH_2)_2O(CH_2)_2{-}Si{-}$$

on 15–20 µm wide-pore silica

Analytical Properties: separation by hydrophobic interaction chromatography, using aqueous salt solutions near pH = 7; used primarily in protein work
Reference: [17]

Name: fluorocarbon polymer phase
Structure: proprietary information of E.I. duPont de Nemours Corp

Analytical Properties: similar separations as obtained using C_3 bonded silica, with a much larger pH stability range than silica-based phases; useful for protein and peptide separations using TFA as a mobile-phase modifier; less mechanical stability than silica-based phases
Reference: [18]

Name: poly (-methyl-L-glutamate) (PMLG)
Structure: partially cross-linked, with long-chain alkyl branches:

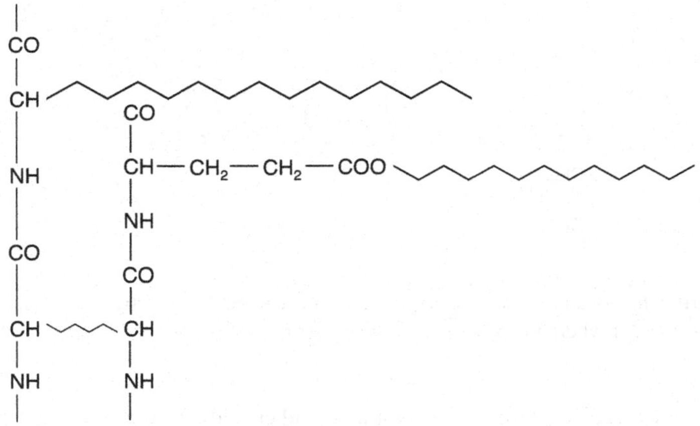

Analytical Properties: separation similar to ODS, but with somewhat higher stability in alkaline solutions; particles are spherical and macroporous
Reference: [19,20]

Name: hydroxyapatite adsorbent
Structure: $Ca_{10}(PO_4)_6(OH)_2$ crystalline, non-stoichiometric mineral rich in surface ions (primarily carbonate)
Analytical Properties: separation of proteins; overcomes some difficulties associated with ion exchange; selectivity and efficiency depend to some extent on particle geometry (that is, sphere, plate, etc.)
Reference: [21–28]

Name: iminodiacetic acid bonded phase
Structure:

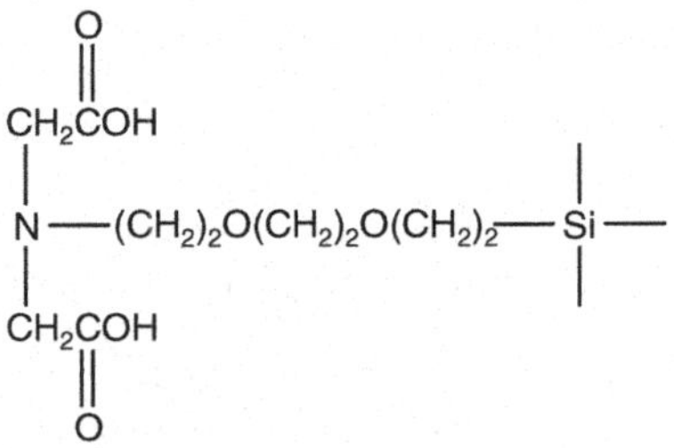

Analytical Properties: separation of proteins by immobilized-metal affinity chromatography (HPIMAC) with Cu(II) or Zn(II) present in the mobile phase
Reference: [28]

Name: Kel-F (polychlorotrifluoroethylene)
Structure: (exact structure is proprietary, 3 M Company)

Analytical Properties: highly inert, even more nonpolar than hydrocarbon phases, with sufficient
mechanical integrity to withstand high pressures; can be functionalized with $-CH_3$, $CH_3(CH_2)^-_3$ and
phenyl (using Grignard reactions) to increase selectivity
Reference: [29]

Name: 2-ethylhexyl carbonate coated or bonded on cholesteryl silica (room temperature liquid crystal)
Structure:

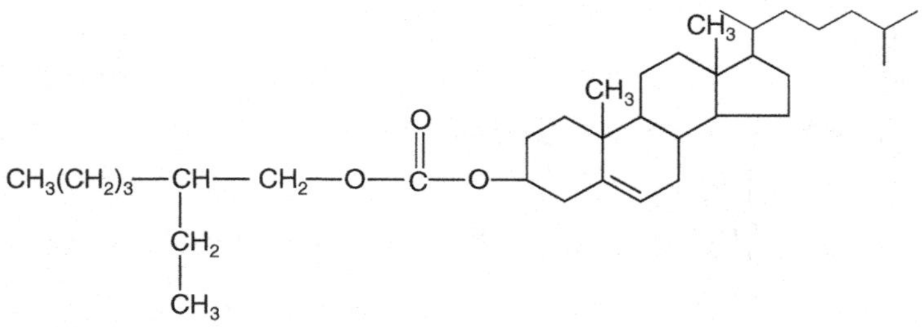

Analytical Properties: has been used for the separation of estrogens and corticoid steroids; liquid crystal
phase retains some order when coated on an active substrate
Reference: [30,31]

Name: cholesteryl-2-ethylhexanoate (room temperature liquid crystal) on silica
Structure:

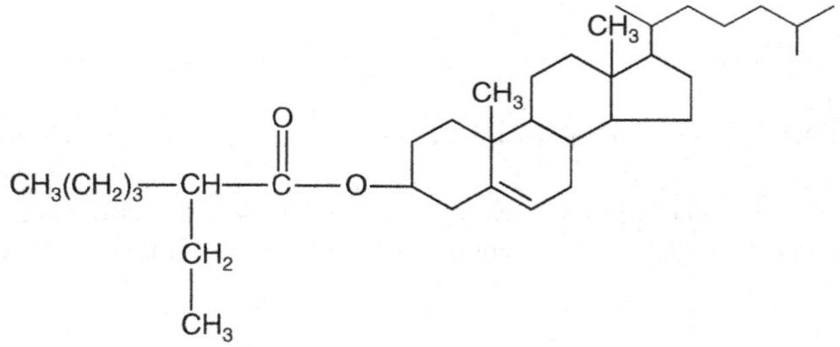

Analytical Properties: has been used for the separation of androstenedione and testosterone
Reference: [30,31]

Name: pentafluorobenzamidopropyl silica gel
Structure:

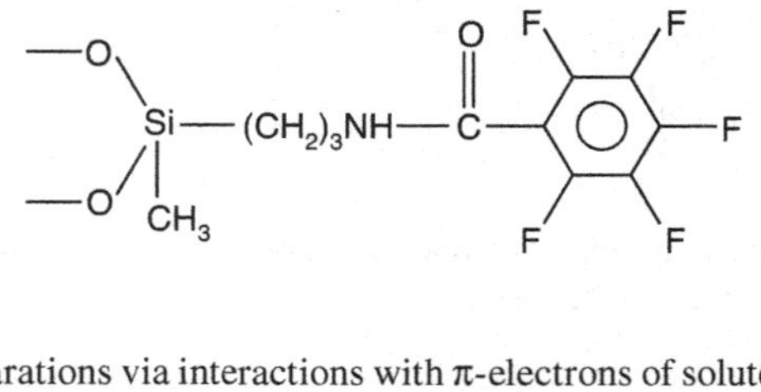

Analytical Properties: separations via interactions with π-electrons of solutes; can be used in both normal
and reverse phase for such π-donor systems as polynuclear aromatic hydrocarbons
Reference: [32]

Name: hydroxymethyl polystyrene
Structure:

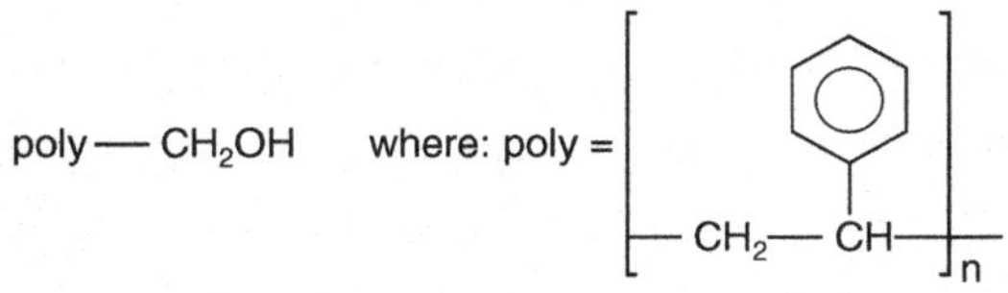

poly — CH₂OH where: poly =

Analytical Properties: separation of polypeptides; usually gives shorter retention times than ODS; hydrophobic interactions not as strong as with ODS
Reference: [33]

Name: polystyrene
Structure:

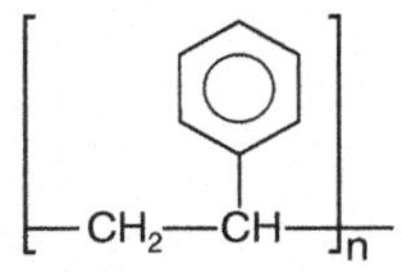

Analytical Properties: separation of polypeptides with results similar to those obtainable with ODS; higher stability at high pH levels (to allow the phase to be washed); stronger hydrophobic interactions than ODS in reverse phase mode
Reference: [34]

Name: polystyrene-divinylbenzene (PS-DVB)
Structure:

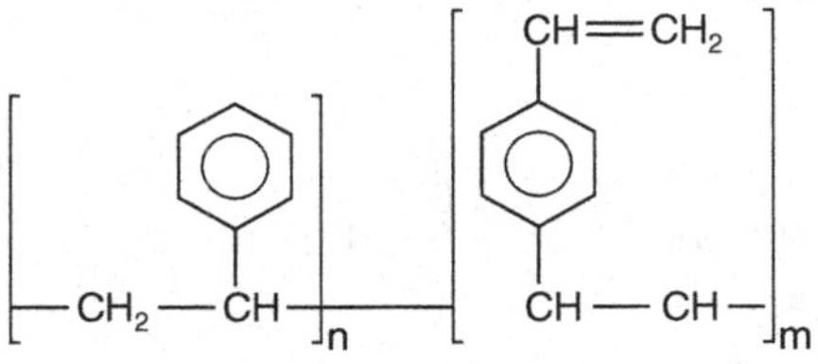

(exact structure is proprietary)
Analytical Properties: useful for the separation of relatively polar compounds such as phenols, carboxylic acids, organic anions, nucleosides, alkylarylketones, chlorophenols, barbiturates, thimine derivatives; good stability under high and low pH; reasonable mechanical integrity at high carrier pressure; compatible with buffered liquid phases
Reference: [35–43]

Name: poly(vinylpyrrolidone) or PVP on silica
Structure:

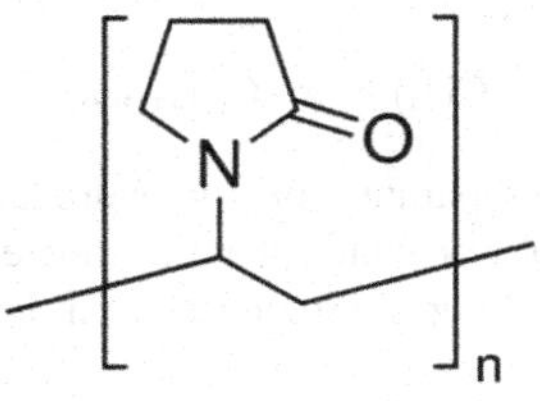

Analytical Properties: separates aromatic and polynuclear aromatic hydrocarbons; can be used in normal phase mode (commonly using n-heptane or n-heptane+dichloromethane liquid phases) or reverse phase mode (commonly using methanol+water, acetonitrile+water, or phosphate-buffered liquid phases)
Reference: [44]

Name: bonded n-propylaniline
Structure:

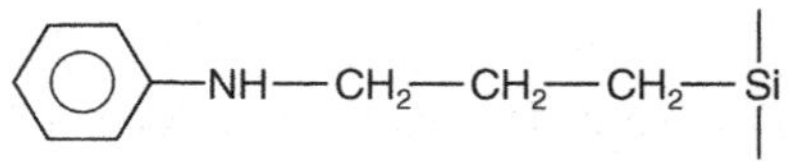

Analytical Properties: selectivity is based on charge transfer interactions; nitroaromatic compounds are separated essentially according to the number of nitro groups, the higher number compounds being most strongly retained when using methanol/water mobile phases
Reference: [45]

Name: n-propylpicrylether bonded phase
Structure:

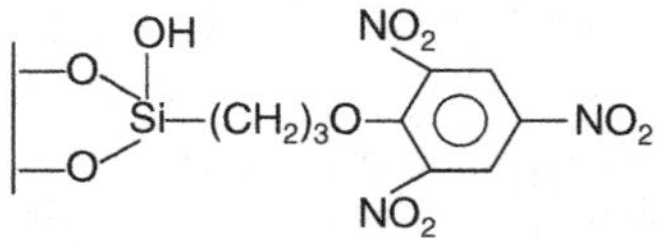

Analytical Properties: separation of aromatic species, including polynuclear aromatic species, by charge transfer interactions
Reference: [46]

Name: 8-quinolinol bonded phases
Structure:

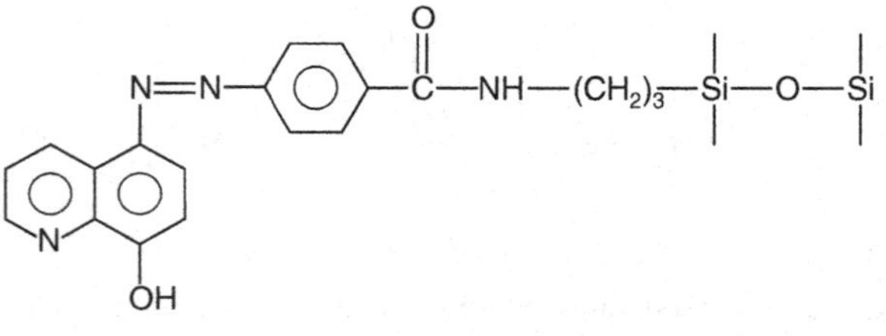

Analytical Properties: separates phenols and EPA priority pollutants; often used with metal ions (such as iron (III)) as chelate ligands; 8-quinolinol has a high affinity for oxygen moieties and will form complexes with upward of 60 metal ions; often with an acidic aqueous mobile phase
Reference: [47–49]

Name: cetyltrimethyl ammonium bromide adsorbed on silica
Structure:

$$(CH_3)_3N^+(n\text{-}C_{16}H_{33})Br^-$$

Analytical Properties: has been used to separate aromatic hydrocarbons, heterocyclic compounds, phenols, and aryl amines using methanol/water/phosphate buffer; extent of adsorption effects retention times; also used as a mobile-phase modifier to provide a dynamically modified silica
Reference: [50–56]

Name: 6,9,12-trioxatridecylmethyl bonded phase
Structure:

$$CH_3(OCH_2CH_2)_3(CH_2)_3Si(CH_3)Cl_2$$

Analytical Properties: phase is very well wetted by water, allowing mobile phases with high water concentration to be used. Somewhat higher efficiency and selectivity than ODS but with similar separation properties
Reference: [57]

Name: ammonium tungstophosphate on silica
Structure: tungstophosphoric acid with ammonium nitrate
Analytical Properties: separation of compounds containing the NH_4^+ group, such as amino acids and peptides; the coated silica also behaves as a reversed phase for the separation of aliphatic and aromatic acids; high selectivity for glycine and tyrosine oligomers
Reference: [58]

CHIRAL STATIONARY PHASES FOR LIQUID CHROMATOGRAPHY

The following table provides information on the properties and application of some of the more specialized stationary phases used to carry out the separation enantiomeric mixtures. In many cases, the phases are not commercially available, and the reader is referred to the appropriate literature citation for details on the synthesis.

REFERENCES

General References

1. Armstrong, D.W., Chiral stationary phases for high performance liquid chromatographic separation of enantiomers: A mini-review, *J. Liq. Chromatogr.*, 7(S-2), 353, 1984.
2. Dappen, R., Arm, H., and Meyer, V.R., Applications and limitations of commercially available chiral stationary phases for high-performance liquid chromatography, *Chromatogr. Rev.* (CHREV 200), 1, 1986.
3. Armstrong, D.W., Chen, S., Chang, C., and Chang, S., A new approach for the direct resolution of racemic beta adrenergic blocking agents by HPLC, *J. Liq. Chromatogr.*, 15(3), 545–556, 1992.
4. Armstrong, D.W., The evolution of chrial stationary phases for liquid chromatography, *J. Chin. Chem. Soc.*, 45, 581–590, 1998.
5. *Cyclobond Handbook: A Guide to Understanding Cyclodextrin Bonded Phases for Chiral LC Separations*, 6th ed., Advanced Separation Technologies, Wippany, NJ, 2002.
6. *Chirobiotic Handbook: A Guide to Using Macrocyclic Glycopeptide Bonded Phases for Chiral LC Separations*, 4th ed., Advanced Separation Technologies, Wippany, NJ, 2002.
7. Busch, K.W. and Busch, M.A., eds., *Chiral Analysis*, Elsevier, Amsterdam, 2006.
8. Teixeira, J., Tiritan, M.E., Pinto, M.M., and Fernandez, C., Chiral stationary phases for liquid chromatography: Recent developments, *Molecules*, 24(5), 865, 2019.

Cited References

1. Erlandsson, P., Hansson, L., and Isaksson, R., Direct analytical and preparative resolution of enantiomers using albumin adsorbed to silica as a stationary phase, *J. Chromatogr.*, 370, 475, 1986.
2. Kuesters, E. and Giron, D., Enantiomeric separation of the beta-blocking drugs pindolol and bipindolol using a chiral immobilized protein stationary phase, HRC&CC, *J. High Resolut. Chromatogr., Chromatogr. Commun.*, 9(9), 531, 1986.
3. Hermannson, J. and Eriksson, M., Direct liquid chromatographic resolution of acidic drugs using a chiral 1-acid glycoprotein column (Enantiopac), *J. Liq. Chromatogr.*, 9(2&3), 621, 1986.
4. Naobumi, O. and Hajimu, K., HPLC separation of amino acid enantiomers on urea derivatives of L-valine bonded to silica gel, *J. Chromatogr.*, 285, 198, 1984.
5. Okamoto, Y., Sakamoto, H., Hatada, K., and Irie, M., Resolution of enantiomers by HPLC on cellulose trans- and cis-tris (4-phenylazophenylcarbamate), *Chem. Lett.*, 983, 1986.
6. Ichid, A., Shibata, T., Okamoto, I., Yuki, Y., Namikoshi, H., and Toga, Y., Resolution of enantiomers by HPLC on cellulose derivatives, *Chromatographia*, 19, 280, 1984.
7. Tagahara, K., Koyama, J., Okatani, T., and Suzuta, Y., Chromatographic resolution of racemic tetrahydroberbeine alkaloids by using cellulose tris (phenylcarbamate) stationary phase, *Chem. Pharm. Bull.*, 34, 5166, 1986.
8. Klemisch, W. and von Hodenberg, A., Separation on crosslinked acetylcellulose, HRC&CC, *J. High Resolut. Chromatogr., Chromatogr. Commun.*, 9(12), 765–767, 1986.
9. Rimboock, K., Kastner, F., and Mannschreck, A., Microcrystallinetribenzoyl cellulose: A high-performance liquid chromatographic sorbent for the separation of enantiomers, *J. Chromatogr.*, 351, 346, 1986.

10. Lindner, K. and Mannschreck, A., Separation of enantiomers by high-performance liquid chromatography on triacetylcellulose, *J. Chromatogr.*, 193, 308–310, 1980.
11. Gubitz, G., Jellenz, W., and Schonleber, D., High performance liquid chromatographic resolution of the optical isomers of D,L-tryptophane, D,L-5-hydroxytryptophan and D,L-dopa on cellulose columns, HRC&CC, *J. High Resolut. Chromatogr., Chromatogr. Commun.*, 3, 31, 1980.
12. Takayanagi, H., Hatano, O., Fujimura, K., and Ando, T., Ligand-exchange high-performance liquid chromatography of dialkyl sulfides, *Anal. Chem.*, 57, 1840, 1985.
13. Armstrong, D.W., U.S. Patent #4,539,399, Assigned to Advanced Separation Technologies Inc., Whippany, NJ, 1985.
14. Armstrong, D.W. and Demond, W., Cyclodextrin bonded phases for the liquid chromatographic separation of optical, geometrical, and structural isomers, *J. Chromatogr. Sci.*, 22, 411, 1984.
15. Armstrong, D.W., Ward, T.J., Armstrong, R.D., and Beesley, T.J., Separation of drug stereoisomers by the formation of β-cyclodextrin inclusion complexes, *Science*, 232, 1132, 1986.
16. Armstrong, D.W., DeMond, W., and Czech, B.P., Separation of metallocene enantiomers by liquid chromatography: Chiral recognition via cyclodextrin bonded phases, *Anal. Chem.*, 57, 481, 1985.
17. Armstrong, D.W., DeMond, W., Alak, A., Hinze, W.L., Riehl, T.E., and Bui, K.H., Liquid chromatographic separation of diastereomers and structural isomers on cyclodextrin-bonded phases, *Anal. Chem.*, 57, 234, 1985.
18. Weaver, D.E. and van Lier R., Coupled β-cyclodextrin and reverse-phase high-performance liquid chromatography for assessing biphenyl hydroxylase activity in hepatic 9000 g supernatant, *Anal. Biochem.*, 154, 590, 1986.
19. Armstrong, D.W., Optical isomer separation by liquid chromatography, *Anal. Chem.*, 59, 84A, 1987.
20. Chang, C.A., Wu, Q., and Tan, L., Normal-phase high-performance liquid chromatographic separations of positional isomers of substituted benzoic acids with amine and b-cyclodextrin bonded-phase columns, *J. Chromatogr.*, 361, 199, 1986.
21. Cline-Love, L. and Arunyanart, M., Cyclodextrin mobile-phase and stationary-phase liquid chromatography, *A.C.S. Symposium Series*, 297, 226, 1986.
22. Feitsma, K., Bosman, J., Drenth, B., and DeZeeuw, R., A study of the separation of enantiomers of some aromatic carboxylic acids by high-performance liquid chromatography on a γ-cyclodextrin-bonded stationary phase, *J. Chromatogr.*, 333, 59, 1985.
23. Fujimura, K., Ueda, T., and Ando, T., Retention behavior of some aromatic compounds on chemically bonded cyclodextrin silica stationary phase in liquid chromatography, *Anal. Chem.*, 55, 446, 1983.
24. Hattori, K., Takahashi, K., Mikami, M., and Watanabe, H., Novel high-performance liquid chromatographic adsorbents prepared by immobilization of modified Cyclodextrins, *J. Chromatogr.*, 355, 383, 1986.
25. Ridlon, C.D. and Issaq, H.J., Effect of column type and experimental parameters on the HPLC separation of dipeptides, *J. Liq. Chromatogr.*, 9(15), 3377, 1986.
26. Sybilska, D., Debowski, J., Jurczak, J., and Zukowski, J., The α- and β-cyclodextrin complexation as a tool for the separation of o-, m- and p-nitro- cis- and trans-cinnamic acids by reversed-phase high-performance liquid chromatography, *J. Chromatgr.*, 286, 163, 1984.
27. Chang, C.A., Wu, Q., and Eastman, M.P., Mobile phase effects on the separations of substituted anilines with β-cyclodextrin-bonded column, *J. Chromatogr.*, 371, 269, 1986.
28. Maguire, J.H., Some structural requirements for resolution of hydantoin enantiomers with β-cyclodextrin liquid chromatography column, *J. Chromatogr.*, 387, 453, 1987.
29. Sinibaldi, M., Carunchio, V., Coradini, C., and Girelli, A.M., High-performance liquid chromatographic resolution of enantiomers on chiral amine bonded silica gel, *Chromatographia*, 18(81), 459, 1984.
30. Gasparrini, D., Misti, D., and Villani, C., Chromatographic optical resolution on trans-1,2-diaminocyclohexane derivatives: Theory and applications, *Chirality*, 4, 447–458, 1992.
31. Zhong, Q., Han, X., He, L., Beesley, T., Trahanovsky, W., and Armstrong, D. Chromatographic evaluation of poly (trans-1,2-cyclohexanediyl-bis acrylamide) as a chiral stationary phase for HPLC, *J. Chromatogr. A*, 1066, 55–70, 2005.
32. Pettersson, C. and Stuurman, H.W., Direct separation of enantiomer of ephedrine and some analogues by reversed-phase liquid chromatography using (+)-di-n-butytartrate as the liquid stationary phase, *J. Chromatogr. Sci.*, 22, 441, 1984.

33. Weems, H.B., Mushtaq, M., and Yang, S.K., Resolution of epoxide enantiomers of polycyclic aromatic hydrocarbons by chiral stationary-phase high-performance liquid chromatography, *Anal. Biochem.*, 148, 328, 1985.

34. Weems, H., Mushtaq, M., Fu, P., and Yang, S., Direct separation of non-k-region monool and diol enantiomers of phenanthrene, benz[a]anthracene, and chrysene by high-performance liquid chromatography with chiral stationary phases, *J. Chromatogr.*, 371, 211, 1986.

35. Yang, S., Mushtaq, M., and Fu, P., Elution order-absolute configuration of k-region dihydrodiol enantiomers of benz[a]anthracene derivatives in chiral stationary phase high performance liquid chromatography, *J. Chromatogr.*, 371, 195–209, 1986.

36. Wainer, I., Applicability of HPLC chiral stationary phases to pharamacokinetic and disposition studies on enantiomeric drugs, *Methodol. Serv. Biochem. Anal Subseries A*, 16, 243, 1986.

37. Vaughan, G.T. and Millborrow, B.V., The resolution by HPLC of RS-[2-^{14}C] Me 1′, 4′-cis-diol of abscisic acid and the metabolism of (–)-R-and-S-abscisic acid, *J. Exp. Bot.*, 35(150), 110–120, 1984.

38. Tambute, A., Gareil, P., Caude, M., and Rosset, R., Preparative separation of racemic tertiary phosphine oxides by chiral high-performance liquid chromatography, *J. Chromatogr.*, 363, 81–93, 1986.

39. Wainer, I. and Doyle, T., The direct enantiomeric determination of (–) and (+)-propranolol in human serum by high-performance liquid chromatography on a chiral stationary phase, *J. Chromatogr.*, 306, 405–411, 1984.

40. Okamoto, Y., Mohri, H., Ishikura, M., Hatada, K., and Yuki, H., Optically active poly (diphenyl-2-pyridylmethyl methacrylate): Asymmetric synthesis, stability of helix, and chiral recognition ability, *J. Polymer Sci.: Polymer Symposium*, 74, 125–139, 1986.

41. Yamazaki, S., Omori, H., and Eon Oh, C., High performance liquid chromatography of alkaline-earth metal ions using reversed-phase column coated with N-n-dodecylimminodiacetic acid, HRC&CC, *J. High Resolut. Chromatogr., Chromatogr. Commun.*, 9, 765, 1986.

42. Gubitz, G. and Mihellyes, S., Direct separation of 2-hydroxy acid enantiomers by high-performance liquid chromatography on chemically bonded chiral phases, *Chromatographia*, 19, 257, 1984.

43. Schulze, J. and Konig, W., Enantiomer separation by high-performance liquid chromatography on silica gel with covalently bound mono-saccharides, *J. Chromatogr.*, 355, 165, 1986.

44. Kip, J., Van Haperen, P., and Kraak, J.C., R-N-(pentafluorobenzoyl) phenylglycine as a chiral stationary phase for the separation of enantiomers by high-performance liquid chromatography, *J. Chromatogr.*, 356, 423, 1986.

45. Gelber, L.R., Karger, B.L., Neumeyer, J.L., and Feibush, B., Ligand exchange chromatography of amino alcohols. Use of schiff bases in enantiomer resolution, *J. Am. Chem. Soc.*, 106, 7729, 1984.

46. Dabashi, Y. and Hara, S., Direct resolution of enantiomers by liquid chromatography with the novel chiral stationary phase derived from (R,R)-tartamide, *Tet. Lett.*, 26(35), 4217, 1985.

47. Facklam, C., Pracejus, H., Oehme, G., and Much, H., Resolution of enantiomers of amino acid derivatives by high-performance liquid chromatography in a silica gel bonded chiral amide phase, *J. Chromatogr.*, 257, 118, 1983.

48. Okamoto, Y., Honda, S., Hatada, K., and Yuki, H., IX. High-performance liquid chromatographic resolution of enantiomers on optically active poly (tri-phenylmethyl methacrylate), *J. Chromatogr.*, 350, 127, 1985.

49. Okamoto, Y. and Hatada, K., Resolution of enantiomers by HPLC on optically active poly (triphenylmethyl methacrylate), *J. Liq. Chromatogr.*, 9 (2&3), 369, 1986.

50. Armstrong, D.W., Tang, Y., Chen, S., Zhou, C., Bagwill, J.R., and Chen, J.R., Macrocyclic antibiotics as a new class of chiral selectors for liquid chromatography, *Anal. Chem.*, 66(9), 1473–1484, 1994.

51. Ekborg-Ott, K.H., Liu, Y., and Armstrong, D.W., Highly enantioselective HPLC separations using the covalently bonded macrocyclic antibiotic ristocetin A chiral stationary phase, *Chirality*, 10, 434–483, 1998.

52. Armstrong, D.W., Liu, Y., and Ekborg-Ott, K.H., A covalently bonded teicoplanin chiral stationary phase for HPLC enantioseparations, *Chirality*, 7(6), 474–497, 1995.

53. Berthod, A., Liu, Y., Bagwill, J.R., and Armstrong, D.W., Facile RPLC enantioresolution of native amino acids and peptides using a teicoplanin chiral stationary phase, *J. Chromatogr. A*, 731,123–137, 1996.

54. Berthod, A., Chen, X., Kullman, J.P., Armstrong, D.W., Gasparrini, F., D'Acquarica, I., Villani, C., and Carotti, A., Role of carbohydrate moieties in chiral recognition on teicoplanin based LC stationary phases, *Anal. Chem.*, 72, 1736–1739, 2000.
55. Aboul-Enein, H.Y. and Serignese, V., Enantiomeric separation of several cyclic imides on a macrocyclic antibiotic (Vancomycin) chiral stationary phase under normal and reverse phase conditions, *Chirality*, 10, 358–361, 1998.
56. Lehotay, J., Hrobonova, J., Krupcik, J., and Cizmarik, J., Chiral separation of enantiomers of amino acid derivatives by HPLC on vancomycin and teichoplanin chiral stationary phases, *Pharmazie*, 53, 863–865, 1998.

Name: bovine serum albumin (covalently fixed to silica gel)

Structure: prolate ellipsoid 14×4 nm, with a molecular mass of 66,500. Amount absorbed is dependent on buffer pH, with the maximum at pH = 4.9.

Analytical Properties: separation of bopindolol and also separation of pindolol after derivatization with isopropyl isocyanate. Separation of DL mixtures of enantiomers; can be used on both the analytical and preparative scales. Changes in pH will cause this phase to leach from the column. Storage at 4°C is recommended

Reference: [1,2]

Name: α-acid glycoprotein

Structure: structure is proprietary (Enantiopac, LKB Co.)

Analytical Properties: separation of the drugs ibuprofen, ketoprofen, naproxen, 2-phenoxypropionic acid, bendroflumethiazide, ethotoin, hexobarbital, disopyramide, and RAC 109

Retention and selectivity of the solutes can be regulated by addition of the tertiary amine N,N,-dimethyloctylamine (DMOA) to the mobile phase.

DMOA decreases retention time and the enantioselectivity of the weaker acids but has opposite effects on the stronger acids.

Reference: [3]

Name: N-(t-butylamino carbonyl-L-valine) bonded silica

Structure:

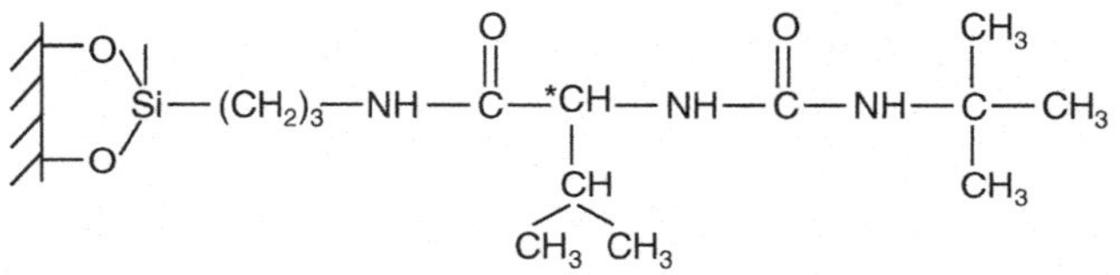

Analytical Properties: separation of amino acid enantiomers; most effective of the L-valine urea derivatives; depends on hydrogen bond interactions usually prepared on LiChro-sorb (10 μm); hexane plus isopropanol modifier has been used as the liquid phase

Reference: [4]

Name: cellulose cis and trans tris (4-phenylazophenyl carbamate) (CPAPC)

Structure:

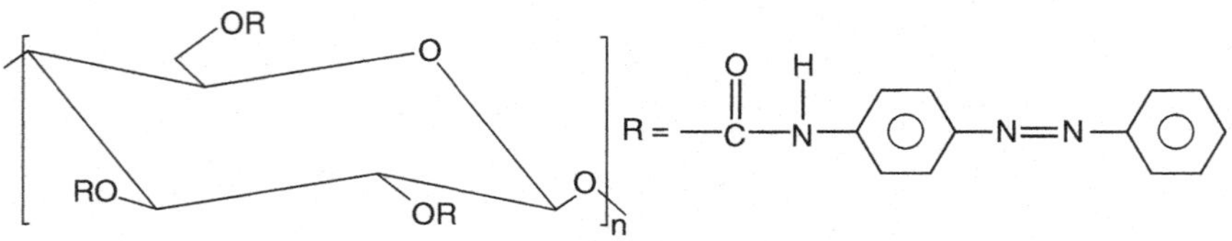

Analytical Properties: trans isomer provides excellent resolution of racemic mixtures such as atropine, pindolol, and flavanone; resolution decreases quickly with increasing cis isomer concentration; the cis/trans equilibrium is controlled by UV radiation, and the phase is adsorbed to silica gel; liquid phase of hexane with 10 % 2-propanol has been found useful

Reference: [5]

Name: cellulose triacetate
Structure:

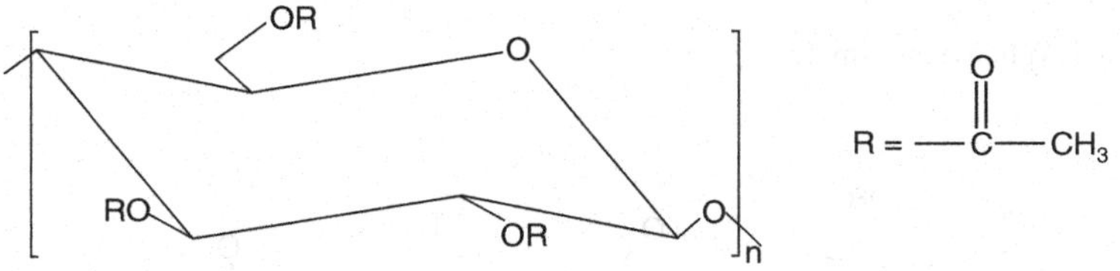

$$R = -\overset{\overset{O}{\|}}{C}-CH_3$$

Analytical Properties: shows chiral recognition for many racemates and is especially effective for substrates with a phosphorus atom at an asymmetric center. However, the degree of chiral recognition is not so high in general
Reference: [6]

Name: cellulose tribenzoate
Structure:

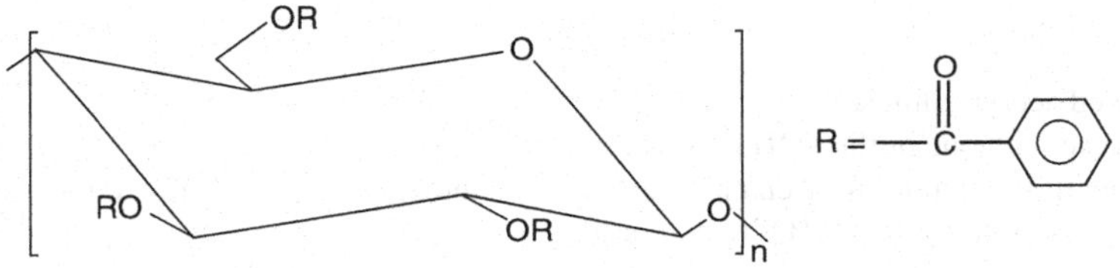

Analytical Properties: demonstrates good chiral recognition for the racemates with carbonyl group(s) in the neighborhood of an asymmetric center
Reference: [6]

Name: cellulose tribenzyl ether
Structure:

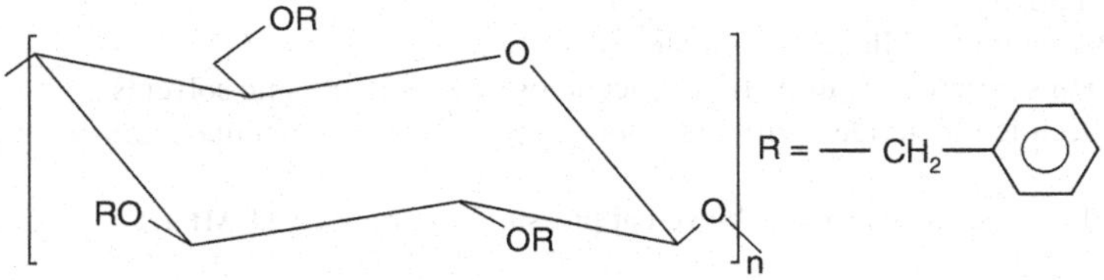

Analytical Properties: effective with protic solvents are used as mobile phases
Reference: [6]

Name: cellulose tricinnamate
Structure:

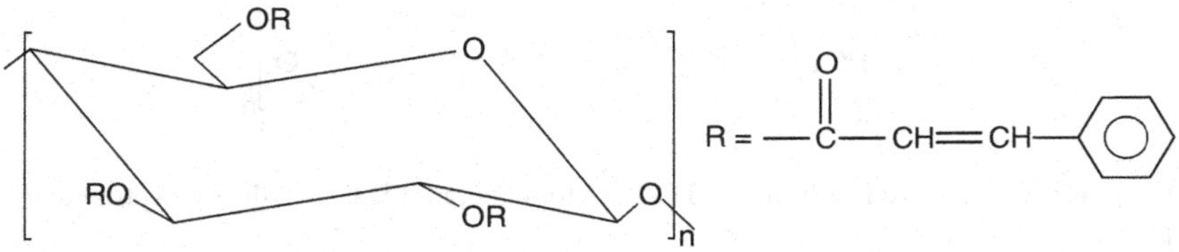

Analytical Properties: shows high chromatographic retention times and a good chiral recognition for many aromatic racemates and barbiturates
Reference: [6]

Name: cellulose tris (phenylcarbamate)
Structure:

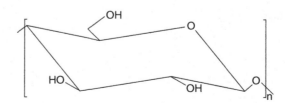

Analytical Properties: separation of racemic mixtures of alkaloids, using ethanol as the eluent; 1-isomers tend to be more strongly retained than d-isomers
good chiral recognition of sulfoxides and high affinity for racemates having an -OH or -NH group, through hydrogen bonding
Reference: [6,7]

Name: cross-linked acetylcellulose
Structure: cellulose with one of the –OH groups acylated
Analytical Properties: separation of enantiomers (such as etozolin, piprozolin, ozolinon, and bunolol) using an ethanol/water, 95/5 (V/V %) liquid phase
Reference: [8]

Name: microcrystalline tribenzoylcellulose
Structure: same structure as cellulose trobenzoate (coated on macroporous silica gel)
Analytical Properties: resolution of trans-1,2, diphenyloxirane, 2-methyl-3-(2′-methylphenyl)-4(3H)-quinazolinone and some aromatic hydrocarbons
Reference: [9]

Name: triacetylcellulose
Structure: same structure as cellulose triacetate.
Analytical Properties: microcrystalline triacetylcellulose swells in organic solvents
separation of racemic thioamides, sulfoxides, organophosphorus compounds, drugs, and amino acids derivatives
separations of these racemates were achieved at pressures at or above 4.9 MPa
Reference: [10]

Name: untreated cellulose (average particle size of 7 μm)
Structure:

Analytical Properties: complete resolution of D,L-tryptophane, and D,L-5-hydroxytryptophane
Reference: [11]

Name: copper (II) 2-amino-1-cyclopentene-1-dithio carboxylate
Structure:

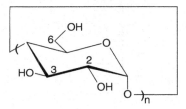

(bonded to silica)
Analytical Properties: separation of dialkyl sulfides when hexane containing methanol or acetonitrile was used as the mobile phase
Reference: [12]

Name: α-cyclodextrin bonded phase
Structure:
Subunit:

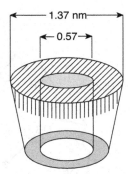

Cavity:

Analytical Properties: α-cyclodextrin (cyclohexamolyose); reversed-phase separation of barbiturates and other drugs and aromatic amino acids. The substrate is composed of 6 glucose units and has a relative molecular mass of 972. The cavity diameter is 0.57 nm, and the substrate has a water solubility of 14.5 g/mL
Reference: [13–28]

Name: β-cyclodextrin bonded phase
Structure:
Subunit:

Cavity:

Ligand:

Note that this structure also illustrates the linkage to silica through one primary –OH group. Either one or two such linkages usually attach the substrate to the silica.

Analytical Properties: β-cyclodextrin (cycloheptamylose); normal-phase separation of positional isomers of substituted benzoic acids; reverse-phase separation of dansyl and napthyl amino acids, several aromatic drugs, steroids, alkaloids, metallocenes, binapthyl crown ethers, aromatics acids, aromatic amines, and aromatic sulfoxides. This substrate has seven glucose units and has a relative molecular mass of 1,135. The inside cavity has a diameter of 0.78 nm, and the substrate has a water solubility of 1.85 g/mL, although this can be increased by derivatization

Reference: [13–28]

Name: β-cyclodextrin, dimethylated bonded phase
Structure:
Cavity:

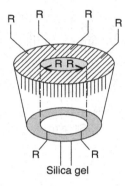

R-ligand:

$$-CH_3$$

Analytical Properties: β-cyclodextrin DM (cycloheptamylose-DM); reversed-phase separation of a variety of structural and geometrical isomers; useful for the separation of analytes that have a carbonyl group off the stereogenic center

Reference: [13–28]

Name: β-cyclodextrin, acylated bonded phase
Structure:
Cavity:

R-Ligand:

$$-COCH_3$$

Analytical Properties: β-cyclodextrin AC (cycloheptamylose-AC); reversed-phase separation of steroids and polycyclic compounds
Reference: [13–28]

Name: β-cyclodextrin, hydroxypropyl ether-modified bonded phase
Structure:
Cavity:

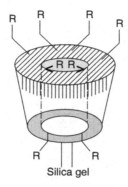

R-Ligand:

$$\begin{array}{c} OH \\ | \\ ---CH_2\underset{*}{C}HCH_3 \end{array}$$

Analytical Properties: β-cyclodextrin SP or RSP (cycloheptamylose-SP, RSP); note that the modifying ligand has a stereogenic center; useful for reversed-phase separation of a variety of analytes, especially for enantiomers that have bulky substituents that are beta to the stereogenic center; can be used for cyclic hydrocarbons and for t-boc amino acids
Reference: [13–28]

Name: β-cyclodextrin, napthylethyl carbamate modified bonded phase
Structure:
Cavity:

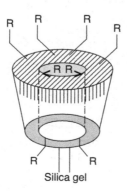

R-Ligand:

$$\text{—CONH}\overset{|}{\underset{*}{\text{CH}}}\text{—}$$

with CH_3 above and naphthyl group attached.

Analytical Properties: β-cyclodextrin SN or RN (cycloheptamylose-SN, RN); note that the modifying ligand has a stereogenic center; useful for normal phase, reversed phase, and polar organic phase separation under specific circumstances; the substrate performs best with normal phase and polar organic mobile phases; the SN modification has shown the highest selectivity
Reference: [13–28]

Name: β-cyclodextrin, 3,5-dimethylphenyl carbamate-modified bonded phase
Structure:
Cavity:

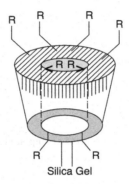

Silica Gel

R-Ligand:

$$\text{—CONH—}$$

with a benzene ring bearing two CH_3 groups at the 3,5-positions.

Analytical Properties: β-cyclodextrin DMP (cycloheptamylose-DMP); useful for normal phase, reversed phase, and polar organic phase separation under specific circumstances; the substrate performs best with normal phase and polar organic mobile phases
Reference: [13–28]

Name: γ-cyclodextrin bonded phase
Structure:
Subunit:

A glucose subunit structure showing positions 6 (OH), 3 (HO), 2 (OH), and ring oxygen with $O\text{—}{)}_n$.

Cavity:

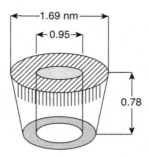

Analytical Properties: γ-cyclodextrin (cyclooctylamalyose), reverse-phase separation of stereoisomers of polycyclic aromatic hydrocarbons. The substrate has 8 glucose units and has a relative molecular mass of 1,297. The cavity has a diameter of 0.59 nm, and the substrate has a water solubility of 23.2 g/mL
Reference: [13–28]

Name: (–)trans-1,2-cyclohexanediamine
Structure:

(bonded to silica gel)
Analytical Properties: resolution of such enantiomeric compounds as 2,2′-dihydroxy-1,1′-binapthyl and trans-1,2-cyclo hexandiol
Reference: [29]

Name: poly (tert-1,2-cyclohexanediyl-bis acrylamide)
Structure:

Analytical Properties: synthetic, covalently bonded chiral stationary phase consisting of a thin, ordered selector layer bonded to the silica surface; high stability and loadability, relatively easy to scale up; reversal of elution order is possible (R,R) to (S,S) configuration; used for resolution of aryloxyacetic acids, alcohols, sulfoxides, selenoxides, phosphinates, t-phospine oxides, benzodiazepines without derivatization; amines, amino acids, amino alcohols, nonsteriod anti-inflammatory drugs with derivatization
Reference: [30,31]

Name: (+)-di-n-butyltartrate
Structure:

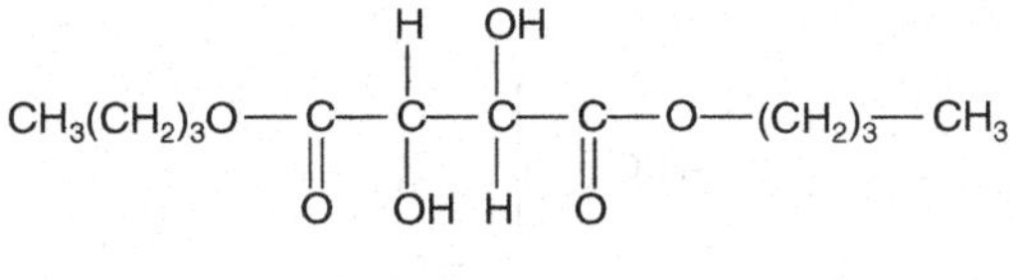

$$CH_3(CH_2)_3O - \underset{\underset{O}{\|}}{C} - \underset{\underset{OH}{|}}{\overset{\overset{H}{|}}{C}} - \underset{\underset{H}{|}}{\overset{\overset{OH}{|}}{C}} - \underset{\underset{O}{\|}}{C} - O - (CH_2)_3 - CH_3$$

(adsorbed on phenyl bonded silica)
Analytical Properties: resolution of ephedrine and norephedrine
Reference: [32]

Name: (S)-N-(3,5-dinitrobenzoyl) leucine or (S)-DNBL
Structure:

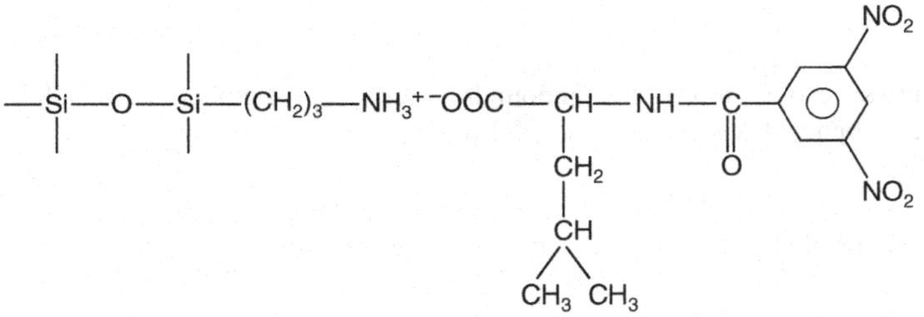

Analytical Properties: resolution of several enantiomers of polycyclic aromatic hydrocarbons, for example, chrysene 5,6-epoxide, dibenz[a,h]anthracene 5,6 epoxide, 7-methyl benz[a]anthracene 5,6-epoxide. Resolution of barbiturates, mephenytoin, benzodiazepinones, and succinimides. Direct separation of some mono-ol and diol enantiomers of phenanthrene, benz[a]anthrene, and chrysene. Ionically bonded to silica gel, this phase provides resolution of enantiomers of cis-dihydroidiols of unsubstituted and methyl- and bromo-substituted benz[a]anthracene derivatives having hydroxyl groups that adopt quasiequatorial-quasiaxial and/or quasiaxial-quasiequatorial conformation
Reference: [33–37]

Name: (R)-N-(3,5-dinitrobenzoyl) phenylglycine
Structure:

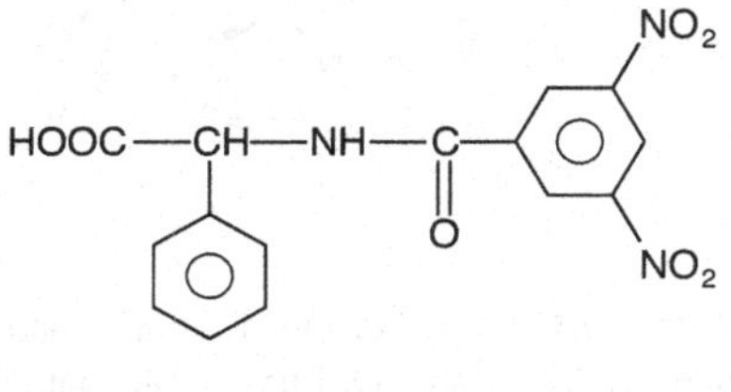

Analytical Properties: ionically bonded to silica, this phase provides good resolution of enantiomeric quasiequatorial trans-dehydrodiols of unsubstituted and methyl- and bromo-substituted benz[a]

anthracene derivatives. Covalently bonded to silica, this phase provides good resolution of enantio-meric pairs of quasidiaxial trans-dihydrodiols of unsubstituted and methyl- and bromo-substituted benz[a]anthracene derivatives. By addition of a third solvent (chloroform) to the classical binary mix-ture (hexane-alcohol) of the mobile phase, resolution of enantiomers of tertiary phosphine oxides is possible

Reference: [33–35,38,39]

Name: poly(diphenyl-2-pyridylmethyl methacrylate) or PD2PyMa
Structure:

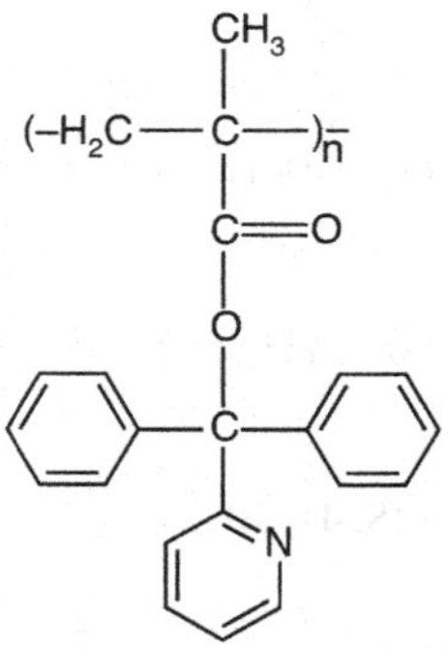

(coated on macroporous silica gel)

Analytical Properties: resolution of such compounds as racemic 1,2-diphenol-ethanol, 2-2′-dihydroxyl-1,1′-dinapthyl, 2,3-diphenyloxirane, and phenyl-2-pyrid-o-toly-1-methanol
Reference: [40]

Name: poly(1,2-diphenylethylenediamine acrylamide) or P-CAP DP
Structure:

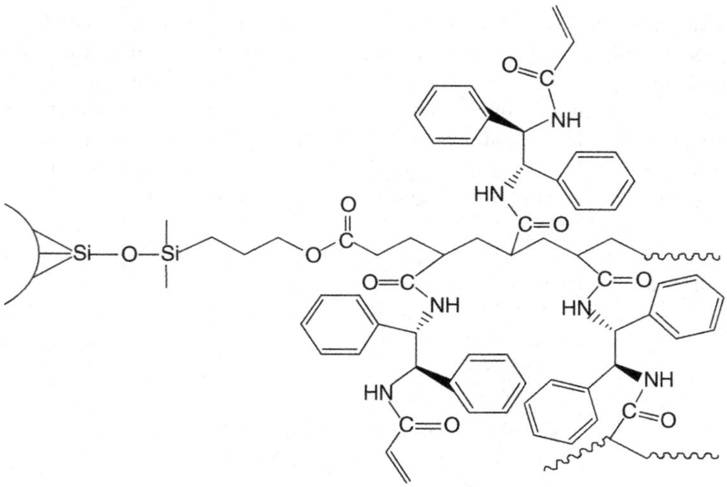

Analytical Properties: synthetic, covalently bonded chiral stationary phase consisting of a thin, ordered selector layer bonded to the silica surface; high stability and loadability, relatively easy to scale up; reversal of elution order is possible (R,R) to (S,S) configuration
Reference: [30,31]

Name: N-n-dodecyliminodiacetic acid (coated on silica gel)
Structure:

Analytical Properties: separation of alkaline earth metal ions
Reference: [40]

Name: L-hydroxyproline
Structure:

(as a fixed ligand on a silica gel; Cu (II) used as complexing agent)

Analytical Properties: chiral phases containing L-hydroxy-proline as a fixed ligand show high enantioselectivity for 2-hydroxy acids; these phases have resolved some aromatic as well as aliphatic 2-hydroxy
acids
Reference: [42]

Name: 1-isothiocyanato-D-glucopyranosides
Structure:

CSP 3

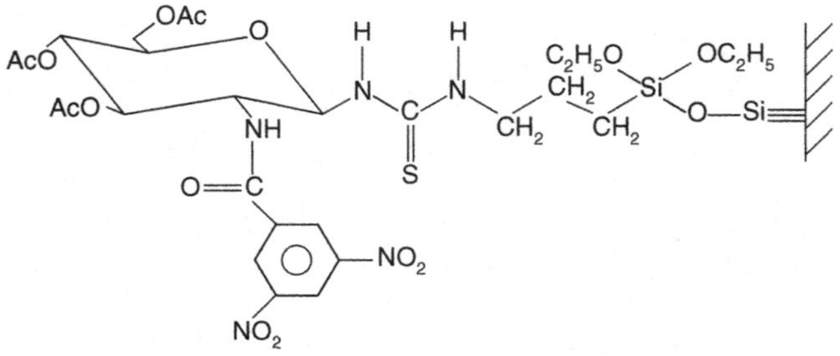

Analytical Properties:

CSP 1 (chiral stationary phase)—Separates some chiral binapthyl derivatives when mixtures of hexane
diethyl ether, dichloromethane, or dioxane are used as the mobile phase.

CSP 2—Separates compounds with carbamate or amide functions. (Mixtures of n-hexane and 2-propanol
can be used as mobile phase.)

CSP 3—Separation of compounds separated by CSP 2. In addition, separation of compounds with car-
banoyl or amide functions and some amino alcohols which have pharmaceutical relevance (β-blockers).

Reference: [43]

Name: N-isopropylamino carbonyl-L-valine bonded silica
Structure:

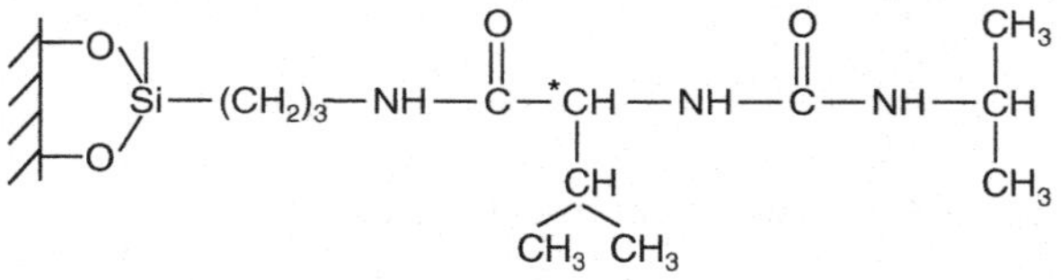

Analytical Properties: separation of amino acid enantiomers; good chiral recognition of N-acetyl amino
acid methyl esters; depends on hydrogen bond interactions; hexane with isopropanol modifier has been
used as the liquid phase; usually prepared on LiChrosorb (10 μm)

Reference: [4]

Name: (R)- or (S)-N-(2-napthyl)alanine
Structure:

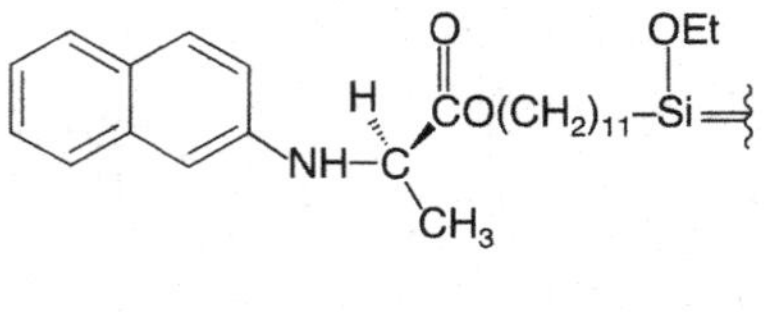

Analytical Properties: high selectivities for a variety of dinitrobenzoyl derivatized compounds
Reference: [19]

Name: (S)-1-(α-napthyl)ethylamine
Structure:

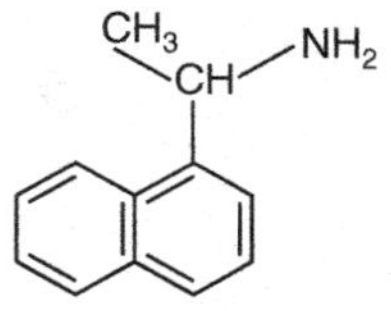

Analytical Properties: separation of 3,5-dinitrobenzoyl derivatives of amino acids; 3,5-dinitroanilide derivatives of carboxylic acids
Reference: [19]

Name: R-N-(pentafluorobenzoyl) phenylglycine
Structure:

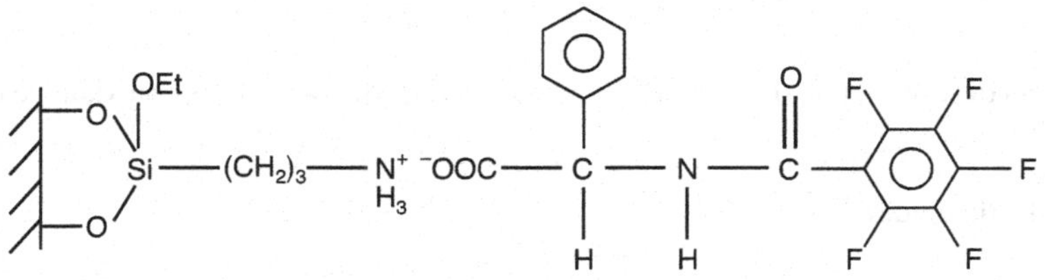

Analytical Properties: higher selectivity for nitrogen-containing racemates than R-N-(3,5-dinitrobenzoyl) phenylglycine. Examples of nitrogen-containing racemates include succinimides, hydantoins, and mandelates
Reference: [44]

Name: N-phenylaminocarbonyl-L-valine bonded silica
Structure:

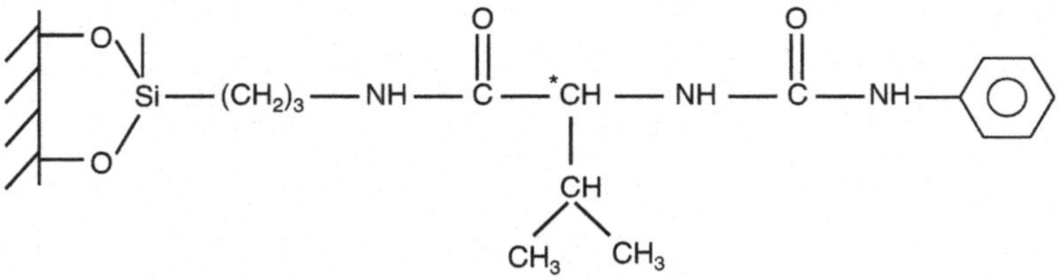

Analytical Properties: separation of amino acid enantiomers by hydrogen bond interactions; usually prepared on LiChrosorb (10:m); hexane plus isopropanol modifier commonly used as liquid phase
Reference: [4]

Name: (L-proline) copper (II)
Structure:

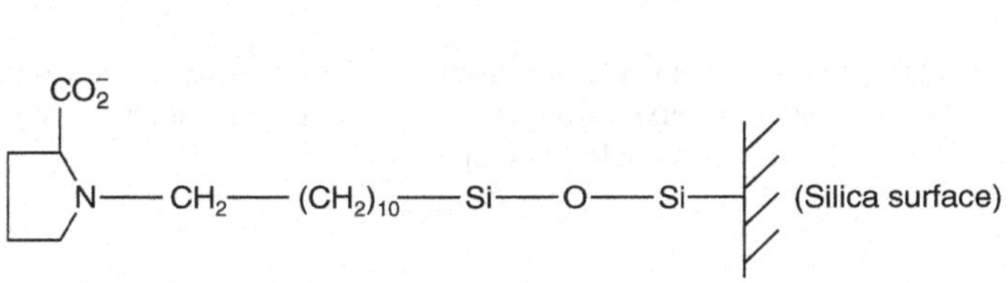

Analytical Properties: separation of primary α-amino alcohols, for example, β-hydroxyphenethylamines and catecholamines
Reference: [45]

Name: derivative of (R,R)-tartramide
Structure:

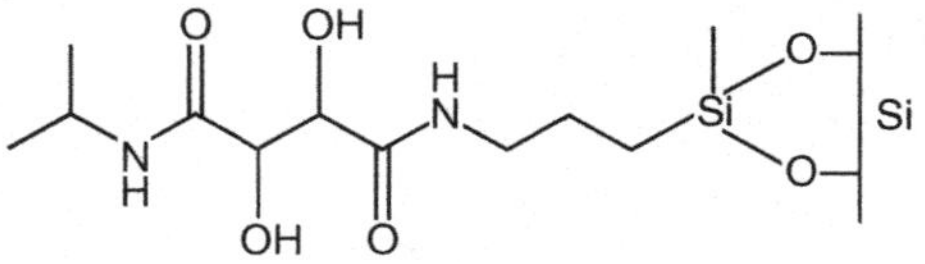

Analytical Properties: resolution of a series of β-hydroxycarboxylic acids as tert-butylamide derivatives
Reference: [46]

Name: tert-butylvalinamide
Structure:

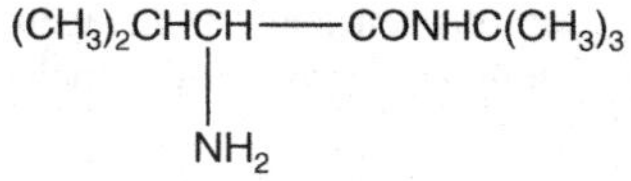

$$(CH_3)_2CHCH\!-\!\!-\!\!-CONHC(CH_3)_3$$
$$|$$
$$NH_2$$

Analytical Properties: resolution of heavier amino acid derivatives
Reference: [47]

Name: (+)-poly(triphenylmethyl methacrylate) or (+)-PTrMA
Structure:

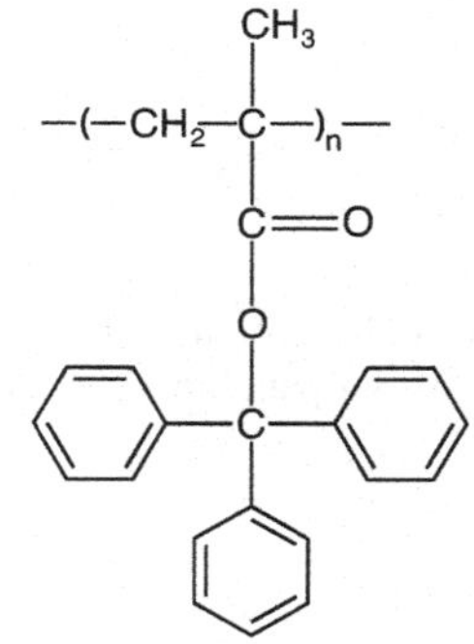

Analytical Properties: resolution of enantiomers such as trans-1,3-cyclohexene dibenzoate, 3,5-pentylene dibenzoate, 3,5-dichlorobenzoate, triacetylacetonates, and racemic compounds having phosphorus as a chiral center. This phase will also resolve achiral compounds
Reference: [46,47]

Name: ristocetin-A bonded phase
Structure:

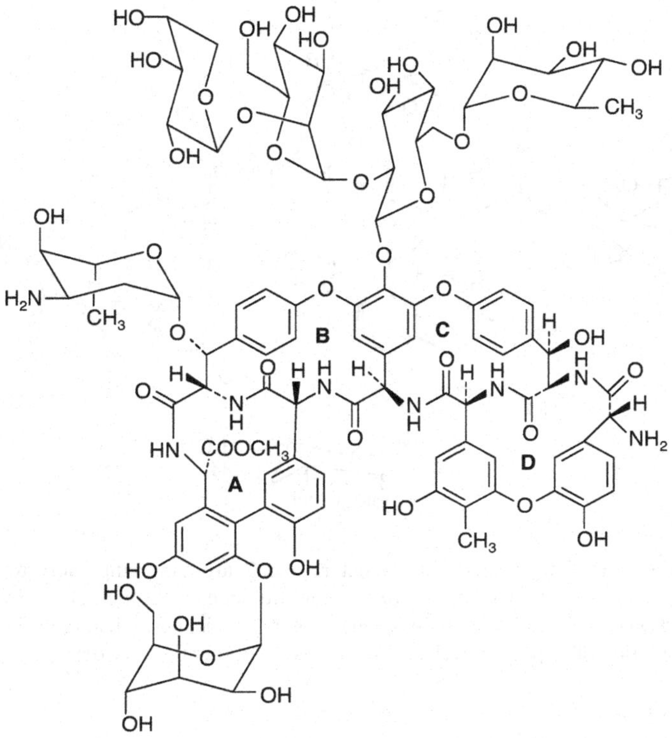

Analytical Properties: substrate has 38 chiral centers and 7 aromatic rings surrounding 4 cavities, making this the most structurally complex of the macrocyclic glycopeptides. Substrate has a relative molecular mass of 2,066. This phase can be used in normal, reverse, and polar organic phase separations; selective for anionic chiral species. With polar organic mobile phases, it can be used for α-hydroxy acids, profens, and N-blocked amino acids; in normal phase mode, it can be used for imides, hydantoins, and N-blocked amino acids, and in reverse phase, it can be used for α-hydroxy and halogenated acids, substituted aliphatic acids, profens, N-blocked amino acids, hydantoins, and peptides
Reference: [50,51]

Name: teicoplanin bonded phase
Structure:

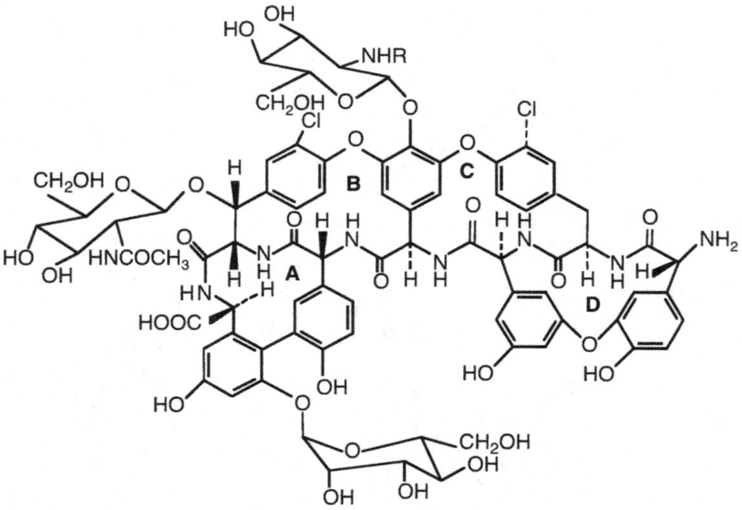

Analytical Properties: substrate has 20 chiral centers and 7 aromatic rings surrounding 4 cavities. Substrate has a relative molecular mass of 1,885. Separation occurs through chiral hydrogen bonding sites, B–B interactions and inclusion complexation in polar organic, normal, and reverse mobile phases. Useful for the resolution of α, β, γ, or cyclic amino acids, small peptides, N-derivatized amino acids
Reference: [52,53]

Name: teicoplanin aglycone bonded phase
Structure:

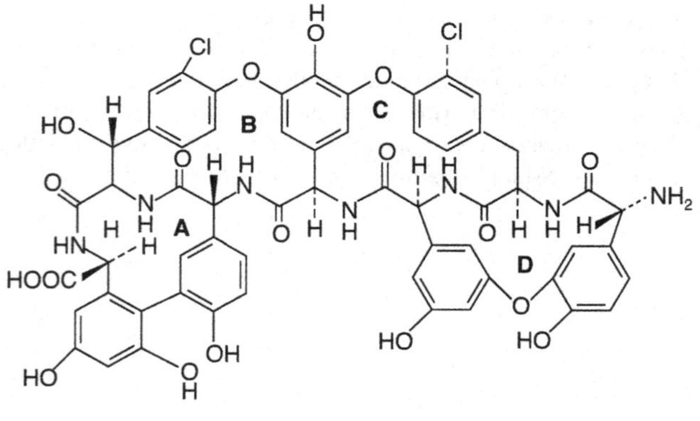

Analytical Properties: substrate has 8 chiral centers and 7 aromatic rings surrounding 4 cavities. Substrate has a relative molecular mass of 1,197. Separation occurs through chiral hydrogen bonding sites, π-π interactions and inclusion complexation in polar organic, normal and reverse mobile phases. Highly selective for amino acids (α, β, γ or cyclic), some N-blocked amines, many neutral cyclic compounds, peptides, diazepines, hydantoins, oxazolidinones, and sulfoxides
Reference: [54]

Name: vancomycin bonded phase
Structure:

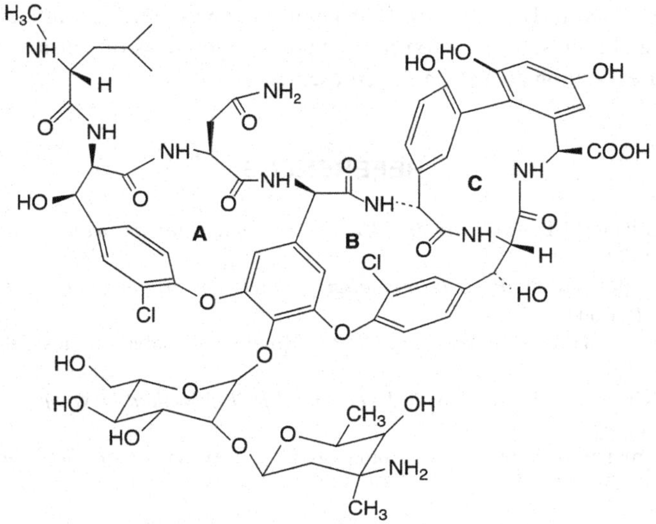

Analytical Properties: substrate has 18 chiral centers and 5 aromatic rings surrounding 3 cavities. Substrate has a relative molecular mass of 1,449, an isoelectric point of 7.2, with pKs of 2.9, 7.2, 8.6, 9.6, 10.4, and 11.7. Separation occurs through chiral hydrogen bonding sites, $\pi-\pi$ interactions, a peptide binding site and inclusion complexation in polar organic, normal and reverse mobile phases. Selective for cyclic amines, amides, acids, esters, and neutral molecules; high sample capacity

Reference: [55,56]

DETECTORS FOR LIQUID CHROMATOGRAPHY

The following table provides some comparative data for the selection and operation of the more common detectors applied to HPLC [1–5]. In general, the operational parameters provided are for optimized systems and represent the maximum obtainable in terms of sensitivity and linearity. In this table, the molar extinction coefficient is represented by ε.

REFERENCES

1. Pryde, A. and Gilbert, M.T., *Applications of High Performance Liquid Chromatography*, Chapman & Hall, London, 1979.
2. Hamilton, R.J. and Sewell, P.A., *Introduction to High Performance Liquid Chromatography*, Chapman & Hall, London, 1977.
3. Ahuja, S., *Trace and Ultratrace Analysis by HPLC*, Chemical Analysis Series, John Wiley & Sons, New York, 1991.
4. Snyder, L.R., Kirkland, J.J., and Glajch, J., *Practical HPLC Method Development*, John Wiley & Sons, New York, 1997.
5. Bruno, T.J., A review of hyphenated chromatographic instrumentation, *Sep. Purif. Meth.*, 29(1), 63–89, 2000.

Detector	Sensitivity	Linearity	Selectivity	Comments
Ultraviolet spectrophotometer	1×10^{-9} g (for compounds of $\varepsilon = 10{,}000\text{--}20{,}000$)	1×10^4	For UV-active functionalities, on the basis of absorptivity	Relatively insensitive to flow and temperature fluctuations; non-destructive, useful with gradient elution; use mercury lamp for 254 nm and quartz-iodine lamp for 350–700 nm; often a diode-array instrument is used to obtain entire UV-vis spectrum
Refractive index detector (RID)	1×10^{-7} g	1×10^4	Universal, dependent on refractive index difference with mobile phase	Relatively insensitive to flow fluctuations but sensitive to temperature fluctuations; non-destructive, cannot be used with gradient elution; solvents must be degassed to avoid bubble formations; laser-based RI detectors offer higher sensitivity.
Fluorometric detector	1×10^{-11} g	1×10^5	For fluorescent species with conjugated bonding and/or aromaticity	Relatively insensitive to temperature and flow fluctuations; non-destructive; can be used with gradient elution; often, chemical derivatization is done on analytes to form fluorescent species; uses deuterium lamp for 190–400 nm or tungsten lamp for 350–600 nm.
Electrochemical detectors Amperometric detector	1×10^{-9} g		Responds to –OH functionalities	Used for aliphatic and aromatic –OH compounds, amines, and indoles; pulsed potential units are most sensitive and can be used with gradient elution and organic mobile phases; senses compounds in oxidative or reductive modes; mobile phases must be highly pure and purged of O_2.
Conductivity detector	1×10^{-9} g	2×10^4	Specific to ionizable compounds	Uses post-column derivatization to produce ionic species; especially useful for certain halogen, sulfur, and nitrogen compounds
Mass spectrometers	Interface dependent	Interface dependent	Universal, within limits imposed by interface.	Complex, expensive devices highly dependent on an efficient interface; electrospray and thermospray interfaces are most common; linear response is difficult to achieve.

ULTRAVIOLET DETECTION OF CHROMOPHORIC GROUPS

The following table is provided to aid the use in the application of ultraviolet spectrophotometric detectors. The data here are used to evaluate the potential of detection of individual chromophoric moieties on analytes [1–3].

REFERENCES

1. Willard, H.H, Merritt, L.L., Dean, J.A., and Settle, F.A., *Instrumental Methods of Analysis*, 7th ed., Wadsworth Publishing Co., Belmont, 1988.
2. Silverstein, R.M., Bassler, G.C., and Morrill, T.C., *Spectrometric Identification of Organic Compounds*, 4th ed., John Wiley & Sons, New York, 1981.
3. Lambert, J.B., Shuruell, H.F., Verbit, L., Cooks, R.G., and Stout, G.H., *Organic Structural Analysis*, MacMillan Publishing Co., New York, 1976.

Chromophore	Functional Group	λ_{max} (nm)	ε_{max}	λ_{max} (nm)	ε_{max}	λ_{max} (nm)	ε_{max}
Ether	-O-	185	1,000				
Thioether	-S-	194	4,600	215	1,600		
Amine	-NH$_2$-	195	2,800				
Amide	-CONH$_2$	<210	—				
Thiol	-SH	195	1,400				
Disulfide	-S-S-	194	5,500	255	400		
Bromide	-Br	208	300				
Iodide	-I	260	400				
Nitrile	-C≡N	160	—				
Acetylide (alkyne)	-C≡C-	175-180	6,000				
Sulfone	-SO$_2$-	180	—				
Oxime	-NOH	190	5,000				
Azido	>C=N-	190	5,000				
Alkene	-C=C-	190	8,000				
Ketone	>C=O	195	1,000	270-285	18-30		
Thioketone	>C=S	205	Strong				
Esters	-COOR	205	50				
Aldehyde	-CHO	210	Strong	280-300	11-18		
Carboxyl	-COOH	200-210	50-70				
Sulfoxide	>S→O	210	1,500				
Nitro	-NO$_2$	210	Strong				
Nitrite	-ONO	220-230	1,000-2,000	300-4,000	10		
Azo	-N=N-	285-400	3-25				
Nitroso	-N=O	302	100				
Nitrate	-ONO$_2$	270 (shoulder)	12				
Conjugated hydrocarbon	-(C=C)$_2$-(acyclic)	210-230	21,000				
Conjugated hydrocarbon	-(C=C)$_3$-	260	35,000				
Conjugated hydrocarbon	-(C=C)$_4$-	300	52,000				
Conjugated hydrocarbon	-(C=C)$_5$-	330	118,000				
Conjugated hydrocarbon	-(C=C)$_2$-(alicyclic)	230-260	3,000-8,000				

(Continued)

CRC HANDBOOK OF BASIC TABLES FOR CHEMICAL ANALYSIS

Chromophore	Functional Group	λ_{max} (nm)	ε_{max}	λ_{max} (nm)	ε_{max}	λ_{max} (nm)	ε_{max}
Conjugated hydrocarbon	C=C–C≡C	219	6,500				
Conjugated system	C=C–C=N	220	23,000				
Conjugated system	C=C–C=O	210–250	10,000–20,000			300–350	Weak
Conjugated system	C=C–NO$_2$	229	9,500				
Benzene		184	46,700	202	6,900	255	170
Diphenyl				246	20,000		
Naphthalene		220	112,000	275	5,600	312	175
Anthracene		252	199,000	375	7,900		
Pyridine		174	80,000	195	6,000	251	1,700
Quinoline		227	37,000	270	3,600	314	2,750
Isoquinoline		218	80,000	266	4,000	317	3,500

Note: ϕ also typically denotes a phenyl group.

DERIVATIZING REAGENTS FOR HPLC

The following table provides a listing of the common derivatizing reagents used in HPLC. Most of these reagents are used to impart a chromophoric or fluorescent group in the sample to enhance the detectability. Occasionally, a derivatization procedure is done in order to enhance selectivity, but this is the exception.

REFERENCES

1. Umagat, H., Kucera, P., and Wen, L.-F., Total amino acid analysis using pre-column fluorescence derivatization, *J. Chromatogr.*, 239, 463, 1982.
2. Nimura, N. and Kinoshita, T., Fluorescent labeling of fatty acids with 9-anthryldiazomethane (ADAM) for high performance liquid chromatography, *Anal. Lett.*, 13(A3), 191, 1980.
3. Lehrfeld, J., Separation of some perbenzoylated carbohydrates by high performance liquid chromatography, *J. Chromatogr.*, 120, 141, 1976.
4. Durst, H.D., Milano, M., Kikta, E.J., Connelly, S.A., and Grushka, E., Phenacyl esters of fatty acids via crown ether catalysts for enhanced ultraviolet detection in liquid chromatography, *Anal. Chem.*, 47(11), 1797, 1975.
5. Farinotti, R., Siard, Ph., Bourson, J., Kirkiacharian, S., Valeur, B., and Mohuzier, G., 4-Bromomethyl-6,7-dimethoxycoumarin as a fluorescent label for carboxylic-acids in chromatographic detection, *J. Chromatogr.*, 269(2), 81, 1983.
6. Lam, S. and Grushka, E., Labeling of fatty acids with 4-bromomethyl-7-methoxycoumarin via crown ether catalyst for fluorimetric detection in high-performance liquid chromatography, *J. Chromatogr.*, 158, 207–214, 1978.
7. Korte, W.D., 9-(chloromethyl) anthracene: A useful derivatizing reagent for enhanced ultraviolet and fluorescence detection of carboxylic acids with liquid chromatography, *J. Chromatogr.*, 243, 153, 1982.
8. Ahnoff, M., Grundevik, I., Arfwidsson, A., Fonselius, J., and Persson, B.-A. Derivatization with 4-chloro-7-nitrobenzofurazan for liquid chromatographic determination of hydroxyproline in collagen hydrolysate, *Anal. Chem.*, 53, 485, 1981.
9. Linder, W., N-Chloromethyl-4-nitro-phthalimide as a derivatizing reagent for HPLC, *J. Chromatogr.*, 198, 367, 1980.
10. Lawrence, J.F., *Organic Trace Analysis by Liquid Chromatography*, Academic Press, New York, 1981.
11. Avigad, G., Dansyl hydrazine as a fluorimetric reagent for thin-layer chromatographic analysis of reducing sugars, *J. Chromatogr.*, 139, 343–347, 1977.
12. Lloyd, J.B.F., Phenanthramidazoles as fluorescent derivatives in the analysis of fatty acids by high-performance liquid chromatography, *J. Chromatogr.*, 189, 359–373, 1980.
13. Frei, R.W. and Lawrence, J.F., eds. *Chemical Derivatization in Analytical Chemistry*, Vol. 1, Plenum Press, New York, 1981.
14. Goto, J., Komatsu, S., Goto, N., and Nambara, T., A new sensitive derivatization reagent for liquid chromatographic separation of hydroxyl compounds, *Chem. Pharm. Bull.*, 29(3), 899, 1981.
15. Musson, D.G. and Sternson, L.A., Conversion of arylhydroxylamines to electrochemically active derivatives suitable for high-performance liquid chromatographic analysis with amperometric detection, *J. Chromatogr.*, 188, 159, 1980.
16. Lankmayr, E.P., Budna, K.W., Müller, K., Nachtmann, F., and Rainer, F., Determination of d-penicillamine in serum by fluorescence derivatization and liquid column chromatography, *J. Chromatogr.*, 222, 249, 1981.
17. Mopper, K., Stahovec, W.L., and Johnson, L., Trace analysis of aldehydes by reversed-phase high-performance liquid chromatography and precolumn fluorigenic labeling with 5,5-dimethyl-1,3-cyclohexanedione, *J. Chromatogr.*, 256, 243, 1983.
18. Lawrence, J.F. and Frei, R.W., *Chemical Derivatization in Liquid Chromatography*, Elsevier, Amsterdam, 1976.
19. Carey, M.A. and Persinger, H.E., Liquid chromatographic determination of traces of aliphatic carbonyl compounds and glycols as derivatives that contain the dinitrophenyl group, *J. Chromatogr. Sci.*, 10, 537, 1972.

20. Fitzpatrick, F.A., Siggia, S., and Dingman Sr., J., High speed liquid chromatography of derivatized urinary 17-keto steroids, *Anal. Chem.*, 44(13), 2211, 1972.

21. Pietrzyk, D.J. and Chan, E.P., Determination of carbonyl compounds by 2-diphenylacetyl-1,3-indandione-1-hydrazone, *Anal. Chem.*, 42(1), 37, 1970.

22. Braun, R.A. and Mosher, W.A., 2-Diphenylacetyl-1,3-indandione 1-hydrazone- a new reagent for carbonyl compounds, *J. Am. Chem. Soc.*, 80, 3048, 1958.

23. Schäfer, M. and Mutschuler, E., Fluorimetric determination of oxprenolol in plasma by direct evaluation of thin-layer chromatograms, *J. Chromatogr.*, 164, 247, 1979.

24. Moye, H.A. and Boning Jr., A.J., A versatile fluorogenic labelling reagent for primary and secondary amines: 9-fluorenylmethyl chloroformate, *Anal. Lett.*, 12(B1), 25, 1979.

25. Lehninger, A.L., *Biochemistry*, 2nd ed., Worth Publishers, New York, 1978.

26. Roos, R.W., Determination of conjugated and esterified estrogens in pharmaceutical tablet dosage forms by high-pressure, normal-phase partition chromatography, *J. Chromatogr. Sci.*, 14, 505, 1976.

27. DeLeenheer, A., Sinsheimer, J.E., and Burckhalter, J.H., Fluorometric determination of primary and secondary aliphatic amines by reaction with 9-isothiocyanatoacridine, *J. Pharm. Sci.*, 62(8), 1370, 1973.

28. Clark, C.R. and Wells, M.M., Precolumn derivatization of amines for enhanced detectability in liquid chromatography, *J. Chromatogr. Sci.*, 16, 332, 1978.

29. Hulshoff, A., Roseboom, H., and Renema, J., Improved detectability of barbiturates in high-performance liquid chromatography by pre-column labelling and ultraviolet detection, *J. Chromatogr.*, 186, 535, 1979.

30. Matthees, D.P. and Purdy, W.C., Napthyldiazomethane as a derivatizing agent for the high-performance liquid chromatography detection of bile acids, *Anal. Chim. Acta.*, 109, 161, 1979.

31. Kuwata, K., Uebori, M., and Yamazaki, Y., Determination of phenol in polluted air as p-nitrobenzeneazophenol derivative by reversed phase high performance liquid chromatography, *Anal. Chem.*, 52, 857, 1980.

32. Nachtmann, F., Spitzy, H., and Frei, R.W., Rapid and sensitive high-resolution procedure for digitalis glycoside analysis by derivatization liquid chromatography, *J. Chromatogr.*, 122, 293, 1976.

33. Knapp, D.R. and Krueger, S., Use of o-p-nitrobenzyl-N,N-diisopropylisourea as a chromogenic reagent for liquid chromatographic analysis of carboxylic acids, *Anal. Lett.*, 8(9), 603–610, 1975.

34. Jupille, T., UV-Visible absorption derivatization in liquid chromatography, *J. Chromatogr. Sci.*, 17, 160–167, 1979.

35. Dunlap, K.L., Sandridge, R.L., and Keller, Jürgen, Determination of isocyanates in working atmospheres by high speed liquid chromatography, *Anal. Chem.*, 48(3), 497–499, 1976.

36. Politzer, I.R., Griffin, G.W., Dowty, B.J., and Laseter, J.L., Enhancement of ultraviolet detectability of fatty acids for purposes of liquid chromatographic-mass spectrometric analyses, *Anal. Lett.*, 6(6), 539–546, 1973.

37. Cox, G.B., Estimation of volatile N-nitrosamines by high-performance liquid chromatography, *J. Chromatogr.*, 83, 471–481, 1973.

38. Borch, R.F., Separation of long chain fatty acids as phenacyl esters by high pressure liquid chromatography, *Anal. Chem.*, 47(14), 2437–2439, 1975.

39. Poole, C.F., Singhawangcha, S., Zlatkis, A., and Morgan, E.D., Polynuclear aromatic boronic acids as selective fluorescent reagents for HPTLC and HPLC, HRC & CC, *J. High Resolut. Chromatogr., Chromatogr. Commun.*, 1, 96–97, 1978.

40. Björkqvist, B. and Toivonen, H., Separation and determination of aliphatic alcohols by high-performance liquid chromatography with U.V. Detection, *J. Chromatogr.*, 153, 265–270, 1978.

41. Munger, D., Sternson, L.A., Repta, A.J., and Higuchi, T., High-performance liquid chromatographic analysis of dianhydrogalactitol in plasma by derivatization with sodium diethyldithiocarbamate, *J. Chromatogr.*, 143, 375–382, 1977.

42. Sugiura, T., Hayashi, T., Kawai, S., and Ohno, T., High speed liquid chromatographic determination of putrescine, spermidine and spermine, *J. Chromatogr.*, 110, 385–388, 1975.

43. Suzuki, Y. and Tani, K., High-speed liquid-chromatography of the aliphatic alcohols as their trityl ether derivatives, *Buneseki Kagaku*, 28, 610, 1979.

Derivatizing Reagent	Structure/Formula	Notes
N-(9-acridinyl) malemide		Used for the precolumn preparation of fluorescent derivatives of thiols Reference: [1]
9-Anthryldiazo-methane		Used for the precolumn preparation of fluorescent derivatives of carboxylic acids; reagent reacts well with fatty acids at room temperature to give intensely fluorescent esters. Reference: [2]
Benzoyl chloride		Used to introduce chromophores into alcohols and amines using pyridine as a solvent; efficient means for the isolation of carbohydrates in complex mixtures. Reference: [3]
Benzyl bromide		Used to introduce chromophores into carboxylic acids. Reference: [4]
4-Bromomethyl-6,7-dimethoxycoumarin		Used for the precolumn preparation of fluorescent derivatives of carboxylic acids using acetone as solvent and with crown ether and alkali as catalysts. Reference: [5]
4-Bromomethyl-7-methoxycoumarin (Br-Mmc)		Used for the precolumn preparation of fluorescent derivatives of carboxylic acids, using a crown ether (18-crown-6) as a catalyst. Reference: [6]
p-Bromophenacyl bromide		Used to introduce chromophores into carboxylic acids; crown ethers are used as phase transfer agents (for example, 18-crown-6 and dicyclohexyl-18-crown-6). Reference: [4]
9-(Chloromethyl) anthracene (9-CIMA)		Used for the precolumn preparation of fluorescent derivatives of carboxylic acids, using cyclohexane as a solvent. Reference: [7]
4-Chloro-7-nitrobenzo-2-oxa-1,3-diazole (NBD-Cl)		Used for the precolumn preparation of fluorescent derivatives of primary and secondary amines, phenols, and thiols (4-chloro-7-nitrobenzofuran). Reference: [8]

(Continued)

Derivatizing Reagent	Structure/Formula	Notes
N-Chloromethyl-4-nitrophthalimide		Used to introduce chromophores into carboxylic acids. Reference: [9]
Dansyl chloride (DnS-Cl)		Used for the precolumn preparation of fluorescent derivatives of primary and secondary amines, phenols, amino acids, and imidazoles. Reference: [10]
Dansyl hydrazine (DnS-H)		Used for the precolumn preparation of fluorescent derivatives of aldehydes and ketones; optimal derivatization of glucose and other sugars occurs at pH 2–3. Reference: [11]
9,10-Diaminophen-anthrene		Used for the precolumn preparation of fluorescent derivatives of carboxylic acids. Reference: [12]
Diazo-4-aminobenzonitrile		Used to introduce chromophores in phenols. Reference: [13]
2,5-Di-n-butylamino-naphthalene-1-sulfonyl chloride (BnS-Cl)		Used for the precolumn preparation of fluorescent derivatives of primary and secondary amines, phenols, amino acids, and imidazoles. Reference: [10]
4-Dimethylamino-1-naphthoyl nitrile		Used for the precolumn preparation of fluorescent derivatives of primary and secondary (but not tertiary) alcohols. Reference: [14]

(Continued)

Derivatizing Reagent	Structure/Formula	Notes
p-Dimethylamino-phenyl isocyanate		Used to introduce chromophores into alcohols. After reaction, excess reagent must be removed as it interferes with ensuing analysis. Reference: [15]
5-Dimethylamino-naphthalene 1-sulfonylaziridine (dansylaziridine)		Used for the precolumn preparation of fluorescent derivatives of thiols; optimum derivatization conditions were found to be pH 8.2 with a minimum of a 2.7-fold molar reagent excess using a reaction time of 1 h at 60 °C. Under these conditions, only free sulfhydryl groups are derivatized. Reference: [16]
5,5-Dimethyl-1,3-cyclohexanedione (dimedone)		Used for the precolumn preparation of fluorescent derivatives of aldehydes using isopropanol as a solvent in the presence of ammonium acetate. Reference: [17]
3,5-Dinitrobenzyl chloride		Used to introduce chromophores into amines (forming phenyl substituted amines), alcohols, glycols, and phenols. References: [18,19]
2,4-Dinitro phenyl-hydrazine		Used to introduce chromophores into aldehydes and ketones in a solution of carbonyl-free methanol; detection of the more common 17-keto steroids as their 2,4-dinitrophenyl derivatives from urine and plasma; suggested potential for clinical use. Reference: [20]
2-Diphenylacetyl-1,3-indandione-1-hydrazone		Used for the precolumn preparation of fluorescent derivatives of aldehydes and ketones; reagent suggested to be especially useful because of its application on the micro level, for the analysis and identification of carbonyl compounds in smog, polluted air, biochemical, and pharmaceutical mixtures. Reagent does not appear to be useful for analysis of sugars. Derivatives are fluorescent in the UV as solids and in solution. References: [21,22]
1-Ethoxy-4-(dichloro-s-triazinyl) naphthalene or EDTN		Used for the precolumn preparation of fluorescent derivatives of primary and secondary alcohols and phenols. Reference: [23]

(Continued)

Derivatizing Reagent	Structure/Formula	Notes
9-Fluorenylmethyl chloroformate (FMOCCl)		Used for the precolumn preparation of fluorescent derivatives of primary and secondary amines in acetone solvent; in the presence of sodium borate, derivation proceeds rapidly under alkaline conditions. Reference: [24]
Fluorescamine or fluram		Used for the precolumn preparation of fluorescent derivatives of primary amines and amino acids by HPLC. Reference: [10]
1-Fluoro-2,4-dinitrobenzene		Used to introduce chromophores into amines using Sanger's procedure. Reference: [25]
p-Iodobenzensulfonyl chloride		Used to introduce chromophores into alcohols and phenols; aids in separation of estrogen derivatives. Reference: [26]
9-Isothiocyanato-acridine		Used for the precolumn preparation of fluorescent derivatives of some primary and secondary amines using toluene as a solvent; only amines with $PK_a \geq 9.33$ have been successfully determined. Reference: [27]
p-Methoxybenzoyl chloride		Used to introduce chromophores into amines using the reagent in tetrahydrofuran. Reference: [28]
2-Naphthacyl bromide (NPB)		Used to introduce chromophores into amines in acetone as a solvent, with cesium bromide as a catalyst; it is suggested that elevated temperatures (up to 80 °C) are necessary for the complete derivatization of compounds containing diisopropylamines. Reference: [29]

(Continued)

Derivatizing Reagent	Structure/Formula	Notes
1-Naphthyldiazo-methane		Used to introduce chromophores into carboxylic acids; reagent is prepared from 1-naphthaldehyde hydrazone by oxidation with Hg(II) oxide, with diethyl ether as a solvent; acetic acid will destroy excess reagent. Reference: [30]
p-nitrobenzenediazonium tetrafluoroborate		Used to introduce chromophores into phenols, suggested derivation takes place in aqueous medium at pH 11.5. Reference: [31]
p-Nitrobenzoyl chloride (4-NBCl)		Used to introduce chromophores into alcohols and amines, using pyridine as the solvent; with silica gel as the stationary phase, relatively low-viscosity, low-polarity solvents can be used for detection of digitalis glycosides by HPLC following derivatization with p-nitrobenzoyl chloride. Reference: [32]
p-Nitrobenzyl-N,N'-diisopropylisourea		Used to introduce chromophores into carboxylic acids, without the need for a base catalyst, and under mild conditions; picomolar concentrations are rendered detectable. Reference: [33]
p-Nitrobenzylhydroxylamine hydrochloride		Used to introduce chromophores into ketones and aldehydes. Reference: [34]
N-p-Nitrobenzyl-N-n-propylamine		Used to introduce chromophores into isocyanates; suggested for use in determining isocyanate levels in air down to 0.2 ppm in a 20 L air sample. Reference: [35]
1-(p-nitrobenzyl)-3-(p-tolyl) triazine		Used to introduce chromophores into carboxylic acids. Reference: [36]
p-Nitrophenyl chloroformate		Used to introduce chromophores into alcohols. Reference: [37]

(Continued)

Derivatizing Reagent	Structure/Formula	Notes
Phenacyl bromide	(phenyl)–C(=O)–CH$_2$–Br	Used to introduce chromophores into carboxylic acids; provides for the subsequent analysis of fatty acid mixtures on the microgram scale using HPLC. Reference: [38]
Phenanthreneboronic acid	B(OH)$_2$ (phenanthrene structure)	Used for the precolumn preparation of fluorescent derivatives of bifunctional compounds. Reference: [39]
Phenyl isocyanate	(phenyl)–N=C=O	Used to introduce chromophores into alcohols; thermal lability of the derivatives can cause problems; can also be used in the presence of water, but more reagent is required in this case. Reference: [40]
o-Phthaldialdehyde (OPT)	benzene with two ortho C(=O)–H groups	Used for the precolumn preparation of fluorescent derivatives of amines and amino acids, in the presence of mercaptoethanol (or ethanethiol) and borate buffer. Reference: [1]
Pyridoxal	pyridine ring with CHO, OH, HOCH$_2$, CH$_3$ substituents	Used for the precolumn preparation of fluorescent derivatives of amino acids. Reference: [18]
Pyruvoyl chloride (2,4-dinitrophenyl hydrazone)	CH$_3$–C(=N–NH–(2,4-dinitrophenyl))–C(=O)–Cl	Used to introduce chromophores into alcohols, amines, ketones, aldehydes, mercaptans, and phenols; aids in separation of estrogen derivatives. Reference: [26]
Sodium diethyldithiocarbamate (DDTC)	(CH$_3$CH$_2$)$_2$N–C(=S)–S$^-$Na$^+$	Used to introduce chromophores into epoxides in the presence of phosphate buffer; dithiocarbamates retain high nucleophilicity and are often water soluble. Reference: [41]
n-Succinimidyl-p-nitrophenylacetate	succinimidyl–N–C(=O)–CH$_2$–(phenyl)–NO$_2$	Used to introduce chromophores into amines; reacts under mild conditions without the need for catalysis. Reference: [34]

(Continued)

Derivatizing Reagent	Structure/Formula	Notes
p-Toluenesulfonyl chloride (TsCl)	CH_3 — benzene ring — SO_2Cl	Used to introduce chromophores into amines; aids in resolution of putrescine, spermidine, and spermine by HPLC; excess TsCl must be removed (by extraction with hexane, for example) before analysis. Reference: [41]
Trityl chloride	three benzene rings — C — Cl	Used to introduce chromophores into alcohols. Reference: [42,43]

CHAPTER 4

Thin-Layer Chromatography

STRENGTH OF COMMON TLC SOLVENTS

The following table contains the common solvents used in thin-layer chromatography (TLC), with a measure of their "strengths" on silica gel and alumina. The solvent strength parameter, $\varepsilon°$, is defined as a relative energy of adsorption per unit area of standard adsorbent [1–3]. It is defined as zero on alumina when pentane is used as the solvent. This series is what was called the eluotropic series in the older literature. For convenience, the solvent viscosity is also provided. Note that the viscosity is tabulated in cP for the convenience of most TLC users. This is equivalent to mPa·s in the SI convention. Additional data on these solvents may be found in the tables on high-performance liquid chromatography.

REFERENCES

1. Snyder, L.R., *Principles of Adsorption Chromatography*, Marcel Dekker, New York, 1968.
2. Willard, H.H., Merritt, L.L., Dean, J.A., and Settle, F.A., *Instrumental Methods of Analysis*, 7th ed., Van Nostrand, New York, Belmont, 1988.
3. Hamilton, R. and Hamilton, S., *Thin Layer Chromatography*, John Wiley & Sons (on behalf of "Analytical Chemistry by Open Learning" London), Chichester, 1987.

Solvent	$\varepsilon°$, (Al$_2$O$_3$)	Viscosity cP, 20°	$\varepsilon°$, (SiO$_2$)
Fluoroalkanes	−0.25	—	
n-Hexane	0.00	0.23	0.00
n-Pentane	0.001	0.23	0.00
2,2,4-Trimethylpentane (isooctane)	0.01	0.54	
n-Heptane	0.01	0.41	
n-Decane	0.04	0.92	
Cyclohexane	0.04	1.00	−0.05
Cyclopentane	0.05	0.47	
Carbon disulfide	0.15	0.37	0.14
Tetrachloromethane (carbon tetrachloride)	0.18	0.97	
1-Chloropentane (n-pentylchloride)	0.26	0.43	
Diisopropyl ether	0.28	0.37	
2-Chloropropane (isopropyl chloride)	0.29	0.33	
Methylbenzene (toluene)	0.29	0.59	
1-Chloropropane (n-propyl chloride)	0.30	0.35	
Chlorobenzene	0.30	0.80	
Benzene	0.32	0.65	0.25
Bromoethane (ethyl bromide)	0.37	0.41	
Diethyl ether (ether)	0.38	0.23	0.38
Trichloromethane (chloroform)	0.40	0.57	
Dichloromethane (methylene chloride)	0.42	0.44	
Tetrahydrofuran	0.45	0.55	
1,2-Dichloroethane	0.49	0.79	
Butanone (methyl ethyl ketone)	0.51	0.43	
Propanone (acetone)	0.56	0.32	0.47
1,4-Dioxane	0.56	1.54	0.49
Ethyl ethanoate (ethyl acetate)	0.58	0.45	0.38
Methyl ethanoate (methyl acetate)	0.60	0.37	
1-Pentanol (n-pentanol)	0.61	4.1	
Dimethyl sulfoxide (DMSO)	0.62	2.24	
Aminobenzene (aniline)	0.62	4.4	
Nitromethane	0.64	0.67	
Cyanomethane (acetonitrile)	0.65	0.37	0.50
Pyridine	0.71	0.94	
2-Propanol (isopropanol)	0.82	2.3	
Ethanol	0.88	1.20	
Methanol	0.95	0.60	
Ethylene glycol	1.11	19.9	
Ethanoic acid (acetic acid)	Large	1.26	
Water	Large	1.00	

MODIFICATION OF THE ACTIVITY OF ALUMINA BY ADDITION OF WATER

The following table describes five different activity grades of commercial alumina used in chromatography [1–4]. The activity grades are defined by the degree of adsorption of azobenzene (called azobenzene number) on the types of hydrated alumina. Those types are prepared by heating commercial alumina to redness, giving grade I, and then adding controlled amounts of water and allowing equilibration in a closed vessel. The azobenzene number decreases with the amount of water added. The R_f value is the ratio of distance traveled by the solute spot to that traveled by the solvent.

REFERENCES

1. Randerath, K., *Thin Layer Chromatography*, Verlag Chemie-Academic Press, Weinheim Bergstr. (Germany), New York, 1968.
2. Gordon, A.J. and Ford, R.A., *The Chemist's Companion: A Handbook of Practical Data, Techniques, and References*, John Wiley & Sons, New York, 1972.
3. Brockmann, H. and Schodder, H., Aluminum oxide with buffered adsorptive properties for purposes of chromatographic adsorption, *Ber.*, 74B, 73, 1941.
4. Dallas, M. S. J., Reproducible R_F values in thin-layer adsorption chromatography, *J. Chromatogr.*, 17, 267, 1965. doi:10.1016/S0021-9673(00)99868-6

Water Added (wt/wt %)	Activity Grade	Azobenzene Number [Maximum Adsorption of Azobenzene (10^{-5} mol/g)]	R_f (p-Aminoazobenzene)
0	I	26	0.00
3	II	21	0.13
6	III	18	0.25
10	IV	13	0.45
15	V	0	0.55

STATIONARY AND MOBILE PHASES

The following table provides a comprehensive guide to the selection of TLC media and solvents for a given chemical family. Mixed mobile phases are denoted with a slash, /, between components, and where available, the proportions are given. Among the references are several excellent texts and reference books [1,2,3,60,99–104], review articles [4–24], and original research papers and reports [25–59,61–98]. A table of abbreviations follows this section.

REFERENCES

1. Krebs, K.G., Heusser, D., and Wimmer, H., *Thin Layer Chromatography, A Laboratory Handbook*, Stahl, E., ed., Springer-Verlag, New York, 1969.
2. Bobbitt, J.B., *Thin Layer Chromatography*, Reinhold, New York, 1963.
3. Touchstone, J.C., *Techniques and Application of Thin Layer Chromatography*, John Wiley & Sons, New York, 1985, 1972.
4. Pataki, G., Paper, thin-layer, and electrochromatography of amino acids in biological material, *Z. Klin. Chem.*, 2, 129, 1964; *Chem. Abs.*, 64, 5425c, 1966.
5. Padley, F.B., Thin-layer chromatography of lipids, Thin-layer Chromatography, *Proceedings Symposium*, Rome, 1963, 87 (Pub. 1964).
6. Honjo, M., Thin-layer chromatography of nucleic acid derivatives, *Kagaku No Ryoiki, Zokan*, 64, 1, 1964.
7. Kazumo, T., Thin-layer chromatography of bile acids, *Kagaku No Ryoiki, Zokan*, 64, 19, 1964.
8. Nakazawa, Y., Thin-layer chromatography of compound lipids, *Kagaku No Ryoiki, Zokan*, 64, 31, 1964.
9. Nishikaze, O., Separation and quantitative analysis of adrenocortical hormone and its metabolite (C_{21}) by thin-layer chromatography, *Kagaku No Ryoiki, Zokan*, 64, 37, 1964.
10. Shikita, M., Kakizazi, H., and Tamaoki, B., Thin-layer chromatography of radioactive substances, *Kagaku No Ryoiki, Zokan*, 64, 45, 1964.
11. Mo, I. and Hashimoto, Y., Method of thin-layer zone electrophoresis, *Kagaku No Ryoiki, Zokan*, 64, 61, 1964.
12. Kinoshita, S., Thin-layer chromatography of sugar esters, *Kagaku No Ryoiki, Zokan*, 64, 79, 1964.
13. Okada, M., Thin-layer chromatography of cardiotonic glycosides, *Kagaku No Ryoiki, Zokan*, 64, 103, 1964.
14. Omoto, T., Thin-layer chromatography of toad toxin, *Kagaku No Ryoiki, Zokan*, 64, 115, 1964.
15. Furuya, C. and Itokawa, H., Thin-layer chromatography of triterpenoids, *Kagaku No Ryoiki, Zokan*, 64, 123, 1964.
16. Zenda, H., Thin-layer chromatography of aconitine-type alkaloids, *Kagaku No Ryoiki, Zokan*, 64, 133, 1964.
17. Hara, S. and Tanaka, H., Thin-layer chromatography of mixed pharmaceutical preparations, *Kagaku No Ryoiki, Zokan*, 64, 141, 1964.
18. Katsui, G., Thin-layer chromatography of vitamins, *Kagaku No Ryoiki, Zokan*, 64, 157, 1964.
19. Fujii, S. and Kamikura, M., Thin-layer chromatography of pigments, *Kagaku No Ryoiki, Zokan*, 64, 173, 1964.
20. Hosogai, Y., Thin-layer chromatography of organic chlorine compounds, *Kagaku No Ryoiki, Zokan*, 64, 185, 1964.
21. Takeuchi, T., Thin-layer chromatography of metal complex salts, *Kagaku No Ryoiki, Zokan*, 64, 197, 1964.
22. Yamakawa, H. and Tanigawa, K., Thin-layer chromatography of organic metal compounds, *Kagaku No Ryoiki, Zokan*, 64, 209, 1964.
23. Takitani, S. and Kawanabe, K., Thin-layer chromatography of inorganic ions (anions), *Kagaku No Ryoiki, Zokan*, 64, 221, 1964.
24. Ibayashi, H., Thin-layer chromatography of steroid hormones and its clinical application, *Kagaku No Ryoiki, Zokan*, 64, 227, 1964.

25. Chilingarov, A.O. and Sobchinskaya, N.M., Quantitative ultramicroanalysis of monoamine dansyl derivatives in biological material, *Lab. Delo*, 1980, 333; *Chem. Abs.*, 93, 109910t, 1980.

26. Heacock, R.A., Nerenberg, C., and Payza, A.N., The chemistry of the "aminochromes" - Part I. The preparation and paper chromatography of pure adrenochrome, *Can. J. Chem.*, 36, 853, 1958.

27. Heacock, R.A. and Powell, W.S., Adrenochrome and related compounds, *Progr. Med. Chem.*, 9, 275, 1972.

28. Heacock, R.A. and Scott, B.D., The chemistry of the "aminochromes" - Part IV. Some new amino-chromes and their derivatives, *Can. J. Chem.*, 38, 516, 1960.

29. Heacock, R.A., The chemistry of adrenochrome and related compounds, *Chem. Rev.*, 59, 181, 1959.

30. Suryaraman, M.G. and Cave, W.T., Detection of some aliphatic saturated long chain hydrocarbon deriv-atives by thin-layer chromatography, *Anal. Chim. Acta*, 30, 96, 1964; *Chem. Abs.*, 60, 7463e, 1964.

31. Knappe, E., Peteri, D., and Rohdewald, I., Thin-layer chromatographic identification of technically important polyhydric alcohols, *Z. Anal. Chem.*, 199, 270, 1964; *Chem. Abs.*, 60, 7464f, 1964.

32. Horak, V. and Klein, R.F.X., Microscale group test for carbonyl compounds, *J. Chem. Ed.*, 62, 806, 1985.

33. Jaminet, F., Paper microchromatography in phytochemical analysis. Application to Congolian Strychnos, *J. Pharm. Belg.*, 8, 339 and 449, 1953; *Chem. Abs.*, 48, 8482c, 1954.

34. Neu, R., A new color method for determining alkaloids and organic bases with sodium tetraphenyl-borate, *J. Chromgr.*, 11, 364, 1963; *Chem. Abs.*, 59, 12181d, 1963.

35. Marini-Bettolo, B.G. and Caggiano, E., Paper chromatography and electrophoresis of tertiary bases, *Liblice, Czech.*, 91, 1961; Chem Abs., 60, 838d, 1964.

36. Knappe, E. and Rohdewald, I., Impregnation of chromatographic thin layers with polyesters. III. Thin-layer chromatographic identification of acetoacetic acid amides, *Z. Anal. Chem.*, 208, 195, 1965; *Chem. Abs.*, 62, 12424f, 1965.

37. Lane, E.S., Thin-layer chromatography of long-chain tertiary amines and related compounds, *J. Chromgr.*, 18, 426, 1965; *Chem. Abs.*, 63, 7630f, 1965.

38. Ashworth, M.R.F. and Bohnstedt, G., Reagent for the detection and determination of N-active hydrogen, *Talanta*, 13, 1631, 1966; *Chem. Abs.*, 66, 25889x, 1967.

39. Heacock, R.A., "The aminochromes," In *Advances in Heterocyclic Chemistry*, Katritsky, A.R., ed., Academic Press, New York, 1965, 205; *Chem. Abs.*, 65, 5432d, 1966.

40. Knappe, E., Peteri, D., and Rohdewald, I., Impregnation of chromatographic thin layers with poly-esters for the separation and identification of substituted 2-hydroxybenzophenones and other ultra-violet absorbers, *Z. Anal. Chem.*, 197, 364, 1963; *Chem. Abs.*, 60, 762g, 1964.

41. Hara, S. and Takeuchi, M., Systematic analysis of bile acids and their derivatives by thin layer chro-matography, *J. Chromgr.*, 11, 565, 1963; *Chem. Abs.*, 60, 838f, 1964.

42. Hauck, A., Detection of caffeine by paper chromatography, *Deut. Z. Gerichtl. Med.*, 54, 98, 1963; *Chem. Abs.*, 60, 838b, 1964.

43. Knappe, E. and Rohdewald, I., Thin-layer chromatography of dicarboxylic acids. IV. Combination of thin-layer chromatographic systems for the identification of individual components in dicarbox-ylic acid mixtures, *Z. Anal. Chem.*, 210, 183, 1965; *Chem. Abs.*, 63, 3600f, 1965.

44. Passera, C., Pedrotti, A., and Ferrari, G., Thin-layer chromatography of carboxylic acids and keto-acids of biological interest, *J. Chromgr.*, 14, 289, 1964; *Chem. Abs.*, 60, 16191f, 1964.

45. Knappe, E. and Peteri, D., Thin-layer chromatography of dicarboxylic acids. I. Separations in the homol-ogous series oxalic to sebacic acids, *Z. Anal. Chem.*, 188, 184, 1962; *Chem. Abs.*, 57, 11836a, 1962.

46. Peteri, D., Thin-layer chromatography of dicarboxylic acids. II. Separation of carbocyclic dicarbox-ylic acids, *Z. Anal. Chem.*, 158, 352, 1962; *Chem. Abs.*, 57, 11836b, 1962.

47. Dutta, S.P. and Baruta, A.K., Separation of cis- and trans-isomers of α,β-unsaturated acids by thin-layer chromatography, *J. Chromgr.*, 29, 263, 1967; *Chem. Abs.*, 67, 96616n, 1967.

48. Dalmaz, Y. and Peyrin, L., Rapid procedure for chromatographic isolation of DOPA, DOPAC, epinephrine, norepinephrine and dopamine from a single urinary sample at endogenous levels, *J. Chromgr.*, 145, 11, 1978; *Chem. Abs.*, 88, 59809c, 1978.

49. Baumgartner, H., Ridl, W., Klein, G., and Preindl, S., Improved radioenzymic assay for the deter-mination of catecholamines in plasma, *Clin. Chim. Acta.*, 132, 111, 1983; *Chem. Abs.*, 99, 99459k, 1983.

50. Hansson, C., Agrup G., Rorsman, H., Rosengren, A.M., and Rosengren, E., Chromatographic separation of catecholic amino acids and catecholamines on immobilized phenylboronic acid, *J. Chromgr.*, 161, 352, 1978; *Chem. Abs.*, 90, 50771d, 1979.

51. Endo, Y. and Ogura, Y., Separation of catecholamines on the phosphocellulose column, *Jap. J. Pharmacol.*, 23, 491, 1973; *Chem. Abs.*, 80, 12002s, 1974.

52. Wada, H., Yamatodani, A., and Seki, T., Systematic determination of amino acids, amines and some nucleotides using dansylchloride, *Kagaku No Ryoiki, Zokan*, 114, 1, 1976; *Chem. Abs.*, 87, 1904f, 1977.

53. Head, R.J., Irvine, R.J., and Kennedy, J.A., The use of sodium borate impregnated silica gel plates for the separation of 3-0-methyl catecholamines from their corresponding catecholamines, *J. Chromgr. Sci.*, 14, 578, 1976; *Chem. Abs.*, 86, 39601x, 1977.

54. Adamec, O., Matis, J., and Galvanek, M., Fractionation and quantitative determination of urinary 17-hydroxycorticosteroids by thin layer chromatography on silica gel, *Steroids*, 1, 495, 1963.

55. Adamec, O., Matis, J., and Galvanek, M., Chromatographic separation of corticoids on a thin-layer of silica gel, *Lancet*, 1962-I, 81; *Chem. Abs.*, 56, 9034d, 1962.

56. Knappe, E. and Rohdewald, I., Thin-layer chromatography of dicarboxylic acids. V. Separation and identification of hydroxy dicarboxylic acids, of di- and tricarboxylic acids of the citrate cycle, and some other dicarboxylic acids of plant origin, *Z. Anal. Chem.*, 211, 49, 1965; *Chem. Abs.*, 63, 7333c, 1965.

57. Snegotskii, V.I. and Snegotskaya, V.A., Thin-layer chromatography of sulfur compounds, *Zavod. Lab.*, 35, 429, 1969; *Chem. Abs.*, 71, 23436b, 1969.

58. Borecky, J., Gasparic, J., and Vecera, M., Identification of organic compounds. XXV. Identification and separation of aliphatic C_1-C_{18} alcohols by paper chromatography, *Chem. Listy*, 52, 1283, 1958; *Chem. Abs.*, 53, 8039h, 1958.

59. Hörhammer, L., Wagner, H., and Hein, H., Thin-layer chromatography of flavonoids on silica gel, *J. Chromgr.*, 13, 235, 1964; *Chem. Abs.*, 60, 13856c, 1964.

60. Mikes, O., ed., *Laboratory Handbook of Chromatographic Methods*, D. Van Nostrand Co., Ltd., London, England, 1966.

61. Wright, J., Detection of humectants in tobacco by thin layer chromatography, *Chem. & Ind.* (London), 1963, 1125.

62. Korte, F. and Vogel, J., Thin-layer chromatography of lactones, lactams and thiolactones, *J. Chromgr.*, 9, 381, 1962; *Chem. Abs.*, 58, 9609c, 1963.

63. Heacock, R.A. and Mahon, M.E., Paper chromatography of some indole - derivatives on acetylated paper, *J. Chromgr.*, 6, 91, 1961.

64. Hackman, R.H. and Goldberg, M., Microchemical detection of melanins, *Anal. Biochem.*, 41, 279, 1971; *Chem. Abs.*, 74, 136114a, 1971.

65. Preussmann, R., Neurath, G., Wulf-Lorentzen, G., Daiber, D., and Hengy, H., Color formation and thin-layer chromatography for N-nitrosocompounds, *Z. Anal. Chem.*, 202, 187, 1964.

66. Preussmann, R., Daiber, D., and Hengy, H., Sensitive color reaction for nitrosamines on thin-layer chromatography, *Nature*, 201, 502, 1964; *Chem. Abs.*, 60, 12663e, 1964.

67. Hranisavljevic-Jakovljevic, M., Pejkovic-Tadic, I., and Stojiljkovic, A., Thin-layer chromatography of isomeric oximes, *J. Chromgr.*, 12, 70, 1963; Chem Abs., 60, 7d, 1964.

68. Abraham, M.H., Davies, A.G., Llewellyn, D.R., and Thain, E.M., The chromatographic analysis of organic peroxides, *Anal. Chim. Acta*, 17, 499, 1957; *Chem. Abs.*, 53, 120b, 1959.

69. Seeboth, H., Thin-layer chromatography analysis of phenols, *Monatsber. Deut. Akad. Wiss. Berlin*, 5, 693, 1963; *Chem. Abs.*, 61, 2489c, 1964.

70. Knappe, E. and Rohdewald, I., Thin-layer chromatographic identification of simple phenols using the coupling products with Fast Red Salt AL, *Z. Anal. Chem.*, 200, 9, 1964; *Chem. Abs.*, 60, 9913g, 1964.

71. Donner, R. and Lohs, K., Cobalt chloride in the detection of organic phosphate ester by paper and especially thin-layer chromatography, *J. Chromgr.*, 17, 349, 1965; *Chem. Abs.*, 62, 13842d, 1965.

72. Engel, J.F. and Barney, J.E., Chromatographic separation of hydrogenation products of dibenz[a,h] anthracene, *J. Chromgr.*, 29, 232, 1967; *Chem. Abs.*, 57, 96617p, 1967.

73. Kucharczyk, N., Fohl, J., and Vymetal, J., Thin-layer chromatography of aromatic hydrocarbons and some heterocyclic compounds, *J. Chromgr.*, 11, 55, 1963; *Chem. Abs.*, 59, 9295g, 1963.

74. Perifoy, P.V., Slaymaker, S.C., and Nager, M., Tetracyanoethylene as a color-developing reagent for aromatic hydrocarbons, *Anal. Chem.*, 31, 1740, 1959; *Chem. Abs.*, 54, 5343e, 1960.

75. Kodicek, E. and Reddi, K.K., Chromatography of nicotinic acid derivatives, *Nature*, 168, 475, 1951; *Chem. Abs.*, 46, 3601g, 1952.

76. Heacock, R.A. and Mahon, M.E., The color reactions of the hydroxyskatoles, *J. Chromgr.*, 17, 338, 1965; *Chem. Abs.*, 62, 13824g, 1965.

77. Martin, H.P., Reversed phase paper chromatography and detection of steroids of the cholesterol class, *Biochim. et Biophys. Acta*, 25, 408, 1957.

78. Lisboa, B.P., Application of thin-layer chromatography to the steroids of the androstane series, *J. Chromgr.*, 13, 391, 1964; *Chem. Abs.*, 60, 13890b, 1964.

79. Lisboa, B.P., Separation and characterization of Δ^5-3-hydroxy-C_{19}-steroids by thin-layer chromatography, *J. Chromgr.*, 19, 333, 1965; *Chem. Abs.*, 63, 16403h, 1965.

80. Lisboa, B.P., Thin-layer chromatography of Δ^4-3-oxosteroids of the androstane series, *J. Chromgr.*, 19, 81, 1965; *Chem. Abs.*, 63, 13619e, 1965.

81. Lisboa, B.P., Thin-layer chromatography of steroids, *J. Pharm. Belg.*, 20, 435, 1965; *Chem. Abs.*, 65, 570c, 1966.

82. Partridge, S.M., Aniline hydrogen phthalate as a spraying reagent for chromatography of sugars, *Nature*, 164, 443, 1949.

83. Grossert, J.S. and Langler, R.F., A new spray reagent for organosulfur compounds, *J. Chromgr.*, 97, 83, 1974: *Chem. Abs.*, 82, 25473n, 1976.

84. Petranek, J. and Vecera, M., Identification of organic compounds. XXIV. Separation and identification of sulfides by paper chromatography, Chem. Listy, 52, 1279, 1958; *Chem. Abs.*, 53, 8039d, 1958.

85. Bican-Fister, T. and Kajganovic, V., Quantitative analysis of sulfonamide mixtures by thin-layer chromatography, *J. Chromgr.*, 16, 503, 1964; *Chem. Abs.*, 62, 8943d, 1965.

86. Bican-Fister, T. and Kajganovic, V., Separation and identification of sulfonamides by thin-layer chromatography, *J. Chromgr.*, 11, 492, 1963; *Chem. Abs.*, 60, 372f, 1964.

87. Reisch, J., Bornfleth, H., and Rheinbay, J., Thin-layer chromatography of some useful sulfonamides, *Pharm. Ztg., Ver. Apotheker-Ztg.*, 107, 920, 1962; *Chem. Abs.*, 60, 372e, 1964.

88. Prinzler, H.W., Tauchmann, H., and Tzcharnke, C., Thin-layer chromatographic separation of organic sulfoxides and dinitrothioethers. Some observations on reproducibility and structural influence. II. Separation of sulfoxide mixtures by one and two-dimensional thin layer chromatography, *J. Chromgr.*, 29, 151, 1967; *Chem Abs.*, 67, 96615m, 1967.

89. Wolski, T., Color reactions for the detection of sulfoxides, *Chem. Anal.* (Warsaw), 14, 1319, 1969; *Chem. Abs.*, 72, 106867q, 1970.

90. Bergstrom, G. and Lagercrantz, C., Diphenylpicrylhydrazyl as a reagent for terpenes and other substances in thin-layer chromatography, *Acta Chem. Scand.*, 18, 560, 1964; *Chem. Abs.*, 61, 2491h, 1964.

91. Dietz, W. and Soehring, K., Identification of thiobarbituric acids in urine by paper chromatography, *Arch. Pharm.*, 290, 80, 1957; *Chem. Abs.*, 52, 4736d, 1958.

92. Curtis, R.F. and Philips, C.T., Thin-layer chromatography of thiophene derivatives, *J. Chromgr.*, 9, 366, 1962; *Chem. Abs.*, 58, 10705c, 1963.

93. Salame, M., Detection and separation of the most important organophosphorus pesticides by thin-layer chromatography, *J. Chromgr.*, 16, 476, 1964; *Chem. Abs.*, 62, 11090b, 1965.

94. Knappe, E. and Rohdewald, I., Thin-layer chromatography of substituted ureas and simple urethanes, *Z. Anal. Chem.*, 217, 110, 1966; *Chem. Abs.*, 64, 16601g, 1966.

95. Fishbein, L. and Fawkes, J., Detection and thin-layer chromatography of sulfur compounds. I. Sulfoxides, sulfones and sulfides, *J. Chromgr.*, 22, 323, 1966; *Chem. Abs.*, 65, 6281e, 1966.

96. Prinzler, H.W., Pape, D., Tauchmann, H., Teppke, M., and Tzcharnke, C., Thin-layer chromatography of organic sulfur compounds, *Ropa Uhlie*, 8, 13, 1966; *Chem. Abs.*, 65, 9710h, 1966.

97. Karaulova, E.N., Bobruiskaya, T.S., and Gal'pern, G.D., Thin-layer chromatography of sulfoxides, *Zh. Analit. Khim.*, 21, 893, 1966; *Chem. Abs.*, 65, 16046f, 1966.

98. Knappe, E. and Yekundi, K.G., Impregnation of chromatographic thin layers with polyesters. II. Separation and identification of lower and middle fatty acids via the hydroxamic acid, *Z. Anal. Chem.*, 203, 87, 1964; *Chem. Abs.*, 61, 5915e, 1964.

99. Spangelberg, B., Poole, C. F., and Weins, C., *Quantitative Thin Layer Chromatography: A Practical Survey*, Springer-Verlag, Berlin, Germany, 2011. ISBN (eBook): 978-3-642- 10729-0; ISBN (Hardcover): 978-3-642-10727-6; ISBN (Softcover): 978-3-642-42347-5.

100. Fried, B. and Sherma, J. *Practical Thin Layer Chromatography: A Multidisciplinary Approach*, Milton, Park, UK, 1996. ISBN-10: 0849326605; ISBN-13: 978-0849326608.

101. Sherma, J. and Fried, B., *Handbook of Thin-Layer Chromatography*, 3rd ed., CRC Press, Boca Raton, FL, 2003. ISBN: 9780824708955.

102. Komsta, L., Waksmundzka-Hajnos, M., and Sherma, J., *Thin Layer Chromatography in Drug Analysis*, CRC Press, Boca Raton, FL, 2013. ISBN: 9781466507159.

103. Kowalska, T. and Sherma J., *Thin Layer Chromatography in Chiral Separation and Analysis*, CRC Press, Boca Raton, FL, 2019. ISBN: 9780367453015.

104. Hahn-Deinstrop, E. *Applied Thin Layer Chromatography: Best Practice and Avoidance of Mistakes*, John & Wiley Sons, New York, 2006. ISBN: 9783527315536; doi:10.1002/ 9783527610259.

Abbreviation	Solvent Name	Abbreviation	Solvent Name
	Abbreviations/Solvent Table		
Ac	Acetone	Et_2O	Diethylether
Ace	Acetate	Foram	Amylformate
AcOH	Acetic acid	HCl	Hydrochloric acid
n-AmOH	n-Amyl alcohol	H_3BO_3	Boric acid
t-AmOH	t-Amyl alcohol	Hex	Hexane
$AmSO_4$	Ammonium sulfate	HForm	Formic acid
i-BuAc	Isobutylacetate	MeCl	Methylene chloride
BuFor	n-Butylformate	MeCN	Acetonitrile
i-BuOH	Isobutanol	MEK	Methylethylketone
n-BuOH	n-Butanol	MeOH	Methanol
$i-Bu_2O$	Diisobutylether	NaAc	Sodium acetate
CCl_4	Carbon tetrachloride	NH_3	Ammonia, aqueous
C_2HCl_3	Trichloroethene	Petet	Petroleum ether
$CHCl_3$	Chloroform	Ph	Phosphate
$(CH_2)_6$	Cyclohexane	PhOH	Phenol
C_6H_6	Benzene	PrAc	Propylacetate
$n-C_6H_{14}$	n-Hexane	PrFor	Propylformate
$n-C_7H_{16}$	n-Heptane	Progl	Propylene glycol
$i-C_8H_{18}$	Isooctane	i-PrOH	Isopropanol
$(ClCH_2)_2$	Dichloroethane	n-PrOH	n-Propanol
DEAE	Diethyl aminoethyl	$i-Pr_2NH$	Diisopropylamine
Diox	Dioxane	$i-Pr_2O$	Diisopropylether
DMF	Dimethylformamide	Py	Pyridine
EtFor	Ethylformate	THF	Tetrahydrofuran
EtOAc	Ethylacetate	Tol	Toluene
EtOH	Ethanol	w	Water
Et_2NH	Diethylamine	m-X	m-Xylene

Family	Stationary Phase	Mobile Phase	Ref.
Adrenaline and derivatives	Alumina (two dimensional)	C_6H_6/EtOAc (60:40) $CHCl_3$/EtOH/Tol (90:6.5:3.1)	[25]
Adrenochromes	Cellulose Whatman #1 (descending)	AcOH (2 %)/w AcOH (2 %)/w	[26,27] [26–29]
Alcohols	Silica gel (G-coated)	EtOAc/Hex	[30]
Alcohols, polyhydric	Alumina or Kieselguhr (impregnated with polyamide) or silica gel	$CHCl_3$/Tol/HForm or n-BuOH/NH_3 or $CHCl_3$	[31]
Aldehydes	Silica gel (G-coated)	EtOAc/Hex	[30]
Aldehydes, 2,4-dinitro-phenylhydrazones	Alumina Alumina IB Silica gel Silica gel IB	C_6H_6 or $CHCl_3$ or Et_2O or C_6H_6/Hex MeCl or Tol/THF (4:1) Hex/EtOAc (4:1 or 3:2) MeCl or Tol/THF (4:1)	[3] [32] [3] [32]
Alkaloids	Alumina Alumina Cellulose (impregnated with formamide) Paper (S&S #2043b) Paper electrophoresis Silica gel	i-BuOH/AcOH or i-BuOH/NH_3 or i-PrOH/AcOH $CHCl_3$ or EtOH or $(CH_2)_6$/$CHCl_3$ (3:7) C_6H_6/n-C_7H_{16}/$CHCl_3$/Et_2NH (6:5:1:0.02) n-BuOH/HCl(25 %)/w (100:26:39) i-BuOH/AcOH or i-BuOH/NH_3 or i-PrOH/HOAc C_6H_6/EtOH (9:1) or $CHCl_3$/Ac/ Et_2NH (5:4:1)	[33] [3] [3] [34] [35] [3]
Amides	Kieselguhr (adipic acid impregnated) Silica gel	i-Pr_2O/Petet/CCl_4/HForm/w i-Pr_2O/Petet/CCl_4/HForm/w	[36] [36]
Amines	Alumina Alumina G Keiselguhr G Silica gel Silica gel (aromatic only)	Ac/n-C_7H_{16} (1:1) i-BuAc or i-BuAc/AcOH Ac/w (99:1) EtOH (95 %)/NH_3 (25 %) (4:1)	[3] [37] [3] [3] [38]
Amino acids	Alumina Cellulose Cellulose (two dimensional) Silica gel	n-BuOH/AcOH/w (3:1:1) or Py/w n-BuOH/AcOH/w (4:1:1) n-BuOH/Ac/NH_3/w (10:10:5:2) followed by i-PrOH/HForm/w (20:1:5) n-BuOH/AcOH/w (3 or 4:1:1) or PhOH/w (3:1) or n-PrOH/NH_3 (34 %) (2:1)	[3] [3] [3] [3]
Aminochromes	Whatman #1 (acid washed)	w or AcOH/w or MeOH/w or EtOH/w or n-BuOH/AcOH/w or i-PrOH/w	[26,39]
Barbiturates	Silica gel	$CHCl_3$/n-BuOH/NH_3 (25 %) (14:8:1)	[3]
Benzophenones, hydroxy	Alumina or cellulose or Kieselguhr (impregnated with adipic acid triethylene glycol polyester) or silica gel	HForm/m-X	[40]
Bile acids	Silica gel	C_6H_6/Et_2O (4:1) or Et_2O/AcOH (99.6:0.4) or $CHCl_3$/MeOH (9:1)	[41]
Caffeine	Chromatography paper	n-BuOH/NH_3 or n-BuOH/HForm	[42]

(Continued)

Family	Stationary Phase	Mobile Phase	Ref.
Carboxylic acids	Kieselguhr/polyethylene glycol	i-Bu$_2$O/HForm/w (90:7:3)	[43]
		i-Pr$_2$O/Petet/CCl$_4$/HForm/w	[43]
	Polyamide powder	(50:20:20:8:1) or MeCN/EtOAc/	[30]
	Silica gel (G-coated)	HForm or BuForm/EtOAc/HForm	[44]
	Silica gel (CaSO$_4$ impregnated)	EtOH/NH$_3$/THF	[43,45,46]
		n-PrOH/NH$_3$ or EtOH/CHCl$_3$/NH$_3$	
	Silica gel/polyethylene glycol M-1000	i-Pr$_2$O/HForm/w (90:7:3)	
Carboxylic acids, unsaturated	Silica gel	CHCl$_3$/MeOH	[47]
Catecholamines	Alumina	HCl (0.025 N)	[48]
	Boric acid gel (neutral pH)	Dilute acids	[48]
	Kieselguhr	Ph buffer (pH=6.2)/EDTA	[49]
	Phenylboronate		[50]
	Phosphocellulose		[51]
Catecholamines, dansyl derivatives	Alumina (two dimensional)	C$_6$H$_6$/EtOAc (60:40) or CHCl$_3$/	[25]
	Amberlite IRC50	EtOH/Tol (90:6.5:3.5)	[52]
Catecholamines, o-methyl derivatives	Silica gel (sodium borate impregnated)		[53]
Corticosteroids	Silica gel	EtOH(5 %)/MeCl or EtOH/CHCl$_3$	[54,55]
Coumarins	Polyamide	MeOH/w (4:1 or 3:2)	[3]
	Silica gel G	Petet/EtOAc (2:1)	[3]
	Silica gel G (impregnated with NaAc)	Tol/EtFor/HForm (5:4:1)	[3]
	Silicic acid (starch bound)	EtOAc/hexanes	[3]
Dicarboxylic acids	Kieselguhr/polyethylene glycol	i-Pr$_2$O/HForm/w (90:7:3)	[43]
		i-Pr$_2$O/Petet/CCl$_4$/HForm/w	[43]
	Polyamide powder	(50:20:20:8:1) or MeCN/EtOAc/	[56]
	Polyamide	HForm (9:1:1) or BuFor/EtOAc/	[43]
	Woelm DC powder	HForm (9:1:1)	[30]
	Silica gel	MeCN/PrFor/PrAc/HForm	
	Silica gel (G-coated)	(45:45:10:10) or i-Pr$_2$O/Petet/	
		CCl$_4$/HForm/w (50:20:20:8:1) or	
		n-AmOH/CCl$_4$/HFor (3:2:1)	
		i-Pr$_2$O/HForm/w (90:7:3)	
		EtOH/NH$_3$/THF	
Diols (see alcohols, polyhydric)			
Disulfides	Alumina	Hex	[57]
Disulfides, 3,5-dinitro-benzoates	Whatman #3 (impregnated with 10 % paraffin oil in cyclohexane, (CH$_2$)$_6$	DMF/MeOH/w or Foram/MeOH/w	[58]
Flavinoids	Paper	n-BuOH/AcOH/w or EtOAc/w or	[60]
	Polyamide	AcOH/w or C$_6$H$_6$/AcOH/w	[3]
	Silica gel	MeOH/H$_2$O	[59]
	Silica gel (impregnated with NaAc)	C$_6$H$_6$/Py/AcOH (36:9:5) Petet/	[3]
		EtOAc (2:1)	[3]
	Silicic acid (starch bound)	Tol/EtForm/HFor (5:4:1)	
		EtOAc/Skellysolve B	
Glycerides	Silica gel G	CHCl$_3$/C$_6$H$_6$ (7:3)	[3]
	Silica gel G (impregnated with silver nitrate)	CHCl$_3$/AcOH (99.5:0.5)	[3]
Glycolipids	Silica gel G	n-PrOH/NH$_3$ (12 %) (4:1)	[3]

(Continued)

Family	Stationary Phase	Mobile Phase	Ref.
Glycols, polyethylene	Paper Silica gel	n-PrOH/EtOAc/w (7:1:2) or n-BuOH/AcOH/w (4:1:5) or t-AmOH/n-PrOH/w (8:2:3) or EtOAc/AcOH/w (9:2:2) Ac or n-BuOH/AcOH/w	[60] [61]
Hydroxamates	Silica gel	i-Pr$_2$O or i-Pr$_2$O/EtOAc (1:4) or i-Pr$_2$O/i-C$_8$H$_{18}$	[62]
Hydroxamic acids	Kieselguhr G (impregnated with diethy-lene glycol or triethylene glycol adipate polyesters)	i-Pr$_2$O/Petet/CCl$_4$/HForm/w (50:20:20:8:1)	[98]
Indoles	Acetylated (ascending) Cellulose (thin layer)	CHCl$_3$/MeOH/w (10:10:6) w or HCl (0.005 N) or n-BuOH/ AcOH/w (12:3:5) or C$_6$H$_6$/AcOH/w (125:72:3)	[63] [64]
α-Ketoacids	Silica gel (CaSO$_4$ impregnated)	EtOH/CHCl$_3$/NH$_3$	[44]
Ketones, 2,4-dinitrophenyl hydrazones	Alumina IB Silica Gel IB	MeCl or Tol/THF (4:1) MeCl or Tol/THF (4:1)	[32] [32]
Lactams	Silica gel	i-Pr$_2$O or i-Pr$_2$O/EtOAc (1:4) or i-Bu$_2$O/i-C$_8$H$_{18}$	[62]
Lactones	Silica gel	i-Pr$_2$O or i-Pr$_2$O/EtOAc (1:4) or i-Bu$_2$O/i-C$_8$C$_{18}$	[62]
Lipids	Alumina Silica gel G Silicic acid	Petet/Et$_2$O (95:5) Petet/Et$_2$O/AcOH (90:10:1) CHCl$_3$/MeOH/w (80:25:3)	[3] [3] [3]
Mercaptans (see thiols)			
Nitrosamines	Silica gel Silica gel	Hex/Et$_2$O/MeCl MeCl/Hex/Et$_2$O (2:3:4) (aliphatic, aromatic); MeCl/Hex/Et$_2$O (5:7:10) (cyclic)	[65] [66]
Nucleotides	Cellulose Cellulose (on DEAE)	AmSO$_4$ (sat'd)/NaAc(1 M)/i-PrOH (80:18:2) HCl (aq)	[3] [3]
Oximes	Silica gel G	C$_6$H$_6$/EtOAc or C$_6$H$_6$/MeOH (abs)	[67]
Peroxides	Silicone filter paper	w/EtOH/CHCl$_3$	[68]
Phenols	Alumina Alumina/AcOH Silica gel A Silica gel G Silica gel/oxalic acid Silica gel/potassium carbonate	Et$_2$O C$_6$H$_6$ CHCl$_3$/AcOH (5:1) or CHCl$_3$/Ac/ AcOH (10:2:1) or C$_6$H$_6$/AcOH (5:1) or Petet (80° C)/ CCl$_4$/AcOH (4:6:1) or CHCl$_3$/Ac/Et$_2$NH (4:2:0.2) C$_6$H$_6$/Diox/AcOH (90:25:4) C$_6$H$_6$ MeCl/EtOAc/Et$_2$NH (92:5:3 or 93:5:2)	[3] [3] [69] [3] [70] [70]
Phosphates, esters	Alumina Silica gel	Hex/C$_6$H$_6$/MeOH (2:1:1) or Hex/ MeOH/Et$_2$O Hex/C$_6$H$_6$/MeOH (2:1:1) or Hex/ MeOH/Et$_2$O	[71] [71]
Phospholipids	Silica gel G	CHCl$_3$/MeOH/w	[3]

(Continued)

Family	Stationary Phase	Mobile Phase	Ref.
Polynuclear aromatics	Alumina	CCl_4	[3]
	Alumina	$C_6H_6/(CH_2)_6$ (15:85)	[72]
	Silica gel	Hex or CH_3CHCl_2 or C_2HCl_3 or CCl_4	[73,74]
Polypeptides	Sephadex G-25	w or NH_3 (0.05 M)	[3]
	Silica gel G	$CHCl_3/MeOH$ (9:1) or $CHCl_3/Ac$ (9:1)	[3]
Pyridines	Whatman #1 (descending)	n-BuOH/w or n-BuOH/w/NH_3 or Ac or i-BuOH/w or MEK/AcOH/w	[75]
Pyridines, quaternary salts (descending)	Whatman #1	Ac/w or $AmSO_4$/Ph buffer (pH=6.8)/n-PrOH(2 %) or n-PrOH	[75]
Purines	Silica gel	Ac/$CHCl_3$/n-BuOH/NH_3 (25 %) (3:3:4:1)	[3]
Pyrrole, tricarboxylic acid	Silica gel	n-BuOH/EtOH/NH_3/w (10:10:1:1)	[64]
Skatoles, hydroxy	Silica gel G	i-Pr_2O or $(ClCH_2)_2$/i-Pr_2NH (6:1)	[76]
Steroids	Alumina	$CHCl_3$EtOH (96:4)	[3]
	Paper	Petet/Tol/MeOH/w or Petet/C_6H_6/ MeOH/w	[60]
	Paper (impregnated with kerosene)	n-PrOH/w	[77]
	Silica gel G	EtOAc/$(CH_2)_6$/EtOH(abs) or EtOAc/ $(CH_2)_6$ or $CHCl_3$/EtOH (abs) or C_6H_6/EtOH or n-C_6H_{14}/EtOAc or EtOAc/n-C_6H_{14}/EtOH(abs)/AcOH or EtOAc/n-C_6H_{14}/AcOH	[78–81]
Sugars	Cellulose	n-BuOH/Py/w (6:4:3) or EtOAc/ Py/w (2:1:2)	[3]
	Kieselguhr G (buffered with 0.02 N NaAc	EtOAc/i-PrOH/w	[3]
	Silica gel (buffered with H_3BO_3)	C_6H_6/AcOH/MeOH (1:1:3)	[3]
	Silica gel (impregnated with sodium bisulfite)	EtOAc/AcOH/MeOH/w (6:1:5:1) or n-PrOH/w (85:15) or i-PrOH/	[3]
	Silica gel G	EtOAc/w (7:1:2) or MEK/AcOH/w (6:1:3)	[82]
	Whatman #1 (descending, two dimensional)	n-PrOH/conc NH_3/w (6:2:1) PhOH or n-BuOH/AcOH	
Sugars, aldoses	Paper	EtOAc/Py/w (2:1:2) or n-BuOH/ AcOH/w (4:1:5) or n-BuOH/	[60]
	Whatman #1	EtOH/H_2O (5:1:4) or EtOAc/ AcOH/w (9:2:2) or EtOAc/ AcOH/ HForm/w or EtOAc/ Py/NaAc (sat'd) PhOH or n-BuOH/AcOH	[82]
Sugars, carbamates	Silica gel	n-BuOH/H_3BO_3 (0.03 M) (9:1)	
Sugars, deoxy	Whatman #1	PhOH or n-BuOH/AcOH	[82]
Sugars, ketoses	Paper	EtOAc/Py/w (2:1:2) or n-BuOH/ AcOH/w (4:1:5) or n-BuOH/	[60]
	Whatman #1	EtOH/H_2O (5:1:4) or EtOAc/ AcOH/w (9:2:2) or w/PhOH (pH=5.5) PhOH or n-BuOH/AcOH	[82]
Sulfides	Alumina	Hex	[75]
	Alumina	$CHCl_3$/MeOH	[96]
	Silica gel	CCl_4 or C_6H_6	[83]
	Silica gel DF-5	Ac/C_6H_6 or Tol/EtOAc	[95]

(*Continued*)

Family	Stationary Phase	Mobile Phase	Ref.
Sulfilimines, p-nitrosobenzene sulfonyl	Whatman #4 (impregnated with formamide)	C_6H_6 or $C_6H_6/(CH_2)_6$	[84]
Sulfonamides	Kieselguhr	$CHCl_3/MeOH$ (9:1) or $CHCl_3/$	[85]
	Silica gel	$MeOH/NH_3$	[86]
	Silica gel (neutral)	Et_2O or $CHCl_3/MeOH$ (10:1)	[87]
	Silica gel (G)	n-BuOH/MeOH/Ac/Et_2NH (9:1:1:1) $CHCl_3/EtOH/n-C_7H_{16}$	[3]
Sulfones	Alumina	Et_2O or Hex/Ac (1:1)	[57]
	Silica gel DF-5	Ac/C_6H_6 or Tol/EtOAc	[95]
Sulfones, esters	Alumina	Et_2O or Hex/Ac (1:1)	[57]
Sulfones, hydroxy-ethyl	Alumina	Hex/w (1:3)	[57]
Sulfoxides	Alumina	C_6H_6/Py (20:1) and Diox	[88]
	Alumina	Ac/CCl_4 (1:4)	[97]
	Silica gel	Ac or EtOAc or $CHCl_3/Et_2O$	[83]
	Silica gel DF-5	Ac/C_6H_6 or Tol/EtOAc	[95]
	Whatman #1	PhOH/w (8:3) or n-BuOH/AcOH/w (9:1:2.5)	[89]
Sulfoxides, hydroxy-ethyl	Alumina	Et_2O or Hex/Ac (1:1) or Hex/Et_2O (1:3)	[57]
Terpenes	Alumina	C_6H_6 or $C_6H_6/Petet$ or $C_6H_6/EtOH$	[3]
	Silica gel G	i-Pr_2O or i-Pr_2O/Ac	[3]
	Silica gel/gypsum	$CHCl_3/C_6H_6$ (1:1)	[90]
	Silicic acid (starch bond)	$n-C_6H_{14}/EtOAc$ (85:15)	[3]
Thiobarbiturates	Paper	n-AmOH/n-BuOH/25 % NH_3 (2:2:1)	[91]
Thiolactones	Silica gel	i-Pr_2O or i-Pr_2O/EtOAc (1:4) or i-Bu_2O/i-C_8H_{18}	[62]
Thiols	Alumina	Hex	[57]
	Alumina (activated)	AcOH/MeCN (3:1)	[96]
	Alumina (5 % cetane impregnated)	AcOH/MeCN (3:1)	[96]
	Silica gel	EtOAc or $CHCl_3$	[83]
Thiophenes	Alumina G	Petet (40 °C–60 °C)	[92]
	Silica gel	MeOH or $C_6H_6/CHCl_3$ (9:1)	[92]
Thiophosphate, esters		Petet or $C_6H_6/CHCl_3$ or Ac or EtOH or EtOAc or MeOH	[93]
Ureas	Acetylated plates	CCl_4/EtOAc/EtOH (100:5:2)	[94]
	Silica gel	CCl_4/MeCl/EtOAc/HOAc (70:50:15:10)	[94]
Urethanes (see ureas)			

TYPICAL STATIONARY AND MOBILE PHASE SYSTEMS USED
IN THE SEPARATION OF VARIOUS INORGANIC IONS

The following table lists a series of stationary and mobile systems that are used in the separation of various inorganic ions [1–8]. The list is far from detailed, and the reader is advised to consult the given references for details.

REFERENCES

1. Kirchner, J.G., *Thin Layer Chromatography*, 2nd ed., Wiley-Interscience, New York, 1978.
2. Bobbitt, J. M., Visualization, *Thin Layer Chromatography,* Reinhold, New York, 1963.
3. Randerath, K., Thin layer chromatography of metals on cellulose behaviour of group III B in the system halogenic acid—alcohol, *Thin Layer Chromatography,* Academic Press, New York, 1963.
4. Randerath, K., Thin layer chromatography of metals on cellulose. II. Behaviour of group IA, IIA, IVB, VB and VIB in the system halogen acid-alcohol, *Thin Layer Chromatography*, 2nd ed., Verlag, Chemie, Weinheim, 1975.
5. Gagliardi, E. and Brodar, B., Thin layer chromatography of metals on cellulose. III. Behavior of the transition metals in the system halogen acid-alcohol, *Chromatographia*, 2, 267, 1969.
6. Gagliardi, E. and Brodar, B., Thin layer chromatography of metals on cellulose. II. Behaviour of group IA, IIA, IVB, VB and VIB in the system halogen acid-alcohol, *Chromatographia*, 3, 7, 1970.
7. Gagliardi, E. and Brodar, B., Thin layer chromatography of metals on cellulose. III. Behavior of the transition metals in the system halogen acid-alcohol, *Chromatographia*, 3, 320, 1970.
8. MacDonald, J.C. ed., *Inorganic Chromatographic Analysis,* John Wiley & Sons, New York.

Stationary Phase	Mobile Phase	Solvent Ratio	Separated Ions
Silica gel G	Butanol/1.5 N HCl/2,5-hehanedione	100:20:0.5	Hydrogen sulfide group
Silica gel G	Acetone/conc. HCl/2,5-hexanedione	100:1:0.5	Ammonium sulfide group
Silica gel G	Water sat'd ethyl acetate/ tributyl phosphate	100:4	U, Ga, Al
Silica gel G	Ethanol/acetic acid	100:1	Alkali metals
Silica gel G	Acetone/1-butanol/conc. NH_4OH/ water	65:25:10:5	Halogens
Silica gel G	Methanol/conc. NH_4OH/10 % trichloroacetic acid/water	50:15:5:30	Phosphates
Dowex 1-cellulose (1:1)	1 M Aqueous sodium nitrate		Halogens
Cellulose	HCl (or HBr)/alcohol mixtures	Variable	Groups IA, IIA, IIIB, IVB, VB, VIB, transition metals
Cellulose	1-Butanol/water/HCl	8:1:1	Fe, Al, Ga, Ti, In
Cellulose	Acetic acid/pyridine/conc. HCl	80:6:20	Ammonium sulfide group
DEAE cellulose	Sodium azide/HCl	Variable	Cd, Cu, Hg
Amberlite CG 400 and CG 120	HCl/HNO_3	Variable	Pb, Bi, Sn, Sb, Cu, Cr, Hg

SPRAY REAGENTS IN THIN-LAYER CHROMATOGRAPHY

The following table lists the most popular spray reagents needed to identify organic compounds on chromatographic plates. These reagents have been thoroughly covered in several books [1–3] and reviews [4–23]. Due to the aerosol nature of the spray and the chemical hazards associated with several of these chemicals, the use of a fume hood is highly recommended. The original references of the spray reagents are given in order to provide information about their results with individual compounds [24–138]. A list and description of some complicated protocols follows this section of the chapter.

Note: $1\ \gamma = 1\ \mu g/cm^2$ on a TLC plate.

REFERENCES

1. Krebs, K.G., Heusser, D., and Wimmer, H., "Spray reagents," In *Thin Layer Chromatography, A Laboratory Handbook*, Stahl, E., ed., Springer-Verlag, New York, 1969, 854.
2. Bobbitt, J.B., "Visualization," In Bobbitt, J.B., *Thin Layer Chromatography*, Reinhold, New York, 1963, Chap. 7.
3. Touchstone, J.C., "Visualization procedures," In Touchstone, J.C. and Sherma, J., *Techniques and Application of Thin Layer Chromatography*, John Wiley & Sons, New York, 1985, 172.
4. Pataki, G., Paper, thin-layer, and electrochromatography of aminoacids in biological material, *Z. Klin. Chem.*, 2, 129, 1964; *Chem. Abs.*, 64, 5425c, 1966.
5. Padley, F.B., Thin-layer chromatography of lipids, Thin-layer Chromatography, *Proceedings Symposium*, Rome, 1963, 87 (Pub. 1964).
6. Honjo, M., Thin-layer chromatography of nucleic acid derivatives, *Kagaku No Ryoiki, Zokan*, 64, 1, 1964.
7. Kazumo, T., Thin-layer chromatography of bile acids, *Kagaku No Ryoiki, Zokan*, 64, 19, 1964.
8. Nakazawa, Y., Thin-layer chromatography of compound lipids, *Kagaku No Ryoiki, Zokan*, 64, 31, 1964.
9. Nishikaze, O., Separation and quantitative analysis of adrenocortical hormone and its metabolite (C_{21}) by thin-layer chromatography, *Kagaku No Ryoiki, Zokan*, 64, 37, 1964.
10. Shikita, M., Kazikazi, H., and Tamaoki, B., Thin-layer chromatography of radioactive substances, *Kagaku No Ryoiki, Zokan*, 64, 45, 1964.
11. Mo, I. and Hashimoto, Y., Method of thin-layer zone electrophoresis, *Kagaku No Ryoiki, Zokan*, 64, 61, 1964.
12. Kinoshita, S., Thin-layer chromatography of sugar esters, *Kagaku No Ryoiki, Zokan*, 64, 79, 1964.
13. Okada, M., Thin-layer chromatography of cardiotonic glycosides, *Kagaku No Ryoiki, Zokan*, 64, 103, 1964.
14. Omoto, T., Thin-layer chromatography of toad toxin, *Kagaku No Ryoiki, Zokan*, 64, 115, 1964.
15. Furnya, C. and Itokawa, H., Thin-layer chromatography of triterpenoids, *Kagaku No Ryoiki, Zokan*, 64, 123, 1964.
16. Zenda, H., Thin-layer chromatography of aconitine-type alkaloids, *Kagaku No Ryoiki, Zokan*, 64, 133, 1964.
17. Hara, S. and Tanaka, H., Thin-layer chromatography of mixed pharmaceutical preparations, *Kagaku No Ryoiki, Zokan*, 64, 141, 1964.
18. Katsui, G., Thin-layer chromatography of vitamins, *Kagaku No Ryoiki, Zokan*, 64, 157, 1964.
19. Fujii, S. and Kamikura, M., Thin-layer chromatography of pigments, *Kagaku No Ryoiki, Zokan*, 64, 173, 1964.
20. Hosogai, Y., Thin-layer chromatography of organic chlorine compounds, *Kagaku No Ryoiki, Zokan*, 64, 185, 1964.
21. Takeuchi, T., Thin-layer chromatography of metal complex salts, *Kagaku No Ryoiki, Zokan*, 64, 197, 1964.

22. Yamakawa, H. and Tanigawa, K., Thin-layer chromatography of organic metal compounds, *Kagaku No Ryoiki, Zokan,* 64, 209, 1964.
23. Ibayashi, H., Thin-layer chromatography of steroid hormones and its clinical application, *Kagaku No Ryoiki, Zokan,* 64, 227, 1964.
24. Beckett, A.H., Beavan, M.A., and Robinson, A.E., Paper chromatography: Multiple spot formation by sympathomimetic amines in the presence of acids, *J. Pharm. Pharmacol.,* 12, 203T, 1960; *Chem. Abs.,* 55, 9785c, 1961.
25. Heacock, R.A. and Scott, B.D., The chemistry of the "aminochromes" - Part IV. Some new aminochromes and their derivatives, *Can. J. Chem.,* 38, 516, 1960.
26. Matthews, J.S., Steroids (CCXXIII) color reagent for steroids in thin-layer chromatography, *Biochim. et Biophys. Acta,* 69, 163, 1963; *Chem. Abs.,* 58, 14043d, 1963.
27. Wasicky, R. and Frehden, O., Spot-plate tests in the examination of drugs (I) aldehyde and amine tests for the recognition of ethereal oils, *Mikrochim. Acta,* 1, 55, 1937; *Chem. Abs.,* 31, 5944, 1937.
28. Lane, E.S., Thin-layer chromatography of long-chain tertiary amines and related compounds, *J. Chromgr.,* 18, 426, 1965; *Chem. Abs.,* 63, 7630f, 1965.
29. Neu, R., A new color method for determining alkaloids and organic bases with sodium tetraphenylborate, *J. Chromgr.,* 11, 364, 1963; *Chem. Abs.,* 59, 12181d, 1963.
30. Zinser, M. and Baumgartel, C., Thin-layer chromatography of ergot alkaloids, *Arch. Pharm.,* 297, 158, 1964; *Chem. Abs.,* 60, 13095f, 1964.
31. Ashworth, M.R.F. and Bohnstedt, G., Reagent for the detection and determination of N-active hydrogen, *Talanta,* 13, 1631, 1966.
32. Whittaker, V.P. and Wijesundera, S., Separation of esters of choline, *Biochem. J.,* 51, 348, 1952; *Chem. Abs.,* 46, 7940g, 1952.
33. Heacock, R.A. and Mahon, M.E., The color reactions of the hydroxyskatoles, *J. Chromgr.,* 17, 338, 1965; *Chem. Abs.,* 62, 13824g, 1965.
34. Micheel, F. and Schweppe, H., Paper chromatographic separation of hydrophobic compounds with acetylated cellulose paper, *Mikrochim. Acta,* 53, 1954; *Chem. Abs.,* 48, 4354i, 1954.
35. Smyth, R.B. and Mckeown, G.G., Analysis of arylamines and phenols in oxidation-type hair dyes by paper chromatography, *J. Chromgr.,* 16, 454, 1964; *Chem. Abs.,* 62, 8930e, 1963.
36. Kawerau, E. and Wieland, T., Aminoacids chromatograms, *Nature,* 168, 77, 1951; *Chem. Abs.,* 46, 382h, 1952.
37. Sturm, A. and Scheja, H.W., Separation of phenolic acids by high voltage electrophoresis, *J. Chromgr.,* 16, 194, 1964; *Chem. Abs.,* 62, 6788b, 1965.
38. Feigl, F., *Spot Tests in Organic Analysis,* 7th ed., Elsevier Pub. Co., Amsterdam, Holland, 1966, p. 251.
39. Curzon, G. and Giltrow, J., A chromatographic color reagent for a group of aminoacids, *Nature,* 172, 356, 1953.
40. Heacock, R.A., Nerenberg, C., and Payza, A.N., The chemistry of the "aminochromes" - Part I. The preparation and paper chromatography of pure adrenochrome, *Can. J. Chem.,* 36, 853, 1958.
41. Heacock, R.A., "The aminochromes," In *Advances in Heterocyclic Chemistry,* Katritsky, ed., Academic Press, New York, 1965, 205; *Chem. Abs.,* 65, 5432d, 1966.
42. Wieland, T. and Bauer, L., Separation of purines and aminoacids, *Angew. Chem.,* 63, 511, 1951; *Chem. Abs.,* 46, 1082h, 1952.
43. Hara, S. and Takeuchi, M., Systematic analysis of bile acids and their derivatives by thin layer chromatography, *J. Chromgr.,* 11, 565, 1963; *Chem. Abs.,* 60, 838f, 1964.
44. Anthony, W.L. and Beher, W.T., Color detection of bile acids using thin layer chromatography, *J Chromgr.,* 13, 570, 1964; *Chem. Abs.,* 60, 13546c, 1964.
45. Hauck, A., Detection of caffeine by paper chromatography, *Deut. Z. Gerichtl. Med.,* 54, 98, 1963; *Chem. Abs.,* 60, 838b, 1964.
46. Suryaraman, M.G. and Cave, W.T., Detection of some aliphatic saturated long chain hydrocarbon derivatives by thin-layer chromatography, *Anal. Chim. Acta,* 30, 96, 1964; *Chem. Abs.,* 60, 7463e, 1964.
47. Passera, C., Pedrotti, A., and Ferrari, G., Thin-layer chromatography of carboxylic acids and keto-acids of biological interest, *J. Chromgr.,* 14, 289, 1964; *Chem. Abs.,* 60, 16191f, 1964.

48. Grant, D.W., Detection of some aromatic acids, *J. Chromgr.*, 10, 511, 1963; *Chem. Abs.*, 59, 5772a, 1963.

49. Roux, D.G., Some recent advances in the identification of leucoanthocyanins and the chemistry of condensed tanins, *Nature*, 180, 973, 1957; *Chem. Abs.*, 52, 5212f, 1958.

50. Abbott, D.C., Egan, H., and Thompson, J., Thin-layer chromatography of organochlorine pesticides, *J. Chromgr.*, 16, 481, 1964; *Chem. Abs.*, 62, 11090c, 1965.

51. Adamec, O., Matis, J., and Galvanek, M., Fractionation and quantitative determination of urinary 17-hydroxycorticosteroids by thin layer chromatography on silica gel, *Steroids*, 1, 495, 1963.

52. French, D., Levine, M.L., Pazur, J.H., and Norberg, E., Studies on the Schardinger dextrins. The preparation and solubility characteristics of alpha, beta and gamma dextrins, *J. Am. Chem. Soc.*, 71, 353, 1949.

53. Knappe, E. and Rohdewald, I., Thin-layer chromatography of dicarboxylic acids. V. Separation and identification of hydroxy dicarboxylic acids, of di- and tricarboxylic acids of the citrate cycle, and some other dicarboxylic acids of plant origin, *Z. Anal. Chem.*, 211, 49, 1965; *Chem. Abs.*, 63, 7333c, 1965.

54. Wright, J., Detection of humectants in tobacco by thin layer chromatography, *Chem. & Ind.* (London), 1125, 1963.

55. Toennies, G. and Kolb, J.J., Techniques and reagents for paper chromatography, *Anal. Chem.*, 23, 823, 1951; *Chem. Abs.*, 45, 8392i, 1951.

56. Kaufmann, H.P. and Sen Gupta, A.K., Terpenes as constituents of the unsaponifiables of fats, *Chem. Ber.*, 97, 2652, 1964; *Chem. Abs.*, 61, 14723b, 1964.

57. Gage, T.B., Douglass, C.D., and Wender, S.H., Identification of flavonoid compounds by filter paper chromatography, *Anal. Chem.*, 23, 1582, 1951; *Chem. Abs.*, 46, 2449c, 1952.

58. Hörhammer, L., Wagner, H., and Hein, K., Thin layer chromatography of flavonoids on silica gel, *J. Chromgr.*, 13, 235, 1964; *Chem. Abs.*, 60, 13856c, 1964.

59. Nakamura, H. and Pisano, J.J., Specific detection of primary catecholamines and their 3-0-methyl derivatives on thin-layer plates using a fluorigenic reaction with fluorescamine, *J. Chromgr.*, 154, 51, 1978; *Chem. Abs.*, 89, 117958x, 1978.

60. Neu, R., Analyses of washing and cleaning agents. XVIII. A new test for polyethylene glycols and their esters, *Chem. Abs.*, 49, 16475c, 1955; Ibid., 54, 2665e, 1960.

61. Korte, F. and Vogel, J., Thin-layer chromatography of lactones, lactams and thiolactones, *J. Chromgr.*, 9, 381, 1962; *Chem. Abs.*, 58, 9609c, 1963.

62. Harley-Mason, J. and Archer, A.A.P.G., p-Dimethylamino-cinnamaldehyde as a spray reagent for indole derivatives on paper chromatograms, *Biochem. J.*, 69, 60, 1958; *Chem. Abs.*, 52, 18600g, 1958.

63. Heacock, R.A. and Mahon, M.E., Paper chromatography of some indole derivatives on acetylated paper, *J. Chromgr.*, 6, 91, 1961.

64. Adams, C.W.M., A perchloric acid-naphthoquinone method for the histochemical localization of cholesterol, *Nature*, 192, 331, 1961.

65. Bennet-Clark, T.A., Tamblah, M.S., and Kefford, N.P., Estimation of plant growth substances by partition chromatography, *Nature*, 169, 452, 1951; *Chem. Abs.*, 46, 6181c, 1952.

66. Gordon, S.A. and Weber, R.P., Estimation of indole acetic acid, *Plant Physiol.*, 26, 192, 1951; *Chem. Abs.*, 45, 4605c, 1951.

67. Dickmann, S.R. and Crockett, A.L., Reactions of xanthydrol: (IV) Determination of tryptophan in blood plasma and proteins, *J. Biol. Chem.*, 220, 957, 1956; *Chem. Abs.*, 49, 7028h, 1956.

68. Mangold, H.K., Lamp, B.G., and Schlenk, H., Indicators for the paper chromatography of lipids, *J. Am. Chem. Soc.*, 77, 6070, 1953; *Chem. Abs.*, 50, 5074f, 1956.

69. Witter, R.F., Marinetti, G.V., Morrison, A., and Heicklin, L., Paper chromatography of phospholipids with solvent mixtures of ketones and acetic acid, *Arch. Biochem. Biophys.*, 68, 15, 1957; *Chem. Abs.*, 51, 12200a, 1957.

70. Martin, H.P., Reversed phase paper chromatography and detection of steroids of the cholesterol class, *Biochim. et Biophys. Acta*, 25, 408, 1957.

71. Preussmann, R., Daiber, D., and Hengy, H., Sensitive color reaction for nitrosamines on thin-layer chromatography, *Nature*, 201, 502, 1964; *Chem. Abs.*, 60, 12663e, 1964.

72. Preussmann, R., Neurath, G., Wulf-Lorentzen, G., Daiber, D., and Hengy, H., Color formation and thin-layer chromatography of N-nitrosocompounds, *Z. Anal. Chem.*, 202, 187, 1964.

73. Hranisavljevic-Jakovljevic, M., Pejkovic-Tadic, I., and Stojiljkovic, A., Thin-layer chromatography of isomeric oximes, *J. Chromgr.*, 12, 70, 1963; *Chem. Abs.*, 60, 7d, 1964.

74. Abraham, M.H., Davies, A.G., Llewellyn, D.R., and Thain, E.M., Chromatographic analysis of organic peroxides, *Anal. Chem. Acta*, 17, 499, 1957; *Chem. Abs.*, 53, 120b, 1959.

75. Knappe, E. and Peteri, D., Thin-layer chromatographic identification of organic peroxides, *Z. Anal. Chem.*, 190, 386, 1962; *Chem. Abs.*, 58, 5021a, 1963.

76. Servigne, Y. and Duval, C., Paper chromatographic separation of mineral anions containing sulfur, *Compt. Rend.*, 245, 1803, 1957; *Chem. Abs.*, 52, 5207b, 1958.

77. Lisboa, B.P., Characterization of Δ^4-3-oxo-C_{21}-steroids on thin-layer chromatography, *J. Chromgr.*, 16, 136, 1964; *Chem. Abs.*, 62, 3409, 1965.

78. Sherma, J. and Hood, L.V.S., Thin-layer solubilization chromatography: (I) phenols, *J. Chromgr.*, 17, 307, 1965; *Chem. Abs.*, 62, 13819b, 1965.

79. Gumprecht, D.L., Paper chromatography of some isomeric monosubstituted phenols, *J. Chromgr.*, 18, 336, 1965; *Chem. Abs.*, 63, 7630h, 1965.

80. Barton, G.M., α,α-Dipyridyl as a phenol-detecting reagent, *J. Chromgr.*, 20, 189, 1965; *Chem. Abs.*, 64, 2724a, 1966.

81. Sajid, H., Separation of chlorinated cresols and chlorinated xylenols by thin-layer chromatography, *J. Chromgr.*, 18, 419, 1965; *Chem. Abs.*, 63, 7630d, 1965.

82. Seeboth, H., Thin-layer chromatography analysis of phenols, *Monatsber. Deut. Akad. Wiss. Berlin*, 5, 693, 1963; *Chem. Abs.*, 61, 2489c, 1964.

83. Burke, W.J., Potter, A.D., and Parkhurst, R.M., Neutral silver nitrate as a reagent in the chromatographic characterization of phenolic compounds, *Anal. Chem.*, 32, 727, 1960; *Chem. Abs.*, 54, 13990d, 1960.

84. Perifoy, P.V., Slaymaker, S.C., and Nager, M., Tetracyanoethylene as a color-developing reagent for aromatic hydrocarbons, *Anal. Chem.*, 31, 1740, 1959; *Chem. Abs.*, 54, 5343e, 1960.

85. Bate-Smith, E.C. and Westall, R.G., Chromatographic behavior and chemical structure (I) naturally occurring phenolic substances, *Biochem. et Biophys. Acta*, 4, 427, 1950; *Chem. Abs.*, 44, 5677a, 1950.

86. Noirfalise, A. and Grosjean, M.H., Detection of phenothiazine derivatives by thin-layer chromatography, *J. Chromgr.*, 16, 236, 1964; *Chem. Abs.*, 62, 10295f, 1965.

87. Schreiber, K., Aurich, O., and Osske, G., Solanum alkaloids (XVIII): Thin-layer chromatography of Solanum steroid alkaloids and steroidal sapogenins, *J. Chromgr.*, 12, 63, 1963; *Chem. Abs.*, 60, 4442h, 1964.

88. Clarke, E.G.C., Identification of solanine, *Nature*, 181, 1152, 1958; *Chem. Abs.*, 53, 7298h, 1959.

89. Donner, R. and Lohs, K., Cobalt chloride in the detection of organic phosphate ester by paper and especially thin-layer chromatography, *J. Chromgr.*, 17, 349, 1965; *Chem. Abs.*, 62, 13842d, 1965.

90. Kucharczyk, N., Fohl, J., and Vymetal, J., Thin-layer chromatography of aromatic hydrocarbons and some heterocyclic compounds, *J. Chromgr.*, 11, 55, 1963; *Chem. Abs.*, 59, 9295g, 1963.

91. Kodicek, E. and Reddi, K.K., Chromatography of nicotinic acid derivatives, *Nature*, 168, 475, 1951; *Chem. Abs.*, 46, 3601g, 1952.

92. Hodgson, E., Smith, E., and Guthrie, F.E., Two-dimensional thin-layer chromatography of tobacco alkaloids and related compounds, *J. Chromgr.*, 20, 176, 1965; *Chem. Abs.*, 64, 3960b, 1966.

93. Stevens, P.J., Thin-layer chromatography of steroids. Specificity of two location reagents, *J. Chromgr.*, 14, 269, 1964; *Chem. Abs.*, 61, 2491b, 1964.

94. Lisboa, B.P., Application of thin-layer chromatography to the steroids of the androstane series, *J. Chromgr.*, 13, 391, 1964; *Chem. Abs.*, 60, 13890b, 1964.

95. Lisboa, B.P., Separation and characterization of Δ^5-3-hydroxy-C_{19}-steroids by thin-layer chromatography, *J. Chromgr.*, 19, 333, 1965; *Chem. Abs.*, 63, 16403h, 1965.

96. Lisboa, B.P., Thin-layer chromatography of Δ^4-3-oxosteroids of the androstane series, *J. Chromgr.*, 19, 81, 1965; *Chem. Abs.*, 63, 13619e, 1965.

97. Neher, R. and Wettstein, A., Steroids (CVII) color reactions; Corticosteroids in the paper chromatogram, *Helv. Chim. Acta*, 34, 2278, 1951; *Chem. Abs.*, 46, 3110d, 1952.

98. Michalec, C., Paper chromatography of cholesterol and cholesterol esters, *Naturwissenschaften*, 42, 509, 1955; *Chem. Abs.*, 51, 5884a, 1957.

99. Scheidegger, J.J. and Cherbuliez, E., Hederacoside A, a heteroside extracted from English ivy, *Helv. Chem. Acta*, 38, 547, 1955; *Chem. Abs.*, 50, 1685g, 1956.

100. Richter, E., Detection of sterols with naphthoquinone-perchloric acid on silica gel layers, *J. Chromgr.*, 18, 164, 1965; *Chem. Abs.*, 63, 7653a, 1965.

101. Lisboa, B.P., Thin-layer chromatography of steroids, *J. Pharm. Belg.*, 20, 435, 1965; *Chem. Abs.*, 65, 570c, 1966.

102. Adachi, S., Thin-layer chromatography of carbohydrates in the presence of bisulfite, *J. Chromgr.*, 17, 295, 1965; *Chem. Abs.*, 62, 13818g, 1965.

103. Bryson, J.L. and Mitchell, T.J., Spraying reagents for the detection of sugar, *Nature*, 167, 864, 1951; *Chem. Abs.*, 45, 8408b, 1951.

104. Sattler, L. and Zerban, F.W., Limitations of the anthrone test for carbohydrates, *J. Am. Chem. Soc.*, 72, 3814, 1950; *Chem. Abs.*, 45, 1039b, 1951.

105. Bacon, J.S.D. and Edelmann, J., Carbohydrates of the Jerusalem artichoke and other Compositae, *Biochem. J.*, 48, 114, 1951; *Chem. Abs.*, 45, 5242b, 1951.

106. Timell, T.E., Glaudemans, C.P.J., and Currie, A.L., Spectrophotometric method for determination of sugars, *Anal. Chem.*, 28, 1916, 1956.

107. Hay, G.W., Lewis, B.A., and Smith, F., Thin-film chromatography in the study of carbohydrates, *J. Chromgr.*, 11, 479, 1963; *Chem. Abs.*, 60, 839b, 1964.

108. Edward, J.T. and Waldron, D.M., Detection of deoxy sugars, glycols and methyl pentoses, *J. Chem. Soc.*, 1952, 3631; *Chem. Abs.*, 47, 1009h, 1953.

109. Johanson, R., New specific reagent for keto-sugars, *Nature*, 172, 956, 1953.

110. Adachi, S., Use of dimedon for the detection of keto sugars by paper chromatography, *Anal. Biochem.*, 9, 224, 1964; *Chem. Abs.*, 61, 13616g, 1964.

111. Sattler, L. and Zerban, F.W., New spray reagents for paper chromatography of reducing sugars, *Anal. Chem.*, 24, 1862, 1952; *Chem. Abs.*, 47, 1543d, 1953.

112. Bailey, R.W. and Bourne, E.J., Color reactions given by sugars and diphenylamine-aniline spray reagents on paper chromatograms, *J. Chromgr.*, 4, 206, 1960; *Chem. Abs.*, 55, 4251c, 1961.

113. Buchan, J.L. and Savage, R.J., Paper chromatography of starch-conversion products, *Analyst*, 77, 401, 1952; *Chem. Abs.*, 48, 8568c, 1954.

114. Schwimmer, S. and Bevenue, A., Reagent for differentiation of 1,4- and 1,6-linked glucosaccharides, *Science*, 123, 543, 1956; *Chem. Abs.*, 50, 8376a, 1956.

115. Partridge, S.M., Aniline hydrogen phthalate as a spraying reagent for chromatography of sugars, *Nature*, 164, 443, 1949.

116. Grossert, J.S. and Langler, R.F., A new spray reagent for organosulfur compounds, *J. Chromgr.*, 97, 83, 1974; *Chem. Abs.*, 82, 25473n, 1976.

117. Snegotskii, V.I. and Snegotskaya, V.A., Thin-layer chromatography of sulfur compounds, *Zavod. Lab.*, 35, 429, 1969; *Chem. Abs.*, 71, 23436b, 1969.

118. Fishbein, L. and Fawkes, J., Detection and thin-layer chromatography of sulfur compounds. I. Sulfoxides, sulfones and sulfides, *J. Chromgr.*, 22, 323, 1966; *Chem. Abs.*, 65, 6281e, 1966.

119. Svoronos, P.D.N., *On the Synthesis and Characteristics of Sulfonyl Sulfilimines Derived from Aromatic Sulfides*, Dissertation, Georgetown University, Washington, D.C., 1980. (Avail. Univ. Microfilms, Order No. 8021272).

120. Petranek, J. and Vecera, M., Identification of organic compounds. XXIV. Separation and identification of sulfides by paper chromatography, *Chem. Listy*, 52, 1279, 1958; *Chem. Abs.*, 53, 8039d, 1958.

121. Bican-Fister, T. and Kajganovic, V., Quantitative analysis of sulfonamide mixtures by thin-layer chromatography, *J. Chromgr.*, 16, 503, 1964; *Chem. Abs.*, 62, 8943d, 1965.

122. Bratton, A.C. and Marshall, Jr., E.K., A new coupling component for sulfanilamide determination, *J. Biol. Chem.*, 128, 537, 1939.

123. Borecky, J., Pinakryptol yellow, reagent for the identification of arenesulfonic acids, *J. Chromgr.*, 2, 612, 1959; *Chem. Abs.*, 54, 16255a, 1960.

124. Pollard, F.H., Nickless, G., and Burton, K.W.C., A spraying reagent for anions, *J. Chromgr.*, 8, 507, 1962; *Chem. Abs.*, 58, 3873b, 1963.

125. Coyne, C.M. and Maw, G.A., Paper chromatography for aliphatic sulfonates, *J. Chromgr.*, 14, 552, 1964; *Chem. Abs.*, 61, 7679d, 1964.

126. Wolski, T., Color reactions for the detection of sulfoxides, *Chem. Anal.* (Warsaw), 14, 1319, 1969; *Chem. Abs.*, 72, 106867q, 1970.

127. Suchomelova, L., Horak, V., and Zyka, J., The detection of sulfoxides, *Microchem. J.*, 9, 196, 1965; *Chem. Abs.*, 63, 9062a, 1965.

128. Thompson, J.F., Arnold, W.N., and Morris, C.J., A sensitive qualitative test for sulfoxides on paper chromatograms, *Nature*, 197, 380, 1963; *Chem. Abs.*, 58, 7351d, 1963.

129. Karaulova, E.N., Bobruiskaya, T.S., and Gal'pern, G.D., Thin-layer chromatography of sulfoxides, *Zh. Analit. Khim.*, 21, 893, 1966; *Chem. Abs.*, 65, 16046f, 1966.

130. Bergstrom, G. and Lagercrantz, C., Diphenylpicrylhydrazyl as a reagent for terpenes and other substances in thin-layer chromatography, *Acta Chem. Scand.*, 18, 560, 1964; *Chem. Abs.*, 61, 2491h, 1964.

131. Urx, M., Vondrackova, J., Kovarik, L., Horsky, O., and Herold, M., Paper chromatography of tetracyclines, *J. Chromgr.*, 11, 62, 1963; *Chem. Abs.*, 59, 9736g, 1963.

132. Dietz, W. and Soehring, K., Identification of thiobarbituric acids in urine by paper chromatography, *Arch. Pharm.*, 290, 80, 1957; *Chem. Abs.*, 52, 4736d, 1958.

133. Prinzler, H.W., Pape, D., Tauchmann, H., Teppke, M., and Tzcharnke, C., Thin-layer chromatography of organic sulfur compound, *Ropa Uhlie*, 8, 13, 1966; *Chem. Abs.*, 65, 9710h, 1966.

134. Curtis, R.F. and Philips, G.T., Thin-layer chromatography of thiophene derivatives, *J. Chromgr.*, 9, 366, 1962; *Chem. Abs.*, 58, 10705c, 1963.

135. Salame, M., Detection and separation of the most important organo-phosphorus pesticides by thin-layer chromatography, *J. Chromgr.*, 16, 476, 1964; *Chem. Abs.*, 62, 11090b, 1965.

136. Siliprandi, D. and Siliprandi, M., Separation and determination of phosphate esters of thiamine, *Biochem. et Biophys. Acta*, 14, 52, 1954; *Chem. Abs.*, 49, 6036f, 1955.

137. Nuernberg, E., Thin-layer chromatography of vitamins, *Deut., Apotheker-Zfg.*, 101, 268, 1961; *Chem. Abs.*, 60, 372, 1964.

138. Mariani, A. and Vicari, C., Determination of vitamin D in the presence of interfering substances, *Chem. Abs.*, 60, 373a, 1964.

139. Waldi, D., *Spray Reagents for Thin Layer Chromatography in Thin Layer Chromatography*, Springer, Berlin, 483–502, 1965.

140. Skipski, V.P., Peterson, R.F., and Barclay, M., *J. Lipid Res.*, 3, 467, 1962.

Family/Functional Group	Test	Result	Ref.
Adrenaline (and derivatives)	2,6-dichloroquinonechloroimide (0.5 % in absolute ethanol)	Variety of colors	[1]
	Potassium ferricyanide (0.6 % in 0.5 % sodium hydroxide	Red spots	[25]
Adrenochromes	4-N,N-Dimethylaminocinnamaldehyde	Blue-green to gray-green spots	[26]
	Ehrlich reagent	Blue-violet to red-violet spots	[26]
	Zinc acetate (20 %)	Blue or yellow fluorescent spots	[26]
Alcohols	Ceric ammonium sulfate (or nitrate)	Yellow/green spots on red background	[1,3]
	2,2-Diphenylpicrylhydrazyl (0.06 % in chloroform)	Yellow spots on purple background after heating (110 °C, 5 min)	[3]
	Vanillin (1 % in conc. sulfuric acid)	Variety of spots after heating (120 °C)—useful only for higher alcohols	[27]
Aldehydes	o-Dianisidine (saturated solution in acetic acid)	Variety of spots	[28]
	2,4-Dinitrophenylhydrazine	Blue colors (saturated ketones); olive green colors (saturated aldehydes); slow developing colors (unsaturated carbonyl compounds)	[1]
	2,4-Diphenylpicrylhydrazyl (0.06 % in chloroform)	Yellow spots on a purple background after heating (110 °C, 5 min)	[3]
	Hydrazine sulfate (1 % in 1 N hydrochloric acid)	Spots under UV (especially after heating)	[3]
Aldehydes, carotenoids	Tollens reagent	Dark spots	[1]
	Rhodamine (1 %–5 % in ethanol)	Variety of spots after treatment with strong alkali (sensitivity 0.03 μg)	[1]
Alkaloids	Bromocresol green (0.05 % in ethanol)	Green spots, especially after exposure to ammonia	[3]
	Chloramine-T (10 % aqueous)	Rose spots after exposure to hydrochloric acid and heat	[4]
	Cobalt (II) thiocyanate		[29]
	4-N,N-Dimethylaminobenzaldehyde (4 % in 1:3 hydrochloric acid/methanol)	Blue spots on a light pink background	[4]
	Iodine/potassium iodide (in 2 N acetic acid)	Characteristic spots for individual alkaloids	[3]
	Kalignost test	Variety of spots	[30]
	Sonnenschein test	Orange/red spots fluorescing under long-wave UV	[1]
	Dragendorff's reagent	Variety of spots	[139]
	Iodine (0.5 %) in chloroform	Variety of colors	[139]
	Iodoplatinate (IP) reagent	Plate may be further sprayed with 5 % ethanolic sulfuric acid or 5 % aqueous sodium nitrite solution to yield a variety of colors (detection up to 0.001 mg)	[139]
	Marquis reagent (especially for morphine, codeine, and thebaine)	Plate is heated (60 °C, 5 min) and evaluated by visible spectrophotometry	[139]
		Reagent sprayed to yield blue-violet color spots	[139]
		Plate evaluated in visible spectrophotometry after spraying	[139]
Alkaloids (ergot or fungal)	4-N,N-Dimethylaminobenzaldehyde/sulfuric acid	Blue spots	[31]

(Continued)

Family/Functional Group	Test	Result	Ref.
Amides	Chlorine/pyrazolinone/cyanide	Red spots turning blue (detection limit 0.5 µg)	[32]
	Hydroxylamine/ferric chloride	Variety of spots	[33]
Amines (all types unless specified)	Alizarin (0.1 % in ethanol)	Violet spots on yellow background	[3]
	Chlorine/pyrazolinon/cyanide	Red spots turning blue (aromatic only)	[32]
	Cobalt (II) thiocyanate	Blue spots on white/pink background	[29]
	Diazotization and α-naphthol coupling	Variety of spots (1° aromatic amines only)	[1]
	Ehrlich reagent	Yellow spots for aromatic amines	[34]
	Fast Blue B Salt	Variety of spots (only for amines that can couple)	[1]
	Glucose/phosphoric acid (4 %)	Variety of spots (aromatic amines only) especially after heating	[35]
	Malonic acid (0.2 %)/salicylaldehyde (0.1 %) (in ethanol)	Yellow spots after heating (120 °C, 15 min)	[3,4]
	1,2-Naphthoquinone-4-sulfonic acid, sodium salt (0.5 % in 1 N acetic acid)	Variety of colors after 30 min (aromatic amines only)	[36]
	Ninhydrin	Red colors when exposed to ammonium hydroxide	[37]
	p-Nitroaniline, diazotized	Variety of colored spots	[38]
	Nitroprusside (2.5 %)/acetaldehyde (5 %)/sodium carbonate (1 %)	Variety of spots (2° aliphatic only)	[39]
	Picric acid (3 % in ethanol)/sodium hydroxide (10 %) (5:1)	Orange spots	[4]
	Potassium iodate (1 %)	Variety of spots for phenylethylamines (after heating)	[3]
	Vanillin-potassium hydroxide	Variety of colors	[40]
	Ninhydrin	Pink spots on white background	[140]
Amino acids	Dehydroascorbic acid (0.1 % in 95 % n-butanol)	Variety of colored spots	[3]
	2,4-Dinitrofluorobenzene	Variety of spots	[1]
	Isatin-zinc acetate	Variety of colors	[1]
	Folin reagent	Variety of colors	[1]
	Ninhydrin	Red colors when exposed to ammonium hydroxide	[36]
	Vanillin/potassium hydroxide	Variety of colors	[40]
Amino alcohols	Alizarin (0.1 % in ethanol)	Violet on yellow background	[3]
Aminochromes	p-N,N-Dimethylaminocinnamaldehyde	Variety of colors	[41]
	Ehrlich reagent	Violet spots	[26,41]
	Ferric chloride (3 %)	Gray-brown spots	[41]
	p-Nitroaniline, diazotized	Red/brown spots	[26,41]
	Sodium bisulfite, aqueous	Yellow fluorescence under UV	[41,42]
Aminosugars	Ninhydrin	Red colors when exposed to ammonium hydroxide	[37]
Ammonium salts, quaternary	Cobalt (II) thiocyanate	Variety of spots	[29]
Anhydrides	Hydroxylamine/ferric chloride	Variety of spots	[33]
Arginine	Sakaguchi reagent	Orange/red spots	[1]

(Continued)

Family/Functional Group	Test	Result	Ref.
Azulenes	4-Dimethylaminobenzaldehyde-acetic acid-phosphoric acid for azulenes	Blue spots (room temperature) that fade to green/yellow shades and can be regenerated with steam	[1]
Barbiturates	Cobalt (II) nitrate (2 %)/lithium hydroxide (0.5 %)	Variety of colors	[1]
	Cupric sulfate/quinine/pyridine	Variety of colors (white, yellow, violet)	[1]
	s-Diphenylcarbazone (0.1 % in ethanol)	Purple spots	[3]
	Ferrocyanide/hydrogen peroxide	Yellow/red colors	[1]
	Fluorescein (0.005 % in 0.5 M ammonia)	Variety of spots under long or short-wave UV	[43]
	Mercurous nitrate (1 %)	Variety of spots	[1]
	Zwikker reagent	Variety of spots	[1]
Bile acids	Anisaldehyde/sulfuric acid	Variety of spots	[4]
	Antimony trichloride (in chloroform)	Variety of spots	[44]
	Perchloric acid (60 %)	Fluorescent spots (long-wave UV) after heating (150 °C, 10 min)	[44]
	Sulfuric acid	Variety of spots	[44,45]
Bromides	Fluorescein/hydrogen peroxide	Non-fluorescent spots	[1]
Caffeine	Chloramine-T	Pink-red spots	[1]
	Silver nitrate (2 % in 10 % sulfuric acid)	Carmine-red spots (limit 2γ)	[46]
Carboxylic acids	Bromcresol blue (0.5 % in 0.2 % citric acid)	Yellow spots on blue background	[3]
	Bromothymol blue (0.2 % in ethanol, pH=7)	Yellow spots upon exposure to ammonia	[47]
	2,6-dichlorophenol/indophenol (0.1 % in ethanol)	Red spots on blue background after heating	[48]
	Hydrogen peroxide (0.3 %)	Blue fluorescence under long-wave UV	[49]
	Schweppe reagent	Dark brown spots	[1]
Carboxylic acids, ammonium salts	Silicomolybdic acid on silica gel	Molybdenum blue formation	[139]
Catechins	p-Toluenesulfonic acid (20 % in chloroform)	Fluorescent spots under long-wave UV	[50]
Catecholamines	Ethylenediamine (50 %)	Spots under short/long-wave UV after heating (50 °C, 20 min)	[1]
Chlorides, alkyl	2,6-Dichlorophenol indophenol (0.2 %)/silver nitrate (3 %) in ethanol	Variety of spots	[1]
	Silver nitrate (0.5 % in ethanol)	Dark spots upon UV irradiation	[51]
	Silver nitrate/formaldehyde	Dark gray spots	[1]
	Silver nitrate/hydrogen peroxide	Dark spots	[1]
Chlorinated insecticides and pesticides	Diphenylamine (0.5 %)/zinc chloride (0.5 %) in acetone	Variety of colors upon heating (200 °C)	[1]
	2-Phenoxyethanol (5 %) in 0.05 % silver nitrate	Variety of spots	[4]
	Silver nitrate/formaldehyde	Dark gray spots	[1]
	o-Toluidine (0.5 %) in ethanol	Green spots under UV (sensitivity 0.5μg)	[4]

(Continued)

Family/Functional Group	Test	Result	Ref.
Choline derivatives	Dipicrylamine (0.2 % in 50 % aqueous acetone)	Red spots on yellow background	[1]
Corticosteroids	Blue tetrazolium (0.05 %)/sodium hydroxide (2.5 M)	Violet spots (limit 1 γ/cm²)	[1,52]
	2,3,5-Triphenyl-H-tetrazolium chloride (2 % in 0.5 NaOH)	Red spots after heating (100 °C, 5 min)	[1]
Coumarins	Benedict reagent	Fluorescent spots under long-wave UV	[1]
	Potassium hydroxide (5 % in methanol)	Variety of spots under long-wave UV	[1]
Dextrins	Iodine/potassium iodide	Blue-black spots (α-dextrins); brown-yellow spots (β- or γ-dextrins)	[53]
Dicarboxylic acids	Bromcresol purple (0.04 % in basic 50 % ethanol, pH=10)	Yellow spots on blue background	[4,54]
Diols (1,2-)	Lead tetraacetate (1 % in benzene)	White spots after heating (110 °C, 5 min) (limit 2 µg)	[55]
Disulfides	Iodine (1.3 % in ethanol)/sodium azide (3.3 % in ethanol)	White spots on brown iodine background	[3]
	Nitroprusside (sodium)	Red spots	[56]
Diterpenes	Antimony (III) chloride/acetic acid	Reddish yellow to blue-violet	[57]
Esters	Hydroxylamine/ferric chloride	Variety of spots	[33]
Flavonoids	Aluminum chloride	Yellow fluorescence on long-wave UV	[58]
	Antimony (III) chloride (10 % in chloroform)	Fluorescence on long-wave UV	[59]
	Benedict's reagent	Fluorescence on long-wave UV (only for o-dihydroxy compounds)	
Flavonoids (cont.)	Lead acetate (basic, 25 %)	Fluorescent spots	[59]
	p-Toluenesulfonic acid (20 % in chloroform)	Fluorescent spots under long-wave UV after heating (100 °C, 10 min)	[4,50]
Fluorescamines	Perchloric acid (70 %)	Blue fluorescent spots	[60]
Glycols, polyethylene	Quercetin/sodium tetraphenylborate	Orange-red spots	[61]
Glycolipids	Diphenylamine (5 % in ethanol) dissolved in 1:1 hydrochloric acid/acetic acid	Blue-gray spots	[1]
Glycosides, triterpene	Liebermann-Burchard reagent	Fluorescence under long-wave UV	[1]
Hydroxamates	Ferric chloride (10 % in acetic acid)	Brown spots	[62]
Hydroxamic acids	Ferric chloride (1 %—5 % in 0.5 N hydrochloric acid)	Red spots	[1]
Imidazoles	p-Anisidine/amyl nitrite	Red/brown spots	[3]

(Continued)

Family/Functional Group	Test	Result	Ref.
Indoles	Chlorine/pyrazolinone/cyanide	Red spots turning blue after a few minutes (limit 0.5 µg)	[32]
	Cinnamaldehyde/hydrochloric acid		[1]
	4-N,N-Dimethylaminocinnamaldehyde	Red spots	[63]
	Ehrlich reagent	Variety of colored spots	[9,34,64]
	Ferric chloride (0.001 M) in 5 % perchloric acid	Purple for indoles; blue for hydroxyindoles	[3]
	Naphthoquinone/perchloric acid	Red spots	[65]
	Perchloric acid (5 %)/ferric chloride (0.001 M)	Orange spots	[66]
	Prochazka reagent	Variety of colored spots	[1]
	Salkowski reagent	Fluorescent (yellow/orange/green) spots under long-wave UV	[67]
Indoles (cont.)	van Urk (or Stahl) reagent	Variety of colored spots	[1]
	Xanthydrol (0.1 % in acidified ethanol)	Variety of colored spots after heating (100 °C)	[68]
Iodides	Sonnenschein test	Variety of spots	[1]
α-Ketoacids	2,6-Dichlorophenol/indophenol (0.1 % in ethanol)	Pink spots upon heating	[4,48]
	o-Phenylenediamine (0.05 % in 10 % trichloroacetic acid or 0.2 % in 0.1 N H_2SO_4/ethanol)	Green fluorescence under long-wave UV after heating (100 °C, 2 min)	[1]
Ketones	o-Dianisidine (saturated solution in acetic acid)	Characteristic spots	[28]
	2,4-Dinitrophenylhydrazine	Yellow-red spots	[3]
Lactones	Hydroxylamine/ferric chloride	Variety of colors	[33]
Lipids	α-Cyclodextrin	Variety of spots (for straight chain lipids)	[53]
	2',7'-Dichlorofluorescein (0.2 %) in ethanol	Spots under long-wave UV	[1,69]
	Fluorescein	Spots after treatment with steam	[1]
	Rhodamine 6G (1 % in acetone)	Spots under long-wave UV	[70]
	Tungstophosphoric acid (20 % in ethanol)	Variety of colored spots after heating	[71]
Mercaptans (see thiols)			
Nitro compounds	4-N,N-Dimethylaminobenzaldehyde/stannous chloride/ hydrochloric acid	Yellow spots	[3]
Nitrosamines	Diphenylamine/palladium chloride	Violet spots after exposure to short-wave UV (limit 0.5 γ)	[1,72]
	Sulfanilic acid (0.5 %)/α-naphthylamine (0.05 %) in 30 % acetic acid	Spraying is preceded by short-wave UV irradiation (3 min); aliphatic nitrosamines yield red/violet spots, while aromatic ones green/blue spots (limit 0.2–0.5 γ)	[1,72,73]
Oximes	Cupric chloride (0.5 %)	Immediate green spots (β-oximes); green-brown spots after 10 min (α-oximes)	[74]

(Continued)

Family/Functional Group	Test	Result	Ref.
Peroxides	Ammonium thiocyanate (1.2 %)/ferrous sulfate (4 %)	Brown-red spots	[74]
	N,N-dimethyl-p-phenylene diammonium dichloride	Purple spots	[76]
	Ferrous thiocyanate	Red-brown spots	[1,75]
	Iodide (potassium)/starch	Blue spots	[1]
Persulfates	Benzidine (0.05 % in 1 N acetic acid)	Blue spots	[77]
Phenols	Anisaldehyde/sulfuric acid	Variety of colors	[1,78]
	p-Anisidine/ammonium vanadate	Variety of spots on pink background	[3]
	Benzidine, diazotized	Variety of colors	[79]
	Ceric ammonium nitrate (46 % in 2 M nitric acid)	Variety of spots	[80]
	α,α′-dipyridyl (0.5 %)/ferric chloride (0.5 %) in ethanol	Variety of spots	[4,81]
	Emerson	Red-orange to pink spots	[1]
	Fast Blue B Salt	Variety of spots	[1]
	Ferric chloride (1 %–5 % in 0.5 N HCl)	Blue-greenish spots	[1]
	Folin–Denis reagent	Variety of spots	[82]
	Gibbs reagent	Variety of colors	[1]
	Millon reagent	Variety of colors after heating	[1]
	Naphthoquinone/perchloric acid	Yellow spots (phenol, catechol); dark blue spots (resorcinol)	[65]
	p-Nitroaniline, diazotized	Variety of colored spots	[38]
	p-Nitrobenzenediazonium fluoroborate	Variety of spots	[84]
	Silver nitrate (saturated in acetone)	Pink to deep green colors	[84]
	Stannic chloride (5 %) in equal volumes of chloroform/acetic acid	Variety of spots after heating (100 °C, 5 min)	[1]
	Tetracyanoethylene (10 % in benzene)	Variety of colors	[85]
	Tollen's (or Zaffaroni) reagent	Dark spots	[86]
	Vanillin (1 % in sulfuric acid)	Variety of colors after heating	[27]
Phenols	Fast Blue Salt (FBS) reagent	Plate sprayed with 0.1 M sodium hydroxide and inspected by visible spectrophotometry	[139]
Phenols, chlorinated	Folin–Denis reagent	Variety of spots	[82]
Phenothiazines	Ferric chloride (5 %)/perchloric acid (20 %)/nitric acid (50 %) (1:9:10)	Variety of colors	[4,87]
	Formaldehyde (0.03 % in phosphoric acid)	Variety of colors	[88,89]
	Palladium (II) chloride (0.5 % pH<7)	Variety of spots	[1]
Phosphates, esters	Cobalt (II) chloride (1 % in acetone or acetic acid)	Blue spots upon warming the plate at 40 °C	[90]
Polynuclear aromatics	Formaldehyde (2 %) in conc. sulfuric acid	Variety of colors	[91]
	Tetracyanoethylene (10 % in benzene)	Variety of colors	[85]
Purines	Fluorescein (0.005 % in 0.5 M ammonia)	Variety of spots under long- or short-wave UV	[43]
Pyrazolones	Ferric chloride (5 %)/acetic acid (2 N) (1:11)	Variety of colors	[4]

(Continued)

Family/Functional Group	Test	Result	Ref.
Pyridines	König reagent	Variety of spots (for free α-position pyridines)	[92,93]
Pyridines, quaternary	König reagent	Blue-white fluorescence under UV	[93]
Pyrimidines	Fluorescein (0.005 % in 0.5 M ammonia)	Variety of spots under long- or short-wave UV	[43]
Pyrones (α- and γ-)	Neu reagent	Fluorescent spots under long-wave UV	[1]
Quinine derivatives	Formic acid vapors	Fluorescent blue spots	[3]
Quinones	5 %–10 % aqueous potassium hydroxide solution	Plates display colors (red, yellow, blue) after spraying and heating	[139]
Sapogenins	Komarowsky reagent	Yellow/pink spots	[94]
	Paraformaldehyde (0.03 % in 85 % phosphoric acid)	Variety of spots	[88]
	Zinc chloride (30 % in methanol)	Fluorescent spots after heating (105 °C, 1 h) in a moisture-free atmosphere	[94]
Saponins	Blood reagent (3.6 % sodium citrate/10 mL in phosphate buffer pH 7.4/30 mL for 90 mL blood sample	White zones formed against reddish background	[139]
	Anisaldehyde–sulfuric (AS) acid reagent	After spraying, plates are heated (100 °C, 10 min) and evaluated in visible or UV (365 nm)	[139]
	Antimony (III) chloride reagent	After spraying, plates are heated (100 °C, 5 min) and evaluated in visible or UV (365 nm)	[139]
Steroids	Anisaldehyde/sulfuric acid	Variety of colors	[95–97]
	Antimony (III) chloride (in acetic acid)	Variety of colors	[57,96]
	Carr-Price reagent	Variety of colors	[1]
	Chlorosulfonic acid/acetic acid	Fluorescence under long-wave UV	[78,95]
	Dragendorff reagent	Variety of spots	[88,89]
	Formaldehyde (0.03 % in phosphoric acid)	Variety of spots	[96]
	Hanes and Isherwood reagent	Variety of spots (only for 3-hydroxy-Δ^5-steroids)	[1]
	Liebermann–Burchard reagent	Fluorescence under long-wave UV	[1,44]
	Perchloric acid (20 %)	Fluorescent spots (long-wave UV) after heating (150 °C, 10 min)	[96]
	Phosphomolybdic acid	Blue color	[95,96,98]
	Phosphoric acid (50 %)	Fluorescent spots after heating (120 °C) (limit 0.005 γ)	[99]
	Phosphotungstic acid (10 % in ethanol)	Variety of spots	[1,100]
	Stannic chloride (5 %) in equal volumes of chloroform/acetic acid (1:1)	Variety of spots after heating (100 °C, 5 min)	[50]
	Sulfuric acid	Variety of spots	[96]
	p-Toluenesulfonic acid (20 % in chloroform)	Fluorescent spots under long-wave UV	[95,96]
	Trichloroacetic acid (50 % aqueous)		
	Zimmermann reagent		

(Continued)

Family/Functional Group	Test	Result	Ref.
Sterols	Antimony (III) chloride (50 % in acetic acid)	Variety of spots	[99]
	Bismuth (III) chloride	Fluorescence under long-wave UV	[1]
	Chlorosulfonic acid/acetic acid	Fluorescence under long-wave UV	[1]
	Liebermann–Burchard reagent	Fluorescence under long-wave UV	[1]
	1,2-Naphthoquinone-4-sulfonic acid/perchloric acid	Pink spots that change to blue upon prolonged heating (cholesterol limit 0.03 γ)	[65,101]
	Phosphoric acid (50 %)	Fluorescent spots after heating (120 °C, 15 min)	[98,102]
	Phosphotungstic acid (10 % in ethanol)	Variety of spots	[99]
	Stannic chloride (5 %) in equal volumes of chloroform/acetic acid	Variety of spots after heating (100 °C, 5 min)	[1]
	Sulfuric acid	Variety of spots	[102]
Sugars	o-Aminodiphenyl (0.3 %)/orthophosphoric acid (5 %)	Brown spots after heating	[103]
	Aniline/phosphoric acid	Variety of colors	[104]
	Anisaldehyde/sulfuric acid	Variety of colors	[1,78]
	Anthrone test	Yellow spots	[105]
	Benzidine/trichloroacetic acid	Red-brown/dark spots	[106]
Sugars (cont.)	Carbazole/sulfuric acid	Violet spots on blue background	[103]
	Lewis–Smith reagent	Brown spots	[107]
	Naphthoquinone/perchloric acid	Pink-brown spots (glucose, mannose, lactose, sucrose)	[65]
	Naphthoresorcinol (0.2 % in ethanol)/phosphoric acid (10:1)	Variety of spots after heating (100 °C, 5–10 min)	[1]
	Naphthoresorcinol (0.1 %)/sulfuric acid (10 %)	Variety of spots after heating (100 °C, 5–10 min)	[1]
	Orcinol reagent	Variety of spots	[1]
	Permanganate, potassium (0.5 % in 1 N sodium hydroxide)	Variety of spots after heating (100 °C)	[108]
	Phenol (3 %)/sulfuric acid (5 % in ethanol)	Brown spots after heating (100 °C, 10 min)	[103]
	Silver nitrate (0.2 % in methanol)/ammonia (saturated)/sodium methoxide (2 % in methanol)	Variety of spots after heating (110 °C, 10 min)	[1]
	Silver nitrate/sodium hydroxide	Variety of spots	[1]
	Sulfuric acid	Variety of spots	[108]
	Thymol (0.5 %) in sulfuric acid (5 %)	Pink spots after heating (120 °C, 20 min)	[103]
Sugars, deoxy	Metaperiodate/p-nitroaniline	Fluorescent (long-wave UV) yellow spots	[109]
Sugars, ketoses	Anthrone test	Bright purple (pentoses); orange-yellow (heptoses); blue fluorescence (aldoses)	[110]
	Dimedone (0.3 %)/phosphoric acid (10 % in ethanol)	Dark gray spots (white light); dark pink fluorescing spots (UV) after heating (110 °C, 15 min)	[1,111]
Sugars, reducing	4-Aminohippuric acid	Fluorescence under long-wave UV	[112]
	Aniline/diphenylamine/phosphoric acid	Variety of colors	[113–115]
	Aniline hydrogen phthalate	Variety of colors (limit 1 μg)	[116]
	p-Anisidine phthalate	Variety of colors	[1]
	3,5-Dinitrosalicylic acid (0.5 % in 4 % sodium hydroxide)	Brown spots (sensitivity 1 μg)	[3]

(Continued)

Family/Functional Group	Test	Result	Ref.
Sulfides	Ceric ammonium nitrate (in 2 M HNO$_3$)	Colorless spots (limit <100 µg/spot)	[117]
	Chloranil (1 %) in benzene	Yellow-brown spots	[119]
	2,3-Dichloro-5,6-dicyano-1,4-benzoquinone (2 %) in benzene	Purple-blue spots changing to orange upon ammonia exposure	[119]
	Gibbs reagent	Yellow-brown spots changing to blue-orange upon exposure to ammonia	[119]
	Iodine vapors	Brown spots	[118]
	Tetracyanoethylene (2 %) in benzene	Orange spots	[119]
	N,2,6-Trichloro-p-benzoquinoneimine (2 %) in ethanol	Brown spots	[119]
Sulfilimines	Potassium permanganate	Colorless spots	[120]
Sulfilimines, p-nitro-benzene-sulfonyl	Tin chloride/4-N,N-dimethylaminobenz-aldehyde	Yellow spots	[121]
Sulfites	Malachite green oxalate	White spots on blue background	[4]
Sulfonamides	Chlorine/pyrazolinone/cyanide	Red spots changing to blue	[32]
	Diazotization and coupling	Variety of spots (limit 0.25 γ)	[122,123]
	Ehrlich	Variety of colors	[124]
	Chloranil (1 %) in benzene	Pink turning to violet or green after heating	[119]
	2,3-Dichloro-5,6-dicyano-1,4-benzoquinone (2 %) in benzene	Lilac-violet turning to yellow-green upon ammonia exposure	[119]
	Gibbs reagent	Violet turning to tan upon exposure to ammonia and heat	[119]
Sulfones	Iodine vapors	Brown spots	[118]
	Tetracyanoethylene (2 %) in benzene	Pink to yellow upon exposure to ammonia and heat	[119]
Sulfonic acids	Pinacryptol yellow (0.1 %)	Yellow-orange spots under long-wave UV	[124]
	silver nitrate/fluorescein	Yellow spots under long-wave UV	[125,126]
Sulfoxides	Acetyl bromide	Yellow-orange spots	[127]
	Ceric ammonium nitrate (40 %) in 2 M nitric acid	Brown spots after heating (especially good for α-polychlorosulfoxides); limit 80 µg/spot	[117]
	Chloranil (1 %) in benzene	Yellow-blue spots	[119]
	2,3-Dichloro-5,6-dicyano-1,4-benzoquinone (2 %) in benzene	Orange-crimson spots	[119]
	Dragendorff reagent	Orange-brown-red spots (limit 30–150 γ)	[128]
	Gibbs reagent	Yellow turning to brown upon ammonia exposure	[168]
	Iodide (sodium)/starch	Brown spots (limits 0.01 µmol/20 µl solution)	[129]
	Iodine vapors	Brown spots	[118,130]
	Tetracyanoethylene (2 %) in benzene	Yellow or crimson turning to white or tan Upon exposure to ammonia	[119]
	N-2,6-Trichloro-p-benzoquinoneimine (2 %) in ethanol	Yellow spots	[119]

(Continued)

Family/Functional Group	Test	Result	Ref.
Terpenes	Anisaldehyde/sulfuric acid	Variety of colors	[1]
	Antimony (V) chloride	Variety of colors	[1]
	Carr-Price reagent	Variety of colors	[1]
	Diphenylpicrylhydrazyl in chloroform	Yellow spots on purple background after heating (110 °C) (limit 1 γ/0.5 cm diameter)	[131]
	Vanillin-phosphoric acid (VPA) reagent	After spraying, plates are heated (100 °C, 10 min) and evaluated in visible or UV (365 nm)	[139]
	Phenol (50 % in carbon tetrachloride)	Variety of spots upon exposure to bromine vapors	[3]
	Vanillin (1 % in 50 % H$_3$PO$_4$)	Variety of spots after heating (120 °C, 20 min)	[4]
Tetracyclines	Ammonium hydroxide	Yellow fluorescence under long-wave UV	[132]
Thioacids	Silver nitrate/ammonium hydroxide/sodium chloride	Yellow-brown spots	[1]
Thiobarbiturates	Cupric sulfate (0.5 %)/diethylamine (3 % in methanol)	Green spots (limit 15 γ)	[3,133]
Thiolactones	Nitroprusside (sodium), basic	Red spots	[62]
Thiols (mercaptans)	Ceric ammonium nitrate (in 2 M nitric acid)	Colorless spots on yellow background (limit <100 μg/spot)	[117]
	Iodine (1.3 % in ethanol)/ethanol	White spots in brown iodine background	[3]
	Nitroprusside (sodium)(3 %)	Red spots	[134]
Thiophenes	Isatin (0.4 % in conc. sulfuric acid)	Variety of colors	[135]
Thiophosphates, esters	Ferric chloride/sulfosalicylic acid	White spots on violet background	[136]
	Palladium (II) chloride (0.5 % in acidified water)	Variety of spots	[1,136]
	Periodic acid (10 % in 70 % perchloric acid)	Variety of spots	[3]
Unsaturated compounds	Fluorescein (0.1 % in ethanol)/bromine	Yellow spots on a pink background upon exposure to bromine vapors	[1]
	Osmium tetroxide vapors	Brown/black spots	[3,95]
Ureas	p-N,N-Dimethylaminobenzaldehyde (1 % in ethanol)	Characteristic spots after exposure to hydrochloric acid	[4]
Vitamin A	Antimony (V) chloride	Variety of colors	[1]
	Carr-Price reagent	Variety of colors	[1]
	Sulfuric (50 % in methanol) followed by heating	Blue spots that turn brown	[1]
Vitamin B1	Dipicrylamine	Characteristic spots	[3]
	Thiochrome	Variety of spots under long-wave UV	[137]
Vitamin B6	N,2,6-Trichloro-p-benzoquinoneimine (0.1 % in ethanol)	Blue spots after exposure to ammonia	[3]
Vitamin B6, acetal	2,6-Dibromo-p-benzoquinone-4-chlorimine (0.4 % in methanol)	Characteristic spots	[138]
Vitamin C	Cacotheline (2 % aqueous)	Purple spot after heating (100 °C)	[3]
	Iodine (0.005 %) in starch (0.4 %)	White spot on blue background	[3]
	Methoxynitroaniline/sodium nitrite	Blue spots on orange background	[3]
Vitamin D	Antimony (V) chloride	Variety of colors	[139]
	Carr-Price reagent	Variety of colors	[1]
	Trichloroacetic (1 % in chloroform)	Variety of spots after heating (120 °C, 5 min)	[1]
Vitamin E	2',7'-Dichlorofluorescein (0.01 % in ethanol)	Spots under long-wave UV light	[1]
	α,α'-dipyridyl (0.5 %)/ferric chloride (0.5 % in ethanol)	Variety of colors	[1]

PROTOCOL FOR REAGENT PREPARATION

The following section gives a summary for the preparation of the major spray reagents listed in the previous section (Spray Reagents in Thin-Layer Chromatography). Reference to the original literature is recommended for any reagents not listed here [1–4].

REFERENCES

1. Krebs, K.G., Heusser, D., and Wimmer, H., *Thin Layer Chromatography, A Laboratory Handbook*, E. Shahl, ed., Springer-Verlag, New York, 1969.
2. Bobbitt, J.B., *Thin Layer Chromatography*, Reinhold, New York, 1963.
3. Touchstone, J.C. and Dobbins, M.F., *Practice of Thin Layer Chromatography*, John Wiley & Sons, New York, 1983.
4. Randerath, K., *Thin-Layer Chromatography*, 2nd ed., Verlag Chemie, GmbH., (in the United States, Academic Press, New York, 1968).

Acetic Anhydride-Sulfuric Acid
See Liebermann–Burchard reagent.

Alizarin
A saturated solution of alizarin in ethanol is sprayed on the moist plate, which is then placed in a chamber containing 25 % ammonium hydroxide solution to yield a variety of colors.

Aluminum Chloride
A 1 % aluminum chloride solution in ethanol is sprayed on the plate, which is then observed under long-wave UV light.

4-Aminoantipyrine-Potassium Ferricyanide
See Emerson reagent.

4-Aminobiphenyl-Phosphoric Acid
See Lewis–Smith reagent.

4-Aminohippuric Acid
A 0.3 % 4-aminohippuric acid solution in ethanol is sprayed on the plate, which is then heated at 140 °C (8 min) and observed under long-wave UV light.

Ammonium Hydroxide
The chromatogram is placed in a chamber containing 25 % ammonium hydroxide, dried and then observed under long-wave UV light.

Aniline-Diphenylamine-Phosphoric Acid
An aniline (1 g)/diphenylamine (1 g)/phosphoric acid (5 mL) solution in acetone (50 mL) is sprayed on the plate, which is then heated at 85 °C (10 min) yielding a variety of colors.

Aniline-Phosphoric Acid
A 20 % aniline solution in n-butanol, saturated with an aqueous (2 N) orthophosphoric acid solution is sprayed on the plate, which is then heated at 105 °C (10 min) yielding a variety of colors.

Aniline Phthalate
An aniline (1 g)/o-phthalic acid (1.5 g) in n-butanol (100 mL) (saturated with water) is sprayed on the plate, which is then heated at 105 °C (10 min) yielding a variety of colors.

Anisaldehyde-Sulfuric Acid

A 1 % anisaldehyde solution in acetic acid (acidified by conc. sulfuric acid) is sprayed on the plate, which is then heated at 105 °C to yield a variety of colors.

p-Anisidine Phthalate

A 0.1 M solution of p-anisidine and phthalic acid in ethanol is sprayed on the plate, which is then heated at 100 °C (10 min) to yield a variety of colors.

Anthrone

A 1 % anthrone solution in 60 % aqueous ethanol solution acidified with 10 mL 60 % phosphoric acid is sprayed on the plate, which is then heated at 110 °C (5 min) to yield yellow spots.

Antimony (III) Chloride

See Carr-Price reagent.

Antimony (III) Chloride-Acetic Acid

A 20 % antimony (III) chloride solution in 75 % chloroform-acetic acid solution is sprayed on the plate, which upon heating at 100 °C (5 min) yields a variety of colors.

Antimony (V) Chloride

A 20 % antimony (V) chloride solution in chloroform or carbon tetrachloride is sprayed on the plate yielding a variety of colors upon heating.

Benedict's Reagent

A solution that is 0.1 M in cupric sulfate, 1.0 M in sodium citrate, and 1.0 M in sodium carbonate is sprayed on the plate which is then observed under long-wave UV light.

Benzidine Diazotized

A 0.5 % benzidine solution in 0.005 % hydrochloric acid is mixed with an equal volume of 10 % sodium nitrite solution in water; the mixture is sprayed on the plate to yield a variety of colors.

Benzidine-Trichloroacetic Acid

A 0.5 % benzidine in (1:1:8) acetic acid/trichloroacetic acid/ethanol is sprayed on the plate to yield red-brown spots upon heating (110 °C) or exposure to unfiltered UV light (15 min).

Bismuth (III) Chloride

A 33 % ethanol solution of bismuth (III) chloride is sprayed on the plate, which upon heating (110 °C) yields fluorescent spots under long-wave UV light.

Carbazole-Sulfuric Acid

A 0.5 % carbazole in ethanol/sulfuric acid (95:5) is sprayed on the plate which yields violet spots (on blue background) after heating at 120 °C (10 min).

Carr-Price Reagent

A 25 % antimony (III) chloride solution in chloroform or carbon tetrachloride is sprayed on the plate which is heated at 100 °C (up to 10 min) to yield a variety of colors.

Ceric Ammonium Sulfate

A 1 % solution of ceric ammonium sulfate in strong acids (phosphoric, nitric) is sprayed on the plate to yield yellow/green spots on a red background, after heating at 105 °C (10 min).

Chloramine-T

A 10 % chloramine-T solution is sprayed on the plate, followed by 1 N hydrochloric acid. The chromatogram is dried and exposed to 25 % ammonium hydroxide and warmed.

Chlorine-Pyrazolinone-Cyanide

An equal volume mixture of 0.2 M 1-phenyl-3-methyl-2-pyrazolin-5-one solution (in pyridine) and 1 M aqueous potassium cyanide solution is sprayed on the plate that has been previously exposed to chlorine vapors. The resulting red spots turn blue after a few minutes.

Chlorosulfonic Acid-Acetic Acid

A 35 % chlorosulfonic acid solution in acetic acid is sprayed on the plate which is then heated at 130 °C (5 min) to produce fluorescence under long-wave UV.

Cinnamaldehyde-Hydrochloric Acid

A 5 % cinnamaldehyde solution in ethanol (acidified with hydrochloric acid) is sprayed on the plate, which is then placed in a hydrochloric acid chamber to yield red spots.

Cobalt (II) Thiocyanate

A ammonium thiocyanate (15 %)/cobalt (II) chloride (5 %) solution in water is sprayed on the plate yielding blue spots.

Cupric Sulfate-Quinine-Pyridine

A solution that is 0.4 % in cupric sulfate, 0.04 % in quinine hydrochloride, and 4 % in pyridine in water is sprayed on the plate followed by a 0.5 % aqueous potassium permanganate solution. A variety of colors (white, yellow, violet) is detected on the chromatogram.

α-Cyclodextrin

A 30 % α-cyclodextrin solution in ethanol is sprayed on the plate, which is further developed in an iodine chamber.

Diazonium

See Fast Blue B Salt.

Diazotization and Coupling Reagent

A 1 % sodium nitrite solution (in 1 M hydrochloric acid) is sprayed on the plate, followed by a 0.2 % α-naphthol solution in 1 M potassium hydroxide and dried.

4-N,N-Dimethylaminobenzaldehyde-Sulfuric Acid

A 0.125 % solution of 4-N,N-dimethylaminobenzaldehyde in 65 % sulfuric acid mixed with 5 % ferric chloride (0.05 mL per 100 mL solution) is sprayed on the plate giving a variety of spots.

4-N,N-Dimethylaminocinnamaldehyde

A 0.2 % solution of 4-N,N-dimethylaminocinnamaldehyde in 6 N HCl/ethanol (1:4) is sprayed on the plate, which is then heated at 105 °C (5 min) revealing a variety of colored spots. Vapors of aqua regia tend to intensify the spots.

2,4-Dinitrofluorobenzene

A 1 % sodium bicarbonate solution in 0.025 M sodium hydroxide is sprayed on the plate followed by a 2,4-dinitrofluorobenzene (10 %) solution in methanol. Heating the plate in the dark (40 °C, 1 h) and further spraying it with diethyl ether yields a variety of spots.

2,4-Dinitrophenylhydrazine

A 0.4 % solution of 2,4-dinitrophenylhydrazine in 2 N hydrochloric acid is sprayed on the plate followed by a 0.2 % solution of potassium ferricyanide in 2 N hydrochloric acid yielding orange/yellow spots.

Dragendorff Reagent

A 1.7 % aqueous solution of basic bismuth nitrate in weak acids (tartaric, acetic) mixed with an aqueous potassium iodide or barium chloride (8 g/20 mL water) solution is sprayed on the plate to yield a variety of spots.

Ehrlich Reagent

A 1 % 4-N,N-dimethylaminobenzaldehyde solution in ethanol is sprayed on the plate, which is dried and then placed in a hydrochloric acid chamber to yield various spots.

Emerson Reagent

A 2 % 4-aminoantipyrine solution in ethanol is sprayed on the plate, followed by an 8 % aqueous potassium ferricyanide solution. The chromatogram is then placed in a chamber containing 25 % ammonium hydroxide.

EP

A 0.3 % solution of 4-N,N-dimethylaminobenzaldelyde in acetic acid/phosphoric acid/water (10:1:4) is sprayed on the plate to yield a variety of spots.

Fast Blue B Salt (Diazonium)

A 0.5 % aqueous solution of Fast Blue B Salt is sprayed on the plate followed by a 0.1 M sodium hydroxide.

Ferric Chloride-Perchloric Acid

A solution made out of 5 mL 5 % aqueous ferric chloride, 45 mL 20 % perchloric acid, and 50 mL 50 % nitric acid is sprayed on the plate to yield a variety of spots.

Ferric Chloride-Sulfosalicylic Acids

The plate is first exposed to a bromine atmosphere and then sprayed with a 0.1 % ethanolic solution of ferric chloride. After air-drying (15 min), the chromatogram is sprayed with a 1 % ethanolic solution of sulfosalicylic acid to yield a variety of spots.

Ferrocyanide-Hydrogen Peroxide

A 0.5 g ammonium chloride is added to a 0.1 % potassium ferrocyanide solution in 0.2 % hydrochloric acid and the resulting solution is sprayed on the plate, which is then dried (100 °C). The chromatogram is further sprayed with 30 % hydrogen peroxide, heated (150 °C, 30 min), and sprayed with 10 % potassium carbonate to yield yellow/red spots.

Ferrous Thiocyanate

A 2:3 mixture of a 4 % aqueous ferrous sulfate and 1.3 % acetone solution of ammonium thiocyanate is sprayed on the plate yielding red-brown spots.

Fluorescein-Hydrogen Peroxide

A 0.1 % fluorescein solution in 50 % aqueous ethanol is sprayed on the plate followed by a 15 % hydrogen peroxide in glacial acetic acid and heated (90 °C, 20 min) yielding non-fluorescent spots.

Folin Reagent

A 0.02 % sodium 1,2-naphthoquinone-4-sulfonate in 5 % sodium carbonate is sprayed on the plate, which is then dried to yield a variety of colors.

Folin–Denis Reagent

A tungstomolybdophosphoric acid solution is sprayed on the plate which is then exposed to ammonia vapors.

Gibbs Reagent

A 0.4 % methanolic solution of 2,6-dibromoquinonechloroimide is sprayed on the plate followed by a 10 % aqueous sodium carbonate yielding a variety of spots.

Glucose-Phosphoric Acid

A 2 % glucose solution in phosphoric acid/water/ethanol/n-butanol (1:4:3:3) is sprayed on the plate followed by heating (115 °C, 10 min) to yield a variety of spots.

Hydroxylamine-Ferric Chloride

A 1:2 mixture of a 10 % hydroxylammonium chloride/10 % potassium hydroxide in aqueous ethanol is sprayed on the plate followed by drying. The chromatogram is then sprayed with an ether solution of ferric chloride in hydrochloric acid to yield a variety of spots.

Iodide (Potassium) Starch

A 1 % potassium iodide solution in 80 % aqueous acetic acid is sprayed on the plate followed by a 1 % aqueous starch solution. A pinch of zinc dust is recommended as an addition to the potassium iodide solution.

Iodide (Sodium) Starch

A solution made by mixing a 5 % starch/0.5 % sodium iodide solution with an equal volume of concentrated hydrochloric acid is sprayed on the plate, which is then exposed to dry sodium hydroxide (desiccator) and evacuated (30–60 min) to yield brown spots.

Iodoplatinate (IP) Reagent

A 1:1 mixture of hydrogen hexachloroplatinate hydrate (0.1 g/100 mL aqueous solution) and potassium iodide (100 mL of 6 % solution)

Isatin-Zinc Acetate

An isatin (1 %)/zinc acetate (1.5 %) solution in isopropanol acidified with acetic acid is sprayed on the plate which is then heated to yield a variety of spots.

Kalignost Reagent

A 1 % solution of sodium tetraphenylborate in aqueous butanone is sprayed on the plate, followed by a 0.015 % methanolic solution of fischtin or quercetin to yield orange-red spots that fluoresce under long-wave UV.

König Reagent

A 2 % p-aminobenzoic acid in ethanolic hydrochloric acid (0.6 M) is sprayed on the plate that has been exposed (1 h) to vapors of cyanogen bromide.

Komarowski Reagent

A 2 % methanolic solution of p-hydroxybenzaldehyde that is 5 % in sulfuric acid is sprayed on the plate which is then heated (105 °C, 3 min) to yield yellow or pink spots.

Lewis–Smith Reagent

o-Aminobiphenyl (0.3 g dissolved in 100 mL of a 19:1 ethanol/phosphoric acid mixture) is sprayed on the plate, which is then heated at 110 °C (15 min).

Liebermann–Burchard Reagent

A freshly prepared mixture of 5 mL acetic anhydride/5 mL conc. sulfuric acid in 50 mL cold absolute ethanol is sprayed on the plate, which is heated at 100 °C (10 min) and observed under long-wave UV light.

Malachite Green Oxalate

A 1 % ethenolic potassium hydroxide solution is sprayed, the plate heated (150 °C, 5 min), and further sprayed with a buffered (pH=7) water/acetone solution of malachite green oxalate to yield white spots on blue background.

Marquis Reagent

A solution of 3 mL formaldehyde in 100 mL sulfuric acid.

Metaperiodate (Sodium)-p-Nitroaniline

A 35 % saturated solution of sodium metaperiodate is sprayed on the plate which is left to dry (10 min). The chromatogram is then sprayed with a 0.2 % p-nitroaniline solution in ethanol/hydrochloric acid (4:1) to yield fluorescing (long-wave UV) yellow spots.

Methoxynitroaniline–Sodium Nitrite

A 0.02 M 4-methoxy-2-nitroaniline solution in 50 % aqueous acetic acid/5 N sulfuric acid is sprayed on the plate, which is dried and re-sprayed with 0.2 % sodium nitrite to yield blue spots on an orange background.

Millon Reagent

A solution of mercury (5 g) in fuming nitric acid (10 mL) diluted with water (10 mL) is sprayed on the plate to yield yellow/orange spots that are intensified by heat (100 °C).

1,2-Naphthoquinone-4-Sulfonic Acid/Perchloric Acid

A 0.1 % 1,2-naphthoquinone-4-sulfonic acid solution in ethanol/perchloric acid/40 % formaldehyde/water (20:10:1:9) is sprayed on the plate which is then heated (70 °C) to yield pink spots that turn to blue on prolonged heating.

Neu Reagent

A 1 % methanolic solution of the β-aminoethylester of diphenylboric acid is sprayed on the plate to yield fluorescent spots under long-wave UV light.

Ninhydrin

A ninhydrin solution (0.3 % in acidified n-butanol or 0.2 % in ethanol) is sprayed on the plate which is then heated (110 °C). The resulting spots are stabilized by spraying with a solution made of 1 mL saturated aqueous cupric nitrate, 0.2 mL 10 % nitric acid, and 100 mL 95 % ethanol, to yield red spots when exposed to ammonium hydroxide (25 %).

p-Nitroaniline, Diazotized

A solution made by mixing 0.1 % aqueous p-nitroaniline/0.2 % aqueous sodium nitrite/10 % aqueous potassium carbonate (1:1:2) is sprayed on the plate to yield colored spots.

p-Nitroaniline, Diazotized (Buffered)

A solution of 0.5 % p-nitroaniline (in 2 N hydrochloric acid), 5 % aqueous sodium nitrite, and 20 % aqueous sodium acetate (10:1:30) is sprayed on the plate to yield a variety of colored spots.

Nitroprusside (Sodium)

A solution made by mixing sodium nitroprusside (1.5 g), 2 N hydrochloric acid (5 mL), methanol (95 mL), and 25 % ammonium hydroxide (10 mL) is sprayed on the plate to yield a variety of colors.

Nitroprusside (Sodium), Basic

A 2 % sodium nitroprusside solution in 75 % ethanol is sprayed on the plate which has already been treated with 1 N sodium hydroxide to yield red spots.

Resorcinol

A mixture consisting of 0.6 % ethanolic orcinol and 1 % ferric chloride in dilute sulfuric acid is sprayed on the plate which is further heated (100 °C, 10 min) to yield characteristic spots.

Phosphate Buffer (pH = 7.4)

A 25.0 mL potassium dihydrogen phosphate solution (27.3 g potassium dihydrogen phosphate dissolved in CO_2 distilled or HPLC water and volume adjusted to 1.0 L) mixed with 39.3 mL 0.1 M sodium hydroxide and volume made up to 100 mL with CO_2 free distilled water.

Prochazka Reagent

A 10 % formaldehyde solution in 5 % hydrochloric acid solution in ethanol is sprayed on the plate which is then heated to yield fluorescent spots (yellow/orange/green) under long-wave UV.

Quercetin-Sodium Tetraphenylborate

A mixture of quercetin (0.015 % in methanol) and sodium tetraphenyl-borate (1 % in n-butanol saturated with water) is sprayed on the plate to yield orange/red spots.

Quinaldine

A 1 %–1.5 % solution of 3,5-diaminobenzoic acid dihydrochloride in 30 % phosphoric acid is sprayed on the plate which is then heated (100 °C, 15 min) to yield fluorescent (green/yellow) spots under long-wave UV or (in case of high concentrations) brown spots in daylight.

Sakaguchi Reagent

A 0.1 % acetone solution of 8-hydroxyquinoline is sprayed on the plate followed by a 0.2 % 0.5 N sodium hydroxide solution to yield orange/red spots.

Salkowski Reagent

A 0.01 M aqueous ferric chloride/35 % perchloric acid solution is sprayed on the plate which is then heated (60 °C, 5 min) to yield a variety of colors intensified when exposed to aqua regia.

Schweppe Reagent

A mixture of 2 % aqueous glucose/2 % ethanolic aniline in n-butanol is sprayed on the plate which is heated (125 °C, 5 min) to yield a variety of spots.

Silver Nitrate-Ammonium Hydroxide-Sodium Chloride

A mixture of silver nitrate (0.05 M)/ammonium hydroxide (5 %) is sprayed on the plate, followed by drying and further spraying with 10 % aqueous sodium chloride to yield yellow/brown spots.

Silver Nitrate-Fluorescein

A mixture of silver nitrate (2 %)/sodium-fluorescein (0.2 %) in 80 % ethanol is sprayed on the plate to yield yellow spots on pink background.

Silver Nitrate-Formaldehyde

The plate is consecutively sprayed with 0.05 M ethanolic silver nitrate, 35 % aqueous formaldehyde, 2 M potassium hydroxide, and, finally, a solution made of equal volumes of hydrogen peroxide (30 %) and nitric acid (65 %). Each spraying is preceded by a 30 min drying, and at the end, the plate is kept in the dark for 12 h before exposing to sunlight to yield dark gray spots.

Silver Nitrate-Hydrogen Peroxide

A 0.05 % silver nitrate solution in water/cellosolve/acetone (1:10:190) (to which a drop of 30 % hydrogen peroxide has been added) is spayed on the plate, which is then treated under unfiltered UV to yield dark spots.

Silver Nitrate-Sodium Hydroxide

A saturated silver nitrate solution is sprayed on the plate followed by a 0.5 M aqueous/methanol solution. Subsequent drying (100 °C, 2 min) yields a variety of spots.

Sonnenschein Reagent

A 2 % ceric sulfate solution in 20 % aqueous trichloroacetic acid (that has been acidified with sulfuric acid) is sprayed on the plate. A variety of colors appears upon heating (110 °C, 5 min).

Stahl Reagent

See van Urk reagent.

Sulfanilic Acid-1-Naphthylamine

A mixture of sulfanilic acid/1-naphthylamine in 30 % acetic acid is sprayed on the plate to yield a variety (violet/green/blue) of colors.

Thiochrome

A 0.3 M aqueous potassium ferricyanide solution that is 15 % in sodium hydroxide is sprayed on the plate yielding a variety of spots under long-wave UV.

Tollens' Reagent

See Zaffaroni reagent.

Vanillin-Potassium Hydroxide

A 2 % solution of vanillin in n-propanol is sprayed on the plate which is heated (100 °C, 10 min) and sprayed again with 1 % ethanolic potassium hydroxide. Reheating yields a variety of colors observed under daylight.

van Urk (Stahl) Reagent

A 0.5 % solution of 4-N,N-dimethylaminobenzaldehyde in concentrated hydrochloric acid/ethanol (1:1) is sprayed on the preheated plate, which is then subjected to aqua regia vapors to yield a variety of colors.

Vanillin-Phosphoric Acid Reagent

1 g vanillin in 100 mL 50 % phosphoric acid (50 %).

Zaffaroni (Tollens') Reagent

A mixture of silver nitrate (0.02 M)/ammonium hydroxide (5 M) is sprayed on the plate which is then heated (105 °C, 10 min) to yield black spots.

Zwikker Reagent

A 1 % cobaltous nitrate in absolute ethanol is sprayed on the plate, which is dried (at room temperature) and exposed to a wet chamber containing 25 % ammonium hydroxide.

CHAPTER 5

Supercritical Fluid Extraction and Chromatography

SOME USEFUL FLUIDS FOR SUPERCRITICAL FLUID EXTRACTION AND CHROMATOGRAPHY

The following table lists some useful carrier and modifier fluids for supercritical fluid extraction and chromatography, along with relevant properties [1–7]. The critical properties are needed to determine successful fluid operating ranges. Where possible, experimental values are provided. In some cases, however, values calculated with a group contribution approach are presented, and these entries are marked with an asterisk. The dipole moment is provided to assess fluid polarity, although these values can be temperature dependent, especially with the more complex fluids. Occasionally, conformations change with temperature, resulting in a change in dipole moment. Data on ultraviolet cutoff are provided to allow the application of UV-Vis monitoring instrumentation. Data are not provided if the only electronic transition is in the very low wavelength range, and the spectrum is largely flat. With respect to the halocarbon fluids, if a commonly used refrigerant designator is available, it is presented with the chemical name. The fluids listed here have been either used or proposed for use in supercritical fluid chromatography or supercritical fluid extraction. The reader should also note that some of these fluids (for example, methanol and toluene) will undergo serious chemical degradation under near critical conditions while in contact with stainless steels and other common materials [4–6].

REFERENCES

1. Bruno, T.J. and Svoronos, P.D.N., *CRC Handbook of Basic Tables for Chemical Analysis*, 3rd ed., CRC Press, Boca Raton, FL, 2011.
2. Bruno, T.J. and Ely, J.F., *Supercritical Fluid Technology, Reviews in Modern Theory and Applications*, CRC Press, Boca Raton, FL, 1991.
3. Bruno, T.J., *CRC Handbook for the Identification and Analysis of Alternative Refrigerants*, CRC Press, Boca Raton, FL, 1995.
4. Bruno, T.J. and Straty, G.C., Thermophysical property measurement on chemically reactive systems - A case study, *J. Res. Nat. Bur. Stds. (U.S.)*, 91(3), 135, 1986.
5. Straty, G.C., Ball, M.J., and Bruno, T.J., Experimental determination of the PVT surface for benzene, *J. Chem. Eng. Data*, 32, 163, 1987.
6. Ashe, W. Mobile phases for supercritical fluid chromatography, *Chromatographia*, 11(7), 411, 1978.
7. Rumble, J.R., ed., *CRC Handbook of Chemistry and Physics*, 100th ed., Taylor & Francis, Boca Raton, FL, 2019.

Inorganic Fluids

Fluid	T_c (°C)	ρ_c (g/mL)	P_c (MPa)	μ (D)	UV Cutoff (nm)
Carbon dioxide	31.98	0.460	7.4	0.0	
Ammonia	132.41	0.235	11.357	1.47	
Nitrous oxide	36.5	0.450	7.375	0.167	
Sulfur dioxide	157.49	0.520	7.884	1.63	
Sulfur hexafluoride	45.6	0.730	3.76	0.0	
Water	374.1	0.40	22.1	1.85	
Xenon	16.6	1.155	5.9		

Hydrocarbon Fluids

Fluid	T_c (°C)	ρ_c (g/mL)	P_c (MPa)	μ (D)	UV Cutoff (nm)
Ethane	32.21	0.203	4.88	0.0	
Propane	96.8	0.220	4.25	$\geq$0.05	
n-Butane	152.1	1.225	3.79	0.0	
n-Pentane	196.6	0.232	3.4	0.0	
n-Hexane	234.2	0.234	3.0	0.08	250
Benzene	288.9	0.304	4.9	0.0	325
Toluene	320.8	0.29	4.2	0.084	325

Alcohols, Ethers, and Ketones

Fluid	T_c (°C)	ρ_c (g/mL)	P_c (MPa)	μ (D)	UV Cutoff (nm)
Methanol	239.5	0.272	8.0	1.70	255
Ethanol	240.8	0.275	6.14	1.7	255
Isopropanol	235.3	0.273	4.8	1.66	255
Dimethyl ether	126.9	0.271	5.31	1.3	
Diethyl ether	193.7	0.265	3.64	1.3	225
Acetone	235	0.278	4.7	2.9	350

Halocarbons

Fluid	T_c (°C)	ρ_c (g/mL)	P_c (MPa)	μ (D)	UV Cutoff (nm)
Fluoromethane, R-41	44.27	0.301	5.88	1.8	285
Difluoromethane, R-32	78.41	0.430	5.83	1.978	240
Trifluoromethane, R-23	25.85	0.526	4.82	1.65	300
Chloromethane, R-40	143.09	0.364	6.72	1.87	240
Chlorodifluoromethane, R-22	96.15	0.521	4.98	1.44	220
Dichlorofluoromethane, R-21	178.5	0.522	5.2	1.24	235
Trichlorofluoromethane, R-11	198	0.554	4.40	NA	
Chlorotrifluoromethane, R-13	28.9	0.578	3.89	0.5	220
Dichlorodifluoromethane, R-12	111.8	0.558	4.12	0.51	245
Fluoroethane, R-161	102.1	0.288	5.02	1.94	210
1,1-Difluoroethane, R-152a	113.3	0.365	4.52	2.262	
1,1,1-Trifluoroethane, R-143a	72.74	0.434	3.77	2.32	
1,1,2-Trifluoroethane, R-143	156.7	0.466	4.52	1.68 (35.9 °C) 1.75 (136.9 °C)	
1,1,1,2-Tetrafluoroethane, R-134a	101.06	0.515	4.06	2.058	
1,1,2,2-Tetrafluoroethane, R-134	118.60	0.542	4.61	0.991 (36 °C) 0.250 (140 C)	300
Pentafluoroethane, R-125	66.1	0.572	3.63	1.54	
1,1-Dichlorotetrafluoro ethane, R-114a	145.5	0.582	3.31		
1-Chloro-1,1-difluoroethane	137.1	0.435	4.12	2.14	265
1,2-Dichlorotetrafluoroethane	145.7	0.582	3.26	0.668 (35 °C) 0.699 (137 °C)	240
2-Chloro-1,1,1,2-tetrafluoroethane, R-124	122.5	0.554	3.63	1.469	230
Chloropentafluoroethane, R-115	79.9	0.596	3.12	0.52	245
1,1,1,2,3,3-Hexafluoropropane, R-236ea	141.1	0.571	3.533	NA	
1,1,1,2,3,3,3-Heptafluoropropane, R-227ea	101.9	0.580	2.93	NA	
2-Chloroheptafluoro propane, R-217ba	127.5*	0.592*	3.12*	NA	210
Bis(difluoromethyl) ether, E-134	147.1	0.522	4.16	1.739 (36 °C) 1.840 (173 °C)	

P-ρ-T SURFACE FOR CARBON DIOXIDE

In earlier editions of this book, we included a tabulation of the P-ρ-T surface for carbon dioxide, the most common fluid used in supercritical extraction and chromatography [1–8]. In this edition, we have withdrawn this table since the best approach is to calculate the density from reference quality equations of state [9,10]. If the user desires the tabular approach, this can be found in the third edition [11].

REFERENCES

1. Schmidt, R. and Wagner, W., A new form of the equation of state for pure substances and its application to oxygen, *Fluid Phase Equilib.*, 19, 175, 1985.
2. Ely, J.F., *National Bureau of Standards*, Boulder, CO, private communication (coefficients for carbon dioxide), 1986.
3. Poling, B.E., Prausnitz, J.M., and O'Connell, J.P., *The Properties of Gases and Liquids*, 5th ed., McGraw-Hill, 2000.
4. Prausnitz, J.M., *Molecular Thermodynamics of Fluid Phase Equilibria*, Prentice-Hall, Englewood Cliffs, 1969.
5. Chao, K.C. and Greenkorn, R.A., *The Thermodynamics of Fluids*, Marcel Dekker, New York, 1975.
6. Jacobson, R.T. and Stewart, R.B., Thermodynamic properties of nitrogen including liquid and vapor phases from 63K to 2000K with pressures to 10,000 bar, *J. Phys. Chem. Ref. Data*, 2(4), 757, 1973.
7. Ely, J.F., *Proceedings of 63rd Gas Processors Association Annual Convention*, pg. 9, 1984.
8. Angus, S., Armstrong, B., and deReuck, K.M., *Carbon Dioxide, International Thermodynamic Tables of the Fluid State*, Pergamon Press, Oxford, 1976.
9. Rowley, R.L., Wilding, W.V., Oscarson, J.L., Zundel, N.A., Marshall, T.L., Daubert, T.E., and Danner, R.P., *DIPPR Data Compilation of Pure Compound Properties*, Design Institute for Physical Properties AIChE, New York, 2008.
10. Lemmon, E. W., Huber, M.L, and McLinden, M.O., REFPROP, Reference fluid thermodynamic and transport properties, NIST Standard Reference Database 23, Version 9.1, National Institute of Standards and Technology, Gaithersburg, MD, 2013.
11. Bruno, T.J. and Svoronos, P.D.N., *CRC Handbook of Basic Tables for Chemical Analysis*, 3rd ed., CRC Press, Boca Raton, FL, 2011.

SOLUBILITY PARAMETERS OF THE MOST COMMON FLUIDS FOR SUPERCRITICAL FLUID EXTRACTION AND CHROMATOGRAPHY

The following table provides the solubility parameters, δ*, for the most common fluids and modifiers used in supercritical fluid extraction and chromatography. The data presented in the first table are for carrier or solvent supercritical fluids at a reduced temperature T_r of 1.02 and a reduced pressure P_r of 2. These values were calculated with the equation of Lee and Kesler [1,2]. The data presented in the second table are for liquid solvents that are potential modifiers [3].

The solubility parameter is defined as the square root of the cohesive energy density. The most common presentation of the solubility parameter is in units of $(cal^{1/2}cm^{-3/2})$. Here, the cohesive energy is expressed per unit volume. A more modern format that is found in some of the literature after 1990 utilizes SI units derived from cohesive pressures. It is possible to convert between the two scales with the following equations:

$$\delta * \left(cal^{1/2}cm^{-3/2}\right) = 0.48888 \times \delta * \left(MPa^{1/2}\right) \tag{1}$$

$$\delta * \left(MPa^{1/2}\right) = 2.0455 \times \delta * \left(cal^{1/2}cm^{-3/2}\right) \tag{2}$$

Thus, as a rough guide, the solubility parameters expressed in the SI system of cohesive pressures are numerically approximately double the values expressed in the older system.

REFERENCES

1. Lee, B.I. and Kesler, M.G., A generalized thermodynamic correlation based on three-parameter corresponding states, *AIChE J.*, 21, 510, 1975.
2. Schoenmakers, P.J. and Vunk, L.G.M., *Advances in Chromatography*, Vol. 30, J.C. Giddings, E. Grushka, and P.R. Brown, eds., Marcel Dekker, New York, 1989.
3. Barton, A., *CRC Handbook of Solubility and Cohesive Energy Parameters*, CRC Press, Boca Raton, FL, 1983.

SOLUBILITY PARAMETERS OF SUPERCRITICAL FLUIDS

In this table, the solubility parameter, δ^*, (in $cal^{1/2}cm^{-3/2}$) was obtained by the methods outlined in Refs. [1] and [2], and the conversion to the pressure scale was done by applying Eq. (2).

Supercritical Fluid	δ^* ($cal^{1/2}cm^{-3/2}$)	δ^* ($MPa^{1/2}$)
Carbon dioxide	7.5	15.3
Nitrous oxide	7.2	14.7
Sulfur hexafluoride	5.5	11.3
Ammonia	9.3	19.0
Xenon	6.1	12.5
Ethane	5.8	11.9
Propane	5.5	11.3
n-Butane	5.3	10.8
Diethyl ether	5.4	11.0

SOLUBILITY PARAMETERS OF LIQUID SOLVENTS

In this table, we provide solubility parameters for some liquid solvents that can be used as modifiers in supercritical fluid extraction and chromatography. The solubility parameters (in $MPa^{1/2}$) were obtained from Ref. [3], and those in $cal^{1/2}cm^{-3/2}$ were obtained by application of Eq. (1), for consistency. It should be noted that other tabulations exist in which these values are slightly different, since they were calculated from different measured data or models. The reader is therefore cautioned that these numbers are for trend analysis and separation design only. For other applications of cohesive parameter calculations, it may be more advisable to consult a specific compilation. This table should be used along with the table on modifier decomposition, since many of these liquids show chemical instability, especially in contact with active surfaces.

Liquid	δ^* ($cal^{1/2}cm^{-3/2}$)	δ^* ($MPa^{1/2}$)
n-Pentane	7.0	14.4
n-Hexane	7.3	14.9
n-Heptane	7.5	15.3
Cyclohexane	8.2	16.8
Benzene	9.1	18.7
Toluene	8.9	18.3
Acetone	9.6	19.7
Methyl ethyl ketone	9.4	19.3
Chloroform	9.1	18.7
Dichloromethane	9.9	20.2
Trichloroethene	9.1	18.7
Methanol	14.5	29.7
Ethanol	12.8	26.2
Diethyl ether	7.5	15.4
Tetrahydrofuran	9.0	18.5
1,4-Dioxane	10	20.5
Water	23.5	48.0

INSTABILITY OF MODIFIERS USED WITH SUPERCRITICAL FLUIDS

Liquid modifiers that are commonly used to increase the effective polarity of supercritical fluids such as carbon dioxide frequently have inherent chemical instabilities that must be considered when designing an analysis, or in the interpretation of results [1–3]. In many cases, such solvents are obtainable with stabilizers added to control the instability or to slow the reaction. Reactive solvents that do not have stabilizers must be used quickly or be given proper treatment. In either case, it is important to understand that the solvents (as they may be used in an analysis) are not necessarily pure materials. The reader is cautioned that many of the other fluids listed earlier in this section are thermally unstable; this table only treats chemical instabilities that are considerable at typical laboratory ambient temperature.

REFERENCES

1. Sadek, P.C., *The HPLC Solvent Guide*, 2nd. Ed., Wiley Interscience, New York, 2002.
2. Bruno, T.J., Straty, G.C., *J. Res. Nat. Bur. Stds. (U.S.)*, 91(3), 135-138, 1986.
3. Ashe, W., Mobile phases for supercritical fluid chromatography, *Chromatographia*, 11(7), 411, 1978.

Solvent	Contaminants, Reaction Products	Stabilizers
Ethers		
Diethyl ether	Peroxides[a]	2 %–3 % (vol/vol) ethanol[b] 1–10 ppm (mass/mass) BHT (1.5 %–3.5 % ethanol) + (0.2 %–0.5 % water) + (5–10 ppm (mass/mass) BHT)
Isopropyl ether	Peroxides[a]	0.01 % (mass/mass) hydroquinone 5–100 ppm (mass/mass) BHT
1,4-Dioxane	Peroxides[a]	25–1,500 ppm (mass/mass) BHT
Tetrahydrofuran	Peroxides[a]	25–250 ppm (mass/mass) BHT
Chlorinated Alkanes		
Chloroform	Hydrochloric acid, chlorine, phosgene (CCl_2O)	0.5 %–1 % (vol/vol) ethanol 50–150 ppm (mass/mass) amylene[c] Various ethanol amylene blends
Dichloromethane	Hydrochloric acid, chlorine, phosgene (CCl_2O)	25 ppm (mass/mass) amylene 25 ppm (mass/mass) cyclohexene 400–600 ppm (mass/mass) methanol Various amylene methanol blends
Alcohols		
Ethanol	Water, numerous denaturants are commonly added	
Methanol	Water; formaldehyde (at elevated temperature)	
Acetone	Diacetone alcohol and higher oligomers	

[a] The peroxide concentration that is usually considered hazardous is 250 ppm (mass/mass). See the treatment of peroxide hazards in Chapter 15 of this book.

[b] Ethanol does not actually stabilize diethyl ether, nor is it a peroxide scavenger, although it was thought to be so in the past. It is still available in chromatographic solvents to preserve the utility of retention relationships and analytical methods.

[c] Amylene is a generic name for 2-methyl-2-butene.

BHT—2,6-di-t-butyl-*p*-cresol.

CHAPTER 6

Electrophoresis

SEPARATION RANGES OF POLYACRYLAMIDE GELS

The following table provides a rough guide to the separation ranges of polyacrylamide gels that have varying gel concentrations, T, expressed in percent, as a function of relative molecular mass [1,2].

REFERENCES

1. Andrews, A.T., *Electrophoresis - Theory Techniques and Biochemical and Clinical Applications*, 2nd ed., Oxford University Press, Oxford, 1986, reproduced with permission.
2. Hames, B.D., ed. "Gel Electrophoresis of proteins: A practical approach," 3rd ed., *The Practical Approach Series*, Oxford University Press, New York, 1998.

T (%)	Optimum Relative Molecular Mass Range
3–5	Above 100,000
5–12	20,000–150,000
10–15	10,000–80,000
15+	Below 15,000

PREPARATION OF POLYACRYLAMIDE GELS

The following table provides in recipe format the typical proportions of reagents needed to prepare 100 mL of the starting material for polyacrylamide gels [1,2]. The factor T is the gel concentration and is related to the ability to separate a given relative molecular mass range. Typically, the tertiary aliphatic amines N,N,N,N-tetramethylethylenediamine (TEMED) or 3-dimethylamino-propionitrile (DMAPN) are used to catalyze the reaction. Note that gelation does not occur readily below T = 2.5 %.

REFERENCES

1. Andrews, A.T., *Electrophoresis - Theory Techniques and Biochemical and Clinical Applications*, 2nd ed., Oxford University Press, Oxford, 1986, reproduced with permission.
2. A guide to polyacrylamide gel electrophoresis and detection, http://www.bio-rad.com/webroot/web/pdf/lsr/literature/Bulletin_6040.pdf, accessed April 2020.

Constituent	Amounts Required for Gels with		
	T = 5 %	T = 7.5 %	T = 10 %
Acrylamide	4.75 g	7.125 g	9.50 g
Bisacrylamide	0.25 g	0.375 g	0.50 g
TEMED or DMAPN	0.05 mL	0.05 mL	0.05 mL
Ammonium persulfate	0.05 g	0.05 g	0.05 g

BUFFER MIXTURES COMMONLY USED FOR POLYACRYLAMIDE GEL ELECTROPHORESIS

The following table provides suggested buffers used for polyacrylamide gel electrophoresis (PAGE). This list is by no means exhaustive; however, these buffers are the most common [1]:

REFERENCE

1. Andrews, A.T., *Electrophoresis - Theory Techniques and Biochemical and Clinical Applications*, 2nd ed., Oxford University Press, Oxford, 1986, reproduced with permission.

Approximate pH Range	Primary Buffer Constituent	pH Adjusted to the Desired Value with
2.4–6.0	0.1 M citric acid	1 M NaOH
2.8–3.8	0.05 M formic acid	1 M NaOH
4.0–5.5	0.05 M acetic acid	1 M NaOH or tris
5.2–7.0	0.05 M maleic acid	1 M NaOH or tris
6.0–8.0	0.05 M KH_2PO_4 or NaH_2PO_4	1 M NaOH
7.0–8.5	0.05 M Na diethyl-barbiturate (Veronal)	1 M HCl
7.2–9.0	0.05 M tris	1 M HCl or glycine
8.5–10.0	0.015 M $Na_2B_4O_7$	1 M HCl or NaOH
9.0–10.5	0.05 M glycine	1 M NaOH
9.0–11.0	0.025 M $NaHCO_3$	1 M NaOH

Note that tris is a buffer made using tris(hydroxymethyl)aminomethane, also abbreviated as THAM.

PROTEINS FOR INTERNAL STANDARDIZATION OF
POLYACRYLAMIDE GEL ELECTROPHORESIS

The following table provides a list of proteins that may be used as internal standards in quantitative applications of polyacrylamide gel electrophoresis. These proteins may be used in isoelectric focusing or in SDS-PAGE. The isoelectric points are reported at 25 °C [1]:

REFERENCE

1. Andrews, A.T., *Electrophoresis - Theory Techniques and Biochemical and Clinical Applications*, 2nd ed., Oxford University Press, Oxford, 1986, reproduced with permission.

Protein	Isoelectric Point (pI at 25 °C)	Relative Molecular Mass
Lysozyme	10.0	14,000
Cytochrome C (horse)	9.3	12,256
Chymotrypsinogen A (ox)	9.0	23,600
Ribonuclease A	8.9	13,500
Myoglobin (sperm whale)	8.2	17,500
Myoglobin (horse)	7.3	17,500
Erythroagglutinin (red kidney bean)	6.5	130,000
Insulin (beef)	5.7	11,466
β-Lactoglobulin B	5.3	36,552
β-Lactoglobulin A	5.1	36,724
Bovine serum albumin	5.1	67,000
Ovalbumin	4.7	45,000
Alkaline phosphatase (calf intestine)	4.4	140,000
α-Lactalbumin	4.3	14,146

CHROMOGENIC STAINS FOR GELS

The following table provides common stain reagents for use in electrophoresis gels [1,2]:

REFERENCES

1. Melvin, M., *Electrophoresis (Analytical Chemistry by Open Learning)*, John Wiley & Sons, Chichester, 1987, reproduced with permission.
2. Hart, C., Schulenberg, B., Steinberg, T.H., Leung, W.-Y., and Patton, W.F., Detection of glycoproteins in polyacrylamide gels on electroblots using Pro-Q Emerald 488 dye, a fluorescent periodate Schiff base stain, *Electrophoresis*, 24, 588, 2003.

Types of Substance Stained	Staining Reagent	Comments
Amino acids, peptides, and proteins	Ninhydrin	Very sensitive stain for amino acids, either free or combined in polypeptides. Used after paper electrophoresis.
Proteins	Amido Black 10B	Binds to cationic groups on proteins. Adsorbs onto cellulose, giving high background staining with paper and dehydration and shrinkage of polyacrylamide gels.
	Coomassie Brilliant Blue	Binds to basic groups on proteins and also by nonpolar interactions. Widely used stain.
	Ponceau S (Ponceau Red)	Used routinely in clinical laboratories for cellulose acetate and starch gels. Very rapid staining reaction which leaves a clear background.
Glycoproteins	Alcian blue	Stains the sugar moiety.
	Emerald 300 Emerald 488	Stains aldehydes on sugar moiety after treatment with periodic acid.
Copper-containing proteins	Alizarin Blue S	Specifically indicates the presence of copper.
Polynucleotides, including RNA and DNA	Acridine orange	Stained product can be assessed quantitatively.
	Pyronine Y (or G)	Gives a permanent staining, so electrophoretogram can be stored for several weeks.
Proteins, lipids, carbohydrates, polynucleotides	Stains-all	Wide applicability, as it forms characteristic colored products with many different types of molecule. Low sensitivity.

FLUORESCENT STAINS FOR GELS

The following table provides common fluorescent stain reagents for use in electrophoresis [1–4]. Note that these agents are typically applied in small amounts before electrophoresis. Other stains are available as proprietary materials, and the reader is advised to consult reviews on staining procedures for additional materials. The reader is advised that anything capable of binding DNA with high affinity is a possible carcinogen; all such stains and dyes must be handled with care. Nitrile laboratory gloves are recommended; latex will not provide adequate protection.

REFERENCES

1. Melvin, M., *Electrophoresis (Analytical Chemistry by Open Learning)*, John Wiley & Sons, Chichester, 1987, reproduced with permission.
2. Williams, L.R., Staining nucleic acids and proteins in electrophoresis gels, *Biotech. Histochemis.* 76(3), 127–132, 2001.
3. Allen, R. and Budowle, B., *Protein Staining and Identification Techniques*, BioTechniques Press, Westborough, MA, 1999.
4. Hames, B.D., *Gel Electrophoresis of Proteins, A Practical Approach*, 3rd ed., Oxford University Press, Oxford, 2002.

Types of Substance Stained	Staining Reagent	Comments
Proteins	Dansyl chloride	Reacts with amine groups
	1-Anilino-8-naphthalene sulfonic acid	Non-fluorescent but gives fluorescent product.
	Fluorescamine	Non-fluorescent (nor are the hydrolysis products) but gives a fluorescent product with a labeled protein.
	2-Methoxy-2,4-diphenyl-2(H)-furanone	Non-fluorescent (nor are the hydrolysis products) but gives a fluorescent product with a labeled protein.
Polynucleotides, including RNA and DNA	Acridine orange	A nucleic acid-selective fluorescent cationic dye; often used in conjunction with ethidium bromide, below.
Double-stranded polynucleotides	Ethidium bromide	Very sensitive. Widely used with agarose gels.

ELECTROPHORESIS SAFETY

In this section, we discuss some of the unique hazards and required safety precautions needed when performing electrophoresis analysis. This section is presented here rather than in Chapter 15 because of the specific nature of the precautions [1]. The hazards associated with electrophoresis include electrical, chemical, and sometimes radiological hazards, and each of these requires consideration and attention.

REFERENCE

1. *Laboratory Safety and Chemical Hygiene Plan*, Queensborough Community College of City University of New York, Bayside, NY, 2012.

Electrical Hazards

Electrophoresis instruments can employ **very high voltage** (up to 2,000 volts) and potentially hazardous current (80 milliamps or more). This high-power output has the potential to cause a fatal electrical shock if not properly handled. Before using any electrophoresis instrument, inspect the power supply to ensure proper working order. Power supplies can be of fixed direct current voltage, variable direct current voltage, and occasionally (but rarely) batteries are used as the power source. Pilot lights should be functional. This is a commonly overlooked item; often a nonfunctional pilot is simply assumed to be a burned-out bulb or LED, and it is ignored. Power cords must be undamaged. Frayed cords must be replaced with a cord of appropriate gauge. See the section on wiring in Chapter 15 of this book for additional guidance. The on/off switch must be fully functional; again, it is common for a nonfunctional switch to be simply bypassed. This is a mistake. Ensure that the on/off switch is easily accessible. If an instrument is operated with a voltage in excess of what is needed for ion migration, distortion of the gel can result, and more importantly for safety, a thermal hazard can result.

The circuit used with an electrophoresis instrument should be equipped with a **Ground Fault Circuit Interrupter** (GFCI), which is identifiable by the test and reset buttons. It is permissible in laboratories for several outlets to be wired into a common GFCI; in such cases, the outlet should be labeled as being GFCI. If a GFCI is not available, it is usually permissible to use a portable GFCI on a short electrical cable. These are available at hardware stores or can be easily constructed. Note, however, that this approach might be prohibited by the institution safety guideline. A three-pronged plug with a ground is required, with at least a 15-ampere circuit. Do not adapt a two-conductor outlet, and do not use an extension cable. Locate the instrument on a nonconductive surface or use a rubber mat to isolate the instrument from any metal surface.

Be sure that the main power switch is turned off before connecting or disconnecting leads to the instrument. One should wear dry nitrile gloves for connecting and disconnecting one lead at a time, using only one hand, with the other lead on a nonconductive surface. An alternative method of connection is to connect both leads simultaneously with two hands, to avoid one lead unexpectedly becoming live. Banana plugs and alligator clips, if used, should be well seated or connected. If alligator clips are used, they should be covered with an insulating sheath. Extraneous equipment should be kept clear of electrophoresis instrumentation, especially if such equipment is conductive or can become a ground. This includes metal plates, spatulas, aluminum foil, and of course jewelry. Avoid touching water piping when manipulating any operating electrophoresis instrument. Do not allow wiring to dangle below the lab bench.

If the gel is light-sensitive, it is possible to drape the instrument with black felt, but this cloth must be kept free of buffer, otherwise it will become a conductor. It is also possible for moisture or leaked buffer on tubing to become conductive and cause an electrical hazard.

Prominently place "Danger – High Voltage" placards on the power supply and buffer tanks. Maintain good housekeeping around the instrument, and ensure that one does not have to reach over the instrument to access reagents or other apparatuses.

Chemical Hazards

Agarose gel electrophoresis often requires the use of **ethidium bromide** (EthBr, or EtBr, an intercalating agent) as a fluorescent tag (orange fluorescence under UV light, highly intensified when bound to DNA). EthBr is a strong mutagen (due to the DNA binding ability) and an irritant. It is critical to use nitrile gloves when handling this reagent. Low concentrations (below 0.4 % mass/mass) are not regulated as a hazardous waste, but institutions may require special handling of this waste. There are substitutes for EthBr that are commercially available and are safer to use. These should be considered where possible.

Acrylamide (prop-2-enamide) is often used to formulate gels; it is highly toxic and a probable carcinogen. It is an irritant to human skin and may possibly initiate skin tumors. In gel preparation, acrylamide forms linear polymers. **Bisacrylamide** (N-[(prop-2-enoylamino)methyl]prop-2-enamide) is often used as a cross-linking agent in gel preparation. The relative ratio of acrylamide and bisacrylamide determines pore size. Based on studies in rats, bisacrylamide is less hazardous than acrylamide, but both must be handled with utmost care.

Phenol, chloroform 2-mercaptoethanol, and also sodium dodecylsulfide (SDS) are also commonly used in electrophoresis. Phenol is corrosive and toxic; chloroform is toxic and a suspected carcinogen. Information on SDS, an anionic detergent applied to protein gels, is covered in the surfactants section in Chapter 14 of this book. 2-Mercaptoethanol (2-sulfanylethan-1-ol) is toxic, causing irritation to the nasal passageways and respiratory tract.

The dyes and buffer components listed in this chapter also have hazardous properties, and these are covered elsewhere in this book.

Personal Protective Equipment

The hazards associated with the use of electrophoresis instrumentation can be minimized by attention to the factors described above and the use of appropriate personal protection. These should include nitrile gloves, the use of a long-sleeved lab coat, safety glasses, long pants, and closed-toed shoes. When using a handheld UV lamp to visualize a gel, use appropriate eye protection.

CHAPTER 7

Electroanalytical Methods

DETECTION LIMITS FOR VARIOUS ELECTROCHEMICAL TECHNIQUES

The following table provides guidance in selection of electrochemical techniques by providing the relative sensitivities of various methods [1–4]. The limit of detection of lead, defined as the minimum detectable quantity (on a mole basis), is used as the basis of comparison [1].

REFERENCES

1. Batley, G.E., *Trace Element Speciation: Analytical Methods and Problems*, CRC Press, Boca Raton, FL, 1989.
2. Ebdon, L., Pitts, L., Cornelis, R., Crews, H., Quevauviller, P., and Donard, O.F.X.D., ed., *Trace Element Speciation for Environment, Food and Health*, Royal Society of Chemistry Cambridge, Cambridge, 2003.
3. Bajger, G., Konieczka, P., and Namieśnik, J., Speciation of trace element compounds in samples of biota from marine ecosystems, *J. Chem. Speciation Bioavailability*, 23(3), 125, 2015, doi: 10.3184/095422911X13032236772803.
4. Svoronos, P.D.N., *Electrochemical Methods, ASM Handbook*, Vol. 10, Materials Characterization, ASM Handbook Committee, ASM International, Materials Park, OH, 2019.

Electrochemical Technique	Limit of Detection for Lead (Mole)
DC polarography (DME)	2×10^{-6}
DC polarography (SMDE)	1×10^{-7}
DP polarography (SMDE)	1×10^{-7}
DP anodic stripping voltammetry (HMDE)	2×10^{-10}
SW anodic stripping voltammetry (HMDE)	1×10^{-10}
DC anodic stripping voltammetry (TMFE)	5×10^{-11}
DP anodic stripping voltammetry (TMFE)	1×10^{-11}
SW anodic stripping voltammetry (TMFE)	5×10^{-12}

DC, direct current; DP, differential pulse; SW, square wave; DME, dropping mercury electrode; SMDE, static mercury drop electrode; HMDE, hanging mercury drop electrode; TMFE, thin mercury film electrode.

VALUES OF (2.3026 RT/F) (IN mV) AT DIFFERENT TEMPERATURES

The following table gives the variation of (2.3026 RT/F) (mV) with temperature (°C) [1]. Electronic pH meters are voltmeters with scale divisions that are equivalent to the value of 2.3026 RT/F (in mV) per pH unit. Generally, an uncertainty of ±0.005 pH unit is feasible when the pH meter is reproducible to 0.2 mV.

REFERENCE

1. Shugar, G.J. and Dean, J.A., *The Chemist's Ready Reference Handbook*, McGraw-Hill Book Company, New York, 1990.

T (°C)	2.3026 RT/F (mV)	T (°C)	2.3026 RT/F (mV)
0	54.199	50	64.120
5	55.191	55	65.112
10	56.183	60	66.104
15	57.175	65	67.096
18	57.770	70	68.088
20	58.167	75	69.080
25	59.159	80	70.073
30	60.152	85	71.065
35	61.144	90	72.057
38	61.739	95	73.049
40	62.136	100	74.041
45	63.128		

POTENTIAL OF ZERO CHARGE (E^{ecm}) FOR VARIOUS ELECTRODE MATERIALS IN AQUEOUS SOLUTIONS AT ROOM TEMPERATURE

The table below lists the potential of zero charge (E^{ecm}) values (in volts) for various electrode materials in aqueous solutions at room temperature (25 °C) [1]. All values are with respect to the normal hydrogen electrode.

REFERENCE

1. Parsons, R., *Handbook of Electrochemical Constants*, Butterworths, London, 1959.

Electrode	E^{ecm} (V)	Solution Composition
Ag	+0.05	0.1 N KNO$_3$
Cd	−0.90	0.0001 N KCl
Ga	−0.60	1 N KCl + 0.1 N HCl
Hg	−0.192	Capillary inactive salts[a]
Ni	−0.06	0.001 N HCl
Pb	−0.69	0.001N KCl
Platinized Pt	+0.11	1 N Na$_2$SO$_4$ + 0.1 N H$_2$SO$_4$
Smooth Pt	+0.27	1 N Na$_2$SO$_4$ + 0.1 N H$_2$SO$_4$
Oxidized Pt	(+0.4)–(0.1)	1 N Na$_2$SO$_4$ + 0.1 N H$_2$SO$_4$
Te	+0.61	1 N Na$_2$SO$_4$
Tl	−0.80	0.001 N KCl
Tl-Hg (satd)	−0.65	1 N KCl
Zn	−0.63	1 N Na$_2$SO$_4$
Graphite	−0.07	0.05N KCl
Activated charcoal	(0.0)–(+0.2)	1 N Na$_2$SO$_4$ + 0.1 N H$_2$SO$_4$

[a] Any salt that is non-reactive with the capillary can be used.

VARIATION OF REFERENCE ELECTRODE POTENTIALS WITH TEMPERATURE

The following table lists the potentials of various (0.1 M KCl calomel, saturated KCl calomel, and 1.0 M KCl Ag/AgCl) electrodes at different temperatures (in °C) [1–3]. The values include the liquid-junction potential.

REFERENCES

1. Bates, R.G., et al, *J. Res. Nat. Bur. Std.*, 45, 418, 1950.
2. Bates, R.G. and Bower, V.E., *J. Res. Nat. Bur. Std.*, 53, 283, 1955.
3. Shugar, G.J. and Dean, J.A., *The Chemist's Ready Reference Handbook*, McGraw-Hill Book Company, New York, 1990.

Temperature (°C)	0.1 M KCl Calomel	Saturated KCl Calomel	1.0 M KCl Ag/AgCl
0	0.3367	0.25918	0.23655
5			0.23413
10	0.3362	0.25387	0.23142
15	0.3361	0.2511	0.22857
20	0.3358	0.24775	0.22557
25	0.3356	0.24453	0.22234
30	0.3354	0.24118	0.21904
35	0.3351	0.2376	0.21565
38	0.3350	0.2355	
40	0.3345	0.23449	0.21208
45			0.20835
50	0.3315	0.22737	0.20449
55			0.20056
60	0.3248	0.2235	0.19649
70			0.18782
80		0.2083	0.1787
90			0.1695

pH VALUES OF STANDARD SOLUTIONS USED IN THE CALIBRATION OF GLASS ELECTRODES

The following table gives the pH values of operational standard solutions recommended for the calibration of glass electrodes at 25 °C and 37 °C [1].

REFERENCE

1. Hibbert, D.B. and James, A.M., *Dictionary of Electrochemistry*, 2nd ed., John Wiley & Sons, New York, 1984.

Standard Solution	pH at	
	25 °C	37 °C
0.1 mol/kg potassium tetraoxalate	1.48	1.49
0.1 mol/dm³ hydrochloric acid + 0.09 mol/dm³ potassium chloride	2.07	2.08
0.05 mol/kg potassium hydrogen phthalate	4.005	4.022
0.10 mol/dm³ acetic acid + 0.10 mol/dm³ sodium acetate	4.644	4.647
0.10 mol/dm³ acetic acid + 0.01 mol/dm³ sodium acetate	4.713	4.722
0.02 mol/kg piperazine phosphate	6.26	6.14
0.025 mol/kg disodium hydrogen phosphate + 0.025 mol/kg potassium dihydrogen phosphate	6.857	6.828
0.05 mol/kg tris(hydroxymethyl)methane hydrochloride + 0.01667 mol/kg tris(hydroxymethyl)methane	7.648	7.332
0.05 mol/kg disodium tetraborate (borax)	9.182	9.074
0.025 mol/kg sodium bicarbonate + 0.025 mol/kg sodium carbonate	9.995	9.889
Saturated calcium hydroxide	12.43	12.05

TEMPERATURE VS. pH CORRELATION OF STANDARD SOLUTIONS
USED FOR THE CALIBRATION OF ELECTRODES

The following table gives the temperature vs. pH correlation of common standard solutions that are used for the calibration of electrodes [1–3]. Such solutions should be stable and easily prepared, whose solutes do not require further purification because of factors such as their hygroscopic nature. It is worth noting that the buffering capacity of these solutions is of little interest.

REFERENCES

1. Hibbert, D.B. and James, A.M., *Dictionary of Electrochemistry*, 2nd ed., John Wiley & Sons, New York, 1984.
2. Koryta, J., Dvorák, J., and Boháckova, V., *Electrochemistry*, Methuen and Co., London, 1970.
3. Robinson, R.H. and Srokes, R.H., *Electrolytic Solutions*, Butterworths, London, 1959.

Temperature (°C)	pH of						
	0.05 M Potassium Tetroxalate	Potassium Hydrogen Tartrate[1]	0.01 M Potassium Tartrate	0.05 M Potassium Hydrogen Phthalate	0.025 M K_2HPO_4 + 0.02 M NaH_2PO_4	0.01 M $Na_2B_4O_7$	$Ca(OH)_2$[a]
0	1.671	—	3.710	4.012	6.893	9.463	13.428
5	1.671	—	3.690	4.005	6.950	9.389	13.208
10	1.669	—	3.671	4.001	6.922	9.328	13.004
15	1.674	—	3.655	4.000	6.896	9.273	12.809
20	1.676	—	3.647	4.001	6.878	9.223	12.629
25	1.681	3.555	3.637	4.005	6.860	9.177	12.454
30	1.685	3.547	3.633	4.011	6.849	9.135	12.296
35	1.693	3.545	3.629	4.019	6.842	9.100	12.135
37	—	—	—	4.022	6.838	9.074	12.05
40	1.697	3.543	3.630	4.030	6.837	9.066	11.985
45	1.704	3.545	3.634	4.043	6.834	9.037	11.841
50	1.712	3.549	3.640	4.059	6.833	9.012	11.704
55	1.719	3.556	3.646	4.077	6.836	8.987	11.575
60	1.726	3.565	3.654	4.097	6.840	8.961	11.454
70	1.74	3.58	—	4.12	6.85	8.93	—
80	1.77	3.61	—	4.16	6.86	8.89	—
90	1.80	3.65	—	4.20	6.88	8.85	—
95	1.81	3.68	—	4.23	6.89	8.83	—

[a] Saturated at 25 °C.

SOLID MEMBRANE ELECTRODES

The following table lists the most commonly used solid membrane electrodes, their applications, and major interferences [1–3]. Often the membrane is composed of a salt (listed first) and a matrix (listed second). Thus, an $AgCl$–Ag_2S electrode involves the finely divided $AgCl$ in an Ag_2S matrix. The salt should be more soluble than the matrix but insoluble enough so that its equilibrium solubility gives a lower anion (Cl^-) activity than that of the sample solution.

REFERENCES

1. Fritz, J.S. and Schenk, G.H., *Quantitative Analytical Chemistry*, 5th ed., Prentice Hall, Englewood Cliffs, 1987.
2. Hall, D.G., Ion selective membrane electrodes: A general limiting treatment of interference effects, *J. Phys. Chem.*, 100, 7230, 1996.
3. Sutter, J., Radu, A., Paper, S., Bakker, E., and Pretsch, E., Solid-contact polymeric membrane electrodes with detection limits in the subnanomolar range, *Anal. Chim. Acta*, 523(1), 53, 2004.

Membrane	Ion Measured	Major Interferences
LaF_3	F^-	OH^-
Ag_2S	S^{-2}, Ag^+	Hg^{+2}
$AgCl$–Ag_2S	Cl^-	Br^-, I^-, S^{-2}, CN^-, NH_3
$AgBr$–Ag_2S	Br^-	I^-, S^{-2}, CN^-, NH_3
AgI–Ag_2S	I^-	S^{-2}, CN^-
$AgSCN$–Ag_2S	SCN^-	Br^-, I^-, S^{-2}, CN^-, NH_3
CdS–Ag_2S	Cd^{+2}	Ag^+, Hg^{+2}, Cu^{+2}
CuS–Ag_2S	Cu^{+2}	Ag^+, Hg^{+2}
PbS–Ag_2S	Pb^{+2}	Ag^+, Hg^{+2}, Cu^{+2}

LIQUID-MEMBRANE ELECTRODES

The following table gives the basic information on several liquid-membrane electrodes [1–3]. The selectivity of a membrane electrode for a given ion is determined primarily by the liquid ion exchanger used. Thus, as the preference of the ion exchanger for a specific ion increases, its selectivity increases. The selectivity is also affected by the organic solvent in which the liquid exchanger is dissolved. In this table, R- may be any organic radical or group.

REFERENCES

1. Durst, R.A., ed., *Ion-Selective Electrodes*, National Bureau of Standards Special Publication, Washington, 314, 70–71, 1969.
2. Frant, M.S. and Ross, J.W., Potassium ion specific electrode with high selectivity for potassium over sodium, *Science*, 167, 987, 1970.
3. Fritz, J.S. and Schenk, G.H., *Quantitative Analytical Chemistry*, 5th ed., Prentice Hall, Englewood Cliffs, 1987.

Ion Measured	Exchange Site	Selectivity Coefficients
K^+	Valinomycin	Na^+, 0.0001
Ca^{+2}	$(RO)_2POO^-$ (dialkylphosphate anion)	Na^+, 0.0016 Mg^{+2}, Ba^{+2}, 0.01 Sr^{+2}, 0.02 Zn^{+2}, 3.2 H^+, 10^7
Ca^{+2} and Mg^{+2}	$(RO)_2POO^-$ (dialkylphosphate anion)	Na^+, 0.01 Sr^{+2}, 0.54 Ba^{+2}, 0.94
Cu^{+2}	$RSCH_2COO^-$ (S-alkyl α-thioacetate anion)	Na^+, K^+, 0.0005 Mg^{+2}, 0.001 Ca^{+2}, 0.002 Ni^{+2}, 0.01 Zn^{+2}, 0.03
NO_3^-	 (1,10-phenanthridinium dication)	F^-, 0.0009 SO_4^{-2}, 0.0006 PO_4^{-3}, 0.0003 Cl^-, CH_3COO^-, 0.006 HCO_3^-, CN^-, 0.02 NO_2^-, 0.06 Br^-, 0.9
ClO_4^-	 (1,10-phenanthridinium dication)	Cl^-, SO_4^{-2}, 0.0002 Br^-, 0.0006 NO_3^-, 0.0015 I^-, 0.012 OH^-, 1.0

STANDARD REDUCTION ELECTRODE POTENTIALS FOR
INORGANIC SYSTEMS IN AQUEOUS SOLUTIONS AT 25 °C

A summary of the potentials, $E°$, in volts (at 25 °C) of the most useful reduction half-reactions is presented below [1–5]. The reactions are arranged in order of decreasing oxidation strength. When comparing two half-reactions, the oxidizing agent of the half-reaction with the higher (more positive) $E°$ will react with the reducing agent of the half-reaction with its lower (less positive) $E°$. Thus, Br_2 (1) ($E° = 1.065$ V) will oxidize H_2O_2 to $O_2(g)$ ($E° = 0.682$ V), but $O_2(g)$ cannot oxidize Br^-. No predictions can be made on the rate of reaction.

If two or more reactions between two substances are possible, the reaction that involves half-reactions which are farthest apart in the table will be most thermodynamically favorable. For instance, in the case of $O_2(g)$ reacting with Cu

$$O_2\left(g\right)+4H^+ +4e^- \rightarrow 2H_2O \ \ (E° = 1.229\,V) \tag{1}$$

$$O_2\left(g\right)+2H^+ +2e^- \rightarrow H_2O_2 \ \ (E° = 0.682\,V) \tag{2}$$

$$Cu^{+2} + 2e^- \rightarrow Cu \ \ (E° = 0.337\,V) \tag{3}$$

The reaction between (1) and (3) will be most favorable. However, if (3) is replaced with (4)

$$Cl_2(g)+2e^- \rightarrow 2Cl^- \ (E° = 1.36\,V) \tag{4}$$

The reactions between (2) and (4) will take place first.

REFERENCES

1. Bard, A.J. and Faulkner, L.R., *Electrochemical Methods*, 2nd ed., John Wiley & Sons, New York, 2001.
2. Day, R.A. and Underwood, A.L., *Quantitative Analysis*, 6th ed., Prentice Hall, 1991.
3. Spreight, J., ed., *Lange's Handbook of Chemistry*, 17th ed., McGraw-Hill Book Co., New York, 2016.
4. Ebbing, D.D. and Gamma, S.D., *General Chemistry*, 9th ed., Haughton Mifflin Co., Boston, 2008.
5. Sugar, G.J. and Dean, J.A., *The Chemist's Ready Reference Handbook*, McGraw-Hill Book Co., New York, 1990.

Half-Reaction		E° (V)
$F_2(g) + 2H^+ + 2e^-$	$\Rightarrow 2HF$	3.06
$O_3 + 2H^+ + 2e^-$	$\Rightarrow O_2 + H_2O$	2.07
$S_2O_8^{2-} + 2e^-$	$\Rightarrow 2SO_4^{2-}$	2.01
$Ag^{2+} + e^-$	$\Rightarrow Ag^+$	2.00
$H_2O_2 + 2H^+ + 2e^-$	$\Rightarrow 2H_2O$	1.77
$MnO_4^- + 4H^+ + 3e^-$	$\Rightarrow MnO_2(s) + 2H_2O$	1.70
$Ce(IV) + e^-$	$\Rightarrow Ce(III)\,(in\,1M\,HClO_4)$	1.61
$H_5IO_6 + H^+ + 2e^-$	$\Rightarrow IO_3^- + 3H_2O$	1.6
$Bi_2O_4 + 4H^+ + 2e^-$	$\Rightarrow 2BiO^+ + 2H_2O$	1.59
$BrO_3^- + 6H^+ + 5e^-$	$\Rightarrow \frac{1}{2}Br_2 + 3H_2O$	1.52
$MnO_4^- + 8H^+ + 5e^-$	$\Rightarrow Mn^{2+} + 4H_2O$	1.51
$PbO_2 + 4H^+ + 2e^-$	$\Rightarrow Pb^{2+} + 2H_2O$	1.455
$Cl_2 + 2e^-$	$\Rightarrow 2Cl^-$	1.36
$CrO_7^{2-} + 14H^+ + 6e^-$	$\Rightarrow 2Cr^{3+} + 7H_2O$	1.33
$MnO_2(s) + 4H^+ + 2e^-$	$\Rightarrow Mn^{2+} + 2H_2O$	1.23
$O_2(g) + 4H^+ + 4e^-$	$\Rightarrow 2H_2O$	1.229
$IO_3^- + 6H^+ + 5e^-$	$\Rightarrow \frac{1}{2}I_2 + 3H_2O$	1.20
$Br_2(liq) + 2e^-$	$\Rightarrow 2Br^-$	1.065
$ICl_2^- + e^-$	$\Rightarrow \frac{1}{2}I_2 + 2Cl^-$	1.06
$VO_2^+ + 2H^+ + e^-$	$\Rightarrow VO^{2+} + H_2O$	1.00
$HNO_2 + H^+ + e^-$	$\Rightarrow NO(g) + H_2O$	1.00
$NO_3^- + 3H^+ + 2e^-$	$\Rightarrow HNO_2 + H_2O$	0.94
$2Hg^{2+} + 2e^-$	$\Rightarrow Hg_2^{2+}$	0.92
$Cu^{2+} + I^- + e^-$	$\Rightarrow CuI$	0.86
$Ag^+ + e^-$	$\Rightarrow Ag$	0.799
$Hg_2^{2+} + 2e^-$	$\Rightarrow 2Hg$	0.79
$Fe^{3+} + e^-$	$\Rightarrow Fe^{2+}$	0.771
$O_2(g) + 2H^+ + 2e^-$	$\Rightarrow H_2O_2$	0.682
$2HgCl_2 + 2e^-$	$\Rightarrow Hg_2Cl_2(s) + 2Cl^-$	0.63
$Hg_2SO_4(s) + 2e^-$	$\Rightarrow 2Hg + SO_4^{2-}$	0.615
$H_3AsO_4 + 2H^+ + 2e^-$	$\Rightarrow HAsO_2 + 2H_2O$	0.581
$Sb_2O_5 + 6H^+ + 4e^-$	$\Rightarrow 2SbO^+ + 3H_2O$	0.559

(Continued)

Half-Reaction		E° (V)
$I_3^- + 2e^-$	$\Rightarrow 3I^-$	0.545
$Cu^+ + e^-$	$\Rightarrow Cu$	0.52
$VO^{2+} + 2H^+ + e^-$	$\Rightarrow V^{3+} + H_2O$	0.361
$Fe(CN)_6^{3-} + e^-$	$\Rightarrow Fe(CN)_6^{4-}$	0.36
$Cu^{2+} + 2e^-$	$\Rightarrow Cu$	0.337
$UO_2^{2+} + 4H^+ + 2e^-$	$\Rightarrow U^{4+} + 2H_2O$	0.334
$BiO^+ + 2H^+ + 3e^-$	$\Rightarrow Bi + H_2O$	0.32
$Hg_2Cl_2(s) + 2e^-$	$\Rightarrow 2Hg + 2Cl^-$	0.2676
$AgCl(s) + e^-$	$\Rightarrow Ag + Cl^-$	0.2223
$SbO^+ + 2H^+ + 3e^-$	$\Rightarrow Sb + H_2O$	0.212
$CuCl_3^{2-} + e^-$	$\Rightarrow Cu + 3Cl^-$	0.178
$SO_4^{2-} + 4H^+ + 2e^-$	$\Rightarrow SO_2(aq) + 2H_2O$	0.17
$Sn^{4+} + 2e^-$	$\Rightarrow Sn^{2+}$	0.154
$S + 2H^+ + 2e^-$	$\Rightarrow H_2S(g)$	0.141
$TiO^{2+} + 2H^+ + e^-$	$\Rightarrow Ti^{3+} + H_2O$	0.10
$S_4O_6^{2-} + 2e^-$	$\Rightarrow 2S_2O_3^{2-}$	0.08
$AgBr(s) + e^-$	$\Rightarrow Ag + Br^-$	0.071
$2H^+ + 2e^-$	$\Rightarrow H_2$	0.00
$Pb^{2+} + 2e^-$	$\Rightarrow Pb$	−0.126
$Sn^{2+} + 2e^-$	$\Rightarrow Sn$	−0.136
$AgI(s) + e^-$	$\Rightarrow Ag + I^-$	−0.152
$Mo^{3+} + 3e^-$	$\Rightarrow Mo$	−0.2
$N_2 + 5H^+ + 4e^-$	$\Rightarrow H_2NNH_3^+$	−0.23
$Ni^{2+} + 2e^-$	$\Rightarrow Ni$	−0.246
$V^{3+} + e^-$	$\Rightarrow V^{2+}$	−0.255
$Co^{2+} + 2e^-$	$\Rightarrow Co$	−0.277
$Ag(CN)_2^- + e^-$	$\Rightarrow Ag + 2CN^-$	−0.31
$Cd^{2+} + 2e^-$	$\Rightarrow Cd$	−0.403
$Cr^{3+} + e^-$	$\Rightarrow Cr^{2+}$	−0.41
$Fe^{2+} + 2e^-$	$\Rightarrow Fe$	−0.440
$2CO_2 + 2H^+ + 2e^-$	$\Rightarrow H_2C_2O_4$	−0.49
$H_3PO_3 + 2H^+ + 2e^-$	$\Rightarrow H_3PO_2 + H_2O$	−0.50
$U^{4+} + e^-$	$\Rightarrow U^{3+}$	−0.61

(Continued)

Half-Reaction		$E°$ (V)
$Zn^{2+} + 2e^-$	$\Rightarrow Zn$	−0.763
$Cr^{2+} + 2e^-$	$\Rightarrow Cr$	−0.91
$Mn^{2+} + 2e^-$	$\Rightarrow Mn$	−1.18
$Zr^{4+} + 4e^-$	$\Rightarrow Zr$	−1.53
$Ti^{3+} + 3e^-$	$\Rightarrow Ti$	−1.63
$Al^{3+} + 3e^-$	$\Rightarrow Al$	−1.66
$Th^{4+} + 4e^-$	$\Rightarrow Th$	−1.90
$Mg^{2+} + 2e^-$	$\Rightarrow Mg$	−2.37
$La^{3+} + 3e^-$	$\Rightarrow La$	−2.52
$Na^+ + e^-$	$\Rightarrow Na$	−2.714
$Ca^{2+} + 2e^-$	$\Rightarrow Ca$	−2.87
$Sr^{2+} + 2e^-$	$\Rightarrow Sr$	−2.89
$K^+ + e^-$	$\Rightarrow K$	−2.925
$Li^+ + e^-$	$\Rightarrow Li$	−3.045

STANDARD REDUCTION ELECTRODE POTENTIALS FOR INORGANIC SYSTEMS IN NON-AQUEOUS SOLUTION AT 25°C

The following table lists some standard electrode potentials (in V) in various solvents. The rubidium ion, which possesses a large radius and shows a low deformability, has a rather low and constant solvation energy in all solvents [1]. As a result, the rubidium electrode is taken as a standard reference electrode in all solvents.

REFERENCE

1. Koryta, J., Dvorák, J., and Boháckova, V., *Electrochemistry*, Methuen and Co., London, 1970.

System	H_2O	CH_3OH	CH_3CN	HCOOH	N_2H_4	NH_3
Li/Li$^+$	−0.03	−0.16	−0.06	−0.03	−0.19	−0.35
Rb/Rb$^+$	0.00	0.00	0.00	0.00	0.00	0.00
Cs/Cs$^+$	+0.06	—	+0.01	−0.01	—	−0.02
K/K$^+$	+0.06	—	+0.01	+0.10	−0.01	−0.05
Ca/Ca^{+2}	+0.14	—	+0.42	+0.25	+0.10	+0.29
Na/Na$^+$	+0.27	+0.21	+0.30	+0.03	+0.18	+0.08
Zn/Zn^{+2}	+2.22	+2.20	+2.43	+2.40	+1.60	+1.40
Cd/Cd^{+2}	+2.58	+2.51	+2.70	+2.70	+1.91	+1.73
Tl/Tl$^+$	+2.64	+2.56	—	—	—	—
Pb/Pb^{+2}	+2.85	+2.74	+3.05	+2.73	+2.36	+2.25
H$_2$/H$^+$	+2.98	+2.94	+3.17	+3.45	+2.01	+1.93
Cu/Cu^{+2}	+3.32	+3.28	+2.79	+3.31	—	+2.36
Cu/Cu$^+$	+3.50	—	+2.89	—	+2.23	+2.34
Hg/Hg^{+2}	+3.78	+3.68	—	+3.63	—	—
Ag/Ag$^+$	+3.78	+3.70	+3.40	+3.62	+2.78	+2.76
Hg/Hg^{+2}	+3.84	—	+3.42	—	—	+2.08
I$^-$/I$_2$	+3.52	+3.30	+3.24	+3.42	—	+3.38
Br$^-$/Br$_2$	+4.04	+3.83	+3.64	+3.97	—	+3.76
Cl$^-$/Cl$_2$	+4.34	+4.16	+3.75	+4.22	—	+3.96

REDOX POTENTIALS FOR SOME BIOLOGICAL HALF-REACTIONS

The following table lists the standard redox potentials of some common biological half-reactions (in V) at 298 K and pH = 7.0 [1].

REFERENCE

1. Hibbert, D.B. and James, A.M., *Dictionary of Electrochemistry*, 2nd ed., John Wiley & Sons, New York, 1984.

Biological System	Half-Cell Reaction	E° (V)
Acetate/pyruvate	$CH_3COOH + CO_2 + 2H^+ + 2e^- \rightarrow CH_3COCOOH + H_2O$	−0.70
Fe^{+3}/Fe^{+2} (ferredoxin)	$Fe^{+3} + e^- \rightarrow Fe^{+2}$	−0.432
H^+/H_2	$2H^+ + 2e^- \rightarrow H_2(g)$	−0.421
$NADP^+/NADPH$	$NADP^+ + 2H^+ + 2e^- \rightarrow NADPH + H^+$	−0.324
$NAD^+/NADH$	$NAD^+ + 2H^+ + 2e^- \rightarrow NADH + H^+$	−0.320
$FAD/FADH_2$	$FAD + 2H^+ + 2e^- \rightarrow FADH_2$	−0.219
Acetaldehyde/ethanol	$CH_3CHO + 2H^+ + 2e^- \rightarrow CH_3CH_2OH$	−0.197
Pyruvate/lactate	$CH_3COCOOH + 2H^+ + 2e^- \rightarrow CH_3CH(OH)COOH$	−0.185
Oxaloacetate/malate	$\begin{array}{c} CH_2COOH \\ \mid \\ O{=}C{-}COOH \end{array} + 2H^+ + 2e^- \rightarrow \begin{array}{c} CH_2COOH \\ \mid \\ HOCHCOOH \end{array}$	−0.166
Methylene blue (ox)(MB)/methylene blue (red)(MBH$_2$)	$MB + 2H^+ + 2e^- \rightarrow MBH_2$	0.011
Fumarate/succinate	$\begin{array}{c} CHCOOH \\ \parallel \\ CHCOOH \end{array} + 2H^+ + 2e^- \rightarrow \begin{array}{c} CH_2COOH \\ \mid \\ CH_2COOH \end{array}$	0.031
Fe^{+3}/Fe^{+2} (myoglobin)	$Fe^{+3} + e^- \rightarrow Fe^{+2}$	0.046
Fe^{+3}/Fe^{+2} (cytochrome b)	$Fe^{+3} + e^- \rightarrow Fe^{+2}$	0.050
Ubiquinone (Ub)/ubihydroquinone (UbH$_2$)	$Ub + 2H^+ + 2e^- \rightarrow UbH_2$	0.10
(Cytochrome c)$^{+3}$/(cytochrome c)$^{+2}$	$Fe^{+3} + e^- \rightarrow Fe^{+2}$	0.254
(Cytochrome a)$^{+3}$/(cytochrome a)$^{+2}$	$Fe^{+3} + e^- \rightarrow Fe^{+2}$	0.29
(Cytochrome f)$^{+3}$/(cytochrome f)$^{+2}$	$Fe^{+3} + e^- \rightarrow Fe^{+2}$	0.365
Cu^{+2}/Cu^+ (hemocyanin)	$Cu^{+2} + e^- \rightarrow Cu^+$	0.540
O_2/H_2O	$O_2(g) + 4H^+ + 4e^- \rightarrow 4H_2O$	0.816

STANDARD EMF OF THE CELL H$_2$/HCL/AgCl, Ag IN VARIOUS AQUEOUS SOLUTIONS OF ORGANIC SOLVENTS AT DIFFERENT TEMPERATURES

The table below lists the standard electromotive force (EMF) values of the cell H$_2$/HCl/AgCl, Ag in water as well as in various aqueous solutions of three common organic solvents, all alcohols, at different temperatures [1–3]. The compositions are given as mass percent of the alcohol in water. All EMF values are expressed in volts.

REFERENCES

1. Koryta, J., Dvorak, J., and Bohackova, V., *Electrochemistry*, Methuen and Co., London 1970.
2. Robinson, R.H. and Stokes, T.H., *Electrolytic Solutions*, Butterworths, London, 1959.
3. Covington, A., ed. *Physical Chemistry of Organic Solvent Systems*, Plenum Press, London, 1973.

°C	100 % Water	10 % aq. Methanol	10 % aq. Ethanol	10 % aq. 2-Propanol	20 % aq. Methanol	20 % aq. Ethanol	20 % aq. 2-Propanol
0	0.23655	0.22762	0.22726	0.22543	0.22022	0.21606	0.21612
5	0.23413	0.22547	0.22527	0.22365	0.21837	0.21486	0.21492
10	0.23142	0.22328	0.22328	0.22158	0.21631	0.21367	0.21336
15	0.22857	0.22085	0.22164	0.21922	0.21405	0.21190	0.21138
20	0.22557	0.21821	0.21901	0.21667	0.21155	0.21013	0.20906
25	0.22234	0.21535	0.21467	0.21383	0.20881	0.20757	0.20637
30	0.21904	0.21220	0.21383	0.21081	0.20567	0.20587	0.20341
35	0.21565	0.20892	0.21082	0.20754	0.20246	0.20275	0.20009
40	0.21208	0.20350	0.20783	0.20410	0.19910	0.19962	0.19652

TEMPERATURE DEPENDENCE OF THE STANDARD
POTENTIAL OF THE SILVER CHLORIDE ELECTRODE

The following table gives the standard potential (in V) of the silver chloride electrode (saturated KCl) at different temperatures (in °C) [1,2]. The uncertainty is ±0.05 mV.

The following correlations for the standard potential of the silver chloride electrode as a function of temperature (where T is temperature in °C) have been reported [3].

$$E°(V) = 0.23695 - 4.8564 \times 10^{-4}\,T - 3.4205 \times 10^{-6}\,T^2 - 5.869 \times 10^{-9}\,T^3 \text{ for } 0 < T < 95°C$$

REFERENCES

1. Conway, B.E., *Theory and Principles of Electrode Process*, Ronald Press, New York, 1965.
2. Koryta, J., Dvorák, J., and Boháckova, V., *Electrochemistry*, Methuen and Co., London, 1970.
3. Bard, A.J., Parson, R., and Jordan, J. *Standard Potentials in Aqueous Solution*, Marcel Dekker, Inc., New York/ Basel, 1985.

Temperature (°C)	E° (V)	Temperature (°C)	E° (V)
0	0.23634	35	0.21563
5	0.23392	40	0.21200
10	0.23126	45	0.20821
15	0.22847	50	0.20437
20	0.22551	55	0.20035
25	0.22239	60	0.19620
30	0.21912		

STANDARD ELECTRODE POTENTIALS OF ELECTRODES OF THE FIRST KIND

The following table lists the standard electrode potentials (in V) of some electrodes of the first kind [1–3]. These are divided into cationic and anionic electrodes. In cationic electrodes, equilibrium is established between atoms or molecules of the substance and the corresponding cations in solution. Examples include metal, amalgam, and the hydrogen electrode. In anionic electrodes, equilibrium is achieved between molecules and the corresponding anions in solution. The potential of the electrode is given by the Nernst equation in the form:

$$E = E° + (RT)/(Z_{\pm}F) \ln a_{\pm},$$

where $E°$ = standard electrode potential (in V)
R = gas constant
T = temperature (in K)
$Z_{\pm}$ = charge, with sign, of the cation (+) or anion (–)
F = Faraday
$a_{\pm}$ = activity of the cation (+) or anion (–)

Electrodes of the first kind differ distinctly from the redox electrodes in that in the latter case both oxidation states can be present in variable concentrations, while in electrodes of the first kind, one of the oxidation states is the electrode material.

REFERENCES

1. Koryta, J., Dvorák, J., and Karan, L., *Principles of Electrochemistry*, 2nd ed., John Wiley & Sons, New York, 1993.
2. Koryta, J., Dvorák, J., and Boháčková, V., *Electrochemistry*, Methuen and Co., London, 1970.
3. Rumble, J.R., ed., *CRC Handbook of Chemistry and Physics*, 100th ed., CRC Taylor & Francis Press, Boca Raton, FL, 2019.

Electrode	$E°$ (V)[a]	Electrode	$E°$ (V)[a]
Li^{+}/Li	–3.0403	Ni^{+2}/Ni	–0.23
Rb^{+}/Rb	–2.98	In^{+}/In	–0.203
Cs^{+}/Cs	–2.92	Sn^{+2}/Sn	–0.1377
K^{+}/K	–2.931	Pb^{+2}/Pb	–0.1264
Ba^{+2}/Ba	–2.912	Cu^{+2}/Cu	+0.3417
Sr^{+2}/Sr	–2.89	Cu^{+}/Cu	+0.52
Ca^{+2}/Ca	–2.868	Te^{+4}/Te	+0.56
Na^{+}/Na	–2.71	Hg^{+2}/Hg	+0.851
Mg^{+2}/Mg	–2.372	Ag^{+}/Ag	+0.7994
Be^{+2}/Be	–1.847	Au^{+3}/Au	+1.42
Al^{+3}/Al	–1.662	$Pt, Se^{-2}/Se$	–0.78
Zn^{+2}/Zn	–0.7620	$Pt, S^{-2}/S$	–0.51
Fe^{+2}/Fe	–0.447	$Pt, OH^{-}/O_2(g)$	+0.401
Cd^{+2}/Cd	–0.4032	$Pt, I^{-}/I_2$	+0.536
In^{+3}/In	–0.3384	$Pt, B^{-}/Br_2$	+1.066
Tl^{+}/Tl	–0.336	$Pt, Cl^{-}/Cl_2(g)$	+1.35793
Co^{+2}/Co	–0.27	$Pt, F^{-}/F_2(g)$	+2.866

[a] All values have been taken from the *CRC Handbook of Chemistry and Physics* and are recalculated to the standard pressure of 1 atm (101.325 kPa).

STANDARD ELECTRODE POTENTIALS OF ELECTRODES OF THE SECOND KIND

The following table lists the standard electrode potentials (in V) of some electrodes of the second kind [1–3]. These consist of three phases. The metal is covered by a layer of its sparingly soluble salt and is immersed in a solution of a soluble salt of the anion. Equilibrium is established between the metal atoms and the solution anions through two partial equilibria: one between the metal and its cation in the sparingly soluble salt and the other between the anion in the solid phase of the sparingly soluble salt and the anion in solution. The silver chloride electrode is preferred for precise measurements.

REFERENCES

1. Koryta, J., Dvorák, J., and Karan, L. *Principles of Electrochemistry*, 2nd ed., John Wiley & Sons, New York, 1993.
2. Koryta, J., Dvorák, J., and Bohácková, V., *Electrochemistry*, Methuen and Co., London, 1970.
3. Rumble, J.R., ed., *CRC Handbook of Chemistry and Physics*, 100th ed., CRC Taylor & Francis Press, Boca Raton, FL, 2019.

Electrode	E°(V)[a]
$PbSO_4$, SO_4^{-2}/Pb, Hg	−0.351
AgI, I^-/Ag	−0.152
AgBr, Br^-/Ag	+0.071
HgO, OH^-/Hg	+0.0975
Hg_2Br_2, Br^-/Hg	+0.140
AgCl, Cl^-/Ag	+0.22216
Hg_2Cl_2, Cl^-/Hg	+0.26791
Hg_2SO_4, SO_4^{-2}/Hg	+0.6123
PbO_2, $PbSO_4$, SO_4^{-2}/Pb	+1.6912

[a] All values have been taken from the *CRC Handbook of Chemistry and Physics* and are recalculated to the standard pressure of 1 atm (101.325 kPa).

POLAROGRAPHIC HALF-WAVE POTENTIALS ($E_{1/2}$) OF INORGANIC CATIONS

The following table lists the polarographic half-wave potentials ($E_{1/2}$, in volts vs. SCE, the standard calomel electrode) of inorganic cations and the supporting electrolyte used during the determination [1–6]. All supporting electrolyte solutions are aqueous unless noted.

REFERENCES

1. Skoog, D.A., West, D.M., Holler, F.J., and Crouch, S.R., *Analytical Chemistry Fundamentals*, 8th ed., Cengage Learning, Boston, MA, 2004.
2. Vogel, A.I., *A Textbook for Quantitative Inorganic Analysis*, 3rd ed., John Wiley & Sons, New York, 1968.
3. Fritz, J.S. and Schenk, G.H., *Quantitative Analytical Chemistry*, 4th ed., Prentice Hall, Englewood Cliffs, 1987.
4. Christian, G.D., Dasgupta, P.K., and Schug, K.A., *Analytical Chemistry*, 7th ed., John Wiley & Sons, New York, 2013.
5. Ewing, G.W., *Instrumental Methods of Analysis*, 5th ed., McGraw Hill, New York, 1985.
6. Meites, L., *Polarographic Techniques*, 2nd ed., Wiley Interscience, New York, 1965.

Supporting Electrolyte

Cation	KCl (0.1F)	NH$_3$ (1F) NH$_4$Cl (1F)	NaOH (1F)	H$_3$PO$_4$ (7.3F)	KCN (1F)	(CH$_3$)$_4$NCl (0.1F)	HCl (1F)	H$_2$SO$_4$ (0.5 M)	0.5 M Tartrate and		Others
									NaOH (0.1F)	pH = 4.5	
Ba^{+2}						−1.94					
Bi^{+3}							−0.09	−0.04	−1.00	−0.23	
Cd^{+2}	−0.64 (−0.60)	−0.81	−0.78	−0.77	−1.18						HNO$_3$ (1.0 F) −0.59 KI (1.0 F) −0.74
Co^{+2}	−1.20	−1.29	−1.46	−1.20	−1.45						Pyridine (0.1 F)/pyridinium (0.1 F) −1.07
Cr^{+3}		−1.43 (to Cr^{+2}) −1.71 (to Cr^0)		−1.02 (to Cr^{+2})	−1.38 (to Cr^{+2})						
Cu^{+2}	+0.04 (to Cu$^+$) −0.22 (to Cu^0)	−0.24 (to Cu$^+$) −0.51 (to Cu^0)	−0.41	−0.09	No reaction					−0.09	
Fe^{+2}	−1.3	−1.49									
Fe^{+3}			−1.12 (to Fe^{+2}) −1.74 (to Fe^0)	+0.06 (to Fe^{+2})					−1.20, −1.73		Ethylenediaminetetraacetic acid (EDTA) (0.1 F)/CH$_3$COONa (2.0 F) −0.17, −1.30 (CH$_3$)$_4$NOH (0.1 M, 50 % C$_2$H$_5$OH) −2.10
K$^+$											
Li$^+$											(CH$_3$)$_4$NOH (0.1 M, 50 % C$_2$H$_5$OH) −2.31
Mn^{+2}	−1.51										H$_2$P$_2$O$_7^{-2}$ (0.2 M), pH=2.2, +0.1
Na$^+$						−2.07					
Ni^{+2}	−1.1	−1.10			−1.36						KSCN (1.0 F) −0.7; pyridine (1.0 F)/HCl, pH=7, −0.78
O$_2$											pH=1–10 (buffered) −0.05 & −0.90
Pb^{+2}	−0.40		−0.75		−0.72				−0.75	−0.48	HNO$_3$ (1 F) −0.40
Sn^{+2}							−0.47				F$^-$ (0.1 F) −0.611; F$^-$ (0.5 F) −0.683
Sn^{+4}			−0.75								HCl (1.0 F)/NH4$^+$ (4.0 F) −0.25 & −0.52
Te$^+$	−0.48	−0.48	−0.48								
Zn^{+2}	−1.00	−1.34	−1.53						−1.15		

POLAROGRAPHIC $E_{1/2}$ RANGES (IN V VS. SCE) FOR THE REDUCTION OF BENZENE DERIVATIVES

The following table lists the polarographic $E_{1/2}$ potential ranges (in V vs. SCE, the standard calomel electrode) obtained at pH = 5–9 in unbuffered media in the reduction of benzene derivatives [1,2].

REFERENCES

1. Zuman, P., *The Elucidation of Organic Electrode Processes*, Academic Press, New York, 1969.
2. Merkel, P.B., Luo, P., Dinnocenzo, J.P., and Farid, S., Accurate oxidation potentials of benzene and biphenyl derivatives via electron-transfer equilibria and transient kinetics, *J. Org. Chem.*, 74(15), 5163–5173, 2009.

Benzene Derivative[a]	Formula[a]	Polarographic $E_{1/2}$ Potential Range[b]
Diaryl alkene	ArCH=CHAr	(−1.8)–(−2.3)
Methyl aryl ester	$ArCOOCH_3$	(−1.0)–(−2.4)
Aryl iodide	ArI	(−1.2)–(−1.9)
Aryl methyl ketone	$ArCOCH_3$	(−1.1)–(−1.8)
Aromatic aldehyde	ArCHO	(−1.1)–(−1.7)
Methyl α,β-unsaturated aryl ketone	$ArCH=CHCOCH_3$	(−1.0)–(−1.6)
Diaryl ketone	ArCOAr	(−0.7)–(−1.4)
Azobenzenes	ArN=NAr	(−0.3)–(−0.8)
Nitroarenes	$ArNO_2$	(−0.3)–(−0.7)
Nitrosoarenes	ArNO	(−0.1)–(−0.4)
Diaryl iodonium salts	Ar_2I^+	(−0.2)–(−0.3)

[a] Ar = aromatic ring.
[b] In V vs. SCE.

VAPOR PRESSURE OF MERCURY

The following table provides data on the vapor pressure of mercury, useful for assessing and controlling the hazards associated with the use of mercury as an electrode [1]. Additional information can be found in Chapter 15.

REFERENCE

1. Rumble, J.R., ed., *CRC Handbook of Chemistry and Physics*, 100th ed., CRC Taylor & Francis Press, Boca Raton, FL, 2019.

Temperature (°C)	Vapor Pressure (mm Hg)	Vapor Pressure (Pa)	Temperature (°C)	Vapor Pressure (mm Hg)	Vapor Pressure (Pa)
0	0.000185	0.0247	28	0.002359	0.3145
10	0.000490	0.0653	30	0.002777	0.3702
20	0.001201	0.1601	40	0.006079	0.8105
22	0.001426	0.1901	50	0.01267	1.689
24	0.001691	0.2254	100	0.273	36.4
26	0.002000	0.2666			

ORGANIC FUNCTIONAL GROUP ANALYSIS OF NON-POLAROGRAPHIC ACTIVE GROUPS

Often an organic functional group is not (or may not be) reduced polarographically at an accessible potential range. In this case, it is necessary to convert this functional group to a derivative whose reduction is feasible within such an accessible potential range. The table below lists the most common functional groups, the reagent needed, and the polarographically active derivative as well as the polarographically active group [1–4]. Such conversions enlarge the number of organic compounds that can be determined by polarography.

REFERENCES

1. Svoronos, P., Horak, V., and Zuman, P., Polarographic study of structure-properties relationship of p-tosyl sulfilimines, *Phosphorus, Sulfur Silicon.*, 42, 139, 1989.
2. Willard, H.H., Merritt, L.L., Jr., Dean, J.A., and Settle, F.A., Jr., *Instrumental Methods of Analysis*, Wadsworth, Belmont, CA, 1988.
3. Zuman, P., Chemical and Engineering News, 94, March 18, 1968.
4. Zuman, P., *Substituent Effects in Organic Polarography*, Plenum, New York, 1967.

Functional Group	Reagent	Polarographically Active Derivative	Active Polarographic Group
Carbonyl (aldehyde, ketone) >C=O	Semicarbazide $H_2NHNCNH_2$ ‖ O	>C=N–NHC–NH$_2$ ‖ O	Semicarbazide, >C=N–N
	Hydroxylamine H_2NOH	>C=N–OH	Oxime, >C=N–OH
Primary amine, R–NH$_2$	Piperonal	–CH=N–R	Azomethine, >C=N–R
	Carbon disulfide, CS$_2$	R–N=C (S–H / \\ S–H)	–N=C (S– / \\ S–)
	Cupric phosphate, Cu$_3$(PO$_4$)$_2$, suspension	[Cu^{+2}–amine] complex	[Cu^{+2}–N]
Secondary amine, R$_2$NH	Nitrous acid, HNO$_2$	R$_2$N–N=O	\\N–N=O Nitroso, /
Primary alcohols, R–CH$_2$OH	Chromic acid, HCrO$_4$	R–CHO	R \\ Aldehyde carbonyl, C=O / R
Secondary alcohols, R–CH$_2$OH	Chromic acid, HCrO$_4$	R$_2$C=O	R \\ Ketone carbonyl, C=O / R

(Continued)

Functional Group	Reagent	Polarographically Active Derivative	Active Polarographic Group
1,2-Diols	Periodic acid, HIO_4	$\begin{matrix} R \\ \diagdown \\ C=O \\ \diagup \\ H \end{matrix}$ $\begin{matrix} R \\ \diagdown \\ C=O \\ \diagup \\ R \end{matrix}$	Aldehyde and/or ketone carbonyl $>C=O$
Carboxylic acid, $R-\overset{\underset{\parallel}{O}}{C}-OH$	Thiourea, $(H_2N)_2C=S$	$RCO_2^- (H_2N)_2 C^+SH$	Thiol, $-S-H$
Phenyl, C_6H_5-, N	Conc. nitric/conc. sulfuric acid, HNO_3/H_2SO_4	$C_6H_5O_2N$	Nitro, $-NO_2$
Sulfides (thioethers), $>S$	Hydrogen peroxide, H_2O_2 or m-chloroperbenzoic acid, $1,3\text{-Cl}-C_6H_4-COOH$	$>S^+ \rightarrow O^-$	$\begin{matrix} R \\ \diagdown \\ \text{sulfoxide, } S^+ - O^- \\ \diagup \end{matrix}$
	Chloramine-T, $p\text{-}CH_3-C_6H_4-SO_2NClNa$	$>S=N-SO_2-C_6H_4-CH_3$	Sulfilimine, $>S=N-$

COULOMETRIC TITRATIONS

The following table lists some common coulometric (also known as constant-current coulometry) titrations [1–4]. Since the titrant is generated electrolytically and reacted immediately, the method gets widespread applications. The generating electrolytic concentrations need to be only approximate, while unstable titrants are consumed as soon as they are formed. The technique is more accurate than methods where visual end points are required, such as in the case of indicators. The unstable titrants in the table below are marked with an asterisk (*).

REFERENCES

1. Christian, G.D., Dasgupta, P.K., and Schug, K.A., *Analytical Chemistry*, 7th ed., John Wiley & Sons, New York, 2013.
2. Christian, G.D., Electrochemical methods for analysis of enzyme systems, in *Advances in Biomedical and Medical Physics*, vol. 4, Levine, S.N., ed., Wiley-Interscience, New York, 1971.
3. Skoog, D.A., West, D.M., Holler, F.J., and Crouch, S.R., *Analytical Chemistry Fundamentals*, 8th ed., Cengage Learning, Boston, MA, 2004.
4. Harris, D.C., *Quantitative Chemical Analysis*, 5th ed., W.H. Freeman, San Francisco, 1998.

Reagent	Generator Electrode Reaction	Typical Generating Electrolyte	Substances Determined
Ag^+	$Ag \rightarrow Ag^+ + e^-$	Ag anode in HNO_3	Br^- Cl^-, thiols
Ag^{+2}	$Ag^+ \rightarrow Ag^{+2} + e^-$		Ce^{+3}, V^{+4}, $H_2C_2O_4$, As^{+3}
*Biphenyl radical anion	$(C_6H_5)_2 + e^- \rightarrow (C_6H_5)_2^-$	Biphenyl/$(CH_3)_4NBr$ in DMF	Anthracene
*Br_2	$2\,Br^- \rightarrow Br_2 + 2e^-$	0.2 M NaBr in 0.1 M H_2SO_4	As^{+3}, Sb^{+3}, U^{+4}, Tl^+, I^-, SCN^-, NH_2OH, N_2H_4, phenols, aromatic amines, mustard gas, olefins, 8-hydroxy-quinoline
*BrO^-	$Br^- + 2\,OH^- \rightarrow BrO^- + H_2O + 2e^-$	1 M NaBr in borate buffer, pH = 8.6	NH_3
Ce^{+4}	$Ce^{+2} \rightarrow Ce^{+4} + 2e^-$	0.1 M $CeSO_4$ in 3 M H_2SO_4	Fe^{+2}, Ti^{+3}, U^{+4}, As^{+3}, I^-, $Fe(SCN)_6^{-4}$
*Cl_2	$2\,Cl^- \rightarrow Cl_2 + 2e^-$		As^{+3}, I^-
*Cr^{+2}	$Cr^{+3} + e^- \rightarrow Cr^{+2}$	$Cr_2(SO_4)_3$ in H_2SO_4	O_2
*$CuCl_3^{-2}$	$Cu^{+2} + 3Cl^- + e^- \rightarrow CuCl_3^-$	0.1 M $CuSO_4$ in 1 M HCl	V^{+5}, Cr^{+6}, IO_3^-
EDTA	$HgNH_3(EDTA)^{-2} + NH_4^+ + 2e^- \rightarrow Hg + 2NH_3 + (HEDTA)^{-3}$	0.02 M Hg^{+2}/EDTA in ammoniacal buffer, pH = 8.5, Hg cathode	Ca^{+2}, Cu^{+3}, Zn^{+2}, Pb^{+2}
EGTA	$HgNH_3(EGTA)^{-2} + NH_4^+ + 2e^- \rightarrow Hg + 2NH_3 + (HEDTA)^{-3}$	0.1 M Hg^{+2}/EGTA in triethanolamine, pH = 8.6, Hg cathode	Ca^{+2} (in the presence of Mg^{+2})
Fe^{+2}	$Fe^{+3} + e^- \rightarrow Fe^{+2}$	Acid solution of $FeNH_4(SO_4)_2$	Cr^{+6}, Mn^{+7}, V^{+5}, Ce^{+4}
I_2	$2I^- \rightarrow I_2 + 2e^-$	0.2 M KI in pH = 8 buffer, pyridine, SO_2, CH_3OH, KI (Karl Fisher titration)	As^{+3}, Sb^{+3}, $S_2O_3^{-2}$, H_2S, H_2O

(Continued)

Reagent	Generator Electrode Reaction	Typical Generating Electrolyte	Substances Determined
H^+	$2H_2O \rightarrow 4H^+ + O_2 + 4e^-$	0.1 M Na_2SO_4 (water electrolysis)	Pyridine
* Mn^{+3}	$Mn^{+2} \rightarrow Mn^{+3} + e^-$	$MnSO_4$ in 2 M H_2SO_4	$H_2C_2O_4$, Fe^{+2}, As^{+3}, H_2O_2
Mo^{+5}	$Mo^{+6} + e^- \rightarrow Mo^{+5}$	0.7 M Mo^{+6} in 4 M H_2SO_4	$Cr_2O_7^{-2}$
* $MV^{+ a}$	$MV^{+2} + e^- \rightarrow MV^+$		Mn^{+3} (in enzymes)
OH^-	$2H_2O + 2e^- \rightarrow H_2 + 2OH^-$	0.1 M Na_2SO_4 (water electrolysis)	HCl
Ti^{+3}	$Ti^{+4} + e^- \rightarrow Ti^{+3}$ or $TiO^{+2} + 2H^+ + e^- \rightarrow Ti^{+3} + H_2O$	3.6 M $TiCl_4$ in 7 M HCl	V^{+5}, Fe^{+3}, Ce^{+4}, U^{+6}
U^{+4}	$UO_2^{+2} + 4H^+ + 2e^- \rightarrow U^{+4} + 2H_2O$	Acid solution of UO_2^{+2}	Cr^{+6}, Ce^{+4}

a MV^+, methyl viologen radical cation; MV^{+2}, methyl viologen radical cation.

CHAPTER 8

Ultraviolet and Visible Spectrophotometry

SOLVENTS FOR ULTRAVIOLET SPECTROPHOTOMETRY

The following table lists some useful solvents for ultraviolet spectrophotometry, along with their wavelength cutoffs and dielectric constants [1–6].

REFERENCES

1. Willard, H.H., Merritt, L.L., Dean, J.A., and Settle, F.A., *Instrumental Methods of Analysis*, 7th ed., Van Nostrand, New York, Belmont, 1988.
2. Strobel, H. A. and Heinemann, W. R., *Chemical Instrumentation: A Systematic Approach*, 3rd ed., John Wiley & Sons, New York, 1989.
3. Dreisbach, R.R., *Physical Properties of Chemical Compounds*, Advances in Chemistry Series, No. 15, American Chemical Society, Washington, DC, 1955.
4. Dreisbach, R.R., *Physical Properties of Chemical Compounds*, Advances in Chemistry Series, No. 22, American Chemical Society, Washington, DC, 1959.
5. Sommer, L., *Analytical Absorption Spectrophotometry in the Visible and Ultraviolet*, Elsevier Science, Amsterdam, 1989.
6. Krieger, P.A., *High Purity Solvent Guide*, Burdick and Jackson, McGaw Park, IL, 1984.

Solvent	Wavelength Cutoff (nm)	Dielectric Constant (20°C)
Acetic acid	260	6.15
Acetone	330	20.7 (25 °C)
Acetonitrile	190	37.5
Benzene	280	2.284
Sec-butyl alcohol (2-butanol)	260	15.8 (25 °C)
n-Butyl acetate	254	
n-Butyl chloride	220	7.39 (25 °C)
Carbon disulfide	380	2.641
Carbon tetrachloride	265	2.238
Chloroform[a]	245	4.806
Cyclohexane	210	2.023
1,2-Dichloroethane	226	10.19 (25 °C)
1,2-Dimethoxyethane	240	
N,N-Dimethylacetamide	268	37.8
N,N-Dimethylformamide	270	36.7

(Continued)

Solvent	Wavelength Cutoff (nm)	Dielectric Constant (20°C)
Dimethylsulfoxide	265	4.7
1,4-Dioxane	215	2.209 (25 °C)
Diethyl ether	218	4.335
Ethanol	210	24.30 (25 °C)
2-Ethoxyethanol	210	
Ethyl acetate	225	6.02 (25 °C)
Methyl ethyl ketone	330	18.5
Glycerol	207	42.5 (25 °C)
n-Hexadecane	200	2.06 (25 °C)
n-Hexane	210	1.890
Methanol	210	32.63 (25 °C)
2-Methoxyethanol	210	16.9
Methyl cyclohexane	210	2.02 (25 °C)
Methyl isobutyl ketone	335	
2-Methyl-1-propanol	230	
N-Methyl-2-pyrrolidone	285	32.0
n-Pentane	210	1.844
n-Pentyl acetate	212	
n-Propyl alcohol	210	20.1 (25 °C)
2-Propanol (sec-propyl alcohol, isopropyl alcohol)	210	18.3 (25 °C)
Pyridine	330	12.3 (25 °C)
Tetrachloroethylene[b]	290	
Tetrahydrofuran	220	7.6
Toluene	286	2.379 (25 °C)
1,1,2-Trichloro-1,2,2-trifluoroethane	231	
2,2,4-Trimethylpentane	215	1.936 (25 °C)
o-Xylene	290	2.568
m-Xylene	290	2.374
p-Xylene	290	2.270
Water		78.54 (25 °C)

[a] Stabilized with ethanol to avoid phosgene formation.
[b] Stabilized with thymol (isopropyl meta-cresol).

ULTRAVIOLET SPECTRA OF COMMON LIQUIDS

The following table presents, in tabular form, the ultraviolet spectra of some common solvents and liquids used in chemical analysis. The data were obtained using a 1.00 cm path length cell, against a water reference [1,2].

REFERENCES

1. Krieger, P.A., *High Purity Solvent Guide*, Burdick and Jackson, McGaw Park, IL, 1984.
2. Sommer, L., *Analytical Absorption Spectrophotometry in the Visible and Ultraviolet*, Elsevier Science, Amsterdam, 1989.

Acetone		Benzene	
Wavelength (nm)	Maximum Absorbance	Wavelength (nm)	Maximum Absorbance
330	1.000	278	1.000
340	0.060	300	0.020
350	0.010	325	0.010
375	0.005	350	0.005
400	0.005	400	0.005

Acetonitrile		1-Butanol	
Wavelength (nm)	Maximum Absorbance	Wavelength (nm)	Maximum Absorbance
190	1.000	215	1.000
200	0.050	225	0.500
225	0.010	250	0.040
250	0.005	275	0.010
350	0.005	300	0.005

2-Butanol		Carbon Tetrachloride	
Wavelength (nm)	Maximum Absorbance	Wavelength (nm)	Maximum Absorbance
260	1.000	263	1.000
275	0.300	275	0.100
300	0.010	300	0.005
350	0.005	350	0.005
400	0.005	400	0.005

n-Butyl Acetate		Chlorobenzene	
Wavelength (nm)	Maximum Absorbance	Wavelength (nm)	Maximum Absorbance
254	1.000	287	1.000
275	0.050	300	0.050
300	0.010	325	0.040
350	0.005	350	0.020
400	0.005	400	0.005

n-Butyl Chloride		Chloroform	
Wavelength (nm)	Maximum Absorbance	Wavelength (nm)	Maximum Absorbance
220	1.000	245	1.000
225	0.300	250	0.300
250	0.010	275	0.005
300	0.005	300	0.005
400	0.005	400	0.005

Cyclohexane		o-Dichlorobenzene	
Wavelength (nm)	Maximum Absorbance	Wavelength (nm)	Maximum Absorbance
200	1.000	295	1.000
225	0.170	300	0.300
250	0.020	325	0.100
300	0.005	350	0.050
400	0.005	400	0.005

Cyclopentane		Diethyl Carbonate	
Wavelength (nm)	Maximum Absorbance	Wavelength (nm)	Maximum Absorbance
200	1.000	256	1.000
215	0.300	265	0.150
225	0.020	275	0.050
300	0.005	300	0.040
400	0.005	400	0.010

Decahydronaphthalene		Dimethyl Acetamide	
Wavelength (nm)	Maximum Absorbance	Wavelength (nm)	Maximum Absorbance
200	1.000	268	1.000
225	0.500	275	0.300
250	0.050	300	0.080
300	0.005	350	0.005
400	0.005	400	0.005

Dimethyl Formamide		2-Ethoxyethanol	
Wavelength (nm)	Maximum Absorbance	Wavelength (nm)	Maximum Absorbance
268	1.000	210	1.000
275	0.300	225	0.500
300	0.050	250	0.200
350	0.005	300	0.005
400	0.005	400	0.005

Dimethyl Sulfoxide		Ethyl Acetate	
Wavelength (nm)	Maximum Absorbance	Wavelength (nm)	Maximum Absorbance
268	1.000	256	1.000
275	0.500	275	0.050
300	0.200	300	0.030
350	0.020	325	0.005
400	0.005	350	0.005

1,4-Dioxane		Diethyl Ether	
Wavelength (nm)	Maximum Absorbance	Wavelength (nm)	Maximum Absorbance
215	1.000	215	1.000
250	0.300	250	0.080
300	0.020	275	0.010
350	0.005	300	0.005
400	0.005	400	0.005

Ethylene Dichloride		n-Hexadecane	
Wavelength (nm)	Maximum Absorbance	Wavelength (nm)	Maximum Absorbance
228	1.000	190	1.000
240	0.300	200	0.500
250	0.100	250	0.020
300	0.005	300	0.005
400	0.005	400	0.005

Ethylene Glycol Dimethyl Ether (Glyme)		n-Hexane	
Wavelength (nm)	Maximum Absorbance	Wavelength (nm)	Maximum Absorbance
220	1.000	195	1.000
250	0.250	225	0.050
300	0.050	250	0.010
350	0.010	275	0.005
400	0.005	300	0.005

n-Heptane		Isobutanol	
Wavelength (nm)	Maximum Absorbance	Wavelength (nm)	Maximum Absorbance
200	1.000	220	1.000
225	0.100	250	0.050
250	0.010	275	0.030
300	0.005	300	0.020
400	0.005	400	0.010

Methanol		Methyl-t-Butyl Ether	
Wavelength (nm)	Maximum Absorbance	Wavelength (nm)	Maximum Absorbance
205	1.000	210	1.000
225	0.160	225	0.500
250	0.020	250	0.100
300	0.005	300	0.005
400	0.005	400	0.005

2-Methoxyethanol		Methylene Chloride	
Wavelength (nm)	Maximum Absorbance	Wavelength (nm)	Maximum Absorbance
210	1.000	233	1.000
250	0.130	240	0.100
275	0.030	250	0.010
300	0.005	300	0.005
400	0.005	400	0.005

2-Methoxyethyl Acetate		Methyl Ethyl Ketone	
Wavelength (nm)	Maximum Absorbance	Wavelength (nm)	Maximum Absorbance
254	1.000	329	1.000
275	0.150	340	0.100
300	0.050	350	0.020
350	0.005	375	0.010
400	0.005	400	0.005

Methyl Isoamyl Ketone		n-Methylpyrrolidone	
Wavelength (nm)	Maximum Absorbance	Wavelength (nm)	Maximum Absorbance
330	1.000	285	1.000
340	0.100	300	0.500
350	0.050	325	0.100
375	0.010	350	0.030
400	0.005	400	0.010

Methyl Isobutyl Ketone		n-Pentane	
Wavelength (nm)	Maximum Absorbance	Wavelength (nm)	Maximum Absorbance
334	1.000	190	1.000
340	0.500	200	0.600
350	0.250	250	0.010
375	0.050	300	0.005
400	0.005	400	0.005

Methyl n-Propyl Ketone		β-Phenethylamine	
Wavelength (nm)	Maximum Absorbance	Wavelength (nm)	Maximum Absorbance
331	1.000	285	1.000
340	0.150	300	0.300
350	0.020	325	0.100
375	0.005	350	0.050
400	0.005	400	0.005

I-Propanol		Pyridine	
Wavelength (nm)	Maximum Absorbance	Wavelength (nm)	Maximum Absorbance
210	1.000	330	1.000
225	0.500	340	0.100
250	0.050	350	0.010
300	0.005	375	0.010
400	0.005	400	0.005

2-Propanol		Tetrahydrofuran	
Wavelength (nm)	Maximum Absorbance	Wavelength (nm)	Maximum Absorbance
205	1.000	212	1.000
225	0.160	250	0.180
250	0.020	300	0.020
300	0.005	350	0.005
400	0.010	400	0.005

Propylene Carbonate		Toluene	
Wavelength (nm)	Maximum Absorbance	Wavelength (nm)	Maximum Absorbance
280	1.000	284	1.000
300	0.500	300	0.120
350	0.050	325	0.020
375	0.030	350	0.050
400	0.020	400	0.005

1,2,4-Trichlorobenzene		2,2,4-Trimethylpentane	
Wavelength (nm)	Maximum Absorbance	Wavelength (nm)	Maximum Absorbance
308	1.000	215	1.000
310	0.500	225	0.100
350	0.050	250	0.020
375	0.010	300	0.005
400	0.005	400	0.005

Trichloroethylene		Water	
Wavelength (nm)	Maximum Absorbance	Wavelength (nm)	Maximum Absorbance
273	1.000	190	0.010
300	0.100	200	0.010
325	0.080	250	0.005
350	0.060	300	0.005
400	0.060	400	0.005

1,1,2-Trichlorotrifluoroethane		o-Xylene	
Wavelength (nm)	Maximum Absorbance	Wavelength (nm)	Maximum Absorbance
231	1.000	288	1.000
250	0.050	300	0.200
300	0.005	325	0.050
350	0.005	350	0.010
400	0.005	400	0.005

TRANSMITTANCE–ABSORBANCE CONVERSION

The following is a conversion table for absorbance and transmittance, assuming no reflection. Included for each pair is the percent error propagated into a measured concentration (using the Beer–Lambert Law), assuming an uncertainty in transmittance of +0.005 [1]. The value of transmittance which will give the lowest percent error in concentration is 3.368. Where possible, analyses should be designed for the low error area.

REFERENCE

1. Kennedy, J.H., *Analytical Chemistry Principles*, Harcourt, Brace and Jovanovich, San Diego, CA, 1984.

Transmittance	Absorbance	Percent Uncertainty
0.980	0.009	25.242
0.970	0.013	16.915
0.960	0.018	12.752
0.950	0.022	10.256
0.940	0.027	8.592
0.930	0.032	7.405
0.920	0.036	6.515
0.910	0.041	5.823
0.900	0.046	5.270
0.890	0.051	4.818
0.880	0.056	4.442
0.870	0.060	4.125
0.860	0.065	3.853
0.850	0.071	3.618
0.840	0.076	3.412
0.830	0.081	3.231
0.820	0.086	3.071
0.810	0.091	2.928
0.800	0.097	2.799
0.790	0.102	2.684
0.780	0.108	2.579
0.770	0.113	2.483
0.760	0.119	2.386
0.750	0.125	2.316
0.740	0.131	2.243
0.730	0.137	2.175
0.720	0.143	2.113
0.710	0.149	2.055
0.700	0.155	2.002
0.690	0.161	1.952
0.680	0.167	1.906
0.670	0.174	1.863
0.660	0.180	1.822
0.650	0.187	1.785

(Continued)

Transmittance	Absorbance	Percent Uncertainty
0.640	0.194	1.750
0.630	0.201	1.717
0.620	0.208	1.686
0.610	0.215	1.657
0.600	0.222	1.631
0.590	0.229	1.605
0.580	0.237	1.582
0.570	0.244	1.560
0.560	0.252	1.539
0.540	0.268	1.502
0.530	0.276	1.485
0.520	0.284	1.470
0.510	0.292	1.455
0.500	0.301	1.442
0.490	0.310	1.430
0.480	0.319	1.419
0.470	0.328	1.408
0.460	0.337	1.399
0.450	0.347	1.391
0.440	0.356	1.383
0.430	0.366	1.377
0.420	0.377	1.372
0.410	0.387	1.367
0.400	0.398	1.364
0.390	0.409	1.361
0.380	0.420	1.359
0.370	0.432	1.358
0.360	0.444	1.359
0.350	0.456	1.360
0.340	0.468	1.362
0.330	0.481	1.366
0.320	0.495	1.371
0.310	0.509	1.376
0.300	0.523	1.384
0.290	0.538	1.392
0.280	0.553	1.402
0.270	0.569	1.414
0.260	0.585	1.427
0.250	0.602	1.442
0.240	0.620	1.459
0.230	0.638	1.478
0.220	0.657	1.500
0.210	0.678	1.525
0.200	0.699	1.553
0.190	0.721	1.584
0.180	0.745	1.619
0.170	0.769	1.659
0.160	0.796	1.704

(*Continued*)

Transmittance	Absorbance	Percent Uncertainty
0.150	0.824	1.756
0.140	0.854	1.816
0.130	0.886	1.884
0.120	0.921	1.964
0.110	0.958	2.058
0.100	1.000	2.170
0.090	1.046	2.306
0.080	1.097	2.473
0.070	1.155	2.685
0.060	1.222	2.961
0.050	1.301	3.336
0.040	1.398	3.881
0.030	1.523	4.751
0.020	1.699	6.387
0.010	2.000	10.852

CORRELATION TABLE FOR ULTRAVIOLET ACTIVE FUNCTIONALITIES

The following table presents a correlation between common chromophoric functional groups and the expected absorptions from ultraviolet spectrophotometry [1–3]. While not as informative as infrared correlations, UV can often provide valuable qualitative information.

REFERENCES

1. Willard, H.H., Merritt, L.L., Dean, J.A., and Settle, F.A., *Instrumental Methods of Analysis*, 7th ed., Wadsworth Publishing Co., Belmont, CA, 1988.
2. Silverstein, R.M. and Webster, F.X., *Spectrometric Identification of Organic Compounds*, 6th ed., Wiley, New York, 1998.
3. Lambert, J.B., Shurvell, H.F., Lightner D.A., Verbit, L., and Cooks, R.G., *Organic Structural Spectroscopy*, Prentice Hall, Upper Saddle River, NJ, 1998.

Chromophore	Functional Group	λ_{max} (nm)	ε_{max}	λ_{max} (nm)	ε_{max}	λ_{max} (nm)	ε_{max}
Ether	–O–	185	1,000				
Thioether	–S–	194	4,600	215	1,600		
Amine	–NH$_2$–	195	2,800				
Amide	–CONH$_2$	<210	—				
Thiol	–SH	195	1,400				
Disulfide	–S–S–	194	5,500	255	400		
Bromide	–Br	208	300				
Iodide	–I	260	400				
Nitrile	–C≡N	160	—				
Acetylide (alkyne)	–C≡C–	175–180	6,000				
Sulfone	–SO$_2$–	180	—				
Oxime	–NOH	190	5,000				
Azido	>C=N–	190	5,000				
Alkene	–C=C–	190	8,000				
Ketone	>C=O	195	1,000	270–285	18–30		
Thioketone	>C=S	205	Strong				
Esters	–COOR	205	50				
Aldehyde	–CHO	210	Strong	280–300	11–18		
Carboxyl	–COOH	200–210	50–70				
Sulfoxide	>S→O	210	1,500				
Nitro	–NO$_2$	210	Strong				
Nitrite	–ONO	220–230	1,000–2,000	300–4,000	10		
Azo	–N=N–	285–400	3–25				
Nitroso	–N=O	302	100				
Nitrate	–ONO$_2$	270 (shoulder)	12				
Conjugated hydrocarbon	–(C=C)$_2$– (acyclic)	210–230	21,000				
Conjugated hydrocarbon	–(C=C)$_3$–	260	35,000				
Conjugated hydrocarbon	–(C=C)$_4$–	300	52,000				

(Continued)

Chromophore	Functional Group	λ_{max} (nm)	ε_{max}	λ_{max} (nm)	ε_{max}	λ_{max} (nm)	ε_{max}
Conjugated hydrocarbon	$-(C=C)_5-$	330	118,000				
Conjugated hydrocarbon	$-(C=C)_2-$ (alicyclic)	230–260	3,000–8,000				
Conjugated hydrocarbon	$C=C-C\equiv C$	219	6,500				
Conjugated system	$C=C-C=N$	220	23,000				
Conjugated system	$C=C-C=O$	210–250	10,000–20,000			300–350	Weak
Conjugated system	$C=C-NO_2$	229	9,500				
Benzene	[benzene ring structure]	184	46,700	202	6,900	255	170
Diphenyl	[two linked benzene rings]			246	20,000		
Naphthalene	[naphthalene structure]	220	112,000	275	5,600	312	175
Anthracene	[anthracene structure]	252	199,000	375	7,900		
Pyridine	[pyridine ring structure, N]	174	80,000	195	6,000	251	1,700
Quinoline	[quinoline structure, N]	227	37,000	270	3,600	314	2,750
Isoquinoline	[isoquinoline structure, N]	218	80,000	266	4,000	317	3,500

Note: ϕ also denotes a phenyl group.

WOODWARD'S RULES FOR BATHOCHROMIC SHIFTS

Conjugated systems show bathochromic shifts in their $\pi \rightarrow \pi^*$ transition bands. Empirical methods for predicting those shifts were originally formulated by Woodward, Fieser, and Fieser [1–4]. This section includes the most important conjugated system rules [1–6]. The reader should consult Refs. [5] and [6] for more details on how to apply the wavelength increment data.

REFERENCES

1. Woodward, R.B., Structure and the absorption spectra of α,β-unsaturated ketones, *J. Am. Chem. Soc.*, 63, 1123, 1941.
2. Woodward, R.B., Structure and absorption spectra. III. Normal conjugated dienes, *J. Am. Chem. Soc.*, 64, 72, 1942.
3. Woodward, R.B., Structure and absorption spectra. IV. Further observations on α,β-unsaturated ketones, *J. Am. Chem. Soc.*, 64, 76, 1942.
4. Fieser, L.F. and Fieser, M., *Natural Products Related to Phenanthrene*, Reinhold, New York, 1949.
5. Silverstein, R.M. and Webster, F.X., *Spectrometric Identification of Organic Compounds*, 6th ed., Wiley, New York, 1998.
6. Lambert, J.B., Shurvell, H.F., Lightner D.A., Verbit, L., and Cooks, R.G., *Organic Structural Spectroscopy*, Prentice Hall, Upper Saddle River, NJ, 1998.

(a) Rules of Diene Absorption

Base value for diene	214 nm
Increments for (each) (in nm)	
Heteroannular diene	+0
Homoannular diene	+39
Extra double bond	+30
Alkyl substituent or ring residue	+5
Exocyclic double bond	+5
Polar groups	
$-OOCR$	+0
$-OR$	+ 6
$-S-R$	+30
Halogen	+ 5
$-NR_2$	+60
λ Calculated	= Total

(b) Rules for Enone Absorption[a]

$$\overset{\delta}{-C}=\overset{\gamma}{C}-\overset{\beta}{C}=\overset{\alpha}{C}-\underset{\underset{O}{\|}}{C}-$$

Base value for acyclic (or six-membered) α,β-unsaturated ketone: 215 nm
Base value for five-membered α,β-unsaturated ketone: 202 nm
Base value for α,β-unsaturated aldehydes: 210 nm
Base value for α,β-unsaturated esters or carboxylic acids: 195 nm
Increments for (each) (in nm)

Heteroannular diene	+0
Homoannular diene	+39
Double bond	+30
Alkyl group	
$\alpha-$	+10
$\beta-$	+12
$\gamma-$ and higher	+18
Polar groups	
−OH	
$\alpha-$	+35
$\beta-$	+30
$\delta-$	+50
−OOCR	
$\alpha,\beta,\gamma,\delta$	+6
−OR	
$\alpha-$	+35
$\beta-$	+30
$\gamma-$	+17
$\delta-$	+31
−SR	
$\beta-$	+85
−Cl	
$\alpha-$	+15
$\beta-$	+12
−Br	
$\alpha-$	+25
$\beta-$	+30
$-NR_2$	
$\beta-$	+95
Exocyclic double bond	+5
λ Calculated	= Total

[a] Solvent corrections should be included. These are: water (−8), chloroform (+1), dioxane (+5), ether (+7), hexane (+11), and cyclohexane (+11). No correction for methanol or ethanol.

(c) Rules for Monosubstituted Benzene Derivatives

Parent Chromophore (benzene): 250 nm

Substituent	Increment
–R	–4
–COR	–4
–CHO	0
–OH	–16
–OR	–16
–COOR	–16

where R is an alkyl group, and the substitution is on C_6H_5-.

Rules for Disubstituted Benzene Derivatives

Parent Chromophore (benzene): 250 nm

Substituent	o–	m–	p–
–R	+3	+3	+10
–COR	+3	+3	+10
–OH	+7	+7	+25
–OR	+7	+7	+25
–O⁻	+11	+20	+78 (variable)
–Cl	+0	+0	+10
–Br	+2	+2	+15
–NH₂	+13	+13	+58
–NHCOCH₃	+20	+20	+45
–NHCH₃	—	—	+73
–N(CH₃)₂	+20	+20	+85

R indicates an alkyl group.

ORGANIC ANALYTICAL REAGENTS FOR THE DETERMINATION OF INORGANIC IONS (WITH DETERMINATIONS BY USE OF UV-VIS SPECTROPHOTOMETRY)

This table is an update of the classical table entitled "Organic Analytical Reagents for the Determination of Inorganic Substances," by G. Ackermann, L. Sommer, and D. Thorburn Burns, which was contained in the *CRC Handbook of Chemistry and Physics* up to the 92nd edition, which was published in 2012. Beginning with the 93rd edition of that handbook, the table has been revised yearly. It has been revamped and considerably expanded for this edition. The many recent advancements in this area have been surveyed, and in this table, specifically for inorganic cations, we present the major tests, reagents, and some guidance as to the expected result or method of observation.

Many of these determinations require the use of UV-Vis spectrometry (spectrophotometry) or the simpler colorimetric methods. For this reason, we have placed this table in this chapter rather than Chapter 13, but we realize there is overlap. For brevity, when a determination calls for a spectrophotometric measurement at a particular wavelength, for example at 500 nm, we denote this as: "spec $\lambda = 500$ nm." No wavelength is specified if it is variable, for example, if there is a variation with pH; here we simply indicate: "spec determination." When a determination calls for spectrofluorometric determination, the excitation and emission wavelengths are provided; thus, specf $\lambda_{ex} = 303.5$ nm and $\lambda_{em} = 353$ nm. Note that often, surfactants (such as cetyltrimethylammonium bromide, abbreviated CTAB) are used in the test. We provide in Chapter 14 a table of the common surfactants and their properties. When relevant, we provide an approximate limit of detection (LOD), and when the uncertainty can vary, this is expressed as a relative standard deviation (RSD). Some of the procedures listed here require the use of hazardous chemicals (carcinogens such as benzene, strong acids such as HF). Appropriate precautions must always be observed.

While a great deal of the information presented here is from the recent literature, the reader is referred to several excellent reviews and monographs for additional information [1–11].

REFERENCES

1. Marczenko, Z., *Separation and Spectrophotometric Determination of Elements*, Ellis Horwood, Chichester, 1986.
2. Sandell, E.B. and Onishi, H., *Photometric Determination of Traces of Metals. General Aspects, Part I*, 4th ed., John Wiley & Sons, New York, 1986.
3. Onishi, H., *Photometric Determination of Traces of Metals. Part IIa: Individual Metals, Aluminium to Lithium*, 4th ed., John Wiley & Sons, New York, 1986.
4. Onishi, H., *Photometric Determination of Traces of Metals. Part IIb: Individual Metals, Magnesium to Zinc*, 4th ed. John Wiley & Sons, New York, 1986.
5. Townshend, A., Burns, D.T., Guilbault, G.G., Lobinski, R., Marczenko, Z., Newman, E., and Onishi, H., ed., *Dictionary of Analytical Reagents*, Chapman and Hall, London, 1993.
6. West, T.S. and Nürnberg, H.W., *The Determination of Trace Metals in Natural Waters*, Oxford, Blackwell, 1988.
7. Savvin, S.B., Shtykov, S.N., Mikhailova, A.V., Organic reagents in spectrophotometric methods of analysis, *Russ. Chem. Rev.*, 75, 341, 2006.
8. Ueno, K., Imamura, T., Cheng, K.L., *Handbook of Organic Analytical Reagents*, CRC Press/Taylor & Francis Group, Boca Raton, FL, 1992.
9. American Chemical Society, Reagent chemicals: specifications and procedures: American Chemical Society specifications, official from January 1, 2006, American Chemican Society, Washington, DC, 2006.
10. Crompton, T.R., *Determination of Anions: A Guide for the Analytical Chemist*, Springer, Berlin, 1996.

11. Svoronos, P.D.N., Bruno, T.J., "Organic analytical reagents for the determination of inorganic ions." In *CRC Handbook of Chemistry and Physics*, 100th ed., CRC Press, Boca Raton, FL, 2019.

I Organic Analytical Reagents for the Determination of Cations

Determination	Reagents	Results
Aluminum	Alizarin Red S	Red color develops; spectrophotometric determination preferred
	Aluminon	Lake pigment stabilized with CTAB
	Chrome Azurol S	Spec $\lambda = 500$ nm, stabilized with CTAB
	Chromazol KS	Spec $\lambda = 625$ nm, stabilized with cetylpyridinium bromide
	Eriochrome cyanine R (ECR) (also known as Mordant Blue 3)+CTAB	Red dye lake (pH = 6) stabilized with CTAB
	ECR (also known as Mordant Blue 3)+N,N-dodecyltrimethylammonium bromide (DTAB)	Spec determination by use of cationic surfactants
	8-Hydroxyquinoline	Produces tris(8-hydroxyquinolinato) aluminum (Alq3), found in organic light-emitting diodes (OLEDs)
	Bromopyrogallol red+CTAB	Spec $\lambda = 627$ nm
	Bromopyrogallol red+nonylphenol tetradecaethylene glycol ether	Spec $\lambda = 612$ nm
	Pyrocatechol violet	Spec $\lambda = 578$ nm after separation of aluminum from the matrix materials by chloroform extraction of its acetylacetone complex (pH = 6.5), from an ammonium acetate–hydrogen peroxide medium
	2,2′,3,4-Tetrahydroxy-3′,5′-disulfoazobenzene	Spec determination of the binary system (pH = 5)
Antimony	Brilliant Green	Isolated as the hexachloroantimonate (V) salt extracted by either toluene or benzene
	Bromopyrogallol red	Used for the determination of antimony (III) using EDTA, cyanide, or fluoride ions as masking agents
	Catechol violet	Ternary complex stabilized with CTAB
	Malachite green (basic green 4)	Isolated as the hexachloroantimonate (V) complex after benzene extraction
	Phenyl fluorene	Sensitive color reaction with antimony (III) using cationic surfactants
	Potassium iodide	Antimony (III)-iodide complex formation in the presence of ascorbic acid
	Pyrocatechol Violet+tridodecylammonium bromide	Spec $\lambda = 530$ nm of the ternary complex with antimony (III)
	Rhodamine B	Ion pair or ion association extraction using toluene or benzene as solvents
	Silver diethyldithiocarbamate	Spec $\lambda = 504$ nm to avoid arsenic interference
	Thiourea	Determination by hydride generation inductively coupled plasma atomic emission spectrometry after reduction of antimony (V) to antimony (III) by thiourea

(Continued)

I Organic Analytical Reagents for the Determination of Cations (*Continued*)

Determination	Reagents	Results
Arsenic	Michler's ketone	Spec $\lambda = 640$ nm trophotometric determination of trace arsenic (V) in water
	Silver diethyldithiocarbamate	Spec at $\lambda = 600$ nm to avoid antimony interference
	Thiourea	Determination by hydride generation inductively coupled plasma atomic emission spectrometry after reduction of arsenic (V) to arsenic (III) by thiourea
Barium	Dimethylsulfonazo-III (DMSA-III)	Spec $\lambda = 662$ nm of the chelate complex
Beryllium	Aluminon	Lake pigment derivative
	Ammonium bifluoride	Derivative detected by fluorescence
	Beryllon II	A resin-phase spectrophotometric method that detects a change in absorbance of the resin phase immobilized with Beryllon II
	Beryllon III	Determination achieved via third-derivative spectrophotometry and decolorization of excess reagent
	Chrome Azurol S	Spec determination with Chrome Azurol S in the presence of EDTA or CTAB
	ECR+CTAB	Spec $\lambda = 590$ nm of ternary complex
	Sulphon Black F	Derivative with a long color-development time
Bismuth	Amberlite XAD-7	Determination system implemented with hydride generation-inductively coupled plasma-atomic emission spectroscopy (HG-ICP-AES) associated with flow injection (FI).
	Sodium azide	Azidodimethylbismuthine precipitate formed by the reaction of the corresponding bismuthine with sodium azide
	Diethyldithiocarbamate	Heterometric micro-determination of lead with sodium diethyldithiocarbamate using a mixture of EDTA, cyanide, and ammonium hydroxide ($\lambda = 400$ nm)
	Dithizone	Orange-red derivative that is extracted in carbon tetrachloride
	Pyrocatechol Violet	Spec determination of bismuth (III) with Pyrocatechol Violet in the presence of Septonex CTAB
	Pyrocatechol Violet+tridodecylammonium bromide	Spec $\lambda = 564$ nm of ternary complex
	Quinolin-8-ol	Determination system implemented with HG-ICP-AES associated with FI
	Thiourea	Determination by HG-ICP-AES
	Xylenol Orange	Derivative used for sol-gel thin films that serve as bismuth (III) sensors
Boron	Azomethine H	Spec $\lambda = 415$ nm
	Carminic acid	Spec $\lambda = 615$ nm

(*Continued*)

I Organic Analytical Reagents for the Determination of Cations (*Continued*)

Determination	Reagents	Results
	Curcumin	Spec $\lambda = 555$ nm of rosocyanine and rubrocurcumin formed by the reaction between borates and curcumin.
	Methylene blue	Spec determination of complex formed between fluoroborate ions and methylene blue after treatment with hydrofluoric and sulfuric acids and extraction with ethylene chloride
Cadmium	2-(5-Bromo-2-pyridylazo)-5-diethylaminophenol (PAR)	Spec determination in the presence of cationic surfactant cetylpyridinium chloride
	Cadion	Determination by β-correction spectrophotometry with Cadion and surfactant Triton-X
	Dithizone	Determination of dithizonate derivative by the extraction–spectrophotometric method
	1-(2-Pyridylazo)-2-naphthol (PAN)	Two-dimensional absorption spec determination of complex in aqueous micellar solutions
	4–(2-Pyridylazo) resorcinol	Preconcentration by cloud point extraction of the complex; determination by ICP optic emission spectrometry. Simultaneous spectrophotometric determination of cadmium and mercury
Calcium	Alizarin S	Red dye used in staining bones for calcium determination
	Chlorophosphonazo III	Spec $\lambda = 667.5$ nm (pH = 2.2)
	Eriochrome Black T+EDTA	Complexometric titration where the initially formed red calcium–Eriochrome Black T color is replaced with the blue calcium–EDTA color at the end point
	Glyoxal-bis(2-hydroxyanil)	Spectrophotometric titration of the complex without preliminary extraction
	Murexide	Spec $\lambda = 506$ nm (pH = 11.3)
	Phthalein purple	High-performance chelation ion chromatography involving dye-coated resins
Cerium	Butaperazine dimaleate propericiazine	Spec determination of the colored complex in a phosphoric acid medium
	Persulfate oxidation to cerium (IV)	Spec $\lambda = 320$ nm
	Propericiazine	Spec determination of the colored complex in a phosphoric acid medium
	Propionyl promazine phosphate (PPP)	Spec $\lambda = 513$ nm of the red-colored radical cation formed upon the reaction of PPP with cerium (IV) in a phosphoric acid medium
	N-Benzoyl-N-phenylhydroxylamine	Spec titration of the cerium (IV) complex (pH = 8–10)
	Sodium triphosphate	Specf $\lambda_{ex} = 303.5$ nm, $\lambda_{em} = 353$ nm of the cerium (III) complex
	8-Hydroxyquinoline	Spec determination of the metal–ligand complex
Chromium	Alizarin S	Lake pigment complex formation

(Continued)

I Organic Analytical Reagents for the Determination of Cations (*Continued*)

Determination	Reagents	Results
	1,5-Diphenylcarbazide	Spec determination during sonication in carbonated aqueous solutions saturated with CCl_4 that produces chlorine radicals.
	3-(2-Pyridyl)-5,6 bis(5-(2 furyl disulfonic acid))-1,2,4-triazine disodium salt (ferene-TM)	Indirect spec $\lambda = 593$ nm in aqueous samples with a chromogen ferene-TM
	4-(2-Pyridylazo) resorcinol (PAR)	Spec determination of the ternary chromium-peroxo-PAR ternary complex
	PAR + hydrogen peroxide	Spec determination of the ternary chromo-peroxo-PAR mixture after ethyl acetate extraction in 0.1 M sulfuric acid
	PAR + xylometazolonium (XMH) chloride	Spec determination of the orange-red anionic complex formed in a heated acetate buffer medium (pH = 4.0–5.5) and extracted with the XMH chloride
	Sulfanilic acid	Spec $\lambda = 360$ nm on the catalytic effect of chromium (VI) in the oxidation of sulfanilic acid by hydrogen peroxide with p-aminobenzoic acid as an activator
Cobalt	8-Hydroxyquinoline	Spec determination of the metal–ligand complex
	p-Nitroso-N,N-dimethylaniline	Spec determination of the binary complex
	Nitroso-R salt	AAS using a continuous online precipitation–dissolution procedure
	1-Nitroso-2-naphthol	Spec determination using nonionic surfactant Triton X-100
	1-Nitroso-2-naphthol	Spec Tween 80 micellar determination
	1-Nitroso-2-naphthol	AAS using a continuous online precipitation–dissolution procedure based on 1-nitroso-2-naphthol
	2-Nitroso-1-naphthol	Spec $\lambda = 530$ nm after isoamyl acetate extraction
	PAN + surfactants	Spec $\lambda = 620$ nm of the cobalt complex in the presence of surfactants (Triton X-100 combined with sodium dodecylbenzene sulfonate (DBS)) and trace of ammonium per sulfate (pH 5.0)
	2,4-Dinitroresorcinol (DNR)	Spec $\lambda = 397$ nm determination
	PAR	Spec determination of complex at both pH = 7.2–7.9 and in 1 M H_2SO_4
	PAR + triethanolamine	Ion-pair reversed-phase high-performance liquid chromatography (HPLC) of the complex
Copper	Bathocuproine disulfonic acid	Spec $\lambda = 470$–550 nm of the bathocuproine disulfonic acid complex after extraction with chloroform and methanol analyzed with partial least squares regression (PLS_2)
	Dithizone	Spec determination of the dithizone complex at pH = 2.3

(Continued)

I Organic Analytical Reagents for the Determination of Cations (*Continued*)

Determination	Reagents	Results
	Neocuproine	Spec determination of the deep orange-red Cu(Neocuproine)$_2^+$ complex color
	Cuprizone	Spec determination $\lambda = 595$ nm of the highly chromogenic copper (III) Cuprizone complex
	p-Nitroso-N,N-dimethylaniline	Spec determination of the binary complex
	1-Nitroso-2-naphthol	Spec determination using nonionic surfactant Triton X-100
	PAR	Preconcentration of copper by cloud point extraction of the complex and determination by ICP optical emission spectrometry.
Europium	1-Nitroso-2-naphthol	Spec Tween 80 micellar determination
	ChromAsurol S	Spec determination of the binary complex
	Arsenazo III	Spec $\lambda = 510$ nm and 655 nm of complex
	PAR + tetradecyldimethylbenzylammonium chloride (TDBA)	Spec $\lambda = 510$ nm of the ion-associate complex extracted with chloroform at pH = 9.7
Gallium	Chrome Azurol S + CTAB	Spec $\lambda = 640$ nm of the ternary complex
	Hematoxylin or its oxidized form + CTAB	Spec determination of the ternary complex of indium with hematoxylin or its oxidized form in the presence of cationic, anionic, and nonionic surfactants such as CTAB
	Pyrocatechol Violet + diphenylguanidine	Spec $\lambda = 345$ nm determination of the ternary complex
	8-Hydroxyquinoline	Spec determination of the gallium complex
	PAN	Spectrofluorimetric determination of gallium (III) with PAN in sodium dodecyl sulfate micellar medium
	Chrome Azurol S + CTAB	Spec $\lambda = 640$ nm of the ternary complex after n-butyl acetate extraction from hydrobromic acid
	PAR	Extraction and spec $\lambda = 510$ nm of gallium with PAR
	Pyrocatechol Violet + tridodecylammonium bromide	Spec $\lambda = 595$ nm of ternary complex
	Rhodamine B	Comparison of the determination of gallium by a Rhodamine B spectrophotometric method and by an AA method based on preliminary solvent extraction
	Xylenol Orange + 8-hydroxyquinoline	Spectrophotometric determination of the ternary complex
Germanium	Brilliant Green + molybdate	Spec $\lambda = 430$ nm of the yellow germanomolybdic acid
	Phenylfluorone	Spec $\lambda = 597$ nm determination of the complex previously extracted with carbon tetrachloride at pH = 3.1

(Continued)

I Organic Analytical Reagents for the Determination of Cations (*Continued*)

Determination	Reagents	Results
Gold	5-(4-Diethylaminobenzylidene) rhodanine	Immobilized 5-(4-dimethylamino-benzylidene) rhodanine serves as a stable solid sorbent for trace amounts of Au(III) ions (pH = 2–4)
	Di(methylheptyl)methyl phosphonate (DMHMP)	Trace Au determination by online preconcentration with Fl atomic absorption spectrometry, using DMHMP as the immobilized phase loaded onto a macroporous resin
	2-Mercaptobenzothiazole	Separation and determination of gold complex matrices employing substoichiometric thermal neutron activation analysis (NAA)
	Molybdate + Nile blue (NB)	A spectrophotometric method based on the reaction of gold(III) with molybdate and NB to form an ion-association complex in the presence of poly(vinyl alcohol)
	Rhodamine B	Aqueous spec determination of Au with Rhodamine B and surfactant
Hafnium	Arsenazo III	Spec determination of the arsenazo III complex in 10 M HCl or H_2SO_4
Indium	Alizarin Red S	Spec $\lambda = 525$ nm determination after extraction in the presence of 1,3-diphenylguanidine
	Bromopyrogallol red	Spec determination of indium after an ether extraction from hydrobromic acid, and benzyl alcohol extraction of its complex with Bromopyrogallol red (pH = 9.0)
	5-Bromine-salicylaldehyde salicyloylhydrazone (5-Br-SASH)	Specf $\lambda_{ex} = 395$ nm, $\lambda_{em} = 461$ nm of the indium: 5-Br-SASH chelate in a water–ethanol (63 %) medium (pH = 4.6)
	2-(5-Bromo-2-pyridylazo)-5-diethyl aminophenol	Spec $\lambda = 558$ nm determination. Interference by Co(II), Ni (II), Ti (III), V(V), and EDTA
	Chrome Azurol S	Spec determination of the binary complex
	Chrome Azurol S + benzyldodecyldimethylammonium bromide	Spec determination of the mixed complex with Chrome Azurol S and benzyldodecyldimethylammonium bromide
	Chrome Azurol S + CTAB	Spec $\lambda = 630$ nm of the ternary complex after n-butyl acetate extraction from hydrobromic acid
	Chrome Azurol S + cationic surfactants	Spec $\lambda = 630$ nm determination using Chrome Azurol S and surfactants such as CTA, CP, or Zephiramine
	4,4′-Dihydroxybenzophenone thiosemicarbazone	Spec $\lambda = 415$ nm of the complex after heating at 45 °C
	Dithizone	Spec determination of the binary complex
	ECR and CTAB	Spec $\lambda = 585$ nm determination. Interference by Be (II), GA (III), Al (III), Fe (III), and V (IV)

(*Continued*)

I Organic Analytical Reagents for the Determination of Cations (*Continued*)

Determination	Reagents	Results
	Hematoxylin or its oxidized form+CTAB	Spec determination of the ternary complex of indium with hematoxylin or its oxidized form in the presence of cationic, anionic, and nonionic surfactants such as CTAB
	2-Hydroxybenzaldehyde isonicotinoylhydrazone	Spec $\lambda = 380$ nm determination after extraction into 1-pentanol
	8-Hydroxyquinoline	Spec determination of the binary complex
	Methylthymol blue+zepheramine	Spec determination of the ternary complex
	2-Oxoguanidine benzoic acid	Spec $\lambda = 545$ nm determination
	PAN	Spec determination of the chelate complex after chloroform extraction (pH = 6); serious interference of Co (II), Fe (III), Ga (III), Ni (II), and V(V) ions
	PAR	Spec $\lambda = 520$ nm after indium extraction from the aqueous phase (pH 5.0–5.5) into chloroform with *N-p*-chlorophenyl-2-furohydroxamic acid and formation of the PAR red chelate. Color develops after 10 min
	Pyrocatechol Violet	Spec determination of the binary complex
	Pyrocatechol Violet+tridodecylammonium bromide	Spec $\lambda = 600$ nm determination of the ternary complex
	Quinalizarin (1,2,5,8-tetrahydroxy anthraquinone)	Spec $\lambda = 565$ nm of the binary 3:1 (quinalizarin:In(III) colored complex in dimethylformamide–water solution
	4,5,6,7-Tetrachlorogallein and cetyl pyridinium chloride	Spec $\lambda = 620$ nm. Color development reached after 30 min
	2,2′,3,4-Tetrahydroxy-3′,5′-disulfoazobenzene	Spec determination of the binary system (pH = 5)
	3,5,7,4′-Tetrahydroxyflavone	Spec $\lambda = 430$ nm determination. Interference by Al(III), Fe (III), Mo (VI), W (V), Sn (IV), Zr (IV), Ti (IV), and V (V)
	2-(2-Thiazolylazo)-p-cresol	Spec $\lambda = 580$ nm determination of the complex. Interference by Cu(II), Ni(II), Fe (II), Co (II) and Ti (IV)
	1-(2-Thiazolylazo)-2-naphthol (TAN)	Spec $\lambda = 575$ nm determination of the complex
	Thiothenoyltrifluoroacetone	Spec $\lambda = 480$ nm after extraction into carbon tetrachloride (pH = 4.5–5)
	Xylenol Orange	Spec $\lambda = 560$ nm determination of the ternary complex
Iridium	1,5-Diphenylcarbazide	Spec determination of the complex (pH = 5.0)
	PAN	Spec $\lambda = 550$ nm of the red complex (pH = 5.1) after chloroform extraction
Iron	Bathophenanthroline	Determination of iron(II) in the presence of 1,000:1 ratio of iron(III) using bathophenanthroline

(*Continued*)

I Organic Analytical Reagents for the Determination of Cations (*Continued*)

Determination	Reagents	Results
	Bathophenantroline–disulfonic acid	Spec λ = 470–550 nm of the bathophenantroline–disulfonic acid complex after extraction with chloroform and methanol
	2,2′-Bipyridine	Spec determination of the iron (II)-2,2′-bipyridine dark red complex
	FerroZine	Spec λ = 562 nm of iron (II)–ferrozine complex after all iron(III) has been reduced by ascorbic acid
	Hematoxylin+CTAB	Spec of the ternary complex. Addition of CTAB shifts λ_{max} from 630 to 640 nm
	1-Nitroso-2-naphthol	Spectrophotometric determination using nonionic surfactant Triton X-100
	1-Nitroso-2-naphthol	Spectrophotometric Tween 40 micellar determination of Fe(III) λ = 446 nm (pH = 1)
	1,10-Phenanthroline (o-Phen)	Spec λ = 508 nm of the Fe(o-Phen)$_3^{+2}$ complex
	1,10-Phenanthroline+Bromothymol Blue	Spec determination of the Fe(o-Phen)$_3^{+2}$ complex in the presence of Bromothymol Blue
	Phenylfluorone	Spec λ = 530 nm of the binary complex (pH = 9.0)
	Phenylfluorone+Triton X	Spec λ = 555 nm of the binary complex sensitized with Triton X-100 (pH = 9.0)
	Xylenol	Spec λ = 560 nm of the binary complex
Lanthanum	Ammonium purpurate	Spec determination of lanthanum with ammonium purpurate as a chromogenic reagent
	Arsenazo III	Spec λ = 652 nm determination of the lanthanum complexation with reagents of the arsenazo III group on the solid phase of fibrous ion exchangers
	ECR	Spec λ = 540 nm of the binary complex
	N-Phenylbenzohydroxamic acid+Xylenol Orange	Solvent extraction followed by the spec λ = 600 nm of the ternary complex (pH 8.8–9.5)
Lead	Dithizone	Orange-red derivative that is extracted in carbon tetrachloride
	Sodium diethyldithiocarbamate	Electrothermal AAS determination of the diethyldithiocarbamate derivative extracted in carbon tetrachloride
	Sodium diethyldithiocarbamate	Heterometric micro-determination of lead diethyldithiocarbamate derivative
	PAR	Spec λ = 520 nm of 4-(2-pyridylazo)-resorcinol:lead (1:1) complex in an ammonia–ammonium chloride medium at pH 10 after extracting the lead in isobutyl methyl ketone
Lithium	1-(o-Arsenophenylazo)-2-naphthol-3, 6-disulfonate (thoron)	Spec λ = 480 nm of thoron–lithium complex in an alkaline acetone medium is measured against the reagent as reference
Magnesium	Chlorophosphonazo III	Spec λ = 669 nm (pH = 7.0)

(*Continued*)

I Organic Analytical Reagents for the Determination of Cations (*Continued*)

Determination	Reagents	Results
	Eriochrome Black T+EDTA	Complexometric titration where the initially formed red magnesium–Eriochrome Black T color is replaced with the blue magnesium–EDTA color at the end point
	8-Hydroxyquinoline	Volumetric, titrimetric, and colorimetric determination
	8-Hydroxyquinoline+butylamine	Spec $\lambda = 380$ nm of the complex after chloroform extraction
	Titan yellow	Spec determination of magnesium by Titan yellow in biological fluids
	Xylidyl blue	Spec determination of magnesium in biological fluids
Manganese	Formaldoxime	Spec $\lambda = 450$ nm of the formaldoxime/ammonia complex (pH = 8.8–8.9) that is stable for 30 min
	2-Hydroxy-4-methoxy acetophenone oxime	Spec determination $\lambda = 410$ nm of binary complex
	PAN	Preconcentration determination using PAN anchored. SiO_2 nanoparticles
Mercury	Dithizone	Spec $\lambda = 500$ nm of dithizone complex after chloroform extraction at pH = 0.3
	Michler's thioketone	Spec $\lambda = 560$ nm of the binary complex in acetate buffer
	4–(2-Pyridylazo) resorcinol	Simultaneous spec determination of cadmium and mercury with 4-(2-pyridylazo)-resorcinol
	Rhodamine 6G	Photoelectrochemical determination of mercury (II) in aqueous solutions using a Rhodamine 6G derivative (RS) and polyaniline (PANI)-coated optical probe in a photoelectrochemical cell
	Xylenol Orange+amine buffer	Spec $\lambda = 590$ nm of Hg(II)/Xylenol Orange complex display a sharp hyperchromic effect in the presence of amine buffers (pH = 7.5)
	Xylenol Orange+citric acid–phosphate buffer	Spec $\lambda = 580$ nm of Hg(II)/Xylenol Orange complex that displays a sharp hypochromic effect upon substituting amine buffers with a citric acid–phosphate (pH = 7.5)
Molybdenum	Bromopyrogallol red+cetylpyridium chloride	Sequential injection analysis (SIA) to the determination of cetylpyridinium chloride based on the sensitized molybdenum–Bromopyrogallol red reaction
	Phenylfluorone	Spec $\lambda = 560$ nm of molybdenum with phenylfluorone (pH of 1.5–3)
	8-Hydroxyquinoline-5-sulfonic acid and phenylfluorone	Diffuse reflection spectrometry with 8-hydroxyquinoline-5-sulfonic acid and phenylfluorone after sorbing on a disk of an anion exchange fibrous material

(*Continued*)

I Organic Analytical Reagents for the Determination of Cations (*Continued*)

Determination	Reagents	Results
	Pyrocatechol Violet	Preconcentration (using basic anion exchanger AV-17-10P) and determination of molybdenum in aqueous solution via diffuse reflection spectroscopy. The colored surface compound to be determined involves
		Mo(VI) sorption on the resin and subsequent treatment of the concentrate obtained with Pyrocatechol Violet.
	Toluene-3,4-dithiol	Spec $\lambda = 415$ nm of molybdenum(VI) with toluene-3,4-dithiol in isobutyl methyl ketone in acidic medium
Neodymium	Semi-Xylenol Orange+cetylpyridinium chloride	Fourth-order derivative spectrophotometric determination of the ternary complex
Neptunium	Arsenazo III	Spec determination after separation by use of the thenoyltrifluoroacetone extraction method and determination in 5 M HNO_3
Nickel	2-(5-Bromo-2-pyridylazo)-5-diethylaminophenol	Spec $\lambda = 520$ and 560 nm of the red–violet complex in water–ethanol (pH = 5.5)
	Dimethylglyoxime+ammonia	Spec $\lambda = 543$ nm of the complex
	Dimethylglyoxime, voltammetry	Nickel voltammetric determination at a chemically modified electrode based on dimethylglyoxime-containing carbon paste
	EDTA	Spec $\lambda = 380$ nm of the complex
	2,2′-Furildioxime	Spec $\lambda = 438$ nm after separation by adsorption of its α-furildioxime complex on naphthalene
	Hematoxylin	Spec $\lambda = 595$ nm of the binary system (pH = 7.8–8.3)
	Hematoxylin+CTAB	Spec $\lambda = 608$ nm of the ternary system (pH = 7.4–8.1)
	1-Nitroso-2-naphthol	Spec Tween 80 micellar determination
	p-Nitroso-N,N-dimethylaniline	Spec determination of the binary complex
	2-(2-Pyridylazo)-2-naphthol	Derivative spectrophotometry $\lambda = 569$ nm determination of nickel complex in Tween 80 micellar solutions
	PAR	Preconcentration of nickel by cloud point extraction of the complex and determination by ICP optic emission spectrometry
	TAN	FI solid-phase spectrophotometry using TAN immobilized on C_{18}-bonded silica ($\lambda = 595$ nm
	Xylenol Orange	Spec determination by mean centering of ratio kinetic profile (pH = 5.3)
Niobium	N-Benzoyl-N-phenylhydroxylamine	Separation and determination of the complex from a tartrate solution at pH > 2
	Bromopyrogallol red	Spec $\lambda = 610$ nm of complex extracted into isopentyl acetate containing di-n-octylmethylamine

(Continued)

I Organic Analytical Reagents for the Determination of Cations (*Continued*)

Determination	Reagents	Results
	o-Hydroxyhydroquinonephthalein (Qnph)+hexadecyltrimethylammonium chloride (HTAC)	Spec λ = 520 nm of the complex in strong acidic media
	PAR+citrate	Determination of the ternary complex by ion-interaction reversed-phase HPLC on a C18 column with a 5 mM citrate buffer (pH = 6.5, λ = 540 nm.)
	Pyrocatechol Violet	Extraction and spec determination of niobium complex in the presence of pyridine and trichloroacetic acid
	Sulfochlorophenol S	Spec λ = 650 nm determination of the complex is extracted into amyl alcohol
	Xylenol Orange	Spec λ = 530 nm of chelate (pH = 5.0)
Osmium	1,5-Diphenylcarbazide	Spec λ = 560 nm of the bluish-violet complex after extraction with isobutyl methyl ketone
	Thiourea	Spec λ = 540 nm of the red-rose complex in acid medium
Palladium	Ammonia+iodide	Thermogravimetric determination of palladium as $Pd(NH_3)_2I_2$
	2,2-Bis-[3-(2-thiazolylazo)-4-4-hydroxyphenyl-propane], TAPHP)	Application of TAPHP immobilized on silica beads to determine the palladium concentration in a trans illuminance configuration (pH = 2) using flow-through spectrophotometric sensing phase
	2-(5-Bromo-2-pyridylazo)-5-diethylaminophenol	Spec determination of the complex in a sulfuric acid medium in the presence of ethanol
	Dithizone	Volumetric determination of palladium with dithizone in acid medium after extraction of its dimethylglyoxime complex with chloroform; PAS determination of the dithizone extraction solution into a thermally thin solid film
	Dithiazone+iodide	Spec determination of palladium with dithizone, using an iodide medium in the presence of sulfite at pH 3–5, to separate platinum
	Dithizone+stannous chloride	Extractive separation and spec determination of palladium in the presence of stannous chloride to separate platinum
	Isonitrosobenzoylacetone	Radiochemical separation and determination of palladium in complex matrices employing substoichiometric thermal NAA
	2-Nitroso-1-naphthol	Spec λ = 370 nm of the violet complex
	PAR	Spec determination of complex at both pH = 7.2–7.9 and in 1 M H_2SO_4
	PAR+diphenylguanidine	Spec determination after extraction of the red Pd(II) chelate with PAR in the presence of *N,N′*-diphenylguanidine into *n*-butanol
	Thioglycolic acid	Spec λ = 384 nm of complex

(Continued)

I Organic Analytical Reagents for the Determination of Cations (*Continued*)

Determination	Reagents	Results
Platinum	N-Phenylbenzimidoylthiourea (PBITU)	Fl analysis spec $\lambda = 345$ nm of the Pd:PBITU complex in 0.2–2.0 M HCl in 10 % (vol/vol) ethanol solution
	Dithizone + iodide	Spec determination of platinum with dithizone, using an iodide medium in the presence of sulfite at pH 3–5, to separate palladium
	Dithizone + stannous chloride	Extractive separation and spectrophotometric determination of platinum in the presence of stannous chloride to separate palladium
	2-Mercaptobenzothiazole	Radiochemical separation and determination of platinum complex matrices employing substoichiometric thermal NAA
Protactinium	Arsenazo III	Spec $\lambda = 680$ nm of the complex after extraction with 7N H_2SO_4 and isoamyl alcohol
	EDTA + tannic acid	Gravimetric analysis of the tannic acid precipitate (pH = 5.0)
Rhenium	2,2′-furildioxime	Spec $\lambda = 532$ nm of the complex
	N,N-Diethyl-N′-benzoylthiourea	Spec $\lambda = 383$ nm of the green complex in hydrochloric acid medium in the presence of tin(II) chloride
	Tin(II)	Spectrophotometric titration for the determination of rhenium using tin(II) as the titrant
Rhodium	1,5-Diphenylcarbazide	Spec determination of the complex after isobutyl alcohol extraction (pH = 5.0)
	o-Methylphenyl thiourea	Spec $\lambda = 320$ nm of the binary complex
	p-Nitrosodimethylaniline	Spec $\lambda = 510$ nm of the cherry red binary complex (pH = 4.4)
	PAN	Spec $\lambda = 598$ nm of the green complex (pH = 5.1) after chloroform extraction
Ruthenium	4-Benzylideneamino-3-mercapto-6-methyl-1,2,4-triazine (4H)-5-one	Spec $\lambda = 620$ nm
	1,10-Phenanthroline	Chemiluminescence determination of chlorpheniramine using tris(1,10-phenanthroline)-ruthenium(II) peroxydisulfate system and SIA; spec determination of the complex after reducing Ru (IV) to Ru(II)
	Thiourea	Spec $\lambda = 640$ nm of ruthenium complex in carbon supported Pt-Ru-Ge catalyst in 5 M HCl
	1,4-Diphenylthiosemicarbazide	Spec determination of the bright red complex
Scandium	Alizarin Red S	Cathodic adsorptive stripping of the scandium–Alizarin Red S complex onto a carbon paste electrode
	Ammonium purpurate	Spec determination of scandium with ammonium purpurate as a chromogenic reagent
	Chrome Azurol S	Spec determination of the binary complex

(*Continued*)

I Organic Analytical Reagents for the Determination of Cations (*Continued*)

Determination	Reagents	Results
	Xylenol Orange	Spec λ = 553 nm of the binary mixture
Selenium	3,3′-Diaminobenzidine	Spec λ = 350 nm of the yellow complex (pH = 3.0)
	2,3-Diaminonaphthaline	Fluorometric determination of the binary complex in water
	Thiourea	Determination of selenium by hydride generation inductively coupled plasma atomic emission spectrometry
Silver	Dithizone	Graphite-furnace atomic-absorption spectrophotometric determination of silver on suspended dithizone particles from acidic sample solutions with sonication to facilitate the separation
	Eosin + 1,10-phenanthroline	Spec λ = 540–555 nm of the 1,10-phenanthroline (PHEN) and Eosin (2,4,5,7-tetrabromofluorescein) association complexes
Strontium	Phthalein purple	High-performance chelation ion chromatography involving dye-coated resins
Tantalum	N-Benzoyl-N-phenylhydroxylamine	Separation and determination of the complex from a tartrate solution at pH < 1.5)
	Crystal violet (CV) + N,N′-diphenylbenzamidine (DPBA)	Spec λ = 600 nm of Ta(V)-F-CV+ cation complex with a benzene solution of DPBA from sulfuric acid solution
	o-Hydroxyhydroquinonephthalein (Qnph) + HTAC	Spec λ = 510 nm of the complex in strong acidic media
	Malachite green	Spec λ = 623 nm of the complex in HF after extraction with benzene or toluene
	Methyl violet	Spec determination of the complex
	PAR + citrate	Determination of the ternary complex by ion-interaction reversed-phase HPLC on a C18 column in 5 mM citrate buffer (pH = 6.5, λ = 540 nm)
	Phenylfluorone	Spectrophotometric determination from HF-HCl solution with methyl isobutyl ketone
	Victoria blue	Spec λ = 600 nm determination of the complex after benzene extraction
Tellurium	Ammonium pyrrolidinedithiocarbamate	Differential determination of tellurium(IV) and tellurium(VI) by AAS with a carbon-tube atomizer
	Bismuthiol II	Spec λ = 330 mm of the yellow complex in acidic medium (pH = 3.5) after chloroform extraction
	Dithizone	Differential determination of tellurium(IV) and tellurium(VI) by AAS with a carbon-tube atomizer
	Sodium diethyldithiocarbamate	Differential determination of tellurium(IV) and tellurium(VI) by AAS with a carbon-tube atomizer
	Thiourea	Determination of selenium by hydride generation inductively coupled plasma atomic emission spectrometry

(*Continued*)

I Organic Analytical Reagents for the Determination of Cations (*Continued*)

Determination	Reagents	Results
Thallium	Brilliant Green+N,N′-diphenyl benzamidine+cetyl chloride	Spec λ = 640 nm determination of the binary complex after toluene extraction
	Dithizone	Spec λ = 505 mm of the complex after chloroform extraction from a citrate-sulfite-cyanide medium at pH = 10.6
	8-Hydroxyquinoline	Spectrophotometric determination of the metal–ligand complex
	Rhodamine B	Fluorimetric determination of thallium (in silicate rocks) with Rhodamine B after separation by adsorption on a crown ether polymer
	Rhodamine B hydroxide	Spec λ = 565 nm via oxidation of Rhodamine B hydrazide by thallium(I) in acidic medium to give a pinkish violet radical cation
Thorium	Arsenazo III	Spec determination of the arsenazo III complex in 10 M HCl or H_2SO_4
	ECR	Spec λ = 540 nm of the binary complex
	Thoron	Spectrophotometric determination of the binary complex in tartaric acid
	Xylenol Orange	Spec λ = 570 nm of the binary complex (pH = 4.0)
	Xylenol Orange+CTAB (CTMAB)	Spec λ = 600 nm of the complex sensitized by CTMAB (pH = 2.5)
Tin	Catechol violet+CTAB (CTMAB)	Spectrophotometric determination of the green complex
	Pyrocatechol Violet+CTAB	Spec λ = 660 nm of the complex (pH = 2.0)
	Gallein	Spec λ = 530 nm determination of complex in acid medium
	Phenylfluorone	Spec λ = 530 nm determination of the complex (pH = 1) in 36 % aqueous ethanol after a preliminary solvent extraction of the tin the iodide form
	Toluene-3,4-dithiol+dispersant	Spec determination of the complex
Titanium	Chromotropic acid	Spec λ = 443 nm of the complex using an FI manifold
	Diantipyrinylmethane	Spec λ = 390 nm of the yellow complex
	3-Hydroxy-2-methyl-1-(4-tolyl)-4-pyridone (HY)	Spec λ = 355 nm of the ternary complex formed in perchloric acid and is extracted by chloroform (see Chapter 15 for information on the safe handling of perchloric acid)
	Methylene blue–ascorbic acid redox reaction	Spec λ = 665 nm of the titanium-methylene blue–ascorbic acid redox reaction
	Tiron	Spec λ = 420 nm of the yellow Tiron derivative (pH = 5.2–5.6)
Tungsten	Cyanate	Spectrophotometric determination of tungsten with thiocyanate after both tungsten(VI) and molybdenum(VI) are extracted into chloroform as benzoin α-oxime complexes

(*Continued*)

I Organic Analytical Reagents for the Determination of Cations (*Continued*)

Determination	Reagents	Results
	Pyrocatechol Violet	Determination of the 2:1 green complex in acid medium which is fixed on a dextran-type anion-exchange resin (Sephadex QAEA-25) by first-derivative solid-phase spectrophotometry ($\lambda = 674$ nm)
	Tetraphenylarsonium chloride + thiocyanate	Gravimetric determination of tungsten with tetraphenylarsonium chloride after its extraction as thiocyanate (pH = 2–4)
	Toluene-3,5-dithiol	Spec $\lambda = 630$ nm of the toluene-3,4-dithiol derivative after extraction with isoamyl acetate
Uranium	Arsenazo III (1,8-dihydroxynaphthalene-3,6-disulfonic acid-2,7-bis[(azo-2)-phenylarsonic acid])	Spec determination of the binary complex
	2-(5-Bromo-2-pyridylazo) diethylaminophenol	Spec $\lambda = 578$ nm of the complex (pH = 7.6)
	Chlorophosphonazo III	Spec $\lambda = 673$ nm of the complex after extraction into 3-methyl-1-butanol from 1.5–3.0 M HCl
	8-Hydroxyquinoline	Spectrophotometric determination of the metal (UO_2 (II))–ligand complex
	2-(2-pyridylazo)-5-diethylaminophenol (PADAP)	Spec $\lambda = 564$ nm of the complex after extraction into methyl isobutyl ketone (pH = 8.2)
	PAN	Spec $\lambda = 560$ nm of the deep red precipitate in ammoniacal solutions extracted with chloroform
	4-(2-Pyridylazo)-resorsinol (PAR)	Spec $\lambda = 530$ nm of the intensely deep red complex after extraction into methyl isobutyl ketone (pH = 8.0)
	2-(2-Thiazolylazo)-p-cresol + surfactants	Spec $\lambda = 588$ nm of the complex (pH = 6.5) with surfactants such as Triton X-100 or N-cetyl-N,N,N-trimethyl ammonium bromide (CTAB)
Vanadium	Chrome azurol S	Spec $\lambda = 598$–603 nm determination of the binary complex
	N-Benzoyl-*N*-phenylhydroxylamine	Spec $\lambda = 416$ nm determination of the violet complex in acidic medium
	3,5-Dinitrocatechol (DNC) + brilliant Green	Extraction-spectrophotometric determination of the system V(V)-3,5-dinitrocatechol (DNC)–Brilliant Green chelate complex.
	8-Hydroxyquinoline (oxine)	Extraction with n-butanol and spec $\lambda = 390$ nm of a ternary complex (vanadium:oxine:n-butanol = 1:2:2).
	8-Hydroxyquinoline-5-sulfonic acid and phenylfluorone	Diffuse reflection spectrometry with 8-hydroxyquinoline-5-sulfonic acid and phenylfluorone after sorbing on a disk of an anion exchange fibrous material
	PAR	Spec determination of the binary complex based on the extraction tetraphenylphosphonium or tetraphenylarsonium chloride
	Xylenol Orange	Spec $\lambda = 490$ nm of chelate (pH = 5.0)

(*Continued*)

I Organic Analytical Reagents for the Determination of Cations (*Continued*)

Determination	Reagents	Results
Yttrium	Alizarin Red S	Spec $\lambda = 550$ nm of the complex
	Ammonium purpurate	Spec determination of yttrium with ammonium purpurate as a chromogenic reagent
	Arsenazo III	Spec $\lambda = 660$ nm of the blue colored complex after preliminary purification with hydroxide and subsequent acidification
	ECR	Spec $\lambda = 540$ nm of the binary complex
	Pyrocatechol Violet	Spec $\lambda = 665$ nm of complex
Zinc	2-(5-Bromo-2-pyridylazo)-5-diethylaminophenol	Spec determination using 2-(5-bromo-2-pyridylazo)-5-diethyl aminophenol in the presence of cationic surfactant cetylpyridinium chloride
	Carbonic anhydrase	Enzymatic determination using carbonic anhydrase after removing zinc by dialysis against dipicolinic acid
	Dithizone	Spec $\lambda = 530$ nm of binary complex after chloroform extraction
	Eriochrome Black T + EDTA	Complexometric titration where the initially formed red zinc–Eriochrome Black T color is replaced with the blue zinc–EDTA color at the end point (pH = 10)
	7-(4-Nitrophenylazo)-8-hydroxyquinoline-5-sulfonic acid (p-NIAZOXS)	Spec $\lambda = 520$ nm of the zinc derivative (pH = 9.2, borax buffer)
	Phenylglyoxal mono(2-pyridyl) hydrazone (PGMPH)	Spec $\lambda = 464$–470 nm of the yellow-orange complex (pH = 7.2–8.5) in 40 % (vol/vol) ethanol
	PAN	Two-dimensional absorption spec determination of complex in aqueous micellar solutions
	PAN	Preconcentration determination using PAN anchored. SiO_2 nanoparticles
	PAR	Spec determination at pH = 7.0; preconcentration of zinc by cloud point extraction of the complex and determination by ICP optical emission spectrometry
	TAN	FI solid-phase spec $\lambda = 595$ nm with TAN immobilized on C18-bonded silica
	Xylenol Orange	Spec determination of 1:1 zinc-Xylenol Orange red–violet complex (pH = 5.8–6.2)
	Xylenol Orange	SIA based on the spec $\lambda = 568$ nm of zinc using Xylenol Orange as a color reagent
	Xylenol Orange	Spec determination by mean centering of ratio kinetic profile (pH = 5.3)
	Xylenol Orange + cetylpyridinium chloride	Spec $\lambda = 580$ nm of the 1:2:4 ratio for the metal:ligand:surfactant ternary complex (pH = 5.0–6.0)
Zirconium	Alizarin Red S	Spec $\lambda = 20$ nm determination of the binary complex after phosphate extraction (pH = 2.5)

(*Continued*)

I Organic Analytical Reagents for the Determination of Cations (*Continued*)

Determination	Reagents	Results
	Arsenazo III (1,8-dihydroxynaphthalene-3,6-disulfonic acid-2,7-bis[(azo-2)-phenylarsonic acid])	Spec determination of the binary complex
	N-p-Chlorophenylbenzohydroxamic acid (N-p-Cl-BHA)+Morin	Spec $\lambda = 420$ nm and fluorimetric determination of the N-p-Cl-BHA greenish yellow complex after extraction with isoamyl alcohol from 0.2 to 0.5 N H_2SO_4 followed by Morin addition; it is critical that the Morin be very pure
	Pyrocatechol Violet	Spec $\lambda = 650$ nm of the blue complex in sulfuric acid solution
	Pyrocatechol Violet+tri-n-octylphosphine oxide (TOPO)	Spec $\lambda = 655$ nm of the blue complex after extraction with tri-n-octylphosphine oxide (TOPO) in cyclohexane
	Morin	Fluorimetric determination of the complex after EDTA addition
	Xylenol Orange	FI spectrophotometric determination in sulfuric acid medium

II Organic Analytical Reagents for the Determination of Anions

Anion	Method	Results/Comments
Arsenate AsO_4^{-3}	X-ray absorption near-edge structure (XANES)	Use of self-consistent multiple-scattering methods that measure the AsO_4^{-3} adsorption on TiO_2
	Dual detection ion chromatography	Simultaneous determination of AsO_4^{-3} and AsO_3^{-3} (LOD = 0.044 mg/L)
	Ion chromatography/inductively coupled plasma/mass spectrometry (MS)	Validation established after ultrasonic-assisted enzymatic extraction
	Ion chromatography	Post-column generation and detection of the AsO_4^{-3} molybdate heteropoly ion
	Surface-enhanced Raman scattering (SERS)	Use of Ag nanostructured multilayer films; LOD = 5 µg/L
	Liquid chromatography–atomic fluorescence detection	Simultaneous determination of six As ions; LOD < 4 µg/L
	LC-ICP/MS	Simultaneous determination of five As ions; LOD < 4 µg/L
	HPLC/hydride generation atomic absorption	Simultaneous determination of six As ions; LOD = 3–9 ng
	Ion exchange determination	Separation of AsO_4^{-3} and AsO_3^{-3} using the weakly basic anion exchanger A63-X4A by difference of respective pKa values
	Titration	As^{+5} is reduced by NH_2NH_2 and titration with Br^-/BrO_3^-
Bromide, Br^-	Ion chromatography using electrochemical detection	Use of Ag working electrode; LOD = 10 ppb (mass/mass)
Carbonate, CO_3^{-2}	Gas chromatography (GC)	Carbonation of salt and use of a GC with methanation reformer and a flame ionization detector (FID)
	GC–MS	Determination of pentafluorobenzyl derivative

(Continued)

II Organic Analytical Reagents for the Determination of Anions (*Continued*)

Anion	Method	Results/Comments
	Fourier transform infrared (FTIR)	CO_3^{-2} measurement in organic matter (875 and 712 cm^{-1})
	Gas measurement	Conversion to CO_2 measured by a barometer or manometer
	Ion chromatography	Measurement in the presence of formyl ions
	Photoacoustic IR spectroscopy	Use of several multi-calibration methods typically in sedimentary rocks
	Thermal optical analysis	Use of semi-continuous carbon elemental analyzer
	Liquid chromatography	Use of a C18 HPLC column, MeOH/AcOH mobile phase and UV detection ($\lambda = 290$ nm)
	Raman spectroscopy	Use of OH stretch band of H_2O as internal standard
	Thermogravimetry	CO_3^{-2} measurement in organic matter after identification of evolved gaseous species
Chlorate, ClO_3^-	FI analysis	Reduction of methylene blue's leuco form to methylene blue followed by optical detection in the visible region
	Spectrophotometry	Reduction to Cl_2 via Tl^{+3} which oxidizes the leuco form of methylene
	FTIR	Absorption band at 973 cm^{-1} ($\lambda = 10,277$ nm)
	Anodic electrochemical dissociation	Determination made during the anodic dissociation of Ti grade VT-1
	Potentiometric titration	Reduction to Cl^- (pH = 7) using As^{+3} (OsO_4 as catalyst)
	Titrimetric analysis	ClO_3^- converted to Cl^- which is titrated against $Hg(NO_3)_2$ (diphenylcarbazone as indicator)
	Titrimetric analysis	Iodometric titration
	Titrimetric analysis	Use of methyl orange (LOD = 10^{-2} μg, mass/mass)
Chromate, CrO_4^{-2}	Potentiometric titration	Selective electrode based on the N,N-butylenebis (salicylideneiminato) copper(II); LOD = 3.0×10^{-6} mol/L
	Spectrophotometry	Determination by oxidation of ascorbic acid ($\lambda = 548$ nm)
	Ion exchange chromatography	Diaminetetraacetic acid (DCTA)/CH_3CN used as eluent
	Ion pair chromatography	Tetrabutylammonium hydroxide (TBAH)/aq. CH_3CN used as eluent
	Capillary electrophoresis	KH_2PO_4 buffer (pH = 7)
	Titration	Iodometric titration
	XANES	LOD = 10 ppm using a 150 μm synchrotron X-ray beam
Cyanide, CN^-	Headspace GC/MS	Use of GC with a silica PLOT column coupled with mass selective detection
	Headspace GC/MS	Use of GC with a nitrogen-phosphorus detector; LOD = 0.05–5 μg/mL

(*Continued*)

II Organic Analytical Reagents for the Determination of Anions (*Continued*)

Anion	Method	Results/Comments
	Headspace GC/MS	Use of headspace solid-phase microextraction (SPME); LOD = 0.006 μg mL
	Headspace GC/MS	Determined as the pentafluorobenzylbromide derivative in the presence of TDMBC
	Headspace GC	Use of GC with thermoionic detection (NPD)
	Headspace GC/MS	Use of GC with the FID
	Headspace GC/MS	Determination after reaction with chloramine-T
		Derivatization by the pyridine–pyazolone method
	Ion chromatography using electrochemical detection	Use of Ag working electrode; LOD = 1μg/L
	Cyclic voltammetry (CV)	Au electrode; LOD = 10^{-5} to 10^{-6} mol/L
	Differential cathodic stripping voltammetry	LOD = 2×10^{-7} M in the presence of 5×10^{-5} M [S^{-2}] and 1×10^{-3} M [Cu^{+2}]
	Differential electrolytic potentiometry (DEP)	DEP coupled with the sequential injection/flow analysis (SIA/FIA) of aqueous solutions; con. range 6–39 μg/mL; LOD = 0.31 μg/mL; RSD = 2.55
	Isotachophoresis	Determination as [$Au(CN)_2^-$], [$Ag(CN)_2^-$], or [$Cu(CN)_3^{-2}$] complexes
	Potentiometric determination	Use of PVC sensors responsive to $Ag(CN)_2^-$
	Spectrofluorometry	Use of naphthoquinone imidazole boronic-based sensors (M-NQB and p-NQB) with CTAB (spec λ = 460 nm)
	Fluorescence	Derivatization with naphthalene 2,3-dicarboxyaldehyde and taurine (specf λ_{ex} = 418 nm, λ_{em} = 460 nm); LOD = 0.002–0.025 μg/mL
	Argentometric method	Argentometric titration/quantification using KCNS ($FeNH_4SO_4$) as indicator
	Capillary electrophoresis/UV spectrophotometry	Conversion of CN^- to SCN^- via rhodanese, separated by capillary electrophoresis, and measured by UV spectrophotometry (spec λ = 200 nm); LOD = 3 μmol)
	Piezoelectric quartz crystal (PQC) sensors	Method makes use of silver-coated sensors, LOD = 1.25×10^{-8} mol/L
	Visible spectrophotometry colorimetric assay	Picrate/resorcinol derivative in bicarbonate (spec λ = 488 nm)
	Visible spectrophotometry colorimetric assay	Lithium/sodium picrate derivative (spec λ = 500/520 nm); LOD = 2.5 μg/100 mg sample
	Visible spectrophotometry colorimetric assay	Isonicotinic acid-barbituric acid derivative after treatment with chloramine-T (spec λ = 600 nm)
	Visible spectrophotometry colorimetric assay	4-Hydroxy-3-(2-oxoindoline-3-ylideneamino)-2-thioxo-2H-1,3-thiazin-5(3H)-one) HOTT derivative (spec λ = 466 nm)

(Continued)

II Organic Analytical Reagents for the Determination of Anions (*Continued*)

Anion	Method	Results/Comments
	Visible spectrophotometry colorimetric assay	Ethanol extraction of the red complex formed on picric acid test strips (spec λ = 500 nm); LOD = 0.5–1.0 µg
	Visible spectrophotometry colorimetric assay	Derivatization with p-nitrobenzaldehyde followed by reaction with o-dinitrobenzene to yield a purple color; LOD = 10 µg/L
	Visible spectrophotometry colorimetric assay	Barbituric or N,N-dimethylbarbituric acid derivative
	Visible spectrophotometry colorimetric assay	Spectrophotometric determination by use of a modified Konig reaction with NaOCl as the chlorinating agent
	Visible spectrophotometry colorimetric assay	Complexation with pyridine or 4-pyridinecarboxylic acid
	Visible spectrophotometry colorimetric assay	Reaction of CN⁻ with Pd-dimethylglyoximate complex that releases dimethylglyoxime, isolated as the Ni derivative
	Visible spectrophotometry colorimetric assay	Determination (spec λ = 570 nm) of pyridine/barbituric acid derivative after chloramine-T derivative
	Visible spectrophotometry colorimetric assay	Measurement of the pyridine–pyrazolone derivative
	Visible spectrophotometry colorimetric assay	Measurement of the isocotinic acid/1-phenyl-3-methyl-5-pyrazolone derivative after reaction with chloramine-T (spec λ = 638 nm)
	Visible spectrophotometry colorimetric assay	Measurement of the derivative with the Hg(II)-p-dimethylaminobenzylidenerhodanine complex (spec λ = 452 nm)
	Visible spectrophotometry colorimetric assay	Measurement of the Hg(II)-diphenylcarbazone derivative (spec λ = 562 nm)
	Visible spectrophotometry colorimetric assay	Measurement of the pyridine–benzidine complex; LOD = 0.02 µg
	Visible spectrophotometry colorimetric assay	Measurement of the pentacyano(N-methylpyrazinium) ferrate (II) complex (spec λ = 658 nm)
	FI analysis	Use of Ag and Pt electrodes to determine CN⁻ as Cu complexes
	FI analysis	Analyzed as the teracyanonickelate (II) complex
	Fluorescent sensors	Fluorescein derivative emits a green light
	Fluorescent sensors	Naphthoquinone-imidazole boronic-based sensors (spec λ = 460 nm); LOD = 1.4 µM
	FTIR	Use of a Horizontal Attenuated Total Reflectance (HATR) accessory; detection range = 0.3–1.0 M
	Chemiluminescence	Measurement of intensity of the reaction with luminal, hemin, and peroxide
	Amperometry/FI analysis	Use of a multimeter as a detection device; LOD = 200 µg/L
	Photoacoustic IR detection	Applied to the determination of HCN and NCCl in air

(Continued)

II Organic Analytical Reagents for the Determination of Anions (*Continued*)

Anion	Method	Results/Comments
	X-ray diffraction	LOD = 0.1 % in waste water analyses
	Coulometric determination	LOD = 10^{-9} mg/100 µL
	HPLC	
	Cavity ringdown spectroscopy	Use of external cavity tunable diode laser
CN^-/SCN^- (only)	Chemical ionization GC/MS	Derivatization with pentafluorobenzyl bromide (PFBBr) via extraction with ethyl acetate and tetrabutylammonium sulfate (TBAS); LOD = 1 µM (CN^-) and <50 nM (SCN^-)
Dichromate, $Cr_2O_7^{-2}$	FTIR	Absorption band at 950 cm^{-1}
	Paper electrophoresis–photo density scanning	Photo density scanning determination after development with diphenyl carbazide
	Cell voltage variation	Applied to the electrochemical synthesis of sodium dichromate from sodium chromate
	Conductimetric determination	Performed in the presence of CrO_4^{-2}
	Adsorptive stripping voltamemtry	Commonly applied to body fluids
	Potentiometric titration	Determination of Pu via the $Fe^{+2}/Cr_2O_7^{-2}$ redox titration
	SIA	$Cr_2O_7^{-2}$ determination based on oxidation of pindolol (spec λ = 640 nm)
	Spectrophotometric analysis	Applied to the determination of pyridine derivatives
		Reduction to Cr^{+3} in organic carbon soil analysis
	Titrimetric analysis	Reduction by Fe^{+3} titration
	Thin-layer chromatography (TLC)	Applied to determination of unsaturated fats and further spectrophotometric analysis (spec λ = 350 nm)
	X-ray absorption fine structure spectroscopy	Applied to the speciation of $CrO_4^{-2}/Cr_2O_7^{-2}$ equilibrium
Hypochlorite, ClO^-	Titrimetric determination	Iodometric titration
	Polarography	Determination by conversion of Br^- to BrO_3^-
	Spectrophotometry	Oxidation of Ir^{+3} to Ir^{+4} (spec λ = 488 nm)
	Spectrophotometry	Measurement of ClO^- absorbance at pH > 7
	Spectrophotometry	Measurement of chloramine formed by addition of aq. NH_3
	Voltametry	Technique compared to iodometric titration
	Potentiometric titration	ClO^- measured by subtraction of BrO^- in a ClO^-/BrO^-
	Potentiometric titration	Reduction to Cl^- (pH < 7, OsO_4 as catalyst)
Iodide, I^-	Ion chromatography using electrochemical detection	Use of Ag working electrode; LOD = 100 ppb
Isocyanates, NCO^-	MS	Use of a proton transfer reaction (PTR), time-of-flight mass spectrometer and high-sensitivity PTR quadruple mass spectrometer
	FTIR	Use of sealed liquid cell for polymer analysis
	HPLC	Analysis of ethanol derivative (spec λ = 245 nm)

(*Continued*)

II Organic Analytical Reagents for the Determination of Anions (Continued)

Anion	Method	Results/Comments
	Capillary electrophoresis	Use of tris(2,2′-bipyridine) ruthenium (II) by electrochemiluminescence
	Potentiometric titration	Titration of carboxylic acid hydrazide derivative with $NaNO_2$
	Spectrophotometric analysis	Conversion of CH_3NCO to CH_3NH_2 and reaction with 1-fluoro-2,4-dinitrobenzene to yield $CH_3NHC_6H_3(NO_2)_2$ (spec $\lambda = 352\,nm$)
Manganate, MnO_4^{-2}	Titrimetric analysis	Oxidation of As^{+3}, Cr^{+3}, Sb^{+3}, Te^{+4}, PO_3^{-3}, various carboxylic acids, trimethylammoniated compounds
	Gravimetric analysis	Precipitation of $BaMnO_4$
	Potentiometric titration	Use of a Pt electrode
	Voltammetry	Use of glassy carbon working electrode
	Spectrophotometry	LOD = 0.185 µg/mL; (spec $\lambda = 635\,nm$)
	Anodic oxidation	Rotating disk electrode
	Gravimetric analysis	Precipitation of $BaMnO_4$
Nitrite, NO_2^-	Ion chromatography	Use of high capacity, OH^--selective, anion exchange column with a guard column in simultaneous determination of BrO_3^-; LOD = 0.05 µg/L; RSD = 3.61 %–4.77 %
	HPLC method	Simultaneous determination of NO_3^-, SO_4^{-2}, and SCN^-; samples are ultrasonically extracted with EtOH/CCl_3COOH and purified by C18 SPE column (highly retentive alkyl-bonded phase); LOD = 0.01–0.02 mg/kg; RSD = 0.87 %–2.6 %
	HPLC method	KOH or NaOH aq. solution used as mobile phase for 10–25 µL sample size
	Quantitative SIA	LOD = 0.002 mg/L
	Diffuse UV-visible reflectance method	Reaction of nitrite with sulfadiazine and o-naphthol; LOD = 0.02 mg/L; RSD = 5 %–8 %
	CV/pulse voltammetry	Electropolymerization of NB at pre-polarized glassy carbon electrode (phosphate buffer, pH = 7.1) leading to a NO_2^- sensor
	CV/pulse voltammetry	Use of a solid-state microelectrode formed through the electrochemical co-deposition of Pt–Fe nanoparticles on an Au microelectrode
	CV/pulse voltammetry	Use of a sol-gel electrode using 3-aminopropyltrimethoxy silane for covalent immobilization of toluidine blue
	Capillary electrophoresis	Simultaneous determination of NO_3^- and $Cr_2O_4^{-2}$
	Capillary zone electrophoresis	Simultaneous determination of NO_3^- and Cl^- by use of a photodiode array (PDA) detector
	UV-visible spectrophotometry	Derivatization with $Zn(OH)_2$/basic fuchsin

(Continued)

II Organic Analytical Reagents for the Determination of Anions (*Continued*)

Anion	Method	Results/Comments
	UV-visible spectrophotometry	Derivatization with Griess reagent (spec $\lambda = 537$ nm); RSD ~1.5 %
	Microfluidic gradient elution moving boundary electrophoresis (GEMBE)	Simultaneous determination of NO_3^- and NH_4^+
	Microfluidic GEMBE	Simultaneous determination of NO_3^-, Cl^-, $Cr_2O_4^{-2}$, NO_3^-, and SO_4^{-2}
	Gasometric analysis	The NO_2^- concentration is quantified by the formation of $N_2(g)$ produced *vis-a-vis* its reaction of with sulfamic acid
	Seeping layer test paper	Derivatization with p-aminobenzenesulfonic acid/N(1-naphthyl) ethylenediamine dihydrochloride
NO_2^- simultaneous NO_2^-/NO_3^- (only)	HPLC	Pre-column derivatization with Griess reagents (spec $\lambda = 540$ nm)
	UV-visible spectrophotometry	Following derivatization with HCl/naphthyl ethylenediamine
		LOD = 0.5–18 mg nitrogen per L $\left(NO_3^-\right)$
		and 0.5–5 mg nitrogen per L $\left(NO_2^-\right)$
	HPLC–UV	LOD = 2.5 mg/kg $\left(NO_3^-\right)$ and 5.0 mg/kg $\left(NO_2^-\right)$
	GC-MS	Quantification of ^{15}N-labeled nitrate and nitrite biological fluids
	Chemiluminescence	LOD = 10 nmol/L; RSD = 3 % (200 nmol/L) to 5 % (400 nmol/L)
Oxalate, $Cr_2O_4^{-2}$	Capillary electrophoresis	Use of acid run buffer to revert electroosmotic flow
	Capillary gas chromatography	LOD = 500 µg by use of an FID
	Chromogenic strip	Test strip equipped with sorghum leaf oxalate oxidase cross-linked with glutaraldehyde and o-toluidine
	FTIR (attenuated)	LOD = 0.07 % of the matrix
	Capillary electrophoresis	Use of acid run buffer to revert electroosmotic flow
	Capillary gas chromatography	LOD = 500 µg by use of an FID
	FI analysis	Oxalate oxidase and peroxidase are immobilized on Au electrode
	GC	Analysis of methylated product
	Ion exchange chromatography method	LOD = 6–9 µg/L
	Isotachophoresis	Detection at $\lambda = 254$ nm; LOD = 20 nmol/L
	LC-tandem MS	System attached to online weak anion exchange chromatography
	Raman spectroscopy	Intensity of peak at 1,462 cm^{-1} is used
	Spectrophotometric analysis	Analysis of catalytic effect of reaction products of methylene blue/$Cr_2O_7^{-2}$
	Spectrophotometric analysis	Analysis of the NaDH → NAD$^+$ reaction with $C_2O_4^{-2}$
	Titrimetric analysis	Titration against MnO_4^-

(*Continued*)

II Organic Analytical Reagents for the Determination of Anions (*Continued*)

Anion	Method	Results/Comments
	Titrimetric analysis	Titration against Ce^{+4}
Perchlorate, ClO_4^-	Spectrophotometric analysis	Analysis of catalytic effect of reaction products of methylene blue/$Cr_2O_7^{-2}$
	Spectrophotometric analysis	Analysis of the NaDH → NAD^+ reaction with $C_2O_4^{-2}$
	Titrimetric analysis	Titration against MnO_4^-
	Titrimetric analysis	Titration against Ce^{+4}
	Spectrophotometry	Yb–Pyrocatechol Violet complex
	Spectrophotometry	Determination of reddish-violet color produced by adding methylene blue/$ZnSO_4$ in presence of NO_3^-
	HPLC/MS	LOD = 0.2 ppb
	Capillary electrophoresis	Analysis typically performed on tobacco plants
	MS	Determination of organic base–phosphate complex
	Ion pair extraction–MS ESI	Use of cationic surfactants
	Titrimetric analysis	Titration against H-tetramethylammonium sulfate with various indicators
	Titrimetric analysis	Back titration of excess $TiCl_3$ against $Fe_2(SO_3)_2$
Phosphate, PO_4^{-3}	FTIR	Use of 1,037 cm^{-1} ($Ca_3(PO_4)_2$) and 1,010 cm^{-1} ($MgNH_4PO_4$)
Phosphide, P^{-3}	Gravimetric	Derivatization as Ag_3PO_3 after formation of Ag_3P and further oxidation with HNO_3
	ESR	Derivatization to GaP
	X-ray electron spectrometry	Derivatization to AlP
	X-ray fluorescence	Derivatization to AlP
	X-ray diffractometry	Derivatization to AlP
	X-ray photoelectron spectroscopy	Derivatization to functionalized GaP
	Fluorescence	Derivatization to Ca_3P_2
Phosphite, PO_3^{-3}	Fluorescence	Enzymatic oxidation to phosphate with NAD^+ as co-substrate to produce fluorescence
	^{31}P NMR	Esterification of OH^- in organic polymers
	Laser Raman spectrometry	Ar-ion laser allows quantitative analysis of $H_2PO_3^-$ and HPO_3^{-2}
	Ion exchange chromatography	Method makes use of conductivity detection
	Visible spectrophotometry	Measurement of 3,5-$(O_2N)_2C_6H_3COOH$ derivative
Silicate, SiO_4	Raman spectroscopy	Use of *ab initio* quantum chemical simulation
	Electron probe microanalysis (EPMA)	Analysis of Na, Mg, Al, Si, K, Ca, Ti, Mn, and Fe salts
	Ion-exclusion chromatography	Use of ascorbate solution as both an eluent and a reducing agent followed by molybdate derivatization

(Continued)

II Organic Analytical Reagents for the Determination of Anions (*Continued*)

Anion	Method	Results/Comments
	Laser-induced plasma spectroscopy	Use of a Q-switched Nd:YAG laser and time-resolved spectroscopy
	Time-of-flight secondary ion mass spectrometry	Use of C60 primary ions
	Laser ablation–inductively coupled plasma mass spectrometry (LA-ICPMS)	No need for internal standard
Sulfide, S^{-2}	Ion chromatography using electrochemical detection	Use of Ag working electrode; LOD = 30 ppb
Sulfite, SO_3^{-2}	Amperometric titration	Use of iodine solution
	Potentiometric titration	Pt wire coated with Hg is balanced against calomel electrode
	Spectrophotometric analysis	Product of reaction with p-rosaniline acid bleached dye and HCHO (spec λ = 560 nm)
	Ion chromatography	
	Titration analysis	Titration against BrO_3^-
	Coulometric determination	LOD = 10^{-11} mg/100 μL
Thiocyanate SCN^-	FI analysis	Formation of the 2-(5-bromo-2-pyridylazo)-5-diethylamino –phenol derivative in $Cr_2O_7^{-2}/H^+$
	FI analysis	Analyzed as the tetracyanonickelate (II) complex

CHAPTER 9

Infrared Spectrophotometry

INFRARED OPTICS MATERIALS

The following table lists the more common materials used for optical components (windows, prisms, etc.) in the infrared region of the electromagnetic spectrum. The properties listed are needed to choose the materials with optimal transmission characteristics [1,2]. The thermal properties are useful when designing experiments for operation at elevated temperatures [3–5]. This listing is far from exhaustive, but these are the most common materials used in instrumentation laboratories.

REFERENCES

1. Gordon, A.J. and Ford, G.A., *The Chemist's Companion*, John Wiley & Sons, New York, 1972.
2. Willard, H.H., Merritt, L.L., Dean, J.A., and Settle, F.A., *Instrumental Methods of Analysis*, 7th ed., Wadsworth, Belmont, 1988.
3. Touloukien, Y.S., Powell, R.W., Ho, C.Y., and Klemens, P.G., *Thermophysical Properties of Matter: Thermal Conductivity of Nonmetallic Solids*, vol. 2, IF - Plenum Data Corp., New York, 1970.
4. Touloukien, Y.S., Kirby, R.K., Taylor, R.E., and Lee, T., *Thermophysical Properties of Matter: Thermal Expansion of Nonmetallic Solids*, vol. 13, IF - Plenum Data Corp., New York, 1977.
5. Wolfe, W.L. and Zissis, G.J., ed., *The Infrared Handbook*, Mir, Moscow, 1995.

Material	Wavelength Range (μm)	Wavenumber Range (cm^{-1})	Refractive Index at 2 μm	Thermal Conductivity w/(m·K) × 10^2	Thermal Expansion ΔL/L, (%)	Notes
Sodium chloride NaCl	0.25–16	40,000–625	1.52	7.61 (273 K) 6.61 (300 K) 4.85 (400 K)	0.448 (400 K) 0.896 (500 K)	Most common material; absorbs water; for aqueous solutions, use saturated NaCl solution as the solvent.
Potassium bromide KBr	0.25–25	40,000–400	1.53	5.00 (275 K) 4.87 (301.5 K) 4.80 (372.2 K)	0.028 (400 K) 0.429 (500 K) 0.846 (600 K)	Useful for the study of C–Br stretch region, useful for solid sample pellets.
Silver chloride AgCl	0.4–23	25,000–435	2.0	1.19 (269.8 K) 1.10 (313.0 K) 1.05 (372.5 K)	0.356 (400 K) 0.729 (500 K) 1.183 (600 K)	Not useful for amines or liquids with basic nitrogen; light sensitive.
Silver bromide AgBr	0.50–35	20,000–286	2.2	0.90 (308.2 K) 0.79 (353.2 K) 0.71 (413.2 K)	0.024 (300 K) 0.109 (325 K) 0.196 (350 K)	Not useful for amines or liquids with basic nitrogen; light sensitive.
Calcium fluoride CaF$_2$	0.15–9	66,700–1,110	1.40	10.40 (237 K) 9.60 (309 K) 4.14 (402 K)	0.214 (400 K) 0.431 (500 K) 0.670 (600 K)	Useful for obtaining high resolution for –OH, N–H, and C–H stretching frequencies.
Barium fluoride BaF$_2$	0.20–11.5	50,000–870	1.46	11.7 (284 K) 10.9 (305 K) 10.5 (370 K)	0.233 (400 K) 0.461 (500 K) 0.698 (600 K)	Shock sensitive, should be handled with care.
Cesium bromide CsBr	1–37	10,000–270	1.67	9.24 (269.4 K) 8.00 (337.5 K) 7.76 (367.5 K)	0.526 (400 K) 1.063 (500 K) 1.645 (600 K)	Useful for C–Br stretching frequencies.
Cesium iodide CsI	1–50	10,000–200	1.74	1.15 (277.7 K) 1.05 (296.0 K) 0.95 (360.7 K)		Useful for C–Br stretching frequencies.

(Continued)

Material	Wavelength Range (μm)	Wavenumber Range (cm⁻¹)	Refractive Index at 2 μm	Thermal Conductivity w/(m·K) × 10²	Thermal Expansion ΔL/L, (%)	Notes
Thallium bromide–thallium iodide TlBr-TlI (KRS-5)	0.5–35	20,000–286	2.37		0.464 (373 K) 1.026 (473 K)	Highly toxic, handle with care; 42 % TlBr, 58 % TlI
Zinc selenide ZnSe	1–18	10,000–555	2.4		0.086 (400 K) 0.175 (500 K) 0.272 (600 K)	Vacuum deposited
Germanium Ge	0.5–11.5	20,000–870	4.0			
Silicon Si	0.20–6.2	50,000–1,613	3.5		0.033 (400 K) 0.066 (500 K) 0.102 (600 K)	
Aluminum oxide (sapphire) Al_2O_3	0.20–6.5	50,000–1,538	1.76	25.1 (293.2 K) 21.3 (323 K) 14.2 (432.2 K)	0.075 (400 K) 0.148 (500 K) 0.225 (600 K)	
Polyethylene	16–300	625–33	1.54			Not useful for many organic compounds.
Mica	200–425	50–23.5				
Fused silica SiO_2	0.2–4.0	50,000–2,500	1.42 (at 3 m)	1.38 (298 K)		Used in near-infrared (NIR) work; can be used with dilute and concentrated acids (except HF), not for use with aqueous alkali; metal ions can be problematic.

INTERNAL REFLECTANCE ELEMENT CHARACTERISTICS

Internal reflectance methods are a common sampling method in infrared spectrophotometry. The following table provides guidance in the selection of elements for reflectance methods [1].

REFERENCE

1. Coleman, P., *Practical Sampling Techniques for Infrared Analysis*, CRC Press, Boca Raton, FL, 1993.

Material	Frequency Range (cm^{-1})	Index of Refraction	Characteristics
Thallium iodide–thallium bromide (KRS-5)	16,000–250	2.4	Relatively soft, deforms easily; warm water, ionizable acids and bases, chlorinated solvents, and amines should not be used with this attenuated total reflection (ATR) element.
Zinc selenide (Irtran-4)	20,000–650	2.4	Brittle; releases H_2Se, a toxic material, if used with acids; water insoluble; electrochemical reactions with metal salts or complexes are possible.
Zinc sulfide (Cleartran)	50,000–770	2.2	Reacts with strong oxidizing agents; relatively inert with typical aqueous, normal acids and bases and organic solvents; good thermal and mechanical shock properties; low refractive index causes spectral distortions at 45 °C.
Cadmium telluride (Irtran-6)	10,000–450	2.6	Expensive; relatively inert; reacts with acids.
Silicon	9,000–1,550	3.5	Hard and brittle; useful at high temperatures to 300 °C; relatively inert.
Germanium	5,000–850	4.0	Hard and brittle; temperature opaque at 125 °C.

WATER SOLUBILITY OF INFRARED OPTICS MATERIALS

The following table provides guidance in the selection of optics materials [1]. Often, the solubility in (pure) water of a particular material is of critical concern.

REFERENCE

1. Coleman, P., *Practical Sampling Techniques for Infrared Analysis*, CRC Press, Boca Raton, FL, 1993.

Material	Formula	Solubility g/100g H_2O at 20 °C
Sodium chloride	NaCl	36.0
Potassium bromide	KBr	65.2
Potassium chloride	KCl	34.7
Cesium iodide	CsI	160 (at 61 °C)
Fused silica	SiO_2	Insoluble
Calcium fluoride	CaF_2	1.51×10^{-3}
Barium fluoride	BaF_2	0.12 (at 25 °C)
Thallium bromide–iodide (KRS-5)	—	$<4.76 \times 10^{-2}$
Silver bromide	AgBr	1.2×10^{-5}
Zinc sulfide	ZnS	Insoluble
Zinc selenide (Irtran-4)	ZnSe	Insoluble
Polyethylene (high density)	—	Insoluble

WAVELENGTH-WAVENUMBER CONVERSION TABLE

The following table provides a conversion between wavelength and wavenumber units, for use in infrared spectrophotometry.

Wavelength (μm)	Wavenumber (cm⁻¹)									
	0	1	2	3	4	5	6	7	8	9
2	5,000	4,975	4,950	4,926	4,902	4,878	4,854	4,831	4,808	4,785
2.1	4,762	4,739	4,717	4,695	4,673	4,651	4,630	4,608	4,587	4,566
2.2	4,545	4,525	4,505	4,484	4,464	4,444	4,425	4,405	4,386	4,367
2.3	4,348	4,329	4,310	4,292	4,274	4,255	4,237	4,219	4,202	4,184
2.4	4,167	4,149	4,232	4,115	4,098	4,082	4,065	4,049	4,032	4,016
2.5	4,000	3,984	3,968	4,953	3,937	3,922	3,006	3,891	3,876	3,861
2.6	3,846	3,831	3,817	3,802	3,788	3,774	3,759	3,745	3,731	3,717
2.7	3,704	3,690	3,676	3,663	3,650	3,636	3,623	3,610	3,597	3,584
2.8	3,571	3,559	3,546	3,534	3,521	3,509	3,497	3,484	3,472	3,460
2.9	3,448	3,436	3,425	3,413	3,401	3,390	3,378	3,367	3,356	3,344
3	3,333	3,322	3,311	3,300	3,289	3,279	3,268	3,257	3,247	3,236
3.1	3,226	3,215	3,205	3,195	3,185	3,175	3,165	3,155	3,145	3,135
3.2	3,125	3,115	3,106	3,096	3,086	3,077	3,067	3,058	3,049	3,040
3.3	3,030	3,021	3,012	3,003	2,994	2,985	2,976	2,967	2,959	2,950
3.4	2,941	2,933	2,924	2,915	2,907	2,899	2,890	2,882	2,874	2,865
3.5	2,857	2,849	2,841	2,833	2,825	2,817	2,809	2,801	2,793	2,786
3.6	2,778	2,770	2,762	2,755	2,747	2,740	2,732	2,725	2,717	2,710
3.7	2,703	2,695	2,688	2,,681	2,674	2,667	2,660	2,653	2,646	2,639
3.8	2,632	2,625	2,618	2,611	2,604	2,597	2,591	2,584	2,577	2,571
3.9	2,654	2,558	2,551	2,545	2,538	2,532	2,525	2,519	2,513	2,506
4	2,500	2,494	2,488	2,481	2,475	2,469	2,463	2,457	2,451	2,445
4.1	2,439	2,433	2,427	2,421	2,415	2,410	2,404	2,398	2,387	2,387
4.2	2,381	2,375	2,370	2,364	2,358	2,353	2,347	2,,342	2,336	2,331
4.3	2,326	2,320	2,315	2,309	2,304	2,299	2,294	2,288	2,283	2,278
4.4	2,273	2,268	2,262	2,257	2,252	2,247	2,242	2,237	2,232	2,227
4.5	2,222	2,217	2,212	2,208	2,203	2,198	2,193	2,188	2,183	2,179
4.6	2,174	2,169	2,165	2,160	2,155	2,151	2,146	2,141	2,137	2,132
4.7	2,128	2,123	2,119	2,114	2,110	2,105	2,101	2,096	2,092	2,088
4.8	2,083	2,079	2,075	2,070	2,066	2,062	2,058	2,053	2,049	2,045
4.9	2,041	2,037	2,033	2,028	2,024	2,020	2,016	2,012	2,008	2,004
5	2,000	1,996	1,992	1,988	1,984	1,980	1,976	1,972	1,969	1,965
5.1	1,961	1,957	1,953	1,949	1,946	1,942	1,938	1,934	1,931	1,927
5.2	1,923	1,919	1,916	1,912	1,908	1,905	1,901	1,898	1,894	1,890
5.3	1,887	1,883	1,880	1,876	1,873	1,869	1,866	1,862	1,859	1,855
5.4	1,852	1,848	1,845	1,842	1,838	1,835	1,832	1,828	1,825	1,821
5.5	1,818	1,815	1,812	1,808	1,805	1,802	1,799	1,795	1,792	1,788
5.6	1,786	1,783	1,779	1,776	1,773	1,770	1,767	1,764	1,761	1,757
5.7	1,754	1,751	1,748	1,745	1,742	1,739	1,736	1,733	1,730	1,727
5.8	1,724	1,721	1,718	1,715	1,712	1,709	1,706	1,704	1,701	1,698
5.9	1,695	1,692	1,689	1,686	1,684	1,681	1,678	1,675	1,672	1,669
6	1,667	1,664	1,661	1,668	1,656	1,653	1,650	1,647	1,645	1,642

(Continued)

Wavelength (μm)	Wavenumber (cm⁻¹)									
	0	1	2	3	4	5	6	7	8	9
6.1	1,639	1,637	1,634	1,631	1,629	1,626	1,623	1,621	1,618	1,616
6.2	1,613	1,610	1,608	1,605	1,603	1,600	1,597	1,595	1,592	1,590
6.3	1,587	1,585	1,582	1,580	1,577	1,575	1,572	1,570	1,567	1,565
6.4	1,563	1,560	1,558	1,555	1,553	1,550	1,548	1,546	1,543	1,541
6.5	1,538	1,536	1,534	1,531	1,529	1,527	1,524	1,522	1,520	1,517
6.6	1,515	1,513	1,511	1,508	1,506	1,504	1,502	1,499	1,497	1,495
6.7	1,493	1,490	1,488	1,486	1,484	1,481	1,479	1,477	1,475	1,473
6.8	1,471	1,468	1,466	1,464	1,462	1,460	1,458	1,456	1,453	1,451
6.9	1,449	1,447	1,445	1,443	1,441	1,439	1,437	1,435	1,433	1,431
7	1,429	1,427	1,425	1,422	1,420	1,418	1,416	1,414	1,412	1,410
7.1	1,408	1,406	1,404	1,403	1,401	1,399	1,397	1,395	1,393	1,391
7.2	1,389	1,387	1,385	1,383	1,381	1,379	1,377	1,376	1,374	1,372
7.3	1,370	1,368	1,366	1,364	1,362	1,361	1,359	1,357	1,355	1,353
7.4	1,351	1,350	1,348	1,346	1,344	1,342	1,340	1,339	1,337	1,335
7.5	1,333	1,332	1,330	1,328	1,326	1,325	1,323	1,321	1,319	1,318
7.6	1,316	1,314	1,312	1,311	1,309	1,307	1,305	1,304	1,302	1,300
7.7	1,299	1,297	1,295	1,294	1,292	1,290	1,289	1,287	1,285	1,284
7.8	1,282	1,280	1,279	1,277	1,276	1,274	1,272	1,271	1,269	1,267
7.9	1,266	1,264	1,263	1,261	1,259	1,258	1,256	1,255	1,253	1,252
8	1,250	1,248	1,247	1,245	1,244	1,242	1,241	1,239	1,238	1,236
8.1	1,235	1,233	1,232	1,230	1,229	1,227	1,225	1,224	1,222	1,221
8.2	1,220	1,218	1,217	1,215	1,214	1,212	1,211	1,209	1,208	1,206
8.3	1,205	1,203	1,202	1,200	1,199	1,198	1,196	1,195	1,193	1,192
8.4	1,190	1,189	1,188	1,186	1,185	1,183	1,182	1,181	1,179	1,178
8.5	1,176	1,175	1,174	1,172	1,171	1,170	1,168	1,167	1,166	1,164
8.6	1,163	1,161	1,160	1,159	1,157	1,156	1,155	1,153	1,152	1,151
8.7	1,149	1,148	1,147	1,145	1,144	1,143	1,142	1,140	1,139	1,138
8.8	1,136	1,135	1,134	1,133	1,131	1,130	1,129	1,127	1,126	1,125
8.9	1,124	1,122	1,121	1,120	1,119	1,117	1,116	1,115	1,114	1,112
9	1,111	1,110	1,109	1,107	1,106	1,105	1,104	1,103	1,101	1,100
9.1	1,099	1,098	1,096	1,095	1,094	1,093	1,092	1,091	1,089	1,088
9.2	1,087	1,086	1,085	1,083	1,082	1,081	1,080	1,079	1,078	1,076
9.3	1,075	1,074	1,073	1,072	1,071	1,070	1,068	1,067	1,066	1,065
9.4	1,064	1,063	1,062	1,060	1,059	1,058	1,057	1,056	1,055	1,054
9.5	1,053	1,052	1,050	1,049	1,048	1,047	1,046	1,045	1,044	1,043
9.6	1,042	1,041	1,040	1,038	1,037	1,036	1,035	1,034	1,033	1,032
9.7	1,031	1,030	1,029	1,028	1,027	1,026	1,025	1,024	1,022	1,021
9.8	1,020	1,019	1,018	1,017	1,016	1,015	1,014	1,013	1,012	1,011
9.9	1,010	1,009	1,008	1,007	1,006	1,005	1,004	1,003	1,002	1,001
10	1,000	999	998	997	996	995	994	993	992	991
10.1	990	989	988	987	986	985	984	983	982	981
10.2	980	979	978	978	977	976	975	974	973	972
10.3	971	970	969	968	967	966	965	964	963	962
10.4	962	961	960	959	958	957	956	955	954	953
10.5	952	951	951	950	949	948	947	946	945	944
10.6	943	943	942	941	940	939	938	937	936	935
10.7	935	934	933	932	931	930	929	929	928	927

(Continued)

Wavelength (µm)	Wavenumber (cm⁻¹)									
	0	1	2	3	4	5	6	7	8	9
10.8	926	925	924	923	923	922	921	920	919	918
10.9	917	917	916	915	914	913	912	912	911	910
11	909	908	907	907	906	905	904	903	903	902
11.1	901	900	899	898	898	897	896	895	894	894
11.2	893	892	891	890	890	889	888	887	887	886
11.3	885	884	883	883	882	881	880	880	879	878
11.4	877	876	876	875	874	873	873	872	871	870
11.5	870	869	868	867	867	866	865	864	864	863
11.6	862	861	861	860	859	858	858	857	856	855
11.7	855	854	853	853	852	851	850	850	849	848
11.8	847	847	846	845	845	844	843	842	842	841
11.9	840	840	839	838	838	837	836	835	835	834
12	833	833	832	831	831	830	829	829	828	827
12.1	826	826	825	824	824	823	822	822	821	820
12.2	820	819	818	818	817	816	816	815	814	814
12.3	813	812	812	811	810	810	809	808	808	807
12.4	806	806	805	805	804	803	803	802	801	801
12.5	800	799	799	798	797	797	796	796	795	794
12.6	794	793	792	792	791	791	790	789	789	788
12.7	787	787	786	786	785	784	784	783	782	782
12.8	781	781	780	779	779	778	778	777	776	776
12.9	775	775	774	773	773	772	772	771	770	770
13	769	769	768	767	767	766	766	765	765	764
13.1	763	763	762	762	761	760	760	759	759	758
13.2	758	757	756	756	755	755	754	754	753	752
13.3	752	751	751	750	750	749	749	748	747	747
13.4	746	746	745	745	744	743	743	742	742	741
13.5	741	740	740	739	739	738	737	737	736	736
13.6	735	735	734	734	733	733	732	732	731	730
13.7	730	729	729	728	728	727	727	726	726	725
13.8	725	724	724	723	723	722	722	721	720	720
13.9	719	719	718	718	717	717	716	716	715	715
14	714	714	713	713	712	712	711	711	710	710
14.1	709	709	708	708	707	707	706	706	705	705
14.2	704	704	703	703	702	702	702	701	701	700
14.3	699	699	698	698	697	697	696	696	695	695
14.4	694	694	693	693	693	692	692	691	691	690
14.5	690	689	689	688	688	687	687	686	686	685
14.6	685	684	684	684	683	683	682	682	681	681
14.7	680	680	679	679	678	678	678	677	677	676
14.8	676	675	675	674	674	673	673	672	672	672
14.9	671	671	670	670	669	669	668	668	668	667

USEFUL SOLVENTS FOR INFRARED SPECTROPHOTOMETRY

The following tables provide the infrared absorption spectra of several useful solvents, along with solvent design properties [1–10]. In most cases, two spectra are provided for each solvent. The first in each set was measured using a double beam spectrophotometer using a neat sample against an air reference. These spectra are presented in both wavenumber (cm^{-1}) and micrometer (μm) scales. The spectra were recorded under high concentration conditions (in terms of path length and attenuation) in order to emphasize the characteristics of each solvent. Thus, these spectra are not meant to be "textbook" examples of infrared spectra. The second spectrum in each set was measured with a Fourier transform instrument. The physical properties listed are those needed most often in designing spectrophotometric experiments [1–10]. The refractive indices are values measured with the sodium-d line. Solvation properties include the solubility parameter, δ, hydrogen bond index, λ, and the solvatochromic parameters, α, β, and π*. The Chemical Abstract Service (CAS) registry numbers and the INChI (International Chemical Identifier) are also provided for each solvent, to allow the reader to easily obtain further information using computerized database services. Note that the heat of vaporization is presented in the commonly used cal/g unit. To convert to the appropriate SI unit (J/g), multiply by 4.184.

We realize that a number of the solvents listed here are not permitted in some academic laboratories. Information on these solvents are presented for users in laboratories equipped to deal with the hazards associated with them.

REFERENCES

1. Larranaga, M.D., Lewis, R.J., and Lewis, R.J., *Hawley's Condensed Chemical Dictionary*, 16th ed., John Wiley & Sons, New York, 2016.
2. Dreisbach, R.R., *Physical Properties of Chemical Compounds*, Advances in Chemistry Series, Number 22, American Chemical Society, Washington, DC, 1959.
3. Jamieson, D.T., Irving, J.B., and Tudhope, J.S., *Liquid Thermal Conductivity—A Data Survey to 1973*, Her Majesty's Stationary Office, Edinburgh, 1975.
4. Lewis, R.J. and Sax, N.I., *Sax's Dangerous Properties of Industrial Materials*, 9th ed., Thompson Publishing, Washington, DC, 1995.
5. Sedivec, V. and Flek, J., *Handbook of Analysis of Organic Solvents*, John Wiley & Sons (Halsted Press), New York, 1976.
6. Epstein, W.W. and Sweat, F.W., Dimethyl sulfoxide oxidations, *Chem. Rev.* 247, 1967.
7. Rumble, J.R., ed., *CRC Handbook for Chemistry and Physics*, 100th. ed., CRC Press, Boca Raton, FL, 2019.
8. Bruno, T.J., and Svoronos, P.D.N., *CRC Handbook of Basic Tables for Chemical Analysis*, 3rd ed., CRC Press, Boca Raton, FL, 2011.
9. NIST Chemistry Web Book, NIST Standard Reference Database Number 69 - (www.webbook.nist.gov/chemistry/), accessed November, 2019.
10. Marcus, Y., The properties of organic liquids that are relevant to their use as solvating solvents, *Chem. Soc. Rev.*, 22(6), 409–416, 1993.

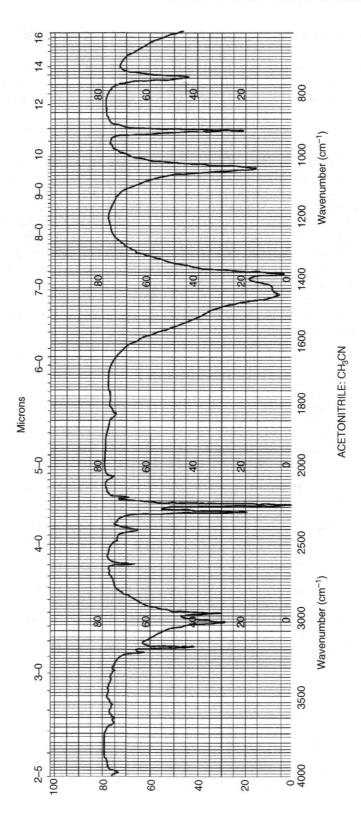

ACETONITRILE: CH$_3$CN

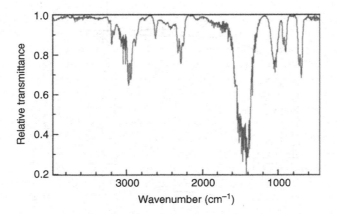

Acetonitrile, CH_3CN

Physical Properties	
Relative molecular mass	41.05
Melting point	−45.7 °C
Normal boiling point	81.6 °C
Refractive index (20 °C)	1.34423
Density (20 °C)	0.7857 g/mL
Viscosity (25 °C)	0.345 mPa·s
Surface tension (20 °C)	29.30 mN/m
Heat of vaporization (at boiling point)	29.75 kJ/mol
Thermal conductivity (20 °C)	0.1762 W/(m·K)
Dielectric constant (20 °C)	38.8
Relative vapor density (air = 1)	1.41
Vapor pressure (20 °C)	0.0097 MPa
Solubility in water[a]	∞
Flash point (OC)	6 °C
Autoignition temperature	509 °C
Explosive limits in air	4.4 %–16 %, vol/vol
CAS registry number	75-05-8
INChI	1S/C2H3N/c1-2-3/h1H3
Exposure limits	40 ppm, 8 h total weighted average (TWA)
Solubility parameter, δ	11.9
Solvatochromic α	0.19
Solvatochromic β	0.4
Solvatochromic π*	0.75

Notes: Highly polar solvent; sweet, ethereal odor; soluble in water; flammable, burns with a luminous flame; highly toxic by ingestion, inhalation, and skin absorption; miscible with water, methanol, methyl acetate, ethyl acetate, acetone, ethers, acetamide solutions, chloroform, carbon tetrachloride, ethylene chloride, and many unsaturated hydrocarbons; immiscible with many saturated hydrocarbons (petroleum fractions); dissolves some inorganic salts such as silver nitrate, lithium nitrate, and magnesium bromide; incompatible with strong oxidants; hydrolyzes in the presence of aqueous bases and strong aqueous acids.

Synonyms: methyl cyanide, acetic acid nitrile, cyanomethane, ethylnitrile.

[a] Forms azeotrope with water (at 16 % mass/mass) that boils at 76 °C.

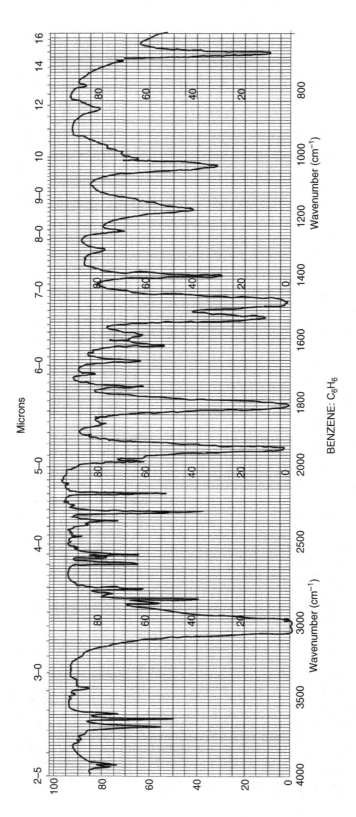

BENZENE: C_6H_6

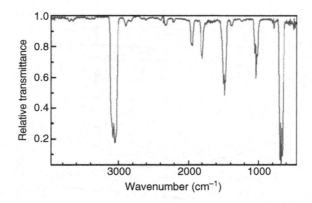

Benzene C$_6$H$_6$

Physical Properties	
Relative molecular mass	78.11
Melting point	5.5 °C
Normal boiling point	80.08 °C
Refractive index (20 °C)	1.5011
(25 °C)	1.4979
Density (20 °C)	0.8790 g/mL
(25 °C)	0.8737 g/mL
Viscosity (25 °C)	0.654 mPa·s
Surface tension (20 °C)	28.87 mN/m
Heat of vaporization (at boiling point)	30.72 kJ/mol
Thermal conductivity (25 °C)	0.1424 W/(m·K)
Dielectric constant (20 °C)	2.284
Relative vapor density (air = 1)	2.77
Vapor pressure (25 °C)	0.0097 MPa
Solubility in water[a]	0.07 %, mass/mass
Flash point (OC)	−11 °C
Autoignition temperature	562 °C
Explosive limits in air	1.4 %–8.0 %, vol/vol
CAS registry number	71-43-2
INChI	1S/C6H6/c1-2-4-6-5-3-1/h1-6H
Exposure limits	10 ppm, 8 h TWA
Solubility parameter, δ	9.2
Hydrogen bond index, λ	2.2
Solvatochromic α	0
Solvatochromic β	0.1
Solvatochromic π*	0.59

Notes: Confirmed human carcinogen; nonpolar, aromatic solvent; sweet odor; very flammable and toxic; confirmed human carcinogen; soluble in alcohols, hydrocarbons (aliphatic and aromatic), ether, chloroform, carbon tetrachloride, carbon disulfide, slightly soluble in water. Incompatible with some strong acids and oxidants, chlorine trifluoride/zinc (in the presence of steam); dimerizes at high temperature to form biphenyl.

Synonyms: cyclohexatriene, benzene, benzol, phenylhydride, 1,3,5-cyclohexatriene. These are the most common, although there are many other synonyms.

[a] Forms azeotrope with ethanol (approximately 65 °C).

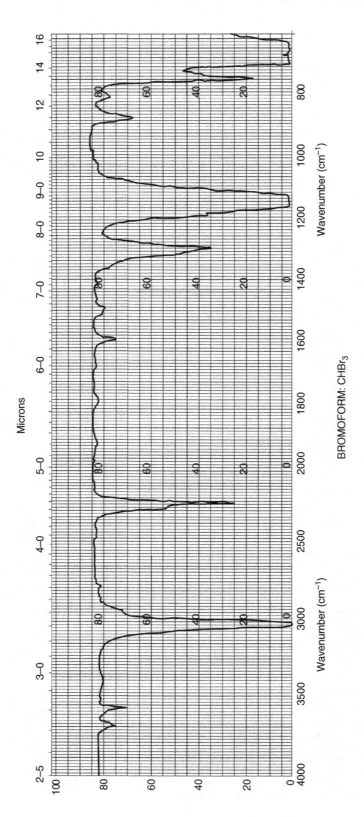

BROMOFORM: CHBr₃

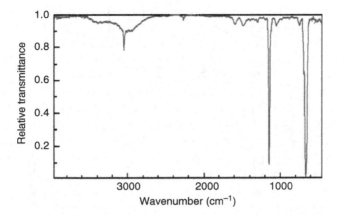

Bromoform CHBr$_3$

Physical Properties	
Relative molecular mass	252.75
Melting point	5.7 °C
Normal boiling point	149.2 °C
Refractive index (20 °C)	1.6005
Density (20 °C)	2.8899 g/mL
Viscosity (25 °C)	1.89 mPa·s
Surface tension (20 °C)	41.53 mN/m
Heat of vaporization (at boiling point)	39.66 kJ/mol
Thermal conductivity (20 °C)	0.0961 W/(m·K)
Dielectric constant (20 °C)	4.39
Relative vapor density (air = 1)	2.77
Vapor pressure (25 °C)	0.0008 MPa
Solubility in water	Slightly
Flash point (OC)	Nonflammable
Autoignition temperature	Not determined
Explosive limits in air	Nonflammable
CAS registry number	75-25-2
INChI	1S/CHBr3/c2-1(3)4/h1H
Exposure limits	0.5 ppm (skin)
Solvatochromic α	0.05
Solvatochromic β	0.05
Solvatochromic π*	0.62

Notes: Moderately polar, weakly hydrogen bonding solvent, dense liquid; gradually decomposes to acquire a yellow color, air and/or light will accelerate this decomposition; nonflammable; commercial product is often stabilized by the addition of 3 %–4 % (mass/mass) alcohols; highly toxic by ingestion, inhalation, and skin absorption; soluble in alcohols, organohalogen compounds, hydrocarbons, benzene, and many oils. Incompatible with many alkali and alkaline earth metals.

Synonyms: tribromomethane.

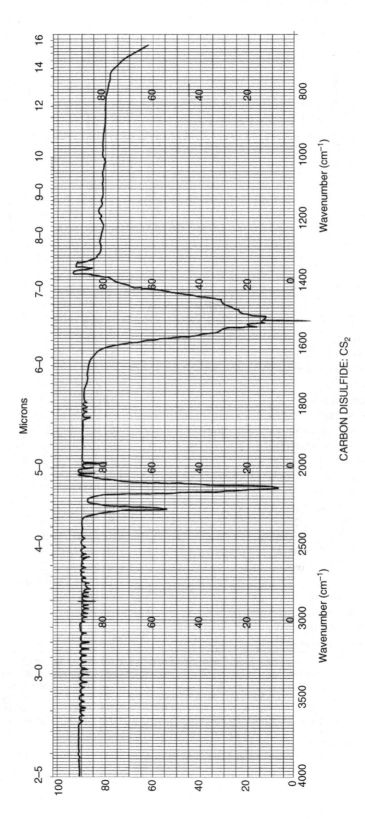

CARBON DISULFIDE: CS$_2$

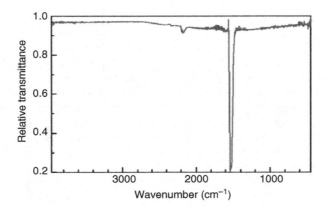

Carbon Disulfide CS$_2$

Physical Properties	
Relative molecular mass	76.14
Melting point	−111 °C
Normal boiling point	46.2 °C
Refractive index (20 °C)	1.6277
(25 °C)	1.6232
Density (20 °C)	1.2631 g/mL
(25 °C)	1.2556 g/mL
Viscosity (20 °C)	0.363 mPa·s
Surface tension (20 °C)	32.25 mN/m
Heat of vaporization (at boiling point)	26.74 kJ/mol
Dielectric constant (20 °C)	2.641
Relative vapor density (air = 1)	2.64
Vapor pressure (25 °C)	0.0448 MPa
Solubility in water (20 °C)	0.29 %, mass/mass
Flash point (OC)	−30 °C
Autoignition temperature	100 °C
Explosive limits in air	1.0 %–50 %, vol/vol
CAS registry number	75-15-0
INChI	1S/CS2/c2-1-3
Exposure limits	20 ppm, 8 h TWA
Solvatochromic α	0
Solvatochromic β	0.07
Solvatochromic π^*	0.61

Notes: Moderately polar solvent, soluble in alcohols, benzene, ethers, and chloroform; slightly soluble in water; very flammable and mobile; can be ignited by friction or contact with hot surfaces such as steam pipes; burns with a blue flame to produce carbon dioxide and sulfur dioxide; toxic by inhalation, ingestion, and skin absorption; strong disagreeable odor when impure; incompatible with aluminum (powder), azides, chlorine, chlorine monoxide, ethylene diamine, ethyleneamine, fluorine, nitrogen oxides, potassium, and zinc and other oxidants; soluble in methanol, ethanol, ethers, benzene, chloroform, carbon tetrachloride, and many oils; can be stored in metal, glass porcelain, and Teflon containers.
Synonyms: carbon bisulfide, dithiocarbon anhydride.

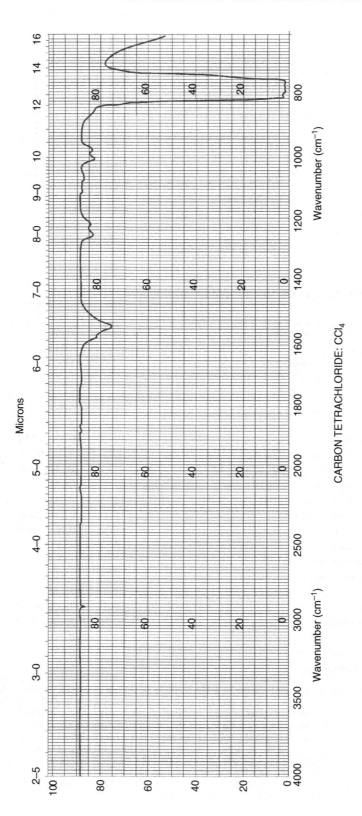

CARBON TETRACHLORIDE: CCl$_4$

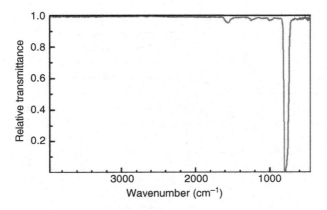

Carbon Tetrachloride CCl₄

Physical Properties	
Relative molecular mass	153.82
Melting point	−22.85 °C
Normal boiling point	76.7 °C
Refractive index (20 °C)	1.4601
(25 °C)	1.457
Density (20 °C)	1.5940 g/mL
(25 °C)	1.5843 g/mL
Viscosity (20 °C)	0.969 mPa·s
Surface tension (20 °C)	26.75 mN/m
Heat of vaporization (at boiling point)	29.82 kJ/mol
Thermal conductivity (20 °C)	0.1070 W/(m·K)
Dielectric constant (20 °C)	2.238
Relative vapor density (air = 1)	5.32
Vapor pressure (25 °C)	0.0122 MPa
Solubility in water (20 °C)	0.08, w/w
Flash point (OC)	Non-combustible
Autoignition temperature	Non-combustible
Explosive limits in air	Non-explosive
CAS registry number	56-23-5
INChl	1S/CCl4/c2-1(3,4)5
Exposure limits	5 ppm (skin)
Solubility parameter, δ	8.6
Hydrogen bond index, λ	2.2
Solvatochromic α	0
Solvatochromic β	0.1
Solvatochromic π*	0.28

Notes: Nonpolar solvent; soluble in alcohols, ethers, chloroform and other halocarbons, benzene, and most fixed and volatile oils, insoluble in water; nonflammable; extremely toxic by inhalation, ingestion, or skin absorption; carcinogenic; incompatible with allyl alcohol, silanes, triethyldialuminum, many metals (for example, sodium).

Synonyms: tetrachloromethane, perchloromethane, methane tetrachloride, Halon-104.

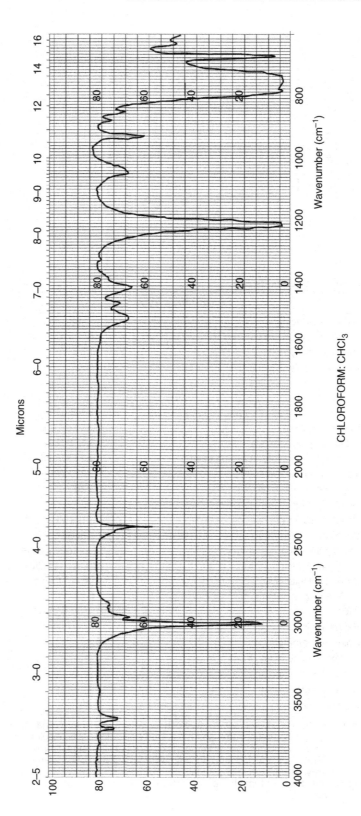

CHLOROFORM: CHCl$_3$

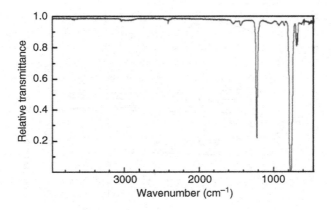

Chloroform CHCl$_3$

Physical Properties	
Relative molecular mass	119.38
Melting point	−63.2 °C
Normal boiling point	61.2 °C
Refractive index (20 °C)	1.4459
(25 °C)	1.4422
Density (20 °C)	1.4892 g/mL
(25 °C)	1.4798 g/mL
Viscosity (20 °C)	0.566 mPa·s
Surface tension	27.2 mN/m
Heat of vaporization (at boiling point)	29.24 kJ/mol
Thermal conductivity (20 °C)	0.1164 W/(m·K)
Dielectric constant (20 °C)	4.806
Relative vapor density (air = 1)	4.13
Vapor pressure (25 °C)	0.0263 MPa
Solubility in water	0.815 %, w/w
Flash point (OC)	Non-combustible[a]
Autoignition temperature	Non-combustible[a]
Explosive limits in air	Non-explosive
CAS registry number	67-66-3
INChI	1S/CHCl3/c2-1(3)4/h1H
Exposure limits	10 ppm, 8 h TWA
Solubility parameter, δ	9.3
Hydrogen bond index, λ	2.2
Solvatochromic α	0.2
Solvatochromic β	0.1
Solvatochromic π*	0.58

Notes: Polar solvent; soluble in alcohols, ether, benzene, and most oils; usually stabilized with methanol to prevent phosgene formation; flammable and highly toxic by inhalation, ingestion, or skin absorption; narcotic; suspected to be carcinogenic; incompatible with caustics, active metals, aluminum powder, potassium, sodium, magnesium.

Synonyms: trichloromethane, methane trichloride.

[a] Although chloroform is nonflammable, it will burn upon prolonged exposure to flame or high temperature.

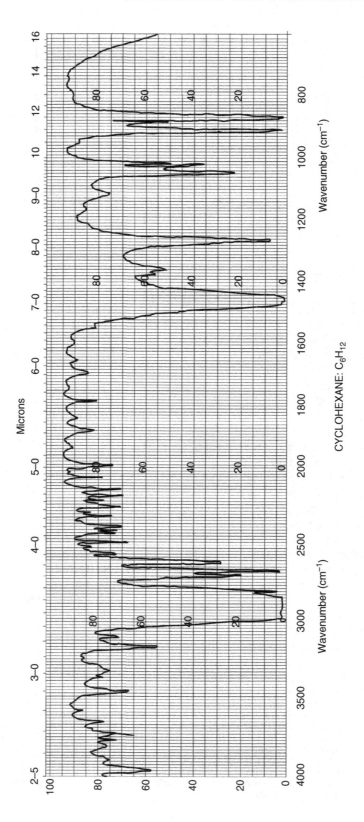

CYCLOHEXANE: C_6H_{12}

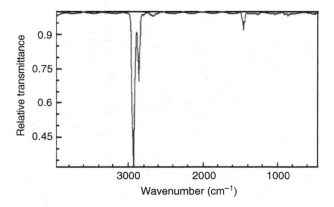

Cyclohexane C$_6$H$_{12}$

Physical Properties	
Relative molecular mass	84.16
Melting point	6.3 °C
Normal boiling point	80.7 °C
Refractive index (20 °C)	1.4235
(25 °C)	1.4235
Density (20 °C)	0.7786 g/mL
(25 °C)	0.7739 g/mL
Viscosity (20 °C)	1.06 mPa·s
Surface tension (20 °C)	24.99 mN/m
Heat of vaporization (at boiling point)	29.97 kJ/mol
Thermal conductivity (20 °C)	0.122 W/(m·K)
Dielectric constant (20 °C)	2.023
Relative vapor density (air = 1)	2.9
Vapor pressure (25 °C)	0.0111 MPa
Solubility in water (20 °C)	<0.01 %, mass/mass
Flash point (OC)	−17 °C
Autoignition temperature	245 °C
Explosive limits in air	1.31 %–8.35 %, vol/vol
CAS registry number	110-82-7
INChI	1S/C6H12/c1-2-4-6-5-3-1/h1-6H2
Exposure limits	330 ppm, 8 h TWA
Solvatochromic α	0
Solvatochromic β	0
Solvatochromic π*	0

Notes: Nonpolar hydrocarbon solvent; mild, gasoline-like odor; soluble in hydrocarbons, alcohols, organic halides, acetone, benzene; flammable; moderately toxic by inhalation, ingestion or skin absorption, may be narcotic at high concentrations; reacts with oxygen (air) at elevated temperatures; decomposes upon heating; incompatible with strong oxidants.
Synonyms: benzene hexahydride, hexamethylene, hexanaphthene, hexahydrobenzene.

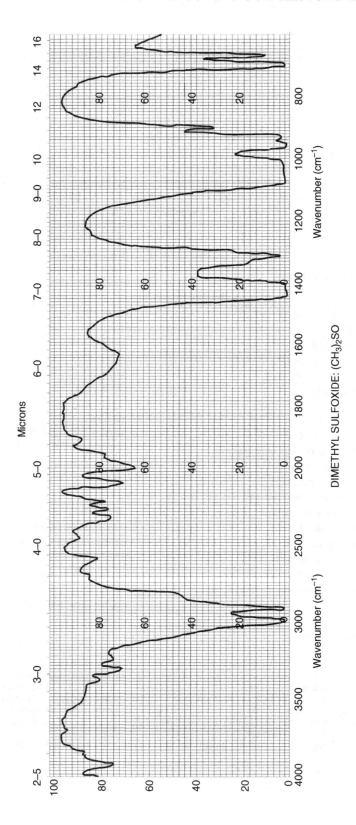

DIMETHYL SULFOXIDE: $(CH_3)_2SO$

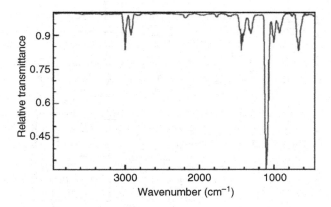

Dimethyl Sulfoxide (CH$_3$)$_2$SO

Physical Properties	
Relative molecular mass	78.13
Melting point	18.5 °C
Normal boiling point	191.9 °C
Refractive index (20 °C)	1.4793
Density (20 °C)	1.1014 g/mL
Viscosity (25 °C)	1.98 mPa·s
Surface tension	43.5 mN/m
Relative vapor density (air = 1)	2.7
Vapor pressure	5.3×10^{-5} MPa
Solubility in water	∞
Flash point (OC)	95 °C
Autoignition temperature	215 °C
Explosive limits in air	26.0 %–28.5 %, vol/vol
CAS registry number	67-68-5
INChI	1S/C2H6OS/c1-4(2)3/h1-2H3
Exposure limits	None established
Solubility parameter, δ	13
Hydrogen bond index, λ	5
Solvatochromic α	0
Solvatochromic β	0.76
Solvatochromic π*	1

Notes: Colorless, odorless (when pure), hygroscopic liquid, powerful aprotic solvent; dissolves many inorganic salts, soluble in water; combustible; readily penetrates the skin; incompatible with strong oxidizers, and many halogenated compounds for example, alkyl halides, aryl halides), oxygen, peroxides, diborane, perchlorates.

Synonyms: DMSO, methyl sulfoxide, sulfinylbismethane.

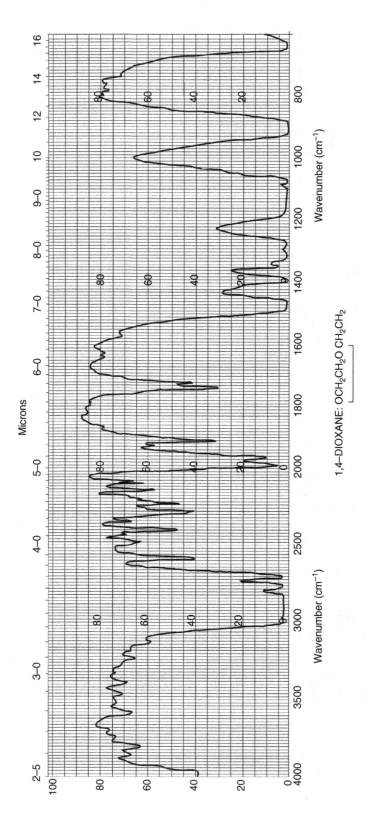

1,4–DIOXANE: OCH$_2$CH$_2$O CH$_2$CH$_2$

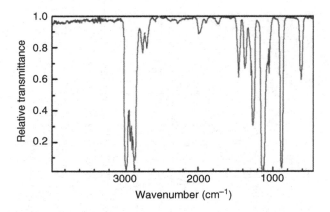

1,4-Dioxane

Physical Properties	
Relative molecular mass	88.11
Melting point	11 °C
Boiling point	101.3 °C
Refractive index (20 °C)	1.4224
(25 °C)	1.4195
Density (20 °C)	1.0338 g/mL
(25 °C)	1.0282 g/mL
Viscosity (20 °C)	1.37 mPa·s
Surface tension (20 °C)	33.74 mN/m
Heat of vaporization (at boiling point)	34.16 kJ/mol
Dielectric constant (20 °C)	2.209
Relative vapor density (air = 1)	3.03
Vapor pressure (25 °C)	0.0053 MPa
Solubility in water	∞
Flash point (OC)	12 °C
Autoignition temperature	180 °C
Explosive limits in air	1.97 %–22.2 %, vol/vol
CAS registry number	123-91-1
INChI	1S/C4H8O2/c1-2-6-4-3-5-1/h1-4H2
Exposure limits	100 ppm (skin)
Solubility parameter, δ	9.9
Hydrogen bond index, λ	5.7
Solvatochromic α	0
Solvatochromic β	0.37
Solvatochromic π^*	0.55

Notes: Moderately polar solvent; soluble in water and most organic solvents; flammable; highly toxic by ingestion and inhalation; absorbed through the skin; may cause central nervous system depression, necrosis of the liver and kidneys; incompatible with strong oxidizers.

Synonyms: diethylene ether, 1,4-diethylene dioxide, diethylene dioxide, dioxyethylene ether.

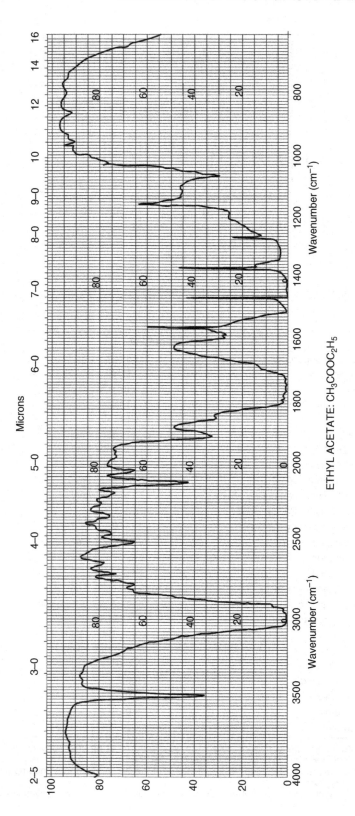

ETHYL ACETATE: $CH_3COOC_2H_5$

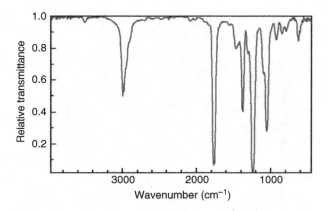

Ethyl Acetate $CH_3COOC_2H_5$

Physical Properties	
Relative molecular mass	88.11
Melting point	−83.58 °C
Boiling point	77.1 °C
Refractive index (20 °C)	1.3723
(25 °C)	1.3698
Density (20 °C)	0.9006 g/mL
(25 °C)	0.8946 g/mL
Viscosity (20 °C)	0.452 mPa·s
Surface tension (20 °C)	23.95 mN/m
Heat of vaporization (at boiling point)	31.94 kJ/mol
Thermal conductivity (20 °C)	0.122 W/(m·K)
Dielectric constant (25 °C)	6.02
Relative vapor density (air = 1)	3.04
Vapor pressure (20 °C)	0.0097 MPa
Solubility in water (20 °C)[a]	3.3 %, mass/mass
Flash point (OC)	−1 °C
Autoignition temperature	486 °C
Explosive limits in air	2.18 %–11.5 %, vol/vol
CAS registry number	141-78-6
INChI	1S/C4H8O2/c1-3-6-4(2)5/h3H2,1-2H3
Exposure limits	440 ppm, 8 h, TWA
Solubility parameter, δ	9.1
Hydrogen bond index, λ	5.2
Solvatochromic α	0
Solvatochromic β	0.43
Solvatochromic $\pi*$	0.55

Notes: Polar solvent; insoluble in water, soluble in alcohols, organic halides, ether, and many oils; flammable; moderately toxic by inhalation and skin absorption; incompatible with strong oxidizers, nitrates, strong alkalis, strong acids.
Synonyms: acedin, acetic ether, acetic ester, vinegar naphtha, acetic acid ethyl ester.
[a] Forms an azeotrope with water at 6.1 %, mass/mass, that boils at 70.4 °C.

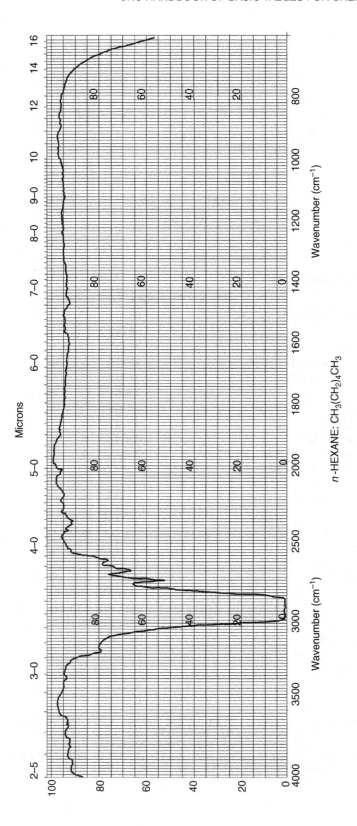

Microns

Wavenumber (cm^{-1})

Wavenumber (cm^{-1})

n-HEXANE: CH$_3$(CH$_2$)$_4$CH$_3$

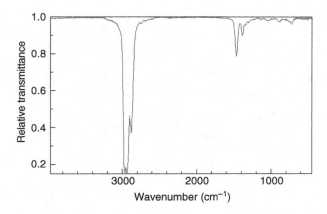

n-Hexane CH₃(CH₂)₄CH₃

Physical Properties	
Relative molecular mass	86.18
Melting point	−95 °C
Normal boiling point	68.72 °C
Refractive index (20 °C)	1.3855
(25 °C)	1.3723
Density (20 °C)	0.6594 g/mL
(25 °C)	0.6548 g/mL
Viscosity (20 °C)	0.31 mPa·s
Surface tension (20 °C)	18.42 mN/m
Heat of vaporization (at boiling point)	28.85 kJ/mol
Thermal conductivity (20 °C)	0.1217 W/(m·K)
Dielectric constant (20 °C)	1.89
Relative vapor density (air = 1)	2.97
Vapor pressure (25 °C)	0.0222 MPa
Solubility in water (20 °C)	0.011 %, mass/mass
Flash point (OC)	−26 °C
Autoignition temperature	247 °C
Explosive limits in air	1.25 %–6.90 %, vol/vol
CAS registry number	110-54-3
INChI	1S/C6H14/c1-3-5-6-4-2/h3-6H2,1-2H3
Exposure limits	500 ppm, 8 h TWA
Solubility parameter, δ	9.3
Hydrogen bond index, λ	2.2
Solvatochromic α	0
Solvatochromic β	0
Solvatochromic π*	0.08

Notes: Nonpolar solvent; soluble in alcohols, hydrocarbons, organic halides, acetone, and ethers; insoluble in water; flammable; moderately toxic by inhalation and ingestion; incompatible with strong oxidizers.
Synonyms: hexane, hexyl hydride.

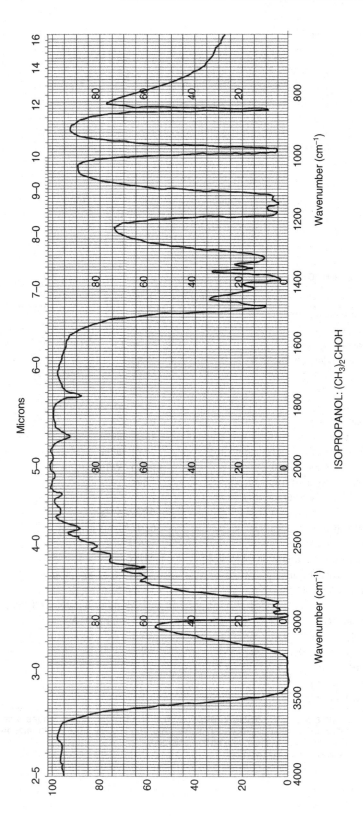

ISOPROPANOL: $(CH_3)_2CHOH$

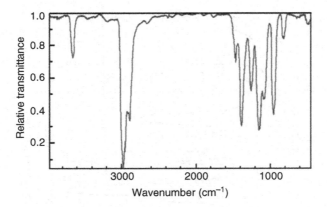

Isopropanol (CH₃)₂CHOH

Physical Properties	
Relative molecular mass	60.1
Melting point	−87.91 °C
Boiling point	82.21 °C
Refractive index (20 °C)	1.3776
(25 °C)	1.375
Density (20 °C)	0.7864 g/mL
(25 °C)	0.7812 g/mL
Viscosity (20 °C)	2.43 mPa·s
Surface tension (20 °C)	21.99 mN/m
Heat of vaporization (at boiling point)	39.85 kJ/mol
Dielectric constant (25 °C)	18.3
Relative vapor density (air = 1)	2.07
Vapor pressure	0.0044 MPa
Solubility in water (20 °C)	∞
Flash point (OC)	16 °C
Autoignition temperature	456 °C
Explosive limits in air	2.02 %–11.8 %, vol/vol
CAS registry number	67-63-0
INChI	1S/C3H8O/c1-3(2)4/h3-4H,1-2H3
Exposure limits	400 ppm (skin)
Solubility parameter, δ	11.5
Hydrogen bond index, λ	8.9
Solvatochromic α	0.76
Solvatochromic β	0.84
Solvatochromic π*	0.48

Notes: Polar solvent; soluble in water, alcohols, ethers, many hydrocarbons, and oils; flammable and moderately toxic by ingestion, inhalation and skin absorption; incompatible with strong oxidizers.

Synonyms: dimethyl carbinol, sec-propyl alcohol, 2-propanol, isopropyl alcohol.

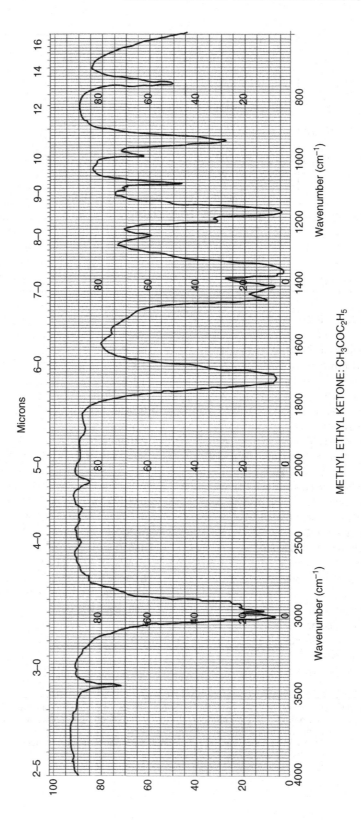

METHYL ETHYL KETONE: $CH_3COC_2H_5$

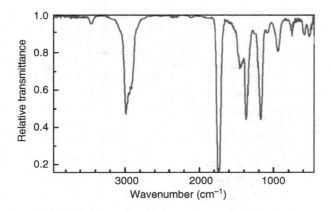

Methyl Ethyl Ketone Methyl Ethyl Ketone $CH_3COC_2H_5$

Physical Properties	
Relative molecular mass	60.1
Melting point	−87.91 °C
Boiling point	82.21 °C
Refractive index (20 °C)	1.3776
(25 °C)	1.375
Density (20 °C)	0.7864 g/mL
(25 °C)	0.7812 g/mL
Viscosity (20 °C)	2.43 mPa·s
Surface tension (20 °C)	21.99 mN/m
Heat of vaporization (at boiling point)	39.85 kJ/mol
Dielectric constant (25 °C)	18.3
Relative vapor density (air = 1)	2.07
Vapor pressure	0.0044 MPa
Solubility in water (20 °C)	∞
Flash point (OC)	16 °C
Autoignition temperature	456 °C
Explosive limits in air	2.02 %–11.8 %, vol/vol
CAS registry number	67-63-0
INChI	1S/C3H8O/c1-3(2)4/h3-4H,1-2H3
Exposure limits	400 ppm (skin)
Solubility parameter, δ	11.5
Hydrogen bond index, λ	8.9
Solvatochromic α	0.76
Solvatochromic β	0.84
Solvatochromic π*	0.48

Notes: Polar solvent; soluble in water, ketones, organic halides, alcohols, ether, and many oils; highly flammable; narcotic by inhalation; incompatible with strong oxidizers, nitrates, nitric acid, reducing agents.
Synonyms: ethyl methyl ketone, 2-butanone, methyl acetone, MEK.

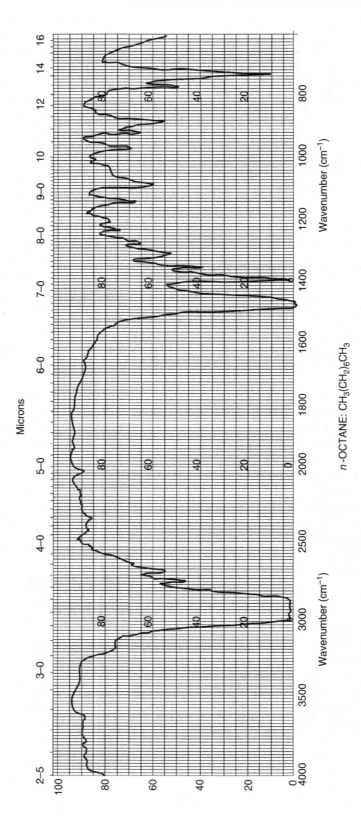

n-OCTANE: CH$_3$(CH$_2$)$_6$CH$_3$

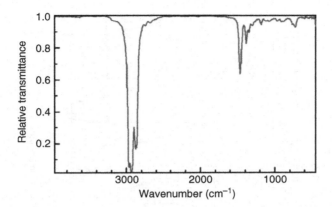

n-Octane $CH_3(CH_2)_6CH_3$

Physical Properties	
Relative molecular mass	114.23
Melting point	−56.7 °C
Boiling point	125.62 °C
Refractive index (20 °C)	1.3944
(25 °C)	1.3951
Density (20 °C)	0.7025 g/mL
(25 °C)	0.6985 g/mL
Viscosity (20 °C)	0.539 mPa·s
Surface tension (20 °C)	21.75 mN/m
Heat of vaporization (at boiling point)	34.41 kJ/mol
Dielectric constant (20 °C)	1.948
Relative vapor density (air = 1)	3.86
Vapor pressure (25 °C)	0.0023 MPa
Solubility in water (20 °C)	~0.002 %, mass/mass
Flash point (CC)	13 °C
Autoignition temperature	232 °C
Explosive limits in air	0.84 %–3.2 %, vol/vol
CAS registry number	111-65-9
INChI	1S/C8H18/c1-3-5-7-8-6-4-2/h3-8H2,1-2H3
Exposure limits	550 ppm, 8 h TWA
Hydrogen bond index, λ	2.2
Solvatochromic α	0
Solvatochromic β	0
Solvatochromic $\pi*$	0.01

Notes: Nonpolar solvent; soluble in alcohol, acetone, and hydrocarbons; insoluble in water; flammable; incompatible with strong oxidizers.
Synonyms: octane.

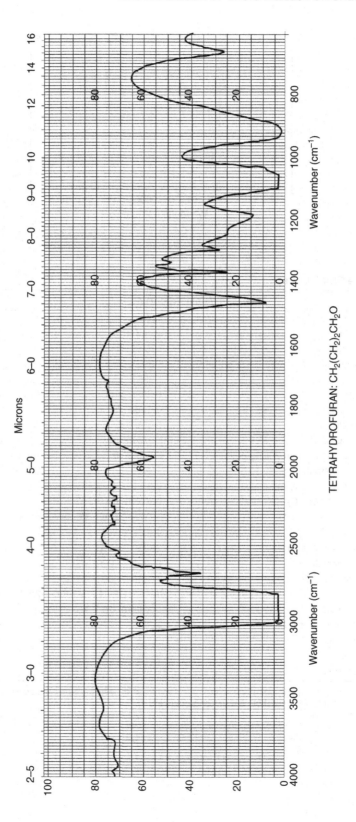

TETRAHYDROFURAN: $CH_2(CH_2)_2CH_2O$

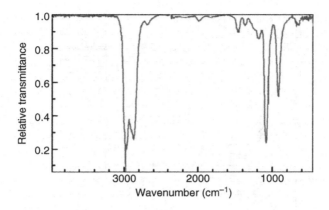

Tetrahydrofuran

Physical Properties	
Relative molecular mass	72.108
Melting point	−65 °C
Normal boiling point	66.0 °C
Refractive index (20 °C)	1.405
(25 °C)	1.404
Density (20 °C)	0.8880 g/mL
(25 °C)	0.8818 g/mL
Viscosity (20 °C)	0.55 mPa·s
Surface tension (20 °C)	26.4 mN/m
Heat of vaporization (at boiling point)	29.81 kJ/mol
Dielectric constant (20 °C)	7.54
Relative vapor density (air = 1)	2.5
Vapor pressure (20 °C)	0.0191 MPa
Solubility in water (20 °C)[a]	∞
Flash point (CC)	−17 °C
Autoignition temperature	260 °C
Explosive limits in air	1.8 %–11.8 %, vol/vol
CAS registry number	109-99-9
INChI	1S/C4H8O/c1-2-4-5-3-1/h1-4H2
Exposure limits	200 ppm, 8 h TWA
Solubility parameter, δ	9.1
Hydrogen bond index, λ	5.3
Solvatochromic α	0
Solvatochromic β	0.55
Solvatochromic π*	0.58

Notes: Moderately polar solvent, ethereal odor; soluble in water and most organic solvents; flammable; moderately toxic; incompatible with strong oxidizers; can form potentially explosive peroxides upon long standing in air; commercially, it is often stabilized against peroxidation with 0.5 %–1.0 % (mass/mass) p-cresol, .05 %–1.0 % (mass/mass) hydroquinone, or 0.01 % (mass/mass) 4,4′-thiobis(6-tert-butyl-m-cresol); can polymerize in the presence of cationic initiators such as Lewis acids or strong proton acids.

Synonyms: THF, tetramethylene oxide, diethylene oxide, 1,4-epoxybutane oxolane, oxacyclopentane.

[a] pH of aqueous solution = 7.

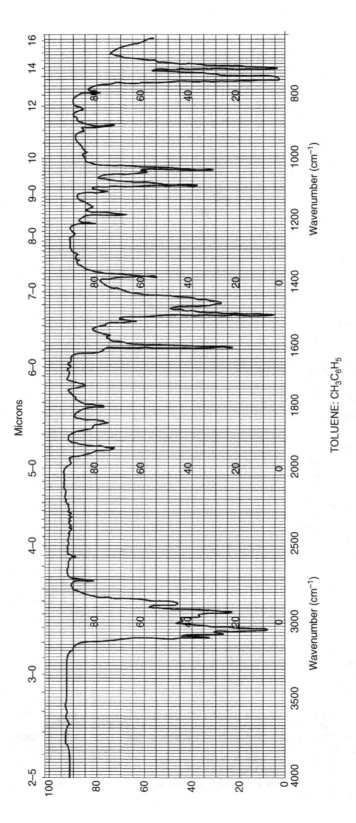

TOLUENE: $CH_3C_6H_5$

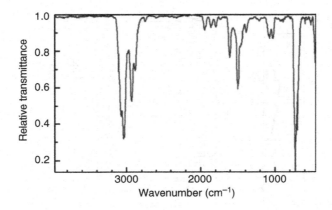

Toluene CH₃C₆H₅

Physical Properties	
Relative molecular mass	92.14
Melting point	−94.5 °C
Normal boiling point	110.6 °C
Refractive index (20 °C)	1.497
(25 °C)	1.4941
Density (20 °C)	0.8669 g/mL
(25 °C)	0.8623 g/mL
Viscosity (20 °C)	0.587 mPa·s
Surface tension (20 °C)	28.52 mN/m
Heat of vaporization (at boiling point)	33.18 kJ/mol
Thermal conductivity (20 °C)	0.1348 W/(m·K)
Dielectric constant (25 °C)	2.379
Relative vapor density (air = 1)	3.14
Vapor pressure (25 °C)	0.0036 MPa
Solubility in water	0.047 %, mass/mass
Flash point (CC)	4 °C
Autoignition temperature	552 °C
Explosive limits in air	1.4 %–7.4 %, vol/vol
CAS registry number	108-88-3
INChl	1S/C7H8/c1-7-5-3-2-4-6-7/h2-6H,1H3
Exposure limits	200 ppm, 8 h TWA
Solubility parameter, δ	8.9
Hydrogen bond index, λ	3.8
Solvatochromic α	0
Solvatochromic β	0.11
Solvatochromic π*	0.54

Notes: Aromatic solvent; sweet pungent odor; soluble in benzene, alcohols, organic halides, ethers, insoluble in water; highly flammable; toxic by ingestion, inhalation, and absorption through the skin, narcotic at high concentrations; incompatible with strong oxidants; decomposes under high heat to form (predominantly) dimethylbiphenyl.
Synonyms: toluol, methylbenzene, methylbenzol, phenylmethane.

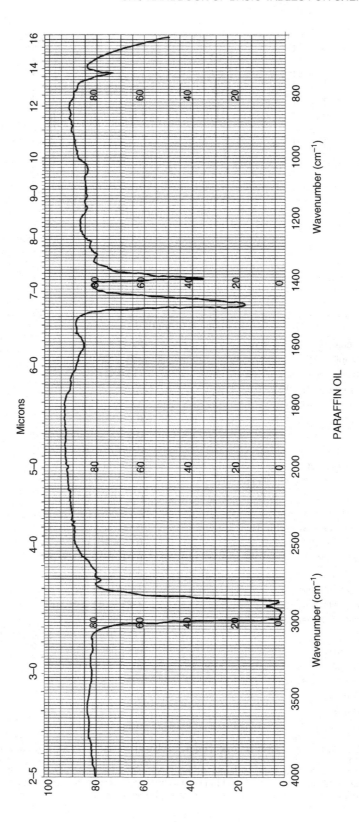

PARAFFIN OIL

Paraffin Oil

Physical Properties	
Relative molecular mass	Variable
Melting point	−20 °C (approximate)
Normal boiling point	315 °C (approximate)
Refractive index (20 °C)	1.472
(25 °C)	1.4697
Specific gravity, 25 °C /25 °C	0.85
Solubility in water	Insoluble
Flash point (OC)	229 °C
Explosive limits in air	0.6 %–6.5 %, vol/vol
CAS registry number	8012-95-1
INChl	NA, not a pure fluid
Exposure limits	50 ppm, 8 h TWA

Notes: Viscous, odorless, moderately combustible liquid used for mull preparation; relatively low toxicity; soluble in benzene, chloroform, carbon disulfide, ethers; incompatible with oxidizing materials and amines.

Synonyms: mineral oil, adepsine oil, lignite oil, nujol.

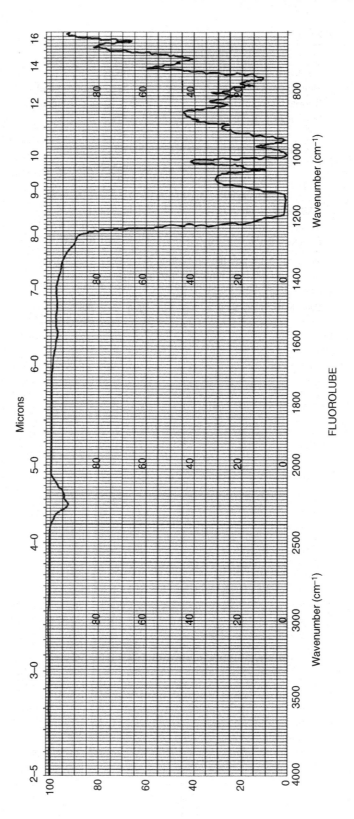

FLUOROLUBE

Fluorolube, Polytrifluorochloroethylene [$-C_2ClF_3-$]

Physical Properties	
Relative molecular mass (monomer)	116.47
Pour point[a]	$-60\,°C\ -13\,°C$
Melting point	$-51\,°C\ -18\,°C$
Acidity (pH)[a]	6.0–7.5
Density (38 °C)[a]	1.865–1.955 g/mL
Viscosity (25 °C)[a]	6–1400 mPa·s
Vapor pressure (93 °C)	0.07–2.2 mmHg
Flash point (OC)	Nonflammable
Autoignition temperature	Nonflammable
Explosive limits in air	Nonflammable
CAS registry number	9002-83-9
INChI	NA, polydisperse fluid
Exposure limits	Not established

Notes: There are six common grades or varieties of this oil, marketed under the name Fluorolube. The properties listed above that are marked with an asterisk depend upon the grade that is used. The primary physical difference between the grades are the viscosities and pour points.

The thermal stability of these materials is dependent on the wetted surfaces. Typical ranges of stability are between 150 °C and 325 °C, but this varies with the wetted surface and residence time. Some metals can accelerate the decomposition into lower molecular mass, more volatile components. It is important to avoid the wetting of metals containing aluminum or magnesium especially in situations in which high friction of galling is possible. Detonation of these fluids is possible under these conditions. Moreover, these fluids can react violently in the presence of sodium, potassium, amines, hydrazine, liquid fluorine, and liquid chlorine.

Since these fluids are essentially transparent from 1,360 to 4,000 cm^{-1} (except for the absorption at 2,321.9 cm^{-1}), they can be used as mulling agents when the bands of paraffin oil obscure or interfere with sample absorptions.

POLYSTYRENE WAVENUMBER CALIBRATION

The following are wavenumber readings assigned to the peaks on the spectrum:

1	–	3,027.1	8	–	1,583.1
2	–	2,924	9	–	1,181.4
3	–	2,850.7	10	–	1,154.3
4	–	1,944	11	–	1,069.1
5	–	1,871	12	–	1,028
6	–	1,801.6	13	–	906.7
7	–	1,601.4	14	–	698.9

Film thickness: 50 μm

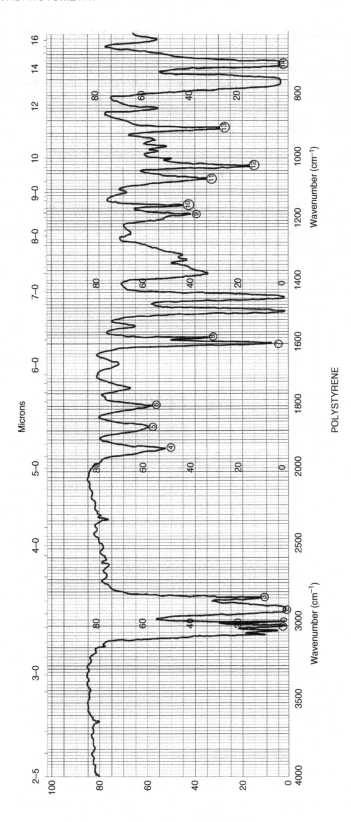

POLYSTYRENE

INFRARED ABSORPTION CORRELATION CHARTS

The following charts provide characteristic infrared absorptions obtained from particular functional groups on molecules [1,2]. These include a general mid-range correlation chart, a chart for aromatic absorptions, and a chart for carbonyl moieties. The general mid-range chart is an adaptation of the work of Professor Charles F. Hammer of Georgetown University, reproduced, with modification and with permission.

REFERENCES

1. Bruno, T.J. and Svoronos, P.D.N., *CRC Handbook of Basic Tables for Chemical Analysis*, 3rd ed., CRC Press, Boca Raton, FL, 2011.
2. Bruno, T.J., and Svoronos, P.D.N., *CRC Handbook of Fundamental Spectroscopic Correlation Charts*, CRC Press, Boca Raton, FL 2006.

Notes:
AR = aromatic
b = broad
sd = solid
sn = solution
sp = sharp
? = unreliable

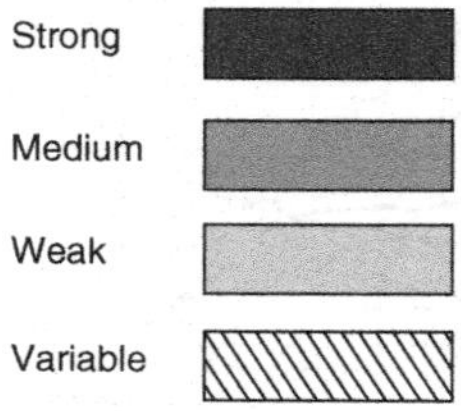

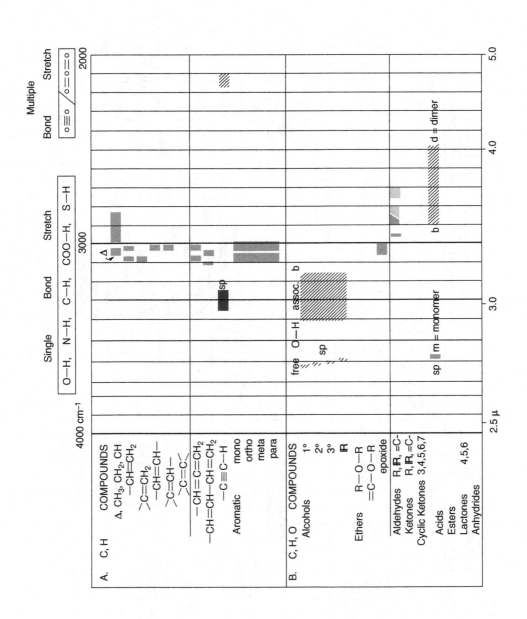

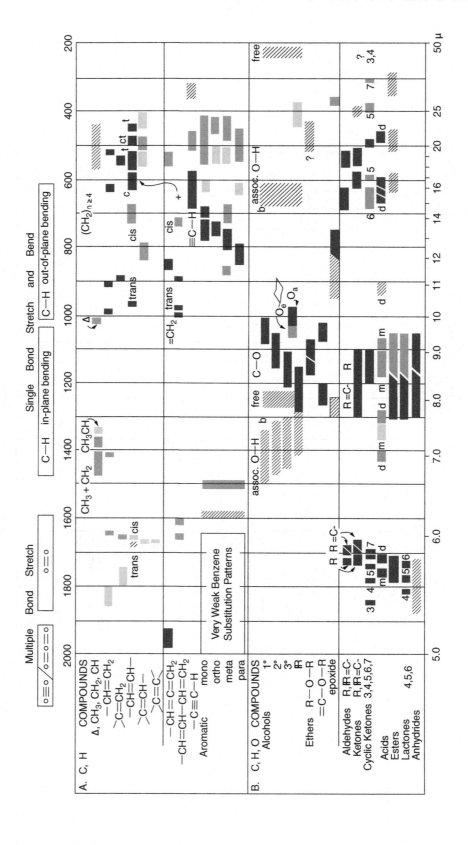

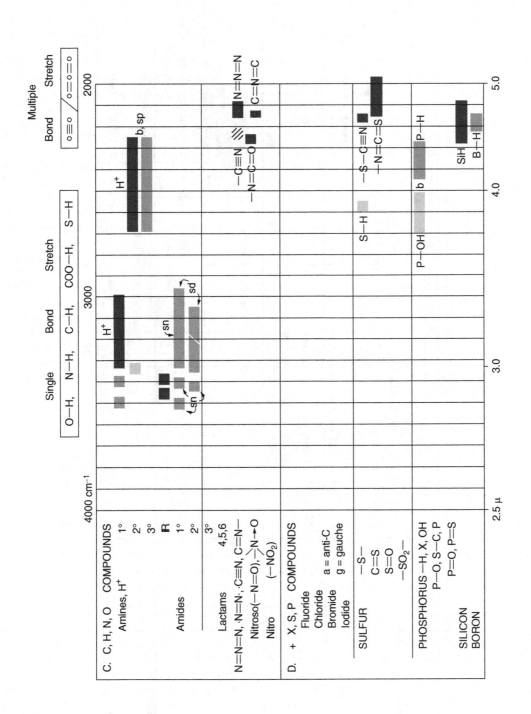

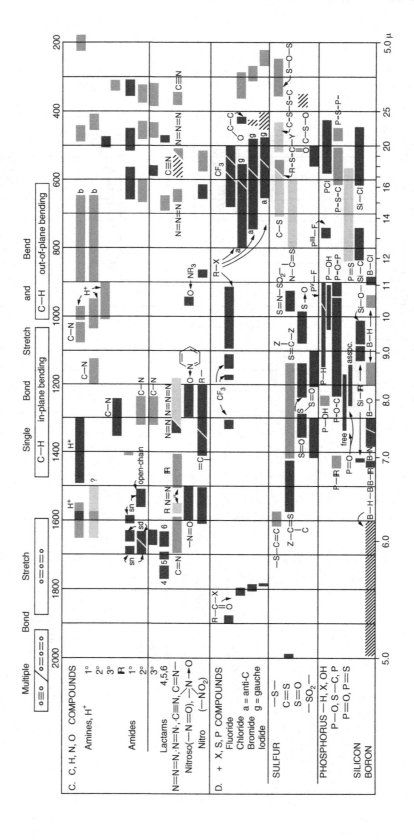

Aromatic Substitution Bands

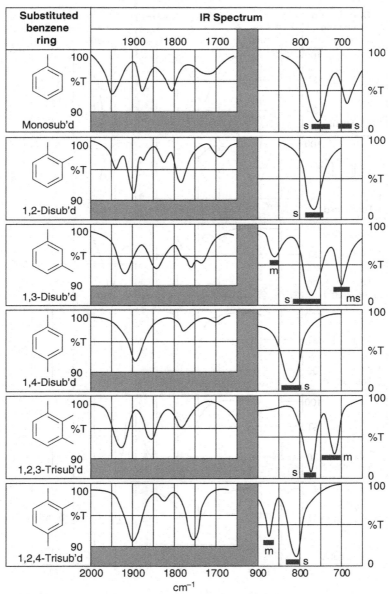

Aromatic Substitution Bands

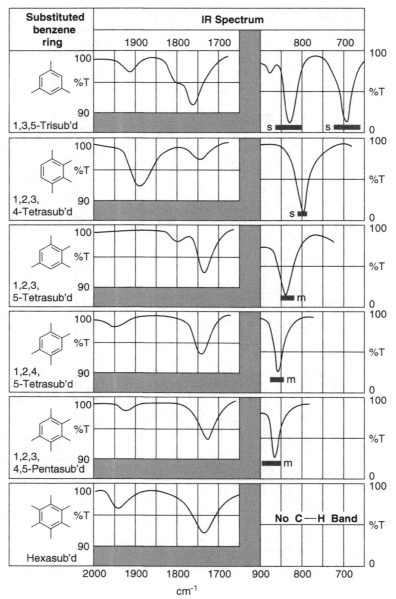

Carbonyl Group Absorptions

Group	Wavenumber, cm^{-1}						
	1850	1800	1750	1700	1650	1600	1550
Acid, Chlorides, Aliphatic		1810–1795					
Acid Chlorides, Aromatic			1785–1765				
Aldehydes, Aliphatic				1740–1718			
Aldehydes, Aromatic				1710–1685			
Amides					1695–1630*		
Amides, typical value, 1°				1684			
Amides, typical value, 2°					1669		
Amides, typical value, 3°					1667		
	5.41	5.56	5.71	5.88	6.06	6.25	6.45
	Wavelength, µm						

* Electron withdrawing groups at the α-position to the carbonyl will raise the wavenumber of the absorption.

Carbonyl Group Absorptions (continued)

Group	Wavenumber, cm^{-1}						
	1850	1800	1750	1700	1650	1600	1550
Anhydrides, acyclic, non-conjugated		1825–1815***	1755–1745**				
Anhydrides, acyclic, conjugated			1780–1770***	1725–1715**			
Anhydrides, ayclic non-conjugated	1870–1845	1800–1775**					
Anhydrides, cyclic conjugated	1860–1850	1780–1760**					
Carbamates				1740–1683			
Carbonates, acyclic			1780–1740				
Carbonates, five-membered ring	1850–1790						
Carbonates, vinyl, typical value			1761				
	5.41	5.56	5.71	5.88	6.06	6.25	6.45
	Wavelength, µm						

** This band is the more intense of the two.
*** Intensity weakens as colinearity is approached.

Carbonyl Group Absorptions (continued)

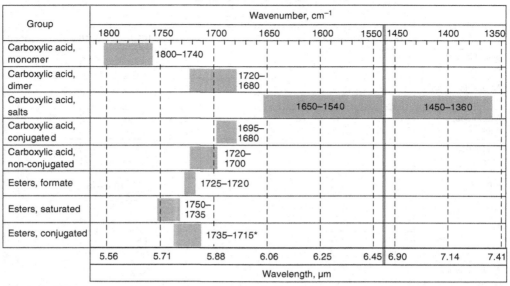

Group	Wavenumber, cm⁻¹
	1800 1750 1700 1650 1600 1550 ‖ 1450 1400 1350
Carboxylic acid, monomer	1800–1740
Carboxylic acid, dimer	1720–1680
Carboxylic acid, salts	1650–1540 ‖ 1450–1360
Carboxylic acid, conjugated	1695–1680
Carboxylic acid, non-conjugated	1720–1700
Esters, formate	1725–1720
Esters, saturated	1750–1735
Esters, conjugated	1735–1715*
	5.56 5.71 5.88 6.06 6.25 6.45 ‖ 6.90 7.14 7.41
	Wavelength, µm

* Electron withdrawing groups in the α-position to the carbonyl will raise the wavenumber adsorption.

Carbonyl Group Absorptions (continued)

Group	Wavenumber, cm⁻¹
	1800 1750 1700 1650 1600 1550 ‖ 1450 1400 1350
Esters, phenyl, typical value	1770
Esters, thiol, non-conjugated	1710–1680
Esters, thiol, conjugated	1700–1640
Esters, vinyl, typical value	1770
Esters, vinylidene, typical value	1764
Ketones, dialkyl	1725–1705
Ketones, α, β- unsaturated	1700–1670
Ketones, α, β, and α', β' conjugated	1680–1640
	5.56 5.71 5.88 6.06 6.25 6.45 ‖ 6.90 7.14 7.41
	Wavelength, µm

Carbonyl Group Absorptions (continued)

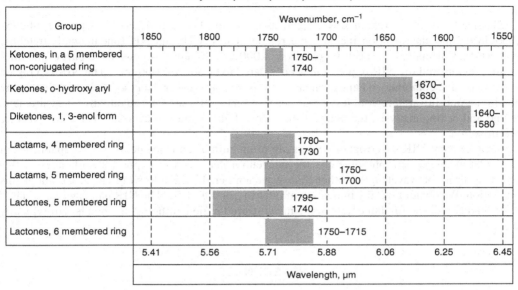

Group	Wavenumber, cm⁻¹
Ketones, in a 5 membered non-conjugated ring	1750–1740
Ketones, o-hydroxy aryl	1670–1630
Diketones, 1, 3-enol form	1640–1580
Lactams, 4 membered ring	1780–1730
Lactams, 5 membered ring	1750–1700
Lactones, 5 membered ring	1795–1740
Lactones, 6 membered ring	1750–1715

1850 1800 1750 1700 1650 1600 1550

5.41   5.56   5.71   5.88   6.06   6.25   6.45

Wavelength, μm

NEAR-INFRARED ABSORPTIONS

Classically, the NIR region was defined as occurring between 0.7 and 3.5 μm, or 14,285–2,860 cm^{-1}. This classification includes the region of CH, OH, and NH fundamental stretching bands [1–5]. Currently, this spectral area, from 4,000 to 2,860 cm^{-1}, is considered part of the mid-infrared region, and the NIR region is now considered to be above 4,000 cm^{-1}. The NIR is a region of overtones and combination bands, which are considerably weaker than the fundamentals that are seen in the mid-infrared region. It is nevertheless a very useful area for quantitative measurement, on line and at line analysis, the analysis of viscous liquids and powders, and even for structure determination.

Because most NIR spectrophotometers are often built as enhancements to the capabilities of ultraviolet-visible spectrophotometers, the convention has been to express absorbances in this region in terms of wavelength, rather than wavenumber. In the following chart, we adopt this convention. We do not give any indication of intensity in this chart. The NIR bands will be related to the intensity of the fundamentals in the mid-infrared region, although the bands will typically be broad.

REFERENCES

1. Colthup, N.B., Daly, L.H., and Wiberley, S.E., *Introduction to Infrared and Raman Spectroscopy*, 3rd ed., Academic Press, Boston, MA, 1990.
2. Conley, R.T., *Infrared Spectroscopy*, Allyn and Bacon, Boston, MA, 1972.
3. Bruno, T.J. and Svoronos, P.D.N., *CRC Handbook of Fundamental Spectroscopic Correlation Charts*, CRC Press, Boca Raton, FL, 2006.
4. Burns, D.A. and Ciurczac, E.W., *Handbook of Near Infrared Analysis*, 3rd ed., CRC Press, Boca Raton, FL, 2007.
5. Pasquini, C.C., Near infrared spectroscopy: fundamentals, practical aspects and analytical applications: a review, *J. Braz. Chem. Soc.*, 14(2), 2003.

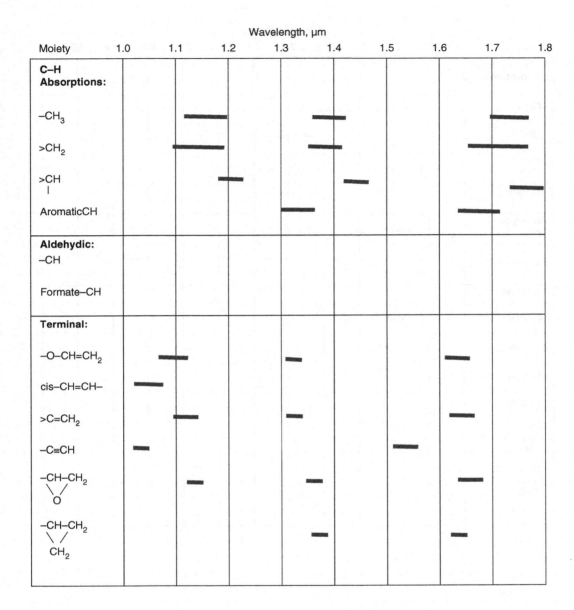

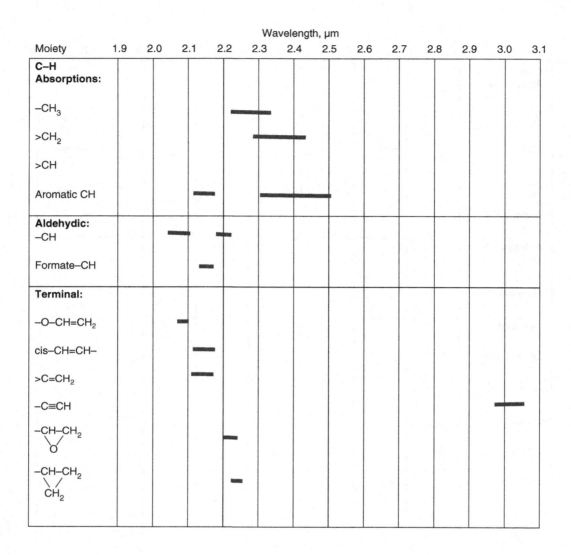

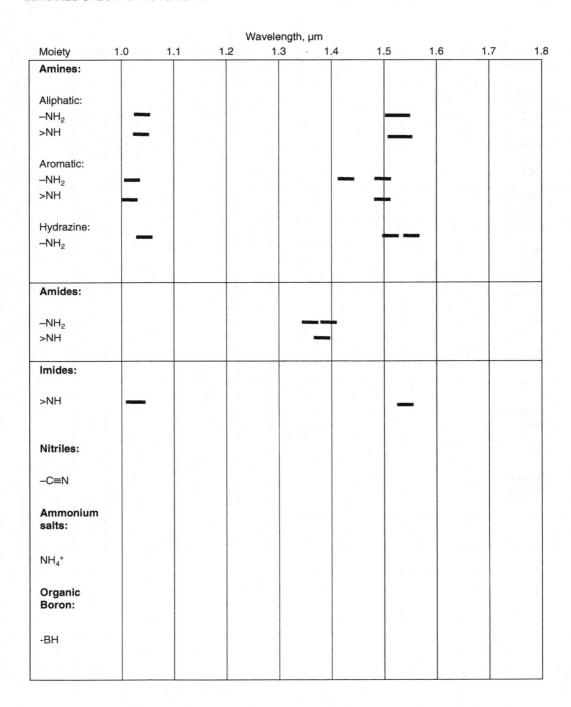

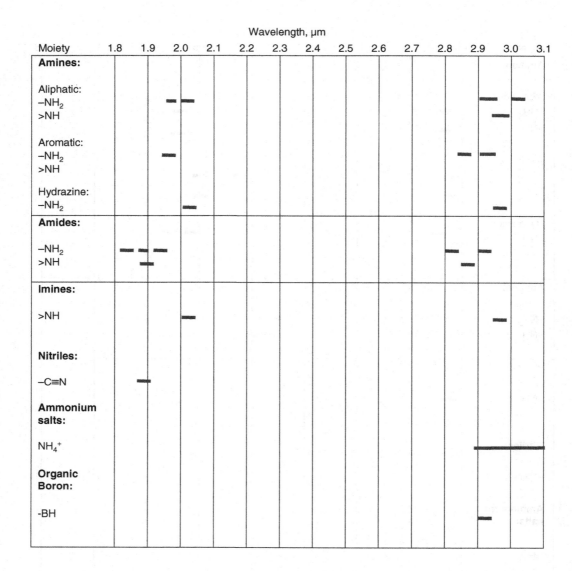

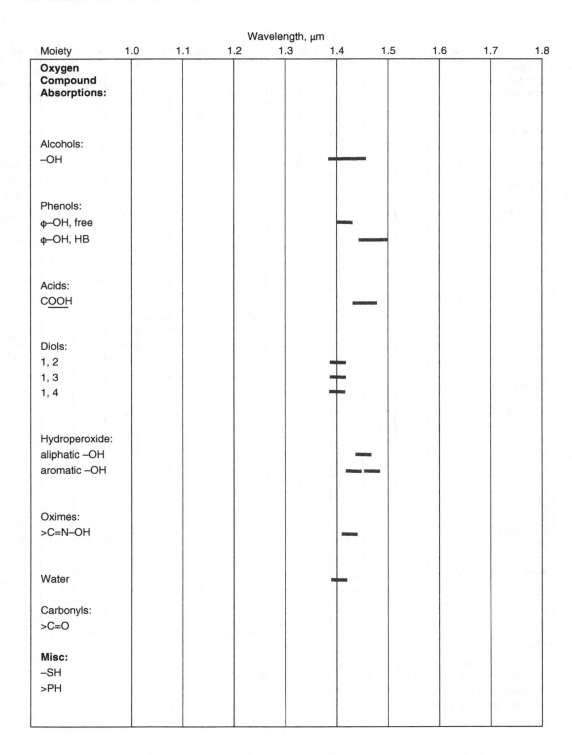

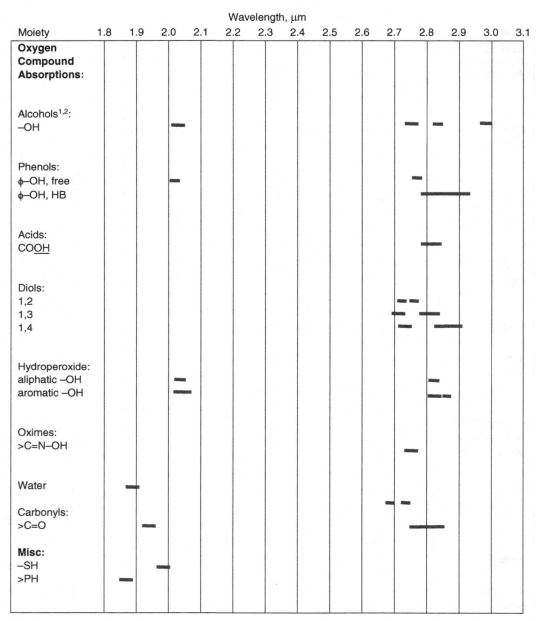

Notes:

1. The dimer bands for alcohols occur between 2.8 and 2.9 μm.
2. The polymer bands of alcohols occur between 2.96 and 3.05 μm.

HB designates the presence of hydrogen bonding.

φ indicates a benzene ring.

INORGANIC GROUP ABSORPTIONS

The following chart provides the infrared absorbance that may be observed from inorganic functional moieties. These have been compiled from a study of the IR absorption of a number of inorganic species [1–3]. It should be understood that the physical state of the sample plays a role in the intensity and position of these bands. These variables include crystal structure, crystallite size, and water of hydration. This chart must therefore be regarded as an approximate guide.

REFERENCES

1. Miller, F.A. and Wilkins, C.H., Infrared spectra and characteristic frequencies of inorganic ions, *Anal. Chem.*, 24(8), 1253–1294, 1952.
2. Bruno, T.J. and Svoronos, P.D.N., *CRC Handbook of Fundamental Spectroscopic Correlation Charts*, CRC Press, Boca Raton, FL, 2006.
3. Nyquist, R. and Kagel, R., *Handbook of Infrared and Raman Spectra of Inorganic Compounds and Organic Salts*, Academic Press, London, 107–108, 1971.

Key:

Strong: ▬▬▬▬

Medium: ▬▬▬▬

Weak: ▬▬▬▬

Note
[a] This water is water of crystallization.

Moiety	3600	3400	3200	3000	2800	2600	2400	2200 cm⁻¹
Water[1]		▬	▬					
Boron:								
BO_2^-								
$B_4O_7^{-2}$								
Carbon								
CO_3^{-2}								
HCO_3^-						▬		
CN^-								
OCN^-								
SCN^-								
Silicon:								
SiO_3^{-2}								
Nitrogen:								
NO_2^-								
NO_3^-								
NH_4^+			▬	▬				
Phosphorus:								
PO_4^{-3}								
HPO_4^{-2}								▬
$H_2PO_4^-$								
Sulfur:								
SO_3^{-2}								
SO_4^{-2}								
HSO_4^-								
$S_2O_3^{-2}$								
$S_2O_5^{-2}$								
$S_2O_8^{-2}$								
Selenium:								
SeO_3^{-2}								
SeO_4^{-2}								
Chlorine:								
ClO_3^-								
ClO_4^-								
Bromine:								
BrO_3^-								

Iodine: IO_3^-									
Vanadium: VO_3^-									
Chromium: CrO_4^{-2} $Cr_2O_7^{-2}$									
Molybdenum: MoO_4^{-2}									
Tungsten: WO_4^{-2}									
Manganese: MnO_4^-									
Iron: $Fe(CN)_6^{-4}$									

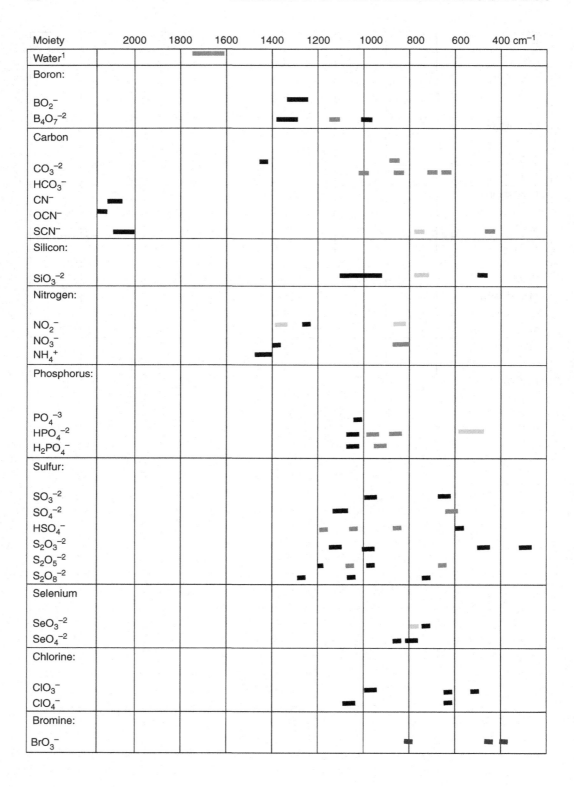

| Moiety | 2000 | 1800 | 1600 | 1400 | 1200 | 1000 | 800 | 600 | 400 cm⁻¹ |

Iodine: IO_3^-										
Vanadium: VO_3^-										
Chromium: CrO_4^{-2} $Cr_2O_7^{-2}$										
Molybdenum: MoO_4^{-2}										
Tungsten: WO_4^{-2}										
Manganese: MnO_4^-										
Iron: $Fe(CN)_6^{-4}$										

INFRARED ABSORPTIONS OF MAJOR CHEMICAL FAMILIES

The following tables provide expected IR absorptions of the major chemical families [1–23]. The ordering of these tables is: hydrocarbons, oxygen compounds, nitrogen compounds, sulfur compounds, silicon compounds, phosphorus compounds, and halogen compounds. In some ways, these data are a more detailed presentation of the spectral correlations.

Abbreviations

s = strong	1° = primary
m = medium	2° = secondary
w = weak	3° = tertiary
vs = very strong	
vw = very weak	
sym = symmetrical	
asym = asymmetrical	

REFERENCES

1. Nakanishi, K. and Solomon, P.H., *Infrared Absorption Spectroscopy*, 2nd ed., Holden-Day, Inc., San Francisco, CA, 1998.
2. Conley, R.T., *Infrared Spectroscopy*, 2nd ed., Allyn and Bacon, Boston, MA, 1972.
3. Silverstein, R.M. and Webster, F.X., *Spectrometric Identification of Organic Compounds*, John Wiley & Sons, New York, 1997.
4. Williams, D.H. and Fleming, I., *Spectroscopic Methods in Organic Chemistry*, McGraw-Hill, London, 1973.
5. Lambert, J.B., Shuzvell, H.F., Verbit, L., Cooks, R.G., and Stout, G.H., *Organic Structural Analysis*, MacMillan Pub. Co., New York, 1976.
6. Meyers, C.Y., *Eighth Annual Report on Research, 1963, Sponsored by the Petroleum Research Fund*, American Chemical Society, Washington, DC, 1964.
7. Kucsman, A., Ruff, F., and Kapovits, I., Bond system of N-acylsulfilimines. I. Infrared spectroscopic investigation of N-sulfonylsulfilimines, *Tetrahedron*, 25, 1575, 1966.
8. Kucsman, A., Kapovits, I., and Ruff, F., Infrared absorption of N-acylsulfilimines, *Acta Chim. Acad. Sci. Hung.*, 40, 75, 1964.
9. Kucsman, A., Ruff, F., and Kapovits, I., Bond system of N-acylsulfilimines. V. IR spectroscopic study on N-(p-nitrophenylsulfonyl) sulfilimines, *Acta Chim. Acad. Sci. Hung.*, 54, 153, 1967.
10. Shah, J.J., Iminosulfuranes (sulfilimines): Infrared and ultraviolet spectroscopic studies, *Can. J. Chem.*, 53, 2381, 1975.
11. Tsujihara, K., Furukawa, N., and Oae, S., Sulfilimine. II. IR, UV and NMR spectroscopic studies, *Bull. Chem. Soc. Japan*, 43, 2153, 1970.
12. Fuson, N., Josien, M.L., and Shelton, E.M., An infrared spectroscopic study of the carbonyl stretching frequency in some groups of ketones and quinones, *J. Am. Chem. Soc.*, 76, 2526, 1954.
13. Davis, F.A., Friedman, A.J., and Kluger, E.W., Chemistry of the sulfur-nitrogen bond. VIII. N-alkylidenesulfinamides, *J. Am. Chem. Soc.*, 96, 5000, 1974.
14. Krueger, P.J. and Fulea, A.O., Rotation about the C–N bond in thioamides: Influence of substituents on the potential function, *Terahedron*, 31, 1813, 1975.
15. Baumgarten, H.E. and Petersen, J.M., Reactions of amines. V. Synthesis of α-aminoketones, *J. Am. Chem. Soc.*, 82, 459, 1960.
16. Gaset, A., Lafaille, L., Verdier, A., and Lattes, A., Infrared spectra of α-aminoketones; configurational study and evidence of an enol form, *Bull. Soc. Chim. Fr.*, 10, 4108, 1968.

17. Cagniant, D., Faller, P., and Cagniant, P., Contribution in the study of condensed sulfur heterocycles. XVII. Ultraviolet and infrared spectra of some alkyl derivatives of thianaphthene, *Bull. Soc. Chim. Fr.*, 2410, 1961.

18. Tamres, M. and Searles, S., Jr., Hydrogen bonding abilities of cyclic sulfoxides and cyclic ketones, *J. Am. Chem. Soc.*, 81, 2100, 1959.

19. George, W.O., Goodman, R.C.W., and Green, J.H.S., The infra-red spectra of alkyl mercapturic acids, their sulphoxides and sulphones, *Spectrochim. Acta.*, 22, 1741, 1966.

20. Cairns, T., Eglinton, G., and Gibson, D.T., Infra-red studies with sulphoxides—part I. The S=O stretching absorptions of some simple sulphoxides, *Spectrochim. Acta*, 20, 31, 1964.

21. Hadzi, D., Hydrogen bonding in some adducts of oxygen bases with acids. Part I. Infrared spectra and structure of crystalline adducts of some phosphine, arsine, and amine oxides, and sulphoxides with strong acids, *J. Chem. Soc.*, 1962, 5128.

22. Currier, W.F. and Weber, J.H., Complexes of sulfoxides. I. Octahedral complexes of manganese (II), iron (II), cobalt (II), nickel (II), and zinc (II), *Inorg. Chem.*, 6, 1539, 1967.

23. Kucsman, A., Ruff, F., and Tanacs, B., IR spectroscopic study of N^{15} labeled acylsulfilimines, *Int. J. Sulfur Chem.*, 8, 505, 1976.

Hydrocarbon Compounds

Family	General Formula	C–H Stretch	C–H Bend	C–C Stretch	C–C Bend
Alkanes, (a) Acyclic	C_nH_{2n+2}				
(i) Straight chain	$CH_3(CH_2)_nCH_3$	3,000–2,840 (s/m): 3,000–2,960 (s) $CH_3–$ (asym) $CH_3–$ (sym): 2,880–2,870 (s) >CH_2 (asym): 2,930–2,920 (s) >CH_2 (sym): 2,860–2,840 (s)	Below 1,500 (w/m/s) $CH_3–$ (asym): 1,460–1,440 (s) $CH_3–$ (sym): 1,380–1,370 (s) >CH_2 (scissoring): ~1,465(s) >CH_2 (rocking): ~720 (s) >CH_2 (twisting and wagging) 1,350–1,150(w)	1,200–800 (w) (Not of practical value for definitive assignment)	Below 500 (Not of practical value for definitive assignment)
(ii) Branched	$R^1–CHR^3$ $\|$ R^2	C–H (3°): ~2,890 (vw)	Gem dimethyl [$(CH_3)_2CH–$]: 1,380, 1,370 (m, symmetrical doublet) Tert-butyl [$(CH_3)_3C–$]: 1,390, 1,370 (m, asymmetrical doublet; latter more intense) $CH_3–$ rocking: 930–920 (w, not reliable)		
(b) Cyclic	$(CH_2)n$	Same as in acyclic alkanes; ring strain increases the wavenumbers up to 3,100 cm^{-1}	>CH_2 (scissoring): lower than in acyclic alkanes (10–15 cm^{-1})		

Hydrocarbon Compounds

Family	General Formula	Wavenumbers (cm^{-1})				Notes
		>C=C< Stretch	>C=C–H Stretch	>C=C–H Bend (in Plane)	>C=C–H Bend (Out of Plane)	
Alkenes (olefins), (I): acyclic	C_nH_{2n}	1,670–1,600	Above 3,000			
(i) Non-conjugated		1,667–1,640 (m)				
(a) Monosubstituted (vinyl)	$R^1CH=CH_2$	1,658–1,648 (m)	3,082–3,000 (m)	1,420–1,415 (m) (scissoring)	~995 (m) ~919 (m)	C–H rocking not dependable for definitive assignment
(b) Disubstituted Cis-	$R^1\!\!>\!\!C\!\!=\!\!C\!\!<\!\!R^2$, H H	1,662–1,652 (m)	3,030–3,015 (m)	~1,406 (m)	715–675 (s) (rocking)	
Trans-	$R^1\!\!>\!\!C\!\!=\!\!C\!\!<\!\!H$, H R^2	1,678–1,668 (w)	3,030–3,020 (m)	1,325–1,275 (m) (deformation)	~965 (s) rocking	
Vinylidene	$R^1\!\!>\!\!C\!\!=\!\!C\!\!<\!\!H$, R^2 H	1,658–1,648 (m)	3,090–3,080 (m) ~2,980 (m)	~1,415 (m)	~890 (s) (rocking)	
(c) Trisubstituted	$R^1\!\!>\!\!C\!\!=\!\!C\!\!<\!\!H$, R^2 R^3	1,675–1,665 (w)	3,090–3,080 (w)	~1,415 (w)	840–800 (m) (deformation)	
(d) Tetrasubstituted	$R^1\!\!>\!\!C\!\!=\!\!C\!\!<\!\!R^3$, R^2 R^4	~1,670 (vw)	—	—	—	>C=C< stretch may not be detected
(ii) Conjugated	>C=C–C=C<	1,610–1,600 (m) (frequently a doublet)	3,050 (vw)		~980 (rocking)	Conjugation of an olefinic >C=C< with an aromatic ring raises the frequency by 20–25 cm^{-1}
(iii) Cumulated	>C=C=C<	2,000–1,900 (m)	3,300 (m)	2,000–1,900 (s) 1,800–1,700 (w)	880–850 (s)	
(II) Cyclic	–C≡C–	1,640–1,560 (variable)			697–625 (w) (wagging)	>C=C< stretch is coupled with C–C stretch of adjacent bonds. Alkyl substitution increases the >C=C< absorption frequency
(III) External exocyclic	$C=CH_2$	1,781–1,650	3,080, 2,995 (m)	~1,300 (w)		>C=C< frequency increases with decreasing ring size

Hydrocarbon Compounds

Family	General Formula	$-C\equiv C-$ Stretch	$-C\equiv C-H$ Stretch	C–H Bend	Notes
			Wavenumbers (cm^{-1})		
Alkynes	C_nH_{2n-2}				
(i) Non-conjugated					
(a) Terminal	$R^1-C\equiv C-H$	2,150–2,100 (m)	3,310–3,200 (m) (sharp)	700–610 (s) 1,370–1,220 (w) (overtone)	$-C\equiv C-H$ stretch peak is narrower than that of –OH or –NH stretch which is broader due to hydrogen bonding
(b) Non-terminal	$R^1-C\equiv C-R^2$	2,260–2,190 (vw)	—	700–610 (s) 1,370–1,220 (w) (overtone)	
(ii) Conjugated					
(a) Terminal	$R^1-C\equiv C-C\equiv C-H$	2,200, 2,040 (doublet)	3,310–3,200 (m) (sharp)	700–610 (s) 1,370–1,220 (w) (overtone)	
(b) Non-terminal	$R^1-C\equiv C-C\equiv C-R^2$	2,200, 2,040 (doublet)	—	700–610 (s) 1,370–1,220 (w) (overtone)	

Hydrocarbon Compounds

Family	General Formula	Wavenumbers (cm⁻¹)			Notes
		H>C=C< Stretch	>C=C< Stretch	>C–H Bend (Out-of-plane)	
Aromatic compounds					All show weak combination and overtone bands between 2,000–16,500 cm⁻¹. See aromatic substitution pattern chart.
(a) Monosubstituted		3,100–3,000	1,600-1,500	770–730 (s) 710–690 (s)	
(b) Disubstituted					
(i) 1,2-		3,100–3,000	1,600–1,500	770–735 (s)	
(ii) 1,3-		3,100–3,000	1,600–1,500	810–750 (s) 710–690 (s)	
(iii) 1,4-		3,100–3,000	1,600–1,500	833–810 (s)	
(c) Trisubstituted				780–760 (s)	
(i) 1,2,3-		3,100–3,000	1,600–1,500	745–705 (m)	
(ii) 1,2,4-		3,100–3,000	1,600–1,500	885–860 (m) 825–805 (s)	
(iii) 1,3,5-		3,100–3,000	1,600–1,500	865–810 (s) 730–765 (m)	
(d) Tetrasubstituted					
(i) 1,2,3,4-		3,100–3,000	1,600–1,500	810–800	
(ii) 1,2,3,5-		3,100-3,000	1,600–1,500	850–840	
(iii) 1,2,4,5-		3,100–3,000	1,600-1,500	870–855	
(e) Pentasubstituted		3,100-3,000	1,600–1,500	~870	
(f) Hexasubstituted		3,100–3,000	1,600–1,500	Below 500	

$H>C=C<$ Stretch is expressed with the inline notation where applicable.

Organic Oxygen Compounds

Family	General Formula	Wavenumbers (cm^{-1})			Notes
		O–H Stretch	>C–O Stretch	–O–H Bend	
Acetals	OR^2 | R^1–C–H | OR^3		1,195–1,060 (s) (three bands) 1,055–1,040 (s) (sometimes obscured)		
Acyl halides	R–C(=O)X X = halogen				See Organic Halogen Compouds
Alcohols	R–OH	3,650–3,584 (s, sharp) for very dilute solutions or vapor-phase spectra. 3,550–3,200 (s, broad) for less dilute solutions where intermolecular hydrogen bonding is likely to occur. Intramolecular hydrogen bonding is responsible for a broad, shallow peak in the range of 3,100–3,050 cm^{-1}.		1,420–1,300 (s)	α-Unsaturation decrease >C–O stretch by 30 cm^{-1}. Liquid spectra of alcohols show a broad out-of-plane bending band (769–650, s).
(i) Primary	R–CH$_2$OH		~1,050	~1,420 (m) and ~1,330 (m) (coupling of O–H in-plane bending and C–H wagging)	
(ii) Secondary	R^1–CHOH | R^2		~1,100	~1,420 (m) and ~1,330 (m) (coupling of O–H in-plane bending and C–H wagging)	
(iii) Tertiary	R^2 | R^1–C–OH | R^3		~1,150	Only one band (1,420–1,330 cm^{-1}), position depending on the degree of hydrogen bonding	

(Continued)

Family	General Formula	>C=O Stretch	–C(=O)H Stretch	Notes
Aldehydes	R–CHO		~2,820 (m), ~2,720 (m) Fermi resonance between C–H stretch and first overtone of the aldehydic C–H bending	
(i) Saturated, aliphatic	R=alkyl	1,720–1,720 (s)		
(ii) Aryl	R=aryl	1,705–1,695 (s)	~2,900 (m), ~2,750 (m) (aromatic)	
(iii) α,β Unsaturated	>C=C–CHO	1,700–1,680 (s)		
(iv) α, β, γ, σ-Unsaturated	>C=C–C=C–CHO	1,680–1,660 (s)		
(v) β-Keto-aldehyde	–C(=O)C–CHO	1,670–1,645 (s) (lowering is possible due to intramolecular hydrogen bonding in enol form)		
(vi) α-Halo-	>C–CHO | X X=halogen	~1,740 (s)		

Wavenumbers (cm^{-1})

Family	General Formula	>C=O Stretch	>C–O Stretch	Notes
Amides				See Organic Nitrogen Compounds
Anhydrides				
(i) Saturated acyclic	>C(=O)O(=O)C<	~1,820 (s) (asym) ~1,760 (m/s) (sym)	1,300–1,050 (s) (one or two bands)	
(ii) Conjugated acyclic	(or Ar–CO)$_2$O	1,795–1,775 (s) (asym) 1,735–1,715 (m/s) (sym)	1,300–1,050 (s) (one or two bands)	
(iii) Cyclic		Ring strain raises band to higher frequencies (up to 1,850 and 1,790 cm^{-1}). Conjugation does not reduce the frequency considerably	1,300–1,175 (s) 950–910 (s)	

Wavenumbers (cm^{-1})

Family	General Formula	>C=O Stretch	>C–O Stretch	–C–H Stretch	–C–H Bend	Notes
			Carboxylic Acids			
(i) Monomer, saturated	R–COOH	~1,760 (s)	~1,420	3,550 (s)	~1,250 (m/s)	700–610 (s) 1,370–1,220 (w) (overtone)
(ii) Monomer, aromatic	Ar–COOH	1,730–1,710	~1,400	3,500 (s)	~1,250 (m/s)	
(iii) Dimer, saturated	R=alkyl	1,720–1,706 (s)	1,315–1,280 (m) (sometimes doublet)	3,300–2,500 (s, broad)	900–860 (m, broad) (out-of-plane)	
(iv) Dimer, α,β-unsaturated (or aromatic)	R=alkenyl	1,700–1,680 (s)	1,315–1,280 (m) (sometimes doublet)	3,300–2,500 (s, broad)	900–860 (m, broad) (out of plane)	

Wavenumbers (cm^{-1})

Family	General Formula	>C=O Stretch	>C–O Stretch	Notes
Carboxylic acids (cont.)				
(v) Salt	R–COO–	1,610–1,550 (s) asym ~1,400 (s) sym.		
Cyanates	R–C≡N→O			See Organic Nitrogen Compounds
Epoxides				~1,250 (s) (ring breathing, sym) 950–810 (s) (asym) 840–810 (s) (C–H bend) 3,050–2,990 (m/s) (C–H stretch)
Esters	R^1–COOR^2			
a) Saturated, aliphatic	R^1,R^2=alkyl	1,750–1,735 (s) α halogen substitution results in an increase in wavenumbers (up to 30 cm−1)	1,210–1,163 (s) [acetates only: 1,240 (s)]	(O–C–C) 1,046–1,031 (s) (1° alcohol) ~1,100 (s) (2° alcohol)
b) Formates	R^1=H, R^2=alkyl	1,730–1,715 (s)	~1,180 (s), ~1,160 (s)	
c) α,β-Unsaturated	>C=C–COOR^2 R^2=alkyl	1,730–1,715 (s)	1,300–1,250 (s) 1,200–1,050 (s)	
d) Benzoate	C$_6$H$_5$–COOR^2 R^2=alkyl	1,730–1,715 (s)	1,310–1,250 (s) 1,180–1,100 (s)	
e) Vinyl	R^1–COOCH=CH$_2$ R^1=alkyl	1,775–1,755 (s)	1,300–1,250 (s) ~1,210 (vs)	
f) Phenyl	R^1–COOC$_6$H$_5$ R^1=alkyl	~1,770 (s)	1,300–1,200 (s) 1,190–1,140 (s)	
g) α-Ketoesters	–C(=O)COOR^2 R^2=alkyl	1,775–1,740 (s)	1,300–1,050 (s) (two peaks)	
h) β-Ketoesters	–C(=O)–C–C(=O)R^2 R^2=alkyl	~1,735 (s) ~1,650 (s) (due to enolization) –C=C–C–OR^2 ‖ O–H....O	1,300–1,050 (s) (two peaks)	
i) Aryl benzoates	R^1–COOR^2 R^1,R^2=aryl	~1,735 (s)	1,300–1,050 (s) (two peaks)	

Wavenumbers (cm⁻¹)

Family	General Formula	$>$C–O–C$<$ Stretch Asymmetrical	$>$C–O–C$<$ Stretch Symmetrical	Notes
Esters	R¹–O–R²	1,150–1,085(s)	Very hard to trace	
a) Aliphatic	R¹,R²=alkyl	Branching off on the carbons adjacent to oxygen creates splitting		
b) Aryl alkyl	R¹=alkyl R²=aryl	1,275–1,200 (s) (high due to resonance)	1,075–1,020 (s)	
c) Vinyl	R¹=vinyl R²=aryl	1,225–1,200 (s) (high due to resonance)	1,075–1,020 (s)	1,660–1,610 (m) ($>$C=CC) ~1,000 (m), 909 (m) ($>$C=C–H) (wagging)
Imides	(R–C=O)₂NH			See Organic Nitrogen Compounds
Isocyanates	R–N=C=O			See Organic Nitrogen Compounds

Ketals	OR³
	  |
	R¹–C–R²
	  |
	OR⁴
		1,190–1,160 (s)
		1,195–1,125 (s)
		1,098–1,063 (s)
		1,055–1,035 (s)		

Ketenes	$>$C=C=O			~2,150 (s) ($>$C=C=O)

Wavenumbers (cm^{-1})

Family	General Formula	>C=O Stretch	>C=C< Stretch	Notes
Ketones	R^{1}C(=O)R^2			>C=O Overtone ~3,400 (w). Solid samples or solutions decrease >C=O stretch (10–20 cm^{-1}). α-Halogenation increases >C=O stretch (0–25 cm^{-1}) >C–H stretch is very weak (3,100–2,900 cm^{-1})
a) Aliphatic, saturated	R^1, R^2=alkyl	1,720–1,710 (s)		
b) α,β-Unsaturated	>C=C–C(=O)R^2 R^2=alkyl	~1,690 (s) (s-cis) ~1,675 (s) (s-trans)	1,650–1,600 (m)	
c) α,β-α^1,β^1-Unsaturated	(>C=C–)$_2$C=O	~1,665 (s)	~1,640 (m)	
d) α,β, γ,δ-Unsaturated	>C=C–C=C–C(=O)R^2 R^2=alkyl	~1,665 (s)	~1,640 (m)	
e) Aryl	R^1=aryl R^2=alkyl	~1,690 (s)	~1,600, 1,500 (m/s) aromatic	
f) Diaryl	R^1, R^2=aryl	~1,665 (s)	~1,600, 1,500 (m/s) (aromatic)	

Wavenumbers (cm^{-1})

Family	General Formula	>C=O Stretch	>C=C< Stretch	Notes
Ketones (cont.)				
g) Cyclic		3-membered: 1,850 (s) 4-membered: 1,780 (s) 5-membered: 1,745 (s) Larger than 6-membered: 1,705 (s)		
h) α-Keto (s-trans)	R^1–C(=O)COR^2	~1,720 (s) (aliphatic) ~1,680 (s) (aromatic)		
i) β-Keto	$R^1COCH_2COR^2$	~1,720 (s) (two bands)	1,640–1,580 (m, broad) due to enol from R^1–C=CH–C–R^2 ... O–H·····O	Shows a shallow broad –OH band (enol form) at 3,000–2,700 cm^{-1}
j) α-Amino ketone hydrochlorides	$R^1COCH_2NH_3^+\ Cl^-$			>C=O decreases 10–15 cm^{-1} with electron deactivating p-substituents
k) α-Amino ketones	R–$COCH_2NR_2$			Strong bands at 3,700–3,600 cm^{-1} (–OH) and 1,700–1,600 cm^{-1} (>C=O) due to the presence of enolic forms

Wavenumbers (cm⁻¹)

Family	General Formula	>C=O Stretch	>C=C< Stretch	>C–O Stretch	Notes
Lactams (cyclic amides)					See Organic Nitrogen Compounds
Lactones (cyclic esters)					
(i) Saturated					
a) α-	x = 4	~1,735 (s)		1,300–1,050 (s, two peaks)	
b) γ-	x = 3	~1,770 (s)		1,300–1,050 (s, two peaks)	
c) β-	x = 2	~1,840 (s)		1,300–1,050 (s, two peaks)	
(ii) Unsaturated, α- to the carbonyl (>C=O)	x = 4	~1,720 (s)		1,300–1,050 (s, two peaks)	
	x = 3	~1,750 (s) (doublet 1,785–1,755 cm⁻¹ when α-hydrogen present)		1,300–1,050 (s, two peaks)	
(iii) Unsaturated, α- to the oxygen	x = 4	~1,760 (s)	~1,685 (s)	1,300–1,050 (s, two peaks)	
	x = 3	~1,790 (s)	~1,660 (s)	1,300–1,050 (s, two peaks)	
(iv) Unsaturated, α- to the carbonyl and α- to the oxygen	x = 4	1,775–1,715 (s, doublet)	1,650–1,620 (s)	1,300–1,050 (s, two peaks)	
	z(α-pyrone; coumarin)		1,570–1,540 (s)	1,300–1,050 (s, two peaks)	

Wavenumbers (cm^{-1})

Family	General Formula	>C=O Stretch	>C=C< Stretch	>C—O Stretch	Notes
Nitramines	$\begin{array}{c}R^1\\ {>}N{-}NO_2\\ R_2\end{array}$				See Organic Nitrogen Compounds
Nitrates	$R{-}NO_3$				See Organic Nitrogen Compounds
Nitro compounds	$R{-}NO_2$				See Organic Nitrogen Compounds
Nitrosamines	$\begin{array}{c}R_1\\ {>}N{-}N{=}O\\ R_2\end{array}$				See Organic Nitrogen Compounds
Nitroso compounds	$\begin{array}{c}R^1{-}N{-}N{=}O\\ \mid\\ R^2\end{array}$				See Organic Nitrogen Compounds

Wavenumbers (cm⁻¹)

Family	General Formula	>C–O Stretch	–O–H Stretch	>C=O Stretch	–O–H Bend	Notes
Peroxides	R^1–O–O–R^2	–C–C–O–		–C(=O)O		
(i): Aliphatic	R^1,R^2=alkyl	890–820 (vw)				
(ii): Aromatic	R^1,R^2=aromatic	~1,000 (vw)				
(iii): Acyl, aliphatic	R^1,R^2=acyl (aliphatic)	890–820 (vw)		1,820–1,810 (s) 1,800–1,780 (s)		
(iv): Acyl, aromatic	R^1,R^2=acyl (aromatic)	~1,000 (vw)		1,805–1,780 (s) 1,785–1,755 (s)		
Peroxyacids	R^1–C(=O)OOH	~1,260 (s)	3,300–3,250 (s, not as broad as in R–COOH)	1,745–1,735 (s) (doublet)	~1,400 (m)	~850 cm⁻¹ (m, –O–O– stretch)
Peroxyacids, anhydride	(R^1–COO)$_2$	(–COO–OOC–)				
(i) Alkyl	R^1=alkyl	1,815 (s), 1,790 (s)				
(ii) Aryl	R^1=aryl	1,790 (s), 1,770 (s)				

Wavenumbers (cm^{-1})

Family	General Formula	>C–O Stretch	–O–H Stretch	–O–H Bend	Notes
Phenols	Ar–OH	~1,230 (m)	~3,610 (m, sharp) (in $CHCl_3$ or CCl_4 solution)	1,410–1,310 (m, broad) (in plane)	
	Ar=aryl		~3,100 (m, broad) (in neat samples)	~650 (m) (out of plane)	
Phosphates	$(R^1O)_3P{=}O$				See Organic Phosphorus Compounds
Phosphinates	$(R^1O)P(=O)H_2$				See Organic Phosphorus Compounds
Phosphine oxides	$R_3P{=}O$				See Organic Phosphorus Compounds
Phosphonates	$(R^1O)_2P(=O)H$				See Organic Phosphorus Compounds
Phosphorus acids	$R_2P(=O)OH$				See Organic Phosphorus Compounds
Pyrophosphates	$(R–P{=}O)_2O$				See Organic Phosphorus Compounds

Wavenumbers (cm⁻¹)

Family	General Formula	>C=O Stretch	>C=C< Stretch	Notes
Quinones				
a) 1,2-		~1,675 (s)	~1,600 (s)	
b) 1,4-		~1,675 (s)	~1,600 (s)	
Silicon compounds				See Organic Silicon Compounds
Sulfates				See Organic Sulfur Compounds
Sulfonamides				See Organic Sulfur Compounds
Sulfonates				See Organic Sulfur Compounds
Sulfones				See Organic Sulfur Compounds
Sulfonyl chlorides				See Organic Sulfur Compounds
Sulfoxides				See Organic Sulfur Compounds

Organic Nitrogen Compounds

Family	General Formula	Wavenumbers (cm^{-1})			Notes
		C–N	N–H	Others	
Amides					
Primary	$R^1\text{-CONH}_2$	1,400 (s) (stretch)	3,520 (m) (stretch) 3,400 (m) (stretch) 1,655–1,620 (m) (bend) 860–666 (m, broad) (wagging)	>C=O (1,650) (s, solid state) (1,690) (s, solution)	Lowering of N–H stretch occurs in solid samples due to hydrogen bonding; higher values arise in dilute samples
Secondary	$R^1\text{-CONHR}^2$	1,400 (s) (stretch)	3,500–3,400 (w) (stretch) 1,570–1,515 (w) (bend) 860–666 (m, broad) (wagging)	>C=O (1,700–1,670) (s, solution); (1,680–1,630) (s, solid state) Band due to interaction of N–H (bend) and (C–N) (stretch) (~1,250) (m, broad)	Lowering of N–H stretch occurs in solid samples due to hydrogen bonding; higher values arise in dilute samples
Tertiary	$R^1\text{-CONR}^2R^3$	1,400 (s) (stretch)	—	>C=O (1,680–1,630) (s); higher values are obtained with electron attracting groups attached to the nitrogen	
Amines					
Primary	$R^1\text{-NH}_2$	1,250–1,020 (m) (for non-conjugated amines) 1,342–1,266 (s) (for aromatic amines)	3,500 (w) (stretch) 3,400 (w) (stretch) 1,650–1,580 (m) (scissoring) 909–666 (m) (wagging)		
Secondary	$R^1\text{-NHR}^2$	1,250–1,020 (m) (for non-conjugated amines) 1,342–1,266 (s) (for aromatic amines)	3,350–3,310 (w) (stretch) 1,515 (vw) (scissoring) 909–666 (m) (wagging)		

(Continued)

Organic Nitrogen Compounds (Continued)

Family	General Formula	Wavenumbers (cm⁻¹)			
		C–N	N–H	Others	Notes
Tertiary	$R^1-NR^2R^3$	1,250–1,020 (m) (for non-conjugated amines) 1,342–1,266 (s) (for aromatic amines)	—		
Amine salts					
Primary	$RNH_3^+X^-$		3,000–2,800 (s) 2,800–2,200 (m) (series of peaks) 1,600–1,575 (m) 1,550–1,504 (m)		
Secondary	$R_2NH_2^+X^-$		3,000–2,700 (s) 2,700–2,250 (m) (series of peaks) 2,000 (w) 1,620–1,560 (m)		
Tertiary	$R_3NH^+X^-$		2,700–2,250 (s)		
Quaternary	$R_4N^+X^-$		—		

Wavenumbers (cm^{-1})

Family	General Formula	C–N	N–H	Others	Notes
Amino acids (alpha)	R^1–CH–COO$^-$ | NH$_2$		3,100–2,600 (s, broad) 2,222–2,000 (s, broad, overtone) 1,610(w) (bend) 1,550–1,485 (s) (bend)	–COO$^-$ (1,600–1,590) (s) –COOH (1,755–1,730) (s)	
	R^1–CH–COO$^-$ | +NH$_3$				
	R^1–CH–COOH | +NH$_3$				
Ammonium ion	NH$_4^+$		3,300–3,040 (s) 2,000–1,709 (m) 1,429 (s)		
Azides	R–N$_3$			2,140 (s) (asym stretch, N$_3$) 1,295 (s) (sym stretch, N$_3$)	
Azocompounds	R^1–N=N–R^2 (trans)	Forbidden in IR but allowed in Raman spectrum (1,576) (w); peak is lowered down to 1,429 cm^{-1} in unsymmetrical p-electron donating substituted azobenzenes			
Azoxy compounds	R–N=N→O				1,310–1,250 (s)

Wavenumbers (cm^{-1})

Family	General Formula	C–N Multiple Bond	Cumulated (–X=C=Y) Double Bond	Notes
Cyanocompounds (nitriles)	R–C≡N	2,260–2,240 (w) (aliphatic) 2,240–2,220 (m) (aromatic, conjugated)		Electronegative elements α- to the C≡N group reduce the intensity of the absorption
Diazonium salts	$R-N≡N^+$			2280–2240 (m) (–N≡N$^+$)
Imides	R–C–NH–C–R (=O, =O)			1,710, 1,700 (>C=O six-membered ring) 1,770, 1,700 (>C=O five-membered ring)
Isocyanates	R–N=C=O		2,273–2,000 (s) (broad) (asym) 1,400–1,350 (w) (sym)	
Isocyanides (isonitriles)	R–N≡C	2,400–2,300 (w) (aliphatic) 2,300–2,200 (w) (aromatic)		
Isonitriles				See isocyanides
Isothiocyanates	R–N=C=S		2,140–2,000 (s) (stretch)	
Ketene	R_1R_2>C=C=O		2,150 (stretch); 1,120	
Ketenimine	R_1R_2>C=C=N–		2,000 (stretch)	

Wavenumbers (cm^{-1})

Family	General Formula	>C–N	>N–O (Asymmetric)	>N–O (Symmetric)	Others	Notes
Lactams					>C=O (s) (stretch) 1,670 (six-membered ring) 1,700 (five-membered ring) 1,745 (four-membered ring) N–H (out-of-plane wagging) (800–700) (broad)	Add ~15 cm^{-1} to every wavenumber in case of a >C=C< in conjugation; amide group is forced into the cis-conformation in rings of medium size.
Nitramines	R^1–N–NO$_2$ $\mid$ R^2		1,620–1,580 (s) (asym) 1,320–1,290 (s) (sym)			
Nitrates	RO–NO$_2$				–N=O 1,660–1,625 (s) (asym) 1,300–1,225 (s) (asym) >N–O 870–833 (s) (stretch) 763–690 (s) (bend)	
Nitriles (cyanocompounds)	R–C≡N					See cyanocompounds
Nitrites	RO–N=O				–N=O stretch 1,680–1,650 (vs) (trans) 1,625–1,610 (vs) (cis) >N–O stretch 850–750 (vs)	

Wavenumbers (cm^{-1})

Family	General Formula	>C–N	>N–O (Asymmetric)	>N–O (Symmetric)	Others	Notes
Nitro compounds	$R-NO_2$		1,615–1,540 (vs) (asym)	1,390–1,320 (vs)	~610 (m) (CNO bend)	Aromatics absorb at lower frequencies than aliphatic
Aliphatic	R–alkyl	870	1,390–1,320 (vs) (sym)			
Aromatic	R–aryl	(Difficult to assign)	1,548–1,508 (s) (asym) 1,356–1,340 (s) (sym)	1,356–1,340 (s)		
Nitrosamines	$R_1R_2N-N=O$				>N–O stretch (1,520–1,500) (s) (vapor) (1,500–1,480) (s) (neat) N–N (1,150–925) (m)	
Nitroso-compounds	R–N=O				N=O stretch 1,585–1,539 (s) (3°, aliphatic) 1,511–1,495 (s) (3°, aromatic)	1°C and 2°C —nitroso-compounds are unstable and rearrange or dimerize.
Pyridines	C_5H_5N				N–H (3,075, 3,030) (s) C–H (out of plane) (920–720) (s) (2,000–1,650) (overtone) C=C ring stretch (1,600, 1,570, 1,500, 1,435)	Characteristic substitution pattern: α-substitution: (795–780), (755–745) β-substitution: (920–880), (840–770), 720.
Sulfilimines	$R^1R^2S=N-R^3$					See Organic Sulfur Compounds.
Sulfonamides	$R-SO_2NH_2$					See Organic Sulfur Compounds.
Thiocyanates	$R-SC{\equiv}N$					See Organic Sulfur Compounds.

Organic Sulfur Compounds

Family	General Formula	Wavenumbers (cm^{-1})				Notes
		>S=O (Asymmetric)	>S=O (Symmetric)	>S=N–	Others	
Disulfides	R^1-S-S-R^2				-S-S- (<500) (w)	
Mercaptans	R-S-H				-S-H (2,600–2,500) (w)	Only significant frequency around that region; lowering of 50–150 cm^{-1} due to hydrogen bonding.
Mercapturic acids	R^2(O=)CNH RSCH$_2$CH HOOC	1,295–1,280 (s) (for sulfones)	1,135–1,100 (s) (for sulfones)		1,025, 970 (>S→O) (for sulfoxides)	Reduction of all >S=O frequencies due to H– bonding with –NH
Sulfates	(RO)$_2$S(=O)$_2$	1,415–1,380 (s)	1,200–1,185 (s)			
Sulfides	R$_1$-S-R^2				R-S- (700–600) (w)	
Sulfilimines	R$_2$S=N-R^1					
(i) N-acyl	R$_2$S=N-COR^1			800 (s)	>C=O (1,625–1,600) (s)	
(ii) N-alkyl	R$_2$S=N-R^1			987–935 (s)		
(iii) N-sulfonyl	R$_2$S=N-SO$_2$R^1	1,280–1,200 (s) 1,095–1,030 (s)	1,160–1,135 (s)	980–901 (s)		
Sulfinamides, N-alkylidene	RS(O)N=CR$_2$				1,520 (amide II band) 1,080 (s, S→O)	
Sulfonamides	R-SO$_2$NH$_2$	1,370–1,335 (s)	1,170–1,155 (s)		>N–H (1°) (3,390–3,330) (s) (3,300–3,247) (s) >N–H (2°) (3,265) (s)	Solid-phase spectra lower wavenumbers by 10–20 cm^{-1}.
Sulfonates	R^1-SO$_2$-OR^2	1,372–1,335 (s)	1,195–1,168 (s)			Electron donating groups on the aryl group cause higher frequency absorption.

(Continued)

Organic Sulfur Compounds *(Continued)*

Family	General Formula	>S=O (Asymmetric)	>S=O (Symmetric)	>S=N–	Others	Notes
		Wavenumbers (cm^{-1})				
Sulfones	$R^1–SO_2–R^2$	1,350–1,300 (s)	1,160–1,120 (s)			Hydrogen bonding reduces the frequency of absorption slightly.
Sulfonic acids (anhydrous)	$R–SO_3H$	1,350–1,342 (s)	1,165–1,150 (s)		–OH (3,300–2,500) (s, broad)	Hydrated sulfonic acids show broad bands at 1,230–1,150 cm^{-1}.
Sulfonic acids, salts	$R–SO_3^-$	ca. 1,175 (s)	ca. 1,055 (s)			
Sulfonyl chlorides	$R–SO_2Cl$	1,410–1,380 (s)	1,204–1,177 (s)			
Sulfoxides cyclic	$R_2S{\rightarrow}O$ $(CH_2)x\ S{\rightarrow}O$				>S→O (1,070–1,030) (s) x = 3 1,192(CCl_4) 1,073($CHCl_3$) x = 4 1,035(CCl_4) 1,020($CHCl_3$) x = 5 1,053(CCl_4) 1,031($CHCl_3$)	Hydrogen bonding reduces the frequency absorption slightly; electronegative substituents increase the >S→O frequency; inorganic complexation reduces the >S→O (up to 50 cm^{-1}).
Thiocarbonyls (not trimerized into cyclic sulfides)	$R^1–C–R^2(H)$ ‖ S				>C=S (1,250–1,020) (s)	
Thiocyanates	$R–S–C{\equiv}N$				–C≡N (2,175–2,140 (s); higher values for aryl thiocyanates	
Thiol esters	$R^1–C–SR^2$ ‖ O				>C=O (1,690) (s) (S-alkyl thioester) (1,710) (s) (S-aryl thioester)	The (+) mesometic effect of sulfur is larger than its (–) inductive effect
Thiols	R–SH					See Mercaptans
Thiophenols	Ar–SH				–S–H (2,600–2,500) (w)	

Organic Silicon Compounds

Family	General Formula	Wavenumbers (cm^{-1})					
		>Si–H Stretch	>Si–H Bend	>C–Si< Stretch	>C–H Bend	>Si–O– Stretch	–OH Stretch
Silanes	R_xSiH_y						
a) Monoalkyl	$R–SiH_3$	2,130–2,100 (s)	890–860 (s)	890–690 (s)	~1,260 (s) (rocking)		
b) Dialkyl	R_2SiH_2	~2,135 (s)	890–860 (s)	820–800 (s)	~1,260 (s) (rocking)		
c) Trialkyl	R_3SiH	2,360–2,150 (s)	890–860 (s)	~840 (s) ~755 (s)	~1,260 (s) (rocking)		
d) Tetraalkyl	R_4Si			890–690 (s)	~1,260 (s) (rocking)		
e) Alkoxy	$R_x^1Si(OR^2)_y$			890–690 (s)	~1,260 (s) (rocking)	1,090–1,080 (s) (doublet)	
Siloxanes	>Si–O–Si<					1,110–1,000 (s) (Si–O–Si)	
a) Disiloxanes						~1,053 (s)	
b) Cyclic trimer						~1,020 (s)	
c) Cyclic tetramer						~1,082 (s)	
Hydroxysilanes	$R_xSi(OH)_y$						~3,680 (s) (confirmed by band at 870–820 cm^{-1})

Organic Phosphorus Compounds

Family	General Formula	Wavenumbers (cm^{-1})				Notes
		>P=O Stretch	>P–H Stretch	>P–O–C< Stretch	–OH Stretch	
Phosphates	O=P(OR)$_3$	1,300–1,100 (s) (doublet)				->P=O stretch can shift up to 65 cm^{-1} due to solvent effect
a) Alkyl		1,285–1,260 (s) (doublet)		~1,050 (s) (alkyl)		
b) Aryl		1,315–1,290 (s) (doublet)		950–875 (s) (aryl)		
Phosphinates	H$_2$P–OR ‖ O	1,220–1,180 (s)	~2,380 (m) ~2,340 (m) (sharp)	~1,050 (s) (alkyl) 950–875 (s) (aryl)		
Phosphine oxides	(R)H–PR$_1$R$_2$ ‖ O					->P=O decreases with complexation
a) Alkyl		1,185–1,150 (s)	2,340–2,280 (m)			
b) Aryl		1,145–1,095 (s)	2,340–2,280 (m)			
Phosphates	H–P(OR)$_2$ ‖ O	1,265–1,230 (s)	2,450–2,420 (m)	~1,050 (s) (alkyl) 950–875 (s) (aryl)		
Phosphorus acids	R^{1}P(=O)OH ｜ R^2	1,240–1,180 (vs)			2,700–2,200 (s, broad) (assoc)	
Phosphorus amides	(RO)$_2$PNR^{1}R^2 ‖ O	1,275–1,200 (s)				
Pyrophosphates	R$_2$P–O–PR$_2$ ‖　‖ O　O	1,310–1,200 (s) (single band)				

Organic Halogen Compounds

Family	General Formula	Wavenumbers (cm^{-1})			
		>C–X Stretch	>CX$_2$ Stretch	–CH$_3$ Stretch	=C–X Stretch
Fluorides	X = F	1,120–1,010	1,350–1,200 (asym) 1,200–1,080 (sym)	1,350–1,200 (asym) 1,200–1,080 (sym)	1,230–1,100
Chlorides	X = Cl	830–500 1,510–1,480 (overtone)	845–795 (asym) ~620 (sym)		
Bromides	X = Br	667–290			
Iodides	X = I	500–200			

COMMON SPURIOUS INFRARED ABSORPTION BANDS

The following table provides some of the common potential sources of spurious infrared absorptions that might appear on a spectrum [1,2].

REFERENCES

1. Bruno, T.J. and Svoronos, P.D.N., *CRC Handbook of Basic Tables for Chemical Analysis*, 3rd ed., CRC Press, Boca Raton, FL, 2011.
2. Bruno, T.J. and Svoronos, P.D.N., *CRC Handbook of Fundamental Spectroscopic Correlation Charts*, CRC Press, Boca Raton, FL, 2006.

Approximate Wavenumber (in cm^{-1})	Wavelength (μm)	Compound or Group	Origin
3,700	2.70	H_2O	Water in solvent (thick layers).
3,650	2.74	H_2O	Water in some quartz windows.
3,450	2.9	H_2O	Hydrogen bonded water, usually in KBr disks.
2,900	3.44	$-CH_3$, $>CH_2$	Paraffin oil, residual from previous mulls.
2,350	4.26	CO_2	Atmospheric absorption or dissolved gas from a dry ice
2,330	4.30	CO_2	bath.
2,300 and 2,150	4.35 and 4.65	CS_2	Leaky cells, previous analysis of samples dissolved in carbon disulfide.
1,996	5.01	BO_2	Metaborate in the halide window.
1,400–2,000	5-7	H_2O	Atmospheric absorption.
1,820	5.52	$COCl_2$	Phosgene, decomposition product in purified $CHCl_3$.
1,755	5.7	Phthalic anhydride	Decomposition product of phthalate esters or resins; paint off-gas product.
1,700–1,760	5.7–5.9	$>C=O$	Bottle cap liners leached by sample.
1,720	5.8	Phthalates	Phthalate polymer plastic tubing.
1,640	6.1	H_2O	Water of crystallization entrenched in sample.
1,520	6.6	CO_2	Leaky cells, previous analysis.
1,430	7.0	CO_3^{-2}	Contaminant in halide window.
1,360	7.38	NO_3^-	Contaminant in halide window.
1,270	7.9	$>SiO-$	Silicone oil or grease.
1,000–1,110	9–10	$->Si-O-Si<-$	Glass; silicones.
980	10.2	SO_4^{-2}	From decomposition of sulfates in KBr pellets.
935	10.7	$(CH_2O)_x$	Deposit from gaseous formaldehyde.
907	11.02	$->C-Cl$	Dissolved R-12 (Freon-12).
837	11.95	NO_3^-	Contaminant in halide window.
823	12.15	KNO_3	From decomposition of nitrates in KBr pellets.
794	12.6	CCl_4 vapor	Leaky cells, from CCl_4 used as a solvent.
788	12.7	CCl_4 liquid	Incomplete drying of cell or contamination, from CCl_4 used as a solvent.
720 and 730	13.7 and 13.9	Polyethylene	Various experimental sources.
728	13.75	$->Si-F$	SiF_4, found in NaCl windows.
667	14.98	CO_3^{-2}	Atmospheric carbon dioxide.
Any	Any	Fringes	If refractive index of windows is too high, or if the cell is partially empty, or the solid sample is not fully pulverized.

DIAGNOSTIC SPECTRA

The interpretation of infrared spectra is often complicated by the presence of spurious absorbances or by instrumental upset conditions that must be recognized. In these cases, it is often helpful to refer to the spectra of common compounds that may be the cause of such difficulties. The following spectra present such diagnostic tools [1]. Carbon dioxide, as an atmospheric constituent, is often present as an unwanted contaminant. Water is also an atmospheric constituent and is also present in many chemical processes. It can also react with certain species such as amines.

REFERENCE

1. NIST Chemistry Web Book, NIST Standard Reference Database Number 69 - (www.webbook.nist. gov/chemistry/), 2018.

Infrared Spectrum of Carbon Dioxide:

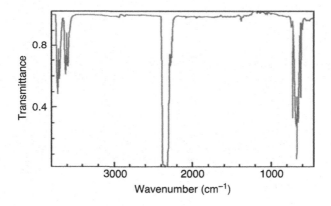

Infrared spectrum of water:

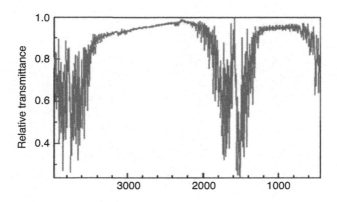

CHAPTER 10

Nuclear Magnetic Resonance Spectroscopy

PROPERTIES OF IMPORTANT NMR NUCLEI

The following table lists the magnetic properties required most often for choosing the nuclei to be used in nuclear magnetic resonance (NMR) experiments [1–14]. The reader is referred to several excellent texts and the literature for guidelines in nucleus selection.

REFERENCES

1. Silverstein, R.M., Bassler, G.C., and Morrill, T.C., *Spectrometric Identification of Organic Compounds*, 5th ed., John Wiley & Sons, New York, 1991.
2. Yoder, C.H. and Shaeffer, C.D., *Introduction to Multinuclear NMR*, Benjamin/Cummings, Menlo Park, CA, 1987.
3. Gordon, A.J. and Ford, R.A., *The Chemist's Companion*, Wiley Interscience, New York, 1971.
4. Silverstein, R.M. and Webster F.X., *Spectrometric Identification of Organic Compounds*, 6th ed., John Wiley & Sons, New York, 1998.
5. Becker, E.D., *High Resolution NMR, Theory and Chemical Applications*, 2nd ed., Academic Press, New York, 1980.
6. Gunther, H., *NMR Spectroscopy: Basic Principles, Concepts and Applications in Chemistry*, John Wiley & Sons, New York, 2003.
7. Rahman, A.-U., *Nuclear Magnetic Resonance*, Springer-Verlag, New York, 1986.
8. Harris, R.K., NMR and the periodic table, *Chem. Soc. Rev.*, 5, 1, 1976.
9. Kitamaru, R., *Nuclear Magnetic Resonance: Principles and Theory*, Elsevier Science, New York, 1990.
10. Lambert, J.B., Holland, L.N., and Mazzola, E.P., *Nuclear Magnetic Resonance Spectroscopy: Introduction to Principles, Applications and Experimental Methods*, Prentice Hall, Englewood Cliffs, NJ, 2003.
11. Bovey, F.A. and Mirau, P.A., *Nuclear Magnetic Resonance Spectroscopy*, 2nd ed., Academic Press, London, 1988.
12. Harris, R.K. and Mann, B.E., *NMR and the Periodic Table*, Academic Press, London, 1978.
13. Hore, P.J. and Hore, P.J., *Nuclear Magnetic Resonance*, Oxford University Press, Oxford, 1995.
14. Nelson, J.H., *Nuclear Magnetic Resonance Spectroscopy*, 2nd ed., John Wiley & Sons, New York, 2003.

Isotope	Natural Abundance	Spin Number I	NMR Frequency[a] at Indicated Field Strength in kG							
			10.000	14.092	21.139	23.487	51.567	93.950	140.925	223.131
$_1H^1$	99.985	1/2	42.5759	60.0000	90.0000	100.0000	220.0000	400.0000	600.0000	950.0000
$_1H^2$	0.015	1	6.53566	9.21037	13.81555	15.35061	33.77134	61.40262	92.10380	145.9830
$_1H^{3*}$	—	1/2	45.4129	63.9980	95.9971	106.6634	234.6595	426.6542	639.9813	1,013.3024
$_6C^{13}$	1.108	1/2	10.7054	15.0866	22.6298	25.1443	55.3174	100.5735	150.8659	2,388.5150
$_7N^{14}$	99.635	1	3.0756	4.3343	6.5014	7.2238	15.924	28.9104	43.3615	68.6557
$_7N^{15}$	0.365	1/2	4.3142	6.0798	9.1197	10.1330	22.2925	40.5306	60.7960	96.2601
$_8O^{17}$	0.037	5/2	5.772	8.134	12.201	13.557	29.825	54.1811	81.3186	128.5801
$_9F^{19}$	100	1/2	40.0541	42.3537	63.5305	94.0769	206.9692	376.2515	564.3781	893.5963
$_{14}Si^{29}$	4.70	1/2	8.4578	11.9191	17.8787	19.8652	43.7035	79.4638	119.1956	188.72
$_{15}P^{31}$	100	1/2	17.235	24.288	36.433	40.481	89.057	161.9828	242.9741	384.7086
$_{16}S^{33}$	0.76	3/2	3.2654	4.6018	6.9026	7.6696	16.8731	30.6826	46.0238	72.8710
$_{16}S^{35*}$	—	3/2	5.08	7.16	10.74	11.932	26.250	47.7267	71.5875	113.3508
$_{17}Cl^{35}$	75.53	3/2	4.1717	5.8790	8.8184	9.7983	21.5562	39.1948	58.7902	93.0876
$_{17}Cl^{36*}$	—	2	4.8931	6.8956	10.3434	11.4927	25.2838	45.9638	68.9432	109.1639
$_{35}Br^{76*}$	—	1	4.18	5.89	8.84	9.82	21.60	39.2768	58.9130	93.2822
$_{35}Br^{79}$	50.54	3/2	10.667	15.032	22.549	25.054	55.119	100.2133	150.3202	238.0064
$_{35}Br^{81}$	49.46	3/2	11.498	16.204	24.305	27.006	59.413	108.0258	162.0386	256.5608
$_{74}W^{183}$	14.40	1/2	1.7716	2.4966	3.7449	4.1610	9.1543	16.6430	24.9646	39.5272

Isotope	Field Value[a] (kg) at Frequency of			Relative Sensitivity		Magnetic Moment $(eh/4BM_c)$	Electric Quadrupole Moment[b] (barns)
	4 MHz	10 MHz	16 MHz	Constant H	Constant ν		
$_1H^1$	0.940	2.349	3.758	1.00	1.00	2.79278	—
$_1H^2$	6.120	15.30	24.48	9.65×10^{-3}	0.409	0.85742	0.0028
$_1H^3*$	0.881	2.202	3.523	1.21	1.07	2.9789	—
$_6C^{13}$	3.736	9.341	14.946	0.0159	0.252	0.7024	—
$_7N^{14}$	13.01	32.51	52.02	1.01×10^{-3}	0.193	0.4036	0.01
$_7N^{15}$	9.272	23.18	37.09	1.04×10^{-3}	0.101	−0.2831	—
$_8O^{17}$	6.93	17.3	27.7	0.0291	1.58	−1.8937	−0.026
$_9F^{19}$	0.999	2.497	3.994	0.834	0.941	2.6288	—
$_{14}Si^{29}$	4.729	11.82	18.92	7.84×10^{-3}	0.199	−0.55477	—
$_{15}P^{31}$	2.321	5.802	9.284	0.0665	0.405	1.1317	—
$_{16}S^{33}$	12.25	30.62	49.0	2.26×10^{-3}	0.384	0.6533	−0.055
$_{16}S^{35}*$	7.87	19.7	31.5	8.50×10^{-3}	0.597	1.00	0.04
$_{17}Cl^{35}$	9.588	23.97	38.35	4.72×10^{-3}	0.490	0.82183	−0.079
$_{17}Cl^{36}*$	0.175	20.44	32.70	0.0122	0.920	1.285	−0.017
$_{35}Br^{76}*$	9.6	24	38	2.52×10^{-3}	0.26	±0.548	±0.25
$_{35}Br^{79}$	3.750	9.375	15.00	0.0794	1.26	2.106	0.31
$_{35}Br^{81}$	3.479	8.697	13.92	0.0994	1.35	2.270	0.26
$_{74}W^{183}$	22.58	56.45	90.31	7.3×10^{-5}	0.042	0.117	—

[a] 1 kG = 10^{-10} T, the corresponding SI unit.
[b] 1 b = 10^{-23} m².
* Nucleus is radioactive.

GYROMAGNETIC RATIO OF SOME IMPORTANT NUCLEI

The following table lists the gyromagnetic ratio, γ, of some important nuclei which are probed in NMR spectroscopy [1–12]. The gyromagnetic ratio is the proportionality constant that correlates the magnetic moment (μ) and the angular momentum, ρ: $\mu = \gamma\rho$.

REFERENCES

1. Carrington, A. and McLaughlin, A., *Introduction to Magnetic Resonance*, Harper and Row, New York, 1967.
2. Levine, I.M., *Molecular Spectroscopy*, John Wiley & Sons, New York, 1975.
3. Becker, E.D., *High Resolution NMR: Theory and Chemical Applications*, Academic Press, New York, 1980.
4. Yoder, C.H. and Shaeffer, C.D., *Introduction to Multinuclear NMR*, Benjamin/Cummings, Menlo Park, CA, 1987.
5. Silverstein, R.M. and Webster F.X., *Spectrometric Identification of Organic Compounds*, 6th ed., John Wiley & Sons, New York, 1998.
6. Rahman, A.-U., *Nuclear Magnetic Resonance*, Springer-Verlag, New York, 1986.
7. Kitamaru, R., *Nuclear Magnetic Resonance: Principles and Theory*, Elsevier Science, New York, 1990.
8. Lambert, J.B., Holland, L.N., and Mazzola, E.P., *Nuclear Magnetic Resonance Spectroscopy: Introduction to Principles, Applications and Experimental Methods*, Prentice Hall, Englewood Cliffs, NJ, 2003.
9. Bovey, F.A. and Mirau, P.A., *Nuclear Magnetic Resonance Spectroscopy*, 2nd ed., Academic Press, London, 1988.
10. Hore, P.J. and Hore, P.J., *Nuclear Magnetic Resonance*, Oxford University Press, Oxford, 1995.
11. Nelson, J.H., *Nuclear Magnetic Resonance Spectroscopy*, 2nd ed., John Wiley & Sons, New York, 2003.
12. Gunther, H., *NMR Spectroscopy: Basic Principles, Concepts and Applications in Chemistry*, John Wiley & Sons, New York, 2003.

Nucleus	γ
$_1H^1$	5.5856
$_1H^2$	0.8574
$_1H^3$	5.9575
$_3Li^7$	2.1707
$_5B^{10}$	0.6002
$_5B^{11}$	1.7920
$_6C^{13}$	1.4044
$_7N^{14}$	0.4035
$_7N^{15}$	−0.5660
$_8O^{17}$	−0.7572
$_9F^{19}$	5.2545
$_{14}Si^{29}$	−1.1095
$_{11}Na^{23}$	1.4774
$_{15}P^{31}$	2.2610
$_{16}S^{33}$	0.4284
$_{17}Cl^{35}$	0.5473
$_{17}Cl^{37}$	0.4555
$_{19}K^{39}$	0.2607
$_{35}Br^{79}$	1.3993
$_{35}Br^{81}$	1.5084
$_{74}W^{183}$	0.2324

CLASSIFICATION OF IMPORTANT QUADRUPOLAR NUCLEI ACCORDING TO NATURAL ABUNDANCE AND MAGNETIC STRENGTH

The following table classifies important quadrupolar nuclei according to their natural abundance and relative magnetic strength [1]. The magnetic strength, while not a commonly recognized physical parameter, is defined as a matter of convenience for classification of nuclei in NMR. It is defined as follows:

Strong: $\gamma/10^7 > 2.5$ rad/(T s)
Medium: 10 rad/(T s) $> \gamma 10^7 > 2.5$ rad/(T s)
Weak: $\gamma 10^7 < 2.5$ rad/(T s)

where the flux density is in units of Teslas (T), and rad refers to 2B. In NMR, one can write:

$$2\pi f = \gamma B,$$

where f is the resonant frequency, γ is the gyromagnetic ratio, and B is the flux density. Thus, for the proton, $\gamma/2\pi = 43$ MHz/T, resulting in a value of $\gamma/10^7 = 4.3$ rad/(T s) and therefore medium magnetic strength.

The less favorable nuclei for a given element are listed in brackets.

REFERENCE

1. Harris, R.K., and Mass, B.E., *NMR and the Periodic Table*, Academic Press, London, 1978.

Magnetic Strength	Natural Abundance		
	High (>90 %)	Medium	Low (<10 %)
Strong	^{7}Li		
Medium	^{9}Be, ^{23}Na, ^{27}Al, ^{45}Sc, ^{51}V, ^{55}Mn, ^{59}Co, ^{75}As, ^{93}Nb, ^{115}In, ^{127}I, ^{133}Cs, ^{181}Ta, ^{209}Bi	[^{10}B], ^{11}B, ^{35}Cl, ^{63}Cu, ^{65}Cu, [^{69}Ga], ^{71}Ga, [^{79}Br], ^{81}Br, [^{85}Rb], ^{87}Rb, ^{121}Sb, [^{123}Sb], ^{137}Ba, ^{139}La, [^{185}Re], ^{187}Re	^{2}H, ^{6}Li, ^{17}O, ^{21}Ne, [^{113}In], [^{135}Ba]
Weak	^{14}N, ^{39}K	^{25}Mg, ^{37}Cl, ^{83}Kr, ^{95}Mo, ^{131}Xe, ^{189}Os, ^{201}Hg	^{33}S, [^{41}K], ^{43}Ca, ^{47}Ti, ^{49}Ti, ^{53}Cr, ^{67}Zn, ^{73}Ge, ^{87}Sr, [^{97}Mo]

CHEMICAL SHIFT RANGES OF SOME NUCLEI

The following table gives an approximate chemical shift range (in ppm) for some of the most popular nuclei. The range is established by the shifts recorded for the most common compounds [1–11].

REFERENCES

1. Yoder, C.H. and Schaeffer, C.D., Jr., *Introduction to Multinuclear NMR*, Benjamin/Cummings Publishing Co., Menlo Park, CA, 1987.
2. Silverstein, R.M., Bassler, G.C., and Morrill, T.C., *Spectrometric Identification of Organic Compounds*, 5th ed., John Wiley & Sons, New York, 1991.
3. Harris, R.U. and Mann, B.E., *NMR and the Periodic Table*, Academic Press, London, 1978.
4. Silverstein, R.M. and Webster F.X., *Spectrometric Identification of Organic Compounds*, 6th ed., John Wiley & Sons, New York, 1998.
5. Gunther, H., *NMR Spectroscopy: Basic Principles, Concepts and Applications in Chemistry*, John Wiley & Sons, New York, 2003.
6. Kitamaru, R., *Nuclear Magnetic Resonance: Principles and Theory*, Elsevier Science, New York, 1990.
7. Lambert, J.B., Holland, L.N., and Mazzola, E.P., *Nuclear Magnetic Resonance Spectroscopy: Introduction to Principles, Applications and Experimental Methods*, Prentice Hall, Englewood Cliffs, NJ, 2003.
8. Bovey, F.A. and Mirau, P.A., *Nuclear Magnetic Resonance Spectroscopy*, 2nd ed., Academic Press, London, 1988.
9. Harris, R.K. and Mann, B.E., *NMR and the Periodic Table*, Academic Press, London, 1978.
10. Hore, P.J. and Hore, P.J., *Nuclear Magnetic Resonance*, Oxford University Press, Oxford, 1995.
11. Nelson, J.H., *Nuclear Magnetic Resonance Spectroscopy*, 2nd ed., John Wiley & Sons, New York, 2003.

Nucleus	Chemical Shift Range (ppm)	Nucleus	Chemical Shift Range (ppm)
^{1}H	15	^{29}Si	400
^{7}Li	10	^{31}P	700
^{11}B	200	^{33}S	600
^{13}C	250	^{35}Cl	820
^{15}N	930	^{39}K	60
^{17}O	700	^{59}Co	14,000
^{19}F	800	^{119}Sn	2,000
^{23}Na	15	^{133}Cs	150
^{27}Al	270	^{207}Pb	10,000

REFERENCE STANDARDS FOR SELECTED NUCLEI

The following table lists the most popular reference standards used when NMR spectra of various nuclei are measured. The standards should be inert, should be soluble in a variety of solvents, and, preferably, should produce one singlet peak that appears close to the lowest frequency end of the chemical shift range. When NMR data are provided, it is always necessary to specify the reference standard employed [1–6].

REFERENCES

1. Yoder, C.H. and Schaeffer, C.D., Jr., *Introduction to Multinuclear NMR*, Benjamin/Cummings Publishing Co., Menlo Park, CA, 1987.
2. Duthaler, R.O. and Roberts, J.D., Steric and electronic effects on ^{15}N chemical shifts of piperidine and decahydroquinoline hydrochlorides, *J. Am. Chem. Soc.*, 100, 3889, 1978.
3. Grim, S.O. and Yankowsky, A.W., On the phosphorus-31 chemical shifts of substituted triarylphosphines, *Phosphorus Sulfur Relat. Elem.*, 3, 191, 1977.
4. Lambert, J.B., Shurrell, H.F., Verbit, L., Cooks, R.G., and Stout, G.H., *Organic Structural Analysis*, MacMillan, New York, 1976.
5. Gunther, H., *NMR Spectroscopy: Basic Principles, Concepts and Applications in Chemistry*, John Wiley & Sons, New York, 2003.
6. Abraham, R.J., Fisher, J. and Loftus, P., *Introduction to NMR Spectroscopy*, John Wiley & Sons, New York, 1988.

Nucleus	Name	Formula
^1H	Tetramethylsilane [TMS]	$(CH_3)_4Si$
	3-(Trimethylsilyl)-1-propane sulfonic acid, sodium salt [DSS][a]	$(CH_3)_3Si(CH_2)_3SO_3Na$
	3-(Trimethylsilyl)-propanoic acid, d_4, sodium salt [TSP]	$(CH_3)_3Si(CD_2)_3CO_2Na$
^2H	Deuterated chloroform [chloroform-d]	$CDCl_3$
^{11}B	Boric acid	H_3BO_3
	Boron trifluoride etherate	$(C_2H_5)_2O.BF_3$
	Boron trichloride	BCl_3
^{13}C	Tetramethylsilane [TMS]	$(CH_3)_4Si$
^{15}N	Ammonium nitrate	NH_4NO_3
	Ammonia	NH_3
	Nitromethane	CH_3NO_2
	Nitric acid	HNO_3
	Tetramethylammonium chloride	$(CH_3)_4NCl$
^{17}O	Water	H_2O
^{19}F	Trichlorofluoromethane [Freon 11, R-11]	CCl_3F
	Hexafluorobenzene	C_6F_6
^{31}P	Trimethylphosphite [methyl phosphite]	$(CH_3O)_3P$
	Phosphoric acid (85 %)	H_3PO_4
^{35}Cl	Sodium chloride	$NaCl$
^{59}Co	Cobalt(III) hexacyanide anion	$[Co(CN)_6]^{-3}$
^{119}Sn	Tetramethyltin	$(CH_3)_4Sn$
^{195}Pt	Platinum(IV) hexacyanide	$[Pt(CN)_6]^{-2}$
	Dihydrogen platinum(IV) hexachloride	H_2PtCl_6
^{183}W	Sodium tungstate (external)	Na_2WO_4

[a] For aqueous solutions (known also as "water-soluble TMS" or 2,2-dimethyl-2-silapentane-5-sulfonate).

¹H AND ¹³C CHEMICAL SHIFTS OF USEFUL SOLVENTS FOR NMR MEASUREMENTS

The following table lists the expected $^1H^{(*H)}$ and $^{13}C^{(*C)}$ chemical shifts for various useful NMR solvents in parts per million (ppm) [1–3]. The table also includes the liquid temperature range (°C) and dielectric constants of these solvents. Slight changes may occur with changes in concentration.

REFERENCES

1. Silverstein, R.M., Bassler, G.C., and Morrill, T.C., *Spectrometric Identification of Organic Compounds*, 5th ed., John Wiley & Sons, New York, 1991.
2. Rahman, A-U., *Nuclear Magnetic Resonance: Basic Principles*, Springer-Verlag, New York, 1986.
3. Abraham, R.J., Fisher, J., and Loftus, P., *Introduction of NMR Spectroscopy*, John Wiley & Sons, Chichester, 1988.

Solvent	Formula	Liquid Temperature Range (°C)	Dielectric Constant (ε)	Chemical Shifts δ_H (ppm)	δ_C (ppm)
Acetone-d_6	$(CD_3)_2CO$	−95 to 56	20.7	2.17	29.2, 204.1
Acetonitrile-d_3	CD_3CN	−44 to 82	37.5	2.00	1.3, 117.7
Benzene-d_6	C_6D_6	6 to 80	2.284	7.27	128.4
Carbon disulfide	CS_2	−112 to 46	2.641	—	192.3
Carbon tetrachloride	CCl_4	−23 to 77	2.238	—	96.0
Chloroform-d_3	$CDCl_3$	−64 to 61	4.806	7.25	76.9
Cyclohexane-d_{12}	C_6D_{12}	6 to 81	2.023	1.43	27.5
Dichloromethane-d_2	CD_2Cl_2	−95 to 40	9.08	5.33	53.6
Difluorobromochloromethane	CF_2BCl	−140 to −25		—	109.2
Dimethylformamide-d_7	$DCON(CD_3)_2$	−60 to 153	36.7	2.9, 3.0, 8.0	31, 36, 132.4
Dimethylsulfoxide-d_6	$(CD_3)_2SO$	19 to 189	46.7	2.62	39.6
1,4-Dioxane-d_8	$C_4D_8O_2$	12 to 101	2.209	3.7	67.4
Hexamethylphosphoramide (HMPA)	$[(CH_3)_2N]_3PO$	7 to 233	30.0	2.60	36.8
Methanol-d_4	CD_3OD	−98 to 65	32.63	3.4, 4.8[a]	49.3
Nitrobenzene	$C_6D_5NO_2$	6 to 211	34.8	8.2, 7.6, 7.5	149, 134, 129, 124
Nitromethane-d_3	CD_3NO_2	−29 to 101	35.87	4.33	57.3
Pyridine-d_5	C_5D_5N	−42 to 115	123	7.0, 7.6, 8.6	124, 136, 150
1,1,2,2-Tetra-chloroethane-d_2	$CDCl_2CDCl_2$	−44 to 146	8.2	5.94	75.5
Tetrahydrofuran-d_8	C_4D_8O	−108 to 66	7.54	1.9, 3.8	25.8, 67.9
1,2,4-Trichlorobenzene	$C_6D_3Cl_3$	17 to 214	3.9	7.1, 7.3, 7.4	133.3, 132.8, 130.7, 130.0, 127.6
Trichlorofluoromethane	$CFCl_3$	−111 to 24	2.3	—	117.6
Vinyl chloride-d_3	$CD_2=CDCl$	−154 to −13		5.4, 5.5, 6.3	126, 117
Trifluoroacetic acid, d	CF_3COOD	−15 to 72	8.6	11.3[1]	114.5, 116.5
Water-d_2	D_2O	0 to 100	78.5	4.7	—

[a] Variable with concentration.

RESIDUAL PEAKS OBSERVED IN THE ^{1}H NMR SPECTRA
OF COMMON DEUTERATED ORGANIC SOLVENTS

The following table lists the residual peaks that are observed in the ^{1}H NMR spectra of common deuterated organic solvents. These peaks are generally attributed to the non-deuterated parent compound that serves as an impurity and are marked with an asterisk (*). In addition, other less significant peaks often arise due to other impurities.

Together with the formula and molecular weight, the table lists the expected chemical shifts, *, multiplicities, and (when possible) the coupling constant, J_{HD}, for every solvent. All spectra are at least 99.5 % deuterium pure [1–5].

REFERENCES

1. Yoder, C.H. and Schaeffer, C.D., Jr., *Introduction to Multinuclear NMR*, Benjamin Cummings, Menlo Park, CA, 1987.
2. Silverstein, R.M. and Webster F.X., *Spectrometric Identification of Organic Compounds*, 6th ed., John Wiley & Sons, New York, 1998.
3. Gunther, H., *NMR Spectroscopy: Basic Principles, Concepts and Applications in Chemistry*, John Wiley & Sons, New York, 2003.
4. Lambert, J.B., Holland, L.N., and Mazzola, E.P., *Nuclear Magnetic Resonance Spectroscopy: Introduction to Principles, Applications and Experimental Methods*, Prentice Hall, Englewood Cliffs, NJ, 2003.
5. Nelson, J.H., *Nuclear Magnetic Resonance Spectroscopy*, 2nd ed., John Wiley & Sons, New York, 2003.

Solvent	Formula	Molecular Weight	δ(mult)[a]	J_{HD}
Acetic acid-d_4	CD_3COOD	64.078	*11.53 (1)	2
			*2.03 (5)	
Acetone-d_6	$(CD_3)_2C=O$	64.117	*2.04 (5)	2.2
			2.78 (1)	
			2.82 (1)	
Acetonitrile-d_3	CD_3CN	44.017	*1.93 (5)	2.5
			2.1–2.15	
			2.2–2.4	
Benzene-d_6	C_6D_6	84.153	*7.12 (b)	
Chloroform-d_3	$CDCl_3$	120.384	1.55[b]	
			1.60[b]	
			7.2	
			*7.24(1)	
Cyclohexane-d_{12}	$(CD_2)_6$	96.236	*1.38(b)	
Deuterium oxide	D_{20}	20.028	*4.63(b)[c]	
			*4.67(b)[d]	
1,2-Dichloroethane-d_4	CD_2ClCD_2Cl	102.985	*3.72(b)	
Dichloromethane-d_2	See methylene chloride-d_2			
Diethylene glycol dimethylether-d_{14}	See diglyme-d_{14}			
Diethylether-d_{10}	$(CD_3CD_2)_2O$	84.185	*3.34(m)	
			*1.07(m)	
Diglyme-d_{14}	$CD_3O(CD_2)_2O(CD_2)_2OCD_3$	148.263	*3.49(b)	
(bis(2-Methoxyethyl) ether)			*3.40(b)	
			*3.22(5)	1.5
N,N-Dimethyl-formamide-d_7	$DCON(CD_3)_2$	80.138	*8.01(b)	
			*2.91(5)	2
			*2.74(5)	2
Dimethylsulfoxide-d_6	$(CD_3)_2SO$		3.3–3.4	
			*2.49(5)	1.7
1,2-Diethoxyethane-d_{10}	See Glyme-d_{10}			
p-Dioxane-d_8	$C_4H_8O_2$	96.156	*3.53(m)	
Ethanol-d_6 (anhydrous)	CD_3CD_2OD	52.106	*5.19(1)	
			*3.55(b)	
			*1.11(m)	
Glyme-d_{10} (dimethoxyethane)	$CD_2OCD_2CD_2OCD_3$	100.184	*3.40(m)	
			*3.22(5)	1.6
Hexamethylphosphoric triamide-d_{18} (HMPT-d_{18})	$[(CD_3)_2N]_3P=O$	197.314	*2.53(m)	
Methanol-d_4	CD_3OH	36.067	*4.78(1)	
			*3.30(5)	1.7
Methylene chloride-d_2	CD_2Cl_2	86.945	*5.32(3)	1
			1.4–1.5(b)	
Nitrobenzene-d_5	$C_6D_5NO_2$	128.143	*8.11(b)	
			*7.67(b)	
			*7.50(b)	
Nitromethane-d_3	CD_3NO_2	64.059	*4.33(5)	2

(*Continued*)

Solvent	Formula	Molecular Weight	δ(mult)[a]	J_{HD}
2-Propanol-d_8	$(CD_3)_2CDOD$	68.146	*5.12(1)	
			*3.89(b)	
			*1.10(b)	
Pyridine-d_5	C_6D_5N	84.133	*8.71(b)	
			*7.55(b)	
			*7.19(b)	
			4.8[b]	
			4.9[b]	
Tetrahydrofuran-d_8	C_4D_8O	80.157	*3.58(b)	
			2.42	
			2.32	
			*1.73(b)	
Toluene-d_8	$C_6D_5CD_3$	100.191	*7.09(m)	
			*7.00(b)	
			*6.98(m)	
			*2.09(5)	2.3
Trifluoroacetic acid-d	CF_3COOD	115.03	*11.50(1)	

[a] Chemical shift, δ, in ppm; mult=multiplicity (indicated by a number); b=broad, m=multiplet.
[b] Two peaks that may often appear as one broad peak.
[c] When DSS, 3-(trimethylsilyl)-1-propane sulfonic acid, sodium salt, is used as a reference standard.
[d] When TSP, sodium-3-trimethylpropionate, is used as a reference standard.

^{1}H NMR CHEMICAL SHIFTS FOR WATER SIGNALS IN ORGANIC SOLVENTS

Often traces of water are encountered in samples whose ^{1}H NMR spectra are being measured. The water signals appear at different chemical shifts depending on the particular solvent used. Listed below are the usual chemical shift positions of the water signal in several common solvents [1,2]. The signal in aprotic solvent solutions is due to the presence of H_2O. On the other hand, the signal observed in protic solvent solutions (in parentheses) is attributed HOD, which is the result of hydrogen exchange with the solvent's deuterium atoms.

REFERENCES

1. Silverstein, R.M., Morrill, T.C., and Bassler G.C., *Spectrometric Identification of Organic Compounds*, 5th ed., John Wiley & Sons, New York, 1991.
2. Notes on NMR Solvents, https://webspectra.chem.ucla.edu/NotesOnSolvents.html, 1997, accessed December 2019.

Solvent	Chemical Shift of H_2O (or HOD)
Acetone	2.8
Acetonitrile	2.1
Benzene	0.4
Chloroform	1.6
Dimethyl sulfoxide	3.3
Methanol	(4.8)
Methylene chloride	1.5
Pyridine	4.9
Water (D_2O)	(4.8)

¹H IN DEUTERATED SOLVENTS

The following table provides the expected chemical shift of solutes that commonly occur as impurities in NMR spectra measured at 298 K. These chemical shifts can vary as a function of (deuterated) solvent. These data are adapted from the compilation of Gottleib et al. [1], as expanded by Fulmer et al. [2]. Only the more common solvents and impurities are considered here. The reader is referred to these sources for additional impurities and solvents. When the same moiety contains multiple protons, bold print indicates the protons under consideration. When protons differ by ring position, the appropriate protons are indicated parenthetically. Note that the chemical shifts of the −OH protons will vary with concentration as a result of hydrogen bonding, so these have not been included.

REFERENCES

1. Gottleib, H.E., Kotlyar, V., and Nudelman, A., NMR chemical shifts of common laboratory solvents as trace impurities, *J. Org. Chem.*, 62, 7512–7515, 1997.
2. Fulmer, G.R., Miller, A.J.M., Sherdan, N.H., Gottleib, H.E., Nudelman, A., Stoltz, B.M., Bercaw, J.E., and Goldberg, K.I., NMR chemical shifts of trace impurities: common laboratory solvents, organics, and gases in deuterated solvents relevant to the organometallic chemist, *Organometallics*, 29, 2176–2179, 2010.

Impurity	Proton	THF-d8	CD_2Cl_2	$CDCl_3$	Toluene-d8	C_6D_6	CD_3OD	D_2O
Acetic acid	CH_3	1.89	2.06	2.10	1.57	1.52	4.87	—
Acetone	CH_3	2.05	2.12	2.17	1.57	1.55	2.15	2.22
Acetonitrile	CH_3	1.95	1.97	2.10	0.69	0.58	2.03	2.06
Benzene	CH	7.31	7.35	7.36	7.12	7.15	7.33	—
Chloroform	CH	7.89	7.32	7.26	6.10	6.15	7.90	—
Cyclohexane	CH_2	1.44	1.44	1.43	1.40	1.40	1.45	—
1,4-Dioxane	CH_2	3.56	3.65	3.71	3.33	3.35	3.66	3.75
Ethanol	CH_3	1.10	1.19	1.25	0.97	0.96	1.19	1.17
	CH_2	3.51	3.66	3.72	3.36	3.39	3.60	3.65
Ethyl acetate	CH_3CO	1.94	2.00	2.05	1.69	1.65	2.01	2.07
	CH_2CH_3	4.04	4.08	4.12	3.87	3.89	4.09	4.14
	CH_2CH_3	1.19	1.23	1.26	0.94	0.92	1.24	1.24
n-Hexane	CH_3	0.89	0.89	0.88	0.88	0.89	0.90	—
	CH_2	1.29	1.27	1.26	1.22	1.24	1.29	—
2-Propanol	CH_3	1.08	1.17	1.22	0.95	0.95	1.50	1.17
(isopropanol)	CH	3.82	3.97	4.04	3.65	3.67	3.92	4.02
Toluene	CH_3	2.31	2.34	2.36	2.11	2.11	2.32	—
	CH (2,4,6)	7.10	7.15	7.17	6.99 (av)	7.02	7.16	—
	CH (3,5)	7.19	7.24	7.25	7.09	7.13	7.16	—

PROTON NMR ABSORPTION OF MAJOR CHEMICAL FAMILIES

The following tables give the region of the expected NMR absorptions of major chemical families. These absorptions are reported in the dimensionless units of ppm vs. the standard compound tetramethylsilane (TMS), which is recorded as 0.0 ppm.

$$
\begin{array}{c}
CH_3 \\
| \\
H_3C-\!Si\!-CH_3 \\
| \\
CH_3
\end{array}
$$

The use of this unit of measure makes the chemical shifts independent of the applied magnetic field strength or the radio frequency. For most proton NMR spectra, the protons in TMS are more shielded than almost all other protons. The chemical shift in this dimensionless unit system is then defined by:

$$
\delta = \frac{v_s - v_r}{v_r} \times 10^6
$$

where v_s and v_r are the absorption frequencies of the sample proton and the reference (TMS) protons (12, magnetically equivalent), respectively. In these tables, the proton(s) whose proton NMR shifts are cited are indicated by underscore. For more detail concerning these conventions, the reader is referred to the general references below [1–11].

REFERENCES

1. Silverstein, R.M. and Webster, F.X., *Spectrometric Identification of Organic Compounds*, 6th ed., Wiley, New York, 1998.
2. Rahman, A.-U., *Nuclear Magnetic Resonance*, Springer Verlag, New York, 1986.
3. Gordon, A. J. and Ford, R.A., *The Chemist's Companion*, Wiley Interscience, New York, 1971.
4. Becker, E.D., *High Resolution NMR, Theory and Chemical Applications*, 2nd ed., Academic Press, New York, 1980.
5. Gunther, H., *NMR Spectroscopy: Basic Principles, Concepts and Applications in Chemistry*, Wiley, New York, 2003.
6. Kitamaru, R., *Nuclear Magnetic Resonance: Principles and Theory*, Elsevier Science, New York, 1990.
7. Lambert, J.B., Holland, L.N., and Mazzola, E.P., *Nuclear Magnetic Resonance Spectroscopy: Introduction to Principles, Applications and Experimental Methods*, Prentice Hall, Englewood Cliffs, NJ, 2003.
8. Bovey, F.A. and Mirau, P.A., *Nuclear Magnetic Resonance Spectroscopy*, 2nd ed., Academic Press, London, 1988.
9. Hore, P.J. and Hore, P.J., *Nuclear Magnetic Resonance*, Oxford University Press, Oxford, 1995.
10. Nelson, J.H., *Nuclear Magnetic Resonance Spectroscopy*, 2nd ed., Wiley, New York, 2003.
11. Abraham, R.J., Fisher, J., and Loftus, P., *Introduction to NMR Spectroscopy*, Wiley, New York, 1988.

Hydrocarbons

Family	δ of Protons (Underlined)			
Alkanes	$\underline{CH_3}$–R	~0.8 ppm		
	–$\underline{CH_2}$–R	~1.1 ppm		
	>$\underline{CH}$–R	~1.4 ppm		
	(Cyclopropane 0.2 ppm)			
Alkenes	$\underline{CH_3}$–C=C<	~1.6 ppm	$\underline{CH_3}$–C–C=C<	~1.0 ppm
	–$\underline{CH_2}$–C=C<	~2.1 ppm	–$\underline{CH_2}$–C–C=C<	~1.4 ppm
	>$\underline{CH}$–C=C<	~2.5 ppm	>$\underline{CH}$–C–C=C<	~1.8 ppm
	>C=C–$\underline{H}$	4.2–6.2 ppm		
Alkynes	$\underline{CH_3}$–C≡C–	~1.7 ppm	$\underline{CH_3}$–C–C≡C–	~1.2 ppm
	–$\underline{CH_2}$–C≡C–	~2.2 ppm	>$\underline{CH_2}$–C–C≡C–	~1.5 ppm
	>$\underline{CH}$–C≡C–	~2.7 ppm	>$\underline{CH}$–C–C≡C–	~1.8 ppm
	R–C≡C–$\underline{H}$	~2.4 ppm		
Aromatics	C_6H_5–G 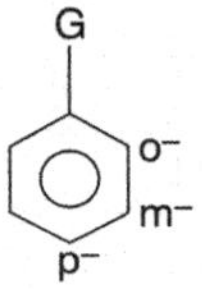	Range: 8.5–6.9 ppm		

G

o–
m–
p–

When $\underline{G}$=electron withdrawing (for example, >C=O, –NO_2, –C≡N), o- and p-hydrogens relative to –G are closer to 8.5 ppm (more downfield)

When $\underline{G}$=electron donating (for example, –NH_2, –OH, –OR, –R), o- and p-hydrogens relative to –G are closer to 6.9 ppm (more upfield)

Organic Oxygen Compounds

Family	Approximate δ of Protons (Underlined)		
Alcohols	C$\underline{H}_3$–OH 3.2 ppm	RC$\underline{H}_2$–OH 3.4 ppm	R$_2$C$\underline{H}$–OH 3.6 ppm
	C$\underline{H}_3$–C–OH 1.2 ppm	RC$\underline{H}_2$–C–OH 1.5 ppm	R$_2$C$\underline{H}$–C–OH 1.8 ppm
	R–$\underline{O}$–$\underline{H}$ (1–5 ppm—depending on concentration)		
Aldehydes	C$\underline{H}_3$–CHO 2.2 ppm	RC$\underline{H}_2$–CHO 2.4 ppm	R$_2$C$\underline{H}$–CHO 2.5 ppm
	C$\underline{H}_3$–C–CHO 1.1 ppm	RC$\underline{H}_2$–C–CHO 1.6 ppm	
Amides	See Organic Nitrogen Compounds		
Anhydrides, acyclic	C$\underline{H}_3$–C(=O)O– 1.8 ppm	RC$\underline{H}_2$–C(=O)O– 2.1 ppm	R$_2$C$\underline{H}$–C(=O)O– 2.3 ppm
	C$\underline{H}_3$–C–C(=O)O– 1.2 ppm	RC$\underline{H}_2$–C–C(=O)O– 1.8 ppm	R$_2$C$\underline{H}$–C–C(=O)O– 2.0 ppm

Anhydrides, cyclic

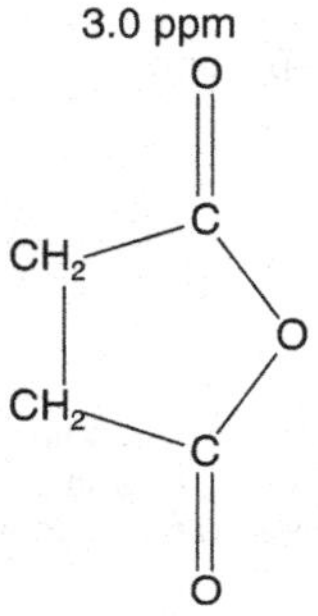

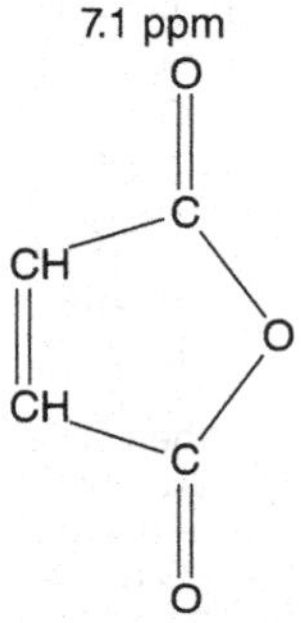

3.0 ppm

7.1 ppm

Carboxylic acids			
CH$_3$–COOH 2.1 ppm	RCH$_2$–COOH 2.3 ppm	R$_2$CH–COOH 2.5 ppm
CH$_3$–C–COOH 1.1 ppm	R-CH$_2$–C–COOH 1.6 ppm	R$_2$CH–C–COOH 2.0 ppm
R-COO–H 11–12 ppm		

Cyclic Ethers   Oxacyclopropane (oxirane)

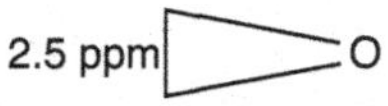

2.5 ppm

Oxacyclobutane (oxetane)

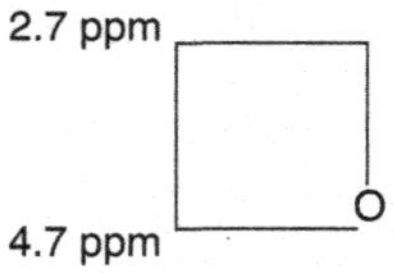

2.7 ppm

4.7 ppm

Oxacyclopentane (tetrahydrofuran)

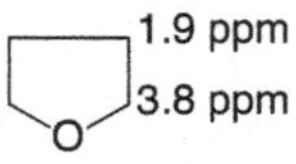

1.9 ppm
3.8 ppm

Oxacyclohexane (tetrahydropyran)

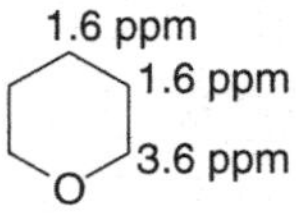

1.6 ppm
1.6 ppm
3.6 ppm

(Continued)

Organic Oxygen Compounds (*Continued*)

Family	Approximate δ of Protons (Underlined)
1,4-Dioxane	3.6 ppm
1,3-Dioxane	1.7 ppm 3.8 ppm 4.7 ppm
Furan	6.3 ppm 7.4 ppm
Dihydropyran	1.9 ppm 4.5 ppm 6.2 ppm

Epoxides		**See Cyclic Ethers**		
Esters		$\underline{CH_3}$–COOR	$R\underline{CH_2}$–COOR	$R_2\underline{CH}$–COOR
	R=alkyl	1.9 ppm	2.1 ppm	2.3 ppm
	R=aryl	2.0 ppm	2.2 ppm	2.4 ppm
		$\underline{CH_3}$–C–COOR	$R\underline{CH_2}$–C–COOR	$R_2\underline{CH}$–C–COOR
		1.1 ppm	1.7 ppm	1.9 ppm
		$\underline{CH_3}$–OOC–R	$R\underline{CH_2}$–OOC–R	$R_2\underline{CH}$–OOC–R
		3.6 ppm	4.1 ppm	4.8 ppm
		$\underline{CH_3}$–C–OOC–R	$R\underline{CH_2}$–C–OOC–R	$R_2\underline{CH}$–C–OOC–R
		1.3 ppm	1.6 ppm	1.8 ppm

Cyclic esters 2.1 ppm ⌐ 4.4 ppm   1.6 ppm
                2.3 ppm   O   1.6 ppm ⌐ 4.1 ppm
                        O   2.3 ppm   O
                                                O

Ethers		$\underline{CH_3}$–O–R	$R\underline{CH_2}$–O–R	$R_2\underline{CH}$–O–R
	R=alkyl	3.2 ppm	3.4 ppm	3.6 ppm
	R=aryl	3.9 ppm	4.1 ppm	4.5 ppm
		$\underline{CH_3}$–C–O–R	$R\underline{CH_2}$–C–O–R	$R_2\underline{CH}$–C–O–R
	R=alkyl	1.2 ppm	1.5 ppm	1.8 ppm
	R=aryl	1.3 ppm	1.6 ppm	2.0 ppm

Isocyanates	See Organic Nitrogen Compounds			
Ketones		$\underline{CH_3}$–C(=O)–	$R\underline{CH_2}$–C(=O) –	$R_2\underline{CH}$–C(=O)
	1.9 ppm	R=alkyl	2.1 ppm	2.3 ppm
	2.4 ppm	R=aryl	2.7 ppm	3.4 ppm
		$\underline{CH_3}$–C(=O)–	$R\underline{CH_2}$–C(=O)–	$R_2\underline{CH}$–C(=O)
	1.1 ppm	R=alkyl	1.6 ppm	2.0 ppm
	1.2 ppm	R=aryl	1.6 ppm	2.1 ppm

Cyclic ketones (n=number of ring carbons)

$(\underline{CH_2})_n$ ⟩=O

	α-Hydrogens	2.0–2.3 ppm (n>5)
		3.0 ppm (n=4)
		1.7 ppm (n=3)
	β-Hydrogens	1.9–1.5 ppm

Lactones	See Esters, cyclic	
Nitro compounds	See Organic Nitrogen Compounds	
Phenols	Ar–O–$\underline{H}$	9–10 ppm (Ar=aryl)

Amides

δ of Proton(s) (Underlined)	Primary R–C(=O)NH$_2$ δ (ppm)	Secondary R–C(=O)NHR$_1$ δ (ppm)	Tertiary R–C(=O)NR$_1$R$_2$ δ (ppm)
(i) N-substitution			
R–C(=O)N–H	5–12	5–12	–
(a) Alpha			
–C(=O)N–CH$_3$	–	~2.9	~2.9
–C(=O)N–CH$_2$	–	~3.4	~3.4
–C(=O)N–CH	–	~3.8	~3.8
(b) Beta			
–C(=O)N–C–CH$_3$	~1.1	~1.1	~1.1
–C(=O)N–C–CH$_2$	~1.5	~1.5	~1.5
–C(=O)N–C–CH	~1.9	~1.9	~1.9
(ii) C-substitution			
(a) Alpha			
CH$_3$–C(=O)N	~1.9	~2.0	~2.1
RCH$_2$–C(=O)N	~2.1	~2.1	~2.1
R$_2$CH–C(=O)N	~2.2	~2.2	~2.2
(b) Beta			
CH$_3$–C–C(=O)N	~1.1	~1.1	~1.1
CH$_2$–C–C(=O)N	~1.5	~1.5	~1.5
>CH–C–C(=O)N	~1.8	~1.8	~1.8

Amines

δ of Proton(s) (Underlined)	Primary R–NH$_2$ δ (ppm)	Secondary RN–HR δ (ppm)	Tertiary RRRN δ (ppm)
(i) Alpha protons			
>N–CH$_3$	~2.5	2.3-3.0	~2.2
>N–CH$_2$	~2.7	2.6-3.4	~2.4
>N–CH<	~3.1	2.9-3.6	~2.8
(ii) Beta protons			
>N–C–CH$_3$			~1.1
>N–C–CH$_2$			~1.4
>N–C–CH<			~1.7

Cyano Compounds (Nitriles)

(i) Alpha Hydrogens δ (ppm)		(ii) Beta Hydrogens δ (ppm)	
CH$_3$–C≡N	~2.1	CH$_3$–C–C≡N	~1.2
–CH$_2$–C≡N	~2.5	–CH$_2$–C–C≡N	~1.6
–CH–C≡N	~2.9	CH–C–C≡N	~2.0

Imides

(i) Alpha Hydrogens δ (ppm)		(ii) Beta Hydrogens δ (ppm)	
$\underline{CH_3}$–C(=O)NHC(=O)–	~2.0	$\underline{CH_3}$–C(=O)C–NH–C(=O)–	~1.2
$\underline{CH_2}$–C(=O)NHC(=O)–	~2.1	$\underline{CH_2}$–C(=O)C–NH–C(=O)–	~1.3
$\underline{CH}$–C(=O)NHC(=O)–	~2.2	–$\underline{CH}$–C(=O)C–NH–C(=O)–	~1.4

Isocyanates

Alpha Hydrogens δ (ppm)	
$\underline{CH_3}$–N=C=O	~3.0
–$\underline{CH_2}$–N=C=O	~3.3
–$\underline{CH}$–N=C=O	~3.6

Isocyanides (Isonitriles)		Isothiocyanates	
Alpha Hydrogens δ (ppm)		Alpha Hydrogens δ (ppm)	
$\underline{CH_3}$–N=C<	~2.9	$\underline{CH_3}$–N=C=S	~3.4
$\underline{CH_2}$–N=C<	~3.3	$\underline{CH_2}$–N=C=S	~3.7
$\underline{CH}$–N=C<	~4.9	>$\underline{CH}$–N=C=S	~4.0

Nitriles δ (ppm)	
–$\underline{CH_2}$–O–N=O	~4.8

Nitro Compounds δ (ppm)					
$\underline{CH_3}$–NO$_2$	~4.1	–$\underline{CH_2}$–NO$_2$	~4.2	–$\underline{CH}$–NO$_2$	~4.4
$\underline{CH_3}$–C–NO$_2$	~1.6	–$\underline{CH_2}$–C–NO$_2$	~2.1	–$\underline{CH}$–C–NO$_2$	~2.5

Organic Sulfur Compounds

Family		δ of Proton(s) (Underlined)				
Benzothiopyrans						
2H–1–	sp^3 C–H	~3.3 ppm	sp^2 C–H	5.8–6.4	Aromatic	~6.8
4H–1–	sp^3 C–H	~3.2 ppm	sp^2 C–H	5.9–6.3	Aromatic	~6.9
2,3,4H–1–	sp^3 C–H	1.9–2.8 ppm	Aromatic			~7.1
Disulfides	C̲H$_3$–S–S–R	~2.4 ppm	C̲H$_3$–C–S–S–R	~1.2 ppm		
	C̲H$_2$–S–S–R	~2.7 ppm	C̲H$_2$–C–S–S–R	~1.6 ppm		
	C̲H–S–S–R	~3.0 ppm	C̲H–C–S–S–R	~2.0 ppm		
Isothiocyanates	C̲H$_3$–N=C=S	~2.4 ppm				
	–C̲H$_2$–N=C=S	~2.7 ppm				
	–C̲H–N=C=S	~3.0 ppm				
Mercaptans (thiols)	C̲H$_3$–S–H	~2.1 ppm	C̲H$_3$–C–S–H	~1.3 ppm		
	–C̲H$_2$–S–H	~2.6 ppm	–C̲H$_2$–C–S–H	~1.6 ppm		
	–C̲H–S–H	~3.1 ppm	–C̲H–C–S–H	~1.7 ppm		
S-methyl salts	+					
	>S–CH$_3$	~3.2 ppm				
Sulfates	(C̲H$_3$–O)$_2$S(=O)$_2$	~3.4 ppm				
	C̲H$_3$–S–	1.8–2.1	C̲H$_3$–C̲H$_2$–S–			1.1–1.2
	R–C̲H$_2$–S–	1.9–2.4	C̲H$_3$–CHR–S–			0.8–1.2
	R–C̲HR–S–	2.8–3.4	C̲H$_3$–CHAr–S–			1.3–1.4
	Ar–C̲H$_2$–S–	4.1–4.2	C̲H$_3$–CR$_2$–S–			1.0
	Ar–C̲HR–S–	3.6–4.2	Ar–C̲H$_2$–CHR–S–			3.0–3.2
	Ar$_2$–C̲H–S–	5.1–5.2	>C=C–C̲H$_2$–CHAr–S–			2.4–2.6
			>C=C–C̲H$_2$–CAr$_2$–S–			2.5
			R$_2$C̲H–CH$_2$–S–			2.6–3.0
			Ar$_2$C̲H–CH$_2$–S–			4.0–4.2
			>C=C–C̲HR–CHAr–S–			2.3–2.4
			>C=C–C̲HR–CAr$_2$–S–			2.8–3.2
Sulfilimines	C̲H$_3$(R)S=N–R^2	~2.5 ppm				
Sulfonamides	C̲H$_3$–SO$_2$NH$_2$	~3.0 ppm				
Sulfonates	C̲H$_3$–SO$_2$–OR	~3.0 ppm				
Sulfones	C̲H$_3$–SO$_2$–R^2	~2.6 ppm				
Sulfonic acids	C̲H$_3$–SO$_3$H	~3.0 ppm				
Sulfoxides	C̲H$_3$–S(=O)R	~2.5 ppm				
	–C̲H$_2$–S(=O)R	~3.1 ppm				
Thiocyanates	C̲H$_3$–S–C≡N	~2.7 ppm				
	–C̲H$_2$–S–C≡N	~3.0 ppm				
	–C̲H–S–C≡N	~3.3 ppm				
Thiols	See mercaptans					

Note: Ar represents aryl.

PROTON NMR CORRELATION CHARTS OF MAJOR FUNCTIONAL GROUPS

The following correlation tables provide the regions of NMR absorptions of major chemical families. These absorptions are reported in the dimensionless units of ppm vs. the standard compound tetramethylsilane (TMS, $(CH_3)_4Si$) which is recorded as 0.0 ppm.

The use of this unit of measure makes the chemical shifts independent of the applied magnetic field strength or the radio frequency. For most proton NMR spectra, the protons in TMS are more shielded than almost all other protons. The chemical shift in this dimensionless unit system is then defined by:

$$\delta = \frac{v_s - v_r}{v_r} \times 10^6$$

where v_s and v_r are the absorption frequencies of the sample proton and the reference (TMS) protons (12, magnetically equivalent), respectively. In these tables, the proton(s) whose proton NMR shifts are cited are indicated by underscore. For more detail concerning these conventions, the reader is referred to the general references below [1–14]. Ref. [15] has a compilation of references for the various nuclei.

Due to the large amount of data, the whole ^1H NMR region is divided into smaller sections 1.0–1.2 ppm range each. This will allow the user to look into specified chemical shift and determine all possibilities for the unknown whose structure is being analyzed.

REFERENCES

1. Silverstein, R.M. and Webster, F.X., *Spectrometric Identification of Organic Compounds*, 6th ed., Wiley, New York, 1998.
2. Rahman, A.-U., *Nuclear Magnetic Resonance*, Springer Verlag, New York, 1986.
3. Gordon, A.J. and Ford, R.A., *The Chemist's Companion*, Wiley Interscience, New York, 1971.
4. Becker, E.D., *High Resolution NMR, Theory and Chemical Applications*, 2nd ed., Academic Press, New York, 1980.
5. Pretsch, E., Bühlmann, P., and Badertscher, M., *Structure Determination of Organic Compounds: Tables of Spectral Data*, 4th Revised and Enlarged English edition, Springer-Verlag, Berlin, 2009.
6. Gunther, H., *NMR Spectroscopy: Basic Principles, Concepts and Applications in Chemistry*, Wiley, New York, 2003.
7. Kitamaru, R., *Nuclear Magnetic Resonance: Principles and Theory*, Elsevier Science, New York, 1990.
8. Lambert, J.B., Holland, L.N., and Mazzola, E.P., *Nuclear Magnetic Resonance Spectroscopy: Introduction to Principles, Applications and Experimental Methods*, Prentice Hall, Englewood Cliffs, NJ, 2003.
9. Bovey, F.A. and Mirau, P.A., *Nuclear Magnetic Resonance Spectroscopy*, 2nd ed., Academic Press, London, 1988.
10. Hore, P.J. and Hore, P.J., *Nuclear Magnetic Resonance*, Oxford University Press, Oxford, 1995.
11. Bruno, T.J. and Svoronos, P.D.N., *Handbook of Basic Tables for Chemical Analysis*, 2nd ed., CRC Press, Boca Raton, FL, 2003.
12. Bruno, T.J. and Svoronos, P.D.N., *CRC Handbook of Fundamental Spectroscopic Correlation Charts*, CRC Press, Boca Raton, FL, 2006.
13. Nelson, J.H., *Nuclear Magnetic Resonance Spectroscopy*, 2nd ed., Wiley, New York, 2003.
14. Abraham, R.J., Fisher, J., and Loftus, P., *Introduction to NMR Spectroscopy*, Wiley, New York, 1988.
15. University of Wisconsin, Organic Structure Determination Bibliography, available online at https://www.chem.wisc.edu/areas/reich/nmr/nmr-biblio.htm, accessed December 2019.

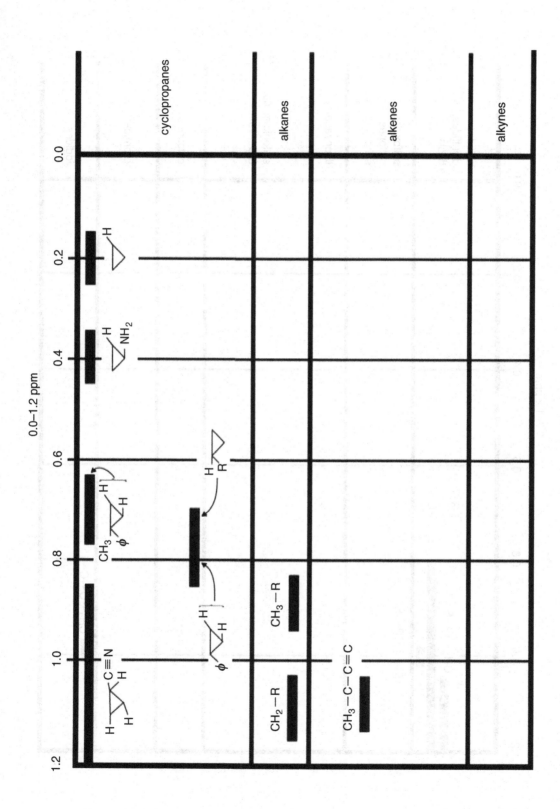

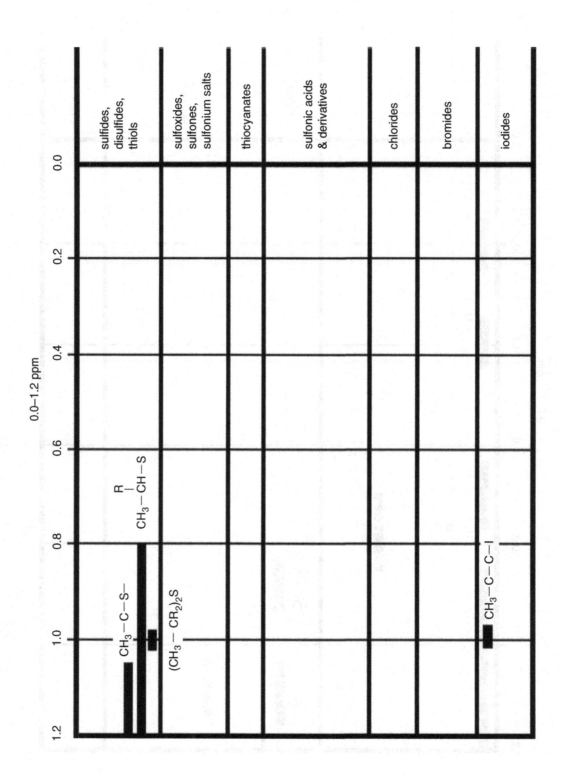

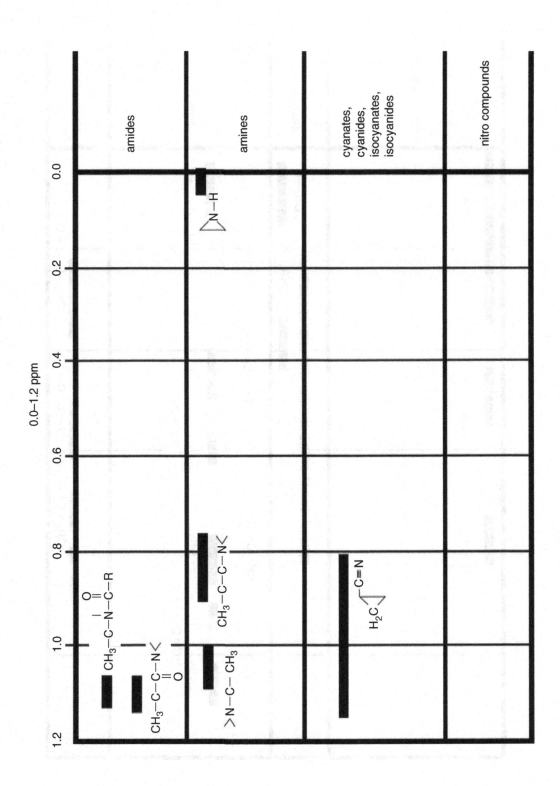

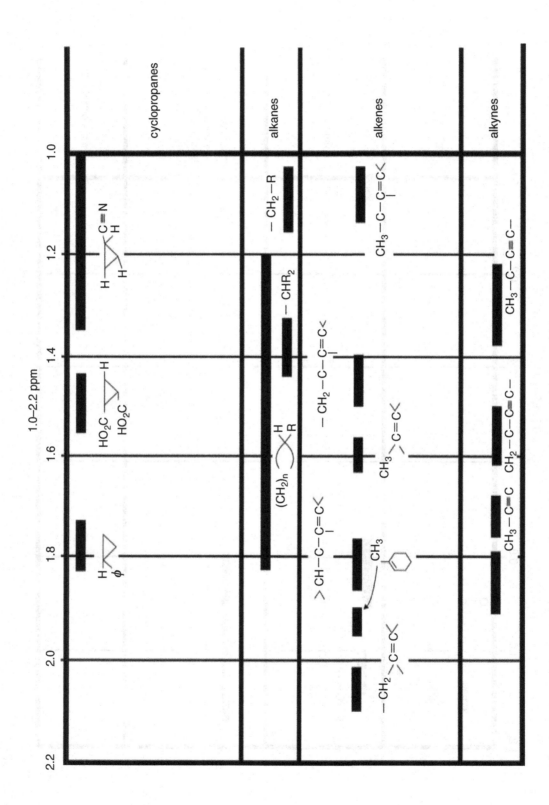

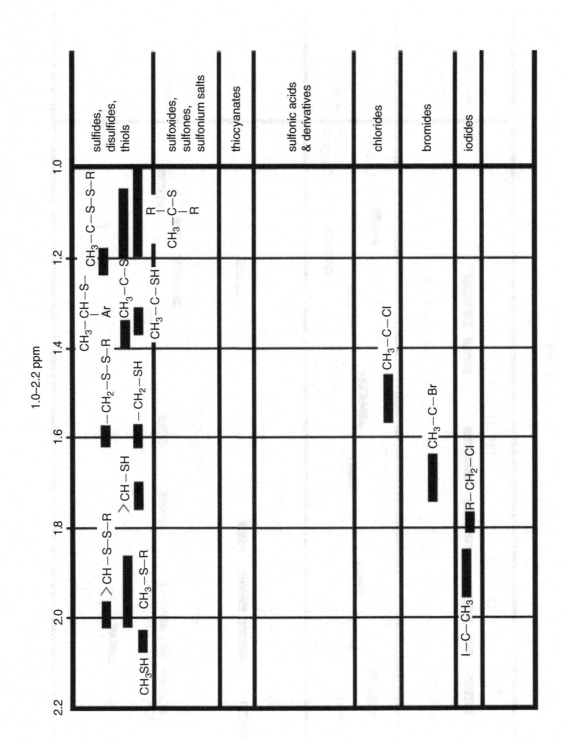

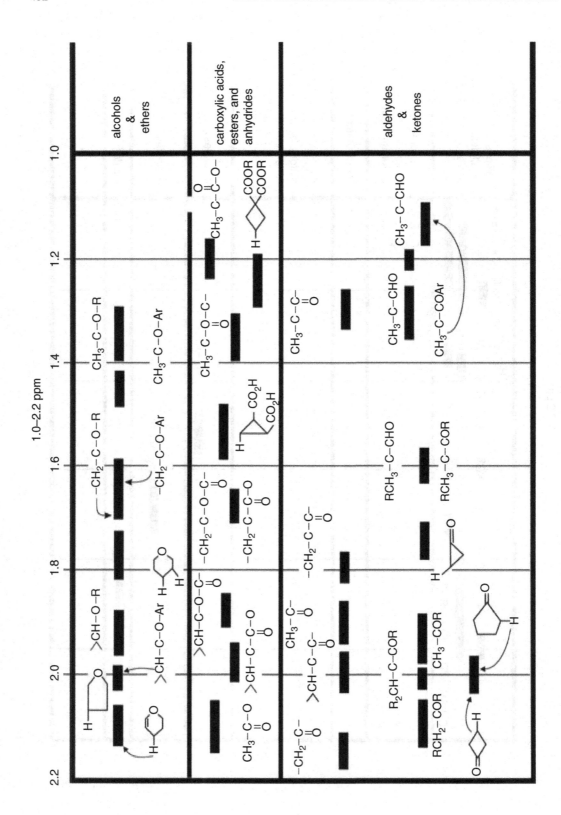

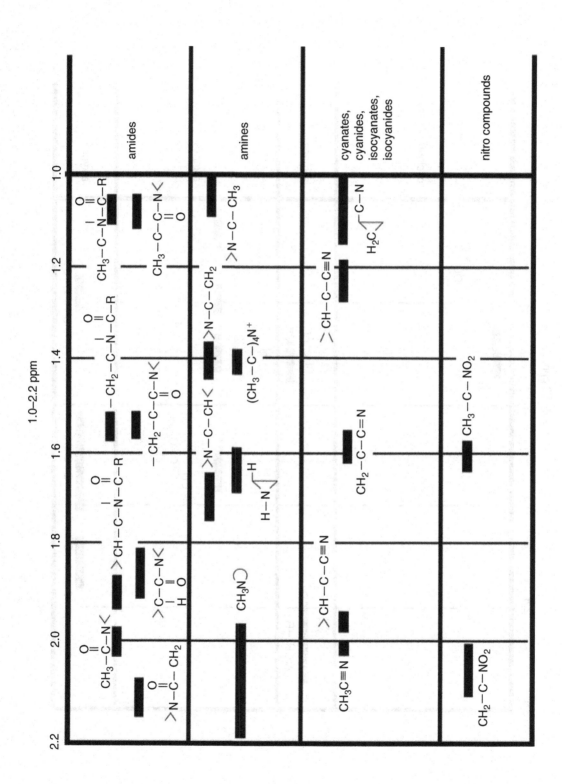

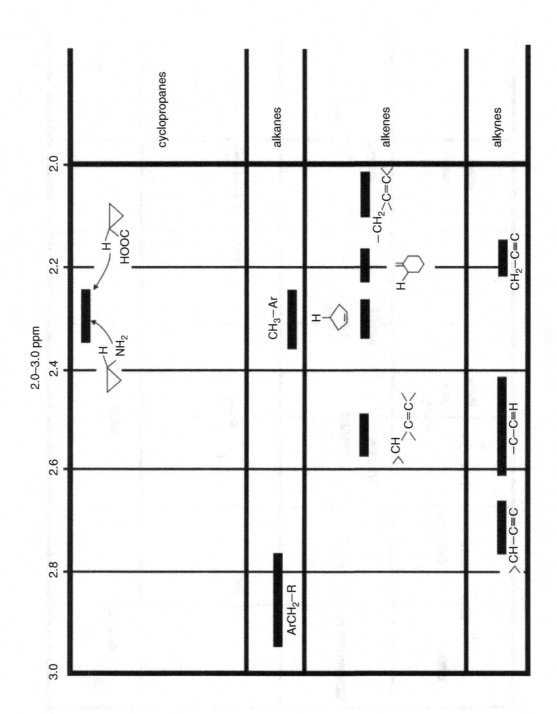

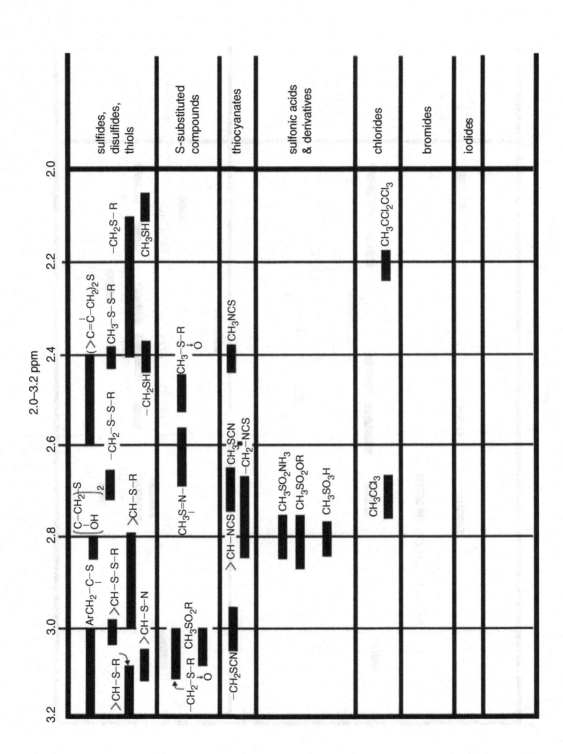

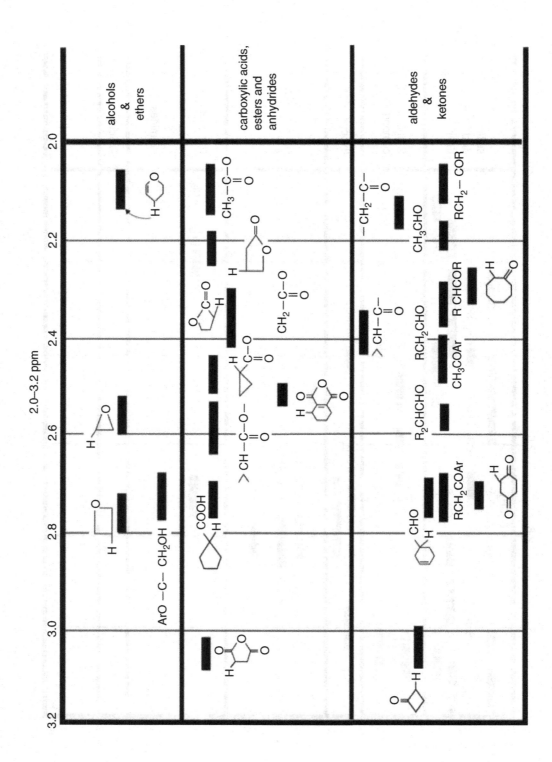

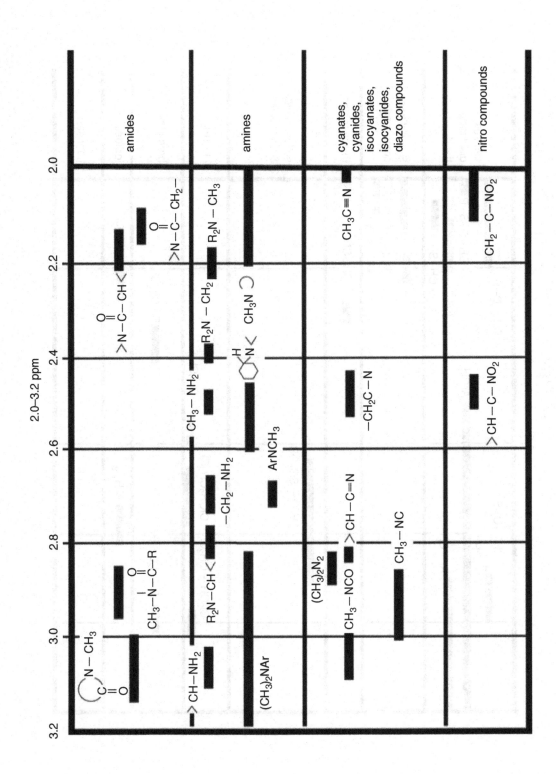

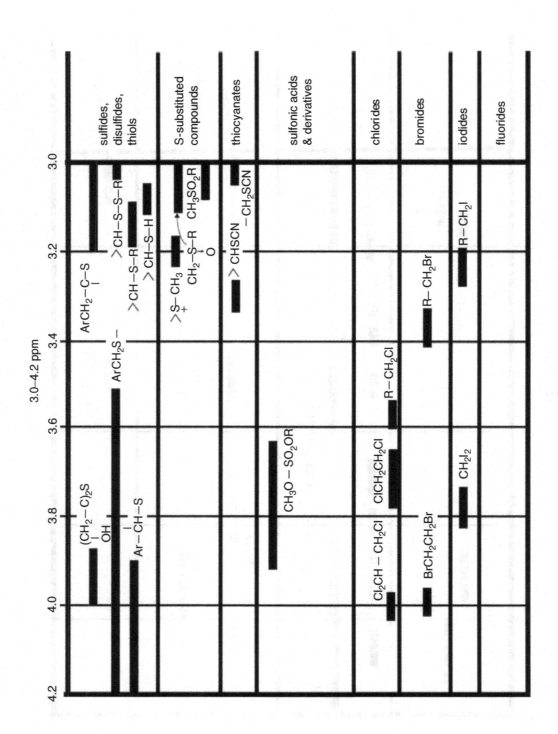

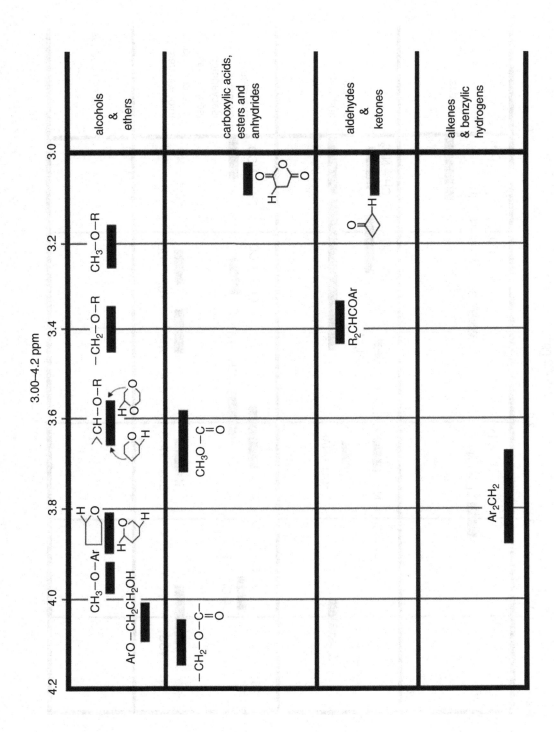

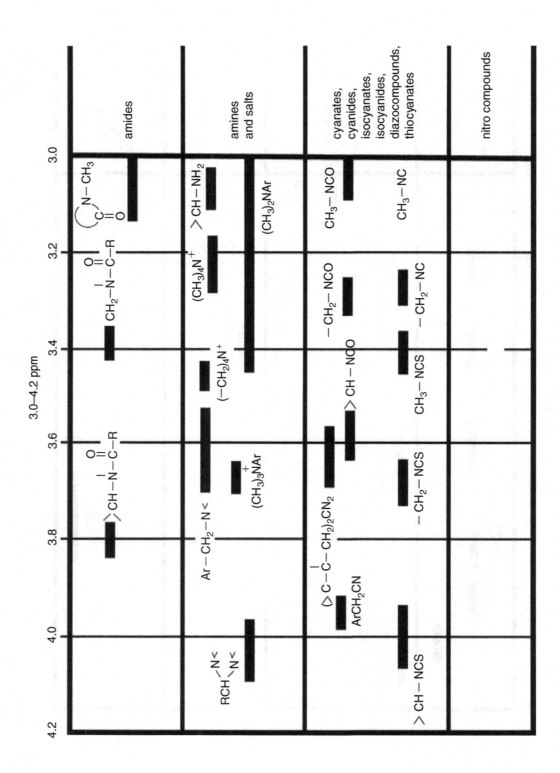

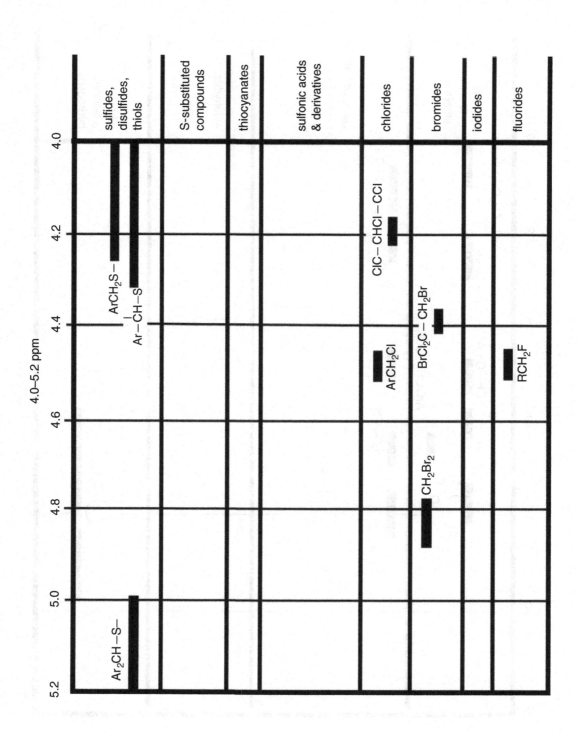

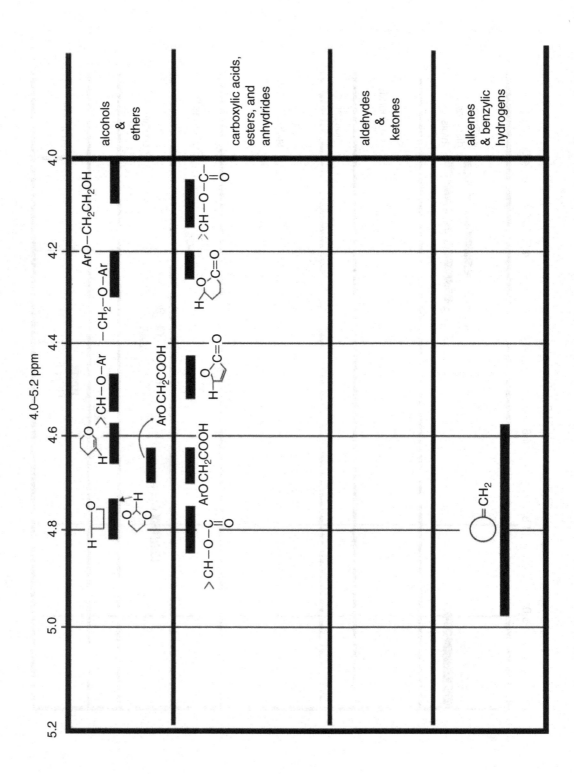

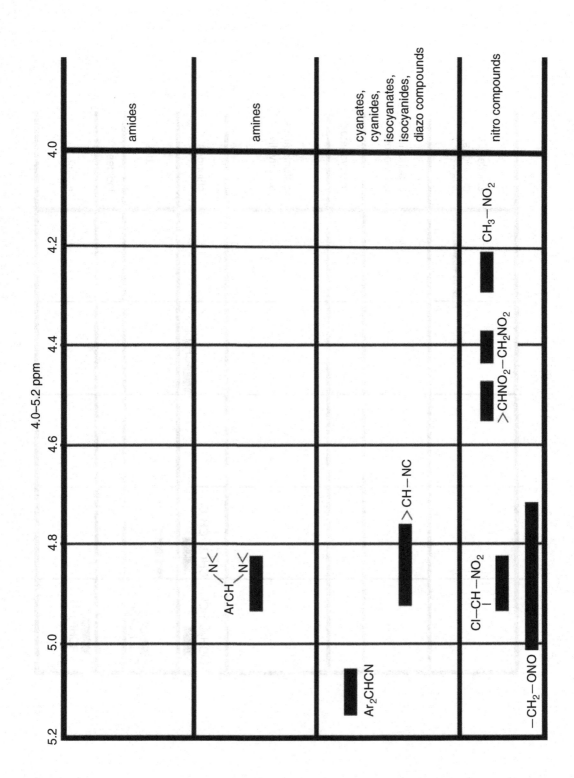

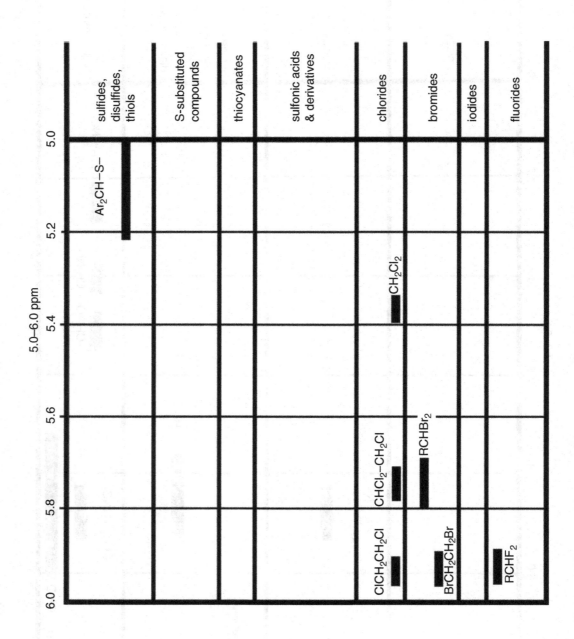

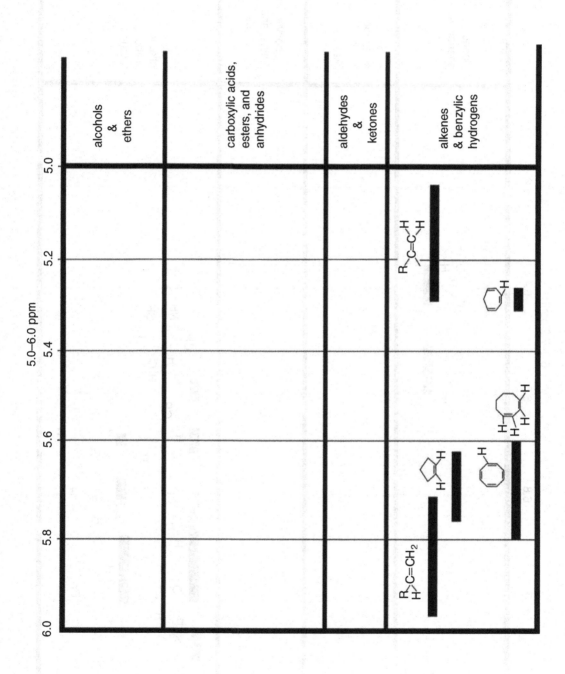

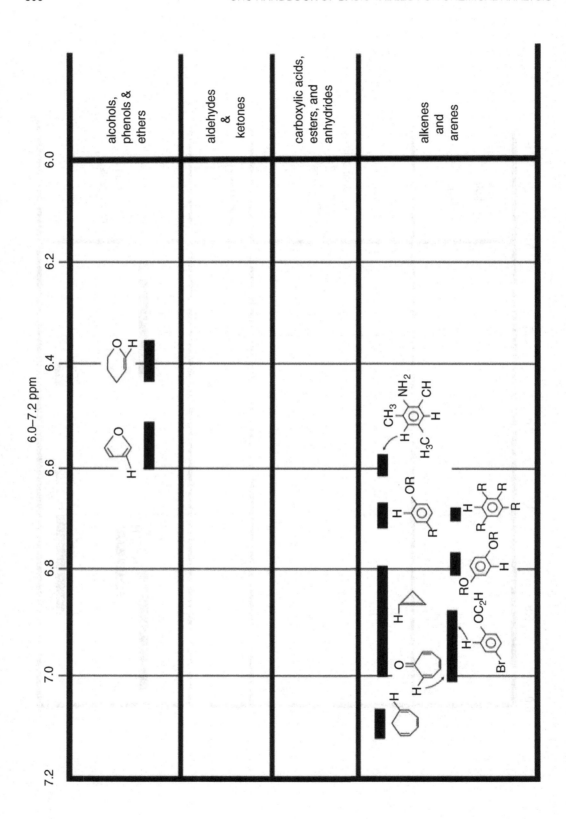

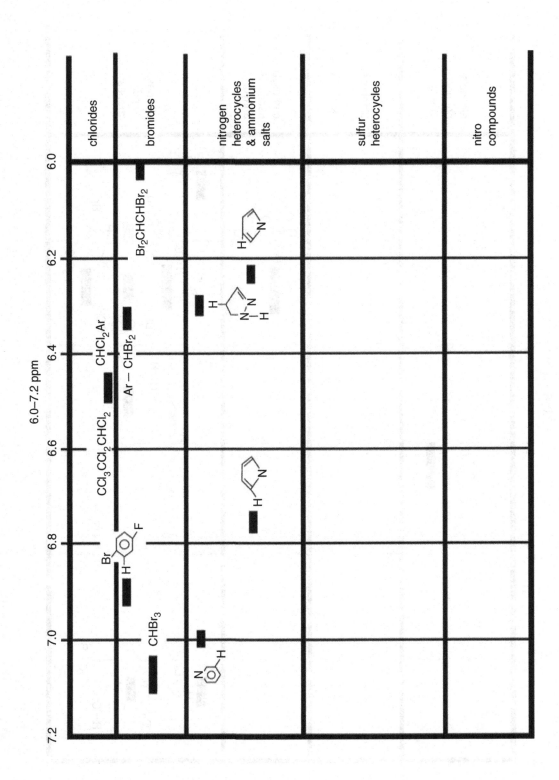

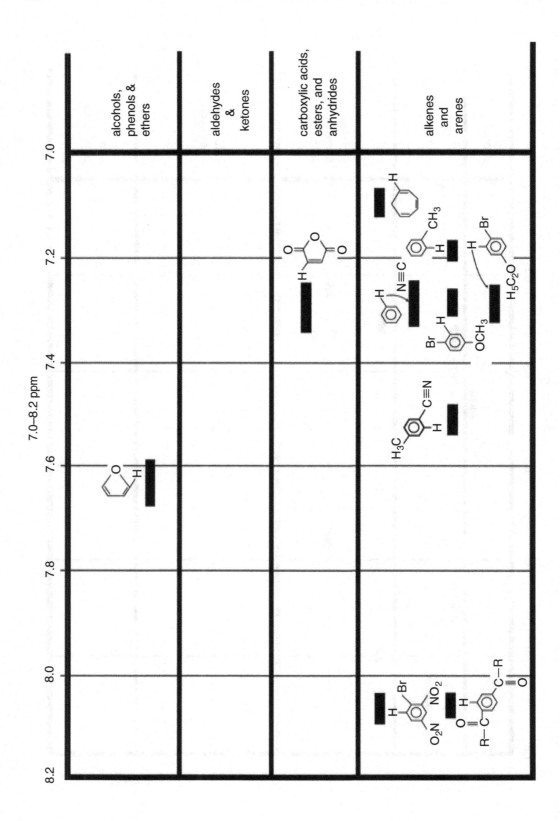

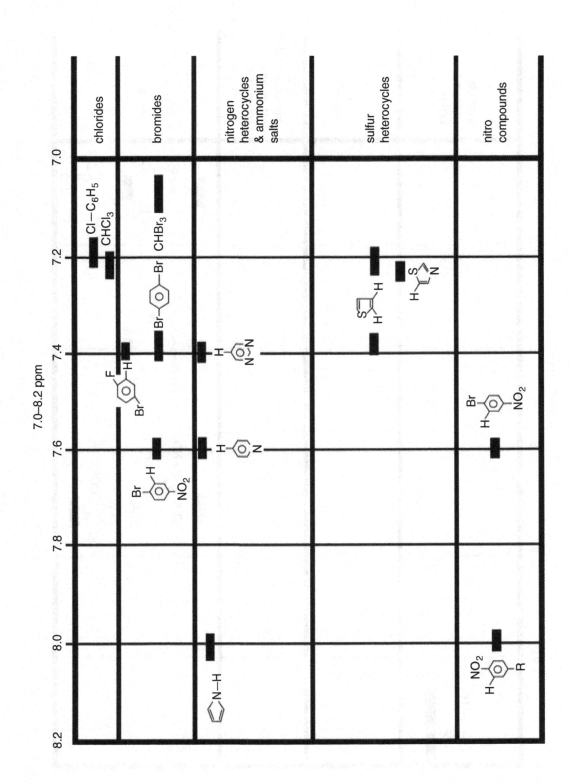

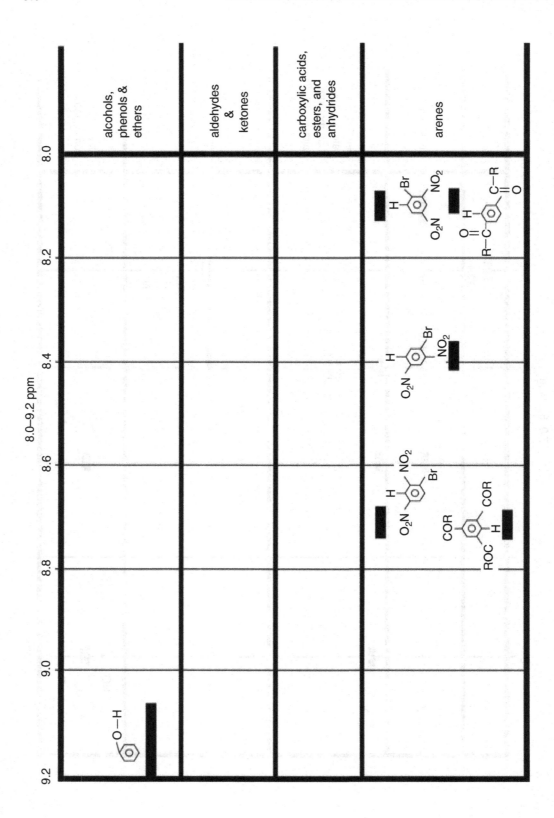

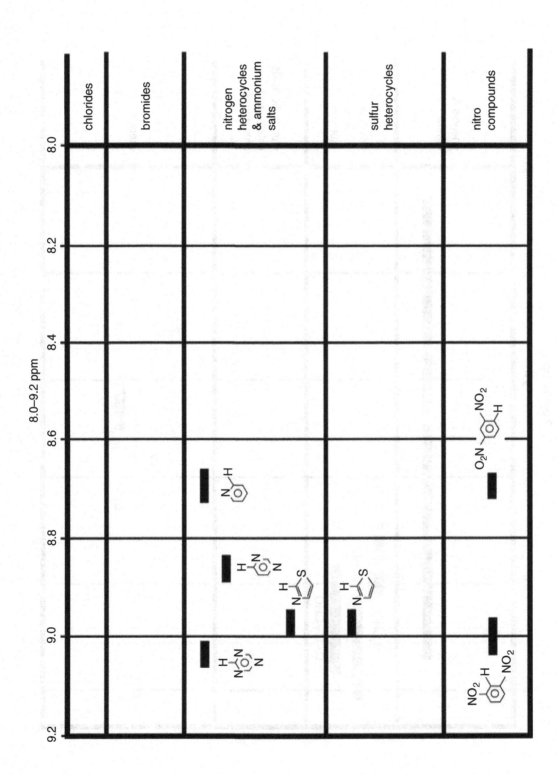

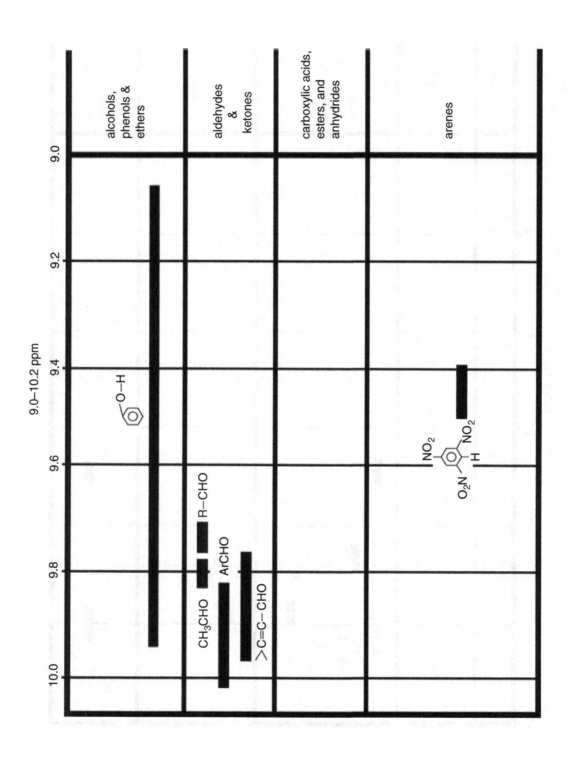

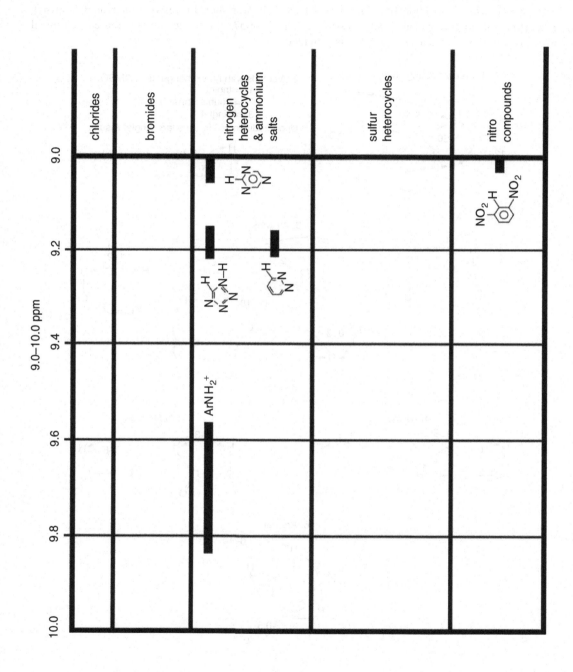

SOME USEFUL ¹H COUPLING CONSTANTS

This section gives the values of some useful proton NMR coupling constants (in Hz). The data are adapted with permission from the work of Dr. C.F. Hammer, Professor Emeritus, Chemistry Department, Georgetown University, Washington, DC 20057. The single numbers indicate a typical average, while in some cases, the range is provided.

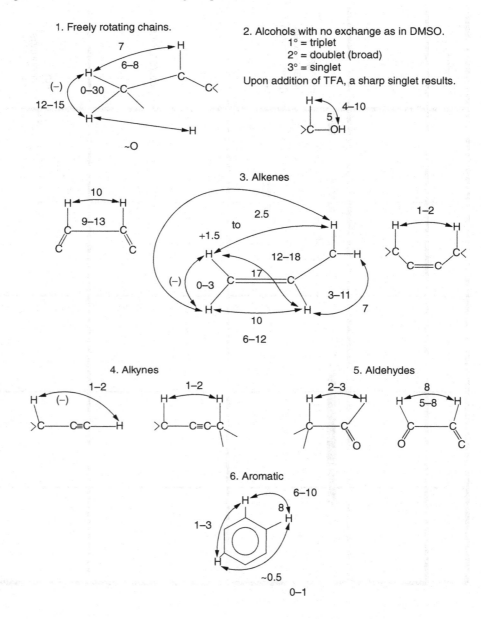

1. Freely rotating chains.

2. Alcohols with no exchange as in DMSO.
   1° = triplet
   2° = doublet (broad)
   3° = singlet
Upon addition of TFA, a sharp singlet results.

3. Alkenes

4. Alkynes

5. Aldehydes

6. Aromatic

^{13}C-NMR ABSORPTIONS OF MAJOR FUNCTIONAL GROUPS

The table below lists the ^{13}C chemical shift ranges (in ppm) with the corresponding functional groups in descending order. A number of typical simple compounds for every family are given to illustrate the corresponding range. The shifts for the carbons of interest are given in parenthesis, either for each carbon as it appears from left to right in the formula, or by the underscore [1–15]. Following the table, correlation charts depicting the ^{13}C chemical shift ranges of various functional groups are presented. The expected peaks attributed to common solvents also appear in the correlation charts.

We provide a list of references that contain many of the ^{13}C chemical shift ranges that appear below [16–44]. This list is certainly not complete and should be augmented with periodic literature searches. Ref. [44] has a compilation of references for the various nuclei.

REFERENCES

1. Yoder, C.H. and Schaeffer, C.D., Jr., *Introduction to Multinuclear NMR: Theory and Application*, Benjamin/Cummings Publishing Co., Menlo Park, CA, 1987.
2. Brown, D.W., A short set of ^{13}C-NMR correlation tables, *J. Chem. Ed.*, 62, 209, 1985.
3. Silverstein, R.M. and Webster F.X., *Spectrometric Identification of Organic Compounds*, 6th ed., John Wiley & Sons, New York, 1998.
4. Becker, E.D., *High Resolution NMR, Theory and Chemical Applications*, 2nd ed., Academic Press, New York, 1980.
5. Gunther, H., *NMR Spectroscopy: Basic Principles, Concepts and Applications in Chemistry*, Wiley, New York, 2003.
6. Kitamaru, R., *Nuclear Magnetic Resonance: Principles and Theory*, Elsevier Science, New York, 1990.
7. Lambert, J.B., Holland, L.N., and Mazzola, E.P., *Nuclear Magnetic Resonance Spectroscopy: Introduction to Principles, Applications and Experimental Methods*, Prentice Hall, Englewood Cliffs, NJ, 2003.
8. Bovey, F.A. and Mirau, P.A., *Nuclear Magnetic Resonance Spectroscopy*, 2nd ed., Academic Press, London, 1988.
9. Harris, R.K. and Mann, B.E., *NMR and the Periodic Table*, Academic Press, London, 1978.
10. Hore, P.J. and Hore, P.J., *Nuclear Magnetic Resonance*, Oxford University Press, Oxford, 1995.
11. Nelson, J.H., *Nuclear Magnetic Resonance Spectroscopy*, 2nd ed., Wiley, New York, 2003.
12. Levy, G.C., Lichter, R.L., and Nelson, G.L., *Carbon-13 Nuclear Magnetic Resonance Spectroscopy*, 2nd ed., Wiley, New York, 1980.
13. Pihlaja, K. and Kleinpeter, E., *Carbon-13 NMR Chemical Shifts in Structural and Stereochemical Analysis*, VCH, New York, 1994.
14. Pouchert, C. and Behnke, J., *Aldrich Library of ^{13}C and ^{1}H FT-NMR Spectra*, Aldrich Chemical Co., Milwaukee, WI, 1993. ISBN-10 0-94163334-9; ISBN-13 978-0-94163334
15. Bruno, T.J. and Svoronos, P.D.N., *Handbook of Basic Tables for Chemical Analysis*, 3rd ed., CRC Press, Boca Raton, FL, 2011.
16. Balci, M., *Basic 1H- and 13C-NMR Spectroscopy*, Elsevier, London, 2005.
17. University of Wisconsin, Organic Structure Determination Bibliography, available online at https://www.chem.wisc.edu/areas/reich/nmr/nmr-biblio.htm, accessed December 2019.

The following bibliography contains information on the NMR absorptions of specific chemical families. This collection is by no means complete, and users should update their own compilations regularly.

Adamantanes

18. Maciel, G.E., Dorn, H.C., Greene, R.L., Kleschick, W.A., Peterson, M.R., Jr., and Wahl, G.H., Jr., ^{13}C Chemical shifts of monosubstituted adamantanes, *Org. Magn. Res.*, 6, 178, 1974.

Amides

19. Jones, R.G. and Wilkins, J.M., Carbon-13 NMR spectra of a series of parasubstituted N, N-dimethylbenzamides, *Org. Magn. Res.*, 11, 20, 1978.

Benzazoles

20. Sohr, P., Manyai, G., Hideg, K., Hankovszky, H., and Lex, L., Benzazoles. XIII. Determination of the E and Z configuration of isomeric 2-(2-benzimidazolyl)-di-and tetra-hydrothiophenes by IR, ^{1}H and ^{13}C NMR spectroscopy, *Org. Magn. Res.*, 14, 125, 1980.

Carbazoles

21. Giraud, J. and Marzin, C., Comparative ^{13}C NMR study of deuterated and undeuterated dibenzo-thiophenes, dibenzofurans, carbazoles, fluorenes, and fluorenones, *Org. Magn. Res.*, 12, 647, 1979.

Chlorinated Compounds

22. Hawkes, G.E., Smith, R.A., and Roberts, J.D., Nuclear magnetic resonance spectroscopy. Carbon-13 chemical shifts of chlorinated organic compounds, *J. Org. Chem.*, 39, 1276, 1974.
23. Mark, V. and Weil, E.D., The isomerization and chlorination of decachlorobi-2,4-cyclopentadien-1-yl, *J. Org. Chem.*, 36, 676, 1971.

Diazoles and Diazines

24. Faure, R., Vincent, E.J., Assef, G., Kister, J., and Metzger, J., Carbon-13 NMR study of substituent effects in the 1,3-diazole and -diazine series, *Org. Magn. Res.*, 9, 688, 1977.

Disulfides

25. Takata, T., Iida, K., and Oae, S., ^{13}C-NMR chemical shifts and coupling constants J_{C-H} of six membered ring systems containing sulfur-sulfur linkage, *Heterocycles*, 15, 847, 1981.
26. Bass, S.W. and Evans, S.A., Jr., Carbon-13 nuclear magnetic resonance spectral properties of alkyl disulfides, thiosulfinates, and thiosulfonates, *J. Org. Chem.*, 45, 710, 1980.
27. Freeman, F. and Angeletakis, C.N., Carbon-13 nuclear magnetic resonance study of the conformations of disulfides and their oxide derivatives, *J. Org. Chem.*, 47, 4194, 1982.

Fluorenes and Fluorenones

28. Giraud, J. and Marzin, C., Comparative 13C NMR study of deuterated and undeuterated dibenzo-thiophenes, dibenzofurans, carbazoles, fluorenes and fluorenones, *Org. Magn. Res.*, 12, 647, 1979.

Furans

29. Giraud, H. and Marzin, C., Comparative ^{13}C NMR study of deuterated and undeuterated dibenzo-thiophenes, dibenzofurans, carbazoles, fluorenes and fluorenones, *Org. Magn. Res.*, 12, 647, 1979.

Imines

30. Allen, M. and Roberts, J.D., Effects of protonation and hydrogen bonding on carbon-13 chemical shifts of compounds containing the >C=N-group, *Can. J. Chem.*, 59, 451, 1981.

Oxathianes

31. Szarek, W.A., Vyas, D.M., Sepulchre, A.M., Gero, S.D., and Lukacs, G., Carbon-13 nuclear magnetic resonance spectra of 1,4-oxathiane derivatives, *Can. J. Chem.*, 52, 2041, 1974.
32. Murray, W.T., Kelly, J.W., and Evans, S.A., Jr., Synthesis of substituted 1,4-oxathianes, mechanistic details of diethoxytriphenylphosphorane—and triphenylphosphine/tetra-chloromethane—promoted cyclodehydrations and ^{13}C NMR spectroscopy, *J. Org. Chem.*, 52, 525, 1987.

Oximes

33. Allen, M. and Roberts, J.D., Effects of protonation and hydrogen bonding on carbon-13 chemical shifts of compounds containing the >C=N-group, *Can. J. Chem.*, 59, 451, 1981.

Polynuclear Aromatics (Naphthalenes, Anthracenes, and Pyrenes)

34. Adcock, W., Aurangzeb, M., Kitching, W., Smith, N., and Doddzell, D., Substituent effects of carbon-13 nuclear magnetic resonance: concerning the π-inductive effect, *Aust. J. Chem.*, 27, 1817, 1974.
35. DuVernet, R. and Boekelheide, V., Nuclear magnetic resonance spectroscopy. Ring-current effects on carbon-13 chemical shifts, *Proc. Nat. Acad. Sci. USA*, 71, 2961, 1974.

Pyrazoles

36. Puar, M.S., Rovnyak, G.C., Cohen, A.I., Toeplitz, B., and Gougoutas, J.Z., Orientation of the sulfoxide bond as a stereochemical probe. Synthesis and ¹H and ¹³C NMR of substituted thiopyrano[4,3-c] pyrazoles, *J. Org. Chem.*, 44, 2513, 1979.

Sulfides

37. Chauhan, M.S., and Still, I.W.J., ¹³C nuclear magnetic resonance spectra of organic sulfur compounds: cyclic sulfides, sulfoxides, sulfones, and thiones, *Can. J. Chem.*, 53, 2880, 1975.
38. Gokel, G.W., Gerdes, H.M., and Dishong, D.M., Sulfur heterocycles. 3. Heterogenous, phase-transfer, and acid catalyzed potassium permanganate oxidation of sulfides to sulfones and a survey of their carbon-13 nuclear magnetic resonance spectra, *J. Org. Chem.*, 45, 3634, 1980.
39. Mohraz, M., Jiam-Qi, W., Heilbronner, E., Solladie-Cavallo, A., and Matloubi-Moghadam, F., Some comments on the conformation of methyl phenyl sulfides, sulfoxides, and sulfones, *Helv. Chim. Acta*, 64, 97, 1981.
40. Srinivasan, C., Perumal, S., Arumugam, N., and Murugan, R., Linear free-energy relationship in naphthalene system-substituent effects on carbon-13 chemical shifts of substituted naphthylmethyl sulfides, *Ind. J. Chem.*, 25A, 227, 1986.

Sulfites

41. Buchanan, G.W., Cousineau, C.M.E., and Mundell, T.C., Trimethylene sulfite conformations: effects of sterically demanding substituents at C-4,6 on ring geometry as assessed by ¹H and ¹³C nuclear magnetic resonance, *Can. J. Chem.*, 56, 2019, 1978.

Sulfonamides

42. Chang, C., Floss, H.G., and Peck, G.E., Carbon-13 magnetic resonance spectroscopy of drugs. Sulfonamides, *J. Med. Chem.*, 18, 505, 1975.

Sulfones (See Also Other Families for the Corresponding Sulfones)

43. Fawcett, A.H., Ivin, K.J., and Stewart, C.D., Carbon-13 NMR spectra of monosulphones and disulphones: substitution rules and conformational effects, *Org. Magn. Res.*, 11, 360, 1978.
44. Gokel, G.W., Gerdes, H.M., and Dishong, D.M., Sulfur heterocycles. 3. Heterogeneous, phase-transfer, and acid catalyzed potassium permanganate oxidation of sulfides to sulfones and a survey of their carbon-13 nuclear magnetic resonance spectra, *J. Org. Chem.*, 45, 3634, 1980.
45. Balaji, T. and Reddy, D.B., Carbon-13 nuclear magnetic resonance spectra of some new arylcyclopropyl sulphones, *Ind. J. Chem.*, 18B, 454, 1979.

Sulfoxides (See Also Other Families for the Corresponding Sulfoxides)

46. Gatti, G., Levi, A., Lucchini, V., Modena, G., and Scorrano, G., Site of protonation in sulphoxides: carbon-13 nuclear magnetic resonance evidence, *J. Chem. Soc. Chem. Comm.*, 251, 1973.
47. Harrison, C.R. and Hodge, P., Determination of the configuration of some penicillin S-oxides by ¹³C nuclear magnetic resonance spectroscopy, *J. Chem. Soc., Perkin Trans. I*, 16, 1772, 1976.

Sulfur Ylides

48. Matsuyama, H., Minato, H., and Kobayashi, M., Electrophilic sulfides (II) as a novel catalyst. V. Structure, nucleophilicity, and steric compression of stabilized sulfur ylides as observed by ^{13}C-NMR spectroscopy, *Bull. Chem. Soc. Jpn*, 50, 3393, 1977.

Thianes

49. Willer, R.L. and Eliel, E.L., Conformational analysis. 34. Carbon-13 nuclear magnetic resonance spectra of saturated heterocycles, 6. Methylthianes, *J. Am. Chem. Soc.*, 99, 1925, 1977.
50. Barbarella, G., Dembech, P., Garbesi, A., and Fara, A., ^{13}C NMR of organosulphur compounds: II. ^{13}C chemical shifts and conformational analysis of methyl substituted thiacyclohexanes, *Org. Magn. Res.*, 8, 469, 1976.
51. Murray, W.T., Kelly, J.W., and Evans, S.A., Jr., Synthesis of substituted 1,4-oxathianes. Mechanistic details of diethoxytriphenyl phosphorane and triphenylphosphine/tetrachloromethane—promoted cyclodehydrations and ^{13}C NMR spectroscopy, *J. Org. Chem.*, 52, 525, 1987.
52. Block, E., Bazzi, A.A., Lambert, J.B., Wharry, S.M., Andersen, K.K., Dittmer, D.C., Patwardhan, B.H., and Smith, J.H., Carbon-13 and oxygen-17 nuclear magnetic resonance studies of organosulfur compounds: the four-membered-ring-sulfone effect, *J. Org. Chem.*, 45, 4807, 1980.
53. Rooney, R.P, and Evans, S.A., Jr., Carbon-13 nuclear magnetic resonance spectra of trans-1-thiadecalin, trans-1,4-dithiadecalin, trans-1,4-oxathiadecalin, and the corresponding sulfoxides and sulfones, *J. Org. Chem.*, 45, 180, 1980.

Thiazines

54. Fronza, G., Mondelli, R., Scapini, G., Ronsisvalle, G., and Vittorio, F., ^{13}C NMR of N-heterocycles. Conformation of phenothiazines and 2,3-diazaphenothiazines, *J. Magn. Res.*, 23, 437, 1976.

Thiazoles

55. Harrison, C.R. and Hodge, P., Determination of the configuration of some penicillin S-oxides by ^{13}C nuclear magnetic resonance spectroscopy, *J. Chem. Soc., Perkin Trans. I*, 16, 1772, 1976.
56. Chang, G., Floss, H.G., and Peck, G.E., Carbon-13 magnetic resonance spectroscopy of drugs. Sulfonamides, *J. Med. Chem.*, 18, 505, 1975.
57. Elguero, J., Faure, R., Lazaro, R., and Vincent, E.J., ^{13}C NMR study of benzothiazole and its nitroderivatives, *Bull. Soc. Chim. Belg.*, 86, 95, 1977.
58. Faure, R., Galy, J.P., Vincent, E.J., and Elguero, J., Study of polyheteroaromatic pentagonal heterocycles by carbon-13 NMR. Thiazoles and thiazolo[2,3-e]tetrazoles, *Can. J. Chem.*, 56, 46, 1978.

Thiochromanones

59. Chauhan, M.S. and Still, I.W.J., ^{13}C nuclear magnetic resonance spectra of organic sulfur compounds: cyclic sulfides, sulfoxides, sulfones and thiones, *Can. J. Chem.*, 53, 2880, 1975.

Thiones

60. Chauhan, M.S. and Still, I.W.J., ^{13}C nuclear magnetic resonance spectra of organic sulfur compounds: cyclic sulfides, sulfoxides, sulfones and thiones, *Can. J. Chem.*, 53, 2880, 1975.

Thiophenes

61. Perjessy, A., Janda, M., and Boykin, D.W., Transmission of substituent effects in thiophenes. Infrared and carbon-13 nuclear magnetic resonance studies, *J. Org. Chem.*, 45, 1366, 1980.
62. Giraud, J. and Marzin, C., Comparative ^{13}C NMR study of deuterated and undeuterated dibenzo-thiophenes, dibenzofurans, carbazoles, fluorenes and flourenones, *Org. Magn. Res.*, 12, 647, 1979.
63. Clark, P.D., Ewing, D.F., and Scrowston, R.M., NMR studies of sulfur heterocycles: III. ^{13}C spectra of benzo[b]thiophene and the methylbenzo[b]thiophenes, *Org. Magn. Res.*, 8, 252, 1976.
64. Osamura, Y., Sayanagi, O., and Nishimoto, K., C-13 NMR chemical shifts and charge densities of substituted thiophenes—the effect of vacant dπ orbitals, *Bull. Chem. Soc. Jpn*, 49, 845, 1976.

65. Balkau, F., Fuller, M.W., and Heffernan, M.L., Deceptive simplicity in ABMX N.M.R. spectra. I. Dibenzothiophen and 9.9'-dicarbazyl, *Aust. J. Chem.*, 24, 2293, 1971.

66. Geneste, P., Olive, J.L., Ung, S.N., El Faghi, M.E.A., Easton, J.W., Beierbeck, H., and Saunders, J.K., Carbon-13 nuclear magnetic resonance study of benzo[b]thiophenes and benzo[b]thiophene S-oxides and S, S-dioxides, *J. Org. Chem.*, 44, 2887, 1979.

67. Benassi, R., Folli, U., Iarossi, D., Schenetti, L., and Tadei, F., Conformational analysis of organic carbonyl compounds. Part 3. A ^{1}H and ^{13}C nuclear magnetic resonance study of formyl and acetyl derivatives of benzo[b]thiophen, *J. Chem. Soc., Perkin Trans. II*, 7, 911, 1983.

68. Kiezel, L., Liszka, M., and Rutkowski, M., Carbon-13 magnetic resonance spectra of benzothiophene and dibenzothiophene, *Spec. Lett.*, 12, 45, 1979.

69. Fujieda, K., Takahashi, K., and Sone, T., The C-13 NMR spectra of thiophenes. II, 2-substituted thiophenes, *Bull. Chem. Soc. Jpn.*, 58, 1587, 1985.

70. Satonaka, H. and Watanabe, M., NMR spectra of 2-(2-nitrovinyl) thiophenes, *Bull. Chem. Soc. Jpn.*, 58, 3651, 1985.

71. Stuart, J.G., Quast, M.J., Martin, G.E., Lynch, V.M., Simmonsen, H., Lee, M.L., Castle, R.N., Dallas, J.L., John B.K., and Johnson, L.R.F., Benzannelated analogs of phenanthro [1,2-b]-[2,1-b]thiophene: Synthesis and structural characterization by two-dimensional NMR and x-ray techniques, *J. Heterocyclic Chem.*, 23, 1215, 1986.

Thiopyrans

72. Senda, Y., Kasahara, A., Izumi, T., and Takeda, T., Carbon-13 NMR spectra of 4-chromanone, 4H-1-benzothiopyran-4-one, 4H-1-benzothiopyran-4-one 1,1-dioxide, and their substituted homologs, *Bull. Chem. Soc. Jpn*, 50, 2789, 1977.

Thiosulfinates and Thiosulfonates

73. Bass, S.W. and Evans, S.A., Jr., Carbon-13 nuclear magnetic resonance spectral properties of alkyl disulfides, thiosulfinates, and thiosulfonates, *J. Org. Chem.*, 45, 710, 1980.

^{13}C NMR Chemical Shift Ranges of Major Functional Groups

δ (ppm)	Group	Family	Example (δ of Underlined Carbon)	
220–165	>C=O	Ketones	$(CH_3)_2\underline{C}O$	(206.0)
			$(CH_3)_2CH\underline{C}OCH_3$	(212.1)
		Aldehydes	$CH_3\underline{C}HO$	(199.7)
		α, β-Unsaturated carbonyls	$CH_3CH=CH\underline{C}HO$	(192.4)
			$CH_2=CH\underline{C}OCH_3$	(169.9)
		Carboxylic acids	$H\underline{C}O_2H$	(166.0)
			$CH_3\underline{C}O_2H$	(178.1)
		Amides	$H\underline{C}ONH_2$	(165.0)
			$CH_3\underline{C}ONH_2$	(172.7)
		Esters	$CH_3\underline{C}O_2CH_2CH_3$	(170.3)
			$CH_2=CH\underline{C}O_2CH_3$	(165.5)
140–120	>C=C<	Aromatic	C_6H_6	(128.5)
		Alkenes	$CH_2=CH_2$	(123.2)
			$CH_2=CHCH_3$	(115.9, 136.2)
			$CH_2=CHCH_2Cl$	(117.5, 133.7)
			$CH_3CH=CHCH_2CH_3$	(124.1, 132.7)
125–115	–C≡N	Nitriles	$CH_3-\underline{C}≡N$	(117.7)
80–70	–C≡C–	Alkynes	$H\underline{C}≡CH$	(71.9)
			$CH_3\underline{C}≡CCH_3$	(73.9)
70–45	>C–O	Esters	$\underline{C}H_3OOCCH_2CH_3$	(57.6, 67.9)
	>C–O	Alcohols	$HO\underline{C}H_3$	(49.0)
			$HO\underline{C}H_2CH_3$	(57.0)
40–20	>C–NH₂	Amines	$\underline{C}H_3NH_2$	(26.9)
			$CH_3\underline{C}H_2NH_2$	(35.9)
30–15	–S–CH₃	Sulfides (thioethers)	$C_6H_5–S–\underline{C}H_3$	(15.6)
30– (–2.3)	>CH–	Alkanes, cycloalkanes	$\underline{C}H_4$	(–2.3)
			$\underline{C}H_3CH_3$	(5.7)
			$\underline{C}H_3\underline{C}H_2CH_3$	(15.8, 16.3)
			$\underline{C}H_3\underline{C}H_2CH_2CH_3$	(13.4, 25.2)
			$\underline{C}H_3\underline{C}H_2\underline{C}H_2CH_2CH_3$	(13.9, 22.8, 34.7)
			Cyclohexane	(26.9)

^{13}C NMR CORRELATION CHARTS OF MAJOR FUNCTIONAL GROUPS

The following correlation tables provide the regions of NMR absorptions of major chemical families.

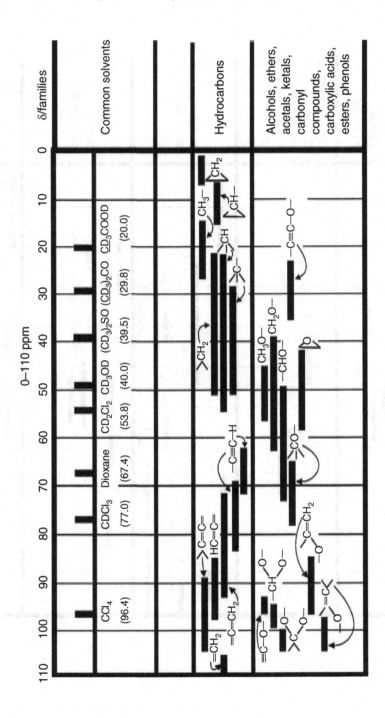

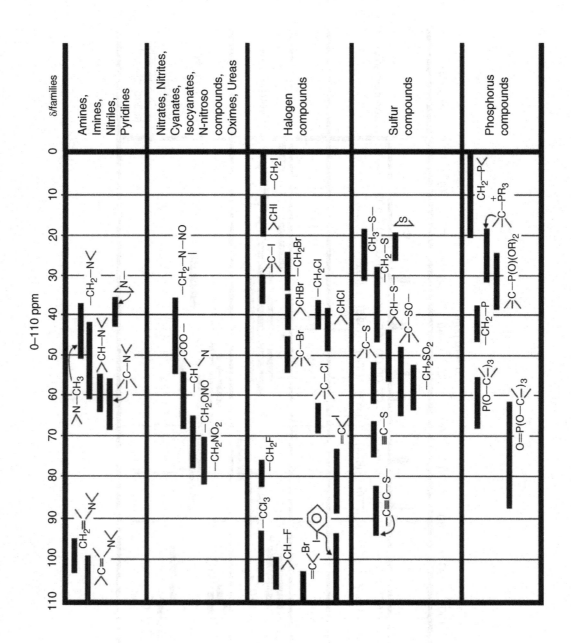

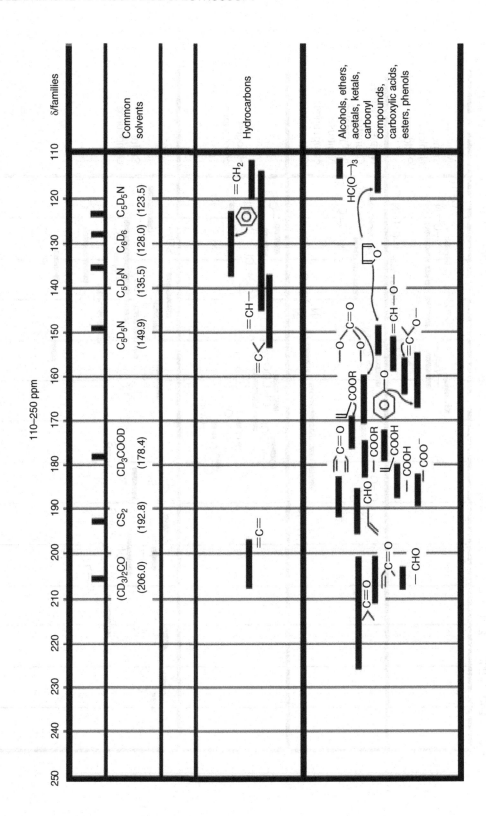

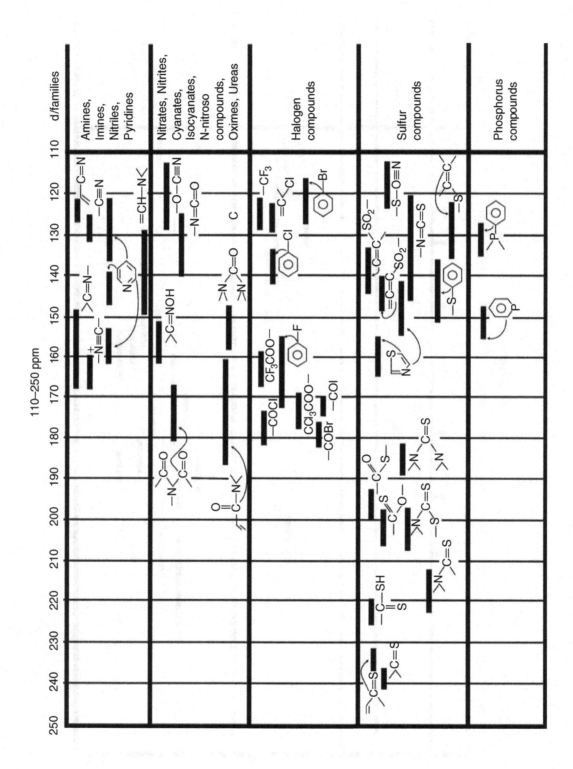

ADDITIVITY RULES IN ^{13}C-NMR CORRELATION TABLES

The wide chemical shift range (~250 ppm) of ^{13}C NMR is responsible for the considerable change of a chemical shift noted when a slight inductive, mesomeric, or hybridization change occurs on a neighboring atom. Following the various empirical correlations in ^{1}H NMR [1–7], D.W. Brown [8] has developed a short set of ^{13}C-NMR correlation tables. This section covers a part of those as adopted by Yoder and Schaeffer [9] and Clerk et al. [10]. The reader is advised to refer to Ref. [8], and should the need for some specific data on more complicated structures arise, additional sources are provided [11–19].

REFERENCES

1. Shoolery, J.N., *Varian Associates Technical Information Bulletin*, 2(3), Palo Alto, CA, 1959.
2. Bell, H.M., Bowles, D.B., and Senese, F., Additive NMR chemical shift parameters for deshielded methine protons, *Org. Magn. Res.*, 16, 285, 1981.
3. Matter, U.E., Pascual, C., Pretsch, E., Pross, A., Simon, W., and Sternhell, S., Estimation of the chemical shifts of olefinic protons using additive increments. II. Compilation of additive increments for 43 functional groups, *Tetrahedron*, 25, 691, 1969.
4. Matter, U.E., Pascual, C., Pretsch, E., Pross, A., Simon, W., and Sternhell, S., Estimation of the chemical shifts of olefinic protons using additive increments. III. Examples of utility in N.M.R. studies and the identification of some structural features responsible for deviations from additivity, *Tetrahedron*, 25, 2023, 1969.
5. Jeffreys, J.A.D., A rapid method for estimating NMR shifts for protons attached to carbon, *J. Chem. Educ.*, 56, 806, 1979.
6. Mikolajczyk, M., Grzeijszczak, S., and Zatorski, A., Organosulfur compounds IX: NMR and structural assignments in α, β-unsaturated sulphoxides using additive increments method, *Tetrahedron*, 32, 969, 1976.
7. Friedrich, E.C. and Runkle, K.G., Empirical NMR chemical shift correlations for methyl and methylene protons, *J. Chem. Educ.*, 61, 830, 1984.
8. Brown, D.W., A short set of ^{13}C-NMR correlation tables, *J. Chem. Educ.*, 62, 209, 1985.
9. Yoder, C.H. and Schaeffer, C.D., Jr., *Introduction to Multinuclear NMR*, Benjamin/Cummings Publishing Co., Menlo Park, CA, 1987.
10. Clerk, J.T., Pretsch, E., and Seibl, J., *Structural Analysis of Organic Compounds by Combined Application of Spectroscopic Methods*, Elsevier, Amsterdam, 1981.
11. Silverstein, R.M. and Webster F.X., *Spectrometric Identification of Organic Compounds*, 6th ed., John Wiley & Sons, New York, 1998.
12. Gunther, H., *NMR Spectroscopy: Basic Principles, Concepts and Applications in Chemistry*, Wiley, New York, 2003.
13. Kitamaru, R., *Nuclear Magnetic Resonance: Principles and Theory*, Elsevier Science, New York, 1990.
14. Lambert, J.B., Holland, L.N., and Mazzola, E.P., *Nuclear Magnetic Resonance Spectroscopy: Introduction to Principles, Applications and Experimental Methods*, Prentice Hall, Englewood Cliffs, NJ, 2003.
15. Bovey, F.A. and Mirau, P.A., *Nuclear Magnetic Resonance Spectroscopy*, 2nd ed., Academic Press, London, 1988.
16. Harris, R.K. and Mann, B.E., *NMR and the Periodic Table*, Academic Press, London, 1978.
17. Nelson, J.H., *Nuclear Magnetic Resonance Spectroscopy*, 2nd ed., Wiley, New York, 2003.
18. Bruno, T.J. and Svoronos, P.D.N., *Handbook of Basic Tables for Chemical Analysis*, 3rd ed., CRC Press, Boca Raton, FL, 2011.
19. Bruno, T.J. and Svoronos, P.D.N., *CRC Handbook of Fundamental Spectroscopic Correlation Charts*, CRC Press, Boca Raton, FL, 2006.

Alkanes

The chemical shift (in ppm) of C^i can be calculated from the following empirical equation:

$$\Delta^i = -2.3 + \Sigma A_i,$$

where ΣA_i is the sum of increments allowed for various substituents depending on their positions (α, β, γ, and δ) relative to the ^{13}C in question, and (-2.3) is the chemical shift for methane relative to tetramethylsilane (TMS).

^{13}C Chemical Shift Increments for A, the Shielding Term for Alkanes and Substituted Alkanes [9,10]

Substituent	Increments			
	A	β	γ	Δ
>C– (sp³)	9.1	9.4	–2.5	0.3
>C=C< (sp²)	19.5	6.9	–2.1	0.4
C≡C– (sp)	4.4	5.6	–3.4	–0.6
C₆H₅	22.1	9.3	–2.6	0.3
–F	70.1	7.8	–6.8	0.0
–Cl	31.0	10.0	–5.1	–0.5
–Br	18.9	11.0	–3.8	–0.7
–I	–7.2	10.9	–1.5	–0.9
–OH	49.0	10.1	–6.2	0.0
–OR	49.0	10.1	–6.2	0.0
–CHO	29.9	–0.6	–2.7	0.0
–COR	22.5	3.0	–3.0	0.0
–COOH	20.1	2.0	–2.8	0.0
–COO⁻	24.5	3.5	–2.5	0.0
–COCl	33.1	2.3	–3.6	0.0
–COOR	22.6	2.0	–2.8	0.0
–OOCR	5.5	6.5	–6.0	
–N<	28.3	11.3	–5.1	
–NH₃⁺	26.0	7.5	–4.6	0.0
[>N<]⁺	30.7	5.4	–7.2	–1.4
–ONO	54.3	6.1	–6.5	–0.5
–NO₂	61.6	3.1	–4.6	–1.0
–CON<	22.0	2.6	–3.2	–0.4
–NHCO–	31.3	8.3	–5.7	0.0
–C≡N	3.1	2.4	–3.3	–0.5
–NC	31.5	7.6	–3.0	0.0
–S–	10.6	11.4	–3.6	–0.4
–S–CO–	17.0	6.5	–3.1	0.0
–SO–	31.1	9.0	–3.5	0.0
–SO₂Cl	54.5	3.4	–3.0	0.0
–SCN	23.0	9.7	–3.0	0.0
–C(=S)N–	33.1	7.7	–2.5	0.6
–C=NOH (syn)	11.7	0.6	–1.8	0.0
–C=NOH (anti)	16.1	4.3	–1.5	0.0
R₁ R₂ R₃Sn				
R₁, R₂, and R₃=organic substituents	–5.2	4.0	–0.3	0.0

Thus, the ^{13}C shift for C^i in 2-pentanol is predicted to be

$$\overset{\beta}{C}H_3 - \overset{\alpha}{C}H_2 - \overset{i}{C}H_2 - \overset{\alpha'}{C}H(OH) - \overset{\beta'}{C}H_3$$

$$\delta^i = (-2.3) + \left[\underset{\alpha}{9.1} + \underset{\beta}{9.4} + \underset{\alpha'}{9.1} + \underset{\beta'}{9.4} + \underset{OH}{10.1} \right] = 44.8 \text{ ppm}$$

Alkenes

For a simple olefin of the type

$$-\overset{\gamma}{C}-\overset{\beta}{C}-\overset{\alpha}{C}-\overset{i}{C}=\overset{\alpha'}{C}-\overset{\beta'}{C}-\overset{\gamma'}{C}$$

$$\delta^i = 122.8 + \Sigma A_i$$

where $A_\alpha = 10.6$, $A_\beta = 7.2$, $A_\gamma = -1.5$, $A_{\alpha'} = -7.9$, $A_{\beta'} = -1.8$, $A_{\gamma'} = 1.5$, and 122.8 is the chemical shift of the sp^2 carbon in ethene.

If the olefin is in the cis- configuration, an increment of -1.1 ppm must be added.

Thus, the ^{13}C shift for C-3 in cis-3-hexene is predicted to be

$$\overset{\beta}{C}H_3 - \overset{\alpha}{C}H_2 - \overset{i}{C}H = CH - \overset{\alpha'}{C}H_2 - \overset{\beta'}{C}H_3$$

$$\delta^i = 122.8 + \left[\underset{(\alpha)}{10.6} + \underset{(\beta)}{7.2} - \underset{(\alpha')}{1.5} + \underset{(\beta')}{7.9} \right] + \underset{(cis)}{(-1.1)} = 130.1 \text{ ppm}$$

Alkynes

For a simple alkyne of the type

$$-\overset{\beta}{C}-\overset{\alpha}{C}-\overset{i}{C}=\overset{\alpha'}{C}-\overset{\beta'}{C}-$$

$$\delta^i = 71.9 + \Sigma A_i$$

where increments A are given in the table below, and 71.9 is the chemical shift of the sp carbon in acetylene [9].

^{13}C Chemical Shift Increments for A, the Shielding Term for Alkynes

Substituents	Increments			
	α	β	α'	β'
C (sp³)	6.9	4.8	-5.7	2.3
–CH₃	7.0		-5.7	
–CH₂CH₃	12.0		-3.5	
–CH(CH₃)₂	16.0		-3.5	
–CH₂OH	11.1		1.9	
–COCH₃	31.4		4.0	
–C₆H₅	12.7		6.4	
–CH=CH₂	10.0		11.0	
–Cl	-12.0		-15.0	

Thus, the ^{13}C shift for C-A in 1-phenyl-1-propyne is predicted to be

$$C_6H_5\underset{B}{-C}\equiv\underset{A}{C}-CH_3$$

$$\delta^i = 71.9 + 7.0 + 6.4 = 85.3\,\text{ppm}$$

while the ^{13}C shift for C-B in the same compound is predicted to be

$$C_6H_5\underset{B}{-C}\equiv\underset{A}{C^i}-CH_3$$

$$\delta^i = 71.9 + 12.7 - 5.7 = 78.9\,\text{ppm}$$

Benzenoid Aromatics

For a benzene derivative, C_6H_5-X, where X = substituent

$$\delta^i = 128.5 + \Sigma A_i$$

where ΣA_i is the sum of increments given below and 128.5 is the chemical shift of benzene [9,10].

^{13}C Chemical Shift Increments for A, the Shielding Term for Benzenoid Aromatics $X-C_6H_5$ Where X = Substituent

Substituent X	C^i	Ortho	Meta	Para
		Increments		
$-CH_3$	9.3	0.8 [9], 0.6 [10]	0.0	-2.9 [9], -3.1 [10]
$-CH_2CH_3$	15.8 [9], 15.7 [10]	-0.4 [9], -0.6 [10]	-0.1	-2.6 [9], -2.8 [10]
$-CH(CH_3)_2$	20.3 [9], 20.1 [10]	-1.9 [9], -2.0 [10]	0.1 [9], 0.0 [10]	-2.4 [9], -2.5 [10]
$-C(CH_3)_3$	22.4 [9], 22.1 [10]	-3.1 [9], -3.4 [10]	-0.2 [9], 0.4 [10]	-2.9 [9], -3.1 [10]
$-CH=CH_2$	7.6	-1.8	-1.8	-3.5
$-C\equiv CH$	-6.1	3.8	0.4	-0.2
$-C_6H_5$	13.0	-1.1	0.5	-1.0
$-CHO$	8.6 [9], 9.0 [10]	1.3 [9], 1.2 [10]	0.6 [9], 1.2 [10]	5.5 [9], 6.0 [10]
$-COCH_3$	9.1 [9], 9.3 [10]	0.1 [9], 0.2 [10]	0.0 [9], 0.2 [10]	4.2
$-CO_2H$	2.1 [9], 2.4 [10]	1.5 [9], 1.6 [10]	0.0 [9], -0.1 [10]	5.1 [9], 4.8 [10]
$-CO_2-$	7.6	0.8	0.0	2.8
$-CO_2R$	2.1	1.2	0.0	4.4
$-CONH_2$	5.4	-0.3	-0.9	5.0
$-CN$	-15.4 [9], -16.0 [10]	3.6 [9], 3.5 [10]	0.6 [9], 0.7 [10]	3.9 [9], 4.3 [10]
$-Cl$	6.2 [9], 6.4 [10]	0.4 [9], 0.2 [10]	1.3 [9], 10.0 [10]	-1.9 [9], -2.0 [10]
$-OH$	26.9	-12.7	1.4	-7.3
$-O-$	39.6 [10]	-8.2 [10]	1.9 [10]	-13.6 [10]
$-OCH_3$	31.4 [9], 30.2 [10]	-14.4 [9], -14.7 [10]	1.0 [9], 0.9 [10]	-7.7 [9], -8.1 [10]
$-OC_6H_5$	29.1	-9.5	0.3	-5.3
$-OC(=O)CH_3$	23.0	-6.4	1.3	-2.3
$-NH_2$	18.7 [9], 19.2 [10]	-12.4	1.3	-9.5
$-NHCH_3$	21.7 [10]	-16.2 [10]	0.7 [10]	-11.8 [10]
$-N(CH_3)_2$	22.4	-15.7	0.8	-11.8
$-NO_2$	20.0 [9], 19.6 [10]	-4.8 [9], -5.3 [10]	0.9 [9], 0.8 [10]	5.8 [9], 6.0 [10]
$-SH$	2.2	0.7	0.4	-3.1
$-SCH_3$	9.9 [10]	-2.0 [10]	0.1 [10]	-3.7 [10]
$-SO_3H$	15.0	-2.2	1.3	3.8

As an example, the ^{13}C shift for the benzene carbon (C^i) carrying the carbonyl in 3,5-dinitroacetophenone, $CH_3C(=O)(C_6H_3)(NO_2)_2$ is predicted to be

$$C^i = 128.5 + 9.1 + 2(0.9) = 132.4 \, ppm$$

¹⁵N CHEMICAL SHIFTS FOR COMMON STANDARDS

The following table lists the ^{15}N chemical shifts (in ppm) for common standards. The estimated uncertainty is less than 0.1 ppm. Nitromethane (according to Levy and Lichter [1]) is the most suitable primary measurement reference but has the disadvantage of lying in the low-field end of the spectrum. Thus, ammonia (which lies in the most upfield region) is the most suitable for routine experimental use [1–8].

REFERENCES

1. Levy, G.C. and Lichter, R.L., *Nitrogen-15 Nuclear Magnetic Resonance Spectroscopy*, John Wiley & Sons, New York, 1979.
2. Lambert, J.B., Shurvell, H.F., Verbit, L., Cooks, R.G., and Stout, G.H., *Organic Structural Analysis*, MacMillan, New York, 1976.
3. Witanowski, M., Stefaniak, L., Szymanski, S., and Januszewski, H., External neat nitromethane scale for nitrogen chemical shifts, *J. Magn. Res.*, 28, 217, 1977.
4. Srinivasan, P.R. and Lichter, R.L., Nitrogen-15 nuclear magnetic resonance spectroscopy. Evaluation of chemical shift references, *J. Magn. Res.*, 28, 227, 1977.
5. Briggs, J.M. and Randall, E.W., Nitrogen-15 chemical shifts in concentrated aqueous solutions of ammonium salts, *Mol. Phys.*, 26, 699, 1973.
6. Becker, E.D., Proposed scale for nitrogen chemical shifts, *J. Magn. Res.*, 4, 142, 1971.
7. Bruno, T.J. and Svoronos, P.D.N., *Handbook of Basic Tables for Chemical Analysis*, 3rd ed., CRC Press, Boca Raton, FL, 2011.
8. Bruno, T.J. and Svoronos, P.D.N., *CRC Handbook of Fundamental Spectroscopic Correlation Charts*, CRC Press, Boca Raton, FL, 2006.

Compound	Formula	Conditions	Chemical Shift (ppm)
Ammonia	NH_3	Vapor (0.5 MPa)	−15.9
		Liquid (25°)	0.0
		Anhydrous liquid (−50°)	3.37
Ammonium nitrate	NH_4NO_3	Aqueous HNO_3	21.60
		Aqueous solution (saturated)	20.68
Ammonium chloride	NH_4Cl	2.9 M (in 1 M HCl)	24.93
		1.0 M (in 10 M HCl)	30.31
		Aqueous solution (saturated)	27.34
Tetraethylammonium chloride	$(C_2H_5)_4N^+Cl^-$	Aqueous solution (saturated)	43.54
		Chloroform solution (saturated)	45.68
		Aqueous solution (0.3 M)	63.94
		Aqueous solution (saturated)	64.39
		Chloroform solution (0.075 M)	65.69
Tetramethyl urea	$[(CH_3)_2N]_2CO$	Neat	62.50
Dimethylformamide (DMF)	$(CH_3)_2NCHO$	Neat	103.81
Nitric acid (aqueous solution)	HNO_3	1 M	375.80
		2 M	367.84
		9 M	365.86
		10 M	362.00
		15.7 M	348.92
Sodium nitrate	$NaNO_3$	Aqueous solution (saturated)	376.53
Ammonium nitrate	NH_4NO_3	Aqueous solution (saturated)	376.25
		5 M (in 2 M HNO_3)	375.59
		4 M (in 2 M HNO_3)	374.68
Nitromethane	CH_3NO_2	1:1 (v/v) in $CDCl_3$	379.60
		0.03M $Cr(acac)_3$ neat	380.23

^{15}N CHEMICAL SHIFTS OF MAJOR CHEMICAL FAMILIES

The following table contains ^{15}N chemical shifts of various organic nitrogen compounds. Chemical shifts are expressed relative to different standards (NH_3, NH_4Cl, CH_3NO_2, NH_4NO_3, HNO_3, etc.) and are interconvertible. Chemical shifts are sensitive to hydrogen bonding and are solvent dependent as seen in the case of pyridine (see note b below). Consequently, the reference as well as the solvent should always accompany chemical shift data. No data are given on peptides and other biochemical compounds. All shifts are relative to ammonia unless otherwise specified. A section of "miscellaneous" data gives the chemical shift of special compounds relative to unusual standards [1–15].

REFERENCES

1. Levy, G.C. and Lichter, R.L., *Nitrogen-15 Nuclear Magnetic Resonance Spectroscopy*, John Wiley & Sons, New York, 1979.
2. Yoder, C.H. and Schaeffer, C.D., Jr., *Introduction to Multinuclear NMR*, Benjamin/Cummings, Menlo Park, CA, 1987.
3. Duthaler, R.O. and Roberts, J.D., Effects of solvent, protonation, and N-alkylation on the ^{15}N chemical shifts of pyridine and related compounds, *J. Am. Chem. Soc.*, 100, 4969, 1978.
4. Duthaler, R.O. and Roberts, J.D., Steric and electronic effects on ^{15}N chemical shifts of saturated aliphatic amines and their hydrochlorides, *J. Am. Chem. Soc.*, 100, 3889, 1978.
5. Kozerski, L. and von Philipsborn, W., ^{15}N chemical shifts as a conformational probe in enaminones: a variable temperature study at natural isotope abundance, *Org. Magn. Res.*, 17, 306, 1981.
6. Duthaler, R.O. and Roberts, J.D., Steric and electronic effects on ^{15}N chemical shifts of piperidine and decahydroquinoline hydrochlorides, *J. Am. Chem. Soc.*, 100, 3882, 1978.
7. Duthaler, R.O. and Roberts, J.D., Nitrogen-15 nuclear magnetic resonance spectroscopy. Solvent effects on the ^{15}N chemical shifts of saturated amines and their hydrochlorides, *J. Magn. Res.*, 34, 129, 1979.
8. Psota, L., Franzen-Sieveking, M., Turnier, J., and Lichter, R.L., Nitrogen nuclear magnetic resonance spectroscopy. Nitrogen-15 and proton chemical shifts of methylanilines and methylanilinium ions, *Org. Magn. Res.*, 11, 401, 1978.
9. Subramanian, P.K., Chandra Sekara, N., and Ramalingam, K., Steric effects on nitrogen-15 chemical shifts of 4-aminooxanes (tetrahydropyrans), 4-amino-thianes, and the corresponding N, N-dimethyl derivatives. Use of nitrogen-15 shifts as an aid in stereochemical analysis of these heterocyclic systems, *J. Org. Chem.*, 47, 1933, 1982.
10. Schuster, I.I. and Roberts, J.D., Proximity effects on nitrogen-15 chemical shifts of 8-substituted 1-nitronaphthalenes and 1-naphthylamines, *J. Org. Chem.*, 45, 284, 1980.
11. Kupce, E., Liepins, E., Pudova, O., and Lukevics, E., Indirect nuclear spin-spin coupling constants of nitrogen-15 to silicon-29 in silylamines, *J. Chem. Soc. Chem. Comm.*, 9, 581, 1984.
12. Allen, M. and Roberts, J.D., Effects of protonation and hydrogen bonding on nitrogen-15 chemical shifts of compounds containing the >C=N-group, *J. Org. Chem.*, 45, 130, 1980.
13. Brownlee, R.T.C., and Sadek, M., Natural abundance ^{15}N chemical shifts in substituted benzamides and thiobenzamides, *Magn. Res. Chem.*, 24, 821, 1986.
14. Dega-Szafran, Z., Szafran, M., Stefaniak, L., Brevard, C., and Bourdonneau, M., Nitrogen-15 nuclear magnetic resonance studies of hydrogen bonding and proton transfer in some pyridine trifluoroacetates in dichloromethane, *Magn. Res. Chem.*, 24, 424, 1986.
15. Lambert, J.B., Shurvell, H.F., Verbit, L., Cooks, R.G., and Stout, G.H., *Organic Structural Analysis*, MacMillan, New York, 1976.

Chemical Shift Range (ppm)	Family	Example (δ)
<930	Nitroso compounds	C_6H_5–NO (913, 930)
608	Sodium nitrite	$NaNO_2$
~500	Azo compounds	C_6H_5–N=N–C_6H_5 (510)
380–350	Nitro compounds	$C_6H_5NO_2$ (370.3); CH_3NO_2 (380.2); 4–F–C_6H_4–NO_2 (368.5); 1,3-$(NO_2)_2C_6H_4$ (365.4)
367	Nitric acid (8.57 M)	HNO_3
360–325	Nitramines	CH_3NHNO_2 (355.6); $CH_3O_2CNHNO_2$ (334.9)
350–300	Pyridines	C_5H_5N (317)[a] (gas); 4–CH_3–C_5H_4N (309.3); 4–NH_2–C_5H_4N (271.5); 4–NC–C_5H_4N (327.9)
~310	Imines (aromatic)	$(C_6H_5)_2C$=NH (308); C_6H_5CH=NCH_3 (318); C_6H_5CH=NC_6H_5 (326)
310.1	Nitrogen (gas)	N_2
250–200	Pyridinium salts	$C_5H_5NH^+$ (215)
260–175	Cyanides (nitriles)	CH_3CN (239.5, 245); C_6H_5CN (258.7); KCN 177.8
~160	Pyrroles	C_4H_4NH (158)
	Isonitriles	CH_3NC (162)
~150	Thioamides	CH_3C(=S)NH_2 (150.2)
120–110	Lactams	HN$(CH_2)_3$C=O (five-membered ring; 114.7) HN$(CH_2)_6$C=O (eight-membered ring; 117.7)
110–100	Amides	$C_6H_5CONH_2$ (100); CH_3CONH_2 (103.4); $CH_3CONHCH_3$ (105.8); $CH_3CON(CH_3)_2$ (103.8); HCONH_2 (108.5)
125–90	Sulfonamides	$CH_3SO_2NH_2$ (95); $C_6H_5SO_2NH_2$ (94.3)
~100	Hydrazines	$C_6H_5NHNHC_6H_5$ (96)
110–60	Ureas	$[H_2N]_2CO$ (75, 82); $[(CH_3)_2N]_2CO$ (63.5); $[C_6H_5NH]_2CO$ (107.7)
100–70	Aminophosphines	$C_6H_5NHP(CH_3)_2$ (71.1)
	Aminophosphine oxides	$C_6H_5NHPO(CH_3)_2$ (86.6)
70–50	Aromatic amines	$C_6H_5NH_2$ (55, 59), (–322.3)[b]; $C_6H_5NH_3^+$ (48), (–326.4)[b], 26.1[c]; p–O_2N–C_6H_4–NH_2 (70)
40–0	Aliphatic amines	CH_3NH_2 (1.3), (–371)[b]; $(CH_3)_2NH$ (–363.3)[b], (–364.9)[e]; 6.7[d]; $(CH_3)_3N$ (–356.9)[b], (–360.7)[e], 13.0[d]
50–10	Isonitriles	CH_3NCO (14.1); C_6H_5NCO (46.5)
65–20	Ammonium salts	NH_4Cl (26.1)[d]; CH_3NH_3Cl (24.5); $(CH_3)_2NH_2Cl$ (26.6); $(CH_3)_3NHCl$ (33.8); $(CH_3)_4NCl$ (44.7)
~15	Isocyanates	CH_3NCO (14.1)

(Continued)

Chemical Shift Range (ppm)	Family	Example (1)
		Miscellaneous
(−130) to (−110) and ~(212)	Imidazoles	N-methylimdazole (−111.4, pyridine N and −215.7, pyrrole N)[b]
(−345) to (−310)[b]	Piperidine, hydrochloride salts	Piperidinium hydrochloride (−344.8); 2-methyl piperdinium hydrochloride (−322.1)[e]
	Decahydroquinolines, hydrochloride salts	Trans-decahydroquinolinium hydrochloride (−322.5); cis-decahydroquinolinium hydrochloride (−328.5)
(−293) to (−280)[f]	Enaminones	$CH_3C(=O)CH=CHNHCH_3$ [(E) − (−294.2); (Z) − (−285.9)]
35–15[g]	4-Aminotetrahydropyrans	2,6-Diphenyl 4-aminotetrahydropyran (34.5)
	4-Aminotetrahydrothiopyrans	2,6-Diphenyl 4-aminotetrahydrothiopyran (33.6)
(−325) to (−310)[c]	1-Naphthylamines	8-Nitro-1-naphthylamine (313.9)
(−350) to (−300)[h]	Silylamines	$HN[Si(CH_3)_3]_2$ (−354.2)

Notes:

[a] Varies with solvent. For instance: cyclohexane (315.5), benzene (312.1), chloroform (304.5), methanol (292.1), water (289), and 2,2,2-trifluoroethanol (277.1). All chemical shifts relative to ammonia [2].

[b] Upfield from external HNO_3 (1 M) (CH_3OH) [4,6,7].

[c] In ppm upfield from external 1 M$D^{15}NO_3$ in D_2O (DMSO) [10].

[d] Downfield from anhydrous liquid ammonia, ±0.2 ppm unless otherwise specified [1].

[e] Upfield from external HNO_3 (1 M) (cyclohexane)[6,7].

[f] Relative to external $CH_3{}^{15}NO_2$[5].

[g] With respect to an external standard of 5 M $^{15}NH_4NO_3$ in 2M HNO_3 ($^{15}NH_4NO_3$=21.6 ppm relative to anhydrous ammonia) [9].

[h] Relative to N (SiH_3) (50 % in $CDCl_3$) [11].

The indicated square bracket numbers following each note above refer to the reference list at the beginning of this table.

^{15}N CHEMICAL SHIFT CORRELATION CHARTS
OF MAJOR FUNCTIONAL GROUPS

The following correlation chart contains ^{15}N chemical shifts of various organic nitrogen compounds. Chemical shifts are often expressed relative to different standards (NH_3, NH_4Cl, CH_3NO_2, NH_4NO_3, HNO_3, etc.) and are interconvertible.

In view of the large chemical shift range (up to 900 ppm), caution in using these correlation charts is of great importance as the chemical shifts are greatly dependent on the inductive, mesomeric, or hybridization effects of the neighboring groups, as well as the solvent used.

Chemical shifts are sensitive to hydrogen bonding and are solvent dependent as seen in case of pyridine. Consequently, the reference as well as the solvent should always accompany chemical shift data. No data are given on peptides and other biochemical compounds. All shifts given in these correlation charts are relative to ammonia unless otherwise specified. A section of "miscellaneous" data gives the chemical shift of special compounds relative to unusual standards [1–16]. Ref. [17] contains a compilation of publications that involve various nuclei.

REFERENCES

1. Levy, G.C. and Lichter, R.L., *Nitrogen-15 Nuclear Magnetic Resonance Spectroscopy*, John Wiley & Sons, New York, 1979.
2. Yoder, C.H. and Schaeffer, C.D., Jr., *Introduction to Multinuclear NMR*, Benjamin/Cummings, Menlo Park, CA, 1987.
3. Duthaler, R.O. and Roberts, J.D., Effects of solvent, protonation, and N-alkylation on the ^{15}N chemical shifts of pyridine and related compounds, *J. Am. Chem. Soc.*, 100, 4969, 1978.
4. Duthaler, R.O. and Roberts, J.D., Steric and electronic effects on ^{15}N chemical shifts of saturated aliphatic amines and their hydrochlorides, *J. Am. Chem. Soc.*, 100, 3889, 1978.
5. Kozerski, L. and von Philipsborn, W., ^{15}N chemical shifts as a conformational probe in enaminones: a variable temperature study at natural isotope abundance, *Org. Magn. Res.*, 17, 306, 1981.
6. Duthaler, R.O. and Roberts, J.D., Steric and electronic effects on ^{15}N chemical shifts of piperidine and decahydroquinoline hydrochlorides, *J. Am. Chem. Soc.*, 100, 3882, 1978.
7. Duthaler, R.O. and Roberts, J.D., Nitrogen-15 nuclear magnetic resonance spectroscopy. Solvent effects on the ^{15}N chemical shifts of saturated amines and their hydrochlorides, *J. Magn. Res.*, 34, 129, 1979.
8. Psota, L., Franzen-Sieveking, M., Turnier, J., and Lichter, R.L., Nitrogen nuclear magnetic resonance spectroscopy. Nitrogen-15 and proton chemical shifts of methylanilines and methylanilinium ions, *Org. Magn. Res.*, 11, 401, 1978.
9. Subramanian, P.K., Chandra Sekara, N., and Ramalingam, K., Steric effects on nitrogen-15 chemical shifts of 4-aminooxanes (tetrahydropyrans), 4-amino-thianes, and the corresponding N, N-dimethyl derivatives. Use of nitrogen-15 shifts as an aid in stereochemical analysis of these heterocyclic systems, *J. Org. Chem.*, 47, 1933, 1982.
10. Schuster, I.I. and Roberts, J.D., Proximity effects on nitrogen-15 chemical shifts of 8-substituted 1-nitronaphthalenes and 1-naphthylamines, *J. Org. Chem.*, 45, 284, 1980.
11. Kupce, E., Liepins, E., Pudova, O., and Lukevics, E., Indirect nuclear spin-spin coupling constants of nitrogen-15 to silicon-29 in silylamines, *J. Chem. Soc.*, Chem. Comm., 9, 581, 1984.
12. Allen, M. and Roberts, J.D., Effects of protonation and hydrogen bonding on nitrogen-15 chemical shifts of compounds containing the >C=N-group, *J. Org. Chem.*, 45, 130, 1980.
13. Brownlee, R.T.C. and Sadek, M., Natural abundance ^{15}N chemical shifts in substituted benzamides and thiobenzamides, *Magn. Res. Chem.*, 24, 821, 1986.
14. Dega-Szafran, Z., Szafran, M., Stefaniak, L., Brevard, C., and Bourdonneau, M., Nitrogen-15 nuclear magnetic resonance studies of hydrogen bonding and proton transfer in some pyridine trifluoroacetates in dichloromethane, *Magn. Res. Chem.*, 24, 424, 1986.

15. Lambert, J.B., Shurvell, H.F., Verbit, L., Cooks, R.G., and Stout, G.H., *Organic Structural Analysis*, MacMillan, New York, 1976.
16. Bruno, T.J. and Svoronos, P.D.N., *Handbook of Basic Tables for Chemical Analysis*, 3rd ed., CRC Press, Boca Raton, FL, 2011.
17. Andersson, H., Carlsson, A.-C.C., Nekoueishahraki, B., Brath, U. and Erdelyi, M., Solvent effects on nitrogen chemical shifts, in *Annual Reports on NMR Spectroscopy*, Chapter 2, Vol. 86, 73–210, 2015.

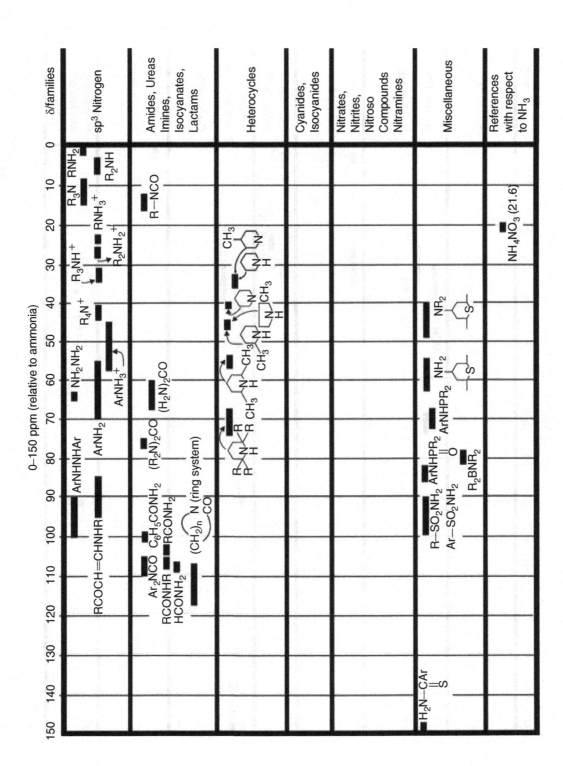

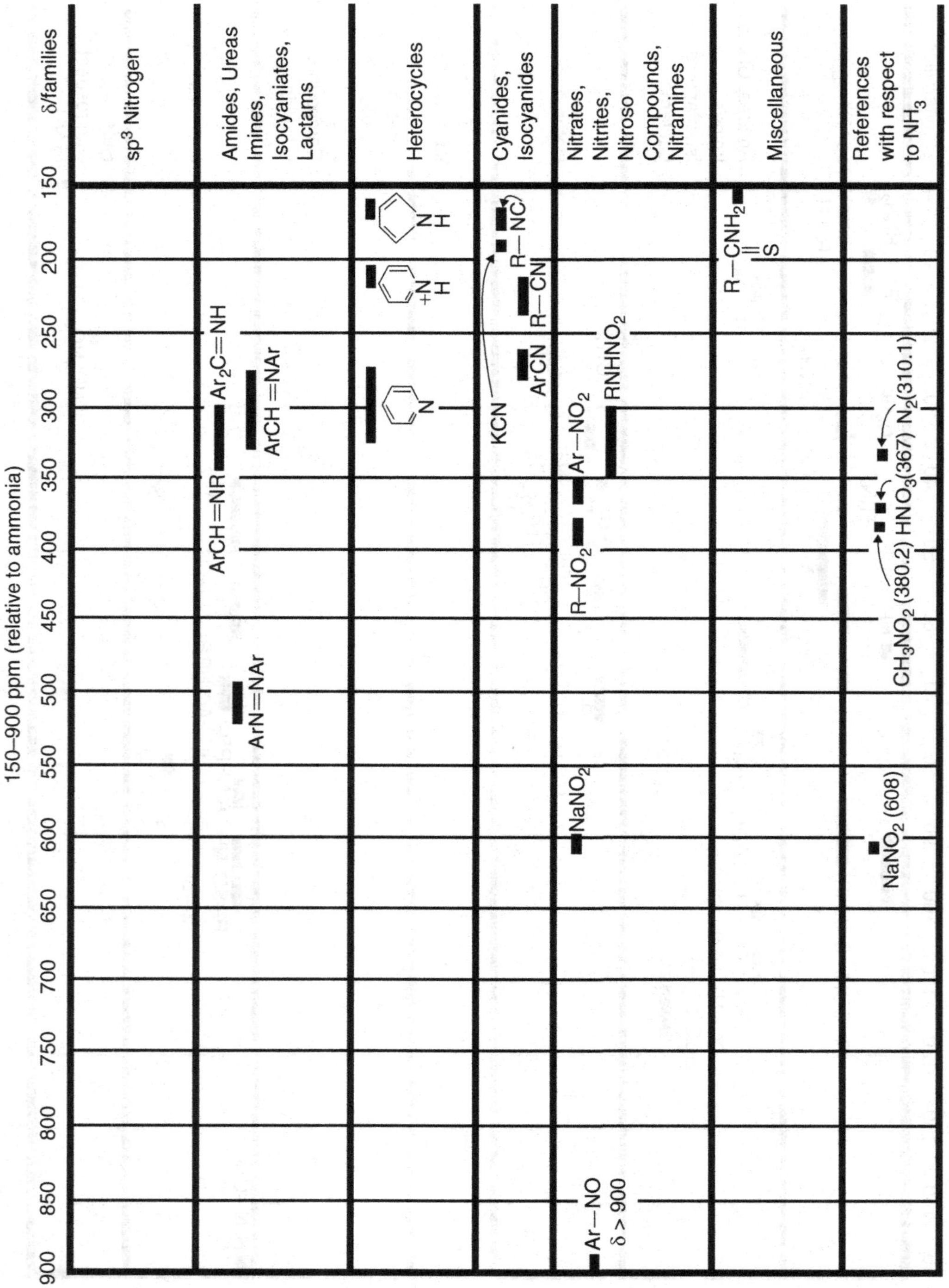

150–900 ppm (relative to ammonia)

δ/families	150	200	250	300	350	400	450	500	550	600	650	700	750	800	850	900

sp³ Nitrogen

Amides, Ureas Imines, Isocyanates, Lactams
- ArCH=NR
- Ar₂C=NH
- ArCH=NAr
- ArN=NAr

Heterocycles

Cyanides, Isocyanides
- R—NC
- R—CN
- ArCN
- KCN

Nitrates, Nitrites, Nitroso Compounds, Nitramines
- RNHNO₂
- Ar—NO₂
- R—NO₂
- NaNO₂
- Ar—NO   δ > 900

Miscellaneous
- R—CNH₂ ‖ S

References with respect to NH₃
- N₂ (310.1)
- HNO₃ (367)
- CH₃NO₂ (380.2)
- NaNO₂ (608)

SPIN–SPIN COUPLING TO ^{15}N

The following table gives representative spin–spin coupling ranges (J_{NH} in Hz) to ^{15}N [1–9].

REFERENCES

1. Levy, G.C. and Lichter, R.L., *Nitrogen-15 Nuclear Magnetic Resonance Spectroscopy*, John Wiley & Sons, New York, 1979.
2. DelBene, J.E. and Bartlett, R.J., N-N spin-spin coupling constants [2hJ(15N-15N)] across N-H—N hydrogen bonds in neutral complexes: to what extent does the bonding at the nitrogens influence 2hJN–N? Communication, *J. Am. Chem. Soc.*, 122, 10480, 2000.
3. Del Bene, J.E., Ajith Perera, S., Bartlett, R.J., Yáñez, M., Elguero, M.J., and Alkorta, I., Two-bond 19F–15N spin–spin coupling constants (2hJFN) across F–H···N hydrogen bonds, *J. Phys. Chem.*, 107, 3121 2003.
4. Duthaler, R.O. and Roberts, J.D., Effects of solvent, protonation, and N-alkylation on the ^{15}N chemical shifts of pyridine and related compounds, *J. Am. Chem. Soc.*, 100, 4969, 1978.
5. Duthaler, R.O. and Roberts, J.D., Steric and electronic effects on ^{15}N chemical shifts of saturated aliphatic amines and their hydrochlorides, *J. Am. Chem. Soc.*, 100, 3889, 1978.
6. Kozerski, L. and von Philipsborn, W., ^{15}N chemical shifts as a conformational probe in enaminones: a variable temperature study at natural isotope abundance, *Org. Magn. Res.*, 17, 306, 1981.
7. Subramanian, P.K., Chandra Sekara, N., and Ramalingam, K., Steric effects on nitrogen-15 chemical shifts of 4-aminooxanes (tetrahydropyrans), 4-amino-thianes, and the corresponding N, N-dimethyl derivatives. Use of nitrogen-15 shifts as an aid in stereochemical analysis of these heterocyclic systems, *J. Org. Chem.*, 47, 1933, 1982.
8. Kupce, E., Liepins, E., Pudova, O., and Lukevics, E., Indirect nuclear spin-spin coupling constants of nitrogen-15 to silicon-29 in silylamines, *J. Chem. Soc. Chem. Comm.*, 9, 581, 1984.
9. Axenrod, T., Structural effects on the one-bond nitrogen-15-proton coupling constant, in *Nucl. Magn. Reson. Spectrosc. Nucl. Other Than Protons*, edited by T. Axenrod and G.A. Webb, Wiley-Interscience, New York, 81–94, 1974

1. ^{15}N–H Coupling Constants

Bond Type	Family	J_{NH}	Example
One-bond	Ammonia	(−61.2)	NH_3
	Amines, aliphatic (1°, 2°)	~(−65)	CH_3NH_2 (−64.5); $(CH_3)_2NH$ (−67.0)
	Ammonium salts	~(−75)	CH_3NH_3Cl (−75.4)
			$(CH_3)_2NH_2Cl$ (−76.1)
			$C_6H_5NH_3^+$ (−76)
	Amines, aromatic (1°, 2°)	(−78) to (−95)	$C_6H_5NH_2$ (−78.5)
			$p-CH_3O-C_6H_4-NH_2$ (−79.4)
			$p-O_2N-C_6H_4-NH_2$ (−92.6)
	Sulfonamides	~(−80)	$C_6H_5SO_2NH_2$ (−80.8)
	Hydrazines	(−90) to (−100)	$C_6H_5NHNH_2$ (−89.6)
	Amides (1°, 2°)	(−85) to (−95)	$HCONH_2$ (−88) (syn); (−92) (anti)
	Pyrroles	(−95) to (−100)	Pyrrole (−96.53)
	Nitriles, salts	~(−135)	$CH_3C≡NH^+$ (−136)
Two-bond	Amines	~(−1)	CH_3NH_2 (−1.0); $(CH_3)_3N$ (−0.85)
	Pyridinium salts	~(−3)	$C_5H_5NH^+$ (−3)
	Pyrroles	~(−5)	C_4H_4NH (−4.52)
	Thiazoles	~(−10)	C_3H_3NS
	Pyridines	~(−10)	C_5H_5N (−10.76)
	Oximes, syn	~(−15)	>C=N–OH (syn)
	Oximes, anti	(−2.5) to (+2.5)	>C=N–OH (anti)
Three-bond	Nitriles, salts	~(2−4)	$CH_3C≡NH^+$ (2.8)
	Amides	~(1−2)	CH_3CONH_2 (1.3)
	Anilines	~(1−2)	$C_6H_5NH_2$ (1.5, 1.8)
	Pyridines	~(0−1)	C_5H_5N (0.2)
	Nitriles	(−1) to (−2)	$CH_3C≡N$ (−1.7)
	Pyridinium salts	~(−4)	$C_5H_5NH^+$ (−3.98)
	Pyrroles	~(−5)	C_4H_4NH (−5.39)

2. ^{15}N–^{13}C Coupling Constants

Bond Type	Family	J_{CH}, H_2	Example
One-bond	Amines, aliphatic	~(−4)	CH_3NH_2 (−4.5); $CH_3(CH_2)_2NH_2$ (−3.9)
	Ammonium salts (aliphatic)	~(−5)	$CH_3(CH_2)_2NH_3^+$ (−4.4)
	Ammonium salts (aromatic)	~(−9)	$C_6H_5NH_3^+$ (−8.9)
	Pyrroles	~(−10)	C_4H_4NH (−10.3)
	Amines, aromatic	(−11) to (−15)	$C_6H_5NH_2$ (−11.43)
	Nitro compounds	(−10) to (−15)	CH_3NO_2 (−10.5); $C_6H_5NO_2$ (−14.5)
	Nitriles	~(−17)	$CH_3C≡N$ (−17.5)
	Amides	~(−14)	$C_6H_5NHCOCH_3$ (−14.3) (CO); (−14.1) (C$_1$)
Two-bond	Amides	7−9	CH_3CONH_2 (9.5)
	Nitriles	~3	$CH_3C≡N$ (3.0)
	Pyridines and N-derivatives	~1−3	C_5H_5N (2.53); $C_5H_5NH^+$ (2.01); C_5H_5NO (1.43)
	Amines, aliphatic	~1−2	$CH_3CH_2CH_2NH_2$ (1.2)
	Nitro compounds, aromatic	~(−1) to (−2)	$C_6H_5NO_2$ (−1.67)
	Amines, aromatic	~(−1) to (−2)	$C_6H_5NH_2$ (−2.68); $C_6H_5NH_3^+$ (−1.5)
	Pyrroles	~(−4)	C_4H_4NH (−3.92)

(Continued)

Bond Type	Family	J_{CH}, H_2	Example
Three-bond	Amides	9	$CH_2=CHCONH_2$ (19)
	Ammonium salts	1–9	$CH_3(CH_2)_2NH_3^+$ (1.3); $C_6H_5NH_3^+$ (2.1)
	Pyridines	~3	C_5H_5N (2.53)
	Amines, aliphatic	1–3	$CH_3(CH_2)_2NH_2$ (1.4)
	Amines, aromatic	~(−1) to (−3)	$C_6H_5NH_2$ (−2.68)
	Nitro compounds	~(−2)	$C_6H_5NO_2$ (−1.67)
	Pyrroles	~(−4)	C_4H_4NH (−3.92)

3. ^{15}N–^{15}N Coupling Constants

Bond Type	Family	J_{NN}, H_2	Example
−N=N−	Azocompounds	12–25	$C_6H_5N=NC(CH_3)_2C_6H_5$ anti (17); syn (21)
>N−N=O	N-nitrosamines	~19	$(C_6H_5CH_2)_2N−N=O$ (19)
>C=N−N<	Hydrazones	~10	$p-O_2NC_6H_4CH=N−NHC_6H_5$ (10.7)
−NH−NH$_2$	Hydrazines	~7	$C_6H_5NHNH_2$ (6.7)

4. ^{15}N–^{19}F Coupling Constants

Bond Type	Family	J_{NF}, H_2	Example
F−N=N	Diflurodiazines transCis	~190 ($^1J_{NF}$)	F-N=N−F (190)
F−N=N		~102 ($^2J_{NF}$)	F-N=N−F (102)
		~203($^1J_{NF}$)	F-N=N−F (203)
		~52($^2J_{NF}$)	F-N=N−F (52)
−C=N−C−F	Fluoropyridines 2-fluoro	(−52.5)	
−C=N−C−C−F	3-Fluoro	(+3.6)	
>N−C=C−F	Fluoroanilines 2-fluoro	0	$1,2−C_6H_4F(NH_2)$
>N−C=C−C−F	3-Fluoro	0	$1,3−C_6H_4F(NH_2)$
>N−C=C−C=C−F	4-Fluoro	1.5	$1,4−C_6H_4F(NH_2)$
+N−C=C−F	Fluoroanilinium salts 2-fluoro	1.4	$1,2−C_6H_4F(NH_3^+)$
+N−C=C−C−F	3-Fluoro	0.2	$1,3−C_6H_4F(NH_3^+)$
+N−C=C−C=C−F	4-Fluoro	0	$1,4−C_6H_4F(NH_3^+)$

^{19}F CHEMICAL SHIFT RANGES

The following table lists the ^{19}F chemical shift ranges (in ppm) relative to neat $CFCl_3$ [1–5].

REFERENCE

1. Yoder, C.H. and Schaeffer, C.D., Jr., *Introduction to Multinuclear NMR: Theory and Application*, Benjamin/Cummings, Menlo Park, CA, 1987.
2. Tonelli, A.E., Schilling, F.C., and Cais, R.E., Fluorine-19 NMR chemical shifts and the microstructure of fluoro polymers, *Macromolecules*, 15(3), 849, 1982. doi:10.1021/ma00231a031.
3. Dungan, C.H. and Van Wazer, I.R., *Compilation of Reported ^{19}F Chemical Shifts 1951 to Mid 1967*, Wiley Interscience, New York, 1970.
4. Emsley, J.W., Phillips, L., and Wray, V., *Fluorine Coupling Constants*, Pergamon, New York, 1977.
5. Dolbier, W.R., *Guide to Fluorine NMR for Organic Chemists*, John Wiley, New York 2009.

Compound Type	Chemical Shift Range (ppm) Relative to Neat $CFCl_3$
F–C(=O)	–70 to –20
CF_3–	+40 to +80
–CF_2–	+80 to +140
>CF–	+140 to +250
Ar–F where Ar = aromatic moiety	+80 to +170

^{19}F CHEMICAL SHIFTS OF SOME FLUORINE-CONTAINING COMPOUNDS

The following table lists the ^{19}F chemical shifts of some fluorine-containing compounds relative to neat $CFCl_3$. All chemical shifts are those of neat samples, and the values pertain to the fluorine present in the molecule [1–3].

REFERENCES

1. Tonelli, A.E., Schilling, F.C., and Cais, R.E., Fluorine-19 NMR chemical shifts and the microstructure of fluoro polymers, *Macromolecules*, 15(3), 849, 1982. doi:10.1021/ma00231a031.
2. Dungan, C.H. and Van Wazer, I.R., *Compilation of Reported ^{19}F Chemical Shifts 1951 to Mid 1967*, Wiley Interscience, New York, 1970.
3. Emsley, J.W., Phillips, L. and Wray, V., *Fluorine Coupling Constants*, Pergamon, New York, 1977.

Compound	Formula	Chemical Shift (ppm)
Fluorotrichloromethane	$CFCl_3$	0.00
Tetrafluoromethane	CF_4	−62.3
Fluoromethane	CH_3F	−271.9
Trifluoromethane	CF_3H	−78.6
Trifluoroalkanes	CF_3R	−60 to −70
Difluoromethane	CF_2H_2	−143.6
Fluoroethane	CH_3CH_2F	−231
Fluoroethene (or vinyl fluoride)	$FCH=CH_2$	−114
1,1-Difluoroethene	$CF_2=CH_2$	−81.3
Tetrafluoroethene	$CF_2=CF_2$	−135
Trifluoroethanoic acid (or trifluoroacetic acid)	CF_3COOH	−78.5
Phenyl trifluoroethanoate (or phenyl trifluoroacetate)	$CF_3COOC_6H_5$	−73.85
Benzyl trifluoroethanoate (or benzyl trifluoroacetate)	$CF_3COOCH_2C_6H_5$	−75.02
Methyl trifluoroethanoate (or methyl trifluoroacetate)	CF_3COOCH_3	−74.21
Ethyl trifluoroethanoate (or trifluoroacetate)	$CF_3COOCH_2CH_3$	−78.7
Hexafluorobenzene	C_6F_6	−164.9
Pentafluorobenzene	C_6F_5H	−113.5
1,4-Difluorobenzene (or p-difluorobenzene)	$p-C_6H_4F_2$	−106.0
(Fluoromethyl)benzene (or benzyl fluoride)	$C_6H_5-CH_2F$	−207
Trifluoromethylbenzene	$C_6H_5-CF_3$	−63.72
Octafluorocyclobutane (or perfluorocyclobutane)	C_4F_8	−135.15
Decafluorocyclopentane (or perfluorocyclopentane)	C_5F_{10}	−132.9
Difluoromethyl ethers	CHF_2OR	−82
Hexafluoropropanone (or hexafluoroacetone)	$(CF_3)_2CO$	−84.6
Fluorine	F_2	+422.92
Chlorotrifluoromethane	CF_3Cl	−28.6
Chlorine trifluoride	ClF_3	+116, −4
Chlorine pentafluoride	ClF_5	+247, +412
Dichlorodifluoromethane	CF_2Cl_2	−8
1,2-Difluoro-1,1,2,2-tetrachloroethane	$CFCl_2-CFCl_2$	−67.8
Fluorotribromomethane	$CFBr_3$	+7.38
Dibromodifluoromethane	CF_2Br_2	+7
Iodine heptafluoride	IF_7	+170

(Continued)

Compound	Formula	Chemical Shift (ppm)
Arsenic trifluoride	AsF_3	−40.6
Arsenic pentafluoride	AsF_5	−66
Boron trifluoride	BF_3	−131.3
Trimethyloxonium tetrafluoroborate	$(CH_3)_2O.BF_3$	−158.3
Triethyloxonium tetrafluoroborate	$(C_2H_5)_2O.BF_3$	−153
Sulfur hexafluoride	SF_6	+57.42
Sulfuryl fluoride (or sulfonyl fluoride)	SO_2F_2	−78.5
Antimony pentafluoride	SbF_5	−108
Selenium hexafluoride	SeF_6	+55
Silicon tetrafluoride	SiF_4	−163.3
Tellurium hexafluoride	TeF_6	−57
Sulfur hexafluoride	SF_6	−57
Xenon difluoride	XeF_2	+258
Xenon tetrafluoride	XeF_4	+438
Xenon hexafluoride	XeF_6	+550
Nitrogen trifluoride	NF_3	+147
Phosphoryl fluoride (or phosphorus oxyfluoride)	POF_3	−90.7
Phosphorus trifluoride	PF_3	−67.5

^{19}F CHEMICAL SHIFT CORRELATION CHART OF SOME FLUORINE-CONTAINING COMPOUNDS

The following correlation chart lists the ^{19}F chemical shifts of some fluorine-containing compounds relative to neat $CFCl_3$. All chemical shifts are those of neat samples and the values pertain to the fluorine present in the molecule [1–4].

REFERENCES

1. Dungan, C.H. and Van Wazer, I.R., *Compilation of Reported ^{19}F Chemical Shifts 1951 to Mid 1967*, Wiley Interscience, New York, 1970.
2. Emsley, J.W., Phillips, L., and Wray, V., *Fluorine Coupling Constants*, Pergamon, New York, 1977.
3. Bruno, T.J. and Svoronos, P.D.N., *Handbook of Basic Tables for Chemical Analysis*, 2nd ed., CRC Press, Boca Raton, FL, 2003.
4. University of Wisconsin, Organic Structure Determination Bibliography, available online at https://www.chem.wisc.edu/areas/reich/nmr/nmr-biblio.htm, accessed December 2019.

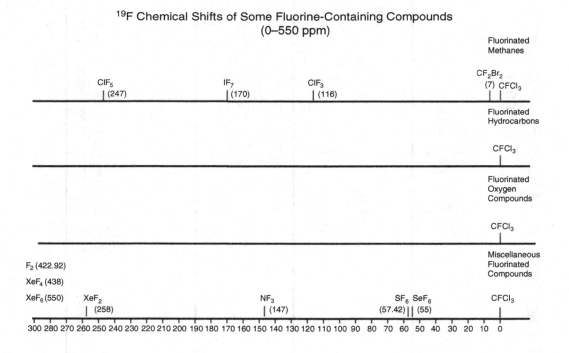

^{19}F Chemical Shifts of Some Fluorine-Containing Compounds (0–550 ppm)

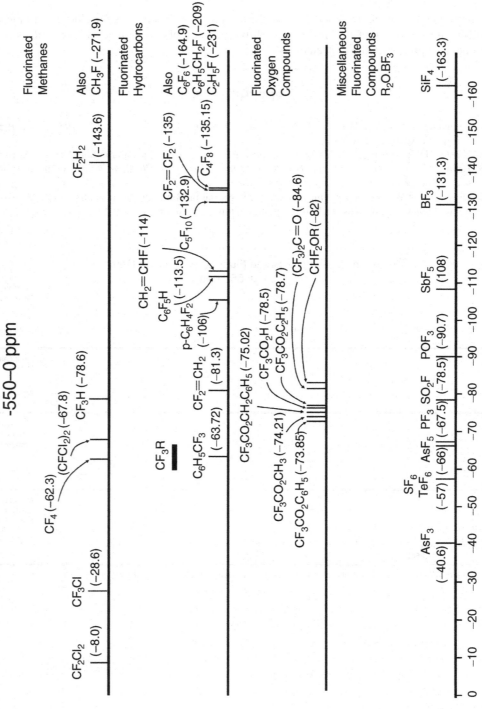

¹⁹F Chemical Shifts of Some Fluorine-Containing Compounds
-550-0 ppm

FLUORINE COUPLING CONSTANTS

The following table gives the most important fluorine coupling constants including J_{FN}, J_{FCF}, and J_{CF} together with some typical examples [1–9]. The coupling constant values vary with the solvent used [3]. The book by Emsley, Phillips, and Wray [1] gives a complete, detailed list of various compounds.

REFERENCES

1. Emsley, J.W., Phillips, L., and Wray, V., *Fluorine Coupling Constants*, Pergamon Press, Oxford, New York, 1977.
2. Lambert, J.R., Shurvell, H.F., Verbit, L., Cooks, R.G., and Stout, G.H., *Organic Structural Analysis*, MacMillan, New York, 1976.
3. Yoder, C.H. and Schaeffer, C.D., Jr., *Introduction to Multinuclear NMR: Theory and Application*, Benjamin/Cummings, Menlo Park, CA, 1987.
4. Schaeffer, T., Marat, K., Peeling, J., and Veregin, R.P., Signs and mechanisms of ^{13}C, ^{19}F spin-spin coupling constants in benzotrifluoride and its derivatives, *Can. J. Chem.*, 61, 2779, 1983.
5. Adcock, W. and Kok, G.B., Polar substituent effects on ^{19}F chemical shifts of aryl and vinyl fluorides: a fluorine-19 nuclear magnetic resonance study of some 1,1-difluoro-2-(4-substituted-bicyclo[2,2,2] oct-1-yl)ethenes, *J. Org. Chem.*, 50, 1079, 1985.
6. Newmark, R.A. and Hill, J.R., Carbon-13-fluorine-19 coupling constants in benzotrifluorides, *Org. Magn. Res.*, 9, 589, 1977.
7. Adcock, W. and Abeywickrema, A.N., Concerning the origin of substituent-induced fluorine-19 chemical shifts in aliphatic fluorides: carbon-13 and fluorine-19 nuclear magnetic resonance study of 1-fluoro-4-phenylbicyclo[2,2,2]octanes substituted in the arene ring, *J. Org. Chem.*, 47, 2945, 1982.
8. Dungan, C.H. and Van Wazer, I.R., *Compilation of Reported ^{19}F Chemical Shifts 1951 to Mid 1967*, Wiley Interscience, New York, 1970.
9. Dolbier, W.R., *Guide to Fluorine NMR for Organic Chemists*, John Wiley & Sons, New York, 2009.

^{19}F–^{1}H Coupling Constants

Fluorinated Family	J_{FH} (Hz)	Example
		(a) Two-Bond
Alkanes	45–80	CH$_3$F (45); CH$_2$F$_2$ (50); CHF$_3$ (79); C$_2$H$_5$F (47); CH$_3$CHF$_2$ (57); CH$_2$FCH$_2$F (48); CH$_2$FCHF$_2$ (54); CF$_3$CH$_2$F (45); CF$_2$HCF$_2$CF$_3$ (52)
Alkyl chlorides	49–65	FCl$_2$CH (53); CF$_2$HCl (63); FCHCl–CHCl$_2$ (49); FCH$_2$–CH$_2$Cl (46)
Alkyl bromides	45–50	FBrCHCH$_3$ (50.5); FH$_2$C–CH$_2$Br (46); FBrCH–CHFBr (49)
Alkenes	45–80	FHC=CHF (cis-71.7; trans-75.1); CH$_2$=CHF (85); CF$_2$=CHF (70.5); FCH$_2$CH=CH$_2$ (47.5)
Aromatics	45–75	Cl–C$_6$H$_4$–CH$_2$F (m-47, p-48); FH$_2$C–C$_6$H$_4$–NO$_2$ (m-47, p-48); FH$_2$C–C$_6$H$_4$–F(m-48, p-48); p–Br–C$_6$H$_4$–OCF$_2$H (73)
Ethers	40–75	FH$_2$COCH$_3$ (74); CF$_2$HCF$_2$OCH$_3$ (46); F$_2$HC–O–CH(CH$_3$)$_2$ (75)
Ketones	45–50	FCH$_2$COCH$_3$ (47); F$_2$HC–COCH$_3$ (54); CH$_3$CH$_2$CHFCOCH$_3$ (50); F$_2$HC–COCH(CF$_3$)$_2$ (54)
Aldehydes	~50	CH$_3$CH$_2$CHFCHO (51)
Esters	45–70	CFH$_2$CO$_2$CH$_2$CH$_3$ (47); CH$_3$CHFCO$_2$CH$_2$CH$_3$ (48)
		(b) Three-Bond
Alkanes	2–25	CF$_2$HCH$_3$ (21); (CH$_3$)$_3$CF (20.4); CH$_3$CHFCH$_2$CH$_2$CH$_3$ (23); CF$_3$CH$_3$ (13)
Alkyl chlorides	8–20	CF$_2$HCHCl$_2$ (8); CF$_2$ClCH$_3$ (15)
Alkyl bromides	15–25	CF$_2$BrCH$_2$Br (22); CF$_2$BrCH$_3$ (16); FC(CH$_3$)$_2$CHBrCH$_3$ (21)
Alkenes	(–5) to 60	CHF=CHF (cis-19.6; trans-2.8)
	J_{HCF} (cisoid)	CH$_2$=CHF (cis-19.6; trans-51.8);
	<20	CHF=CF$_2$ (cis-(–4.2); trans-12.5);
	J_{HCF} (transoid)	CH$_2$=CF$_2$ (cis-0.6; trans-33.8)
	>20	
Alcohols	5–30	CF$_3$CH$_2$OH (8); FCH$_2$CH$_2$OH (29); CH$_3$CHFCH$_2$OH (23.6, 23.6); CF$_3$CH(OH)CH$_3$ (7.5); CF$_3$CH(OH)CF$_3$ (6); FC(CH$_3$)$_2$C(OH)(CH$_3$)$_2$ (23)
Ketones	5–25	CH$_3$CH$_2$CHFCOCH$_3$ (24); FC(CH$_3$)$_2$COCH$_3$ (21); (CF$_3$)$_2$CHCOCH$_3$ (8); CF$_2$HCOCH(CF$_3$)$_2$ (7)
Aldehydes	10–25	(CH$_3$)$_2$CFCHO (22)
Esters	10–25	CH$_3$CHFCO$_2$CH$_2$CH$_3$ (23); (CH$_3$CH$_2$)$_2$CFCO$_2$CH$_3$ (16.5)

^{19}F–^{19}F Coupling Constants

Carbon	J_{FCF} (Hz)	Examples
		(a) Two-Bond
Saturated (sp³)	140–250	$CF_3CF_2^a$, $^bCFHCH_3$ ($J_{ab}=270$); CF_2^a, $^bBrCHFSO_2F$ ($J_{ab}=188$); $CH_3O–CF_2^a$, $^bCFHSO_2F$ ($J_{ab}=147$); $CH_3O–CF_2^a$, ^bCFHCl ($J_{ab}=142$); $CH_3S–CF_2^a$, ^bCFHCl ($J_{ab}=222$)
Cycloalkanes	150–240	$F_2C(CH_2)_2$ (150) (three-membered); $F_2C(CH_2)_3$ (200) (four-membered); $F_2C(CH_2)_4$ (240) (five-membered); $F_2C(CH_2)_5$ (228) (six-membered)
Unsaturated (sp²)	≤100	$CF_2=CH_2$ (31,36); $CF_2=CHF$ (87); $CF_2=CBrCl$ (30); $CF_2=CHCl$ (41); $CF_2=CFBr$ (75); $CF_2=NCF_3$ (82); $CF_2=CFCN$ (27); $CF_2=CFCOF$ (7); $CF_2=CFOCH_2CF_3$ (102); $CF_2=CBrCH_2N(CF_3)_2$ (30); $CF_2=CFCOCF_2CF_3$ (12); $CF_2=CHC_6H_5$ (33); $CF_2=CH(CH_2)_5CH_3$ (50); $CF_2=CH–Ar[Ar=aryl]$ (50)
		(b) Three-Bond
Saturated (sp³)	0–16	CF_3CH_2F (16); CF_3CF_3 (3.5); CF_3CHF_2 (3); CH_2FCH_2F (10–12); $CF_2^aHCF^bHCF_2H$ ($J_{ab}=13$); $CF_2HCF_2^aCH_2F$ ($J_{ab}=14$); $CF_3^aCF_2^bCF^cHCH_3$ ($J_{ab}<l$; $J_{bc}=15$); $CF_3^aCF^bHCF_2^cH$ ($J_{ab}=12$; $J_{bc}=12$); $CF_3^aCF_2^bC\equiv CF_3$ ($J_{ab}=3.3$); $CF_3^aCF_2^bC\equiv CCF_3$ ($J_{ab}=3.3$); $CF_3^aCF_2^bC\equiv CCl$ ($J_{ab}=10$); $CF_3CF_2COCH_2CH_3$ (1); $FCH_2CFHCO_2C_2H_5$ (−11.6); $CF_3^aCF_2^bCF_2^cCOOH$ ($J_{ab}<l$; $J_{bc}<l$); $(CF_3^a)_2CF^bS(O)OC_2H_5$ ($J_{ab}=8$)
Unsaturated (sp²)	>30	$FCH=CHF$[cis (−18.7); trans (−133.5)]; $CF_2=CHBr$ (34.5); $CF_2=CHCl$ (41); $CF_2=CH_2$ (37)

^{13}C–^{19}F Coupling Constants

Fluorinated Family	J_{CF}, (Hz)	Examples
		One-Bond
Alkanes	150–290	CH_3F (158); CH_2F_2 (237); CHF_3 (274); CF_4 (257); CF_3CF_3 (281); CF_3CH_3 (271); $(CH_3)_3CF$ (167); $(C^aF_3^b)_2C^cF_2^d$ [$J_{ab}=285$; $J_{cd}=265$]
Alkenes	250–300	$CF_2=CD_2$ (287); $CF_2=CCl_2$ (−289); $CF_2=CBr_2$ (290); ClFC=CHCl [cis (−300); trans (−307)]; ClFC=CClF [cis (290); trans (290)]
Alkynes	250–260	$C^aF_3^bC\equiv CF$ [$J_{ab}=259$]; $CF_3C\equiv CCF_3$ (256)
Alkyl chlorides	275–350	$CFCl_3$ (337); CF_2Cl_2 (325); CF_3Cl (299); $CF_3(CCl_2)_2CF_3$ (286); CF_3CH_2Cl (274); $CF_3CCl=CCl_2$ (274); $CF_2=CCl_2$ (−289); CF_3CCl_3 (283)
Alkyl bromides	290–375	$CFBr_3$ (372); CF_2Br_2 (358); CF_3Br (324); CF_3CH_2Br (272); $CF_2=CBr_2$ (290)
Acyl fluorides	350–370	$HCOF$ (369); CH_3COF (353)
Carboxylic acids	245–290	CF_3COOH (283); CF_2HCO_2H (247)
Alcohols	~275	CF_3CH_2OH (278)
Nitriles	~250	CF_2HCN (244)
Esters	~285	$CF_3CO_2CH_2CH_3$ (284)
Ketones	~290	CF_3COCH_3 (289)
Ethers	~265	$(CF_3)_2O$ (265)

³¹P NMR ABSORPTIONS OF REPRESENTATIVE COMPOUNDS

³¹P is considered to be a medium sensitivity nucleus that has the advantage of yielding sharp lines over a very wide chemical shift range. Its sensitivity is much less than that of ¹H, but it is superior to that of ¹³C [1].

The following charts provide information on the characteristic values for the ³¹P spectra of representative phosphorus-containing compounds [1–4]. The list is far from complete but gives an insight on the spectra of both organic and inorganic compounds. All data are presented in a correlation chart form. The reference in each case is 85 % (mass/mass) phosphoric acid. The first chart provides the general chemical shift range of the various phosphorus families and is followed by more detailed charts that provide representative compounds for each of the families. These families are classified according to the coordination number around phosphorus, which ranges from 2 to 6.

Since this section only gives a general information on ³¹P NMR spectroscopy, the reader is advised to consult Refs. [2] and [3] which include a large, detailed amount of updated spectral information and numerous references.

REFERENCES

1. Kuhl, O., Phosphorus-31 NMR Spectroscopy: A Concise Introduction for the Synthetic Organic and Organometallic Chemist. Springer, Berlin/Heidelberg, 2008.
2. Quin, L.D. and Verkade, J.G., eds., *Phosphorus-31 NMR Spectral Properties in Compounds: Characterization and Structural Analysis.* John Wiley & Sons, New York, 1994.
3. Tebby, J.C., ed. *Handbook of Phosphorus-31 Nuclear Magnetic Resonance Data.* CRC Press, Boca Raton, FL, 1991.
4. Koo, I.S., Ali, D., Park, Y., Wardlaw, D.M., and Buncel, E. Theoretical study of ³¹P NMR chemical shifts for organophosphorus esters, their anions and o, o-dimethylthiophosphorate anion with metal complexes, *Bull. Korean Chem. Soc.*, 29(11), 2252, 2008.

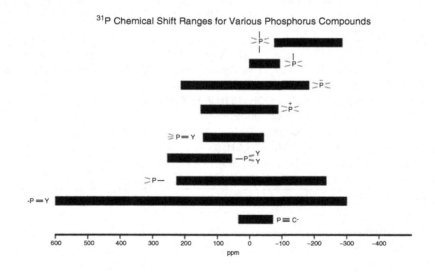

³¹P Chemical Shift Ranges for Various Phosphorus Compounds

Monocoordinated Phosphorus Compounds (C ≡ P)

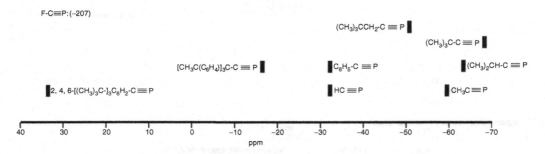

Dicoordinated Phosphorus Compounds (X=P−)

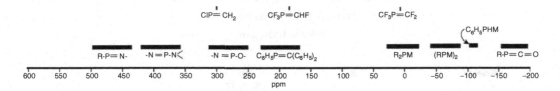

Tricoordinated Phosphorus Compounds (⪫P)
(+240) - 0 ppm

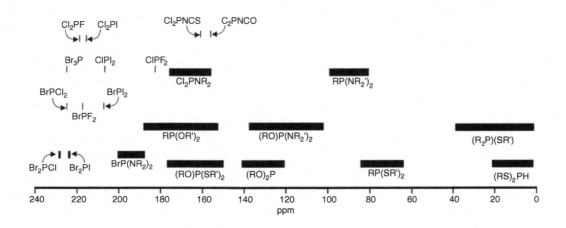

Tricoordinated Phosphorus Compounds (⩾P)
0–(−240) ppm

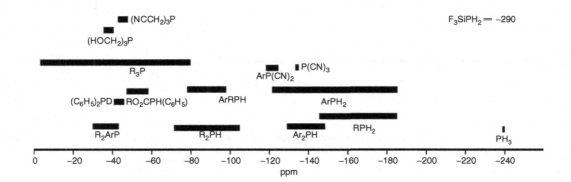

Tetracoordinated Phosphorus Compounds
(phosphonium salts) (⩾P⁺–X⁻)
140–(−140) ppm

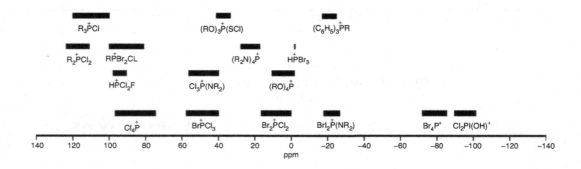

Tetracoordinated Phosphorus Compounds (⪢P═)

Abbreviation: R, R'=Alkyl

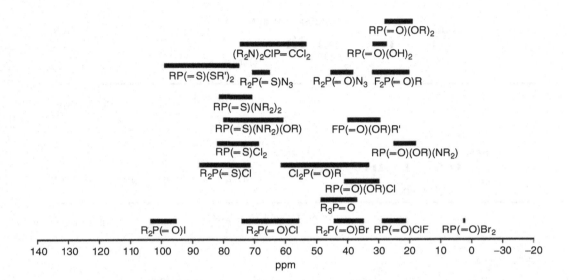

Pentacoordinated Phosphorus Compounds (>P<)

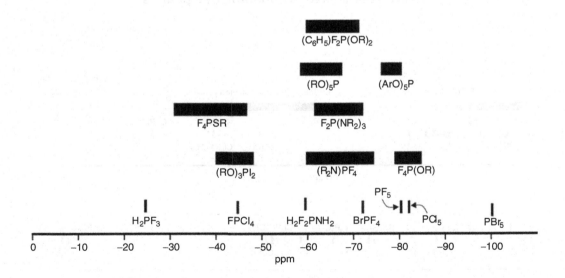

Hexacoordinated Phosphorus Compounds

$$F_3P(X)_x(Y)_{3-x} / Cl_3P(X)_x(Y)_{3-x}\ x \leq 3$$

$Cl_3P(NCS)_3 = -271$

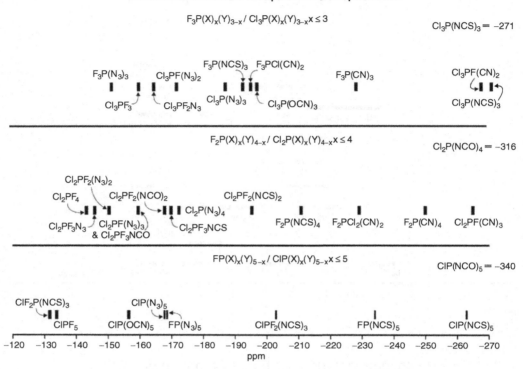

$$F_2P(X)_x(Y)_{4-x} / Cl_2P(X)_x(Y)_{4-x}\ x \leq 4$$

$Cl_2P(NCO)_4 = -316$

$$FP(X)_x(Y)_{5-x} / ClP(X)_x(Y)_{5-x}\ x \leq 5$$

$ClP(NCO)_5 = -340$

Hexacoordinated Phosphorus Compounds

Abbreviation: R = alkyl
Q = quinoline

F_5PX Cl_5PX

$Cl_5P(Pyr)$ (Pyr = substituted pyridine)

$Cl_5PNCO = -280$
$Cl_5PNCS = -287$
$Cl_6P = (-287)-(-307)$
$Cl_5P(CN) = -310$

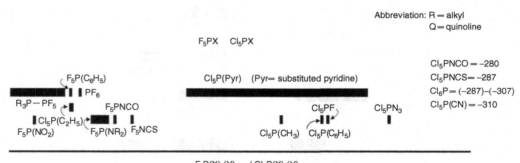

$$F_4P(X)_x(Y)_{2-x} / Cl_4P(X)_x(Y)_{2-x}$$

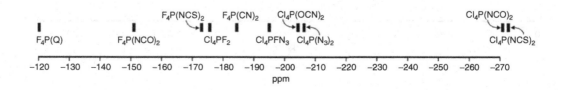

^{29}Si NMR ABSORPTIONS OF MAJOR CHEMICAL FAMILIES

The following correlation tables provide the regions of ^{29}Si NMR absorptions of some major chemical families. These absorptions are reported in the dimensionless units of ppm vs. the standard compound tetramethylsilane (TMS, $(CH_3)_4Si$), which is recorded as 0.0 ppm.

^{29}Si NMR (natural abundance 4.7 %) is a low sensitivity nucleus that has a wide chemical shift range that is useful in determining the identity of certain silicon-containing compounds, many of which are important in biological systems. It should be noted that when taking such spectra, there is always a background signal that is attributed to the glass comprising the measurement tube. Modifying the probe, which can be costly, or a simple adjustment in the pulse sequence, often overcomes the background signals. The ^{29}Si sensitivity is approximately 7.85×10^{-2} (at constant field) and 0.199 (at constant frequency) of that of ^{1}H.

For more detail concerning the chemical shifts, the reader is referred to the general references below and the literature cited therein [1–8].

REFERENCES

1. Williams, E.A. and Cargioli, J.D., Silicon-29 NMR spectroscopy, *Ann. Rep. NMR Spectr.*, 9, 221, 1979; 15, 235, 1983.
2. Schraml, J. and Bellama J.M., ^{29}Si nuclear magnetic resonance, in *Determination of Organic Structures by Physical Methods*, edited by F.C. Nachod and J.J. Zuckerman, 6, 203, Academic Press, New York, 1976.
3. Williams, E.A., NMR spectroscopy of organosilicon compounds, in *Chemistry and Physics of DNA-Ligand Interactions*, edited by N.R. Kallenboch, Adenine Press, New York, 511, 1990.
4. Schraml, J., ^{29}Si NMR spectroscopy of trimethyl silyl tags, in *Progress in NMR Spectroscopy*, edited by J.W. Emsley, J. Feeneym, and L.H. Sutcliffe, 22, 289, 1990.
5. Günther, H., Magnetic resonance in chemistry, in *Encyclopedia of Nuclear Magnetic Resonance, Vol. 9. Advances in NMR*, Wiley, Chichester, 2002.Mason, J., ed., *Multinuclear NMR*, Plenum Press, New York, 1987.
6. Takeuchi, Y. and Takayama, T., ^{29}Si NMR spectroscopy of organosilicon compounds, in *The Chemistry of Organic Silicon Compounds*, 2, John Wiley & Sons, 267, 2003. doi:10.1002/0470857250.ch6.
7. Takayama, T. and Ando, I., Silicon-29 NMR chemical shifts of organosilicons as studied by the FPT CNDO/2 method, *Bull. Chem Soc. Jpn.*, 60, 3125, 1987.
8. Bruno, T.J. and Svoronos, P.D.N., Handbook of Basic Tables for Chemical Analysis, 3rd ed., CRC Press, Boca Raton, FL, 2011.

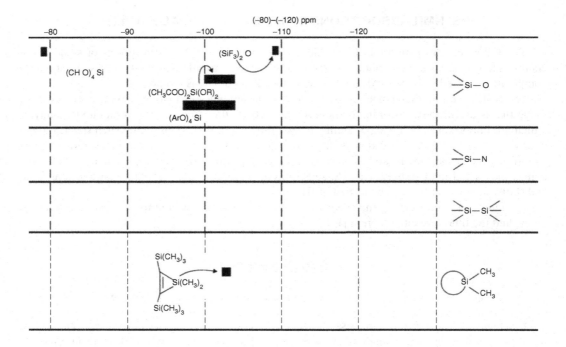

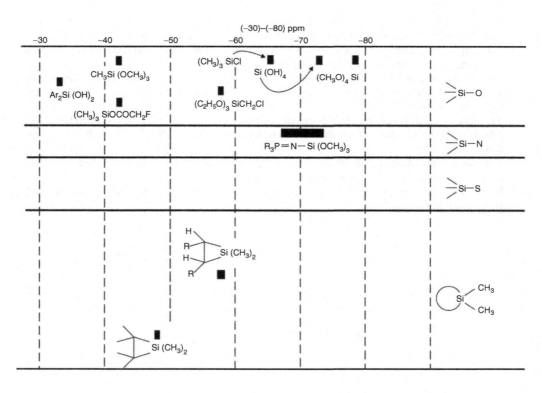

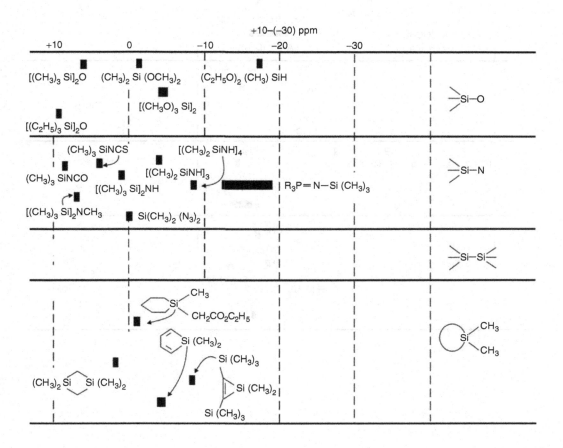

10–50 ppm

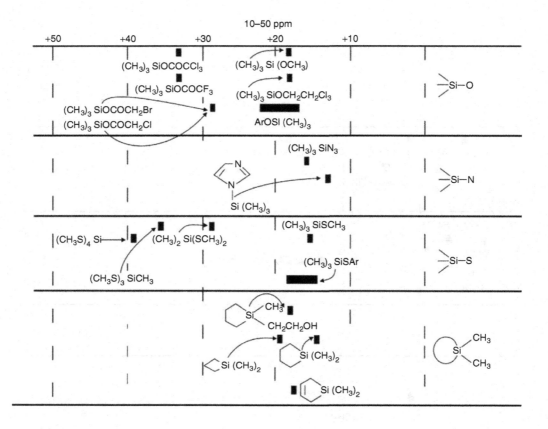

^{119}Sn NMR ABSORPTIONS OF MAJOR CHEMICAL FAMILIES

The following correlation tables provide the regions of ^{119}Sn NMR absorptions of some major chemical families. These absorptions are reported in the dimensionless units of ppm vs. the standard compound tetramethylstanate $(CH_3)_4Sn$, which is recorded as 0.0 ppm.

^{119}Sn NMR is a low sensitivity nucleus that has a relatively wide chemical shift range, which is useful in determining the identification of certain tin-containing compounds. The ^{119}Sn sensitivity is approximately 5.18×10^{-2} (at constant field) that of ^{1}H. Its abundance (8.59 %) is slightly higher than that of ^{117}Sn (7.68 %) and much higher than that of ^{115}Sn (0.34 %).

For more detail concerning the chemical shifts, the reader is referred to the general references below and the literature cited therein [1–9]. The reader should be aware that there is a great deal of chemical shift variation when tin compounds are measured in different solvents. Moreover, many tin compounds are difficult to dissolve in common solvents.

REFERENCES

1. Petrosyan, V.S., NMR spectra and structures of organotin compounds, *Progr. NMR Spectr.*, 11, 115, 1978.
2. Smith, P.J., Chemical shifts of Sn-119 nuclei in organotin compounds,, *Ann. Rep. NMR Spectr.*, 8, 292, 1978.
3. Wrackmeyer, B., Tin-119 NMR parameters, *Ann. Rep. NMR Spectr.*, 16, 73, 1985.
4. Hari, R. and Geanangel, R.A., Tin-119 NMR in coordination chemistry, *Coord. Chem. Rev.*, 44, 229, 1982.
5. Wrackmeyer, B., Multinuclear NMR and Tin chemistry, *Chem. Br.*, 26, 48, 1990.
6. Kaur, A. and Sandhu, G.K., Use of ^{119}Sn Mossbauer and ^{119}Sn NMR spectroscopies in the study of organotin complexes, *J. Chem. Sci.*, 2, 1, 1986.
7. Günther, H., Magnetic resonance in chemistry, in *Encyclopedia of Nuclear Magnetic Resonance, Vol. 9. Advances in NMR*, Wiley, Chichester, 2002.Mason, J., ed., *Multinuclear NMR*, Plenum Press, New York, 1987.
8. Web elements periodic table. http://www.webelements.com/webelements/elements/text/Sn/nucl.html, accessed December 2019.
9. Bruno, T.J. and Svoronos, P.D.N., Handbook of Basic Tables for Chemical Analysis, 3rd ed., CRC Press, Boca Raton, FL, 2011.

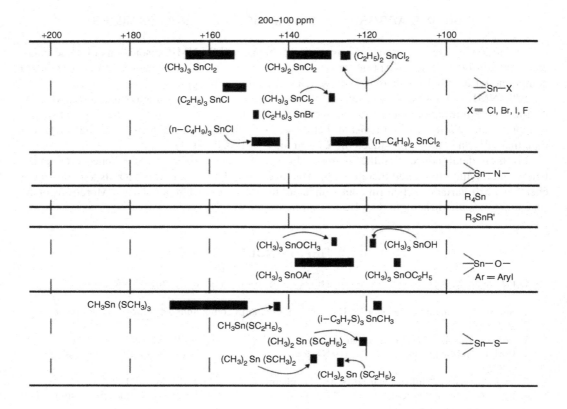

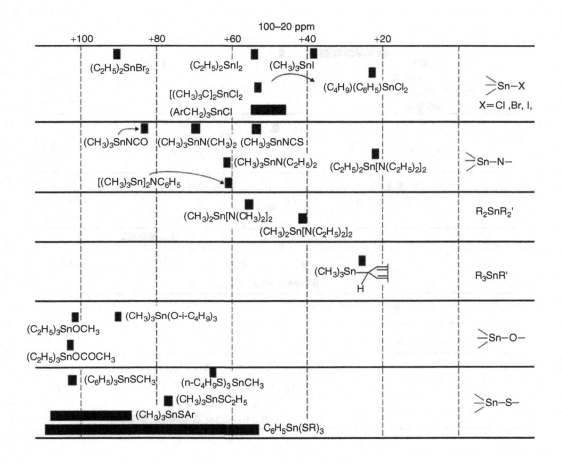

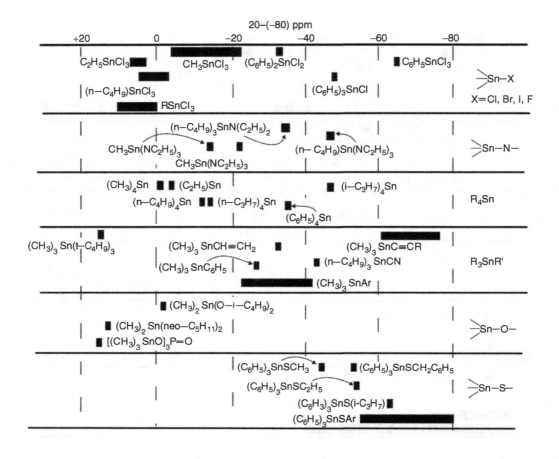

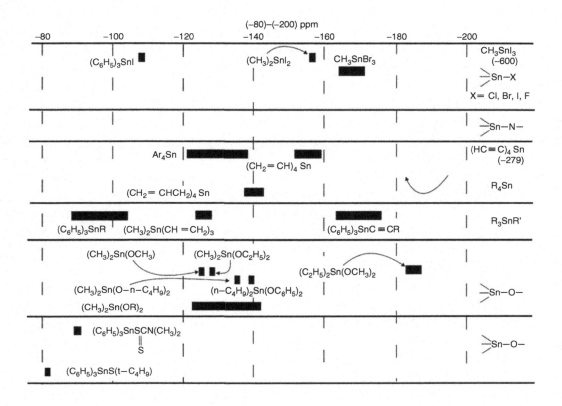

CHAPTER 11

Mass Spectroscopy

NATURAL ABUNDANCE OF IMPORTANT ISOTOPES

The following table lists the atomic masses and relative percent concentrations of naturally occurring isotopes of importance in mass spectroscopy [1–5].

REFERENCES

1. deHoffmann, E. and Stroobant, V., *Mass Spectrometry: Principles and Applications*, 2nd and 3rd ed., Wiley Interscience, Chichester, UK, 2007.
2. Johnstone, R.A.W. and Rose, M.E., *Mass Spectrometry for Chemists and Biochemists*, Cambridge University Press, Cambridge, 1996.
3. Rumble, J. ed., *CRC Handbook for Chemistry and Physics*, 100th ed., CRC Press, Boca Raton, FL, 2019.
4. McLafferty, F.W. and Turecek, F., *Interpretation of Mass Spectra*, 4th ed., University Science Books, Mill Valley, CA, 1993.
5. Watson, J.T. *Introduction to Mass Spectrometry*, 3rd ed., Lippincott-Raven, Philadelphia, PA, 1997.

Element	Total # of Isotopes	More Prominent Isotopes (Mass, Percent Abundance)		
Hydrogen	3	^{1}H (1.00783, 99.985)	^{2}H (2.01410, 0.015)	
Boron	6	^{10}B (10.01294, 19.8)	^{11}B (11.00931, 80.2)	
Carbon	7	^{12}C (12.00000, 98.9)	^{13}C (13.00335, 1.1)	
Nitrogen	7	^{14}N (14.00307, 99.6)	^{15}N (15.00011, 0.4)	
Oxygen	8	^{16}O (15.99491, 99.8)		^{18}O (17.9992, 0.2)
Fluorine	6	^{19}F (18.99840, ≈100.0)		
Silicon	8	^{28}Si (27.97693, 92.2)	^{29}Si (28.97649, 4.7)	^{30}Si (29.97376, 3.1)
Phosphorus	7	^{31}P (30.97376, ≈100.0)		
Sulfur	10	^{32}S (31.972017, 95.0)	^{33}S (32.97146, 0.7)	^{34}S (33.96786, 4.2)
Chlorine	11	^{35}Cl (34.96885, 75.5)		^{37}Cl (36.96590, 24.5)
Bromine	17	^{79}Br (78.9183, 50.5)		^{81}Br (80.91642, 49.5)
Iodine	23	^{127}I (126.90466, ≈100.0)		

RULES FOR DETERMINATION OF MOLECULAR FORMULA

The following rules are used in the mass spectroscopic determination of the molecular formula of an organic compound [1–5]. These rules should be applied to the molecular ion peak and its isotopic cluster. The molecular ion, in turn, is usually the highest mass in the spectrum. It must be an odd-electron ion and must be capable of yielding all other important ions of the spectrum via a logical neutral species loss. The elements that are assumed to possibly be present on the original molecule are carbon, hydrogen, nitrogen, the halogens, sulfur, and/or oxygen. The molecular formula that can be derived is not the only possible one, and consequently, information from nuclear magnetic resonance spectrometry and infrared spectrophotometry is necessary for the final molecular formula determination.

Modern mass spectral databases allow the automated searching of very extensive mass spectral libraries [6]. This has made the identification of compounds by mass spectrometry a far more straightforward task. One must understand, however, that such databases are no substitute for the careful analysis of each mass spectrum, and that the results of database match-up are merely suggestions.

REFERENCES

1. Lee, T.A., *A Beginner's Guide to Mass Spectral Interpretation*, Wiley, New York, 1998.
2. McLafferty, F.W., *Interpretation of Mass Spectra*, 4th ed., University Science Books, Mill Valley, CA, 1993.
3. Shrader, S.R., *Introductory Mass Spectrometry*, Allyn and Bacon, Boston, FL, 1971.
4. Smith, R.M., *Understanding Mass Spectra: A Basic Approach*, 2nd ed., Wiley Interscience, New York, 2004.
5. Watson, J.T. and Watson, T.J., *Introduction to Mass Spectrometry*, Lippincott, Williams and Wilkins, Philadelphia, PA, 1998.
6. NIST Standard Reference Database 1A, NIST/EPA/NIH Mass Spectral Library with Search Program: (Data Version: NIST vol 17, Software Version 2.3), accessed December 2019.

Rule 1

An odd molecular ion value suggests the presence of an odd number of nitrogen atoms; an even molecular ion value is due to the presence of zero or an even number of nitrogen atoms. Thus, $m/z = 141$ suggests 1, 3, 5, 7, etc., nitrogen atoms, while $m/z = 142$ suggests 0, 2, 4, 6, etc., nitrogen atoms.

Rule 2

The maximum number of carbons (N_C^{max}) can be calculated from the formula

$$\left(N_C^{max}\right) = \frac{\text{Relative intensity of M}+1 \text{ peak}}{\text{Relative intensity of M}^+ \text{ peak}} \times \frac{100}{1.1}$$

where M + 1 is the peak one unit above the value of the molecular ion (M^+). This rule gives the *maximum* number of carbons but not necessarily the *actual* number. If, for example, the relative intensities of M^+ and M + 1 are 100 % and 9 %, respectively, then the maximum number of carbons is

$$\left(N_C^{max}\right) = (9/100)\times(100/1.1) = 8$$

In this case, there is a possibility for seven, six, etc., carbons but not for nine or more.

Rule 3

The maximum numbers of sulfur atoms ((N_S^{max})) can be calculated from the formula

$$\left(N_S^{max} \right) = \frac{\text{Relative intensity of M} + 2 \text{ peak}}{\text{Relative intensity of M}^+ \text{peak}} \times \frac{100}{4.4}$$

where M + 2 is the peak two units above that of the molecular ion M$^+$.

Rule 4

The actual number of chlorine and/or bromine atoms can be derived from the Chlorine–Bromine Combination Isotope Intensities table later in this chapter.

Rule 5

The residual mass difference should be only oxygen and hydrogen atoms. These rules assume the absence of phosphorus, silicon, or any other elements.

NEUTRAL MOIETIES EJECTED FROM SUBSTITUTED
BENZENE RING COMPOUNDS

The following table lists the most common substituents encountered in benzene rings and the neutral particles lost and observed on the mass spectrum [1]. Complex rearrangements are often encountered and are enhanced by the presence of one or more heteroatomic substituent(s) in the aromatic compound. All neutral particles that are not the product of rearrangement appear in parentheses and are produced alongside the species that are formed via rearrangement. Prediction of the more abundant moiety is not easy, as it is seriously affected by factors that dictate the nature of the compound. These include the nature and the position of any other substituents, as well as the stability of any intermediate(s) formed. Correlations of the data with the corresponding Hammett σ constants have been neither consistent nor conclusive.

REFERENCE

1. Rose, M.E. and Johnstone, R. A. W., *Mass Spectroscopy for Chemists and Biochemists*, Second edition, Cambridge University Press, Cambridge, 1996.

Substituent	Neutral Moiety(s) Ejected after Rearrangement	Mass Lost
NO_2	NO, CO, (NO_2)	30, 28, (48)
NH_2	HCN	27
$NHCOCH_3$	C_2H_2O, HCN	42, 27
CN	HCN	27
F	C_2H_2	26
OCH_3	CH_2O, CHO, CH_3	29, 30, 15
OH	CO, CHO	28, 29
$SO_2 NH_2$	SO_2, HCN	64, 27
SH	CS, CHS (SH)	44, 45, (33)
SCH_3	CS, CH_2S, SH, (CH_3)	44, 46, 33, (15)

ORDER OF FRAGMENTATION INITIATED BY THE PRESENCE OF A SUBSTITUENT ON A BENZENE RING

The following table lists the relative order of ease of fragmentation that is initiated by the presence of a substituent in the benzene ring in mass spectroscopy [1]. The ease of fragmentation decreases from top to bottom. The substituents marked with an asterisk (*) are very similar in their ease of fragmentation. Particularly in the case of disubstituted benzene rings, the order of fragmentation at the substituent linkage may be easily predicted using this table. As a rule of thumb, the more complex the size of the substituent, the easier its decomposition. For instance, in all chloroacetophenone isomers (1,2-, 1,3-, or 1,4-), the elimination of the methyl radical occurs before the loss of chlorine. On the other hand, under normal mass conditions all bromofluorobenzenes (1,2-, 1,3-, and 1,4-) easily lose the bromine but not the fluorine. Deuterium labeling studies have indicated that any rearrangement of the benzene compounds occurs in the molecular ion and before fragmentation.

REFERENCE

1. Rose, M.E. and Johnstone, R. A. W., *Mass Spectroscopy for Chemists and Biochemists*, Second edition, Cambridge University Press, Cambridge, 1996.

Substituent	Neutral Moiety Eliminated	Mass
$COCH_3$	CH_3	15
CO_2CH_3	OCH_3	31
NO_2	NO_2	46
*I	I	127
*OCH_3	CH_2O, CHO	30, 29
*Br	Br	79 (81)
OH	CO, CHO	28, 29
CH_3	H	1
Cl	Cl	35 (37)
NH_2	HCN	27
CN	HCN	27
F	C_2H_2	26

CHLORINE–BROMINE COMBINATION ISOTOPE INTENSITIES

Due to the distinctive mass spectral patterns caused by the presence of chlorine and bromine in a molecule, interpretation of a mass spectrum can be much easier if the results of the relative isotopic concentrations are known. The following table provides peak intensities (relative to the molecular ion (M^+) at an intensity normalized to 100 %) for various combinations of chlorine and bromine atoms, assuming the absence of all other elements except carbon and hydrogen [1–4]. The mass abundance calculations were based upon the most recent atomic mass data [1].

REFERENCES

1. Rumble, J., ed., *CRC Handbook for Chemistry and Physics*, 100th ed., CRC Press, Boca Raton, FL, 2019.
2. McLafferty, F.W. and Turecek, F., *Interpretation of Mass Spectra*, 4th ed., University Science Books, Mill Valley, CA, 1993.
3. Silverstein, R.H., Bassler, G.C., and Morrill, T.C., *Spectroscopic Identification of Organic Compounds*, 6th ed., John Wiley & Sons, New York, 1998.
4. Williams, D.H. and Fleming, I., *Spectroscopic Methods in Organic Chemistry*, 4th ed., McGraw-Hill, London, 1989.

Relative Intensities of Isotope Peaks for Combinations of Bromine and Chlorine (M^+ = 100 %)

		Br_0	Br_1	Br_2	Br_3	Br_4
Cl_0	P + 2		98.0	196.0	294.0	390.8
	P + 4			96.1	288.2	574.7
	P + 6				94.1	375.3
	P + 8					92.0
Cl_1	P + 2	32.5	130.6	228.0	326.1	424.6
	P + 4		31.9	159.0	383.1	704.2
	P + 6			31.2	187.4	564.1
	P + 8				30.7	214.8
	P + 10					30.3
Cl_2	P + 2	65.0	163.0	261.1	359.3	456.3
	P + 4	10.6	74.4	234.2	490.2	840.3
	P + 6		10.4	83.3	312.8	791.6
	P + 8			10.2	91.7	397.5
	P + 10				9.8	99.2
	P + 12					10.1
Cl_3	P + 2	97.5	195.3	294.0	393.3	489
	P + 4	31.7	127.0	99.7	609.8	989
	P + 6	3.4	34.4	159.4	473.8	1,064
	P + 8		3.3	37.1	193.9	654
	P + 10			3.2	39.6	229
	P + 12				3.0	42
	P + 14					3.2

(Continued)

Relative Intensities of Isotope Peaks for Combinations of Bromine and Chlorine (M$^+$ = 100 %)
(Continued)

		Br$_0$	Br$_1$	Br$_2$	Br$_3$	Br$_4$
Cl$_4$	P + 2	130.0	228.3	326.6	4.2	522
	P + 4	63.3	190.9	414.9	735.3	1,149
	P + 6	13.7	75.8	263.1	670.0	1,388
	P + 8	1.2	14.4	88.8	347.1	1,002
	P + 10		1.1	15.4	102.2	443
	P + 12			1.3	16.2	117
	P + 14				0.7	17
Cl$_5$	P + 2	162.6	260.7	358.9		
	P + 4	105.7	265.3	520.8		
	P + 6	34.3	137.9	397.9		
	P + 8	5.5	39.3	174.5		
	P + 10	0.3	5.8	44.3		
	P + 12		0.3	5.7		
	P + 14			0.5		
Cl$_6$	P + 2	195.3				
	P + 4	158.6				
	P + 6	68.8				
	P + 8	16.6				
	P + 10	2.1				
	P + 12	0.1				
Cl$_7$	P + 2	227.8				
	P + 4	222.1				
	P + 6	120.3				
	P + 8	39.0				
	P + 10	7.5				
	P + 12	0.8				
	P + 14	0.05				

REFERENCE COMPOUNDS UNDER ELECTRON IMPACT CONDITIONS IN MASS SPECTROMETRY

The following table lists the most popular reference compounds for use under electron impact conditions in mass spectrometry. For accurate mass measurements, the reference compound is introduced and ionized concurrently with the sample and the reference peaks are resolved from sample peaks. Reference compounds should contain as few heteroatoms and isotopes as possible. This is to facilitate the assignment of reference masses and minimize the occurrence of unresolved multiplets within the reference spectrum [1]. An approximate upper mass limit should assist in the selection of the appropriate reference [1,2].

REFERENCES

1. Chapman, J.R., *Computers in Mass Spectrometry*, Academic Press, London, 1978.
2. Chapman, J.R., *Practical Organic Mass Spectrometry*, 2nd ed., John Wiley & Sons, Chichester, 1995.

Reference Compound	Formula	Upper Mass Limit
Perfluoro-2-butyltetrahydrofuran	$C_8F_{16}O$	416
Decafluorotriphenyl phosphine (ultramark 443; DFTPP)	$(C_6F_5)_3P$	443
Heptacosafluorotributylamine (perfluorotributylamine; heptacosa; PFTBA)	$(C_4F_9)_3N$	671
Perfluoro kerosene, low-boiling (perfluoro kerosene-L)	$CF_3(CF_2)_nCF_3$	600
Perfluoro kerosene, high-boiling (perfluoro kerosene-H)	$CF_3(CF_2)_nCF_3$	800–900
Tris (trifluoromethyl)-s-triazine	$C_3N_3(CF_3)_3$	285
Tris (pentafluoroethyl)-s-triazine	$C_3N_3(CF_2CF_3)_3$	435
Tris (heptafluoropropyl)-s-triazine	$C_3N_3(CF_2CF_2CF_3)_3$	585
Tris (perfluoroheptyl)-s-triazine	$C_3N_3[(CF_2)_6CF_3]_3$	1,185
Tris (perfluorononyl)-s-triazine	$C_3N_3[(CF_2)_8CF_3]_3$	1,485
Fluoroalkoxy cyclotriphosphazine (mixture, ultramark 1621)	$P_3N_3[OCH_2(CF_2)_nH]_6$	~2,000
Fomblin diffusion pump fluid (ultramark F-series; perfluoropolyether)	$CF_3O[CF(CF_3)CF_2O]_m$ $(CF_2O)_nCF_3$	≥3,000

MAJOR REFERENCE MASSES IN THE SPECTRUM OF HEPTACOSAFLUOROTRIBUTYLAMINE (PERFLUOROTRIBUTYLAMINE)

The following list tabulates the major reference masses (with their relative intensities and formulas) of the mass spectrum of heptacosafluorotributylamine [1]. This is one of the most widely used reference compounds in mass spectrometry.

REFERENCE

1. Chapman, J.R., *Practical Organic Mass Spectrometry*, 2nd ed., John Wiley & Sons, Chichester, 1995.

Mass	Relative Intensity	Formula	Mass	Relative Intensity	Formula
613.9647	2.6	$C_{12}F_{24}N$	180.9888	1.9	C_4F_7
575.9679	1.7	$C_{12}F_{22}N$	175.9935	1.0	C_4F_6N
537.9711	0.4	$C_{12}F_{20}N$	168.9888	3.6	C_3F_7
501.9711	8.6	$C_9F_{20}N$	163.9935	0.7	C_3F_6N
463.9743	3.8	$C_9F_{18}N$	161.9904	0.3	C_4F_6
425.9775	2.5	$C_9F_{16}N$	149.9904	2.1	C_3F_6
413.9775	5.1	$C_8F_{16}N$	130.9920	31	C_3F_5
375.9807	0.9	$C_8F_{14}N$	118.9920	8.3	C_2F_5
325.9839	0.4	$C_7F_{12}N$	113.9967	3.7	C_2F_4N
313.9839	0.4	$C_6F_{12}N$	111.9936	0.7	C_3F_4
263.9871	10	$C_5F_{10}N$	99.9936	12	C_2F_4
230.9856	0.9	C_5F_9	92.9952	1.1	C_3F_3
225.9903	0.6	C_5F_8N	68.9952	100	CF_3
218.9856	62	C_4F_9	49.9968	1.0	CF_2
213.9903	0.6	C_4F_8N	30.9984	2.3	CF

COMMON FRAGMENTATION PATTERNS OF
FAMILIES OF ORGANIC COMPOUNDS

The following table provides a guide to the identification and interpretation of commonly observed mass spectral fragmentation patterns for common organic functional groups [1–9]. It is of course highly desirable to augment mass spectroscopic data with as much other structural information as possible. Especially useful in this regard will be the confirmatory information of infrared and ultraviolet spectrophotometry as well as nuclear magnetic resonance spectrometry.

REFERENCES

1. Bowie, J.H., Williams, D.H., Lawesson, S.O., Madsen, J.O., Nolde, C., and Schroll, G., Studies in mass spectrometry - XV. Mass spectra of sulphoxides and sulphones. The formation of C–C and C–O bonds upon electron impact, *Tetrahedron*, 22, 3515, 1966.
2. Johnstone, R.A.W. and Rose, M.E., *Mass Spectrometry for Chemical and Biochemists*, Cambridge University Press, Cambridge, 1996.
3. Lee, T.A., *A Beginner's Guide to Mass Spectral Interpretation*, Wiley, New York, 1998.
4. McLafferty, F.W. and Turecek, F., *Interpretation of Mass Spectra*, 4th ed., University Science Books, Mill Valley, CA, 1993.
5. Pasto, D.J. and Johnson, C.R., *Organic Structure Determination*, Prentice-Hall, Englewood Cliffs, NJ, 1969.
6. Silverstein, R.M., Bassler, G.C., and Morrill, T.C., *Spectroscopic Identification of Organic Compounds*, 6th ed., John Wiley & Sons, Chichester, 1998.
7. Smakman, R. and deBoer, T.J., The mass spectra of some aliphatic and alicyclic sulphoxides and sulphones, *Org. Mass Spec.*, 3, 1561, 1970.
8. Smith, R.M. *Understanding Mass Spectra: A Basic Approach*, Wiley, New York, 1999.
9. Watson, T.J. and Watson, J.T., *Introduction to Mass Spectrometry*, Lippincott, Williams and Wilkins, Philadelphia, PA, 1997.

Family	Molecular Ion Peak	Common Fragments; Characteristic Peaks
Acetals		Cleavage of all C–O, C–H, and C–C bonds around the original aldehydic carbon.
Alcohols	Weak for 1° and 2°; not detectable for 3°; strong for benzyl alcohols	Loss of 18 (H_2O—usually by cyclic mechanism); loss of H_2O and olefin simultaneously with four (or more) carbon-chain alcohols; prominent peak at $m/z = 31 (CH_2\ddot{O}H)^+$ for 1° alcohols; prominent peak at $m/z = (RCH\ddot{O}H)^+$ for 2°, and $m/z = (R_2C\ddot{O}H)^+$ for 3° alcohols.
Aldehydes	Low intensity	Loss of aldehydic hydrogen (strong M-1 peak, especially with aromatic aldehydes); strong peak at $m/z = 29 (HC\equiv O^+)$; loss of chain attached to alpha carbon (beta cleavage); McLafferty rearrangement via beta cleavage if gamma hydrogen is present.
Alkanes		Loss of 14 mass units (CH_2).
a. Chain	Low intensity	
b. Branched	Low intensity	Cleavage at the point of branch; low intensity ions from random rearrangements.
c. Alicyclic	Rather intense	Loss of 28 mass units ($CH_2=CH_2$) and side chains.
Alkenes (olefins)	Rather high intensity (loss of π-electron) especially in case of cyclic olefins	Loss of units of general formula C_nH_{2n-1}; formation of fragments of the composition C_nH_{2n} (via McLafferty rearrangement); retro Diels–Alder fragmentation.
Alkyl halides	Abundance of molecular ion F<Cl<Br<I; intensity decreases with increase in size and branching	Loss of fragments equal to the mass of the halogen until all halogens are cleaved off.
a. Fluorides	Very low intensity	Loss of 20 (HF); loss of 26 (C_2H_2) in case of fluorobenzenes.
b. Chlorides	Low intensity; characteristic isotope cluster	Loss of 35 (Cl) or 36 (HCl); loss of chain attached to the gamma carbon to the carbon carrying the Cl.
c. Bromides	Low intensity; characteristic isotope cluster	Loss of 79 (Br); loss of chain attached to the gamma carbon to the carbon carrying the Br.
d. Iodides	Higher than other halides	Loss of 127 (I).
Alkynes	Rather high intensity (loss of π-electron)	Fragmentation similar to that of alkenes.
Amides	Rather high intensity	Strong peak at $m/z = 44$ indicative of a 1° amide ($O=C=NH^+_2$); base peak at $m/z = 59$ ($CH_2=C(OH)$ N^+H_2); possibility of McLafferty rearrangement; loss of 42 (C_2H_2O) for amides of the form $RNHCOCH_3$ when R is aromatic ring.
Amines	Hardly detectable in case of acyclic aliphatic amines; high intensity for aromatic and cyclic amines	Beta cleavage yielding >C=N$^+$<; base peak for all 1° amines at $m/z = 30$ ($CH_2=N^+H_2$); moderate M-1 peak for aromatic amines; loss of 27 (HCN) in aromatic amines; fragmentation at alpha carbons in cyclic amines.
Aromatic hydrocarbons (arenes)	Rather intense	Loss of side chain; formation of RCH=CHR' (via McLafferty rearrangement); cleavage at the bonds beta to the aromatic ring; peaks at $m/z = 77$ (benzene ring; especially monosubstituted), 91 (tropyllium); the ring position of alkyl substitution has very little effect on the spectrum.
Carboxylic acids	Weak for straight-chain monocarboxylic acids; large if aromatic acids	Base peak at $m/z = 60$ ($CH_2=C(OH)_2$) if α-hydrogen is present; peak at $m/z = 45$ (COOH); loss of 17 (–OH) in case of aromatic acids or short-chain acids.
Disulfides	Rather low intensity	Loss of olefins (m/z equal to R–S–S–H$^+$); strong peak at $m/z = 66$ (HSSH$^+$).

(Continued)

Family	Molecular Ion Peak	Common Fragments; Characteristic Peaks
Phenols	Highly intense peak (base peak* generally)	Loss of 28 (C=O) and 29 (CHO); strong peak at $m/z = 65$ ($C_5H_5^+$).
Sulfides (thioethers)	Rather low-intensity peak but higher than that of corresponding ether	Similar to those of ethers (–O– substituted by –S–); aromatic sulfides show strong peaks at $m/z = 109$ ($C_6H_5S^+$); 65 ($C_5H_5^+$); 91 (tropyllium ion).
Sulfonamides	Rather intense	Loss of $m/z = 64$ ($SONH_2$) and $m/z = 27$ (HCN) in case of benzenesulfonamide.
Esters	Rather weak intensity	Base peak at m/z equal to the mass of $R–C\equiv O^+$; peaks at m/z equal to the mass of $^+O\equiv C–OR'$, the mass of OR' and R'; McLafferty rearrangement possible in case of (a) presence of a beta hydrogen in R' (peak at m/z equal to the mass of $R–C(^+OH)OH$), and (b) presence of a gamma hydrogen in R (peak at m/z equal to the mass of $CH_2=C(^+OH)OR$); loss of 42 ($CH_2=C=O$) in case of benzyl esters; loss of ROH via the ortho effect in case of o–substituted benzoates).
Ketones	Rather high-intensity peak	Loss of R groups attached to the >C=O (alpha cleavage); peak at $m/z = 43$ for all methyl ketones (CH_3CO^+); McLafferty rearrangement via beta cleavage if gamma hydrogen is present; loss of $m/z = 28$ (C=O) for cyclic ketones after initial alpha cleavage and McLafferty rearrangement.
Mercaptan (thiols)	Rather low intensity but higher than that of corresponding alcohol	Similar to those of alcohols (–OH substituted by –SH); loss of $m/z = 45$ (CHS) and $m/z = 44$ (CS) for aromatic thiols.
Nitriles	Unlikely to be detected except in case of acetonitrile (CH_3CN) and propionitrile (C_2H_5CN)	M + 1 ion may appear (especially at higher pressures); M – 1 peak is weak but detectable ($R–CH=C=N^+$); base peak at $m/z = 41$ ($CH_2=C=N^+H$); McLafferty rearrangement possible; loss of HCN is case of cyanobenzenes.
Nitrites	Absent (or very weak at best)	Base peak at $m/z = 30$ (NO^+); large peak at $m/z = 60$ ($CH_2=O^+NO$) in all unbranched nitrites at the alpha carbon; absence of $m/z = 46$ permits differentiation from nitro compounds.
Nitro compounds	Seldom observed	Loss of 30 (NO); subsequent loss of CO (in case of aromatic nitrocompounds); loss of NO_2 from molecular ion peak.
Sulfones	High intensity	Similar to sulfoxides; loss of mass equal to RSO_2; aromatic heterocycles show peaks at M-32 (sulfur), M-48 (SO), M-64 (SO_2).
Sulfoxides	High intensity	Loss of 17 (OH); loss of alkene (m/z equal to $RSOH^+$); peak at $m/z = 63$ ($CH_2=SOH)^+$; aromatic sulfoxides show peak at $m/z = 125$ ($^+S–CH=CHCH=CHC=O$), 97($C_5H_5S^+$), 93(C_6H_5OH); aromatic heterocycles show peaks at M-16 (oxygen), M-29(COH); M-48(SO)

* The base peak is the most intense peak in the mass spectrum and is often the molecular ion peak, M^+.

COMMON FRAGMENTS LOST

The following table gives a list of neutral species that are most commonly lost when measuring the mass spectra of organic compounds. The list is suggestive rather than comprehensive and should be used in conjunction with other sources [1–4]. The listed fragments include only combination of carbon, hydrogen, oxygen, nitrogen sulfur, and the halogens.

REFERENCES

1. Hamming, M. and Foster, N., *Interpretation of Mass Spectra of Organic Compounds*, Academic Press, New York, 1972.
2. McLafferty, F.W. and Turecek, F., *Interpretation of Mass Spectra*, 4th ed., University Science Books, Mill Valley, CA, 1993.
3. Silverstein, R.M., Bassler, G.C., and Morrill, T.C., *Spectroscopic Identification of Organic Compounds*, 6th ed., John Wiley & Sons, New York, 1996.
4. Bruno, T.J., *CRC Handbook for the Analysis and Identification of Alternative Refrigerants*, CRC Press, Boca Raton, FL, 1995.

Mass Lost	Fragment Lost	Mass Lost	Fragment Lost
1	H·	51	·CHF$_2$
15	CH$_3$·	52	C$_4$H$_4$·, C$_2$N$_2$
17	OH·	54	CH$_2$=CHCH=CH$_2$
18	H$_2$O	55	CH$_2$=CH–CH·CH$_3$
19	F·	56	CH$_2$=CH–CH$_2$CH$_3$; CH$_3$CH=CHCH$_3$; CO (2 moles)
20	HF	57	C$_4$H$_9$·
26	HC≡CH; ·C≡N	58	·NCS; (CH$_3$)$_2$C=O; (NO and CO)
27	CH$_2$–CH·; HC≡N	59	CH$_3$OC=O·; CH$_3$CONH$_2$; C$_2$H$_3$S·
28	CH$_2$=CH$_2$; C=O; (HCN and H·)	60	C$_3$H$_7$OH
29	CH$_3$CH$_2$·; H–·C=O	61	CH$_3$CH$_2$S·; (CH$_2$)$_2$S·H
30	·CH$_2$NH$_2$; HCHO; NO	62	[H$_2$S and CH$_2$=CH$_2$]
31	CH$_3$O·; ·CH$_2$OH; CH$_3$NH$_2$	63	·CH$_2$CH$_2$Cl
32	CH$_3$OH; S	64	S$_2$·, SO$_2$·, C$_5$H$_4$·
33	HS·	68	CH$_2$=CHC(CH$_3$)=CH$_2$
34	H$_2$S	69	CF$_3$·; C$_5$H$_9$·
35	Cl·	71	C$_5$H$_{11}$·
36	HCl$_2$H$_2$O	73	CH$_3$CH$_2$OC·=O
37	H$_2$Cl	74	C$_4$H$_9$OH
38	C$_3$H$_2$·; C$_2$N; F$_2$	75	C$_6$H$_3$
39	C$_3$H$_3$; HC$_2$N	76	C$_6$H$_4$; CS$_2$
40	CH$_3$C≡CH	77	C$_6$H$_5$; HCS$_2$
41	CH$_2$=CHCH$_2$·	78	C$_6$H$_6$·,H$_2$CS$_2$·, C$_5$H$_4$N
42	CH$_2$=CHCH$_3$; CH$_2$=C=O; (CH$_2$)$_3$; NCO; NCNH$_2$	79	Br·; C$_5$H$_5$N
43	C$_3$H$_7$·; CH$_3$C=O·; CH$_2$=CH–O·; HCNO	80	HBr
44	CH$_2$=CHOH; CO$_2$; N$_2$O; CONH$_2$; NHCH$_2$CH$_3$	85	·CClF$_2$
45	CH$_3$CHOH; CH$_3$CH$_2$O·; CO$_2$H; CH$_3$CH$_2$NH$_2$	100	CF$_2$=CF$_2$
46	CH$_3$CH$_2$OH; ·NO$_2$	119	CF$_3$CF$_2$·
47	CH$_3$S·	122	C$_6$H$_5$CO$_2$H
48	CH$_3$SH; SO; O$_3$	127	I·
49	·CH$_2$Cl	128	HI

IMPORTANT PEAKS IN THE MASS SPECTRA OF COMMON SOLVENTS

The following table gives the most important peaks that appear in the mass spectra of the most common solvents, which may be found as an impurity in organic samples. The solvents are classified in ascending order, based upon their M+ peaks. The highest intensity peaks are indicated with (100 %) [1–4].

REFERENCES

1. Clere, J.T., Pretsch, E., and Seibl, J., *Studies in Analytical Chemistry I. Structural Analysis of Organic Compounds by Combined Application of Spectroscopic Methods*, Elsevier, Amsterdam, 1981.
2. McLafferty, F.W. and Turecek, F., *Interpretation of Mass Spectra*, 4th ed., University Science Books, Mill Valley, CA, 1993.
3. Pasto, D.J. and Johnson, C.R., *Organic Structure Determination*, Prentice Hall, Englewood Cliffs, NJ, 1969.
4. Smith, R.M., *Understanding Mass Spectra: A Basic Approach*, Wiley, New York, 1999.

Solvents	Formula	M+	Important Peaks (m/z)
Water	H_2O	18 (100 %)	17
Methanol	CH_3OH	32	31 (100 %), 29, 15
Acetonitrile	CH_3CN	41 (100 %)	40, 39, 38, 28, 15
Ethanol	CH_3CH_2OH	46	45, 31 (100 %), 27, 15
Dimethylether	CH_3OCH_3	46 (100 %)	45, 29, 15
Acetone	CH_3COCH_3	58	43 (100 %), 42, 39, 27, 15
Acetic acid	CH_3CO_2H	60	45, 43, 18, 15
Ethylene glycol	$HOCH_2CH_2OH$	62	43, 33, 31 (100 %), 29, 18, 15
Furan	C_4H_4O	68 (100 %)	42, 39, 38, 31, 29, 18
Tetrahydrofuran	C_4H_8O	72	71, 43, 42 (100 %), 41, 40, 39, 27, 18, 15
n-Pentane	C_5H_{12}	72	57, 43 (100 %), 42, 41, 39, 29, 28, 27, 15
Dimethylformamide (DMF)	$HCON(CH_3)_2$	73 (100 %)	58, 44, 42, 30, 29, 28, 18, 15
Diethylether	$(C_2H_5)_2O$	74	59, 45, 41, 31 (100 %), 29, 27, 15
Methylacetate	$CH_3CO_2CH_3$	74	59, 43 (100 %),42, 32, 29, 28, 15
Carbon disulfide	CS_2	76 (100 %)	64, 44, 32
Benzene	C_6H_6	78 (100 %)	77, 52, 51, 50, 39, 28
Pyridine	C_5H_5N	79 (100 %)	80, 78, 53, 52, 51, 50, 39, 26
Dichloromethane	CH_2Cl_2	84	86, 51, 49 (100 %), 48, 47, 35, 28
Cyclohexane	C_6H_{12}	84	69, 56, 55, 43, 42, 41, 39, 27
n-Hexane	C_6H_{14}	86	85, 71, 69, 57 (100 %), 43, 42, 41, 39, 29, 28, 27
p-Dioxane	$C_4H_8O_2$	88 (100 %)	87, 58, 57, 45, 43, 31, 30, 29, 28
Tetramethylsilane (TMS)	$(CH_3)_4Si$	88	74, 73, 55, 45, 43, 29
1,2-Dimethoxy ethane	$(CH_3OCH_2)_2$	90	60, 58, 45 (100 %), 31, 29
Toluene	$C_6H_5CH_3$	92	91 (100 %), 65, 51, 39, 28
Chloroform	$CHCl_3$	118	120, 83, 81 (100 %), 47, 35, 28
Chlorodorm-d$_1$	$CDCl_3$	119	121, 84, 82 (100 %), 48, 47, 35, 28
Carbon tetrachloride	CCl_4	152 (not seen)	121, 119, 117 (100 %), 84, 82, 58.5, 47, 35, 28
Tetrachloroethene	$CCl_2=CCl_2$	164 (not seen)	168, 166 (100 %), 165, 164, 131, 129, 128, 94, 82, 69, 59, 47, 31, 24

REAGENT GASES FOR CHEMICAL IONIZATION MASS SPECTROMETRY

The following tables provide guidance in the selection and optimization of reagents in high-pressure chemical ionization mass spectrometry, as applied with gas chromatography or as a stand-alone technique [1–3]. The first table provides data on positive ion reagent gases, which are called Bronsted acid reagents. Here, we provide the proton affinity (PA) of the conjugate base and the hydride ion affinity (the enthalpy of the reaction of the positive ion with H^-). The second table provides data on negative ion reagent gases, which are called Bronsted base reagents. Here, we provide the PA of the negative ion and the electron affinity of the base.

REFERENCES

1. Harrison, A.G., *Chemical Ionization Mass Spectrometry*, CRC Press, Boca Raton, FL, 1992.
2. Message, G.M., *Practical Aspects of Gas Chromatography/Mass Spectrometry*, John Wiley & Sons (Wiley Interscience), New York, 1984.
3. Karasek, F.W. and Clement, R.E., *Basic Gas Chromatography – Mass Spectrometry*, Elsevier, Amsterdam, 1988.

Reagent Gas	Reactant Ion (s)	PA (kJ/mol)	PA (kcal/mol)	HIA (kJ/mol)	HIA (kcal/mol)	Comments
		Positive Ion Reagent Gases for Chemical Ionization Mass Spectrometry				
H_2	H_3^+	423.7	101.2	1,260	300	General-purpose reagent gas.
$N_2 + H_2$	N_2H^+	494.9	118.2	1,180	282	
$CO_2 + H_2$	CO_2H^+	547.6	130.8	1,130	270	
$N_2O + H_2$	N_2OH^{+2}	581.1	138.8	1,090	261	Significant signals observed for NO^+.
$CO + H_2$	HCO^+	596.2	142.4	1,080	258	
CH_4	CH_5^+	551.0	131.6	1,130	269	Most widely used reagent gas; usually used initially for most work; degree of fragmentation is relatively large; background spectrum is often large; can produce a large number of addition ions and quasimolecular ions.
	$C_2H_5^+$	680.8	162.6	1,130	271	
H_2O	$H^+(H_2O)_x$ x is pressure dependent	697.1	166.5	980	234	Used for alcohols, ketones, esters, and amines.
CH_3OH	$H^+(CH_3OH)_x$ x is pressure dependent	761.6	181.9	917	219	
C_3H_8	$C_3H_7^+$	751.5	179.5	1,050	250	Uncommon reagent gas.
i-C_4H_{10}	$C_4H_9^+$	820.2	195.9	976	233	General-purpose reagent gas; fragmentation pattern is similar to that produced by ammonia.
NH_3	$H^+(NH_3)_x$ x is pressure dependent	854.1	204.0	825	197	
		Negative Ion Reagent Gases for Chemical Ionization Mass Spectrometry				
H_2	H^-	1,675	400	72.9	17.4	H^- ion is difficult to form in good yields; sometimes used for analysis of alcohols.
NH_3	NH_2^-	1,691	404	75.4	18.0	General-purpose gas, used for the analysis of esters.
N_2O	OH^-	1,637	391	177	42.2	Most common negative ion reagent gas used; often used as a mixture with N_2O, to eliminate O^- signal; sometimes used as a N_2O/He/N_2O, 1:1:1 mixture; used with CH_4 for simultaneous ± ion work.
CH_3NO_2	CH_3O^-	1,595	381	152	36.2	Almost as strong a base as OH^-; used as a 1 % mixture in CH_4.
O_2	O_2^-	1,478	353	42.3	10.1	Used in the analysis of alcohols.
$C_2Cl_3F_3$ (R-113)	Cl^-	1,394	333	349	83.4	Cl^- is a weak Bronsted base useful for acidic compounds.
CH_2Br_2	Br^-	1,357	324	325	77.6	Br^- is a weak Bronsted base (weaker than Cl^-) which reacts with analytes that have a moderately acidic hydrogen.

PROTON AFFINITIES OF SOME SIMPLE MOLECULES

The following table gives the PAs of some simple molecules. For the occurrence of proton transfer (or reaction) between a reactant ion and a sample molecule, the reaction must be exothermic. Thus,

$$\Delta H \text{ reaction} = PA(\text{reactant gas}) - PA(\text{sample}) < 0$$

The more exothermic the reaction, the greater the degree of fragmentation. Endothermic reactions do not yield a protonated form of a sample; therefore, the sample compound cannot be recorded. One can choose the proper reactant gas that will give the correct fragmentation pattern of a desired compound out of a mixture of compounds [1–3]. Chapman [3] lists positive ion chemical ionization applications by reagent gas and by compounds analyzed. The values are provided in kcal/mol for convenience; to convert to the appropriate SI unit (kJ/mol), multiply by 4.1845.

REFERENCES

1. Field, F.H., Chemical ionization mass spectrometry, *Accts. Chem. Res.*, 1, 42, 1968.
2. Harrison, A.G., *Chemical Ionization Mass Spectrometry*, 2nd ed., CRC Press, Boca Raton, FL, 1992.
3. Chapman, J.R., *Practical Organic Mass Spectrometry*, 2nd ed., John Wiley & Sons, Chichester, 1995.

Family	Typical Examples (PA in kcal/mol)
Alcohols	CH_3OH (184.9); CH_3CH_2OH (190.3); $CH_3CH_2CH_2OH$ (191.4); $(CH_3)_3COH$ (195.0); CF_3CH_2OH (174.9)
Aldehydes	$HCHO$ (177.2); CH_3CHO (188.9); CH_3CH_2CHO (191.4); $CH_3CH_2CH_2CHO$ (193.3)
Alkanes	CH_4 (130.5); $(CH_3)_3CH$ (195)
Alkenes	$H_2C=CH_2$ (163.5); $CH_3CH=CH_2$ (184.9); $(CH_3)_2C=CH_2$ (196.9); trans-$CH_3CH=CHCH_3$ (182.0)
Aromatics, substituted C_6H_5-G	G= –H (182.8); –Cl (181.7); –F (181.5); –CH_3 (191.2); –C_2H_5 (192.2); –$CH_2CH_2CH_3$ (191.0); –$CH(CH_3)_2$ (191.4); –$C(CH_3)_3$ (191.6); –NO_2 (193.8); –OH (196.2); –CN (196.3); –CHO (200.3); –OCH_3 (200.6); –NH_2 (211.5)
Amines	1°: NH_3 (205.0); CH_3NH_2 (214.1); $C_2H_5NH_2$ (217.1); $CH_3CH_2CH_2NH_2$ (218.5); $CH_3CH_2CH_2CH_2NH_2$ (219.0) 2°: $(CH_3)_2NH$ (220.5); $(C_2H_5)_2NH$ (225.1); $(CH_3CH_2CH_2)_2NH$ (227.4) 3°: $(CH_3)_3N$ (224.3); $(C_2H_5)_3N$ (231.2); $(CH_3CH_2CH_2)_3N$ (233.4)
Carboxylic acids	HCO_2H (182.8); CH_3CO_2H (190.7); $CH_3CH_2CO_2H$ (193.4); CF_3CO_2H (176.0)
Dienes	$CH_2=CHCH=CH_2$ (193); E–$CH_2=CHCH=CHCH_3$ (201.8); E–$CH_2=CHC(CH_3)=CHCH_3$ (205.7); cyclopentadiene (200.0)
Esters	HCO_2CH_3 (190.4); $HCO_2C_2H_5$ (194.2); $HCO_2CH_2CH_2CH_3$ (195.2); $CH_3CO_2CH_3$ (198.3); $CH_3CO_2C_2H_5$ (201.3); $CH_3CO_2CH_2CH_2CH_3$ (202.0)
Ethers	$(CH_3)_2O$ (193.1); $(C_2H_5)_2O$ (200.4); $(CH_3CH_2CH_2)_2O$ (202.9); $(CH_3CH_2CH_2CH_2)_2O$ (203.9); tetrahydrofuran (199.6); tetrahydropyran (200.7)
Ketones	CH_3COCH_3 (197.2); $CH_3COC_2H_5$ (199.4)
Nitriles (cyanocompounds)	HCN (178.9); CH_3CN (190.9); C_2H_5CN (192.8); $CH_3CH_2CH_2CN$ (193.8)
Sulfides	$(CH_3)_2S$ (200.7); $(C_2H_5)_2S$ (205.6); $[(CH_3)_2CH]_2S$ (209.3)
Thiols	H_2S (176.6); CH_3SH (188.6); C_2H_5SH (192.0); $[(CH_3)_2CH]_2SH$ (194.7)

PROTON AFFINITIES OF SOME ANIONS

The following table lists the PAs of some common anions (X^-). Since the reaction of an anion (X^-) with a proton (H^+)

$$X^- + H^+ \rightarrow H - X$$

is exothermic, it can be used to generate other anions that possess a smaller PA value by the addition of the corresponding neutral species [1,2].

REFERENCES

1. Chapman, J.R., *Practical Organic Mass Spectrometry*, 2nd ed., John Wiley & Sons, Chichester, 1995.
2. Harrison, A.G., *Chemical Ionization Mass Spectrometry*, 2nd ed., CRC Press, Boca Raton, FL, 1992.

Anion	PA (kJ/mol)
NH_2^-	1,689
H^-	1,676
OH^-	1,636
$O^{\cdot-}$	1,595
CH_3O^-	1,583
$(CH_3)_2CHO^-$	1,565
$-CH_2CN$	1,556
F^-	1,554
$C_5H_5^-$	1,480
$O_2^{\cdot-}$	1,465
CN^-	1,462
Cl^-	1,395

DETECTION OF LEAKS IN MASS SPECTROMETER SYSTEMS

The following tables provide guidance for troubleshooting possible leaks in the vacuum systems of mass spectrometers, especially those operating in electron impact mode. Leak testing is commonly done by playing a stream of a pure gas against a fitting, joint, or component that is suspected of being a leak source. If in fact the component is the source of a leak, one should be able to note the presence of the leak detection fluid on the mass spectrum. Here we present the mass spectra of methane tetrafluoride, 1,1,1,2-tetrafluoroethane (R-134a), n-butane, and acetone [1,2]. Methane tetrafluoride, 1,1,1,2-tetrafluoroethane, and n-butane are handled as gases, while acetone is handled as a liquid. Typically, n-butane is dispensed from a disposable lighter, and acetone is dispensed from a dropper. Care must be taken when using acetone or a butane lighter for leak checking because of the flammability of these fluids.

REFERENCES

1. Bruno, T.J., *CRC Handbook for the Analysis and Identification of Alternative Refrigerants*, CRC Press, Boca Raton, FL, 1994.
2. NIST Chemistry Web Book, NIST Standard Reference Database Number 69 - March, 2018 Release, accessed December 2019.

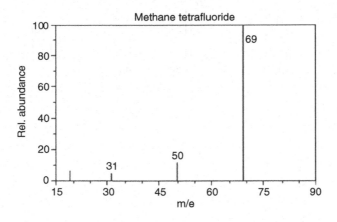

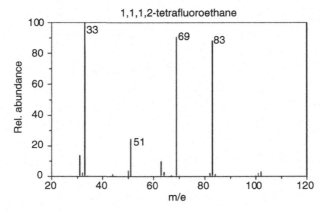

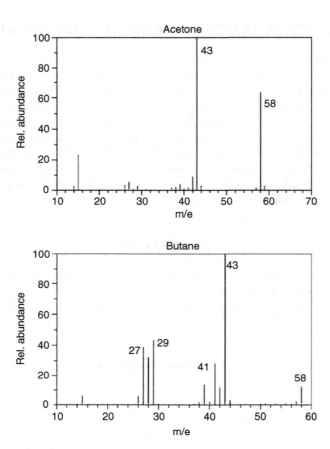

COMMON SPURIOUS SIGNALS OBSERVED IN MASS SPECTROMETERS

The following table provides guidance in the recognition of spurious signals (m/z peaks) that will sometimes be observed in measured mass spectra [1]. Often, the occurrence of these signals can be predicted by the recent history of the instrument or the method being used. This is especially true if the mass spectrometer is interfaced to a gas chromatograph.

REFERENCE

1. Maintaining your GC-MS System Agilent Technologies, Applications Manual, 2001, available online at www.agilent.com/chem, accessed December 2019.

Ions Observed (m/z)	Possible Compound	Possible Source
13, 14, 15, 16	Methane*	Chlorine reagent gas
18	Water*	Residual impurity; outgassing of ferrules; septa and seals.
14, 28	Nitrogen*	Residual impurity, outgassing of ferrules; septa and seals; leaking seal.
16, 32	Oxygen*	Residual impurity; outgassing of ferrules; septa and seals; leaking seal.
44	Carbon dioxide*	Residual impurity, outgassing of ferrules; septa and seals; leaking seal; note it may be mistaken for propane in a sample.
31, 51, 69, 100, 119, 131, 169, 181, 214, 219, 264, 376, 414, 426, 464, 502, 576, 614	Perfluorotributyl amine (PFTBA) and related ions	This is a common tuning compound; may indicate a leaking valve.
31	Methanol	Solvent; can be used as a leak detector.
41, 43, 55, 57, 69, 71, 85, 99	Hydrocarbons	Mechanical pump oil; fingerprints
43, 58	Acetone	Solvent; can be used as a leak detector.
78	Benzene	Solvent; can be used as a leak detector.
91, 92	Toluene	Solvent; can be used as a leak detector.
105, 106	Xylenes	Solvent; can be used as a leak detector.
151, 153	Trichloroethane	Solvent; can be used as a leak detector.
69	Mechanical pump fluid, PFTBA	Back diffusion of mechanical pump fluid; possible leaking valve of tuning compound vial.
73, 147, 207, 221, 281, 295, 355, 429	Dimethylpolysiloxane	Bleed from a column or septum, often during high-temperature program methods in GC-MS
77, 94, 115, 141, 168, 170, 262, 354, 446	Diffusion pump fluid	Back diffusion from diffusion pump, if present.
149	Phthalates	Plasticizer in vacuum seals, gloves.
X–14 peaks	Hydrocarbons	Loss of a methylene group indicates a hydrocarbon sample.

* It is possible to operate the analyzer to ignore these common background impurities. They will be present to contribute to poor vacuum if these impurities result from a significant leak.

MASS RESOLUTION REQUIRED TO RESOLVE COMMON SPECTRAL INTERFERENCES ENCOUNTERED IN INDUCTIVELY COUPLED PLASMA MASS SPECTROMETRY (ICP-MS)

The table below lists some common spectral interferences that are encountered in inductively coupled plasma mass spectrometry (ICP-MS) as well as the resolution that is necessary to analyze them [1]. The resolution is presented as a dimensionless ratio. As an example, the mass of the polyatomic ion $^{15}N^{16}O^+$ would be 15.000108 + 15.994915 = 30.995023. This would interfere with $^{31}P^+$ at a mass of 30.973762. The required resolution would be RMM/ΔRMM or 30.973762/0.021261 = 1457. One should bear in mind that as resolution increases, the sensitivity decreases with subsequent effects on the price of the instrument. Note that small differences exist in the published exact masses of isotopes, but for the calculation of the required resolution, these differences are trivial. Moreover, recent instrumentation has provided rapid, high-resolution mass spectra with an uncertainty of <0.01 %.

REFERENCE

1. Gregoire, D.C., Analysis of geological materials by inductively coupled plasma mass spectrometry, *Spectroscopy*, 14, 14–19, 1999.

Polyatomic Ion	Interfered Isotope (Natural Abundance %)	Required Resolution
$^{14}N_2^+$	$^{14}Si^+$ (92.21)	958
$^{15}N^{16}O^+$	$^{31}P^+$ (100)	1,457
$^{40}Ar^{12}O^+$	$^{52}Cr^+$ (83.76)	2,375
$^{32}S^{16}O^+$	$^{48}Tl^+$ (73.94)	2,519
$^{35}Cl^{16}O^+$	$^{51}V^+$ (99.76)	2,572
$^{40}Ar^{35}Cl^+$	$^{75}As^+$ (100)	7,775
$^{40}Ar_2^+$	$^{80}Se^+$ (49.82)	9,688

CHAPTER 12

Atomic Absorption Spectrometry

INTRODUCTION FOR ATOMIC SPECTROMETRIC TABLES

The tables presented in this section are designed to aid in the area of atomic spectrometric methods of analysis. The following conventions for abbreviation are recommended by the International Union of Pure and Applied Chemistry [1].

Atomic Emission Spectrometry—AES
Atomic Absorption Spectrometry—AAS
Flame Atomic Emission Spectrometry—FAES
Flame Atomic Absorption Spectrometry—FAAS
Electrothermal Atomic Absorption Spectrometry—EAAS
Inductively Coupled Plasma Atomic Emission Spectrometry—ICP-AES

Other variations such as cold vapor and hydride generation are not abbreviated but spelled out, for example, cold vapor AAS and hydride generation FAAS. These abbreviations are used whenever appropriate throughout the section.

Several of these tables have appeared in Parsons' handbook [2] in one form or another. They have been updated to the extent possible, and the wavelength values have been made to conform to those in the National Standard Reference Data System–National Bureau of Standards (NSRDS-NBS) 68 [3] wherever possible.

As several of the tables cite the same references, all cited references will be listed at the end of this introduction instead of being repeated at the end of each table. These tables were originally prepared by Parsons for the first edition of this book [1–18].

REFERENCES

1. Commission on spectrochemical and other optical procedures for analysis, nomenclature, symbols, units and their usage in spectrochemical analysis—I. General atomic emission spectroscopy; II. Data interpretation; and III. Analytical flame spectroscopy and associated procedures, Slavin, W., Manning, D.C., The L'vov platform for furnace atomic absorption analysis, *Spectrochim. Acta, Part B*, 33, 219, 1978.
2. Parsons, M.L., Smith, B.W., and Bentley, G.E., *Handbook of Flame Spectroscopy*, Plenum Press, New York, 1975.
3. Reader, J., Corliss, C.H., Weise, W.L., and Martin, G.A., *Wavelengths and Transition Probabilities for Atoms and Atomic Ions*, NSRDS-NBS 68, U.S. Government Printing Office, Washington, D.C., 1980.
4. Smith, B.W. and Parsons, M.L., Preparation of standard solutions: critically selected compounds, *J. Chem. Ed.*, 50, 679, 1973.

5. Dean, J.A. and Rains, T.C., *Flame Emission and Atomic Absorption Spectrometry*, Vol. 2, Marcel Dekker, New York, 1971, pp. 327–341.

6. Thermo Jarrell Ash Corp., *Guide to Analytical Values for TJA Spectrometers*, Waltham, MA, 1987.

7. Anderson, T.A. and Parsons, M.L., ICP emission spectra III: the spectra for the Group IIIA elements and spectral interferences due to Group IIA and IIIA elements, *Appl. Spectrosc.*, 38, 625, 1984; Parsons, M.L., Forster, A., and Anderson, D., *An Atlas of Spectral Interferences in ICP Spectroscopy*, Plenum Press, New York, 1980.

8. Park, D.A., *Further Investigations of Spectra and Spectral Interferences Due to Group A Elements in ICP Spectroscopy: Groups IVA and VA*, Ph.D. thesis, Arizona State University, Tempe; Parsons, M.L., unpublished data, Los Alamos National Laboratory, Los Alamos, NM, 1987.

9. Perkin-Elmer Corp., *Mercury/Hydride System*, Report No. 1876/6.79, Norwalk, CT, 1987.

10. Lovett, R.J., Welch, D.L., and Parsons, M.L., On the importance of spectral interferences in atomic absorption spectroscopy, *Appl. Spectrosc.*, 29, 470, 1975.

11. Layman, L., Palmer, B., and Parsons, M.L., Unpublished data taken with the Los Alamos National Laboratory FTS Facility, Los Alamos, NM, 1987.

12. Sneddon, J., Background correction techniques in atomic spectroscopy, *Spectroscopy*, 2(5), 38, 1987.

13. Wittenberg, G.K., Haun, D.V., and Parsons, M.L., The use of free-energy minimization for calculating beta factors and equilibrium compositions in flame spectroscopy, *Appl. Spectrosc.*, 33, 626, 1979.

14. Parsons, M.L., Smith, B.W., and McElfresh, P.M., On the selection of analysis lines in atomic absorption spectrometry, *Appl. Spectrosc.*, 27, 471, 1973.

15. Parker, L.R., Jr., Morgan, S.L., and Deming, S.N., Simplex optimization of experimental factors in atomic absorption spectrometry, *Appl. Spectrosc.*, 29, 429, 1975.

16. Parsons, M.L. and Winefordner, J.D., Optimization of the critical instrumental parameters for achieving maximum sensitivity and precision in flame-spectrometric methods of analysis, *Appl. Spectrosc.*, 21, 368, 1967.

17. Wiese, W.L., Smith, M.W., and Glennon, B.M., *Atomic Transition Probabilities: Vol. I Hydrogen Through Neon*, NSRDS-NBS 4, U.S. Government Printing Office, Washington, D.C., 1966.

18. Bruno, T.J. and Svoronos, P.D.N., *CRC Handbook of Basic Tables for Chemical Analysis*, 3rd ed., CRC Press, Boca Raton, FL, 2011.

STANDARD SOLUTIONS—SELECTED COMPOUNDS AND PROCEDURES

The compounds selected for this table were chosen using a rather stringent set of criteria, including stability, purity, ease of preparation, availability, high molecular mass, and toxicity. It is very important to have a compound that is pure and can be dried, weighed, and dissolved with comparative ease. The list of compounds provided here meets those goals as much as possible. No attempt was made to include all compounds that meet these criteria nor are the compounds in this list trivial to dissolve; some require a rather long time and/or vigorous conditions.

In this table, the significant figures in all columns represent the accuracy with which the atomic masses of the elements are known.

Table 12.1 was compiled from Refs. 4 and 5.

Table 12.1   Standard Solutions—Selected Compounds and Procedures

Element	Compound	Relative Formula Mass	Weight for 1,000 µg/L (PPM)-g/L	Solvent	Note
Aluminum	Al-metal	26.982	1.0000	Hot dil. HCl-2M	APS
Antimony	$KSbOC_4H_4O_6$	324.92	2.6687	Water	f
	* 1/2 H_2O (antimony potassium tartarate)				
	Sb-metal	121.75	1.0000	Hot aq. reg.	
Arsenic	As_2O_3	197.84	1.3203	1:1 NH_3	PS, c, NIST
Barium	$BaCO_3$	197.35	1.4369	Dil. HCl	h
	$BaCl_2$	208.25	1.5163	Water	g
Beryllium	Be-metal	9.0122	1.0000	HCl	c
	$BeSO_4$ * $4H_2O$	177.135	19.6550	Water+acid	i
Bismuth	Bi_2O_3	465.96	1.1148	HNO_3	
	Bi metal	208.980	1.00000	HNO_3	
Boron	H_3BO_3	61.84	5.720	Water	PS, NIST, m
Bromine	KBr	119.01	1.4894	Water	APS
Cadmium	CdO	128.40	1.1423	HNO_3	
	Cd metal	112.40	1.0000	Dil. HCl	
Calcium	$CaCO_3$	100.09	2.4972	Dil. HCl	h
Cerium	$(NH_4)_2Ce(NO_3)_4$	548.23	3.9126	Water	
Cesium	Cs_2SO_4	361.87	1.3614	Water	
Chlorine	NaCl	58.442	1.6485	Water	PS
Chromium	$K_2Cr_2O_7$	294.19	2.8290	Water	PS, NIST
	Cr metal	51.996	1.0000	HCl	
Cobalt	Co metal	58.933	1.0000	HNO_3	APS
Copper	Cu metal	63.546	1.0000	Dil. HNO_3	APS
	CuO	69.545	1.2517	Hot HCl	APS
	$CuSO_4$ * $5H_2O$	249.678	3.92909	Water	
Dysprosium	Dy_2O_3	373.00	1.477	Hot HCl	e
Erbium	Er_2O_3	382.56	1.1435	Hot HCl	e
Europium	Eu_2O_3	351.92	1.1579	Hot HCl	e
Fluorine	NaF	41.988	2.2101	Water	j
Gadolinium	Gd_2O_3	362.50	1.1526	Hot HCl	e
Gallium	Ga metal	69.72	1.000	Hot HNO_3	k
Germanium	GeO_2	104.60	1.4410	Hot 1 M NaOH or 50 g oxalic acid + water	
Gold	Au metal	196.97	1.0000	Hot aq. reg.	APS, NIST
Hafnium	Hf metal	178.49	1.0000	Hf, fusion	1

(Continued)

Table 12.1 (*Continued*)   Standard Solutions—Selected Compounds and Procedures

Element	Compound	Relative Formula Mass	Weight for 1,000 µg/L (PPM)-g/L	Solvent	Note
Holmium	Ho_2O_3	377.86	1.1455	Hot HCl	e
Indium	In_2O_3	277.64	1.2090	Hot HCl	
	In metal	114.82	1.0000	Dil. HCl	
Iodine	KIO_3	214.00	1.6863	Water	PS
Iridium	Na_3IrCl_6	473.8	2.466	Water	
Iron	Fe-metal	55.847	1.0000	Hot HCl	APS
Lanthanum	La_2O_3	325.82	1.1728	Hot HCl	e
Lead	$Pb(NO_3)_2$	331.20	1.5985	HCl	APS, NIST
Lithium	Li_2CO_3	73.890	5.3243	Dil. HCl	APS, h
Lutetium	Lu_2O_3	397.94	1.1372	Hot HCl	e
Magnesium	MgO	40.311	1.6581	HCl	
	Mg metal	24.312	1.0000	Dil. HCl	
Manganese	$MnSO_4 * H_2O$	169.01	3.0764	Water	o
Mercury	$HgCl_2$	271.50	1.3535	Water	c
	Hg metal	200.59	1.0000	$5 M HNO_3$	
Molybdenum	MoO_3	143.94	1.5003	$1 M NaOH or 2 M HN_3$	
Neodymium	Nd_2O_3	336.48	1.1664	HCl	e
Nickel	Ni metal	58.71	1.000	Hot HNO_3	APS
Niobium	Nb_2O_5	265.81	1.4305	HF, fusion	p, q
	Nb metal	92.906	1.0000	$HF+H_2SO_4$	q
Osmium	Os metal	190.20	1.0000	Hot H_2SO_4	d
Palladium	Pd metal	106.40	1.0000	Hot HNO_3	
Phosphorus	KH_2PO_4	136.09	4.3937	Water	
	$(NH_3)_2HPO_4$	209.997	6.77983	Water	
Platinum	K_2PtCl_4	415.12	2.1278	Water	APS, NIST
	Pt metal	195.05	1.0000	Hot aq. reg.	
Potassium	KCl	74.555	1.9067	Water	PS, NIST
	$KHC_6H_4O_4$	204.22	5.2228	Water	PS, NIST
	(Potassium hydrogen phthalate)				
	$K_2Cr_2O_7$	294.19	3.7618	Water	PS, NIST
Praseodymium	Pr_6O_{11}	1,021.43	1.20816	HCl	e
Rhenium	Re metal	186.2	1.000	HNO_3	
	$KReO_4$	289.3	1.554	Water	
Rhodium	Rh metal	102.91	1.0000	Hot H_2SO_4	
Rubidium	Rb_2SO_4	267.00	1.5628	Water	
Ruthenium	RuO_4	165.07	1.6332	Water	
Samarium	Sm_2O_3	348.70	2.3193	Hot HCl	e
Scandium	Sc_2O_3	137.91	1.5339	Hot HCl	
Selenium	Se metal	78.96	1.000	Hot HNO_3	
	SeO_2	110.9	1.405	Water	
Silicon	Si metal	28.086	1.0000	NaOH, conc.	
	SiO_2	60.085	2.1393	HF	
Silver	$AgNO_3$	169.875	1.57481	Water	APS, r
	Ag metal	107.870	1.0000	HNO_3	
Sodium	NaCl	58.442	2.5428	Water	PS
	$Na_2C_2O_4$	134.000	2.91432	Water	PS, NIST
	(sodium oxalate)				

(Continued)

Table 12.1 (*Continued*) Standard Solutions—Selected Compounds and Procedures

Element	Compound	Relative Formula Mass	Weight for 1,000 µg/L (PPM)-g/L	Solvent	Note
Strontium	$SrCO_3$	147.63	1.6849	Dil. HCl	APS, h
Sulfur	K_2SO_4	174.27	5.4351	Water	
	$(NH_4)_2SO_4$	114.10	3.5585	Water	
Tantalum	Ta_2O_5	441.893	1.22130	HF, fusion	p, q
	Ta metal	180.948	1.0000	$HF+H_2SO_4$	q
Tellurium	TeO_2	159.60	1.2507	HCl	
Terbium	Tb_2O_3	365.85	1.1512	Hot HCl	e
Thallium	Tl_2CO_3	468.75	1.1468	Water	APS, c
	$TlNO_3$	266.37	1.3034	Water	
Thorium	$Th(NO_3)_4$ * $4H_2O$	552.118	2.37943	HNO_3	
Thulium	Tm_2O_3	385.87	1.1421	Hot HCl	e
Tin	Sn metal	118.69	1.0000	HCl	
	SnO	134.69	1.1348	HCl	
Titanium	Ti metal	47.90	1.000	1:1 H_2SO_4	APS
Tungsten	Na_2WO_4 * $2H_2O$	329.86	1.7942	Water	s
	Na_2WO_4	293.83	1.5982	Water	f
Uranium	UO_2	270.03	1.1344	HNO_3	PS, NIST
	U_3O_6	842.09	1.1792	HNO_3	
	$UO_2(NO_3)_2$ * $6H_2O$	502.13	2.1095	Water	
Vanadium	V_2O_5	181.88	1.78521	Hot HCl	
	NH_4VO_3	116.98	2.2963	Dil. HNO_3	
Ytterbium	Yb_2O_3	394.08	1.1386	Hot HCl	e
Yttrium	Y_2O_3	225.81	1.2700	Hot HCl	e
Zinc	ZnO	81.37	1.245	HCl	APS
	Zn metal	65.37	1.000	HCl	APS, NIST
Zirconium	Zr metal	91.22	1.000	HF, fusion	1
	$ZrOCl_2$ * $8H_2O$	322.2	3.533	HCl	

Notes:
PS = Primary standard.
APS = Compounds which approach primary standard quality.
NIST = These compounds are sold as primary standards by the NIST Standard Reference Materials Program, 100 Bureau Drive, Gaithersburg, MD 20899-3460 (www.nist.gov, accessed January 2020).
c = Highly toxic.
d = Very highly toxic.
e = The rare-earth oxides, because they absorb CO_2 and water vapor from the atmosphere, should be freshly ignited prior to weighing.
f = Loses water at 110 °C. Water is only slowly regained, but rapid weighing and desiccator storage are required.
g = Drying at 250 °C, rapid weighing, and desiccator storage are required.
h = Add a quantity of water, then add dilute acid and swirl until the CO_2 has ceased to bubble out, then dilute.
I = Dissolve in water, then add 5 mL of concentrated HCl and dilute.
j = Sodium fluoride solutions will etch glass and should be freshly prepared.
k = Because the melting point is 29.6 °C, the metal may be warmed and weighed as a liquid.
l = Zr and Hf compounds were not investigated in the laboratory of Ref. 5.
m = Boric acid may be weighed directly from the bottle. It loses 1 H_2O at 100 °C, but it is difficult to dry to a constant mass.
n = Several references suggest that the addition of acid will help stabilize the solution.
o = This compound may be dried at 100 °C without losing the water of hydration.
p = Nb and Ta are slowly soluble in 40 % HF. The addition of H_2SO_4 accelerates the dissolution process.
q = Dissolve in 20 mL hot HF in a platinum dish, add 40 mL H_2SO_4 and evaporate to fumes, dilute with 8M H_2SO_4.
r = When kept dry, silver nitrate crystals are not affected by light. Solutions should be stored in brown bottles.
s = Sodium tungstate loses both water molecules at 110 °C. The water is not rapidly regained, but the compound should be kept in a desiccator after drying and should be weighed quickly once it is removed.

LIMITS OF DETECTION TABLES FOR COMMON
ANALYTICAL TRANSITIONS IN AES AND AAS

The following five tables present the common transitions for analysis and the detection limits for AES and AAS on the basis of source, where appropriate for the specific atom cell indicated. The detection limits are from the literature cited and are given in parts per billion (ppb) or nanograms per milliliter of aqueous solution. The limits of detection (LODs) are generally defined as a signal-to-noise ratio of two or three. This generally relates to a concentration that produces a signal of two or three times the standard deviation of the measurement. These are measured in dilute aqueous solution and represent the best that the system was capable of measuring. In most cases, the detection limit in real samples will be one or two orders of magnitude higher, or worse, than those stated here. The type designation is I for free atom and II for single ion. In all cases, NO means that no observation was made for the situation indicated, NA means that either AES or AAS was observed but no detection limit was reported.

In all cases where possible, the wavelengths of the transitions were made to conform with Ref. 3; any wavelength below 200 nm is the wavelength given in vacuum, all others are in air.

Limits of Detection for the Air–Hydrocarbon Flame[a]

Element	Symbol	Wavelength (nm)	Type	LOD-AAS (ppb)
Antimony	Sb	217.581	I	100
		231.147	I	100
Bismuth	Bi	223.061	I	50
Calcium	Ca	22.673	I	2
Cesium	Cs	455.5276	I	600
		852.1122	I	50
Chromium	Cr	357.869	I	5
Cobalt	Co	240.725	I	5
Copper	Cu	324.754	I	50
		327.396	I	50
Gallium	Ga	287.424	I	70
Gold	Au	242.795	I	20
Indium	In	303.936	I	50
Iridium	Ir	208.882	I	15,000
		2,639.71	I	2,000
Iron	Fe	248.3271	I	5
Lead	Pb	283.3053	I	10
Lithium	Li	670.776	I	5
Magnesium	Mg	285.213	I	0.3
Manganese	Mn	279.482	I	2
		403.076	I	2
Mercury	Hg	253.652	I	500
Molybdenum	Mo	313.259	I	30
Nickel	Ni	232.003	I	5
Osmium	Os	290.906	I	17,000
Palladium	Pd	244.791	I	2,000
		247.642	I	30
Platinum	Pt	265.945	I	100
Potassium	K	766.490	I	5
Rhodium	Rh	343.489	I	30
Rubidium	Rb	420.180	I	NA
		780.027	I	5
Ruthenium	Ru	349.894	I	300
		372.803	I	3,000
Selenium	Se	196.09	I	100
		203.98	I	2,000
Silver	Ag	328.068	I	5
		338.289	I	200
Sodium	Na	330.237	I	NA
		588.9950	I	2
		589.5924	I	2
Strontium	Sr	407.771	II	NA
		460.733	I	10
Tellurium	Te	214.281	I	100
Thallium	Tl	276.787	I	30
		377.572	I	2,400
Tin	Sn	224.605	I	30
Zinc	Zn	213.856	I	2

[a] Flames formed from air combined with the lighter hydrocarbons, such as methane, propane, butane, or natural gas, behave in a very similar fashion with similar temperatures, similar chemical properties, etc.
These data were taken from Ref. 2.

Limits of Detection for the Air–Acetylene Flame

Element	Symbol	Wavelength (nm)	Type	LOD-AES (ppb)	LOS-AAS (ppb)
Aluminum	Al	308.2153	I	NO	700
		309.2710	I	NO	500
		396.1520	I	NA	600
Antimony	Sb	206.833	I	NA	50
		217.581	I	NA	40
		231.147	I	3,000	40
		259.805	I	NA	NO
Arsenic	As	193.759	I	10,000	140
Barium	Ba	455.403	II	NA	NO
		553.548	I	NA	NO
Bismuth	Bi	223.061	I	3,000	25
Boron	B	249.677	I	NA	NO
Cadmium	Cd	228.8022	I	500	1
		326.1055	I	NA	NA
Calcium	Ca	393.366	II	NO	5,000
		396.847	II	NO	5,000
		422.673	I	0.5	0.5
Cesium	Cs	455.5276	I	NA	NO
		852.1122	I	NA	8
Chromium	Cr	357.869	I	NA	3
		425.435	I	NA	200
Cobalt	Co	240.725	I	NO	4
		352.685	I	NA	125
Copper	Cu	324.754	I	NA	1
		327.396	I	NA	120
Gallium	Ga	287.424	I	NO	50
		294.364	I	NA	50
		417.204	I	NA	1,500
Germanium	Ge	265.1172	I	7,000	
Gold	Au	242.795	I	NA	6
		267.595	I	NA	90
Indium	In	303.936	I	NA	30
		325.609	I	NA	20
		451.131	I	NA	200
Iodine	I	183.038	I	NO	8,000
		206.163	I	2,500,000	NO
Iridium	Ir	208.882	I	NO	600
		2,639.71	I	NO	2,500
Iron	Fe	248.3271	I	NO	5
		371.9935	I	NA	700
Lead	Pb	217.000	I	NO	9
		283.3053	I	NA	240
		368.3462	I	NA	NO
Lithium	Li	670.776	I	NA	0.3
		451.857	I	NO	NA
Magnesium	Mg	279.553	II	NO	NA
		280.270	II	NO	NA
		285.213	I	NA	0.1
Manganese	Mn	279.482	I	NA	2
		403.076	I	NA	600
Mercury	Hg	253.652	I	NA	140
Molybdenum	Mo	313.259	I	NO	20
		379.825	I	80,000	900
		390.296	I	100	1,600

(*Continued*)

Limits of Detection for the Air–Acetylene Flame (*Continued*)

Element	Symbol	Wavelength (nm)	Type	LOD-AES (ppb)	LOS-AAS (ppb)
Nickel	Ni	232.003	I	NO	2
		352.454	I	NA	350
Niobium	Nb	309.418	II	NO	NA
Osmium	Os	290.906	I	NA	1,200
Palladium	Pd	244.791	I	NO	20
		247.642	I	NO	20
		340.458	I	NA	660
		363.470	I	NA	300
Phosphorus	P	213.547	I	NO	30,000
Platinum	Pt	214.423	I	NO	350
		265.945	I	NA	50
Potassium	K	766.490	I	NA	1
Rhenium	Re	346.046	I	NO	800
Rhodium	Rh	343.489	I	NA	2
		369.236	I	NA	70
Rubidium	Rb	420.180	I	NA	NO
		780.027	I	NA	0.3
Ruthenium	Ru	349.894	I	NA	400
		372.803	I	NA	250
Selenium	Se	196.09	I	NA	50
		203.98	I	50,000	10,000
Silver	Ag	328.068	I	NA	1
		338.289	I	NA	70
Sodium	Na	330.237	I	NO	NA
		588.9950	I	NA	1
		589.5924	I	NA	0.2
Strontium	Sr	407.771	II	NA	400
		421.552	II	NO	NA
		460.733	I	NA	2
Sulfur	S	180.7311	I	NO	30,000
Tellurium	Te	214.281	I	500	30
		238.578	I	NO	NA
Thallium	Tl	276.787	I	NA	30
		377.572	I	NA	1,200
		535.046	I	NA	12,000
Tin	Sn	224.605	I	NO	10
		235.484	I	2,000	600
		283.999	I	NA	1,000
		326.234	I	NA	NO
Tungsten	W	255.135	I	90,000	3,000
		400.875	I		
Uranium	U	591.539	I	NA	NO
Vanadium	V	318.540	I	NA	NO
		437.924	I	300	NO
Ytterbium	Yb	398.799	I	NO	80
Zinc	Zn	213.856	I	7,000	1
Zirconium	Zr	351.960	I	NO	NA

These data were taken from Refs. 2 and 6.

CRC HANDBOOK OF BASIC TABLES FOR CHEMICAL ANALYSIS

Limits of Detection for the Nitrous Oxide–Acetylene Flame

Element	Symbol	Wavelength (nm)	Type	LOD-AES (ppb)	LOD-AAS (ppb)
Aluminum	Al	308.2153	I	NA	NO
		309.2710	I	NA	20
		396.1520	I	3	900
Barium	Ba	553.548	I	1	8
Beryllium	Be	234.861	I	100	1
Boron	B	208.891	I	NO	NA
		208.957	I	NO	24,000
		249.677	I	NO	700
		249.773	I	NO	1,500
Cadmium	Cd	326.1055	I	800	NO
Calcium	Ca	422.673	I	0.1	1
Cesium	Cs	455.5276	I	600	NO
		852.1122	I	0.02	NO
Chromium	Cr	425.435	I	1	NO
Cobalt	Co	352.685	I	200	NO
Copper	Cu	324.754	I	30	NO
		327.396	I	3	NO
Dysprosium	Dy	353.170	II	NO	800
		404.597	I	20	500
		421.172	I	NO	50
Erbium	Er	337.271	II	NO	100
		400.796	I	20	40
Europium	Eu	459.403	I	0.2	30
Gadolinium	Gd	368.413	I	NO	2,000
		440.186	I	1,000	NO
Gallium	Ga	417.204	I	5	NO
Germanium	Ge	265.1172	I	400	50
Gold	Au	267.595	I	500	NO
Hafnium	Hf	307.288	I	NO	2,000
Holmium	Ho	345.600	II	NO	3,000
		405.393	I	10	400
		410.384	I	NO	40
Indium	In	303.936	I	NO	1,000
		325.609	I	NO	700
		451.131	I	1	3,500
Iridium	Ir	208.882	I	NO	500
Iron	Fe	371.9935	I	10	NO
Lanthanum	La	408.672	II	NO	7,500
		550.134	I	4,000	2,000
Lead	Pb	368.3462	I	0.2	NO
Lithium	Li	670.776	I	0.001	NO
Lutetium	Lu	261.542	II	NO	3,000
		451.857	I	400	NO
Magnesium	Mg	285.213	I	1	NO
Manganese	Mn	403.076	I	1	NO
Mercury	Hg	253.652	I	10,000	NO
Molybdenum	Mo	313.259	I	10	25
		379.825	I	300	NO
		390.296	I	10	NO
Neodymium	Nd	463.424	I	200	600
		492.453	I		700

(Continued)

Limits of Detection for the Nitrous Oxide–Acetylene Flame (*Continued*)

Element	Symbol	Wavelength (nm)	Type	LOD-AES (ppb)	LOD-AAS (ppb)
Nickel	Ni	352.454	I	20	NO
Niobium	Nb	334.906	I	NO	1,000
		405.894	I	60	5,000
Osmium	Os	290.906	I	NO	80
		442.047	I	2,000	NA
Palladium	Pd	363.470	I	40	NO
Phosphorus	P	177.499	I	NO	30,000
		213.547	I	NO	29,000
Platinum	Pt	265.945	I	2,000	2,000
Potassium	K	766.490	I	0.01	NO
Praseodymium	Pr	495.137	I	500	2,000
Rhenium	Re	364.046	I	200	200
Rhodium	Rh	343.489	I	NO	700
		369.236	I	10	1,400
Rubidium	Rb	780.027	I	8	NO
Ruthenium	Ru	372.803	I	300	NO
Samarium	Sm	429.674	I	NO	500
		476.027	I	50	14,000
Scandium	Sc	391.181	I	10	20
Selenium	Se	196.09	I	100,000	NO
Silicon	Si	251.6113	I	3,000	20
		288.1579	I	NO	NA
Silver	Ag	328.068	I	2	NO
Sodium	Na	588.9950	I	0.01	NO
		589.5924	I	0.01	NO
Strontium	Sr	469.733	I	0.1	50
Tantalum	Ta	271.467	I	NO	800
		474.016	I	4,000	NO
Terbium	Tb	432.643	I	NA	600
Thallium	Tl	377.572	I	50	NO
		535.046	I	2	
Thorium	Th	324.4448	I	NO	181,000
		491.9816	II	10,000	NO
Thulium	Tm	371.791	I	4	10
Tin	Sn	224.605	I	NO	3,000
		235.484	I	NO	90
		283.999	I	100	NO
Titanium	Ti	334.941	II	NO	NA
		364.268	I	NA	10
		365.350	I	30	500
Tungsten	W	255.135	I	NO	500
		400.875	I	200	7,500
Uranium	U	358.488	I	NO	7,000
Vanadium	V	318.540	I	200	20
		437.924	I	7	100
Ytterbium	Yb	398.799	I	0.2	5
Yttrium	Y	410.238	I	NO	50
Zinc	Zn	213.856	I	10,000	NO
Zirconium	Zr	351.960	I	1,200	NO
		360.119	I	3,000	1,000

These data were taken from Refs. 2 and 6.

Limits of Detection for Graphite Furnace AAS[a]

Element	Symbol	Wavelength (nm)	Type	LOD (ppb)
Aluminum	Al	308.2153	I	NA
		309.2710	I	0.01
		396.1520	I	600
Antimony	Sb	206.833	I	NA
		217.581	I	0.08
		231.147	I	NA
Arsenic	As	189.042	I	NA
		193.759	I	0.12
Barium	Ba	553.548	I	0.04
Beryllium	Be	234.861	I	0.003
Bismuth	Bi	223.061	I	0.01
Cadmium	Cd	228.8022	I	0.0002
Calcium	Ca	422.673	I	0.01
Chromium	Cr	357.869	I	0.004
Cobalt	Co	240.725	I	8
Copper	Cu	324.754	I	0.005
		327.396	I	NA
Erbium	Er	400.796	I	0.3
Gadolinium	Gd	440.186	I	0.3
Gallium	Ga	287.424	I	0.01
Germanium	Ge	265.1172	I	0.1
Gold	Au	242.795	I	0.01
Holmium	Ho	345.600	II	NA
		405.393	I	NA
Indium	In	303.936	I	0.02
Iodine	I	183.038	I	40,000
Iridium	Ir	208.882	I	0.5
Iron	Fe	248.3271	I	0.01
		371.9935	I	NA
Lanthanum	La	550.134	I	0.5
Lead	Pb	217.000	I	0.007
		283.3053	I	NA
Lithium	Li	670.776	I	0.01
Magnesium	Mg	285.213	I	0.0002
Manganese	Mn	279.482	I	0.0005
		403.076	I	NA
Mercury	Hg	253.652	I	0.2
Molybdenum	Mo	313.259	I	0.03
Nickel	Ni	232.003	I	0.05
Osmium	Os	290.906	I	2
Palladium	Pd	247.642	I	0.05
Phosphorus	P	177.499	I	NA
		213.547	I	20
		253.561	I	NA
Platinum	Pt	265.945	I	0.2
Potassium	K	766.490	I	0.004
Rhenium	Re	346.046	I	10
Rhodium	Rh	343.489	I	0.1
Rubidium	Rb	780.027	I	NA
Selenium	Se	196.09	I	0.05

(Continued)

Limits of Detection for Graphite Furnace AAS[a] (Continued)

Element	Symbol	Wavelength (nm)	Type	LOD (ppb)
Silicon	Si	251.6113	I	0.6
Silver	Ag	328.068	I	0.001
Sodium	Na	588.9950	I	0.004
Strontium	Sr	460.733	I	0.01
Sulfur	S	180.7311	I	NA
		182.0343	I	NA
		216.89		NA
Tellurium	Te	214.281	I	0.03
Thallium	Tl	276.787	I	0.01
Tin	Sn	235.484	I	0.03
		283.999	I	NA
Titanium	Ti	364.268	I	0.3
		365.350	I	NA
Uranium	U	358.488	I	30
Vanadium	V	318.540	I	0.4
Ytterbium	Yb	398.799	I	0.01
Yttrium	Y	410.238	I	10
Zinc	Zn	213.856	I	0.001

[a] The detection limits for the graphite furnace AAS are calculated using 100 μL of sample. In graphite furnace AAS, additional chemicals are often added to aid in determining certain elements. Walter Slavin has published an excellent guide to these issues and has provided an excellent bibliography: Slavin, W., *Graphite Furnace Source Book*, Perkin-Elmer Corp., Ridgefield, CT, 1984; and Slavin, W. and Manning, D.C., Furnace interferences, a guide to the literature, *Progress Anal. At. Spectrosc.*, 5, 243, 1982.

Limits of Detection for ICP-AES

Element	Symbol	Wavelength (nm)	Type	LOD (ppb)	References
Aluminum	Al	167.0787	II	1	6
		308.2153	I	0.4	7
		309.2710	I	0.02	8
		396.1520	I	0.2	7
Antimony	Sb	206.833	I	10	7
		217.581	I	15	7
		231.147	I	61	7
		259.805	I	107	7
Arsenic	As	189.042	I	136	8
		193.759	I	2	7
		197.262	I	76	7
		234.984	I	90	7
Barium	Ba	455.403	II	0.001	8
		493.409	II	0.3	7
		553.548	I	2	7
Beryllium	Be	234.861	I	0.003	7
		313.042	II	0.1	6
		313.107	II	0.01	8
Bismuth	Bi	223.061	I	0.03	8
		289.798	I	10	7
Boron	B	208.891	I	5	8
		208.957	I	3	8
		249.677	I	0.1	8
		249.773	I	2	8
Bromine	Br	470.486	II	NA	8
		827.244	I	NA	8
Cadmium	Cd	214.441	II	0.1	8
		226.502	II	0.05	8
		228.8022	I	0.08	8
		326.1055	I	3	8
Calcium	Ca	364.441	I	0.5	8
		393.366	II	0.0001	8
		396.847	II	0.002	8
		422.673	I	0.2	8
Carbon	C	193.0905	I	40	6
		247.856	I	100	8
Cerium	Ce	394.275	II	2	8
		413.765	II	40	6
		418.660	II	0.4	7
Chlorine	Cl	413.250	II	NA	7
		837.594	I	NA	8
Chromium	Cr	205.552	II	0.009	8
		267.716	II	0.08	8
		357.869	I	0.1	8
		425.435	I	5	8
Cobalt	Co	228.615	II	0.3	8
		238.892	II	0.1	7
Copper	Cu	213.5981	II	7	8
		324.754	I	0.01	8
		327.396	I	0.06	8
Dysprosium	Dy	353.170	II	1	8
Erbium	Er	337.271	II	1	8
		400.796	I	1	7
Europium	Eu	381.967	II	0.06	7
Fluorine	F	685.603	I	NA	8

(Continued)

Limits of Detection for ICP-AES (*Continued*)

Element	Symbol	Wavelength (nm)	Type	LOD (ppb)	References
Gadolinium	Gd	342.247	II	0.4	7
Gallium	Ga	287.424	I	78	7
		294.364	I	3	8
		417.204	I	0.6	8
Germanium	Ge	199.8887	I	0.6	8
		209.4258	I	11	8
		265.1172	I	4	7
Gold	Au	242.795	I	2	8
		267.595	I	0.9	7
Hafnium	Hf	277.336	II	2	8
		339.980	II	5	6
Holmium	Ho	345.600	II	1	6
		389.102	II	0.9	8
Hydrogen	H	486.133	I	NA	8
		656.2852	I	NA	7
Indium	In	230.605	II	30	8
		303.936	I	15	8
		325.609	I	15	6
		451.131	I	30	7
Iodine	I	183.038	I	NA	7
		206.163	I	10	8
Iridium	Ir	224.268	II	0.6	8
		2,639.71	I	0.6	8
Iron	Fe	238.204	II	0.004	8
		259.9396	II	0.09	7
		371.9935	I	0.3	7
Lanthanum	La	333.749	II	2	6
		408.672	II	0.1	8
Lead	Pb	217.000	I	30	8
		220.3534	II	0.6	8
		283.3053	I	2	7
		368.2462	I	20	8
Lithium	Li	670.776	I	0.02	7
Lutetium	Lu	261.542	II	0.1	7
		451.857	I	8	7
Magnesium	Mg	279.553	II	0.003	7
		280.270	II	0.01	7
		285.231	I	0.2	7
Manganese	Mn	257.610	II	0.01	7
		403.076	I	0.6	7
Mercury	Hg	184.905	II	1	7
		194.227	II	10	6
		253.652	I	1	7
Molybdenum	Mo	202.030	II	0.3	8
		313.259	I	NA	8
		379.825	I	0.2	7
		390.296	I	80	8
Neodymium	Nd	401.225	II	0.3	7
Nickel	Ni	221.648	II	2	8
		232.003	I	6	8
		352.454	I	0.2	7
Niobium	Nb	309.418	II	0.2	7
Nitrogen	N	174.2729	I	1,000	8
		821.634	I	27,000	8

(*Continued*)

Limits of Detection for ICP-AES (Continued)

Element	Symbol	Wavelength (nm)	Type	LOD (ppb)	References
Osmium	Os	225.585	II	4	8
		290.906	I	6	8
Oxygen	O	426.825	I	NA	8
		777.194	I	NA	8
Palladium	Pd	340.458	I	2	8
		363.470	I	1	8
Phosphorus	P	177.499	I	NA	8
		213.547	I	16	6
		253.561	I	15	7
Platinum	Pt	214.423	I	16	6
		265.945	I	0.9	7
Potassium	K	766.490	I	5	8
Praseodymium	Pr	390.805	II	0.3	8
		422.535	II	10	7
Rhenium	Re	197.3	?	6	7
		221.426	II	4	6
Rhodium	Rh	233.477	II	30	7
		343.489	I	8	6
		369.236	I	7	8
Rubidium	Rb	420.180	I	38,000	8
		780.027	I	100	6
Ruthenium	Ru	240.272	II	8	6
		349.894	I	NA	8
		372.803	I	60	7
Samarium	Sm	359.260	II	0.5	8
		373.912	II	2	7
Scandium	Sc	361.384	II	0.1	8
Selenium	Se	196.09	I	0.1	8
		203.98	I	0.03	8
Silicon	Si	251.6113	I	2	7
		288.1579	I	10	7
Silver	Ag	328.068	I	0.8	8
		338.289	I	7	8
Sodium	Na	330.237	I	100	8
		588.9950	I	0.1	7
		589.5924	I	0.5	8
Strontium	Sr	407.771	II	0.2	6
		421.552	II	0.1	8
		460.733	I	0.4	8
Sulfur	S	180.7311	I	15	6
		182.0343	I	30	7
		216.89		NA	7
Tantalum	Ta	226.230	II	15	8
		240.063	II	13	6
		296.513	II	5	7
Tellurium	Te	214.281	I	0.7	8
		238.578	I	2	8
Terbium	Tb	350.917	II	0.1	7
		367.635	II	1.5	8
Thallium	Tl	190.864	II	4	8
		276.787	I	27	6
		377.572	I	17	8
Thorium	Th	283.7295	II	8	6
		401.9129	II	1.3	8

(Continued)

Limits of Detection for ICP-AES (*Continued*)

Element	Symbol	Wavelength (nm)	Type	LOD (ppb)	References
Thulium	Tm	313.126	II	0.9	6
		346.220	II	0.2	7
Tin	Sn	189.991	II	0.05	8
		235.484	I	9	8
		283.999	I	10	8
		326.234	I	0.5	8
Titanium	Ti	334.941	II	0.1	8
		365.350	I	230	8
		368.520	II	0.2	8
Tungsten	W	207.911	II	7	8
		276.427	II	0.8	7
		400.875	I	3	7
Uranium	U	263.553	II	70	6
		385.957	II	2	7
Vanadium	V	309.311	II	0.06	7
		311.062	II	0.06	7
		437.924	I	0.2	7
Ytterbium	Yb	328.937	II	0.01	8
		369.419	II	0.02	7
Yttrium	Y	371.030	II	0.04	7
		377.433	II	0.1	8
Zinc	Zn	202.548	II	0.6	8
		213.856	I	0.07	8
Zirconium	Zr	343.823	II	0.06	7

These data were taken from Refs. 7 and 8.

DETECTION LIMITS BY HYDRIDE GENERATION AND COLD VAPOR AAS

In addition to the AAS methods in flames or graphite furnaces, the elements listed below are detected and determined at extreme sensitivity by introduction into a flame or a hot quartz cell by AAS.

Element	Wavelength[a] (nm)	LOD[b] (ppb)
Antimony, Sb	217.581	0.1
Arsenic, As	193.759	0.02
Bismuth, Bi	223.061	0.02
Mercury, Hg	313.652	0.02
Selenium, Se	196.09	0.02
Tellurium, Te	214.281	0.02
Tin, Sn	235.484	0.5

[a] It has been assumed that the transitions used for these detection limits were the most sensitive cited for AAS.
[b] The detection limits are based on 50 mL sample solution volumes.
These data were taken from Ref. 9.

SPECTRAL OVERLAPS

In FAES and FAAS, the analytical results will be totally degraded if there is a spectral overlap of an analyte transition. This can result from an interfering matrix element with a transition close to that of the analyte. This table presents a list of those overlaps that have been observed and those which are predicted to happen. In many cases, the interferant element has been present in great excess when compared to the analyte species. Therefore, if the predicted interferant element is a major component of the matrix, a careful investigation for spectral overlap should be made. Excitation sources other than flames were not covered in this study.

A. Observed Overlaps

Analyte Element	Wavelength (nm)	Interfering Element	Wavelength (nm)
Aluminum	308.2153	Vanadium	308.211
Antimony	217.023	Lead	217.000
Antimony	231.147	Nickel	231.096
Cadmium	228.8022	Arsenic	228.812
Calcium	422.673	Germanium	422.6562
Cobalt	252.136	Indium	252.137
Copper	324.754	Europium	324.755
Gallium	403.299	Manganese	403.307
Iron	271.9027	Platinum	271.904
Manganese	403.307	Gallium	403.299
Mercury	253.652	Cobalt	253.649
Silicon	250.690	Vanadium	250.690
Zinc	213.856	Iron	213.859

B. Predicted Overlaps

Analyte Element	Wavelength (nm)	Interfering Element	Wavelength (nm)
Boron	249.773	Germanium	249.7962
Bismuth	202.121	Gold	202.138
Cobalt	227.449	Rhenium	227.462
Cobalt	242.493	Osmium	242.497
Cobalt	252.136	Tungsten	252.132
Cobalt	346.580	Iron	346.5860
Cobalt	350.228	Rhodium	350.252
Cobalt	351.348	Iridium	351.364
Copper	216.509	Platinum	216.517
Gallium	294.417	Tungsten	294.440
Gold	242.795	Strontium	242.810
Hafnium	295.068	Niobium	295.088
Hafnium	302.053	Iron	302.0639
Indium	303.936	Germanium	303.9067
Iridium	208.882	Boron	208.891
Iridium	248.118	Tungsten	248.144
Iron	248.3271	Tin	248.339
Lanthanum	370.454	Vanadium	370.470
Lead	261.3655	Tungsten	261.382
Molybdenum	379.825	Niobium	379.812

(*Continued*)

B. Predicted Overlaps (*Continued*)

Analyte Element	Wavelength (nm)	Interfering Element	Wavelength (nm)
Osmium	247.684	Nickel	247.687
Osmium	264.411	Titanium	264.426
Osmium	271.464	Tantalum	271.467
Osmium	285.076	Tantalum	285.098
Osmium	301.804	Hafnium	301.831
Palladium	363.470	Ruthenium	363.493
Platinum	227.438	Cobalt	227.449
Rhodium	350.252	Cobalt	350.262
Scandium	298.075	Hafnium	298.081
Scandium	298.895	Ruthenium	298.895
Scandium	393.338	Calcium	393.366
Silicon	252.4108	Iron	252.4293
Silver	328.068	Rhodium	328.055
Strontium	421.552	Rubidium	421.553
Tantalum	263.690	Osmium	263.713
Tantalum	266.189	Iridium	266.198
Tantalum	269.131	Germanium	269.1341
Thallium	291.832	Hafnium	291.858
Thallium	377.572	Nickel	377.557
Tin	226.891	Aluminum	226.910
Tin	266.124	Tantalum	266.134
Tin	270.651	Scandium	270.677
Titanium	264.664	Platinum	264.689
Tungsten	265.654	Tantalum	265.661
Tungsten	271.891	Iron	271.9027
Vanadium	252.622	Tantalum	252.635
Zirconium	301.175	Nickel	301.200
Zirconium	386.387	Molybdenum	386.411
Zirconium	396.826	Calcium	396.847

These data were taken from Ref. 10.

RELATIVE INTENSITIES OF ELEMENTAL TRANSITIONS FROM HOLLOW CATHODE LAMPS

In AAS, the hollow cathode lamp (HCL) is the most important excitation source for most of the elements determined. However, sufficient light must reach the detector for the measurement to be made with good precision and detection limits. For elements in this table with intensities of less than 100, HCLs are probably inadequate, and other sources such as electrodeless discharge lamps should be investigated.

Element	Fill Gas	Wavelength (nm)	Relative Emission Intensity[a]
Aluminum	Ne	309.2710	1,200
		309.2839	800
		396.1520	
Antimony	Ne	217.581	250
		231.147	250
Arsenic	Ar	193.759	125
		197.262	125
Barium	Ne	553.548	400
		350.111	200
Beryllium	Ne	234.861	2,500
Bismuth	Ne	223.061	120
		306.772	400
Boron	Ar	249.773	400
Cadmium	Ne	228.8022	2,500
		326.1055	5,000
Calcium	Ne	422.673	1,400
Cerium	Ne	520.012	8
		520.042	8
		569.699	
Chromium	Ne	357.869	6,000
		425.435	5,000
Cobalt	Ne	240.725	1,000
		345.350	1,500
		352.685	1,300
Copper	Ne	324.754	7,000
		327.396	6,000
Dysprosium	Ne	404.597	2,000
		418.682	2,000
		421.172	2,500
Erbium	Ne	400.796	1,600
		386.285	1,600
Europium	Ne	459.403	1,000
		462.722	950
Gadolinium	Ne	368.413	350
		407.870	700
Gallium	Ne	287.424	400
		417.204	1,100
Germanium	Ne	265.1172	500
		265.1568	250
		259.2534	
Gold	Ne	242.795	750
		267.595	1,200

(Continued)

Element	Fill Gas	Wavelength (nm)	Relative Emission Intensity[a]
Hafnium	Ne	307.288	300
		286.637	200
Holmium	Ne	405.393	2,000
		410.384	2,200
Indium	Ne	303.936	500
		410.176	500
Iridium	Ne	263.971	400
Iron	Ne	248.3271	400
		371.9935	2,400
Lanthanum	Ne	550.134	120
		392.756	45
Lead	Ne	217.000	200
		283.3053	1,000
Lithium	Ne	670.776	700
Lutetium	Ar	335.956	30
		337.650	25
		356.784	15
Magnesium	Ne	285.213	6,000
		202.582	130
Manganese	Ne	279.482	3,000
		280.106	2,200
		403.076	14,000
Mercury	Ar	253.652	1,000
Molybdenum	Ne	313.259	1,500
		317.035	800
Neodymium	Ne	463.424	300
		492.453	600
Nickel	Ne	232.003	1,000
		341.476	2,000
Niobium	Ne	405.894	400
		407.973	360
Osmium	Ar	290.906	400
		301.804	200
Palladium	Ne	244.791	400
		247.642	300
		340.458	3,000
Phosphorus	Ne	215.547	30
		213.618	20
		214.914	
Platinum	Ne	265.945	1,500
		299.797	1,000
Potassium	Ne	766.490	6
		404.414	300
Praseodymium	Ne	495.137	100
		512.342	70
Rhenium	Ne	346.046	1,200
		346.473	900
Rhodium	Ne	343.489	2,500
		369.236	2,000
		350.732	200
Rubidium	Ne	780.027	1.5
		420.180	80
Ruthenium	Ar	349.894	600
		392.592	300

(Continued)

Element	Fill Gas	Wavelength (nm)	Relative Emission Intensity[a]
Samarium	Ne	429.674	600
		476.027	800
Scandium	Ne	391.181	3,000
		390.749	2,500
		402.040	1,800
		402.369	2,100
Selenium	Ne	196.09	50
		203.98	50
Silicon	Ne	251.6113	500
		288.1579	500
Silver	Ar	328.068	3,000
		338.289	3,000
Sodium	Ne	588.9950	2,000
		330.237	40
		330.298	
Strontium	Ne	460.733	1,000
Tantalum	Ar	271.467	150
		277.588	100
Tellurium	Ne	214.281	60
		238.578	50
Terbium	Ne	432.643	110
		432.690	90
		431.883	60
		433.841	
Thallium	Ne	276.787	600
		258.014	50
Thulium	Ne	371.791	40
		409.419	50
		410.584	70
Tin	Ne	224.605	100
		286.332	250
Titanium	Ne	364.268	600
		399.864	600
Tungsten	Ne	255.100	200
		255.135	1,400
		400.875	
Uranium	Ne	358.488	300
		356.659	200
		351.461	200
		348.937	150
Vanadium	Ne	318.314	600
		318.398	200
		385.537	
		385.584	
Ytterbium	Ar	398.799	2,000
		346.437	800
Yttrium	Ne	407.738	500
		410.238	600
		414.285	300
Zinc	Ne	213.856	2,500
		307.590	2,500

[a] The most intense line is the Mn 403.076 transition with a relative intensity of 14,000.
These data were obtained using Westinghouse HCLs and a single experimental setup. No correction has been made for the spectral response of the monochromator/photomultiplier tube (PMT) system.
These data were taken from Ref. 2.

Inert Gases

In AAS, the excitation source inert gas emission offers a potential background spectral interference. The most common inert gases used in HCLs are Ne and Ar. The data taken for this table and the other tables in this book on lamp spectra are from HCLs; however, electrodeless discharge lamps emit very similar spectra. The emission spectra for Ne and Ar HCLs and close lines that must be resolved for accurate analytical results are provided in the following four tables. This information was obtained for HCLs and flame atom cells and should not be considered with respect to plasma sources. In the "Type" column, an "I" indicates that the transition originates from an atomic species and an "II" indicates a singly ionized species.

NEON HOLLOW CATHODE LAMP (HCL) SPECTRUM

Wavelength (nm)	Type	Relative Intensity[a]
323.237	II	5.4
330.974	II	2.8
331.972	II	8.7
332.374	II	28
332.916	II	1.7
333.484	II	5.2
334.440	II	17
335.502	II	3.5
336.060	II	1.7
336.9908	I	7.8
336.9908	II	17
337.822		
339.280	II	8.3
341.7904	I	16
344.7703	I	12
345.4195	I	15
346.0524	I	6.6
346.6579	I	12
347.2571	I	12
349.8064	I	2.9
350.1216	I	3.8
351.5191	I	3.6
352.0472	I	61
356.850	II	7.8
357.461	II	5.9
359.3526	I	19
360.0169	I	3.5
363.3665	I	3.6
366.407	II	1.9
369.421	II	3.5
370.962	II	4.9
372.186	II	3.1
404.264	I	1.4
533.0778	I	1.6
534.920	I	1.6
540.0562	I	3.3
576.4419	I	2.3
585.2488	I	100
588.1895	I	8.7
594.4834	I	14
597.4627	I	2.6
597.5534		
602.9997	I	2.8
607.4338	I	11
609.6163	I	15
614.3063	I	20
616.3594	I	5.2

[a] These data are referenced to the Ne transition at 585.2488 nm which has been assigned the value of 100.

These data were taken with a Copper HCL operated at 10 mA. The Cu 324.7 nm transition was a factor of 2.9 more intense than the 585.249 nm Ne transition. The spectrum was taken with an IP28 PMT. The relative intensities were not corrected for the instrumental/PMT response.

These data were taken from Ref. 2.

NEON LINES WHICH MUST BE RESOLVED FOR
ACCURATE AAS MEASUREMENTS

Analyte Element	Wavelength (nm)	Neon Line (nm)	Required Resolution (nm)[a]
Chromium	357.869	357.461	0.20
Chromium	359.349	359.3526	0.002
Chromium	360.533	360.0169	0.26
Copper	324.754	323.237	0.75
Dysprosium	404.597	404.264	0.17
Gadolinium	371.357	370.962	0.20
Gadolinium	371.748	372.186	0.22
Lithium	670.776	335.502 in second order is 671.004	0.11
Lutetium	335.956	336.060	0.05
Niobium	405.894	404.264	0.82
Rhenium	346.046	346.0524	0.003
Rhenium	346.473	346.6579	0.11
Rhenium	345.188	345.4195	0.12
Rhodium	343.489	344.7703	0.64
Rhodium	369.236	369.421	0.09
Ruthenium	372.803	372.186	0.31
Scandium	402.369	404.264	0.94
Silver	338.289	337.822	0.23
Sodium	588.995	588.1895	0.40
Sodium	589.592	588.1895	0.70
Thulium	371.792	372.186	0.19
Titanium	337.145	336.9808 and	0.08
Titanium	364.268	336.9908	0.45
Titanium	365.350	363.3665 366.407	0.53
Uranium	356.660	356.850	0.09
Uranium	358.488	359.3526	0.43
Ytterbium	346.436	346.6579	0.11
Zirconium	351.960	352.0472	0.04
Zirconium	360.119	360.0169	0.05

[a] The monochromator settings must be at least one-half of the separation of the analyte and interferant transition.

These data were taken from Ref. 10.

ARGON HOLLOW CATHODE LAMP SPECTRUM

Wavelength (nm)	Type	Relative Intensity[a]
294.2893	II	3.5
297.9050	II	1.9
329.3640	II	1.5
330.7228	II	1.5
335.0924	II	2.2
337.6436	II	2.2
338.8531	II	1.8
347.6747	II	3.7
349.1244	II	2.0
349.1536	II	7.2
350.9778	II	7.0
351.4388	II	4.0
354.5596	II	11
354.5845	II	12
355.9508	II	16
356.1030	II	1.8
357.6616	II	11
358.1608	II	3.9
358.2355	II	8.5
358.8441	II	1.2
360.6522	I	2.0
362.2138	II	1.3
363.9833	II	3.3
371.8206	II	5.5
372.9309	II	1.3
373.7889	II	9.8
376.5270	II	5.1
376.6119	II	6.8
377.0520	II	1.7
378.0840	II	4.0
380.3172	II	5.5
380.9456	II	1.8
383.4679	I	2.6
385.0581	II	1.2
386.8528	II	7.0
392.5719	II	9.9
392.8623	II	6.9
393.2547	II	3.2
394.6097	II	14
394.8979	I	5.2
397.9356	II	5.6
399.4792	II	6.2
401.3857	II	4.3
403.3809	II	2.5
403.5460	II	2.3
404.2894	II	1.6
404.4418	I	9.0
405.2921	II	21
407.2005	II	34
407.2385	II	5.5
407.6628	II	2.0
407.9574	II	4.4
408.2387	II	3.2
410.3912	II	10

(Continued)

Wavelength (nm)	Type	Relative Intensity[a]
413.1724	II	61
415.6086	II	2.4
415.8590	I	1.4
416.4180	I	4.4
418.1884	I	6.9
419.0713	I	9.1
419.1029	I	8.9
419.8317	I	38
420.0674	I	38
421.8665	II	2.2
422.2637	II	4.3
422.6988	II	5.6
422.8158	II	12
423.7220	II	15
425.1185	I	2.3
425.9326	I	42
426.6286	I	11
426.6527	II	7.4
427.2169	I	18
427.7528	II	100
428.2898	II	2.5
430.0101	I	13
430.0650	II	3.3
430.9239	II	6.6
433.1200	II	17
433.2030	II	4.2
433.3561	I	12
433.5338	I	5.8
434.5168	I	3.7
434.8064	II	1.5
435.2205	II	5.4
236.2066	II	3.3
436.7832	II	9.6
437.0753	II	32
437.1329	II	6.5
437.5954	II	10
437.9667	II	20
438.5057	II	6.7
440.0097	II	5.7
440.0986	II	15
442.6001	II	1.6
443.0189	II	1.2
443.0996	II	5.4
443.3838	II	5.3
443.9461	II	5.3
444.8879	II	7.9
447.4759	II	19
448.1811	II	33
451.0733	I	20
452.2323	I	2.0
453.0552	II	3.2
454.5052	II	1.3
457.9350	II	1.4
458.9898	II	47
459.6097	I	1.8
460.9567	II	1.3
462.8441	I	1.4

(Continued)

Wavelength (nm)	Type	Relative Intensity[a]
463.7233	II	5.5
465.7901	II	1.9
470.2316	I	2.5
472.6868	II	43
473.2053	II	9.7
473.5906	II	1.3
476.4865	II	1.5
480.6020	II	36
484.7812	II	1.6
486.5910	II	1.3
487.9864	II	58
488.9042	II	11
490.4752	II	3.5
493.3209	II	4.1
496.5080	II	28
500.9334	II	5.5
501.7163	II	12
506.2037	II	5.9
509.0495	II	2.9
514.1783	II	5.7
514.5308	II	3.7
516.2285	I	3.8
516.5773	II	1.8
518.7746	I	3.8
522.1271	I	1.2
545.1652	I	1.7
549.5874	I	3.1
555.8702	I	4.0
557.2541	I	1.9
560.6733	I	4.9
565.0704	I	1.7
588.8584	I	1.9
591.2085	I	4.1
592.8813	I	1.4
603.2127	I	4.1
604.3223	I	1.6
611.4923	II	2.2
617.2278	II	1.1
696.5431	I	3.2
706.7218	I	1.7
738.3980	I	1.2
750.3869	I	2.7

[a] These data are referenced to the Ar transition at 427.7528 nm which has been assigned the value of 100.
These data were taken from an Ar filled Ga HCL at the Los Alamos Fourier transform spectrometer facility [11].

CLOSE LINES FOR BACKGROUND CORRECTION

In AAS, it is possible to make background corrections in many cases by measuring a normally non-absorbing transition near the analytical transition. This table presents a list of suitable transitions for such a background measurement. It is often desirable to check the background absorbance by more than one method even if there is a built-in background measurement by some other means such as the continuum or Zeeman methods. In the table below, the first two columns give the analyte element and wavelength of the analytical transition, and the last two columns give the transition useful for the background measurement and its source. If the source is Ne and the HCL is Ne filled, the same HCL can be used for the background measurement; if not, a different HCL must be placed in the spectrometer to make the measurement.

These data were taken from Ref. 12.

Element	Analysis Line (nm)		Background Line (nm)		Source
Aluminum	309.2711	I	306.614	I	Al
Antimony	217.581	I	217.919	I	Sb
Arsenic	231.147	I	231.398	I	Ni
Barium	193.759	I	191.294	II	As
	553.548	I	540.0562	I	Ne
			553.305	I	Mo
			557.742	I	Y
Beryllium	234.861	I	235.484	I	Sn
Bismuth	223.061	I	226.502	II	Cd
Bromine	306.772	I	306.614	I	Al
Cadmium	148.845	I	149.4675	I	N
	228.8022	I	226.502	II	Cd
Calcium	422.673	I	421.9360	I	Fe
Cesium	852.1122	I	423.5936	I	Fe
Chromium	357.869	I	854.4696	I	Ne
			352.0472	I	Ne
			358.119	I	Fe
Cobalt	204.206	I	238.892	II	Co
Copper	324.754	I	242.170	I	Sn
Dysprosium	421.172	I	324.316	I	Cu
			421.645	II	Fe
			421.096	I	Ag
Erbium	400.796	I	394.442	I	Er
Europium	459.403	I	460.102	I	Cr
Gallium	287.424	I	283.999	I	Sn
Gold	242.795	I	283.690	I	Cd
			242.170	I	Sn
Indium	303.936	I	306.614	I	Al
Iodine	183.038	I	184.445	I	I
Iron	248.3271	I	249.215	I	Cu
Lanthanum	550.134	I	550.549	I	Mo
			548.334	I	Co
Lead	283.3053	I	280.1995	I	Pb
Lithium	217.000	I	283.6900	I	Cd
Magnesium	670.791	I	220.3534	II	Pb
	285.213	I	671.7043	I	Ne
			283.690	I	Cd
			283.999	I	Sn

(Continued)

Element	Analysis Line (nm)		Background Line (nm)		Source
Manganese	279.482	I	282.437	I	Cu
Mercury	253.652	I	280.1995	I	Pb
Molybdenum	313.259	I	249.215	I	Cu
Nickel	232.003	I	312.200	II	Mo
			232.138	I	Ni
Palladium	247.642	I	249.215	I	Cu
Phosphorus	213.618	I	213.856	I	Zn
Potassium	766.490	I	769.896	I	K
Rhodium	343.489	I	767.209	I	Ca
			350.732	I	Rh
			352.0472	I	Ne
Rubidium	780.027	I	778.048	I	Ba
Ruthenium	249.894	I	352.0472	I	Ne
Selenium	196.09	I	199.51	I	Se
Silicon	251.6113	I	249.215	I	Cu
Silver	328.068	I	332.374	II	Ne
			326.234	I	Sn
Sodium	588.9950	I	588.833	I	Mo
Strontium	460.733	I	460.500	I	Ni
Tellurium	214.281	I	213.856	I	Zn
Thallium	276.787	I	217.581	I	Sb
			280.1995	I	Pb
Tin	224.605	I	226.502	II	Cd
Titanium	286.332	I	283.999	I	Sn
Uranium	364.268	I	361.939	I	Ni
	365.350	I	361.939	I	Ni
	358.488	I	358.119	I	Fe
Vanadium	318.398	I	324.754	I	Cu
Zinc	318.540	I	324.754	I	Cu
	213.856	I	212.274	II	Zn

BETA VALUES FOR THE AIR–ACETYLENE AND
NITROUS OXIDE–ACETYLENE FLAMES

Beta values represent the fraction of free atoms present in the hot flame gases of the flame indicated. These values have been taken from various sources and were either experimentally measured or calculated from thermodynamic data using the assumption of local thermodynamic equilibrium in the flame. These values do not have very good agreement within each element; however, the values do provide an indication of the probable sensitivity of the particular flame.

These data were taken from Refs. 2 and 13.

Element	Symbol	Beta A/AC Flame	Beta N/AC Flame
Aluminum	Al	<0.0001	0.13
		<0.00005	0.29
		0.0005	0.97*
			0.5
Antimony	Sb	0.03	
Arsenic	As	0.0002	
Barium	Ba	0.0009	0.074
		0.002	0.074
		0.003	0.98
		0.0018	
Beryllium	Be	0.0004	0.095
		0.00006	0.98
			0.98
Bismuth	Bi	0.17	0.35
Boron	B	<0.0006	0.0035
		<0.000001	0.2
Cadmium	Cd	0.38	0.56
		0.50	0.60
		0.80	
Calcium	Ca	0.066	0.34
		0.14	0.52*
		0.05*	0.98
		0.018	
Cesium	Cs	0.02	0.0004
		0.0057	
Chromium	Cr	0.071	0.63
		0.13	1.02
		0.53	1.00
		0.042	
Cobalt	Co	0.023	0.11
		0.28	0.25
		0.41	
Copper	Cu	0.4	0.49
		0.82	0.66
		0.98	1.00*
Gallium	Ga	0.16	0.73
		0.16	
Germanium	Ge	0.001	
Gold	Au	0.21	0.16
		0.40	0.27
		0.63	

(Continued)

Element	Symbol	Beta A/AC Flame	Beta N/AC Flame
Indium	In	0.10	0.37
		0.67	0.93
		0.67	
Iridium	Ir	0.1	
Iron	Fe	0.38	0.83
		0.66	0.91
		0.84	1.00
		0.66	
Lead	Pb	0.44	0.84
		0.77	
Lithium	Li	0.21	0.34*
		0.26*	0.96*
		0.20*	0.041
		0.08	0.91*
Magnesium	Mg	0.59	0.88
		1.05	0.99
		0.62	0.92
			0.99*
Manganese	Mn	0.45	0.37
		0.93	0.77
		1.0	
Mercury	Hg	0.04	
Molybdenum	Mo	0.03	
Nickel	Ni	1	
Palladium	Pd	1	
Platinum	Pt	0.4	
Potassium	K	0.7*	0.12*
		0.25	0.0004
		0.45	0.17*
		0.59*	
Rhodium	Rh	1	
Rubidium	Rb	0.16	
Ruthenium	Ru	0.3	
Selenium	Se	0.0001	
Silicon	Si	<0.001	0.55
		<0.0000001	0.12
			0.36
Silver	Ag	0.66	0.57
		0.70	
Sodium	Na	0.63	0.32
		1.00	0.97*
		1.00*	0.012
		0.56	0.80
Strontium	Sr	0.068	0.26
		0.10	0.57
		0.13	0.99
		0.021	
Tantalum	Ta		0.045
Thallium	Tl	0.36	0.55
		0.52	
Tin	Sn	<0.0001	0.35
		0.043	0.82
		0.078	
		0.061	

(Continued)

Element	Symbol	Beta A/AC Flame	Beta N/AC Flame
Titanium	Ti	<0.001	0.11
			0.33
			0.49
Tungsten	W	0.004	0.71
Vanadium	V	0.0004	0.32
		0.015	0.99
		0.000001	
Zinc	Zn	0.66	0.49
		0.45	

* Ionization has been suppressed for these measurements/calculations.

LOWER ENERGY LEVEL POPULATIONS (IN PERCENT) AS A FUNCTION OF TEMPERATURE

It is possible to calculate the relative number of atoms in the ground energy level(s) using the following equation:

$$\%\text{Atoms (ith level)} = n_i/n_t * 100 = g_i/Z * \exp(-E_i/kT)$$

where n_i is the number of atoms in the ith level per unit volume of atoms cell, n_t is the total number of atoms per unit volume of atom cell, g_i is the statistical weight for energy level i, Z is the electronic partition function, E_i is the energy of the ith level, k is the Boltzmann constant, and T is the absolute temperature. Of course all of the data must be in consistent units.

In utilizing these data, it should be remembered that, other things being equal, the larger the percentage of atoms in the ground or lower level of a transition, the larger the absorption signal from that transition should be. For example, a transition with 100 % of the atoms in the ground state should be ten times more sensitive than one with 10 %. Also, these data refer to the percent of atoms in the atomic state only; therefore, this information should be used in conjunction with the beta values table.

These data were taken from Ref. 12.

Element	Energy Level (cm⁻¹)	Percent Population at Temperature (°C)		
		2,000	2,500	3,000
Aluminum	0.0	35.1	34.8	34.5
	112.040	64.9	65.2	65.5
Antimony	0.0	99.7	98.7	97.0
	8,512.100	0.2	0.7	1.6
	9,854.100	0.1	0.5	1.7
Arsenic	0.0	99.9	99.5	98.5
	10,592.500	0.0	0.2	0.6
	10,914.600	0.1	0.3	0.8
Barium	0.0	98.0	92.6	82.6
	9,033.985	0.4	1.5	3.3
	9,215.518	0.6	2.3	5.0
	9,596.551	0.7	2.6	5.8
	11,395.382	0.1	0.7	1.8
Beryllium	0.0	100.0	100.0	100.0
Bismuth	0.0	100.0	99.8	99.5
Boron	0.0	33.5	33.5	33.5
	16.0	66.5	66.5	66.5
Cadmium	0.0	100.0	100.0	100.0
Calcium	0.0	100.0	99.8	99.3
Cesium	0.0	99.9	99.5	98.3
	11,178.240	0.0	0.2	0.5
	111,732.350	0.0	0.2	0.7
Chromium	0.0	98.6	95.9	91.5
	7,593.160	0.3	0.9	1.7
	7,750.780	0.1	0.2	0.3
	7,810.820	0.1	0.2	0.3
	7,927.470	0.1	0.4	0.9
	8,095.210	0.2	0.7	1.3
	8,307.570	0.3	0.8	1.7

(*Continued*)

Element	Energy Level (cm⁻¹)	Percent Population at Temperature (°C)		
		2,000	2,500	3,000
Cobalt	0.0	51.8	45.6	40.9
	816.000	23.1	22.8	22.1
	1,406.840	11.3	12.2	12.5
	1,809.330	5.6	6.4	6.9
	3,482.820	4.2	6.1	7.7
	4,142.660	2.1	3.4	4.5
	4,690.180	1.1	1.8	2.6
	5,075.830	0.5	1.0	1.4
	7,442.410	0.2	0.5	0.9
Copper	0.0	99.9	99.4	98.3
	11,202.565	0.1	0.5	1.9
Gallium	0.0	47.5	44.6	42.6
	826.240	52.5	55.4	57.4
Germanium	0.0	20.6	18.2	16.6
	557.100	41.4	39.7	38.7
	1,409.900	37.4	40.5	42.4
	7,125.260	0.6	1.5	2.7
Gold	0.0	99.5	98.5	96.4
	9,161.300	0.4	1.5	3.5
Hafnium	0.0	73.9	64.0	55.9
	2,356.680	19.0	23.1	25.3
	4,567.640	5.0	8.3	11.3
	5,521.780	0.3	0.5	0.8
	5,638.620	1.3	2.5	3.7
	6,572.550	0.4	0.9	1.4
Indium	0.0	71.0	64.1	59.1
	2,212.560	29.0	35.9	40.9
Iridium	0.0	85.1	77.2	69.9
	2,834.980	11.1	15.1	18.0
	4,078.940	1.8	3.0	4.0
	5,784.620	0.8	1.7	2.6
	6,323.910	0.7	1.6	2.7
	7,106.610	0.4	1.0	1.9
Iron	0.0	46.2	43.4	41.0
	415.933	26.7	26.6	26.1
	704.003	15.5	16.1	16.3
	888.123	8.1	8.7	8.9
	978.074	2.5	2.7	2.9
	6,928.280	0.4	1.0	1.8
	7,376.775	0.2	0.6	1.2
	7,728.071	0.1	0.4	0.8
	7,985.795	0.1	0.2	0.5
Lanthanum	0.0	42.5	34.2	28.3
	1,053.200	29.9	28.0	25.6
	2,668.200	6.2	7.4	7.9
	3,010.010	7.3	9.1	10.0
	3,494.580	6.9	9.2	10.6
	4,121.610	5.5	8.0	9.8
	7,011.900	0.4	0.9	1.5
	7,231.360	0.1	0.3	0.4
	7,490.460	0.2	0.5	0.8
	7,679.940	0.3	0.6	1.1
Lead	0.0	98.7	95.7	90.8
	7,819.350	1.1	3.2	6.4
	10,650.470	0.2	1.1	2.8
Lithium	0.0	100.0	99.9	99.8
Magnesium	0.0	100.0	100.0	100.0

(Continued)

Element	Energy Level (cm⁻¹)	Percent Population at Temperature (°C)		
		2,000	2,500	3,000
Manganese	0.0	100.0	100.0	99.8
Mercury	0.0	100.0	100.0	100.0
Molybdenum	0.0	99.9	99.4	98.1
Nickel	0.0	39.5	36.4	34.2
	204.786	26.6	25.2	24.1
	879.813	11.7	12.2	12.4
	1,332.153	11.8	13.2	14.0
	1,713.080	3.9	4.5	5.0
	2,216.519	4.5	5.7	6.6
	3,409.925	1.9	2.8	3.7
Niobium	0.0	7.5	6.5	5.7
	154.190	13.4	11.8	10.6
	391.990	17.0	15.4	14.2
	695.250	18.2	17.3	16.4
	1,050.260	17.6	17.7	17.3
	1,142.790	6.6	6.7	6.6
	1,586.900	7.2	7.8	8.0
	2,154.110	6.4	7.5	8.1
	2,805.360	5.0	6.4	7.5
	4,998.170	0.2	0.4	0.5
	5,297.920	0.3	0.6	0.9
	5,965.450	0.3	0.6	1.0
Osmium	0.0	86.5	78.3	70.4
	2,740.490	6.7	9.0	10.5
	4,159.320	3.4	5.6	7.5
	5,143.920	2.6	5.0	7.3
	5,766.140	0.5	0.9	1.5
	6,092.790	0.1	0.3	0.4
	8,742.830	0.2	0.5	1.1
Palladium	0.0	91.7	80.3	67.2
	6,464.110	6.2	13.8	21.6
	7,754.990	1.8	4.7	8.3
	10,093.940	0.2	0.7	1.6
	11,721.770	0.1	0.5	1.2
Platinum	0.0	47.0	43.8	41.5
	775.900	19.2	20.0	20.4
	823.700	33.4	35.1	36.0
	6,140.000	0.1	0.2	0.3
	6,567.5000	0.3	0.7	1.3
Potassium	0.0	100.0	99.8	99.4
Rhenium	0.0	99.9	99.5	98.3
Rhodium	0.0	69.0	60.0	53.6
	1,529.970	18.1	19.9	20.6
	2,598.030	6.3	8.1	9.3
	3,309.860	3.8	5.4	6.6
	3,472.680	2.2	3.3	4.1
	5,657.970	0.5	0.9	1.4
	5,690.970	0.9	1.8	2.8
Rubidium	0.0	100.0	99.8	99.3
Ruthenium	0.0	62.4	55.2	49.3
	1,190.640	21.7	22.8	22.8
	2,091.540	8.8	10.6	11.5
	2,713.240	4.0	5.3	6.1
	3,105.490	1.8	2.5	3.0
	6,545.030	0.5	1.0	1.8
	7,483.070	0.2	0.6	1.1
	8,084.120	0.1	0.3	0.7
	9,183.660	0.0	0.1	0.3

(Continued)

Element	Energy Level (cm⁻¹)	Percent Population at Temperature (°C)		
		2,000	2,500	3,000
Scandium	0.0	42.9	42.2	41.5
	168.340	57.0	57.4	57.4
Selenium	0.0	85.0	80.5	76.8
	1,989.490	12.2	15.4	17.8
	2,534.350	2.7	3.7	4.6
	9,576.080	0.1	0.3	0.8
Silicon	0.0	12.3	11.9	11.6
	77.150	34.8	34.2	33.5
	223.310	52.2	52.3	52.1
	6,298.810	0.7	1.6	2.8
Silver	0.0	100.0	100.0	100.0
Sodium	0.0	100.0	100.0	99.9
Strontium	0.0	100.0	99.8	98.9
Tantalum	0.0	64.9	53.8	45.0
	2,010.00	23.0	25.4	25.8
	3,963.920	7.5	11.0	13.5
	5,621.040	2.9	5.3	7.6
	6,049.420	0.4	0.8	1.2
	9,253.430	0.1	0.4	0.8
	9,705.380	0.1	0.4	0.9
Technetium	0.0	27.2	25.7	24.0
	170.132	33.8	32.6	31.0
	386.873	37.1	37.0	35.9
	6,556.860	0.1	0.4	0.6
	6,598.830	0.2	0.6	1.0
	6,661.000	0.3	0.8	1.4
	6,742.790	0.4	1.0	1.7
	6,843.000	0.4	1.1	2.0
	7,255.290	0.1	0.4	0.7
Tungsten	0.0	28.4	20.5	15.9
	1,670.300	25.6	23.6	21.5
	2,951.290	23.8	26.3	27.1
	3,325.530	13.0	15.2	16.2
	4,830.000	6.2	8.9	11.0
	6,219.330	2.9	5.2	7.3
Vanadium	0.0	14.1	12.6	11.5
	137.380	19.2	17.5	16.2
	323.420	22.4	20.9	19.7
	553.020	23.7	22.9	22.1
	2,112.320	1.5	1.9	2.1
	2,153.200	3.0	3.7	4.1
	2,220.130	4.3	5.3	6.0
	2,311.370	5.4	6.7	7.6
	2,424.890	6.2	7.8	9.0
Yttrium	0.0	49.3	47.2	45.4
	530.360	50.5	52.2	52.8
Zinc	0.0	100.0	100.0	100.0
Zirconium	0.0	34.2	29.1	25.1
	570.410	31.7	29.3	26.7
	1,240.840	25.2	25.6	24.9
	4,186.110	1.7	2.6	3.4
	4,196.850	0.3	0.5	0.7
	4,376.280	0.9	1.4	1.8
	4,870.530	0.6	1.1	1.5
	5,023.410	0.9	1.6	2.3
	5,101.680	0.9	1.5	2.2
	5,249.070	1.1	2.0	2.8
	5,540.540	1.1	2.2	3.2
	5,888.930	1.1	2.2	3.3
	8,057.300	0.2	0.5	1.0

CRITICAL OPTIMIZATION PARAMETERS FOR AES/AAS METHODS

In most multiparameter instrumental techniques, the parameters can be classified into two types: independent and dependent. Independent parameters can be optimized independently from all other parameters and can therefore be subjected to a univariate approach, that is, the variable can be adjusted until the largest signal-to-noise ratio (SNR) is obtained and set at that value for the best instrumental performance. This is the simplest situation and can be handled in a very straight-forward manner.

Dependent parameters are an entirely different matter. Most dependent parameters have optimum values that depend on the value of the other parameters. If the value of any variable is changed, then the optimum for the parameter under question will be different.

The following table lists the parameters for FAAS, EAAS, and FAES which are both dependent and independent. A "yes" in any column indicates that the listed parameter is appropriate for that technique. If an optimization is necessary when independent parameters are involved, it is important to use a systematic approach which permits one to vary all parameter values to develop the optimum for each. If the variables are simply varied one at a time, false optimum values and poor results will be obtained. Experimental design techniques are required for good results; one of the best approaches is the SIMPLEX technique which has been fully discussed in the literature [15].

A. Independent Parameters

Parameter	FAAS	EAAS	FAES
Excitation source power	Yes	Yes	na
Photomultiplier voltage[a]	Yes	Yes	Yes
Readout gain[b]	Yes	Yes	Yes
Noise suppression setting[c]	Yes	Yes	Yes

na = not applicable.
[a] The PMT voltage does not affect the SNR unless extreme voltages are used. It will specify the level of signal which is observed.
[b] The gain does not affect the SNR until electronic noise becomes important. It also specifies the level of signal which is observed.
[c] This specifies the frequency response of the system and is accompanied by a time requirement. More noise filtering requires a long measurement.

B. Dependent (Interdependent) Parameters

Parameter	FAAS	EAAS	FAES
Oxidant gas flow rate	Yes	na	Yes
Fuel-to-oxidant ratio	Yes	na	Yes
Sheath gas flow rate[a]	Yes	Yes	Yes
Solution flow rate[b]	Yes	na	Yes
Sample size	na	Yes	na
Height of optical measurement	Yes	Yes	Yes
Monochromater slit setting	Yes	Yes	Yes
Burner variables[c]	Yes	na	Yes
Furnace variables[d]	na	Yes	na

na = not applicable.
[a] Most commercial burners do not use a sheath gas; however, there is always the possibility of a sheath gas in EAAS.
[b] This is important if the sample solution flow rate is controlled by a pump rather than by the oxidant gas flow rate.
[c] Some burners have additional variables such as bead position and nebulizer position.
[d] The timing cycle and temperature are always critical variables for the graphite furnaces.
This information was taken from Ref. 16.

FLAME TEMPERATURES AND REFERENCES ON
TEMPERATURE MEASUREMENTS

Flame Type	Experimental Measurement Range (K)	Calculated Stoichiometric Temperature (K)	Typical[a] (K)
Hydrocarbon/air	1,900–2,150	2,228	2,000
Acetylene/air	2,360–2,600	2,523	2,450
Acetylene/nitrous oxide	2,830–3,070	3,148	2,950
Hydrogen/air	2,100–2,300	2,373	2,300
Hydrogen/oxygen	2,500–2,900	3,100	2,800
Acetylene/oxygen	2,900–3,300	3,320	3,100

[a] This value represents the value most often cited for flames used in analytical spectroscopy. These data were taken from Ref. 2.

REFERENCES WHICH DISCUSS THE TECHNIQUES
OF TEMPERATURE MEASUREMENT

Gaydon, A.G. and Wolfhard, H.G., *Flames, Their Structure, Radiation, and Temperature*, Chapman and Hall Ltd., London, 1970.

Fristrom, R.M. and Westenberg, A.A., *Flame Structure*, McGraw-Hill, New York, 1965.

Tourin, R.H., *Spectroscopic Gas Temperature Measurement*, Elsevier Publishing Co., Amsterdam, 1966.

Gaydon, A.G. and Wolfhard, H.G., The spectrum-line reversal method of measuring flame temperature, *Proc. Phys. Soc. (London)*, 65A, 19, 1954.

Browner, R.F. and Winefordner, J.D., Measurement of flame temperatures by a two-line atomic absorption method, *Anal. Chem.*, 44, 247, 1972.

Omenetto, N., Benetti, P., and Rossi, G., Flame temperature measurements by means of atomic fluorescence spectrometry, *Spectrochim. Acta*, 27B, 253, 1972.

Herzfield, C.M., *Temperature, Its Measurement and Control in Science and Industry*, Vol. III, Part 2, Ed. by I. Dahl, Reinhold, New York, 1962.

Alkemade, C.Th.J., Hollander, Tj., Snelleman, W., and Zeegers, P.J.Th., *Metal Vapours in Flames*, Pergamon Press, New York, 1982.

FUNDAMENTAL DATA FOR THE COMMON TRANSITIONS

To the extent possible, the fundamental data for the transition commonly used with the methods discussed in this section are given in this table. The transition in nm, the type of transition (I indicates atomic and II indicates ionic), the lower and upper energy levels, E-low and E-high, in cm^{-1}, the statistical weight, g(i), of the lower level (i), the transition probability, A(ji), in s^{-1}, and the merit and reference for the transition probability are listed. In some cases, the g(i) and the A(ji) were only available in the multiplied form, and in these cases, the "gA = xx" format was used. If a blank appears, no information was available for that specific column.

Element	Symbol	Wavelength (nm)	Type	E-low (cm^{-1})	E-high (cm^{-1})	g(i)	A(ji) $10^5 s^{-1}$	Merit[a]	References
Aluminum	Al	167.0787	II	0	32,435	4	0.63	C	3
		308.2153	I	112	32,437	6	0.73	C	3
		309.2710	I	112	25,348	2	0.98	C	3
		396.1520	I						
Antimony	Sb	206.833	I	0	48,332	6	42	E	2
		217.581	I	0	45,945	4	13.8	E	2
		231.147	I	0	43,249	2	3.75	E	2
		259.805	I	8,512	46,991	2	32	E	2
Arsenic	As	189.042	I	0	52,898	6	2.0	D	3
		193.759	I	0	51,610	4	2.0	D	3
		197.262	I	10,592	50,694	2	2.0	D	3
		234.984	I	18,186	53,136	4	3.1	D	3
		286.044	I		53,136	2	0.55	D	3
Barium	Ba	455.403	II	0	21,952	4	1.17	A	3
		493.409	II	0	20,262	2	0.955	B	3
		553.548	I	0	18,080	3	1.15	B	3
Beryllium	Be	234.861	I	0	42,565	3	5.56	B	3
		313.042	II	0	31,935	4	1.14	B	3
		313.107	II	0	31,929	2	1.15	B	3
Bismuth	Bi	223.061	I	0	44,817	4	0.25	D	3
		289.798	I	11,418	45,916	2	1.53	C	3
Boron	B	208.891	I	0	47,857	4	0.28	D	3
		208.957	I	16	47,857	6	0.33	D	3
		249.677	I	0	40,040	2	0.84	C	3
		249.773	I	16	40,040	2	1.69	C	3
Bromine	Br	470.486	II		115,176	7	1.1	D	3
		827.244	I						
Cadmium	Cd	214.441	II	0	46,619	4	2.8	C	3
		226.502	II	0	44,136	2	3.0	C	3
		228.8022	I	0	43,692	3	0.24	D	3
		326.1055	I	0	30,656	3	0.004	C	3
Calcium	Ca	364.441	I	15,316	42,747	7	0.355	C	3
		393.366	II	0	25,414	4	1.47	C	3
		396.847	II	0	25,192	2	1.4	C	3
		422.673	I	0	23,652	3	2.18	B	3
Carbon	C	193.0905	I	21,648	61,982	3	3.7	D	3
		247.856	I		61,982	3	0.18	D	3
Cerium	Ce	394.275	II	6,913	32,269	gA = 19		E	2
		413.765	II	4,166	28,327	gA = 4.8		E	2
		418.660	II	6,968	30,847	gA = 18		E	2
Cesium	Cs	455.5276	I	0	21,946	4	0.019	C	3
		852.1122	I	0	11,732	4	0.32	E	2

(Continued)

Element	Symbol	Wavelength (nm)	Type	E-low (cm⁻¹)	E-high (cm⁻¹)	g(i)	A(ji) 10⁵s⁻¹	Merit[a]	References
Chlorine	Cl	413.250	II		153,259	5	1.6	D	3
		837.594	I						
Chromium	Cr	205.552	II	0	48,632	gA = 9.1		E	2
		267.716	II	12,304	49,646	gA = 132		E	2
		357.869	I	0	27,935	gA = 8.3		E	2
		425.435	I	0	23,499	9	0.315	B	3
Cobalt	Co	228.615	II	3,350	47,078	gA = 169		E	2
		238.892	II	3,350	45,198	gA = 278		E	2
		240.725	I	0	41,529	12	3.08	E	2
		352.685	I	0	28,346	10	0.12	C	3
Copper	Cu	213.5981	II	0	30,784	4	1.39	B	3
		324.754	I	0	30,784	4	1.39	B	3
		327.396	I	0	30,535	2	1.37	B	3
Dysprosium	Dy	353.170	II	0	28,307	gA = 19		E	2
		404.597	I	0	24,709	15	1.5	D	3
		421.172	I	0	23,737	19	2.08	C	3
Erbium	Er	337.271	II	0	29,641	gA = 13		E	2
		400.796	I	0	24,943	15	26	D	3
Europium	Eu	381.967	II	0	26,173	gA = 4.8		E	2
		459.403	I	0	21,761	10	1.4	D	3
Fluorine	F	685.603	I		116,987	8	0.42	D	3
Gadolinium	Gd	342.247	II	1,935	31,146	gA = 19		E	2
		368.413	I	0	27,136	gA = 12		E	2
		440.186	I	1,719	24,430	gA = 4.2		E	2
Gallium	Ga	287.424	I	0	34,782	4	1.2	C	3
		294.364	I	826	34,788	6	1.4	C	3
		417.204	I	826	24,789	2	0.92	C	3
Germanium	Ge	199.8887	I	1,410	51,438	5	0.55	C	3
		209.4258	I	1,410	49,144	7	0.97	C	3
		265.1172	I	1,410	39,118	5	2.0	C	3
Gold	Au	242.795	I	0	41,174	4	1.5	D	3
		267.595	I	0	37,359	2	1.1	D	3
Hafnium	Hf	277.336	II	6,344	42,391	gA = 14		E	2
		307.288	I	0	32,533	gA = 3.2		E	2
		339.980	II	0	29,405	gA = 1.1		E	2
Holmium	Ho	345.600	II	637	26,331				
		389.102	II	0	24,660				
		405.393	I	0	24,361				
		410.384	I						
Hydrogen	H	486.133	I	82,259	102,824	32	0.084	A	17
		656.2852	I	82,259	97,492	18	0.441	A	17
Indium	In	230.605	II	0	43,349	gA = 0.032		E	2
		303.936	I	0	32,892	gA = 7.1		E	2
		325.609	I	2,213	32,915	6	1.3	D	3
		451.131	I	2,213	24,373	2	1.02	C	3
Iodine	I	183.038	I		56,093	4	2.71	C	3
		206.163	I						
Iridium	Ir	208.882	I	0	47,858	12	28	E	2
		224.268	II	0	44,576	10	0.56	E	2
		263.971	I	0	37,872				
Iron	Fe	238.204	II	0	41,968	gA = 92		E	2
		248.3271	I	0	40,257	11	4.9	C	3
		259.9396	II	0	38,459	10	2.22	C	3
		371.9935	I	0	26,875	11	0.163	B	3

(Continued)

Element	Symbol	Wavelength (nm)	Type	E-low (cm⁻¹)	E-high (cm⁻¹)	g(i)	A(ji) $10^5 s^{-1}$	Merit[a]	References
Lanthanum	La	333.749	II	3,250	33,204	gA = 3.5		E	2
		408.672	II	0	24,463	5	0.20	E	2
		550.134	I	0	18,172	4	0.08	E	2
Lead	Pb	217.000	I	0	46,068	3	1.5	D	3
		220.3534	II	14,081	59,448	gA = 5.7		E	2
		283.3053	I	0	35,287	3	0.58	D	3
		368.3462	I	7,819	34,960	1	1.5	D	3
Lithium	Li	670.776	I	0	1,494	4	0.372	B	3
Lutetium	Lu	261.545	II	0	38,223	gA = 5.8		E	2
		451.857	I	0	22,125	4	0.21	B	3
Magnesium	Mg	279.553	II	0	35,761	4	4.0	C	3
		280.270	II	0	35,669	2	2.6	C	3
		285.213	I	0	35,051	3	5.3	D	3
Manganese	Mn	257.610	II	0	38,807	9	8.89	E	2
		279.482	I	0	35,770	8	3.7	C	3
		403.076	I	0	24,802	8	0.19	C	3
Mercury	Hg	184.905	II	0	39,412	3	0.13	D	3
		194.227	II						
		253.652	I						
Molybdenum	Mo	202.030	II	0	49,481	gA = 24		E	2
		313.259	I	0	31,913	9	1.09	E	2
		379.825	I	0	26,321	9	0.49	E	2
		390.296	I	0	25,614	5	0.42	E	2
Neodymium	Nd	401.225	II	5,086	30,002	20	0.55	D	3
		463.424	I	0	21,572	gA = 2.0		E	2
		492.453	I	0	20,301	gA = 2.0		E	2
Nickel	Ni	221.648	II	0	53,496	12	5.5	D	3
		232.003	I	205	43,090	11	6.9	C	3
		352.454	I		28,569	5	1.0	C	3
Niobium	Nb	309.418	II	4,146	36,455	13	1.1	E	2
		334.906	I	2,154	32,005	10	0.45	E	2
		405.894	I	1,050	25,680	12	0.65	E	2
Nitrogen	N	174.2729	I						
		821.634	I						
Osmium	Os	225.585	II	0	44,315	11	1.0	E	2
		290.906	I	0	34,365	9	0.034	E	2
		442.047	I	0	22,616				
Oxygen	O	436.825	I		86,631	7	0.34	B	3
		777.194	I						
Palladium	Pd	244.791	I	0	40,839	3	0.28	E	2
		247.642	I	0	40,369	3	0.37	E	2
		340.458	I	6,564	35,928	9	1.33	E	2
		363.470	I	6,564	34,069	5	1.24	E	2
Phosphorus	P	177.499	I	0	56,340	6	2.17	C	3
		213.547	I	11,362	58,174	4	0.211	C	3
		253.561	I	18,748	58,174	4	0.20	C	3
Platinum	Pt	214.423	I	0	46,622	7	5.14	E	2
		265.945	I	0	37,591	9	0.91	E	2
Potassium	K	766.490	I	0	13,043	4	0.387	B	3
Praseodymium	Pr	390.805	II	0	23,660	gA = 1.4		E	2
		422.535	II	0	20,190				
		495.137	I						

(Continued)

Element	Symbol	Wavelength (nm)	Type	E-low (cm^{-1})	E-high (cm^{-1})	g(i)	A(ji) 10^{5}s^{-1}	Merit[a]	References
Rhenium	Re	197.3	II	0	45,148	gA = 15		E	2
		221.426	I	0	28,890				
		346.046							
Rhodium	Rh	233.477	II	16,885	59,702	gA = 44		E	2
		343.489	I	0	29,105	12	0.34	E	2
		369.236	I	0	27,075	8	0.35	E	2
Rubidium	Rb	420.180	I	0	23,793	4	0.018	C	3
		780.027	I	0	12,817	4	0.370	B	3
Ruthenium	Ru	240.272	II	9,152	50,758	gA = 247		E	2
		349.894	I	0	28,572	13	0.46	E	2
		372.803	I	0	26,816	11	0.42	E	2
Samarium	Sm	359.260	II	3,053	30,880	gA = 6.3		E	2
		373.912	II	326	27,063	gA = 21		E	2
		429.674	I	4,021	27,288	gA = 3.3		E	2
		476.027	I	812	21,813				
Scandium	Sc	361.384	II	178	27,841	9	0.14	D	3
		391.181	I	168	25,725	8	1.37	C	3
Selenium	Se	196.09	I	0	50,997	2	100	E	2
		203.98	I	1,989	50,997	2	65	E	2
Silicon	Si	251.6113	I	223	39,955	5	1.21	C	3
		288.1579	I	6,299	40,992	3	1.89	C	3
Silver	Ag	328.068	I	0	30,473	4	1.4	B	3
		338.2068	I	0	29,552	2	1.3	B	3
Sodium	Na	330.237	I	0	30,273	4	0.028	C	3
		588.9950	I	0	16,973	4	0.622	A	3
		589.5924	I	0	16,956	2	0.618	A	3
Strontium	Sr	407.771	II	0	24,517	4	1.42	C	3
		421.552	II	0	23,715	2	1.27	C	3
		460.733	I	0	21,698	3	2.01	B	3
Sulfur	S	180.7311	I		55,331	3	3.8	C	3
		182.0343	I		55,331	3	2.2	C	3
		216.89							
Tantalum	Ta	226.230	II	2,642	46,831	gA = 35		E	2
		240.063	II	6,187	47,830	gA = 516		E	2
		271.467	I	0	36,826	6	1.17	E	2
		296.513	II	0	33,715	gA = 7.8		E	2
		474.016	I	9,976	31,066	4	0.028	E	2
Tellurium	Te	214.281	I	0	46,653	3	38	E	2
		238.578	I	4,751	46,653	3	5.47	E	2
Terbium	Tb	350.917	II	0	28,488	gA = 7.2		E	2
		367.635	II	1,016	28,209				
		432.643	I	0	23,107				
Thallium	Tl	190.864	II	0	36,118	4	1.26	C	3
		276.787	I	0	26,478	2	0.625	B	3
		377.572	I	7,793	26,478	2	0.705	B	3
		535.046	I						
Thorium	Th	283.7295	II	6,214	41,448	gA = 0.12		E	2
		324.4448	I	0	30,813	gA = 0.66		E	2
		401.9129	II	0	24,874	gA = 0.08		E	2
		491.9816	II	6,168	26,489				
Thulium	Tm	313.126	II	0	31,927	gA = 4.6		E	2
		346.220	II	0	28,875	gA = 2.5		E	2
		371.791	I	0	26,889	gA = 8.3		E	2

(Continued)

Element	Symbol	Wavelength (nm)	Type	E-low (cm^{-1})	E-high (cm^{-1})	g(i)	A(ji) 10^5 s^{-1}	Merit[a]	References
Tin	Sn	189.991	II	0	44,509	3	1.6	D	3
		224.605	I	1,692	44,145	5	1.7	D	3
		235.484	I	3,428	38,629	5	1.7	D	3
		283.999	I	8,613	39,257	3	2.7	D	3
		326.234	I						
Titanium	Ti	334.941	II	393	30,241	12	1.3	D	3
		364.268	I	170	27,615	9	0.67	C	3
		365.350	I	387	27,750	11	0.66	C	3
		368.520	II	4,898	32,026				
Tungsten	W	207.911	II	6,147	54,229	gA = 93		E	2
		255.135	I	0	39,183	7	1.17	E	2
		276.427	II	0	36,165	gA = 6.9		E	2
		400.875	I	2,951	27,890	9	0.20	E	2
Uranium	U	263.553	II	0	27,887	15	0.10	B	3
		358.488	I	289	26,191	gA = 2.6		E	2
		385.957	II	0	16,900	gA = 0.12		E	2
		591.539	I						
Vanadium	V	309.311	II	3,163	35,483	13	1.8	D	3
		311.062	II	2,809	34,947	9	1.5	D	3
		318.540	I	553	31,937	12	1.4	D	3
		437.924	I	2,425	25,254	12	1.2	D	3
Ytterbium	Yb	328.937	II	0	30,392	4	1.8	C	3
		369.419	II	0	27,062	2	1.4	C	3
		398.799	I	0	25,068	3	1.76	C	3
Yttrium	Y	362.094	I	530	28,140	4	1.55	E	2
		371.030	II	1,450	28,394	8	0.64	E	2
		377.433	II	1,045	27,532				
		410.238	I	530	24,900				
Zinc	Zn	202.548	II	0	49,355	4	3.3	C	3
		213.856	I	0	46,745	3	7.09	B	3
Zirconium	Zr	343.823	II	763	29,840	gA = 13		E	2
		351.960	I	0	28,404	7	0.71	E	2
		360.119	I	1,241	29,002	11	0.91	E	2

[a] The key for the merit of the A(ji) values follows that given in Ref. 3 as follows: A = within 3 %; B = within 10 %; C = within 25 %; and D = within.

ACTIVATED CARBON AS A TRAPPING SORBENT FOR TRACE METALS

Activated carbon is commonly used to preconcentrate samples of heavy metals before spectrometric analysis [1]. This material is typically used by passing the sample through a thin layer (50–150 mg) of the activated carbon that is supported on a filter disk. It can also be used by shaking 50–150 mg of activated carbon in the solution containing the heavy metal and then filtering the sorbent out of the solution.

REFERENCE

1. Alfasi, Z.B. and Wai, C.M., *Preconcentration Techniques for Trace Elements*, CRC Press, Boca Raton, FL, 1992.

Matrices	Trace Metals	Complexing Agents	Determination Methods
Water	Ag, Bi, Cd, Co, Cu, Fe, In, Mg, Mn, Ni, Pb, Zn	(NaOH; pH 7–8)	AAS
Water	Ag, As, Ca, Cd, Ce, Co, Cu, Dy, Fe, La, Mg, Mn, Nb, Nd, Ni, Pb, Pr, Sb, Sc, Sn, U, V, Y, Zn	8-Quinolinol	spark source mass spectrometry (SSMS), X-ray fluorescence (XRF)
Water	Ba, Co, Cs, Eu, Mn, Zn	APDC, DDTC, PAN, 8-quinolinol	XRF
Water	Hg, methyl mercury	—	AAS
Water	Hg (halide)	—	AAS
Water	Hg (halide)	—	AAS
Water	U	L-Ascorbic acid	INAA
HNO_3, water, Al, KCl	Ag, Bi, Cd, Cu, Hg, Pb, Zn	Dithizone	AAS
Mn, MnO_3, Mn salts	Bi, Cd, Co, Cu, Fe, In, Ni, Pb, Tl, Zn	Ethyl xanthate	AAS
Co, $Co(NO_3)_2$	Ag, Bi	APDC	AAS
Ni, $Ni(NO_3)_2$	Ag, Bi	APDC	AAS
Mg, $Mg(NO_3)_2$	Ag, Cu, Fe, Hg, In, Mn, Pb, Zn	(pH 8.1–9)	AAS
Al	Cd, Co, Cu, Ni, Pb	Thioacetamide	AAS
Ag, $TINO_3$	Bi, Co, CU, Fe, In, Pb	Xenol orange	AAS
Cr salts	Ag, Bi, Cd, Co, Cu, In, Ni, Pb, Tl, Zn	HAHDTC	AAS
Co, In, Pb, Ni, Zn	Ag, Bi, Cu, Tl	DDTC	AAS
Se	Cd, Co, Cu, Fe, Ni, Pb, Zn	DDTC	AAS
$NaClO_4$	Ag, Bi, Cd, Co, Cu, Fe, Hg, In, Mn, Ni, Pb	(pH 6)	AAS

APDC, ammonium pyrrolidinecarbodithioate; DDTC, diethyldithiocarbamate; HAHDTC, hexamethyleneammonium hexaethylenedithiocarbamate; PAN, 1-(2-pyridylazo)-2-naphthol.

REAGENT-IMPREGNATED RESINS AS TRAPPING SORBENTS FOR TRACE MINERALS

Reagent-impregnated resins can be used as a trapping sorbents for the preconcentration of heavy metals [1]. These materials can be used in the same way as activated carbons.

REFERENCE

1. Alfasi, Z.B. and Wai, C.M., *Preconcentration Techniques for Trace Elements*, CRC Press, Boca Raton, FL, 1992.

Reagents	Adsorbents	Metals
TBP	Porous polystyrene DVB resins	U
YBP	Levextrel (polystyrene DVB resins)	U
DEHPA	Levextrel	Zn
DEHPA	XAD-2	Zn
Alamine 336	XAD-2	U
LIX-63	XAD-2	Co, Cu, Fe, Ni, etc.
LIX-64N, -65N	XAD-2	Cu
Hydroxyoximes	XAD-2	Cu
Kelex 100	XAD-2	Co, Cu, Fe, Ni
Kelex 100	XAD-2,4,7,8,11	Cu
Dithizone, STTA	Polystyrene DVB resins	Hg
Dithizone (acetone)	XAD-1,2,4,7,8	Hg, methyl mercury
DMABR	XAD-4	Au
Pyrocatechol violet	XAD-2	In, Pb
TPTZ	XAD-2	Co, Cu, Fe, Ni, Zn

DEHPA, di-ethylhexyl phosphoric acid; DMABR, 5-(4-dimethylaminobenzylidene)-rhodanine; LIX 63, aliphatic α-hydroxyoxime; LIX 64N, a mixture of LIX 65N with approximately 1 % (vol/vol) of LIX-63; LIX 65N, 2-hydroxy-5-nonylbensophenoneoxime; STTA, monothiothenolytrifluoroacetone; TBP, tributyl phosphate; TPTZ, 2,4,6-tri(2-pyridyl)-1,3,5-triazine.

REAGENT-IMPREGNATED FOAMS AS A TRAPPING
SORBENTS FOR INORGANIC SPECIES

Reagent-impregnated foams can be used as a trapping sorbents for the preconcentration of heavy metals [1]. These materials can be used in the same way as activated carbons.

REFERENCE

1. Alfasi, Z.B. and Wai, C.M., *Preconcentration Techniques for Trace Elements*, CRC Press, Boca Raton, FL, 1992.

Matrices	Elements	Conc.	Foam Type	Reagents	Determination Methods
Water	^{131}I, ^{203}Hg	Traces	Polyether	Alamine 336	Radiometry
Natural water					
Water	Bi, Cd, Co, Cu, Fe, Hg, Ni, Pb, Sn, Zn	Traces	Polyether	Amberlite LA-2	Spectroph., AAS
Water	Co, Fe, Mn	Traces to: g/L	Polyether	PAN	Radiometry
Natural water	Cd	μg/L	Polyether	PAN	AAS
Water	Au, Hg	μg/L	Polyether	PAN	NAA
Water	Ni	Traces to :g/L	—	DMG, α-benzyldioxime	Spectroph., AAS
Water	Cr	μg/L	Polyether	DPC	Colorimetry
Water	Hg, methyl-Hg, phenyl-Hg	μg/L	Polyether	DADTC	Radiometry
Natural water	Sn	Traces	Polyether	Toluene-3,4-dithiol	Spectroph.
Water	Cd, Co, Fe, Ni	Traces	Polyether	Aliquot	Spectroph.
Water	Th	Traces	Polyether	PMBP HDEHP-TBP	Radiometry Spectroph.
Water	PO_4^{3-}	Traces		Amine-molybdate-TBP	Colorimetry

AAS, atomic absorption spectrometry; DADTC, diethylammonium diethyldithiocarbamate; DMG, dimethylglyoxime; DPC, 1,5-diphenylcarbazide; HDEHP, bis-[2-ethylhexyl]phosphate; NAA, neutron activation analysis; PAN, 1-(2-pyridylazo)-2-naphthol; PMBP, 1-phenyl-3-methyl-4-bensoyl-pyrazolone-5; Spectroph., spectrophotometry; TBP, tributyl phosphate.

CHAPTER 13

Qualitative Tests

PROTOCOL FOR CHEMICAL TESTS

The following section gives a suggested protocol for the chemical tests used in the identification of organic compounds. Variations of the procedures are possible, but these protocols have been used successfully for most organic identifications [1–10].

REFERENCES

1. Vogel, A.I., Furniss, B.S., and Tatchell, A.R., *Vogel's Textbook of Practical Organic Chemistry,* John Wiley and Sons, New York, 1989.
2. Shriner, R.L., Hermann, C.K.F., Morrill, T.C., Fuson, R.C. and D.Y. Curtin, *The Systematic Identification of Organic Compounds, A Laboratory Manual,* John Wiley and Sons, New York, 1998.
3. Vogel, A.I., *Elementary Practical Organic Chemistry,* Part 2, John Wiley and Sons, New York, 1966.
4. Pasto, D.J. and Johnson, C.R., *Laboratory Text for Organic Chemistry,* Prentice Hall, Englewood Cliffs, NJ, 1979.
5. Roberts, R.M., Gilbert, J.C., Rodewald, L.B., and Wingrove, A.S., *Modern Experimental Organic Chemistry,* Saunders, New York, 1985.
6. Uamm, O., *Qualitative Organic Analysis,* John Wiley and Sons, New York, 1932.
7. Behforout, M., "Getting the acid out of your 2,4-DNPH", *J. Chem. Ed.* 63, 723, 1986.
8. Durst, H.D. and Gokel, G.W., *Experimental Organic Chemistry,* McGraw-Hill, New York, 1987.
9. Fieser, L.F. and Freser, M., *Reagents for Organic Synthesis,* John Wiley and Sons, New York, 1968.
10. Svoronos, P., Sarlo, E. and Kulawiec, R., *Experiments in Organic Chemistry,* McGraw-Hill, Dubuque, 1997.

Aluminum chloride–chloroform test: To a mixture of 2 mL chloroform and 0.2–0.4 *dry* aluminum chloride in a test tube, add 5–10 drops of your unknown aromatic compound. A color formation will indicate the presence of a benzene ring.

Basic hydrolysis test: Reflux 0.1 g of the compound in 5 mL of a 10 % sodium hydroxide solution.

Benedict's test: Add 5–10 drops of your unknown to 1–2 mL of the Benedict's reagent and heat. A positive test for reducing sugars will change the blue copper(II) color of the reagent with subsequent precipitation of the red copper(I) oxide.

Bromine test: The compound to be tested is treated with a few drops of 1 %–5 % Br_2/CCl_4 solution. A positive test is indicated by decolorization of the bromine color.

Ceric ammonium nitrate test: To 1–2 mL 5 % ceric ammonium nitrate, add ten drops of the compound to be tested. A change to an orange/red color is indicative of an alcohol (detection limit 100 mg— compounds tested $C_1–C_{10}$).

Dichromate test (Jones Test): Add ten drops of the alcohol to be tested to a mixture of 1 mL 1 % $Na_2Cr_2O_7$ and five drops conc. H_2SO_4. A blue-green solution is positive test for a 1° or 2° alcohol; 3° alcohols do not react and, therefore, the solution stays orange. (Detection limit 20 µg—compounds tested C_1–C_8). Slight heating may be necessary for water-immiscible alcohols. Extensive heat gives a positive test also for tertiary alcohols which is due to the water elimination of the alcohol and oxidation of the formed alkene.

2,4-Dinitrophenylhydrazine (2,4-DNP) test: Add ten drops of the compound to be tested to 1 mL of the 2,4-DNP reagent. A yellow to orange-red precipitate is considered a positive test. The crystals can be purified by washing them with 5 % $NaHCO_3$, then with water, and finally recrystallized from ethanol. The 2,4-DNP reagent can be prepared by dissolving 1 g 2,4-dinitrophenylhydrazine in 5 mL conc. H_2SO_4 and then mixing it with 8 mL of water and 20 mL 95 % ethanol. The solution should be filtered before reacting it with the unknown compound. (Detection limit 20 µg—compounds tested C_1–C_8).

Fehling's test: The test is similar to the Benedict's test (see above).

Ferric chloride test: Add ten drops of 3 % aqueous $FeCl_3$ solution to 1 mL of a 5 % aqueous ethanol solution of the compound in question. Phenols give red, blue, purple, or green colorations. The same test can be done by using chloroform as a solvent (detection limit 50 µg).

Hinsberg test: To 0.5 mL of the amine (0.5 g, if solid) in a test tube add 1 mL of benzenesulfonyl chloride and 8 mL 10 % NaOH. Stopper the tube and shake for 3–5 minutes. Remove the stopper and warm the tube with shaking in a hot water bath (70 °C) for about 1 minute. No reaction is indicative of a 3° amine; the amine becomes soluble upon acidification (pH = 2–4) with 10 % HCl. If a precipitate is present in the alkaline solution, dilute with 5–8 mL H_2O and shake. If the precipitate does not dissolve, the original amine is probably a 2° one. If the solution is clear, acidify (pH = 4) with 10 % HCl. The formation of a precipitate is indicative of a 1° amine. (Detection limit 100 mg—compounds tested C_1–C_{10}).

Iodoform test: The reagent calls for the mixture of 10 g I_2 and 20 g KI in 100 mL water. The reagent is then added dropwise to a mixture of ten drops of the compound in question in 2 mL of water (or dioxane, to facilitate the solubility) and 1 mL 10 % aqueous NaOH solution until a *persistent* brown color remains (even when heating in a hot water bath at 60 °C). A yellow precipitate is indicative of iodoform (CHI_3) formation and is characteristic of a methyl ketone, acetaldehyde, or an alcohol of the general formula $CH_3CH(R)OH$ (R = alkyl, hydrogen). Aldols, $RC(=O)CH_2CH(OH)R'$, may also give a positive iodoform test by a retro aldol condensation first yielding $RC(=O)CH_3$ and RCHO (detection limit 100 mg). In this case, at least one of the products should be a methyl ketone or acetaldehyde.

Molisch test: The reagent is made by preparing a solution of 95 % (vol/vol) of 1-naphthol in ethanol. The reagent is added to the test solution, which is then acidified with sulfuric acid. The development of a purple color at the interface of the test solution with the reagent mixture is indicative of a carbohydrate.

Lucas test: The reagent is made by dissolving 16 g anhydrous $ZnCl_2$ in 10 mL concentrated hydrochloric acid and cooling to avoid HCl loss. Add 10–15 drops of the *anhydrous* alcohol to 2 mL of the reagent; 3° alcohols form an emulsion that appears as two layers (due to the water-insoluble alkyl halide) almost immediately; 2° alcohols form this emulsion after 2–5 minutes, while 1° alcohols react after a very long time (if at all). Some secondary alcohols (for example, isopropyl) may not *visually* form the layers because of the low-boiling alkyl halide which may evaporate. Allyl alcohols and most benzyl alcohols also yield results that are identical to the results obtained for 3° alcohols.

Permanganate test: The compound to be tested is treated with 10–15 drops of 1 % $KMnO_4$ solution. A positive test is indicated by the decolorization of the solution and subsequent formation of a black (MnO_2) precipitate.

Silver nitrate test: The compound to be tested is treated with a few drops of 1 % alcoholic silver nitrate. A white precipitate indicates a positive reaction. This could be due to either silver chloride (reaction with a reactive alkyl halide), silver alkynide (reaction with a terminal alkyne), or the silver salt of a carboxylic acid (reaction with a carboxylic acid).

Sodium fusion test: Treat 100 mg of the compound to be analyzed with a fresh tiny piece of sodium metal of the size of a small pea in a 4-inch test tube. The test tube is warmed gently until melting of the sodium metal and decomposition (indicated by charring) of the compound occurs. When it appears that all the volatile material has been decomposed, the test tube is strongly heated until the residue acquires a red color. After 3 minutes of constant heating, the mixture is left to cool to room temperature, then a few drops of methanol are added. If no smoke appears, then excess of sodium metal is not present and incomplete conversion of the elements (nitrogen, sulfur, and halides) is very likely. Addition of another tiny piece of sodium metal and repetition of the heating process is necessary. If smoke appears, then the red hot test tube is plunged in a small beaker containing 10–15 mL *distilled water* and covered with a watch glass or a wire gauze. The test tube might shatter, and therefore, having the small beaker placed inside a larger one is recommended. The contents of the test tube together with the broken glass are ground in a mortar using a pestle, then transferred to the small beaker and heated for a few minutes. The solution is then filtered and the solution divided into two larger portions and 1 mL part.

Detection of nitrogen: To one of the two larger portions, add ten drops of 6 M NaOH (pH adjusted to 13), five drops of saturated $Fe(NH_4)_2(SO_4)_2$ solution, and five drops of 30 % KF solution. The mixture is then boiled for 30 seconds and immediately acidified with 6 M H_2SO_4 with stirring until the colloidal iron hydroxides are dissolved. The formation of a blue color is indicative of the presence of nitrogen.

Detection of sulfur: To the 1 mL part, add ten drops of 6 M acetic acid and two to three drops of 5 % lead(II) acetate solution. A black precipitate is indicative of sulfur presence.

Detection of halogens: To the other larger part add 10 % H_2SO_4 (dropwise) until the solution is acidic. Boil off the solution to one-third its volume to secure evaporation of H_2S and HCN gases. Formation of a precipitate upon addition of a few drops of 10 % $AgNO_3$ solution is indicative of the presence of a halogen: white for a chloride (which is soluble in 6 M NH_4OH), pale yellow for a bromide (which is only slightly soluble in 6 M NH_4OH), and canary yellow for iodide (which is insoluble in 6 M NH_4OH). Should the color of the precipitate be difficult to provide satisfactory identification of the halogen, proceed as follows: the working solution which has been acidified with 10 % H_2SO_4 and boiled down is treated with four to five drops 0.1 N $KMnO_4$ solution, with enough oxalic acid added to discharge the color of excess permanganate and 0.5 mL carbon disulfide. Color formation in the carbon disulfide layer indicates the presence of bromine (red brown) or iodine (purple). Chlorine's presence cannot be detected by color formation. Should the compound to be tested carry both bromine and iodine, the identification is difficult (red-brown to purple carbon disulfide layer). In this case, addition of a few drops of allyl alcohol decolorizes bromine but does not decolorize iodine.

Tollen's test: The reagent should be freshly prepared by mixing two solutions (A and B). Solution A is a 10 % aqueous $AgNO_3$ solution, and solution B is a 10 % aqueous NaOH solution. When the test is required, 1 mL of solution A and 1 mL of solution B are mixed, and the silver oxide thus formed is dissolved by dropwise addition of 10 % aqueous NH_4OH. To the clear solution, ten drops of the compound to be tested are added. A silver mirror is indicative of the presence of an aldehyde. The reagent mixture (A + B) is to be prepared *immediately prior to use*, otherwise explosive silver fulminate will form. The silver mirror is usually deposited on the walls of the test tube either immediately or after a short warming period in a hot water bath. This is to be disposed of immediately by dissolving it in dilute HNO_3. (Detection limit 50 mg—compounds tested C_1–C_6.)

FLOW CHARTS FOR CHEMICAL TESTS

The following flow charts and notes provide a step by step process for the identification of functional groups which may be present in an unknown sample [1–12]. These are meant to augment and confirm information obtainable using instrumental methods of analysis. It will usually be necessary to use gas or liquid chromatography before these "wet" chemical tests are performed in order to determine the number of components present in a given sample. Since many of these tests require the use of sometimes toxic compounds, the strictest rules of laboratory safety must be observed at all times. The use of a fume hood is often required. The book by Feigl et al. is an excellent guide for spot tests [11].

Note
ppt=precipitate; conc=concentrated; dil=dilute.

REFERENCES

1. Pasto, D.J. and Johnson, C.R., *Laboratory Text for Organic Chemistry*, Prentice Hall, Englewood Cliffs, NJ, 1979.
2. Svoronos, P., Sarlo, E., and Kulawiec, R., *Experiments in Organic Chemistry*, McGraw-Hill, Dubuque, 1997.
3. Roberts, R.M., Gilbert, J.C., Rodewald, L.B., and Wingrove, A.S., *Modern Experimental Organic Chemistry*, Sanders, New York, 1985.
4. Hodgman, C.D., Weast, R.C., and Selby, S.M., *Tables for Identification of Organic Compounds*, Chemical Rubber Publishing Co., Cleveland, OH, 1950.
5. Kamm, O., *Qualitative Organic Analysis*, John Wiley and Sons, New York, 1932.
6. Vogel, A.I., Furniss, B.S., and Tatchell, A.R., *Vogel's Textbook of Practical Organic Chemistry*, John Wiley and Sons, New York, 1989.
7. Shriner, R.L., Hermann, C.K.F., Morrill, T.C., Fuson, R.C. and D.Y. Curtin, *The Systematic Identification of Organic Compounds, A Laboratory Manual* John Wiley and Sons, New York, 1998.
8. Vogel, A.I., *Elementary Practical Organic Chemistry*, Part 2, John Wiley and Sons, New York, 1966.
9. Behforouz, M., "Getting the acid out of your 2,4-DNPH", *J. Chem. Educ.*, 63, 723, 1986.
10. Durst, H.D. and Gokel, G.W., *Experimental Organic Chemistry*, McGraw-Hill, New York, 1987.
11. Feigl, F. Anger, V., and Oesper, R.E., *Spot Tests in Organic Analysis*, Elsevier, Amsterdam, 1966.
12. Ott, L., Classical wet analytical chemistry, ASM Handbook, Vol 10, *Materials Characterization*, ASM Handbook Committee, ASM International, Materials Park, OH, 2019.

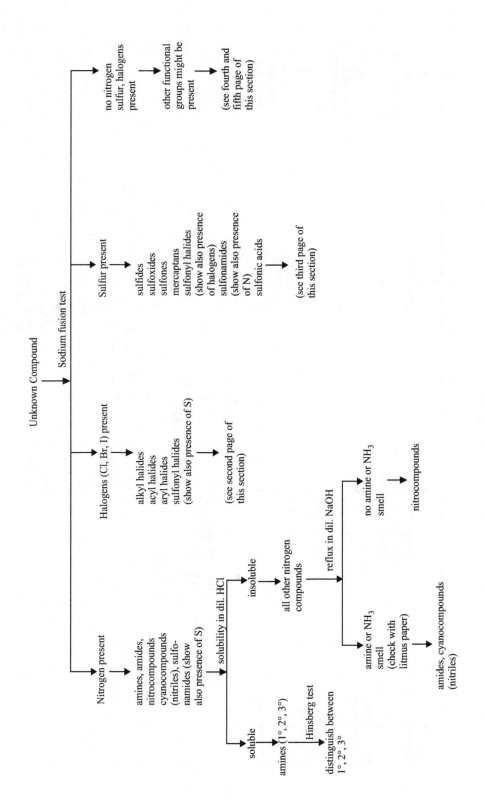

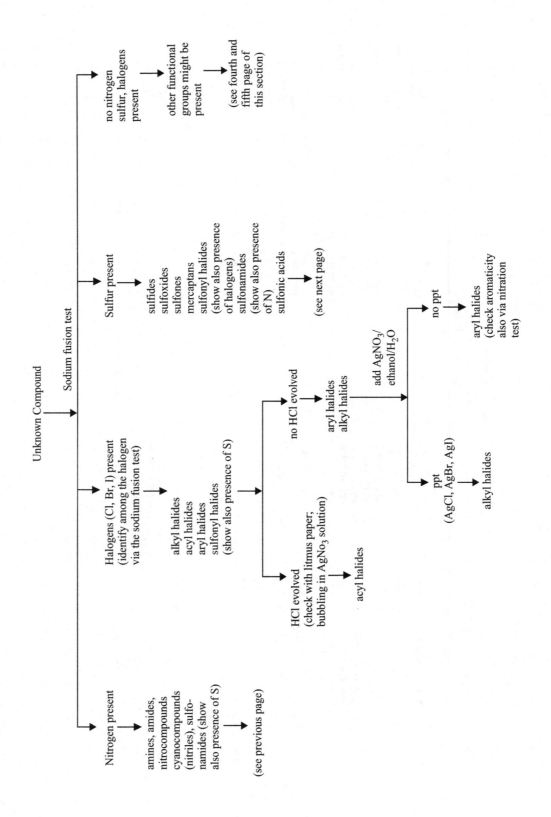

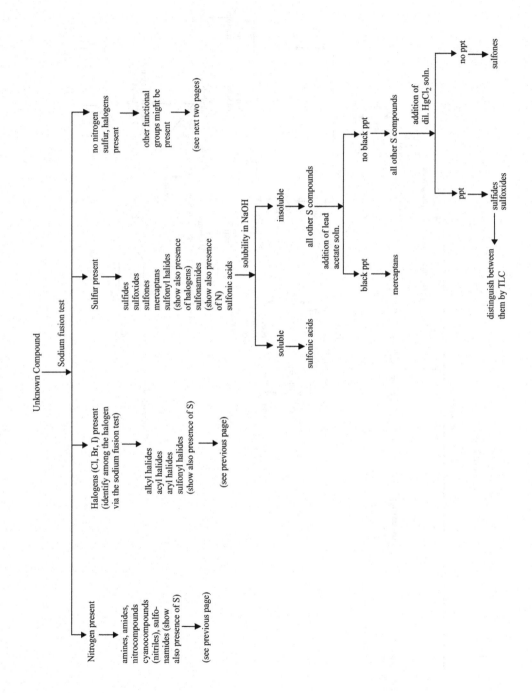

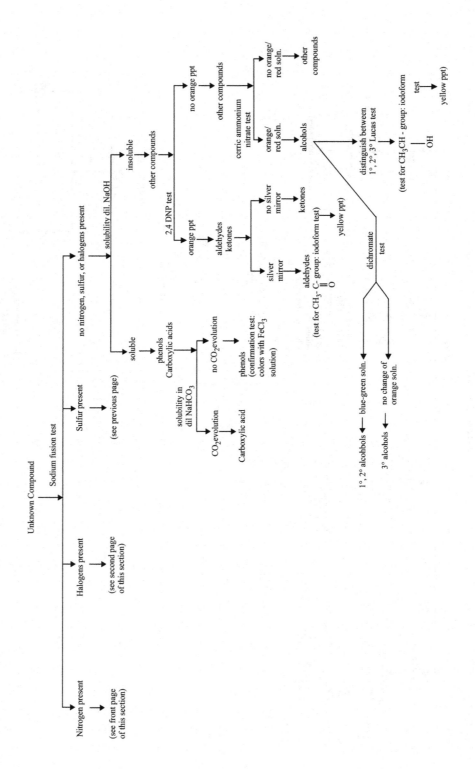

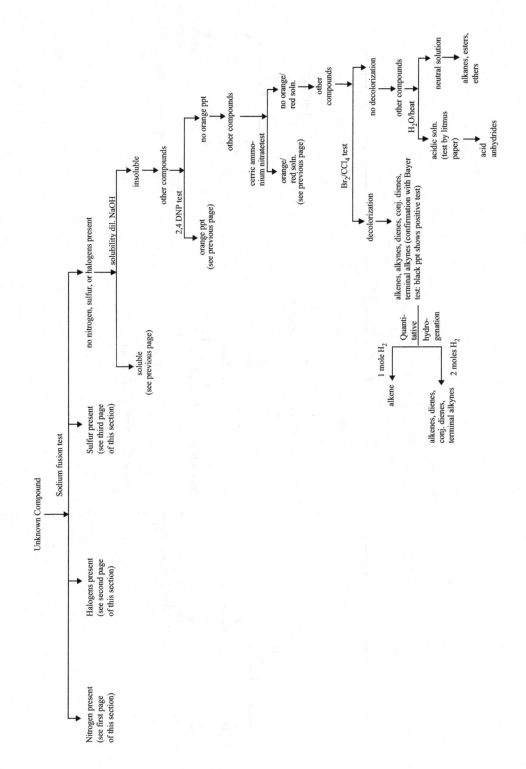

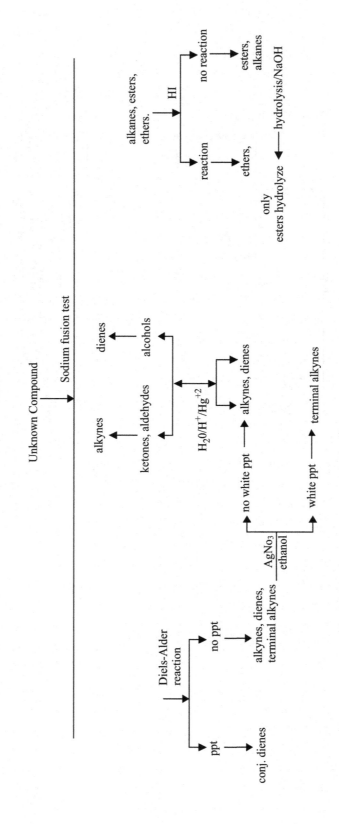

ORGANIC FAMILIES AND CHEMICAL TESTS

The following section gives the major organic families and their most important confirmatory chemical tests. This part serves as a complement "Organic Group Qualitative Tests" section [1–8].

Note
ppt. = precipitate.

REFERENCES

1. Pasto, D.J. and Johnson, C.R., *Laboratory Text for Organic Chemistry*, Prentice Hall, Englewood Cliffs, NJ, 1979.
2. Roberts, R.M., Gilbert, J.C., Rodewald, L.B., and Wingrove A.S., *Modern Experimental Organic Chemistry,* Saunders, New York, 1985.
3. Kamm, O., *Qualitative Organic Analysis,* John Wiley and Sons, New York, 1932.
4. Shriner, R.L., Hermann, C.K.F., Morrill, T.C., Fuson, R.C. and D.Y. Curtin, *The Systematic Identification of Organic Compounds, A Laboratory Manual* , John Wiley and Sons, New York, 1998.
5. Vogel, A.I., *Elementary Practical Organic Chemistry*, Part 2, John Wiley and Sons, New York, 1966.
6. Durst, H.D. and Gokel, G.W., *Experimental Organic Chemistry*, McGraw-Hill, New York, 1987.
7. Fieser, L.F. and Freser, M., *Reagents for Organic Synthesis*, John Wiley and Sons, New York, 1968.
8. Svoronos, P., Sarlo, E., and Kulawiec, R., *Experiments in Organic Chemistry*, McGraw-Hill, Dubuque, IA, 1997.

Organic Families and Chemical Tests

Family	Test	Notes
Alcohols	Ceric ammonium nitrate	Positive for all alcohols
	Dichromate test	Positive for 1° and 2° alcohols; negative for 3° alcohols
	Iodoform test	
	Lucas test	Positive for all alcohols of the general formula CH₃CH(OH)R
		Immediate reaction for 3°, allylic or benzylic alcohols; slower reaction (2–5 min) for 2°; no reaction for 1° alcohols
Aldehydes	Benedict's test	Positive for all aldehydes
	Dichromate test	Positive for all aldehydes
	2,4-Dinitrophenylhydrazine (2,4-DNP)	Positive for all aldehydes (and ketones)
	Fehling's test	Positive for all aldehydes
	Iodoform test	Positive only for acetaldehyde
	Oxime	Positive for all aldehydes (and ketones)
	Permanganate test	Positive for all aldehydes
	Semicarbazone	Positive for all aldehydes (and ketones)
	Tollen's test	Positive for all aldehydes
Alkanes	No test	
Alkenes	Bromine test	Positive for all alkenes
	Permanganate test	Positive for all alkenes
	Solubility in conc. sulfuric acid	All alkenes dissolve
Alkynes	Bromine test	Positive for all alkynes
	Permanganate test	Positive for all alkynes
	Silver nitrate	Positive for all terminal alkynes only
	Sodium metal addition	Positive for all terminal alkynes only
	Sulfuric acid	Positive for all alkynes
Amides	Basic (reflux) hydrolysis	All amides yield ammonia or the corresponding amine detected by odor or by placing wet blue litmus paper on top of the condenser
Amines	Diazotization	All 1° amines give red azodyes with β-naphthol
	Hinsberg test	Distinguishes between 1°, 2°, or 3°
	Solubility in dilute HCl	All amines are soluble
Arenes	Aluminum chloride–chloroform	Positive for all arenes
Aryl halides	Aluminum chloride–chloroform	Positive for all aryl halides
Carboxylic acids	Solubility in dilute sodium bicarbonate	All carboxylic acids are soluble
	Solubility in dilute sodium hydroxide	All carboxylic acids are soluble
Ketones	2,4-Dinitrophenylhydrazine (2,4-DNP)	Positive for all ketones (and aldehydes)
	Hydrazine	Positive for all ketones (and aldehydes)
	Iodoform	Positive for methyl ketones
	Oxime	Positive for all ketones (and aldehydes)
	Semicarbazone	Positive for all ketones (and aldehydes)
Nitriles	Basic hydrolysis	Positive for all nitriles
Phenols	Acetylation	Ppt. of a characteristic melting point
	Benzoylation	Ppt. of a characteristic melting point
	Sulfonation	Ppt. of a characteristic melting point
	Ferric chloride test	Variety of colors characteristic of the individual phenol
	Solubility in aqueous base	Most phenols are soluble in dilute sodium hydroxide but insoluble in dilute sodium bicarbonate. Phenols with strong electron withdrawing groups (for example, picric acid) are soluble in sodium bicarbonate
Sulfonamides	Basic (reflux) hydrolysis	Positive for all sulfonamides
	Sodium fusion test	Presence of sulfur and nitrogen
Sulfonic acids	Sodium fusion test	Presence of sulfur
	Solubility in aqueous base	Most sulfonic acids are soluble in dilute sodium hydroxide and generate carbon dioxide with sodium bicarbonate

INORGANIC GROUP QUALITATIVE PRECIPITATION TESTS

The following tables list some simple chemical tests which will indicate the presence or absence of a given inorganic cation or anion [1–4]. For most of these tests, the anion or cation must be present at a relatively high concentration; the approximate lower bound is 0.05 % unless otherwise specified. It may therefore be necessary to concentrate more dilute samples before successful results can be obtained. Often centrifuging solutions to clearly confirm the formation of a precipitate is necessary. These tests should be used in conjunction with other methods such as the chromatographic methods or spectrometry and spectrophotometry (primarily atomic absorption and atomic emission). Since many of these tests require the use of potentially toxic compounds, the strictest rules of laboratory safety must be observed at all times. The use of a fume hood is strongly recommended. All of the test reagents specified in this section are assumed to be in aqueous solution, unless otherwise designated. The reader is referred to the work of Svehla and Suehla [3] for details on reagent preparation.

Note: In previous editions of this book, we suggested the destruction of the cyanide ion by treatment with chlorine. It is now illegal in most jurisdictions in the United States to pretreat chemical waste prior to submitting such waste to authorized collection. We have therefore removed the destruction procedure from this edition. Users who require this reaction sequence are urged to consult the third edition.

REFERENCES

1. Barber, H.H. and Taylor, T.I., *Semimicro Qualitative Analysis,* Harper Brothers, New York, 1953.
2. Bruno, T.J. and Svoronos, P.D.N., *Basic Tables for Chemical Analysis*, National Bureau of Standards Technical Note 1096, April 1986.
3. Svehla, G. and Suehla, G., *Vogels Qualitative Inorganic Analysis*, 7th ed., Addison Wesley, New York, 1996.
4. De, A.K., *Separation of Heavy Metals*, Pergamon Press, New York, 1961.

Tests for Anions

Acetates, CH_3COO^-

1. Sulfuric acid, dilute	Evolution of acetic acid (vinegar-like odor); concentrated sulfuric acid also evolves sulfur dioxide under mild heating.
2. Dry ethanol and concentrated sulfuric acid	Evolves ethyl acetate (fruity odor) upon heating; dry isoamyl alcohol may be substituted for ethanol.
3. Silver nitrate	Formation of white precipitate of silver acetate which is soluble in dilute ammonia solution.
4. Iron(III) chloride	Deep red coloration (coagulates on boiling forming a brownish-red precipitate).

Benzoates, $C_6H_5COO^-$ (or $C_7H_5O_2^-$)

1. Dilute sulfuric acid	Formation of white precipitate of benzoic acid.
2. Dilute hydrochloric acid	Formation of a crystalline precipitate melting between 121 °C and 123 °C.
3. Silver nitrate	White precipitate of silver benzoate from cold solutions, soluble in hot water and also in dilute ammonia solution.
4. Iron(III) chloride	Buff-colored (light yellow-red) precipitate of iron(III) benzoate from neutral solution, soluble in hydrochloric acid.

Borates, BO_3^{3-}, $B_4O_7^{2-}$, BO_2^-

1. Concentrated sulfuric acid	Upon heating solution, white fumes of boric acid are evolved.
2. Silver nitrate	White precipitate of silver metaborate, soluble in dilute ammonia solution and in acetic acid.
3. Barium chloride	White precipitate of barium metaborate which is soluble in excess reagent, dilute acids as well as ammonium salt solutions.

Bromates, BrO_3^-

1. Concentrated sulfuric acid	Evolution of red bromine vapors even when cold.
2. Silver nitrate	White precipitate of silver bromate which is soluble in dilute ammonia.
3. Sodium nitrite	Brown color develops after addition and subsequent acidification with dilute nitric acid.

Note: Bromates are reduced to bromides by sulfur dioxide, hydrogen sulfide, or sodium nitrite solution.

Bromides, Br^-

1. Concentrated sulfuric acid	Reddish-brown coloration, followed by reddish-brown vapors (hydrogen bromide+bromine) evolution.
2. Manganese dioxide+sulfuric acid	Reddish-brown bromine vapors evolve upon mild heating.
3. Silver nitrate	Pale-yellow, curdy, precipitate of silver bromide, slightly soluble in ammonia solution; insoluble in nitric acid.
4. Lead acetate	White crystalline precipitate of lead bromide which is soluble in hot water.

Special tests: The addition of an aqueous solution of chlorine (or sodium hypochlorite) will liberate free bromine, which may be isolated in a layer of carbon tetrachloride or carbon disulfide.

Carbonates, CO_3^{2-}

1. Hydrochloric acid	Decomposition with effervescence and evolution of carbon dioxide (odorless).
2. Barium chloride	White precipitate of barium carbonate which is soluble in HCl (calcium chloride may be substituted for barium chloride).
3. Silver nitrate	Gray precipitate of silver carbonate.
4. Magnesium sulfate	White precipitate is formed, which can be dissolved by the addition of dilute acetic acid.

Special tests: Effervescence with all acids, producing carbon dioxide which makes limewater cloudy.

(Continued)

Tests for Anions (*Continued*)

Chlorates, ClO$_3^-$

1. Concentrated sulfuric acid	Liberates chlorine dioxide gas (green); solids decrepitate (crackle explosively) when warmed; *large quantities may result into a violent explosion.*
2. Concentrated HCl	Chlorine dioxide gas evolved, imparts yellow color to acid.
3. Manganese(II) sulfate + phosphoric acid	Violet coloration due to diphosphatomanganate formation; peroxydisulfate nitrates, bromates, iodates, periodates, react similarly.
4. Heat of neat sample	Decomposition and formation of gaseous oxygen.

Chlorides, Cl$^-$

1. Concentrated sulfuric acid	Evolution of hydrogen chloride gas (pungent odor).
2. Silver nitrate	White precipitate of silver chloride which is soluble in ammonia solution (reprecipitate with HNO$_3$).
3. Lead acetate	White precipitate of lead bromide, soluble in boiling water.

Special tests: 1. MnO$_2$ + H$_2$SO$_4$ evolves Cl$_2$ gas.
2. An aqueous solution of chlorine + carbon disulfide produces no coloration.

Chromates, CrO$_4^{2-}$, Dichromates, Cr$_2$O$_7^{2-}$

1. Barium chloride	Pale-yellow precipitate of barium chromate, soluble in dilute mineral acids, insoluble in water and in acetic acid.
2. Silver nitrate	Brownish-red precipitate of silver chromate, soluble in dilute nitric acid and in ammonia solution; insoluble in acetic acid.
3. Lead acetate	Yellow precipitate of lead chromate, soluble in dilute nitric acid, insoluble in acetic acid.
4. Hydrogen peroxide	Deep-blue coloration in acidic solution, which quickly turns green, with the subsequent liberation of oxygen.
5. Hydrogen sulfide	Dirty yellow deposit of sulfur is produced in acidic solutions.

Citrates, C$_6$H$_5$O$_7^{3-}$

1. Concentrated sulfuric	Evolution of carbon dioxide and carbon monoxide (highly poisonous).
2. Silver nitrate	White precipitate of silver citrate which is soluble in dilute ammonia solution.
3. Cadmium acetate	White gelatinous precipitate of cadmium citrate, practically insoluble in boiling water, soluble in warm acetic acid.
4. Pyridine + acetic anhydride, 3:1 (vol/vol)	A red brown color develops upon addition to the reagent mixture.

Cyanates, OCN$^-$

1. Sulfuric acid, concentrated and dilute	Vigorous effervescence, due largely to evolution of carbon dioxide, with concentrated acid producing a more dramatic effect.
2. Silver nitrate	Curdy white precipitate of silver cyanate.
3. Copper sulfate–pyridine	Lilac-blue precipitate (interference by thiocyanates). Reagent is prepared by adding two to three drops of pyridine to 0.25 M CuSO$_4$ solution.

Cyanides, CN$^-$

1. Cold dilute HCl	Liberation of hydrogen cyanide (odor of bitter almond; caution—highly toxic).
2. Silver nitrate	White precipitate of silver cyanide.
3. Concentrated sulfuric acid, hot	Liberation of carbon monoxide (caution).
4. Mercury(I) nitrate	Gray precipitate of mercury.
5. Copper sulfide	Formation of colorless tetracyanocuprate(I) ions. This test can be done on a section of filter paper.

(*Continued*)

Tests for Anions (*Continued*)

Dithionites, $S_2O_4^{2-}$

1. Dilute sulfuric acid	Orange coloration which disappears quickly, accompanied by evolution of sulfur dioxide gas and deposition of pale-yellow sulfur.
2. Concentrated sulfuric acid	Fast evolution of sulfur dioxide and precipitation of pale-yellow sulfur.
3. Silver nitrate	Black precipitate of silver.
4. Copper sulfate	Red precipitate of copper.
5. Mercury(II) chloride	Gray precipitate of mercury.
6. Methylene blue	Decolorization in cold solution.
7. Potassium hexacyanoferrate(II) and iron(II) sulfate	White precipitate of dipotassium iron(II) hexacyanoferrate(II); turns from white to Prussian blue.

Fluorides, F^-

1. Concentrated sulfuric acid	Evolution of hydrogen fluoride dimer.
2. Calcium chloride	White, slimy precipitate of calcium fluoride, slightly soluble in dilute hydrochloric acid.
3. Iron(III) chloride	White precipitate.

Special tests: HF etches glass (only visible after drying).

Formates, $HCOO^-$

1. Dilute sulfuric acid	Formic acid is evolved (pungent odor).
2. Concentrated sulfuric acid	Carbon monoxide (highly poisonous, odorless, colorless) is evolved on warming.
3. Ethanol and concentrated H_2SO_4, heat	Ethyl formate evolved (pleasant odor).
4. Silver nitrate	White precipitate of silver formate in neutral solutions, forming a black deposit of elemental silver upon mild heating.
5. Iron(III) chloride	Red coloration due to complex formation.
6. Mercury(II) chloride	White precipitate of calomel produced on warming; upon boiling, a black deposit of elemental mercury is produced.

Hexacyanoferrate(II) Ions, $[Fe(CN)_6]^{4-}$

1. Silver nitrate	White precipitate of silver hexacyanoferrate(II).
2. Iron(III) chloride	Prussian blue is formed in neutral or acid conditions, which is decomposed by alkali bases.
3. Iron(II) sulfate (aq)	White precipitate of potassium iron(II) hexacyanoferrate which turns blue by oxidation.
4. Copper sulfate	Brown precipitate of copper hexacyanoferrate(II).
5. Thorium nitrate	White precipitate of thorium hexacyanoferrate(III).

Hexacyanoferrate(III) Ions $[Fe(CN)_6]^{3-}$

1. Silver nitrate	Orange-red precipitate of silver hexacyanoferrate(III), which is soluble in ammonia solution but not in nitric acid.
2. Iron(II) sulfate	Dark-blue precipitate in neutral or acid solution (Prussian or Turnbull's blue).
3. Iron(III) chloride	Brown coloration.
4. Copper sulfate	Green precipitate of copper(II) hexacyanoferrate(III).
5. Concentrated hydrochloric acid	Brown precipitate of hexacyanoferric acid.

(*Continued*)

Tests for Anions (*Continued*)

Hexafluorosilicates (Silicofluorides), [SiF$_6$]$^{2-}$

1. Barium chloride — White, crystalline precipitate of barium hexafluorosilicate, insoluble in dilute HCl, slightly soluble in water.
2. Potassium chloride — White gelatinous precipitate of potassium hexafluorosilicate, slightly soluble in water.
3. Ammonia solution — Gelatinous precipitate of silica acid.

Hydrogen Peroxide, H$_2$O$_2$

1. Potassium iodide and starch — If sample is previously acidified by dilute sulfuric acid, a deep-blue coloration occurs due to the production of iodine complexation with starch.
2. Potassium permanganate — Decolorization, evolution of oxygen.
3. Titanium(IV) chloride — Orange-red coloration; very sensitive test.

Special test: $4H_2O_2 + PbS \rightarrow PbSO_4\downarrow + 4H_2O$.
Black lead sulfide reacts to produce white lead sulfate.

Special test: A reagent prepared from p-hydroxyphenylacetic acid (HPPA), 7.6 mg, hematin (typically from pig albumin), 1.0 mg, in 100 mL of 0.1 M KOH (aq) will produce a fluorescent dimer (6,6′-dihydroxy-3,3′-biphenyl acetic acid) with hydrogen peroxide. This test is extremely sensitive.

Hypochlorites, OCl$^-$

1. Dilute hydrochloric acid — Yellow coloration, followed by chlorine gas evolution.
2. Lead(II) acetate or nitrate — Brown lead(IV) oxide forms upon heating.
3. Cobalt nitrate — Black precipitate of cobalt(II) hydroxide.
4. Mercury — On shaking slightly acidified solution of a hypochlorite with Hg, a brown precipitate of mercury(II) chloride is formed.

Hypophosphites, H$_2$PO$_2^-$

1. Silver nitrate — White precipitate of silver hypophosphite.
2. Mercury(II) chloride — White precipitate of calomel in cold solution, which darkens upon warming.
3. Copper(II) sulfate — Red precipitate of copper(I) oxide forms upon warming.
4. Potassium permanganate — Immediate decolorization under cold conditions.

Iodates, IO$_3^-$

1. Silver nitrate — White, curdy precipitate of silver iodate, soluble in dilute ammonia solution.
2. Barium chloride — White precipitate of barium iodate, sparingly soluble in hot water or dilute nitric acid; insoluble in ethanol and methanol.
3. Mercury(II) nitrate — White precipitate of mercury(II) iodate.

Iodides, I$^-$

1. Concentrated sulfuric acid — Produces hydrogen iodide and iodine.
2. Silver(I) nitrate — Yellow precipitate of silver(I) iodide which is slightly soluble in ammonia solution and insoluble in dilute nitric acid.
3. Lead acetate — Yellow precipitate of lead iodide, soluble in excess hot water.
4. Potassium dichromate and concentrated sulfuric acid — Liberation of iodine.
5. Sodium nitrite — Liberation of iodine.
6. Copper sulfate — Brown precipitate.
7. Mercury(II) chloride — Scarlet precipitate of mercury(II) iodide.

Special tests: 1. $MnO_2 + H_2SO_4$ produces I_2.
2. Cl_2 (aq)/CS_2 produces I_2 in CS_2 (purple).
3. Starch paste+Cl_2 (aq), deep blue coloration.

(*Continued*)

Tests for Anions (*Continued*)

Lactates, $CH_3CH(OH)COO^-$

1. Potassium permanganate solution	The odor of acetaldehyde is observed upon the addition of dilute potassium permanganate solution, followed by acidification with dilute sulfuric acid and heating.

Metaphosphates, PO_3^-

1. Silver nitrate	White precipitate, soluble in dilute nitric acid, in dilute ammonia solution, and in dilute acetic acid.
2. Albumin and dilute acetic acid.	Coagulation.
3. Zinc sulfate solution	White precipitate on warming; soluble in dilute acetic acid.

Nitrates, NO_3^-

1. Concentrated sulfuric acid	Solid nitrate with concentrated sulfuric acid evolves reddish-brown vapors of nitrogen dioxide+nitric acid vapors when heated.

Special tests: 1. Add iron(II) sulfate, shake, then add concentrated sulfuric acid; produces brown ring.
2. White precipitate is formed upon addition of nitron reagent ($C_{20}H_{16}N_4$); test is not specific to only nitrates, however; see table of Precipitation Reagents.

Nitrites, NO_2^-

1. Dilute hydrochloric acid	Cautious addition of acid to a solid nitrite in cold gives a transient pale (of nitrous acid or the anhydride) blue liquid and consequent evolution of brown fumes of nitrogen dioxide.
2. Silver nitrate	White precipitate of silver nitrite.
3. Iron(II) sulfate solution (25 %, acidified with either acetic or sulfuric acid)	A brown ring forms at the junction of the two liquids due to the formation of a complex.
4. Acidified potassium permanganate	Decolorization with no gas evolution.
5. Ammonium chloride (solid)	Boiling with excess of solid reagent causes nitrogen to be evolved.
6. Concentrated sulfuric acid	Liberates brown nitrogen dioxide gas.

Special tests: Acidified solutions of nitrites liberate iodine from potassium iodide.

Orthophosphates, PO_4^{3-}

1. Silver nitrate	Yellow precipitate of silver orthophosphate, soluble in dilute ammonia and in dilute nitric acid.
2. Barium chloride	White precipitate of barium hydrogen phosphate, soluble in dilute mineral acids and acetic acid.
3. Magnesium nitrate reagent or magnesia mixture	White crystalline precipitate of magnesium ammonium phosphate, soluble in acetic acid and mineral acids, practically insoluble in 2.5 % ammonia solution.
4. Ammonium molybdate	Addition of 2–3 mL excess reagent to approximately 0.5 mL sample gives yellow precipitate of ammonium phosphomolybdate which is soluble in ammonia solution and in solutions of caustic alkalis. Large quantities of hydrochloric acid interfere.
5. Iron(III) chloride	Yellowish-white precipitate of iron(III) phosphate, soluble in mineral acids, insoluble in dilute acidic acid.
6. Ammonium molybdate quinine	Yellow precipitate of unknown composition, reducing agents interfere.

Note: The orthophosphates are salts of orthophosphoric acid, H_3PO_4, and are simply referred to as phosphates.

(*Continued*)

Tests for Anions (*Continued*)

Oxalates, (COO)$_2^{2-}$

1. Silver nitrate

White precipitate of silver oxalate that is soluble in ammonia solution and dilute nitric acid.

2. Calcium chloride

White precipitate of calcium oxalate that is insoluble in dilute acetic acid, oxalic acid, and in ammonium oxalate solution; soluble in dilute hydrochloric acid and in dilute nitric acid.

3. Potassium permanganate

Decolorization upon warming to 60°C –70°C, in acidified solution.

Perchlorates, ClO$_4^-$

1. Potassium chloride

White precipitate of potassium perchlorate, insoluble in alcohol.

Special tests: 1. Neutral ClO$_4^-$ + cadmium sulfate in concentrated ammonia produces [Cd(NH$_3$)$_4$](ClO$_4$)$_2$ (white precipitate).
   2. Cautious heating of solids evolves oxygen.

Peroxydisulfates, S$_2$O$_8^{2-}$

1. Water

On boiling, decomposes into the sulfate, free sulfuric acid, and oxygen.

2. Silver nitrate

Black precipitate of silver peroxide.

3. Barium chloride

On boiling or standing for some time, a precipitate of barium sulfate is formed.

4. Manganese(II) sulfate

Brown precipitate of hydrate complex in neutral or alkaline test solution.

Phosphites, HPO$_3^{2-}$

1. Silver nitrate

White precipitate of silver phosphite, which yields black metallic silver on standing.

2. Barium chloride

White precipitate of barium phosphite, soluble in dilute acids.

3. Mercury(II) chloride

White precipitate in cold solutions, which yields gray metallic mercury on warming.

4. Copper sulfate

Light-blue precipitate which dissolves in hot acetic acid.

5. Lead(II) acetate

White precipitate of lead(II) hydrogen phosphite.

Pyrophosphates, P$_2$O$_7^{4-}$

1. Silver nitrate

White precipitate, soluble in dilute nitric acid and in dilute acetic acid.

2. Copper sulfate

Pale-blue precipitate.

3. Magnesia mixture or magnesium reagent

White precipitate, soluble in excess reagent but reprecipitated on boiling.

4. Cadmium acetate and dilute acetic acid

White precipitate.

5. Zinc sulfate

White precipitate, insoluble in dilute acetic acid; soluble in dilute ammonia solution yielding a white precipitate on boiling.

Salicylates, C$_6$H$_4$(OH)COO$^-$ (or C$_7$H$_5$O$_3^-$)

1. Concentrated sulfuric acid

Evolution of carbon monoxide and sulfur dioxide (poisonous).

2. Concentrated sulfuric acid and methanol

0.5 g sample + 3 mL reagent + heat evolves methyl salicylate (odor of wintergreen).

3. Dilute hydrochloric acid

Crystalline precipitate of salicylic acid.

4. Silver nitrate

Heavy crystalline precipitate of silver salicylate (which is soluble in boiling water and recrystallizes upon cooling).

5. Iron(III) chloride

Violet-red coloration which clears upon the addition of dilute mineral acids.

(*Continued*)

Tests for Anions (*Continued*)

Silicates, SiO_3^{2-}

1. Dilute hydrochloric acid	Gelatinous precipitate of metasilicic acid, insoluble in concentrated acids, soluble in water and dilute acids.
2. Ammonium chloride or ammonium carbonate	Gelatinous precipitate.
3. Silver nitrate	Yellow precipitate of silver silicate, soluble in dilute acids as well as ammonia solution.
4. Barium chloride	White precipitate of barium silicate which is soluble in dilute nitric acid.

Succinates, $C_4H_4O_4^{2-}$

1. Silver nitrate	White precipitate of silver succinate, soluble in dilute ammonia solution.
2. Iron(III) chloride	Light-brown precipitate of iron(III) succinate.
3. Barium chloride	White precipitate of barium succinate.
4. Calcium chloride	Slow precipitation of calcium succinate.

Sulfates, SO_4^{2-}

1. Barium chloride	White precipitate of barium sulfate, insoluble in warm dilute hydrochloric acid and in dilute nitric acid, slightly soluble in boiling hydrochloric acid.
2. Lead acetate	White precipitate of lead sulfate; soluble in hot concentrated sulfuric acid, ammonium acetate, ammonium tartrate, and sodium hydroxide.
3. Silver nitrate	White precipitate of silver sulfate.
4. Mercury(II) nitrate	Yellow precipitate of mercury(II) sulfate.

Sulfides, S^{2-}

1. Dilute hydrochloric acid or sulfuric acid	Hydrogen sulfide gas is evolved and detected by odor or lead acetate paper.
2. Silver nitrate	Black precipitate of silver sulfide, soluble in hot, dilute nitric acid.
3. Lead acetate	Black precipitate of lead sulfide.
4. Sodium nitroprusside solution ($Na_2[Fe(CN)_5NO]$)	Transient purple color in the presence of solutions of alkalis.

Special tests: Catalysis of iodine–azide reaction. Solution of sodium azide, (NaN_3) and iodine reacts with a trace of a sulfide to evolve nitrogen. Thiosulfates and thiocyanates act similarly and therefore must be absent.

Sulfites, SO_3^{2-}

1. Dilute hydrochloric acid	Decomposition (which becomes more rapid on warming) and evolution of sulfuric dioxide (odor of burning sulfur).
2. Barium chloride strontium chloride	White precipitate of the respective sulfite, the precipitate being soluble in dilute hydrochloric acid.
3. Silver nitrate	At first no change; upon addition of more reagent, white crystalline precipitate at silver sulfite forms, which darkens to metallic silver upon heating.
4. Potassium permanganate solution acidified with dilute sulfuric acid	Decolorization (Fuchsin test).
5. Potassium dichromate in with dilute sulfuric acid	Green color formation.
6. Lead acetate or lead nitrate solution	White precipitate of lead sulfite.
7. Zinc and sulfuric acid	Hydrogen sulfide gas evolved, detected by holding lead acetate paper to mouth of test tube.
8. Concentrated sulfuric acid	Evolution of sulfur dioxide gas.
9. Sodium nitroprusside–zinc sulfate	Red compound of unknown composition.

(*Continued*)

Tests for Anions (Continued)

Tartrates, $C_4H_4O_6^{2-}$

1. Concentrated sulfuric acid	When sample is heated, the evolution of carbon monoxide, carbon dioxide, and sulfur dioxide (burned sugar odor) results.
2. Silver nitrate	White precipitate of silver tartrate.
3. Calcium chloride	White precipitate of calcium tartrate, soluble in dilute acetic acid, dilute mineral acids, and in cold alkali solutions.
4. Potassium chloride	White precipitate, the reaction is: $C_4H_4O_6^{2-} + K^+ + CH_3COOH \rightarrow C_4H_5O_6 K\downarrow + CH_3COO^-$

Special test: One drop 25 % iron(II) sulfate, two to three drops hydrogen peroxide: produces deep violet-blue color (Fenton's test).

Thiocyanates, SCN⁻

1. Sulfuric acid	In cold solution, yellow coloration is produced; upon warming, violent reaction occurs and carbonyl sulfide is released. In basic solution, carbonyl sulfide is hydrolyzed to hydrogen sulfide.
2. Silver nitrate	White precipitate of silver thiocyanate.
3. Copper sulfate	First a green coloration, then black precipitate of copper(II) thiocyanate is formed.
4. Mercury(II) nitrate	White precipitate of mercury(II) thiocyanate.
5. Iron(III) chloride	Blood-red coloration due to complex formation.
6. Dilute nitric acid	Upon warming, red coloration is observed, with nitrogen oxide and hydrogen cyanide (poisonous) being evolved.
7. Cobalt nitrate	Blue coloration due to complex ion formation.

Thiosulfates, $S_2O_3^{2-}$

1. Iodine solution	Decolorized; a colorless solution of tetrathionate ions is formed.
2. Barium chloride	White precipitate of barium thiosulfate.
3. Silver nitrate	White precipitate of silver thiosulfate.
4. Lead(II) acetate or nitrate solution	First no change; on further addition of reagent, a white precipitate of lead thiosulfate forms.
5. Iron(III) chloride solution	Dark-violet coloration due to complex formation.
6. Nickel ethylenediamine nitrate $[Ni(NH_2(CH_2)_2NH_2)_3]$ $(NO_3)_2$	Violet complex precipitate forms; hydrogen sulfide and ammonium sulfide interfere.

Special tests: Blue Ring Test: When solution of thiosulfate mixed with ammonium molybdate solution is poured slowly down the side of a test tube which contains concentrated sulfuric acid, a blue ring is formed temporarily at the contact zone.

TESTS FOR CATIONS

This table provides a summary of the common tests for cations, primarily in aqueous solution. The cations are grouped according to the usual convention of reactivity to a set of common reagents.

Abbreviations
>   conc—concentrated
>   dil—dilute
>   g—gaseous
>   EtOH—ethanol.

Group I: Pb, Ag, Hg(I)

All members are precipitated by dilute HCl to give lead chloride ($PbCl_2$), silver chloride (AgCl), or mercury(I) chloride (Hg_2Cl_2).

Lead(II), Pb^{+2}

1. Potassium chromate	Yellow precipitate of lead(II) chromate.
2. Potassium iodide	Yellow precipitate of lead(II) iodide.
3. Sulfuric acid, dilute	White precipitate of lead(II) sulfate.
4. Hydrogen sulfide gas	Black precipitate of lead(II) sulfide.
5. Potassium cyanide	White precipitate of lead(II) cyanide.
6. Tetramethyldiaminodiphenyl methane	Blue oxidation product (presence of Bi, Ce, Mn, Th, Co, Ni, Fe, and Cu may interfere).
7. Gallocyanine	Deep violet precipitate, unknown composition (Bi, Cd, Cu, and Ag may interfere).
8. Diphenylthiocarbazone	Brick-red complex in neutral or ammoniacal solution.

Silver(I), Ag^+

1. Potassium chromate	Reddish brown precipitate of silver chromate.
2. Potassium iodide	Yellow precipitate of silver iodide.
3. Hydrogen sulfide gas	Black precipitate of silver sulfide.
4. Disodium hydrogen phosphate	Yellow precipitate of silver phosphate.
5. Sodium carbonate	Yellow-white precipitate of silver carbonate, forming the brown oxide upon heating.
6. p-Dimethylaminobenzylidene-rhodanine	Reddish-violet precipitate in acidic solution.
7. Ammonia solution	Brown precipitate of silver oxide, dissolving in excess to form Ag_3N, which is explosive.

Mercury(I), Hg_2^{+2}

1. Potassium carbonate	Red precipitate of mercury(I) chromate.
2. Potassium iodide	Green precipitate of mercury(I) iodide.
3. Dilute sulfuric acid	White precipitate of mercury(I) sulfate.
4. Elemental copper, aluminum, or zinc	Amalgamation occurs.
5. Hydrogen sulfide	Black precipitate (in neutral or acid medium) of mercury(I) sulfide and mercury.
6. Ammonia solution	Black precipitate of $HgO \cdot Hg(NH_2)(NO_3)$.
7. Diphenylcarbizide (1 % in ethanol, with 0.2 M nitric acid	Violet-colored complex results (high sensitivity and selectivity).
8. Potassium cyanide	Mercury(I) cyanide solution, with a precipitation of elemental mercury (mercury(II) interferes).

Group II: Hg(II), Cu, Bi, Cd, As(III), As(V), Sb(III), Sb(V), Sn(II), Sn(IV)

All members show no reaction with HCl; all form a precipitate with H_2S.

Bismuth(III), Bi^{3+}

1. Potassium iodide	Black precipitate of bismuth(III) iodide.
2. Potassium chromate	Yellow precipitate of bismuth(III) chromate.
3. Ammonia solution	White precipitate of variable composition, approximate formula: $Bi(OH)_2NO_3$.
4. Pyrogallol (10 %)	Yellow precipitate of bismuth(III) pyrogallate.
5. 8-Hydroxyquinoline (5 %) + potassium iodide (6 M)	Red precipitate of the tetraiodobismuthate salt (characteristic in the absence of Cl⁻, F⁻, and Br⁻).
6. Sodium hydroxide	White precipitate of bismuth(III) hydroxide.

Copper(II), Cu^{+2}

1. Potassium iodide	Brown precipitate of copper(I) iodide, colored brown due to I_3^-.
2. Potassium cyanide	Yellow precipitate of copper(II) cyanide; which then decomposes.
3. Potassium thiocyanate	Black precipitate of copper(II) thiocyanate, which then decomposes.
4. α-Benzoin oxime (or cupfon), 5 % in EtOH	Green precipitate of the α-benzoin oxime salt derivative.
5. Salicylaldoxime (1 %)	Greenish-yellow precipitate of the copper complex.
6. Rubeanic acid (0.5 %) (dithiooxamide)	Black precipitate of the rubeanate salt.

Cadmium(II), Cd^{2+}

1. Ammonia solution	White precipitate of cadmium hydroxide which dissolves in excess ammonia.
2. Potassium cyanide	White precipitate of cadmium cyanide which dissolves in excess potassium cyanide.
3. Sodium hydroxide	White precipitate of cadmium hydroxide which is insoluble in excess sodium hydroxide.
4. Dinitro-p-diphenyl carbizide	Brown precipitate with cadmium hydroxide.

Arsenic(III), As^{+3}

1. Silver nitrate	Yellow precipitate of silver arsenite in neutral solution.
2. Copper(II) sulfate	Green precipitate of copper(II) arsenite (or $Cu_3(AsO_3)_2 \cdot xH_2O$).
3. Potassium triiodide ($KI + I_2$)	Decolorization due to oxidation.
4. Tin(II) chloride + concentrated hydrochloric acid	Black precipitate forms in the presence of excess reagent.

Arsenic(V), As^{+5}

1. Silver nitrate	Brownish-red precipitate of silver arsenate from neutral solutions.
2. Ammonium molybdate	Yellow precipitate (in presence of excess reagent) of ammonium arsenomolybdate $(NH_4)_3AsMo_{12}O_4$.
3. Potassium iodide + concentrated hydrochloric acid	Iodine formation.

Small amounts of As(III) or As(V) can be identified by the response to the Marsh, Gutzeit, or Fleitmann tests (see references at the beginning of this section).

Antimony(III), Sb^{+3}

1. Sodium hydroxide	White precipitate of the hydrated oxide $Sb_2O_3 \cdot xH_2O$.
2. Elemental zinc or tin	Black precipitate of antimony.
3. Potassium iodide	Yellow color of $[SbI_6]^{3-}$ ion.
4. Phosphomolybdic acid, $H_3[PMo_{12}O_{40}]$	Blue color produced; Sn(II) interferes; 0.2 µg sensitivity.

(Continued)

Group II: Hg(II), Cu, Bi, Cd, As(III), As(V), Sb(III), Sb(V), Sn(II), Sn(IV) (Continued)

Antimony(V), Sb⁺⁵

1. Water	White precipitate of basic salts and ultimately antimonic acid, H_3SbO_4.
2. Potassium iodide	Formation of iodine as a floating precipitate.
3. Elemental zinc or tin	Black precipitate of antimony (in the presence of hydrochloric acid).

Small amounts of antimony can be identified using Marsh's test and/or Gutzeit's test (see references at the beginning of this section for details).

Tin(II), Sn⁺²

1. Mercury(II) chloride	White precipitate of mercury(I) chloride (in an excess of tin ions, precipitate turns gray).
2. Bismuth nitrate	Black precipitate of bismuth metal.
3. Cacotheline (nitro derivative of brucine, $C_{21}H_{21}O_7N_3$)	Violet coloration with stannous salts. The following interfere: strong reducing agents (hydrogen sulfide, dithionites, sulfites, and selenites); also, U, V, Te, Hg, Bi, Au, Pd, Se, and Sb.
4. Diazine green (dyestuff formed by coupling diazotized safranine with N,N-dimethylaniline)	Color change blue→violet→red.

Tin(IV), Sn⁺⁴

1. Iron powder	Reduces Sn(IV) to Sn(II).

Group III: Fe(II) and (III), Al³⁺, Cr(III) and (VI), Ni²⁺, Co²⁺, Mn(II) and (VII), and Zn²⁺

All members are precipitated by H_2S in the presence of ammonia and ammonium chloride, or ammonium sulfide solutions.

Iron(II), Fe⁺²

1. Ammonia solution	Precipitation of iron(II) hydroxide. If large amounts of ammonium ion are present, precipitation does not occur.
2. Ammonium sulfide	Black precipitate of iron(II) sulfide.
3. Potassium cyanide (POISON)	Yellowish-brown precipitate of iron(II) cyanide, soluble in excess reagent, forming the hexacyanoferrate(II) ion.
4. Potassium hexacyanoferrate(II) solution	In complete absence of air, white precipitate of potassium iron(II) hexacyanoferrate. If air is present, a pale blue precipitate is formed.
5. Potassium hexacyanoferrate(III) solution	Dark-blue precipitate, called Turnbull's blue.
6. α,α′-Dipyridyl	Deep red bivalent cation $[Fe(C_5H_4N)_2]^{2+}$ formed with iron(II) salts in mineral acid solution; sensitivity: 0.3 μg.
7. Dimethylglyoxime (DMG)	Red, iron(II) dimethylglyoxime; nickel, cobalt, and large quantities of copper salts interfere; sensitivity: 0.04 μg.
8. o-Phenanthroline (0.1 wt% in water)	Red coloration due to the complex cation $[Fe(C_{12}H_8N_2)_3]^{2+}$, in slightly acidic conditions.

Iron(III), Fe⁺³

1. Ammonia solution	Reddish-brown gelatinous precipitate of iron(III) hydroxide.
2. Ammonium sulfide	Black precipitate mixture of iron(II) sulfide and sulfur.
3. Potassium cyanide	When added slowly, reddish-brown precipitate of iron(III) cyanide is formed, which dissolves in excess potassium cyanide to yield a yellow solution.
4. Potassium hexacyanoferrate(III)	A brown coloration is produced due to the formation of iron(III) hexacyanoferrate(III).

(Continued)

Group III: Fe(II) and (III), Al³⁺, Cr(III) and (VI), Ni²⁺, Co²⁺, Mn(II) and (VII), and Zn²⁺ *(Continued)*

5. Disodium hydrogen phosphate	Yellowish-white precipitate of iron(III) phosphate.
6. Sodium acetate solution	Reddish-brown coloration caused by complex formation.
7. Cupferron ($C_6H_5N(NO)ONH_4$) aqueous solution, freshly prepared	Reddish-brown precipitate formed in the presence of hydrochloric acid.
8. Ammonium thiocyanate + dilute acid	Deep red coloration of iron(III) thiocyanate complex.
9. 7-Iodo-8-hydroxyquinoline-5-sulfonic acid (ferron)	Green or greenish-blue coloration in slightly acidic solutions; sensitivity: 0.5 μg.

Cobalt(II), Co⁺²

1. Ammonia solution	In the absence of ammonium salts, small amounts of $Co(OH)NO_3$ precipitate which is soluble in excess aqueous ammonia.
2. Ammonium sulfide	Black precipitate of cobalt(II) sulfide, from neutral or alkaline solutions.
3. Potassium cyanide (POISON)	Reddish-brown precipitate of cobalt(II) cyanide, which dissolves in excess.
4. Potassium nitrite	Yellow precipitate of potassium hexacyanocobaltate(III), $K_3[Co(NO_2)_6]$.
5. Ammonium thiocyanate (crystals)	Gives blue coloration when added to neutral or acid solution of cobalt, due to a complex formation (Vogel's reaction); sensitivity: 0.5 μg.
6. α-Nitroso-β-naphthol (1 % in 50 % acetic acid)	Red-brown (chelate) precipitate, extractable using carbon tetrachloride; sensitivity: 0.05 μg.

Nickel(II), Ni⁺²

1. Ammonia solution	Green precipitate of nickel(II) hydroxide which dissolves in excess ammonia.
2. Potassium cyanide (poison)	Green precipitate of nickel(II) cyanide which dissolves in excess potassium cyanide.
3. Dimethylglyoxime, DMG ($C_4H_8O_2N_2$)	Red precipitate of nickel–DMG chelate complex in ammoniacal solution; sensitivity 0.16 μg.

Manganese(II), Mn⁺²

1. Ammonia solution	Partial precipitation of white manganese(II) hydroxide.
2. Ammonium sulfide	Pink precipitate of manganese(II) sulfide, which is soluble in mineral acids.
3. Sodium phosphate (in the presence of ammonia or ammonium ions).	Pink precipitate of manganese ammonium phosphate, $Mn(NH_4)PO_4 \cdot 7H_2O$, which is soluble in acids.

Aluminum(III), Al³⁺

1. Ammonia	White gelatinous precipitate of aluminum hydroxide.
2. Sodium hydroxide	White gelatinous precipitate of aluminum hydroxide which is soluble in excess sodium hydroxide.
3. Ammonium sulfide	White precipitate of aluminum sulfide.
4. Sodium acetate	Upon boiling with excess reagent, a precipitate of basic aluminum acetate, $Al(OH)_2CH_3COO$, is formed.
5. Sodium phosphate	White gelatinous precipitate of aluminum phosphate.
6. "Aluminon" (a solution of the ammonium salt of aurine tricarboxylic acid)	Bright-red solution.
7. Quinalizarin, alizarin-S, and alizarin	Red precipitate or "lake."

(Continued)

Group III: Fe(II) and (III), Al³⁺, Cr(III) and (VI), Ni²⁺, Co²⁺, Mn(II) and (VII), and Zn²⁺ (*Continued*)

Chromium(III), Cr⁺³

1. Ammonia solution	Gray-green to gray-blue gelatinous precipitate of chromium(III) hydroxide.
2. Sodium carbonate	Precipitate of chromium(III) hydroxide.

Zinc(II), Zn²⁺

1. Ammonia solution	White precipitate of zinc hydroxide which is soluble in excess ammonia
2. Disodium hydrogen phosphate	White precipitate of zinc phosphate which is soluble in dilute acids.
3. Potassium hexacyanoferrate(II)	White precipitate of variable composition, which is soluble in sodium hydroxide.
4. Ammonium tetrathiocyanato-mercurate(II)–copper sulfate, slightly acidic	Solution is treated with five drops of 0.25 M copper(II) sulfate solution followed by 2 mL ammonium tetrathiocyanato-mercurate to give a violet precipitate.

Group IV: Ba²⁺, Sr²⁺, and Ca²⁺

All members of this group react with ammonium carbonate.

Barium(II), Ba⁺²

1. Ammonium carbonate	White precipitate of barium carbonate which is soluble in dilute acids.
2. Ammonium oxalate	White precipitate of barium oxalate which is soluble in dilute acids.
3. Dilute sulfuric acid	Heavy, white, finely divided precipitate of barium sulfate.
4. Saturated calcium sulfate (or strontium sulfate)	White precipitate of barium sulfate.
5. Sodium rhodizonate	Red-brown precipitate; sensitivity: 0.25 µg.

Strontium(II), Sr⁺²

1. Ammonium carbonate	White precipitate of strontium carbonate.
2. Dilute sulfuric acid	White precipitate of strontium sulfate.
3. Saturated calcium sulfate	White precipitate of strontium sulfate.
4. Potassium chromate	Yellow precipitate of strontium chromate.
5. Ammonium oxalate	White precipitate of strontium oxalate that is soluble in mineral acids.

Calcium(II), Ca⁺²

1. Ammonium carbonate	White precipitate of calcium carbonate.
2. Dilute sulfuric acid	White precipitate of calcium sulfate.
3. Ammonium oxalate	White precipitate of calcium oxalate that is soluble in mineral acids.
4. Potassium chromate	Yellow precipitate of strontium chromate that is soluble in mineral acids.
5. Sodium rhodizonate	Red-brown precipitate; sensitivity: 4 µg.

Group V: Mg, Na, K, and NH₄⁺

No common reaction or reagent.

Magnesium(II), Mg²⁺

1. Ammonia solution, sodium hydroxide	Partial precipitation of white magnesium hydroxide.
2. Ammonium carbonate	White precipitate of magnesium carbonate, only in the absence of ammonia salts.
3. Oxine + ammoniacal ammonium chloride solution	Yellow precipitate of $Mg(C_9H_6NO)_2 \cdot 4H_2O$.
4. Quinalizarin	Blue precipitate, or blue-colored solution, which can be cleared by a few drops of bromine water.

Sodium(I), Na⁺

1. Uranyl magnesium acetate solution (in 30 v/v ETOH)	Yellow precipitate of sodium magnesium uranyl acetate.
2. Uranyl zinc acetate solution	Yellow precipitate of sodium zinc uranyl acetate; sensitivity: 12.5 µg Na.

Potassium(I), K⁺

1. Sodium hexanitrocobaltate(III) ($Na_3[Co(NO_2)_6]$)	Yellow precipitate of potassium hexanitrocobaltate(III); insoluble in acetic acid.
2. Tartaric acid solution (sodium acetate buffered)	White precipitate of potassium hydrogen tartrate.
3. Perchloric acid	White precipitate of potassium perchlorate.
Note: Perchloric acid is a powerful oxidizing agent that must be handled carefully. See the Materials Compatibility table in Chapter 3 and the incompatibilities table in Chapter 15.	
4. Dipicrylamine	Orange-red complex precipitate (NH₄⁺ interferes); sensitivity: 3 µg K.
5. Sodium tetraphenylboron + acetic acid	White precipitate of potassium tetraphenylboron.

Ammonium, NH₄⁺

1. Sodium hydroxide	Evolution of ammonia gas upon heating.
2. Potassium tetraiodomercurate (Nessler's reagent)	Brown-yellow color or brown precipitate of mercury(II) amidoiodide; high sensitivity; all other metals (except Na and K) interfere.
3. Tannic acid–silver nitrate	Precipitate of black elemental silver, from neutral solution; very sensitive.
4. p-Nitrobenzene-diazonium chloride	Red-colored solution results in the presence of sodium hydroxide; sensitivity: 0.7 µg NH₄⁺.

Note: Ammonium ions will cause a similar reaction to that of potassium in the presence of: sodium hexanitrocobaltate(III) sodium hydrogen tartrate.

ORGANIC PRECIPITATION REAGENTS FOR INORGANIC IONS

The following table lists the most important organic reagents used for precipitating various inorganic species from solution [1,2]. Many of these reagents are subject to the serious disadvantage caused by lack of selectivity. Thus, many of the listed reagents will precipitate more than one species. The selectivity of some of the reagents can be controlled to a certain extent by adjustment of pH, reagent concentrations, and the use of masking reagents. The first two factors, pH and concentration, are the most critical. A number of these reagents form rather large, bulky complexes. While this can serve to enhance sensitivity (especially for gravimetric procedures), it can also impose rather stringent concentration limits. The reader is referred to several excellent "recipe" texts for further guidance [3–10].

REFERENCES

1. Kennedy, J.H., *Analytical Chemistry*, 2nd ed., Saunders College Publishing, New York, 1990.
2. Christian, G.D., *Analytical Chemistry*, 6th ed., John Wiley and Sons, New York, 2003.
3. Barber, H.H. and Taylor, T.I., *Semimicro Qualitative Analysis*, Harper and Brothers, New York, 1953.
4. Greenfield, S. and Clift, M., *Analytical Chemistry of the Condensed Phosphates*, Pergamon Press, Oxford, 1975.
5. Ryabchikov, D.I. and Gol'Braikh, E.K., *The Analytical Chemistry of Thorium*, The MacMillan Company, New York, 1963.
6. Jungreis, E., *Spot Test Analysis*, Wiley Interscience, New York, 1985.
7. Jungreis, E., *Spot Test Analysis: Clinical, Environmental, Forensic and Geochemical Applications*, 2nd ed., John Wiley and Sons, New York, 1997.
8. Svehla, G. and Suehla, G., *Vogel's Qualitative Inorganic Analysis*, Addison-Wesley, New York, 1996.
9. Skoog, D.A., West, D.M. and Holler, F.J., *Fundamentals of Analytical Chemistry,* 7th ed., Saunders College Publishing, Philadelphia, PA, 1996.
10. Harris, D.C., *Quantitative Chemical Analysis*, 5th ed., Freeman, New York, 1997.

ORGANIC PRECIPITATION REAGENTS

Reagent	Structure/Formula	Applications and Notes
Alizarin-S (sodium alizarin sulfonate)	*(structure of alizarin-S anthraquinone with OH, OH, SO_3Na groups)*	Will precipitate Al in ammoniacal solution; high sensitivity.
Ammonium nitroso-phenylhydroxylamine (cupferron)	$C_6H_5\text{-}N(N=O)O^-$ NH_4^+ *(structure with N=O, N, $O^-NH_4^+$)*	Will precipitate Fe(III), V(V), Ti(IV), Zr(IV), Sn(IV), and U(IV) in the presence of moderate acidity; will also precipitate rare earths.
Anthranilic acid	$o\text{-}H_2N\text{-}C_6H_4COOH$ *(structure with NH_2 and COOH)*	Will precipitate Cu(II), Cd(II), Ni(II), Co(II), Pb(II), and Zn(II) in acetic acid or nearly neutral solution.
α-Benzoin oxime (cupron)	*(structure with OH, NOH, C, C, H and two phenyl rings)*	Will precipitate Cu(II) in the presence of NH_3 or tartarate; will precipitate Mo(VI), W(VI) in acidic medium.
Dimethylglyoxime (DMG)	$[CH_3C(=NOH)]_2$ $CH_3-C=N-OH$ $CH_3-C=N-OH$	Will precipitate Ni(II) in the presence of NH_3 or buffered acetate; Pd(II) in HCl solution; the addition of tartaric acid to the reagent will mask Fe(III) and Cr(III) interferences; Pd(II) and Bi(III) will also precipitate.
Dimethyl oxalate	$(CH_3)_2C_2O_4$ $H_3C-O-C-C-O-CH_3$ *(with two C=O)*	Will precipitate Ac, Am, Ca, and Th ions as well as rare earth metals.
Dimethyl sulfate	$(CH_3O)_2SO_2$ $CH_3O-S-OCH_3$ *(with two S=O)*	Will precipitate Ba, Ca, Pb(II), and Sr ions; particular care must be taken while using this reagent since it is a powerful methylating agent.

(Continued)

Reagent	Structure/Formula	Applications and Notes
Dipicrylamine (hexanitro-diphenylamine)		Will precipitate K^+ (as well as NH_4, Na, and Li ions but with much less sensitivity).
Dithio-oxamide (rubeanic acid)	$[H_2N-C(=S)]_2$ 	Will precipitate Cu(II) (black), Ni (blue), and Co (brown); high sensitivity.
Gallocyanine		Will precipitate lead (deep violet precipitate of uncertain composition); high sensitivity.
8-Hydroxyquinoline (oxine)		Will precipitate Al(III) at pH 4–5 and Mg(II) in the presence of NH_3; will precipitate Be, Bi, Cd, Cu, Ga(I), Hf, Fe, In, Mg, Hg(II), Nb, Pd, Sc, Ta, Ti, Th, U, Zn, Zr, and W ions at pH 4–5 or in the presence of NH_3. Use of pH control provides a measure of selectivity.
Nitron reagent (in acetic acid)		Will precipitate nitrate, bromide, iodide, nitrite, chromate chlorate, perchlorate, thiocyanate, oxalate, and picrate anions.
α-Nitroso-β-naphthol		Will precipitate Co(II), Ni^{2+}, Fe(III), and Pd(II) in a weakly acidic solution.
Oxalic acid	HOOC-COOH 	Will precipitate Ca^{+2}; high concentrations of Mg will interfere.
p-Dimethylamino-benzylidenerhodanine (0.3 % in acetone)		Will precipitate Ag(I), Hg(II), Au(II), Pt(II), and Pd(II) under slightly acidic conditions.

(Continued)

Reagent	Structure/Formula	Applications and Notes
Picrolonic acid	O_2N—CH—C—CH_3 | | CO N \\ / N | (4-NO_2-phenyl)	Will precipitate Ca^{+2}; subject to interference from many other cations, however.
Pyrogallol	1,2,3-$C_6H_3(OH)_3$	Will precipitate Bi(III) and Sb(III); high sensitivity; note that this reagent is also used as an antioxidant for vegetable oils and derived products such as biodiesel fuel.
Quinalizarin	[anthraquinone with OH groups]	Will precipitate Al^{+3}.
Salicylaldoxime (1 % in acetic acid)	1,2-$C_6H_4(OH)CH{=}NOH$	Will precipitate Cu(II), with interference by Pd(II) and Au(II).
Sodium diethylthiocarbamate	$(C_2H_5)N{-}C({=}S)S^-Na^+$	Useful for the precipitation of many metals.
Sodium dihydroxy-tartrate osazone	[C_6H_5—NH—N=C—COONa ; C_6H_5—NH—N=C—COONa]	Will precipitate Ca^{+2}; subject to interference from many other cations, however.
Sodium rhodizonate	[CO—CO—C—ONa ; CO—CO—C—ONa]	Will precipitate Ba^{+2} and Sr^{+2}; subject to interference by all H_2S reactive cations.
Sodium tetraphenylboron	$(C_6H_5)_4B^-Na^+$	Used to precipitate K, Rb, Cs, Tl, Ag, Hg(I), Cu(I), NH_4^+, RNH_3^+, $R_2NH_2^+$, R_3NH^+, and R_4N^+, in cold acidic solution; selectivity is high for K^+ and NH_4^+.

(Continued)

Reagent	Structure/Formula	Applications and Notes
Tetraphenyl arsonium chloride	$(C_6H_5)_4AsCl$	Will precipitate $Cr_2O_7^{2-}$, MnO_4^-, ReO_4^-, MoO_4^{2-}, WO_4^{2-}, ClO_4^-, and I_3^- in acidic solution.

Thioacetamide	$CH_3C(=S)NH_2$	Used to provide a source of H_2S for the precipitation of As, Bi, Cd, Cu, Hg, Mn, Mo, Pb, Sb, and Sn with heating in acid medium.

Urea	$(C_2H_5O)_3PO$	Will precipitate Al, Fe(III), Ga, Sn, Th, and Zn ions.

Triethyl phosphate		Will precipitate Hf and Zr ions.

Trichloroacetic acid		Will precipitate Ba, Ra, and the rare earths.

CHAPTER 14

Solution Properties

We note that guidance and data on solutions and solvents specific to different analytical methods is provided throughout this book. This section deals with information that does not fit easily in the other sections.

PHYSICAL PROPERTIES OF LIQUID WATER

The table below provides data on the most important properties of pure water under different temperatures. These properties are density (g/mL), molar volume (mL/mol), vapor pressure (in kPa and mm Hg), static dielectric constant, and dynamic viscosity (mPa·s). The properties other than the vapor pressure are evaluated at a pressure of 101.325 kPa or the vapor pressure, whichever is higher.

The properties were computed by a software implementation [1] of standards adopted by the International Association for the Properties of Water and Steam (IAPWS) [2].

REFERENCES

1. Harvey, A.H., Peskin, A.P., and Klein, S.A., *NIST/ASME Steam Properties*, NIST Standard Reference Database 10, Version 3.0, National Institute of Standards and Technology, Gaithersburg, MD, 2013.
2. Documentation of IAPWS standards is available at www.iapws.org, accessed December 2019.

Temperature (°C)	Density (g/mL)	Molar Volume (mL/mol)	Vapor Pressure (mm Hg)	Vapor Pressure (kPa)	Dielectric Constant	Viscosity (mPa·s)
0	0.99984	18.0181	4.584	0.6112	87.903	1.792
5	0.99997	18.0159	6.545	0.8726	85.916	1.518
10	0.99970	18.0206	9.212	1.228	83.975	1.306
15	0.99910	18.0314	12.794	1.706	82.078	1.138
18	0.99860	18.0405	15.487	2.065	80.960	1.053
20	0.99821	18.0476	17.546	2.339	80.223	1.002
25	0.99705	18.0686	23.776	3.170	78.408	0.8901
30	0.99565	18.0940	31.855	4.247	76.634	0.7973
35	0.99403	18.1234	42.221	5.629	74.898	0.7193
40	0.99222	18.1566	55.391	7.385	73.201	0.6530
45	0.99021	18.1933	71.968	9.595	71.540	0.5961
50	0.98804	18.2334	92.646	12.352	69.916	0.5468
55	0.98569	18.2768	118.22	15.762	68.328	0.5040
60	0.98320	18.3232	149.61	19.946	66.774	0.4664
65	0.98055	18.3726	187.83	25.042	65.256	0.4333
70	0.97776	18.4250	234.02	31.201	63.770	0.4039
75	0.97484	18.4802	289.49	38.595	62.318	0.3777
80	0.97179	18.5382	355.63	47.414	60.898	0.3543
85	0.96861	18.5991	434.03	57.867	59.509	0.3333
90	0.96531	18.6627	526.40	70.182	58.152	0.3144
95	0.96189	18.7291	634.61	84.608	56.825	0.2973
100	0.95835	18.7982	760.69	101.42	55.527	0.2817

REFRACTIVE INDEX OF WATER

The following table provides the refractive index of water at various temperatures [1].

REFERENCE

1. Rumble, J.R., ed, *CRC Handbook for Chemistry and Physics*, 100th ed., CRC Taylor and Francis Press, Boca Raton, FL, 2019.

Temperature (°C)	Refractive Index (η, Na d-line)
15	1.33341
20	1.33299
30	1.33192
40	1.33051
50	1.32894
60	1.32718
70	1.32511
80	1.32287
90	1.32050
100	1.31783

APPROXIMATE pK$_a$ VALUES OF COMPOUNDS USEFUL IN BUFFER SYSTEMS

The following table provides the pK values of acids and bases needed to make the most popular buffers [1–4]. The approximate composition of buffers can be calculated from the equation:

$$pH = pK_a + \log\{[salt]/[acid]\}$$

Note that the quantities in square brackets denote concentrations, and the logarithmic quantity refers to the common (base 10) logarithm.

REFERENCES

1. Ramette, R.W., *Chemical Equilibrium and Analysis*, Addison-Wesley, Reading, MA, 1981.
2. Skoog, D.A., West, D.M., Holler, F.J., and Crouch, S.R., *Fundamentals of Analytical Chemistry*, 9th ed., Cengage Learning, Florence, KY, 2020.
3. Serjeant, E.P., Dempsey, B., eds., *Ionization Constants of Organic Acids in Solution*, IUPAC Chemical Data Series, No. 23, Pergamon Press, Oxford, 1979.
4. Harris, D.C., *Quantitative Chemical Analysis*, 10th ed., MacMillan Publishing, New York, 2020.

pKa	Compound	Formula
1.45	(K_1) oxalic acid	$HOOC–COOH$
1.9	(K_1) maleic acid	$HOOC–CH=CH–COOH$ (cis)
2.12	(K_1) phosphoric acid	H_3PO_4
2.35	(K_1) glycine	$H_2NCH_2CO_2H$
2.83	(K_1) malonic acid	$CH_2(COOH)_2$
2.95	(K_1) phthalic acid	$C_6H_4–1,2–(CO_2H)_2$
3.22	(K_1) citric acid	$HOC(COOH)(CH_2COOH)_2$
3.46	(K_1) isophthalic acid	$C_6H_4–1,3–(CO_2H)_2$
3.51	(K_1) terephthalic acid	$C_6H_4–1,4–(CO_2H)_2$
3.66	(K_1) β,β′-dimethyl glutamic acid	$[HO_2CCH_2–]_2C(CH_3)_2$
3.83	(K_1) glycolic acid	$HOCH_2COOH$
4.14	(K_2) oxalic acid	$HOOC–COOH$
4.21	(K_1) succinic acid	$HO_2CCH_2CH_2CO_2H$
4.46	(K_2) isophthalic acid	$C_6H_4–1,3–(CO_2H)_2$
4.76	(K_1) acetic acid	CH_3CO_2H
4.82	(K_2) terephthalic acid	$C_6H_4–1,4–(CO_2H)_2$
4.84	(K_2) citric acid	$HOC(COOH)(CH_2COOH)_2$
5.41	(K_2) phthalic acid	$C_6H_4–1,2–(CO_2H)_2$
5.64	(K_2) succinic acid	$HO_2CCH_2CH_2CO_2H$
5.67	(K_2) malonic acid	$CH_2(COOH)_2$
6.07	(K_2) maleic acid	$HOOC–CH=CH–COOH$ (cis)
6.4	(K_1) carbonic acid	H_2CO_3
6.15	(K_1) cacodylic acid	$(CH_3)_2As(O)OH$
6.2	(K_2) β,β′-dimethyl glutaric acid	$[HO_2CCH_2]_2C(CH_3)_2$
6.33	(K_2) maleic acid	$HO_2CCH=CHCO_2H$
6.39	(K_3) citric acid	$HOC(COOH)(CH_2COOH)_2$
7.21	(K_2) phosphoric acid	H_3PO_4
8.07	(K_1) tris-(hydroxymethyl)-aminomethane	$(HOCH_2)_3CNH_2$
8.67	(K_1) 2-amino-2-methyl-1,3-propanediol	$(HOCH_2)_2C(CH_3)NH_2$
9.23	(K_1) boric acid	H_3BO_3
9.78	(K_2) glycine	$H_2NCH_2CO_2H$
10.33	(K_2) carbonic acid	H_2CO_3
12.32	(K_3) phosphoric acid	H_3PO_4

PREPARATION OF BUFFERS

The following table gives the necessary information for preparing various buffers at different pHs. These buffers are suitable for use either in enzymatic or histochemical studies [1–4]. The uncertainty of the tables is within ±0.05 pH at 23 °C and the pH values do not change considerably even at 37 °C. The recommended mixture of the various solutions is given under the corresponding pH with the "final solution volume" indicated. This assumes addition of water to the necessary dilution. A list of stock solutions follows the buffer/pH table. The approximate composition of buffers can be calculated from the equation:

$$pH = pK + \log\left\{[salt]/[acid]\right\}$$

Note that the quantities in square brackets denote concentrations, and the logarithmic quantity refers to the common (base 10) logarithm.

REFERENCES

1. Colowick, S.P., and Kaplan, N.O., ed., *Methods in Enzymology*, vol. 1, Academic Press, New York, 1955.
2. Perrin, D.D., and Dempsey, B., *Buffers for pH and Metal Ion Control*, Chapman and Hall, London, 1974.
3. Sergeant, E.P., and Dempsey, B., ed., *Ionization Constants of Organic Acids in Solution*, IUPAC Chemical Data Series No. 23, Pergamon Press, Oxford, 1979.
4. Stoll, V.S. and Blanchard, J.S., Buffers: Principles and practice. *Meth. Enzmol.* **182**, 24–38 (1990).

Buffer	pH					Final Solution Volume
	1.0	1.1	1.2	1.3	1.4	
Hydrochloric acid/ potassium chloride	50.0 mL A + 97.0 mL B	50.0 mL A + 78.0 mL B	50.0 mL A + 64.5 mL B	50.0 mL A + 51.0 mL B	50.0 mL A + 41.5 mL B	200 mL

Buffer	pH					Final Solution Volume
	1.5	1.6	1.7	1.8	1.9	
Hydrochloric acid/ potassium chloride	50.0 mL A + 33.3 mL B	50.0 mL A + 26.3 mL B	50.0 mL A + 20.6 mL B	50.0 mL A + 16.6 mL B	50.0 mL A + 13.2 mL B	200 mL

Buffer	pH					Final Solution Volume
	2.0	2.1	2.2	2.3	2.4	
Hydrochloric acid/ potassium chloride	50.0 mL A + 10.6 mL B	50.0 mL A + 8.4 mL B	50.0 mL A + 6.7 mL B			200 mL
Glycine/hydrochloric acid			50.0 mL C + 44.0 mL B		50.0 mL C + 32.4 mL B	200 mL
Potassium phthalate/ hydrochloric acid			50.0 mL D + 46.7 mL B		50.0 mL D + 39.6 mL B	200 mL

Buffer	pH					Final Solution Volume
	2.5	2.6	2.7	2.8	2.9	
Glycine/hydrochloric acid		50.0 mL C + 24.2 mL B		50.0 mL C + 16.8 mL B		200 mL
Potassium phthalate/ hydrochloric acid		50.0 mL D + 33.0 mL B		50.0 mL D + 26.4 mL B		200 mL
Aconitate	20.0 mL E + 15.0 mL F	20.0 mL E + 18.0 mL F	20.0 mL E + 21.0 mL F	20.0 mL E + 24.6 mL F	20.0 mL E ± 28.0 mL F	200 mL
Citrate/phosphate		44.6 mL G + 5.4 mL K		42.2 mL C + 7.8 mL K		100 mL

Note: A list of the stock solutions follows the buffer table.

(Continued)

Buffer	pH					Final Solution Volume
	3.0	3.1	3.2	3.3	3.4	
Glycine/hydrochloric acid	50.0 mL C + 11.4 mL B		50.0 mL C + 8.2 mL B		50.0 mL C + 6.4 mL B	200 mL
Potassium phthalate/ hydrochloric acid	50.0 mL D + 20.3 mL B		50.0 mL D + 14.7 mL B		50.0 mL D + 9.9 mL B	200 mL
Aconitate	20.0 mL E + 32.0 mL F	20.0 mL E + 36.0 mL F	20.0 mL E+ 40.0 mL F	20.0 mL E + 44.0 mL F	20.0 mL E + 48.0 mL F	200 mL
Citrate	46.5 mL G + 3.5 mL H		43.7 mL G + 6.3 mL H		40.0 mL G + 10.0 mL H	100 mL
Citrate/phosphate	39.8 mL G + 10.2 mL K		37.7 mL G + 12.3 mL K		35.9 mL G + 14.1 mL K	100 mL

Buffer	pH					Final Solution Volume
	3.5	3.6	3.7	3.8	3.9	
Glycine/hydrochloric acid		50.0 mL C + 5.0 mL B				200 mL
Potassium phthalate/ hydrochloric acid		5.0 mL D + 6.0 mL B		50.0 mL D + 2.63 mL B		200 mL
Aconitate	20.0 mL E + 52.0 mL F	20.0 mL E + 56.0 mL F	20.0 mL E + 60.0 mL F	20.0 mL E + 64.0 mL F	20.0 mL E + 68.0 mL F	200 mL
Citrate		37.0 mL G + 13.0 mL H		35.0 mL G + 15.0 mL H		100 mL
Acetate		46.3 mL I + 3.7 mL J		44.0 mL I + 6.0 mL J		100 mL
Citrate/phosphate		33.9 mL G + 16.1 mL K		32.3 mL G + 17.7 mL K		100 mL
Succinate				25.0 mL L + 7.5 mL F		100 mL

Note: A list of the stock solutions follows the buffer table.

(Continued)

Buffer	pH 4.0	4.1	4.2	4.3	4.4	Final Solution Volume
Aconitate	20.0 mL E + 72.0 mL F	20.0 mL E + 76.0 mL F	20.0 mL E + 79.6 mL F	20.0 mL E + 83.0 mL F	20.0 mL E + 86.6 mL F	200 mL
Citrate	33.0 mL G + 17.0 mL H		31.5 mL G + 18.5 mL H		28.0 mL G + 22.0 mL H	100 mL
Acetate	41.0 mL I + 9.0 mL J		36.8 mL I + 13.2 mL J		30.5 mL I + 19.5 mL J	100 mL
Citrate/phosphate	30.7 mL G + 19.3 mL K		29.4 mL G + 20.6 mL K		27.8 mL G + 22.2 mL K	100 mL
Succinate	25.0 mL L + 10.0 mL F		25.0 mL L + 13.3 mL F		25.0 mL L + 16.7 mL F	100 mL
Potassium phthalate/sodium hydroxide			50.0 mL D + 3.7 mL F		50.0 mL D + 7.5 mL F	200 mL

Buffer	pH 4.5	4.6	4.7	4.8	4.9	Final Solution Volume
Aconitate	20.0 mL E + 90.0 mL F	20.0 mL E + 93.6 mL F	20.0 mL E + 97.0 mL F	20.0 mL E + 100.0 mL F	20.0 mL E + 103.0 mL F	200 mL
Citrate		25.5 mL G + 24.5 mL H		23.0 mL G + 27.0 mL H		200 mL
Acetate		25.5 mL I + 24.5 mL J		20.0 mL I + 30.0 mL J		100 mL
Citrate/phosphate		26.7 mL G + 23.3 mL K		25.2 mL G + 24.8 mL K		100 mL
Succinate		25.0 mL L + 20.0 mL F		25.0 mL L + 23.5 mL F		100 mL
Potassium phthalate/sodium hydroxide		50.0 mL D + 12.2 mL F		50.0 mL D + 17.7 mL F		200 mL

Note: A list of the stock solutions follows the buffer table.

(Continued)

Buffer	pH					Final Solution Volume
	5.0	5.1	5.2	5.3	5.4	
Aconitate	20.0 mL E + 105.6 mL F	20.8 mL E + 108.0 mL F	20.0 mL E + 110.6 mL F	20.0 mL E + 113.0 mL F	20.0 mL E + 116.0 mL F	200 mL
Citrate	20.5 mL G + 29.5 mL H		18.0 mL G + 32.0 mL H		16.0 mL G + 34.0 mL H	100 mL
Acetate	14.8 mL I + 35.2 mL J		10.5 mL I + 39.5 mL J		8.8 mL I + 41.2 mL J	100 mL
Citrate/phosphate	24.3 mL G + 25.7 mL K		23.3 mL G + 26.7 mL K		22.2 mL G + 27.8 mL K	100 mL
Succinate	25.0 mL L + 26.7 mL F		25.0 mL L + 30.3 mL F		25.0 mL L + 34.2 mL F	100 mL
Potassium phthalate/sodium hydroxide	50.0 mL D + 23.9 mL F		50.0 mL D + 30.0 mL F		50.0 mL D + 35.5 mL F	200 mL
Maleate			50.0 mL M + 7.2 mL F		50.0 mL M + 10.5 mL F	200 mL
Cacodylate	50.0 mL N + 47.0 mL B		50.0 mL N + 45.0 mL B		50.0 mL N + 43.0 mL B	200 mL
tris-Maleate			50.0 mL Q + 7.0 mL F		50.0 mL Q + 10.8 mL F	200 mL

Note: A list of the stock solutions follows the buffer table.

(Continued)

Buffer	pH					Final Solution Volume
	5.5	5.6	5.7	5.8	5.9	
Aconitate	20.0 mL E + 119.0 mL F	20.0 mL E + 122.6 mL F	20.0 mL E + 126.0 mL F			200 mL
Citrate		13.7 mL G + 36.3 mL H		11.8 mL G + 38.2 mL H		100 mL
Acetate		4.8 mL I + 45.2 mL J				100 mL
Citrate/phosphate		21.0 mL G + 29.0 mL K		19.7 mL G + 30.3 mL K		100 mL
Succinate		25.0 mL L + 37.5 mL F		25.0 mL L + 40.7 mL F		100 mL
Potassium phthalate/ sodium hydroxide		50.0 mL D + 39.8 mL F		50. mL D + 43.0 mL F		200 mL
Maleate		50.0 mL M + 15.3 mL F		50.0 mL M + 20.8 mL F		200 mL
Cacodylate		50.0 mL N + 39.2 mL B		50.0 mL N + 34.8 mL B		200 mL
Phosphate			93.5 mL O + 6.5 mL P	92.0 mL O + 8.0 mL P	90.0 mL O + 10.0 mL P	200 mL
tris-Maleate		50.0 mL Q + 15.5 mL F		50.0 mL Q + 20.5 mL F		200 mL

Note: A list of the stock solutions follows the buffer table.

(Continued)

CRC HANDBOOK OF BASIC TABLES FOR CHEMICAL ANALYSIS

Buffer	pH 6.0	6.1	6.2	6.3	6.4	Final Solution Volume
Citrate	9.5 mL G + 41.5 mL H		7.2 mL G + 42.8 mL H			100 mL
Citrate/phosphate	17.9 mL G + 32.1 mL K		16.9 mL G + 33.1 mL K		15.4 mL G + 34.6 mL K	100 mL
Succinate	25.0 mL L + 43.5 mL F					100 mL
Potassium phthalate/ sodium hydroxide	50.0 mL D + 45.5 mL F					200 mL
Maleate	50.0 mL M + 26.9 mL F		50.0 mL M + 33.0 mL F		50.0 mL M + 38.0 mL F	200 mL
Cacodylate	50.0 mL N + 29.6 mL B		50.0 mL N + 23.8 mL B		50.0 mL N + 18.3 mL B	200 mL
Phosphate	87.7 mL O + 12.3 mL P	85.0 mL O + 15.0 mL P	81.5 mL O + 18.5 mL P	77.5 mL O + 22.5 mL P	73.5 mL O + 26.5 mL P	200 mL
tris-Maleate	50.0 mL Q + 26.0 mL F		50.0 mL Q + 31.5 mL F		50.0 mL Q + 37.0 mL F	200 mL

Buffer	pH 6.5	6.6	6.7	6.8	6.9	Final Solution Volume
Barbital				50.0 mL R + 45.0 mL B		200 mL
Citrate/phosphate		13.6 mL G + 36.4 mL K		9.1 mL G + 40.9 mL K		100 mL
Maleate		50.0 mL M + 41.6 mL F		50.0 mL M + 44.4 mL F		200 mL
Cacodylate		40.0 mL N + 13.3 mL B		50.0 mL N + 9.3 mL B		200 mL
Phosphate	68.5 mL O + 31.5 mL P	62.5 mL O + 37.5 mL P	56.5 mL O + 43.5 mL P	51.0 mL O + 49.0 mL P	45.0 mL O + 55.0 mL P	200 mL
tris-Maleate		50.0 mL Q + 42.5 mL F		50.0 mL Q + 45.0 mL F		200 mL

Note: A list of the stock solutions follows the buffer table.

(Continued)

Buffer	pH					Final Solution Volume
	7.0	7.1	7.2	7.3	7.4	
Barbital	50.0 mL R + 43.0 mL B		50.0 mL R + 39.0 mL B		50.0 mL R + 32.5 mL B	200 mL
Tris			50.0 mL S + 44.2 mL B		50.0 mL S + 41.4 mL B	200 mL
Citrate/phosphate	6.5 mL G + 43.6 mL K					100 mL
Cacodylate	50.0 mL N + 6.3 mL B		50.0 mL N + 4.2 mL B		50.0 mL N + 2.7 mL B	200 mL
Phosphate	39.0 mL O + 61.0 mL P	33.0 mL O + 67.0 mL P	28.0 mL O + 72.0 mL P	23.0 mL O + 77.0 mL P	19.0 mL O + 81.0 mL P	200 mL
tris-Maleate	50.0 mL Q + 48.0 mL F		50.0 mL Q + 51.0 mL F		50.0 mL Q + 54.0 mL F	200 mL

Buffer	pH					Final Solution Volume
	7.5	7.6	7.7	7.8	7.9	
Barbital		50.0 mL R + 27.5 mL B		50.0 mL R + 22.5 mL B		200 mL
Tris		50.0 mL S + 38.4 mL B		50.09 mL S + 32.5 mL B		200 mL
Boric acid/borax		50.0 mL T + 2.0 mL U		50.0 mL T + 3.1 mL U		200 mL
Ammediol				50.0 mL V + 43.5 mL B		200 mL
Phosphate	16.0 mL O + 84.0 mL P	13.0 mL O + 87.0 mL P	10.5 mL O + 90.5 mL P	8.5 mL O + 91.5 mL P	7.0 mL O + 93.0 mL P	200 mL
tris-Maleate		50.0 mL Q + 58.0 mL F		50.0 mL Q + 63.5 mL F		200 mL

Note: A list of the stock solutions follows the buffer table.

(Continued)

Buffer	pH					Final Solution Volume
	8.0	8.1	8.2	8.3	8.4	
Barbital	50.0 mL R + 17.5 mL B		50.0 mL R + 12.7 mL B		50.0 mL R + 9.0 mL B	200 mL
Tris	50.0 mL S + 26.8 mL B		50.0 mL S + 21.9 mL B		50.0 mL S + 16.5 mL B	200 mL
Boric acid/borax	50.0 mL T + 4.9 mL U		50.0 mL T + 7.3 mL U		50.0 mL T + 11.5 mL U	200 mL
Ammediol	50.0 mL V + 41.0 mL B		50.0 mL V + 37.7 mL B		50.0 mL V + 34.0 mL B	200 mL
Phosphate	5.3 mL O + 94.7 mL P					200 mL
tris-Maleate	50.0 mL Q + 69.0 mL F		50.0 mL Q + 75.0 mL F		50.0 mL Q + 81.0 mL F	200 mL

Buffer	pH					Final Solution Volume
	8.5	8.6	8.7	8.8	8.9	
Barbital		50.0 mL R + 6.0 mL B		50.0 mL R + 4.0 mL B		200 mL
Tris		50.0 mL S + 12.2 mL B		50.0 mL S + 8.1 mL B		200 mL
Boric acid/borax		50.0 mL T + 17.5 mL U	50.0 mL T + 22.5 mL U	50.0 mL T + 30.0 mL U	50.0 mL T + 42.5 mL U	200 mL
Ammediol		50.0 mL V + 29.5 mL B		50.0 mL V + 22.0 mL B		200 mL
Glycine/sodium hydroxide		50.0 mL C + 4.0 mL F		50.0 mL C + 6.0 mL F		200 mL
tris-Maleate		50.0 mL Q + 86.5 mL F				200 mL

Note: A list of the stock solutions follows the buffer table.

(Continued)

Buffer	pH 9.0	pH 9.1	pH 9.2	pH 9.3	pH 9.4	Final Solution Volume
Barbital	50.0 mL R + 2.5 mL B		50.0 mL R + 1.5 mL B			200 mL
Tris	50.0 mL S + 5.0 mL B					200 mL
Boric acid/borax	50.0 mL T + 59.0 mL U	50.0 mL T + 83.0 mL U	50.0 mL T + 115.0 mL U			200 mL
Ammediol	50.0 mL V + 16.7 mL B		50.0 mL V + 12.5 mL B		50.0 mL V + 8.5 mL B	200 mL
Glycine/sodium hydroxide	50.0 mL C + 8.8 mL F		50.0 mL C + 12.0 mL F		50.0 mL C + 16.8 mL F	200 mL
Borax/sodium hydroxide			50.0 mL U (pH = 9.28)		50.0 mL U + 11.0 mL F	200 mL
Carbonate/bicarbonate			4.0 mL W + 46.0 mL X	7.5 mL W + 42.5 mL X	9.5 mL W + 40.5 mL X	200 mL

Buffer	pH 9.5	pH 9.6	pH 9.7	pH 9.8	pH 9.9	Final Solution Volume
Ammediol		50.0 mL V + 5.7 mL B		50.0 mL V + 3.7 mL B		200 mL
Glycine/sodium hydroxide		50.0 mL C + 22.4 mL F		50.0 mL C + 27.2 mL F		200 mL
Borax/sodium hydroxide	50.0 mL U + 17.6 mL F	50.0 mL U + 23.0 mL F	50.0 mL U + 29.0 mL F	50.0 mL U + 34.0 mL F	50.0 mL U + 38.0 mL F	200 mL
Carbonate/bicarbonate	13.0 mL W + 37.0 mL X	16.0 mL W + 34.0 mL X	19.5 mL W + 30.5 mL X	22.0 mL W + 28.0 mL X	25.0 mL W + 25.0 mL X	200 mL

Note: A list of the stock solutions follows the buffer table.

(Continued)

Buffer	pH					Final Solution Volume
	10.0	10.1	10.2	10.3	10.4	
Ammediol	50.0 mL V + 2.0 mL B					200 mL
Glycine/sodium hydroxide	50.0 mL C + 32.0 mL F				50.0 mL C + 45.5 mL F	200 mL
Borax/sodium hydroxide	50.0 mL U + 43.0 mL F	50.0 mL U + 46.0 mL F				200 mL
Carbonate/bicarbonate	27.5 mL W + 22.5 mL X	30.0 mL W + 20.0 mL X	33.0 mL W + 17.0 mL X	35.5 mL W + 14.5 mL X	38.5 mL W + 11.5 mL X	200 mL

Buffer	pH					Final Solution Volume
	10.5	10.6	10.7	10.8	10.9	
Glycine/sodium hydroxide		50.0 mL C + 45.5 mL F				200 mL
Carbonate/bicarbonate	40.5 mL W + 9.5 mL X	42.5 mL W + 7.5 mL X	45.0 mL W + 5.0 mL X			200 mL

STOCK SOLUTIONS

A = 0.2 M potassium chloride (14.91 g in 1,000 mL)

B = 0.2 M hydrochloric acid

C = 0.2 M glycine (15.01 g in 1,000 mL)

D = 0.2 M potassium acid phthalate (40.84 g in 1,000 mL)

E = 0.5 M aconitic acid, 1-propene-1,2,3-tricarboxylic acid, (87.05 g in 1,000 mL)

F = 0.2 M sodium hydroxide

G = 0.1 M citric acid (21.01 g in 1,000 mL)

H = 0.1 M sodium citrate dihydrate (29.41 g in 1,000 mL); avoid using any other hydrated salt.

I = 0.2 M acetic acid

J = 0.2 M anhydrous sodium acetate (16.4 g in 1,000 mL) or 0.2 M sodium acetate trihydrate (27.2 g in 1,000 mL)

K = 0.2 M dibasic sodium phosphate heptahydrate (53.65 g in 1,000 mL) or 0.2 M dibasic sodium phosphate dodecahydrate (71.7 g in 1,000 mL)

L = 0.2 M succinic acid (23.6 g in 1,000 mL)

M = 0.2 M sodium maleate (8.0 g NaOH + 23.2 g maleic acid or 19.6 g maleic anhydride in 1,000 mL)

N = 0.2 M sodium cacodylate (42.8 g sodium cacodylate trihydrate in 1,000 mL)

O = 0.2 M monobasic sodium phosphate (27.8 g in 1,000 mL)

P = 0.2 M dibasic sodium phosphate (53.65 g dibasic sodium phosphate heptahydrate or 71.7 g dibasic sodium phosphate dodecahydrate in 1,000 mL).

Q = 0.2 M tris acid maleate (24.2 g tris (hydroxymethyl) amino methane + 23.2 g maleic acid or 19.6 g maleic anhydride in 1,000 mL)

R = 0.2 M sodium barbital (veronal) (41.2 g in 1,000 mL)

S = 0.2 M tris (hydroxymethyl) aminomethane (24.2 g in 1,000 mL)

T = 0.2 M boric acid (12.4 g in 1,000 mL)

U = 0.05 M borax (19.05 g in 1,000 mL)

V = 0.2 M 2-amino-2-methyl-1,3-propanediol (21.03 g in 1,000 mL)

W = 0.2 M anhydrous sodium carbonate (21.2 g in 1,000 mL)

X = 0.2 M anhydrous sodium bicarbonate (16.8 g in 1,000 mL).

INDICATORS FOR ACIDS AND BASES

The following table lists the most common indicators together with their pH range and colors in acidic and basic media. Since the color change is not instantaneous at the pKa value, a pH range is given where a mixture of colors is present. This pH range, which varies between indicators, generally falls between the pKa with a spread or uncertainty of 1 pH unit. All solutions are either aqueous or ethanol/aqueous (% ethanol, vol/vol) [1–3]. Ref. 4 lists the exact quantities needed for the indicator solutions.

REFERENCES

1. Lange, N.A. *Lange's Handbook of Chemistry,* 8th ed., Handbook Publishers, New York, 1952.
2. Kolthoff, I.M., and V.A. Stenger (translated in English by N. H. Furman), *Volumetric Analysis*, 2nd ed., Interscience Publishers, New York, 1942.
3. Sabnis, R.W., *Handbook of Acid-Base Indicators*, CRC Press, Taylor and Francis Group, Boca Raton, FL, 2008.
4. http://www2.csudh.edu/oliver/che230/textbook/ch02.htm, accessed January 2020.

Indicator	pH Range	Solvent	Acid	Base
Gentian violet (crystal violet)	0.0–2.0	Aqueous	Yellow	Blue-violet
Thymol blue	1.2–2.8	Aqueous	Red	Yellow
Pentamethoxy red	1.2–2.3	70 % ethanol	Red-violet	Colorless
Tropeolin OO	1.3–3.2	Aqueous	Red	Yellow
2,4-Dinitrophenol	2.4–4.0	50 % ethanol	Colorless	Yellow
Methyl yellow	2.9–4.0	90 % ethanol	Red	Yellow
Methyl orange	3.1–4.4	Aqueous	Red	Orange
Bromophenol blue	3.0–4.6	Aqueous	Yellow	Blue-violet
Tetrabromphenol blue, sodium salt	3.0–4.6	Aqueous	Yellow	Blue
Congo red	3.0–5.0	Aqueous	Blue-violet	Red
Alizarin sodium sulfonate	3.7–5.2	Aqueous	Yellow	Violet
α-Naphthyl red	3.7–5.0	70 % ethanol	Red	Yellow
p-Ethoxychrysoidine	3.5–5.5	Aqueous	Red	Yellow
Bromocresol green, sodium salt	4.0–5.6	Aqueous	Yellow	Blue
Methyl red, sodium salt	4.4–6.2	Aqueous	Red	Yellow
Bromocresol purple	5.2–6.8	Aqueous	Yellow	Purple
Chlorophenol red	5.4–6.8	Aqueous	Yellow	Red
Bromophenol blue, sodium salt	6.2–7.6	Aqueous	Yellow	Blue
p-Nitrophenol	5.0–7.0	Aqueous	Colorless	Yellow
Azolitmin	5.0–8.0	Aqueous	Red	Blue
Bromothymol blue, sodium salt	6.0–7.6	70 % ethanol	Yellow	Blue
Phenol red, sodium salt	6.4–8.0	Aqueous	Yellow	Red
Neutral red	6.8–8.0	70 % ethanol	Red	Yellow
Rosolic acid	6.8–8.0	90 % ethanol	Yellow	Red
Cresol red, sodium salt	7.2–8.8	Aqueous	Yellow	Red
α-Naphtholphthalein	7.3–8.7	70 % ethanol	Rose	Green
Tropeolin OOO	7.6–8.9	Aqueous	Yellow	Rose-red
Thymol blue, sodium salt	8.0–9.6	Aqueous	Yellow	Blue
Phenolphthalein	8.0–10.0	70 % ethanol	Colorless	Red
α-Naphtholbenzein	9.0–11.0	70 % ethanol	Yellow	Blue
Thymolphthalein	9.4–10.6	90 % ethanol	Colorless	Blue
Nile blue	10.1–11.1	Aqueous	Blue	Red
Alizarin yellow R	10.0–12.0	Aqueous	Yellow	Lilac
Salicyl yellow	10.0–12.0	90 % ethanol	Yellow	Orange-brown
Diazo violet	10.1–12.0	Aqueous	Yellow	violet
Tropeolin O	11.0–13.0	Aqueous	Yellow	Orange-brown
Nitramine	11.0–13.0	70 % ethanol	Colorless	Orange-brown
Poirrier's blue	11.0–13.0	Aqueous	Blue	Violet-pink
Trinitrobenzoic acid	12.0–13.4	Aqueous	Colorless	Orange-red

DIELECTRIC CONSTANTS OF INORGANIC SOLVENTS

The dielectric constant (ε) of a substance is a macroscopic property that measures the reduction of the strength of the electric field that surrounds a charged particle when immersed in that substance, as compared to the field strength around the same particle when placed in vacuum. As a result, the higher the value of the dielectric constant of the substance, the greater the tendency of the charged particle to ionize. Although the dielectric constant gives only one of several quantitative measures of the polarity of the substance, it is nonetheless a useful property in describing solvents as polar or nonpolar. The table below lists the dielectric constants of some inorganic solvents at a specific temperature [1–3].

REFERENCES

1. Rumble, J.R., ed., *CRC Handbook for Chemistry and Physics*, 100th ed., CRC Press, Boca Raton, FL, 2019.
2. Lowry, T.H. and Richardson, K.S., *Mechanism and Theory in Organic Chemistry*, Harper Collins Publishers, New York, 1987.
3. Parsons, R., *Handbook of Electrochemical Constants*, Butterworths, London, 1959.

Name	Formula	Dielectric Constant (ε)	Temp. (°C)
Aluminum bromide	$AlBr_3$	3.38	100
Ammonia	NH_3	16.9	25
Argon	Ar	1.53	−191
Arsenic trichloride	$AsCl_3$	12.8	20
Arsine	AsH_3	2.50	−100
Boron tribromide	BBr_3	2.58	0
Bromine	Br_2	3.09	20
Chlorine	Cl_2	2.10	−50
Deuterium	D_2	1.277	−253
Deuterium oxide (see water-d_2)			
Dinitrogen tetroxide	N_2O_4	2.4	18
Fluorine	F_2	1.54	−202
Germanium tetrachloride	$GeCl_4$	2.43	25
Helium	He	1.055	−271
Hydrogen bromide	HBr	7.0	−70
Hydrogen chloride	HCl	12.0	−113
Hydrogen cyanide	HCN	106.8	25
Hydrogen fluoride	HF	17.0	−73
		83.6	25
Hydrogen iodide	HI	3.39	−50
Hydrogen sulfide	H_2S	9.26	−85.5
Hydrogen peroxide	H_2O_2	84.2	0
Iodine	I_2	11.1	118
Iodine pentafluoride	IF_5	36.2	25
Hydrazine	N_2H_4	51.7	25
Lead tetrachloride	$PbCl_4$	2.78	20
Mercury (II) bromide	$HgBr_2$	9.8	240
Nitrosyl bromide	NOBr	13.4	15
Nitrosyl chloride	NOCl	18.2	12
Phosphorus trichloride	PCl_3	3.43	25
Phosphorus pentachloride	PCl_5	2.8	160
Phosphoryl chloride	$POCl_3$	13.3	22
Selenium	Se	5.40	250
Seleninyl chloride	$SeOCl_2$	26.2	20
Silicon tetrachloride	$SiCl_4$	2.40	20
Sulfur	S	3.52	118
Sulfur dioxide	SO_2	14.1	20
Sulfuric acid	H_2SO_4	101	25
Sulfuryl chloride	SO_2Cl_2	9.2	20
Thionyl bromide	$SOBr_2$	9.06	20
Thionyl chloride	$SOCl_2$	9.25	20
Thiophosphoryl chloride	$PSCl_3$	5.8	22
Titanium tetrachloride	$TiCl_4$	2.80	20
Water	H_2O	80.22	20
		78.41	25
Water-d_2	D_2O	77.94	25

DIELECTRIC CONSTANTS OF METHANOL–WATER MIXTURES FROM 5 °C TO 55 °C

The table below lists the value of dielectric constant of methanol–water mixtures as a function of their respective w/w % composition at various temperatures [1]. This information is useful for a variety of chromatographic and extractive applications.

REFERENCE

1. Parsons, R., *Handbook of Electrochemical Constants*, Butterworths, London, 1959.

Methanol	5 °C	15 °C	25 °C	35 °C	45 °C	55 °C
			Weight %			
0	85.76	83.83	78.30	74.83	71.51	68.35
10	81.68	77.83	74.18	70.68	67.32	64.08
20	77.38	73.59	69.99	66.52	63.24	60.06
30	72.80	69.05	65.55	62.20	58.97	55.92
40	67.91	64.31	60.94	57.72	54.62	51.69
50	62.96	59.54	56.28	53.21	50.29	47.53
60	57.92	54.71	51.67	48.76	46.02	43.42
70	52.96	49.97	47.11	44.42	41.83	39.38
80	48.01	45.24	42.60	40.08	37.70	35.46
90	42.90	40.33	37.91	35.65	33.53	31.53
95	39.98	37.61	35.38	33.28	31.29	29.43
100	36.88	34.70	32.66	30.74	28.02	27.21

COMMON DRYING AGENTS FOR ORGANIC LIQUIDS

The following table gives the suggested common agents for drying various organic liquids. Those squares marked "X" are the best combination of organic family/drying agent. Those marked "never" are the worst combinations, primarily due to possible chemical reactions. For instance, alcohols and sodium metal react vigorously. Consequently, one should look for other drying agents. Those that are blank might be efficient, but are not recommended for use, unless the suggested drying agents are not available. Some combinations do not give efficient results due to complexation (marked "d") [1–6].

REFERENCES

1. Vogel, A.I., *A Textbook of Practical Organic Chemistry*, Longmans, Green and Co., London, 1951.
2. Brewster, R.Q., Vanderwerf, C.A., and McEwen, W.E., *Unitized Experiments in Organic Chemistry*, D. Van Nostrand Co., New York, 1977.
3. Gordon, A.J. and Ford, R.A., *The Chemist's Companion; a Handbook of Practical Data, Techniques and References*, John Wiley and Sons, New York, 1972.
4. Bruno, T.J., and Svoronos, P.D.N., *Basic Tables for Chemical Analysis*, NBS Technical Note 1096, U.S. Dept. of Commerce, National Bureau of Standards, Washington, DC, 1986.
5. Bruno, T.J., and Svoronos, P.D.N., *CRC Handbook of Basic Tables for Chemical Analysis*, 2nd ed., CRC Press, Boca Raton, FL, 2003.
6. Sarlo, E., Svoronos, P.D.N., and Kulawiec, R., *Organic Chemistry Laboratory Manual*, 2nd ed., W.C. Brown, Dubuque, 1997.

Family	Na_2CO_3[a]	K_2CO_3[a]	$MgSO_4$[b]	$CaSO_4$[c]	Na_2SO_4[c]	$CaCl_2$[d]
Alcohols		X	X	X		d
Aldehydes			X	X	X	d
Alkyl halides						X
Amines						d
Anhydrides						
Aryl halides						X
Carboxylic acids	Never	Never	X	X	X	e
Esters			X		X	d
Ethers				X		X
Hydrocarbons, aromatic	X	X		X	Poor	X
Hydrocarbons, saturated	X	X		X	Poor	X
Hydrocarbons, unsaturated	X	X			Poor	X
Ketones		X	X	X	X	d
Nitriles						

Family	Na	P_2O_5	NaOH (Solid)	KOH (Solid)	CaO	CaH_2	$LiAlH_4$
Alcohols	Never	Never	Never	Never	X	g	Never
Aldehydes	Never	Never	Never	Never		Never	Never
Alkyl halides	Never	X	Never	Never		X	Never
Amines	Never	Never	X	X	X	f	Never
Anhydrides		X					
Aryl halides	Never	X				X	X
Carboxylic acids	Never	X	Never	Never	Never	Never	Never
Esters			Never	Never		X	Never
Ethers	X	X			X	X	X
Hydrocarbons, aromatic	X	X				X	X
Hydrocarbons, saturated	X	X				X	X
Hydrocarbons, unsaturated		X				X	X
Ketones	Never	Never	Never	Never	Never	Never	Never
Nitriles		X	Never	Never			Never

[a] Excellent in salting out.
[b] Best all-purpose drying agent.
[c] High capacity, but slow reacting.
[d] Forms complexes.
[e] Lime (common impurity) reacts with acidic hydrogen.
[f] Only for 3° amines (R_3N).
[g] Only for C_4 and higher alcohols.

COMMON RECRYSTALLIZATION SOLVENTS

The following table gives a list of solvents (and their useful properties) in order of decreasing polarity (on the basis of eluotropic series number, ε) and the organic compounds they are capable of recrystallizing. In choosing a solvent, one should consider the following criteria: (a) low toxicity, (b) low cost, (c) ease of separation of the solvent from the crystals (relatively high degree of volatility), (d) the ability to dissolve the crystals while hot, but not while cold, with impurities being either soluble or insoluble both in hot and cold, and (e) the boiling point of the solvent should be lower than the melting point of the compound. While not all of these factors may be optimized with each application, an attempt should be made to achieve optimization of as many as possible. For the same compound, a variety of recrystallizing solvents can be employed based on the type of impurities that are present [1–4].

REFERENCES

1. Gordon, A.J., and Ford, R.A., *The Chemist's Companion: A Handbook of Practical Data, Techniques, and References*, John Wiley and Sons, New York, 1972.
2. Roberts, R.M., Gilbert, J.C., Rodewald, L.B., and Wingrove, A.S., *An Introduction to Modern Experimental Organic Chemistry*, Holt, Rinehart, and Winston, New York, 1969.
3. Sarlo, E., Svoronos P.D.N., and Kulawiec, R., *Organic Chemistry Laboratory Manual*, 2nd ed., W.C. Brown, Dubuque, 1997.
4. Rumble, J.R., ed., *CRC Handbook of Chemistry and Physics*, 100th ed., CRC Press/Taylor & Francis Group, Boca Raton, FL, 2019.

Solvent	bp^a(°C)	$\varepsilon^{b,c}$	Flammability^d	Toxicity	Good For	Second Solvent in Mixture^e	Comments
Water	100	78.5^c	0	0	Amides, salts, some carboxylic acids	Methanol, ethanol, acetone, dioxane, acetonitrile	Difficult to remove from crystals
Acetic acid	117.9	6.15^b	1	2	Amides, some carboxylic acids, some sulfoxides	Water	Difficult to remove from crystals
Acetonitrile	81.6	37.5^b	3	3	Some carboxylic acids, hydroquinones	Water, ether, benzene	
Methanol	64.5	32.63^c	3	1	Nitro compounds, esters, bromo compounds, some sulfoxides, sulfones, and sulfilimines, anilines	Water, ether, benzene	
Ethanol	78.24	24.30^c	3	0	Same as methanol	Water, ethyl acetate, hydrocarbons, methylene chloride	
Acetone	56	20.7^c	3	1	Nitro compounds, osazones	Water, ether, hydrocarbons	
2-Methoxyethanol(methyl cellosolve)	124	123^c	2	2	Carbohydrates	Water, ether, benzene	
Pyridine	115.2	123^c	3	3	Quinones, thiazoles, oxazoles	Water, methanol	Difficult to remove from crystals
Methyl acetate	56.7	6.68	4	2	Esters, carbonyl compounds, sulfide derivatives, carbinols	Water, ether	
Ethyl acetate	77.1	6.02^c	3	1	Same as methyl acetate	Water, ether, chloroform, methylene chloride	
Methylene chloride (dichloromethane)	39.6	9.08	0	2	Low-melting compounds	Ethanol, hydrocarbons	Easily removed
Ether (diethyl ether)	34.5	4.34^b	4	2	Low-melting compounds	Acetone, acetonitrile, methanol, ethanol, acetate esters	Easily removed, can create peroxides

(Continued)

Solvent	bp^a(°C)	$\varepsilon^{b,c}$	Flammability^d	Toxicity	Good For	Second Solvent in Mixture^e	Comments
Chloroform	61.2	4.81^b	0	4	Polar compounds	Ethanol, acetate esters, hydrocarbons	Easily removed; suspected carcinogen^f
1,4-Dioxane	102	2.21^c	3	2	Amides	Water, hydrocarbons, benzene	Can form complexes with ethers
Carbon tetrachloride	76.7	2.24^b	0	4	Acid chlorides, anhydrides	Ether, benzene, hydrocarbons	Can react with strong organic bases; suspected carcinogen^f
Toluene	110.60	2.38^c	3	2	Aromatics, hydrocarbons	Ether, ethyl acetate, hydrocarbons	A little difficult to remove from crystals
Benzene	80.1	2.28^b	3	3	Aromatics, hydrocarbons, molecular complexes, sulfides, ethers	Ether, ethyl acetate, hydrocarbons	Carcinogen^f
Ligroin (naphtha solvent)	90–110	—	3	1	Hydrocarbons, aromatic heterocycles	Ethyl acetate, benzene, methylene chloride	
Petroleum ether (ACS)	35–60	—	4	1	Hydrocarbons	Any solvent less polar than ethanol	Easy to separate
n-Pentane	36.06	1.84^b	4	1	Hydrocarbons	Any solvent less polar than ethanol	Easy to separate
n-Hexane	68.72	1.89^b	4	1	Hydrocarbons	Any solvent less polar than ethanol	
Cyclohexane	80.7	2.02^b	4	1	Hydrocarbons	Any solvent less polar than ethanol	
n-Heptane	98.38	1.91^c	4	1	Hydrocarbons	Any solvent less polar than ethanol	

^{a} Normal boiling point (°C).
^{b} Dielectric constant (20°C).
^{c} Dielectric constant (25°C).
^{d} Scale varies from 4 (highly flammable, highly toxic) to 0 (not flammable, not toxic).
^{e} Second solvent used to facilitate dissolving the crystals in a solvent mixture.
^{f} See Carcinogen Table in Chapter 15.

SURFACE ACTIVE CHEMICALS (SURFACTANTS)

The following table provides the structure and properties of common surface active chemicals or surfactants [1–3] and is based on a table prepared for the *96th Edition of the CRC Handbook of Chemistry and Physics*, updated yearly and for this book [4]. These reagents are used industrially to decrease surface tension in many different applications, and they are used in the laboratory in many analytical and biochemical procedures (as well as in consumer applications). In this way, the table complements many of the tables presented in Chapters 8 and 13, in which these surfactants are incorporated in reagents. The table is arranged alphabetically according to chemical name, although the most common generic name or abbreviation is provided in bold type (unless the chemical name is also the most common name). The surfactant class is provided (anionic, cationic, or nonionic). Anionic surfactants are molecules in which the hydrophilic moiety is a negatively charged group such as a sulfonate, sulfate, or carboxylate. Cationic surfactants are molecules in which the hydrophilic moiety is a positively charged group such as a quaternary ammonium ion. Nonionic surfactants are molecules in which the hydrophilic moiety is uncharged, such as an ethoxylate. Another classification exists, though not represented here, called amphoteric surfactants, in which the ionic character is pH dependent.

Where available, the melting and normal boiling temperatures are provided; if decomposition occurs before this state point is reached, this is indicated by the notation "dec." The measured density of either the liquid or a solution is provided, along with temperature, where possible. Predicted density, while available for some surfactants in which measurements are unavailable, is not provided here.

Where available, the critical micelle concentration (CMC), the concentration of the surfactant above which micelles spontaneously form in water, is provided. The preferred unit is mM, but in the case of mixtures, this is provided as a percent or ppm, usually on the basis of mass (mass/mass). The CMC is dependent on temperature and the ionic strength of the solution, and is also dependent to some extent on the measurement technique. For anionic and cationic surfactants, the CMC is reduced by increasing ionic strength, but temperature has a minor effect. For nonionic surfactants, the CMC is relatively insensitive to ionic strength but increases with increase in temperature. For the values provided here, the temperature is provided where possible. If not provided, the temperature is to be regarded as ambient.

Where available, the aggregation number is provided [1, 5, 6]. This is the mean number of surfactant molecules present in a micelle after the CMC has been reached. It is measured by use of luminescent probes by varying surfactant concentration and is dependent on temperature and the concentration of any organic or ionic species present. Finally, the hydrophilic lipophilic balance (HLB) is provided where available. This is a fit-for-purpose property rather than a fundamental property which has evolved in definition and determination since being devised, a matter beyond the scope of this chapter. The HLB is a numerical scale between 0 and 20 descriptive of the tendency of a surfactant to be either hydrophilic or hydrophobic. A relatively high HLB (greater than ten) indicates hydrophilicity or good water or polar solvent solubility. A relatively low HLB (lower than ten) indicates lipophilicity or good solubility in nonpolar solvents such as oils or organics. The HLB changes with concentration and mixture composition.

REFERENCES

1. Surfactant Basics: Dow Performance Materials and Chem Answer Center, Answer 1760, 2015.
2. Lange, R.K., *Surfactants: A Practical Handbook*, Hanser Publishers, Munich, 1999.
3. Porter, M.R., *Handbook of Surfactants*, Blackie Academic and Professional, London, 1994.

4. Bruno, T.J., and Svoronos, P.D.N., Surface active chemicals, in Rumble, J.R., Ed., *CRC Handbook of Chemistry and Physics*, CRC Press, Boca Raton, FL, 2019.

5. Alargova, R.G., I.I. Kochijashky, M.L. Sierra, and R. Zana, Micelle aggregation numbers and surfactants in aqueous solutions: a comparison between the results from steady state and time resolved fluorescence quenching, *Langmuir*, 14, 5412–5418, 1998.

6. Technical Information Surface Chemistry, Nouryon (Akzo Nobel), https://surfacechemistry.nouryon.com/, accessed January 2020.

Name: abietic acid (CAS No: 514-10-3)
Class: anionic surfactant
Synonyms: abietinic acid; sylvic acid; abieta-7,13-dien-18-oic acid (1R,4aR,4bR,10aR)-7-isopropyl-1,4a-dimethyl-1,2,3,4,4a,4b,5,6,10,10a-decahydrophenanthrene-1-carboxylic acid.
Chemical Structure:

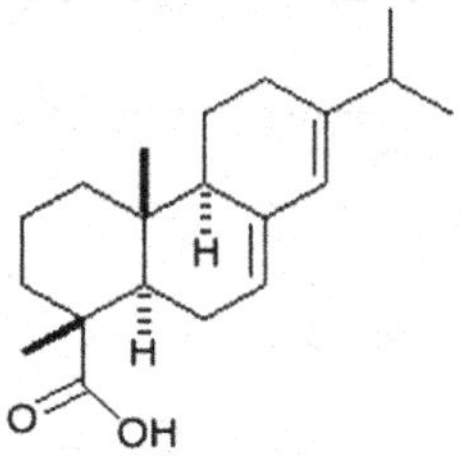

Characteristics and Properties:
Molecular formula: $C_{20}H_{30}O_2$
Molecular mass: 302.45 g/mol
Normal boiling temperature: 439.5 °C
Melting temperature: 173.5 °C
Density: 1.06 g/mL at 25 °C
CMC: 2 mM
Water solubility: insoluble
Soluble in: ethanol, acetone, ether, chloroform, and benzene.
Notes: natural product is primary component of resin acid of pine wood; used as bow rosin; used at high pH.

Name: cetyltrimethylammonium bromide (CAS No: 57-09-0)
Class: cationic surfactant
Synonyms: CTAB, CTABr, alkyltrimethylammonium bromide; HTAB; hexadecyl palmityltrimethylammonium bromide.
Chemical Structure:

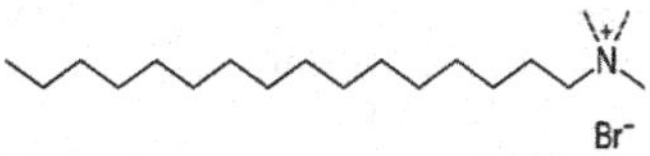

Characteristics and Properties:
Molecular formula: $CH_3(CH_2)_{15}N(CH_3)_3$
Molecular mass: 364.45 g/mol
Normal boiling temperature: dec.
Melting temperature: 237 °C–243 °C, dec.
CMC: 0.92–1.00 mM at 25 °C
Aggregation number: 75–170
HLB: ~10
Water solubility: 0.3 g/100 mL at 20 °C
Soluble in: alcohols, slightly soluble in acetone; insoluble in diethyl ether, benzene.
Notes: aid in high molecular mass DNA isolation and polymerase chain reaction (PCR) analysis; titrant for perchlorate; phase transfer catalyst in arene and heterocycle reductions.

Name: deoxycholic acid (CAS No: 83-44-3)
Class: anionic surfactant
Synonyms: (3α,5β,12α,20R)-3,12-Dihydroxycholan-24-oic acid; cholanoic acid; deoxycholate; deoxycholic, Pyrochol; Septochol (l).
Chemical Structure:

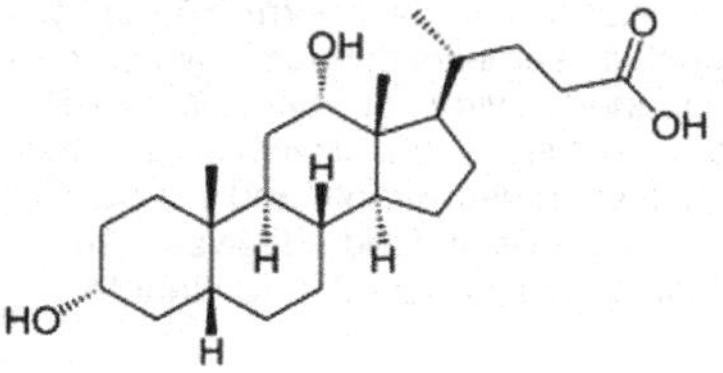

Characteristics and Properties:
Molecular formula: $C_{24}H_{40}O_4$
Molecular mass: 392.57 g/mol
Normal boiling temperature: dec.
Melting temperature: 174 °C–176 °C
CMC: 5 mM
Aggregation number: 22
HLB: 17.6
Water solubility: 0.24 mg/mL at 15 °C
Soluble in: ethanol (1 g/100mL at 20 °C), acetone, ether, chloroform.
Notes: readily dialyzable; used in a modified procedure to recover 40 %–80 % of a protein; $pK_a = 6.58$; $\lambda_{max} = 310$ nm; specific rotation +55°; forms molecular coordination compounds; complexes with fatty acids; sodium salt is much more soluble (333 mg/mL in water at 15 °C) but its aqueous solutions precipitate at pH < 5.

Name: dioctyl sodium sulfosuccinate (CAS No. 577-11-7)
Class: anionic surfactant
Synonyms: docusate; sodium dioctyl sulfosuccinate; dioctyl sulfosuccinate sodium salt; sodium 1,4-bis(2-ethylhexoxy)-1,4-dioxobutane-2-sulfonate; 1,4-bis (2-ethylhexyl) sodium sulfosuccinate; sulfosuccinic acid 1,4-bis (2-ethylhexyl) ester sodium salt; sulfobutanedioic acid 1,4-bis (2-ethylhexyl) ester sodium salt; dioctyl sodium sulfosuccinate; Colace; Ex-Lax; Senokot; Comfolax.
Chemical Structure:

Characteristics and Properties:
Molecular formula: $C_{20}H_{37}NaO_7S$
Molecular mass: 444.56 g/mole
Normal boiling temperature: dec.
Melting temperature: 153 °C–157 °C
Density: 1.1 g/cm³
Water solubility: 15 g/L at 25 °C; 23 g/L at 40 °C; 30 g/L at 50 °C; 55 g/L at 70 °C
Soluble in ethanol (20 g/L), chloroform (300 g/L), diethyl ether (300 g/L), petroleum ether (unlimited), glycerol, carbon tetrachloride, xylene, petroleum ether, acetone, vegetable oils.
Notes: stable in acid and neutral solutions; hydrolyzes in basic media; used as emulsifier, dispersant, wetting agent, and laxative to treat constipation and earwax removal.

Name: n-dodecyl-β-D-maltoside (CAS No: 69227-93-6)
Class: nonionic surfactant
Synonyms: DDM, n-dodecyl-beta-D-maltopyranoside; dodecyl 4-O-alpha-D-glucopyranosyl-beta-D-glucopyranoside; dodecyl B-D-maltopyranoside.
Chemical Structure:

Characteristics and Properties:
Molecular formula: $C_{24}H_{46}O_{11}$
Molecular mass: 510.62 g/mol
Normal boiling temperature: dec.
Melting temperature: 224 °C–226 °C
Density: 1.28 g/mL at 25 °C
CMC: 0.15–0.17 mM at 25 °C
Aggregation number: 70–140
HLB: ~10
Water solubility: soluble.
Notes: used to aid in the solubilization and isolation of hydrophobic membrane proteins to preserve their activity.

Name: ethoxylated tall oil (CAS No: 65071-95-6)
Class: nonionic surfactant
Synonyms: Ethofat242/25, Renex; Industrol TO 16HR; OKM; OKM (surfactant); OKM 10; OKM 12; OKM 50; OKM 60; OKM75; T 13; T 13 (emulsifier); Teric T 2.
Chemical Structure:
mixture

Characteristics and Properties:
Molecular formula: mixture
Molecular mass: 945 g/mol typical
Normal boiling temperature: dec.
Density: 1.08 g/mL at 25 °C
CMC: Typically 4 mM at 25 °C
HLB: 12.2
Water solubility: soluble.
Notes: derived from tall oil obtained from paper processing, can be used at wide range of pH.

Name: glyceryl laurate (CAS No. 27215-38-9)
Class: nonionic surfactant
Synonyms: monolaurin; glycerol monolaurate; glycerol α-monolaurate; glycerin monolaurate; glycerol 1-laurate; glycerol 1-monolaurate; dodecanoic acid 2,3-dihydroxypropyl ester; 1-lauroyl-glycerol; monolauroylglycerin; 2,3-dihydroxypropyl dodecanoate; 2,3-dihydroxypropyl laurate; dodecanoic acid α-monoglyceride; 1-mono-dodecanoyl glycerol; Lauricidin R; 1-monomyristin; Imwitor 312; Lauricidin 802; Dimodan ML 90; Monomuls L 90; Monomuls 90L 12.
Chemical Structure:

Characteristics and Properties:
Molecular formula: $C_{15}H_{30}O_4$
Molecular mass: 274.40 g/mol
Normal boiling temperature: 397 °C
Melting temperature: 63.2 °C
Water solubility: 12.67 mg/L (at 25 °C)
Soluble in: methanol and chloroform (50 mg/L).
Notes: white fluffy semi-solid; has shown both antibacterial and antiviral activity in vitro; used against some infections, but clinical studies are incomplete; used in deodorants and in some cases as an emulsifier in food additives; used as a methane mitigation agent in ruminants.

Name: octyl-β-D-1-thioglucopyranoside (CAS No 85618-21-9)
Class: nonionic
Synonyms: (1S)-octyl-β-D-thioglucoside; n-octyl-β-D-
thioglucoside; OTG; (2R,3S,4S,5R,6R)-2-(hydroxymethyl)-6-
octylsulfanyl-oxane-3,4,5-triol;
(2R,3S,4S,5R,6R)-2-(hydroxymethyl)-6-(octylthio)tetrahydro-2H-
pyran-3,4,5-triol; octylthioglycoside; octyl thioglycoside;
octyl-β-D-thioglucopyranoside.
Chemical Structure:

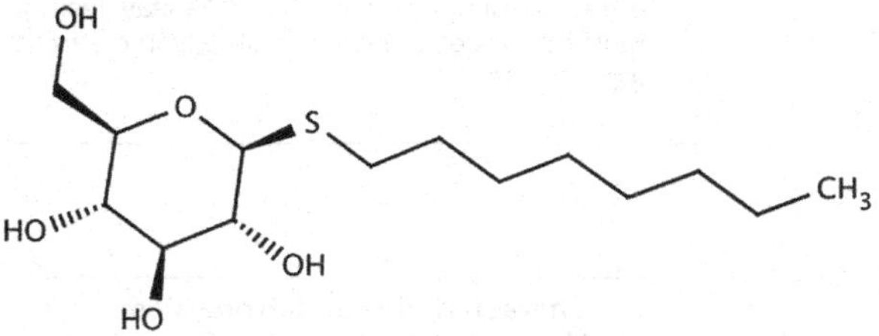

Characteristics and Properties:
Molecular formula: $C_{14}H_{28}O_5S$
Molecular mass: 308.43 g/mol
Normal boiling temperature: dec.
Melting temperature: 125 °C–131 °C
CMC: 9 mM
Water solubility: slightly
Soluble in: ethanol (50 mg/mL).
Notes: colorless powder; used for cell
lysis and for solubilizing proteins without
denaturing them; dialyzable.

Name: perfluorobutanesulfonic acid (CAS: 375-73-5)
Class: anionic surfactant
Synonyms: FC-98; PFBS; nonaflate;
1,1,2,2,3,3,4,4,4-nonafluorobutanesulfonic acid;
1,1,2,2,3,3,4,4,4-nonafluorobutane-1-sulfonic acid;
nonafluorobutanesulfonic acid; nonafluoro-1-butanesulfonic
acid; nonafluorobutane-1-sulfonic acid;
nonafluorobutanesulfonic acid; perfluorobutane sulfonate
Chemical Structure:

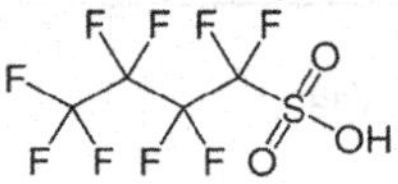

Characteristics and Properties:
Molecular formula: $C_4HF_9O_3S$
Molecular mass: 300.10 g/mol
Normal boiling temperature: 211 °C
Melting temperature: 76 °C–84 °C
Density: 1.811 g/mL (at 25 °C)
Water solubility: reacts violently.
Notes: used as a replacement for
perfluoroctanesulfonic acid stain repellent.

Name: Polyethylene glycol tert-octylphenyl ether
(CAS No: 9002-93-1)
Class: Nonionic surfactant
Synonyms: Octoxynol-9, **Triton X-100**, TX-100.
Chemical Structure:

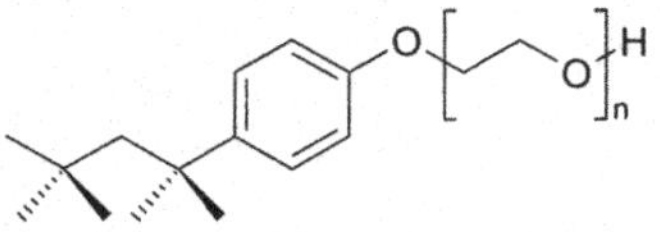

Characteristics and Properties:
Molecular formula: $C_{14}H_{22}O(C_2H_4O)_n$
(n = 9–10)
Molecular mass: 647 g/mol typical
Normal boiling temperature: 270 °C
Melting temperature: 6 °C
Density: 1.07 g/mL at 20 °C
CMC: 0.24–0.27 mM at 25 °C
Aggregation number: 140
HLB: 13.5
Water solubility: fully miscible
Soluble in: benzene, toluene, xylene, trichloroethylene,
ethylene glycol, ethyl ether, ethanol, isopropanol, and
ethylene dichloride.
Notes: a nonionic surfactant and emulsifier widely used
for solubilizing membrane proteins and lysing cells; no
antimicrobial properties; absorbs in the UV, so can
interfere with protein quantitation.

Name: polyoxyethylene (20) sorbitan monolaurate
(CAS No: 9005-64-5)
Class: nonionic surfactant
Synonyms: Polysorbate 20; PEG(20) sorbitan
monolaurate; Alkest TW 20; **Tween 20**.
Chemical Structure:

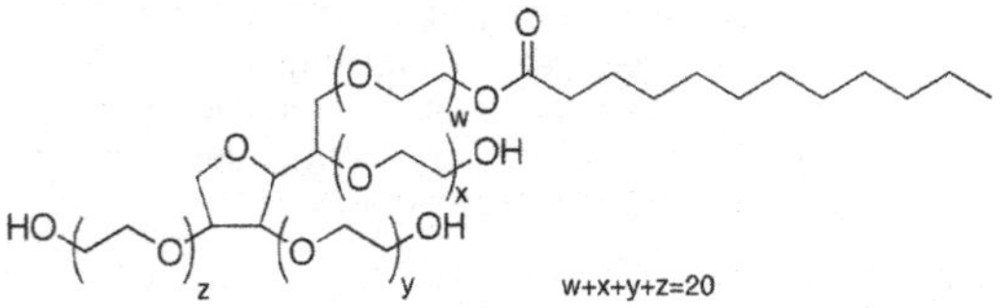

w+x+y+z=20

Characteristics and Properties:
Molecular formula: $C_{58}H_{114}O_{26}$
Molecular mass: 1,227.54 g/mol
Normal boiling temperature: dec.
Density: 1.06 g/mL at 25 °C
CMC: 0.06 mM at 25 °C
Aggregation number: 62
HLB: 16.7
Water solubility: soluble.
Notes: washing agent in immunoassay; lysing
mammalian cells; used industrially in cleaning
applications.

Name: polyoxyethylene (20) sorbitan monooleate
(CAS No: 9005-65-6)
Class: nonionic surfactant
Synonyms: Polysorbate 80; Alkest TW 80; **Tween 80**
Chemical Structure:

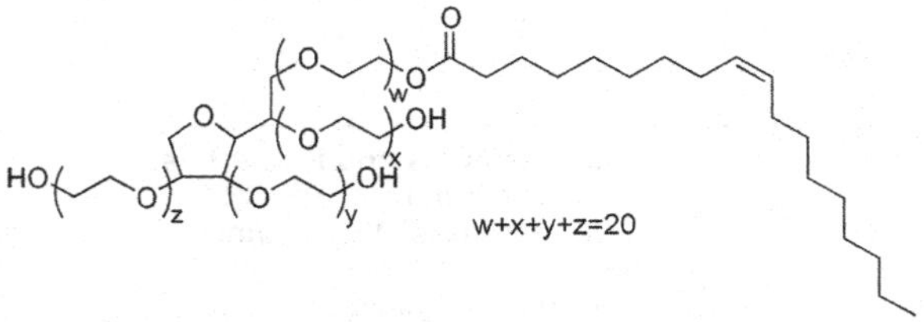

w+x+y+z=20

Characteristics and Properties:
Molecular formula: $C_{64}H_{124}O_{26}$
Molecular mass: 1,310 g/mol
Normal boiling temperature: dec.
Density: 1.06–1.09 g/mL at 25 °C
CMC: 0.01 mM at 25 °C
HLB: 13.4–15
Water solubility: soluble.
Notes: used to stabilize aqueous drug
formulations; emulsifier for drugs;
surfactant used in cosmetics and soaps,
and in mouthwash, oily liquid.

Name: sodium dodecyl sulfate (CAS No: 151-21-3)
Class: anionic surfactant
Synonyms: Sodium lauryl sulfate, **SDS**, sodium
monododecyl sulfate; sodium lauryl sulphate;
sodium monolauryl sulfate; sodium dodecanesulfate;
sodium coco-sulfate; dodecyl alcohol, hydrogen
sulfate, sodium salt; n-dodecyl sulfate sodium;
sulfuric acid monododecyl ester sodium salt.
Chemical Structure:

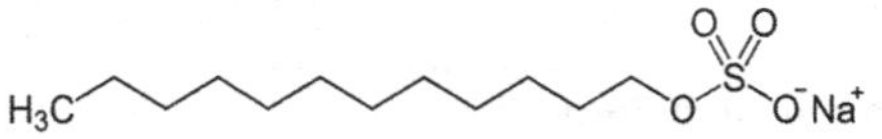

Characteristics and Properties:
Molecular formula: $CH_3(CH_2)_{11}SO_4Na$
Molecular mass: 288.37 g/mol
Normal boiling temperature: dec.
Melting temperature: 206 °C
Density: 1.06 g/mL at 25 °C
CMC: 8.2 mM at 25 °C
Aggregation number: 62
HLB: 40
Water solubility: 200 mg/mL at 20 °C
Soluble in: ethanol.
Notes: aid in lysing cells during DNA extraction and
for unraveling proteins in SDS-PAGE; low pK_a; can
be used in pH as low as 4; can be prone to foaming.

Name: sodium lauroyl sarcosinate (CAS No. 137-16-6)
Class: ionic (amphiphilic) surfactant
Synonyms: sarkosyl; sarcosyl NL; sodium
N-lauroylsarcosinate; sarcosyl; sarcosyl NL; maprosyl
30; sodium [dodecanoyl(methyl)amino]acetate;
n-lauroylsarcosine, sodium salt; N-methyl-N-(1-
oxododecyl)glycine, sodium salt; sodium-n-lauriyl
sarcosinate.
Chemical Structure:

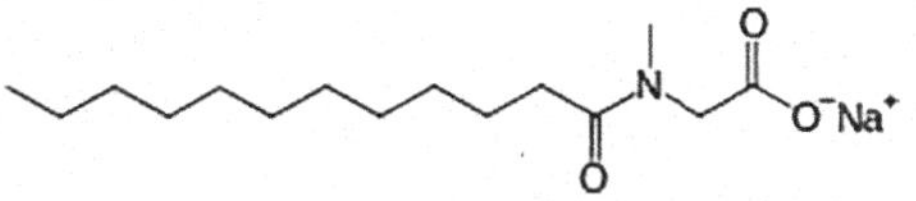

Characteristics and Properties:
Molecular formula: $C_{15}H_{28}NaNO_3$
Molecular mass: 293.38 g/mol
Normal boiling temperature: dec.
Melting point: 140 °C
Water solubility: soluble (7.0 %–8.6 %).
Notes: white powder (94 %); pH 7.5–8.5 (10 % sol);
foam stabilizer; lubricant; corrosion inhibitor;
bacteriostat; used in emulsion polymerization.

Name: Tergitol 15-S-9 (CAS No. 68131-40-8)
Class: nonionic surfactant
Synonyms: secondary alcohol ethoxylate
Chemical Structure:

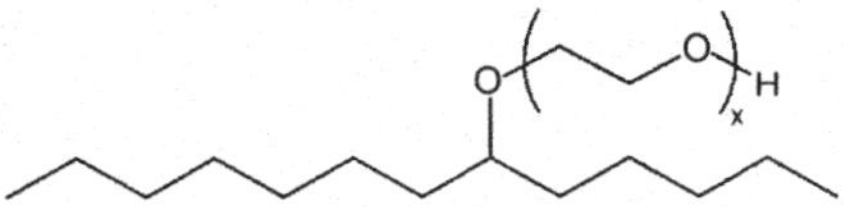

Characteristics and Properties:
Molecular formula: $C_{12-24}H_{25-29}O[CH_2CH_2O]_xH$
Molecular mass: 230–378 g/mol
Density: 1.006 g/mL at 20 °C
CMC: 52 ppm (mass/mass) at 25 °C
Water solubility: soluble
Soluble in: chlorinated solvents and most polar organic
solvents (ethanol, acetone).
Notes: pale yellow liquid; pH (1 % aqueous solution) 7.1;
chemically stable in the presence of dilute acids, bases
and salts; compatible with anionic, cationic, and other
nonionic surfactants; used in chemical analysis,
high-performance cleaners and agrochemicals,
member of the large Tergitol family of surfactants.

CHAPTER **15**

Tables and Guidelines for Laboratory Safety

Many if not most institutions now have comprehensive safety guidelines and procedures that are laid out as conditions of employment. This section is not meant to replace any local policies or standard operating procedures, but rather it is meant to augment and support such policies. We also provide useful guidance in the safe use of equipment in the laboratory; it is presented as much as a how-to, rather than a how-not-to guide.

MAJOR CHEMICAL INCOMPATIBILITIES

The following chemicals react, sometimes violently, in certain chemical environments [1–9]. Incompatibilities may cause fires, explosions, or the release of toxic gases. Extreme care must be taken when working with these materials. This list is not inclusive, and the reader is urged to consult multiple sources for more specific information. When using any chemicals, thorough reading of the Material Safety Data Sheets (MSDS) [10] is strongly recommended.

REFERENCES

1. Dean, J. A., ed., *Lange's Handbook of Chemistry*, 16th ed., McGraw-Hill Book Co., New York, 2005.
2. Fieser, L.F. and Fieser, M., *Reagents for Organic Synthesis*, John Wiley and Sons, New York, 1967.
3. Gordon, A.J. and Ford, R.A., *The Chemist's Companion: A Handbook of Practical Data, Techniques and References*, John Wiley and Sons, New York, 1972.
4. Shugar, G.J. and Dean, J.A., *The Chemist's Ready Reference Handbook*, McGraw-Hill Book Company, New York, 1990.
5. Svoronos, P., Sarlo, E., and Kulawiec, R., *Organic Chemistry Laboratory Manual*, 2nd ed., McGraw-Hill Book Co., New York, 1997.
6. Pohanish, R.P. and Greene S.A., *Wiley Guide to Chemical Incompatibilities*, 3rd ed., John Wiley and Sons, New York, 2009.
7. Pohanish, R.P., *Sittig's Handbook of Toxic and Hazardous Chemicals and Carcinogens*, Noyes Publications, Bracknell, Berkshire, UK, 2007.
8. Pohanish, R.P., *Rapid Guide to Hazardous Chemicals in the Environment*, Noyes Publications, Bracknell, Berkshire, UK, 1997.
9. Pohanish, R.P., *HazMat Data: For First Response, Transportation, Storage, and Security*, Wiley-Interscience, New York, NY, 2004.
10. http://www.ilpi.com/MSDS/, accessed January 2020.

Chemical	Incompatible Chemicals
Acetic acid	Strong acids (chromic, nitric, perchloric), peroxides.
Acetylene	Air, copper, halogens (chlorine, bromine, iodine).
Alkali metals	Acids, water, hydroxy compounds, polychlorinated hydrocarbons (for example, CCl_4), halogens, carbon dioxide, oxidants.
Ammonia, anhydrous	Halogens (bromine, chlorine, iodine), hydrofluoric acid, liquid oxygen, calcium or sodium hypochlorite, heavy metals (silver, gold mercury), nitric acid.
Ammonium nitrate	Metal powders, chlorates, nitrites, sulfur, sugar, flammable and combustible organics, acids, sawdust.
Anilines	Concentrated acids (nitric, chromic), oxidizing agents (chromium (III) ions, peroxides, permanganate).
Carbon, activated	Oxidizing agents, unsaturated oils.
Carboxylic acids	Metals (alkalis), organic bases, ammonia.
Chlorates	Flammable and combustible organic compounds, finely powdered metals, manganese dioxide, ammonium salts.
Chromic acid	Anilines, 1° or 2° alcohols.
Halogens (chlorine, bromine)	Finely powdered metals, diethyl ether, hydrogen, unsaturated organic compounds, carbide salts, acetylene, alkali metal.
Copper	Oxidizing agents.
Ether (including diethyl)	Peroxides (especially after long exposure of ether to air; see the other tables in this chapter dealing with this compound).
Fluorine	Reactive (as a strong oxidizing agent) to a certain degree with most compounds, but it can sometimes cause a violent reaction.
Hydrocarbons (saturated)	Halogens (especially fluorine) in the presence of ultraviolet light and peroxides.
Hydrocarbons (unsaturated)	Halogens, concentrated strong acids, peroxides.
Hydrofluoric acid	Ammonia, glass.
Hydrogen peroxide	Metals, alcohols, potassium permanganate, thiols, flammable and combustible materials.
Iodine	Acetaldehyde, antimony, unsaturated hydrocarbons, ammonia and some amines.
Mercury	Some metals, ammonia, terminal alkynes (see the other tables in this chapter dealing with this element).
Nitric acid (concentrated)	Flammable liquids, unsaturated organics, lactic acid, coal, ammonia, powdered metals, wood, alcohols, electron-rich aromatic rings (phenols, analines).
Perchloric acid	Some organics, acetic anhydride, metals, alcohols, wood and its derivatives (see the other tables in this chapter dealing with this compound).
Permanganates, general	Aldehydes, alcohols, unsaturated hydrocarbons.
Peroxides	Flammable liquids, metals, aldehydes, alcohols, impact, hydrocarbons (unsaturated).
Picric acid	Dryness and impact, alkali metals, oxidizing agents, concentrated bases.
Potassium, metal	See alkali metals.
Potassium permanganate	Hydrochloric acid, glycerol, hydrogen peroxide, sulfuric acid, wood.
Silver salts, organic	Dryness and prolonged air exposure.
Sodium, metal	See alkali metals.
Sulfuric acid, concentrated	Electron-rich aromatic rings (phenols, anilines), unsaturated hydrocarbons potassium permanganate, chlorates, perchlorates.

PROPERTIES OF HAZARDOUS SOLIDS

The following table lists some of the more important properties of hazardous room temperature solids commonly used in the analytical laboratory [1,2]. The flash points were determined with the open cup method.

REFERENCES

1. Turner, C.F. and McCreery, J.W., *The Chemistry of Fire and Hazardous Materials*, Allyn and Bacon, Boston, MA, 1981.
2. Schnepp, R., *Hazardous Materials: Awareness and Operations, Jones and Bartlett Learning*, Burlington, MA, 2019.

Name	Formula	Specific Gravity (at 20 °C)	Melting Point, °C	Boiling Point, °C	Flash Point, °C	Autoignition Point, °C	Ignition/Explosion Mechanism	Fire Suppression Media
Acetyl peroxide	$(CH_3CO)_2O_2$	1.2	30	63.1	—	—	Heat, shock	a, c
Aluminum (finely divided)	Al	2.7	660.323	2,519	—	—	Mixing with iron oxides	a
Aluminum chlorate	$Al(ClO_3)_3$	—	—	—	—	—	Heat, impact agents, reducing agents	a
Aluminum chloride	$AlCl_3$	2.4	192.6	180 (sb)	—	—	Heat, moisture	a
Ammonium nitrate	NH_4NO_3	1.7	169.7	200–260 (dec)	—	—	Heat	b
Ammonium nitrite	NH_4NO_2	1.7	dec	—	70	—	Heat, shock, impact	a
Ammonium perchlorate	NH_4ClO_4	—	dec	—	—	—	Shock, impact	—
Antimony, gray	Sb	6.7	630.628	1,587	—	—	Heat, water	a
Antimony trisulfide	Sb_2S_3	4.6	—	—	—	—	Heat, strong organic acids, oxidizers	b
Antimony pentasulfide	Sb_2S_5	4.1	—	—	—	—	Heat, strong oxidizers, acids	a
Barium	Ba	3.6	727	≈1,845	—	—	Heat	a
Beryllium	Be	1.87	1,287	2,468	—	—	Heat, friction	a
Cadmium	Cd	8.6	321.069	767	—	—	Heat	a
Calcium hypochlorite	$Ca(ClO)_2{\cdot}4H_2O$	—	dec	—	—	—	Heat, contact with combustible material, acid	—
Camphor	$C_{10}H_{16}O$	1.0	176	409 (sub)	66	466	High conc. in air	a, c
Cesium	Cs	1.9	28.5	671	—	—	Water	f
Iodine	I_2	4.9	113.7	184.4	—	—	Heat	c
Lithium	Li	0.53	180.50	1,342	—	—	Water, inorganic acids	a
Magnesium	Mg	1.75	650	1,090	—	—	Water	a
Phosphorus, red	P_4	2.2	579.2	431 (sub)	—	260	Heat, oxidizers	b, e
Phosphorus, white	P_4	1.82	44.15	280.5	Ambient	30	Heat, oxidizers, dry atmosphere	b

(Continued)

Name	Formula	Specific Gravity (at 20 °C)	Melting Point, °C	Boiling Point, °C	Flash Point, °C	Autoignition Point, °C	Ignition/Explosion Mechanism	Fire Suppression Media
Phosphorus pentachloride	PCl_5	4.7	167 (triple point)	260 (sub)	—	—	Moist air, heat	a, c
Phosphorus pentasulfide (diphosphorus pentasulfide)	P_2S_5	2.03	285	515	142	287	Water, acids	—
Potassium	K	0.89	285	63.5	759	—	Water, acids	a
Potassium chlorate	$KClO_3$	2.3	357	Dec	—	—	Charcoal, sulfur, and phosphorus	b
Potassium nitrate	KNO_3	2.1	334	400 (dec)	—	—	Fiction, contact with organics	b
Potassium nitrite	KNO_2	1.9	438	537 (dec)	—	—	Friction, impact	a
Sodium	Na	0.97	97.794	882.940	—	—	Moisture	a
Sodium hydride	NaH	0.9	800	—	—	—	Water, oxidizers	a
Sodium nitrate	$NaNO_3$	2.3	306.5	—	—	—	Contact with organics	b
Sodium nitrite	$NaNO_2$	2.17	284	>320 (dec)	—	—	Contact with organics	b
Sodium styrene sulfonate	$C_8H_7SO_3Na$	—	225 (dec)	—	—	462	Hot surfaces, sparks	a, b, c,
Sulfur (monoclinic)	S/S_8	2.07	115.21	444.61	207	232	Heat in the presence of oxygen	a, d
Triphenylboron	$(C_6H_5)_3B$	—	136	347	—	220	Water (produces benzene), heat	a, b, c, d

dec, decomposes; a, dry chemical extinguisher; b, H_2O, c, CO_2; d, foam; e, wet sand; f, chlorinated hydrocarbons; sb, sublimes.

NANOMATERIAL SAFETY GUIDELINES

The classification of nanomaterials includes nanoobjects and nanoparticles; nano-objects are materials with at least one dimension (length, width, height, and /or diameter) that is between 1 and 100 nm (1×10^{-9} m), nanoparticles are materials in which all three dimensions are on this scale. [Note that the ASTM (American Society for Testing and Materials) definition allows for two dimensions between 1 and 100 nm.] Beyond scale, nanomaterials can be classified as natural, incidental, and engineered, depending on origin. Natural nanomaterials include volcanic products, viruses, sea spray, and mineral aerosols and are ubiquitous in nature at appreciable concentrations. Incidental nanomaterials include metal vapors produced during welding, sandblasting dust and other industrial effluent, cooking smoke, and diesel engine particulates. The environmental health and safety aspects of natural nanomaterials have received some study, and among the incidental nanomaterials, welding vapors and diesel fuel particulates have received extensive study. In recent years, however, there has been a great emphasis on engineered nanomaterials, and it is this class, which includes metal nanoparticles, nanorods, nanowires, nanotubes, Buckyballs, nanocapsules, and quantum dots that are the main concern here. Study of the environmental health and safety risks of engineered nanomaterials remains an active area of research that is receiving increasing attention due to the widespread use of these materials in numerous applications ranging from medicine to energy storage. While much is still unknown regarding the fate and toxicity of this class of materials, here, we provide some general guidelines for the safe handling of nanomaterials. We begin with some simplified definitions or terms used in nanotechnology, needed for understanding of these safety guidelines as well as those provided elsewhere [1,2].

REFERENCES

1. Bruno, T.J. and Smith, B.L., *CRC Handbook of Chemistry and Physics*, 100th Ed., Boca Raton, FL, 2019.
2. Schnepp, R., *Hazardous Materials: Awareness and Operations*, Jones and Bartlett Learning, Burlington, MA, 2019.

DEFINITIONS

Aerodynamic diameter: An indirect measure of particle diameter defined as the diameter of a sphere with a density of 1,000 kg/m^3, having the same settling velocity of a particle of interest.

Agglomerate: A group of particles (which may include nanoparticles) held together in a loose cluster by weak forces that may include van der Waals forces, surface tension, and electrostatic forces. Agglomerates are often re-suspendable.

Aggregate: A heterogeneous particle held together with relatively strong forces such that the particle is not easily disassembled. Aggregates are typically not re-suspendable.

Buckyballs: Spherical carbon (C_{60}) fullerenes.

Fullerenes: Molecules composed entirely of carbon, usually in the form of a hollow sphere, ellipsoid, or tubes.

Graphene: A one-atom thick sheet of carbon.

Multi-walled carbon nanotube: Multiple sheets of sheet graphene wrapped into a tube of nanoscale dimensions.

Nanoaerosol: A collection of nanomaterials suspended in a gas.

Nanocolloid: A nanomaterial suspended in a gel or other semi-solid substance.

Nanocomposite: A solid material composed of two or more nanomaterials having different physical characteristics.

Nanohydrosol: A nanomaterial suspended in a solution.

Nanotube: A seamless tube with a diameter on the order of nanometers.

Nanowire: A wire of dimensions on the order of nanometers.

Quantum dot: A nanomaterial or nanoparticle that confines the motion of conduction band electrons, valence band holes, or excitons (pairs of conduction band electrons and valence band holes) in all three spatial directions.

Single-walled carbon nanotube: A single sheet graphene wrapped into a tube of nanoscale dimensions.

Ultrafine particle: A (usually) airborne particle with a diameter less than 100 nm.

SAFETY ISSUES AND EXPOSURE ROUTES

The unique safety issues posed by nanomaterials result from the potential of deep penetration into tissue, the potential of passing through the blood–brain barrier, and the possible ability to translocate between organs. Biological effects result from the size, shape, polarity/charge, adsorptive capacity and surface composition, and the ability to bind biological proteins and receptors. Nanomaterials have a higher reactivity than the parent compounds, often having catalytic effects and often presenting greater flammability or explosion risks. For example, bulk elemental gold is considered inert, but gold nanoparticles below 5 nm are catalytic toward a number of oxidation reactions.

The most obvious exposure route of nanomaterials is respiratory; particles depositable in the air exchange region of the lungs are considered respirable. Ingestion can occur from unintentional hand-to-mouth transfer. Finally, nanoparticles can be absorbed through skin or cuts/abrasions to the skin.

Guidelines for Safe Handling of Nanomaterials

The safe handling of nanomaterials will generally follow the usual laboratory safety grid:

Elimination: A change in the experimental design to avoid the hazard
Substitution: The use of a surrogate of lower hazard
Engineering controls: The use of enclosures, fume hoods, etc.
Administrative: Adherence to standard procedures and protocols
Personal protective equipment (PPE): The last line of safety, including gloves, clothing, and respirators.

Clearly, elimination and substitution are most useful for nanoparticles of incidental origin. Research with engineered nanoparticles must make use of engineering controls, administrative controls, and PPE.

The use of non-regenerating general ventilation systems, such as fume hoods, and high-efficiency particulate air (HEPA) dust collection systems are critical to safe handling. Where possible, installation of ultra-low particulate air (ULPA) filters should be used, since they are widely viewed as being more effective for engineered nanomaterials. The lab in which nanomaterials are handled should be under negative pressure relative to the surroundings (corridor, service galley, etc.). The entry on fume hoods and/or biological safety cabinets (BSCs) in this section provides additional information on fume hood selection and operation.

Where possible, manipulations should be conducted in solution (or in a liquid phase) to minimize the potential of aerosol formation.

Solution-phase nanomaterials should be handled wearing gloves. Gloves should be compatible with the solvent used to disperse the nanomaterials in solution. In general, nitrile gloves are recommended, and double gloving is advisable for heavy or prolonged usage. Gloves with cuffs, clothing with full sleeves to protect wrists, or a laboratory coat is recommended. Liquid- or solution-phase nanomaterial manipulations are best conducted in a fume hood or BSC, especially when employing nanomaterials dispersed in open containers, in solvents with known health risks, or when higher risk activities such as sonication, agitation, and vortex mixing are involved. Manipulations conducted in closed containers need not be performed in a fume hood or BSC; however, the container should be opened inside of such an enclosure since aerosols could be released upon opening. Disposable bench covers should be used where spillage is possible. Any spillage should be cleaned up immediately. Contaminated gloves should be removed and replaced immediately.

Manipulation of dry nanomaterials must be performed in a fume hood or BSC. Transport of dry nanomaterials from place to place in the lab must be done in closed containers.

For manipulations of air-sensitive nanomaterials, a glove box or glove bag is required.

Hand washing must be done after manipulations of nanomaterials. Work areas should be cleaned after completion of tasks. Adequate consideration should be given to tasks involved with the maintenance of equipment or instrumentation used in work on nanomaterials. Such maintenance should be done with the assumption of the presence of nanomaterials.

Waste Disposal

Though the fate and toxicity of nanomaterials remains largely unknown and is still an area of active investigation, nanomaterials and any by-products from their synthesis should be treated as potentially hazardous waste. Nanomaterials should be properly disposed of based on their nanomaterial and solvent compositions. Nanomaterials containing heavy metals should be treated accordingly and separately from other waste streams. When possible, it is advisable to collect or fully dissolve nanomaterials that are present in solution to limit the volume of waste generated. Often nanomaterials can be aggregated or precipitated from solution with an anti-solvent and filtered off to be recycled or collected as solid waste. This is called "crashing out" in laboratory vernacular. Alternatively, adsorbents such as activated charcoal can often be used to remove certain types of nanomaterials from solution upon filtration and collection of the adsorbent following exposure to the nanomaterial-containing solution. Other types of nanomaterials such as metal oxides can often be dissolved completely with strong acids.

COMPOUNDS THAT ARE REACTIVE WITH WATER

The following is a partial listing of families of compounds that are known to have reactivity with water [1,2]. Depending upon the specific compound, the reaction can be rapid and even violent, or simply slow hydrolysis.

REFERENCES

1. Bretherick, L., Urben, P.G., and Pitt, M.J., *Bretherick's Handbook of Reactive Chemical Hazards: An Indexed Guide to Published Data,* 6th Ed., Butterworths, London, 1999.
2. Schnepp, R., *Hazardous Materials: Awareness and Operations, Jones and Bartlett Learning,* Burlington, MA, 2019.

Acid anhydrides
Acyl halides
Alkali metals
Alkylaluminum derivatives
Alkylmagnesium derivatives
Alkyl-nonmetal halides
Complex anhydrides
Metal halides
Metal oxides
Nonmetal halides and their oxides
Nonmetal oxides.

PYROPHORIC COMPOUNDS—COMPOUNDS THAT ARE REACTIVE WITH AIR

The following listing provides the classes of compounds, with some examples, which can undergo spontaneous reaction upon exposure to air [1,2]. In some cases, the reaction is vigorous, while in others, the reaction is more subdued or will only occur if other conditions (such as temperature, humidity, or a reactive surface) are present. The reader is advised to check the literature for more specific information.

REFERENCES

1. Bretherick, L., Urben, P.G., Pitt, M.J., *Bretherick's Handbook of Reactive Chemical Hazards: An Indexed Guide to Published Data,* 6th Ed., Butterworths, London, 1999.
2. Pyrophoric Materials, https://www.chem.tamu.edu/safety, Texas A&M University, Accessed January 2020.

Alkali metals* (sodium, potassium, potassium/sodium alloy, lithium/tin alloys)
Alkylaluminum derivatives (diethylaluminum hydride)
Alkylated metal alkoxides (diethylethoxyaluminum)
Alkylboranes
Alkylhaloboranes (bromodimethyl borane)
Alkylhalophosphines
Alkylhalosilanes
Alkyl metals
Alkyl nonmetal hydrides
Boranes (diborane)
Carbonyl metals (pentacarbonyl iron, octacarbonyl dicobalt, nickel carbonyl)
Complex acetylides
Complex hydrides (diethylaluminum hydride)
Finely divided metals*(calcium, zirconium)
Haloacetylene derivatives
Hexamethylnitrato dialuminum salts
Metal hydrides (germane, sodium hydride, lithium aluminum hydride)
Nonmetal hydrides
Some nonmetal (organic) halides (dichloro(methyl)silane)
Spent hydrogenation catalysts (can be especially hazardous because of adsorbed hydrogen; for example, Raney nickel)
White phosphorus*(slang term: willie pete).

---

* Note that the reactivity depends on the particle size and the ease at which oxides are formed on the metal surface.

VAPOR PRESSURE OF MERCURY

The following table provides data on the vapor pressure of mercury, useful for assessing and controlling the hazards associated with use of mercury, for example, as an electrode [1].

REFERENCE

1. Rumble, J., ed., *CRC Handbook of Chemistry and Physics*, 100th Ed. CRC Press, Boca Raton, FL, 2019.

Temperature (°C)	Vapor Pressure (mm Hg)	Vapor Pressure (Pa)	Temperature (°C)	Vapor Pressure (mm Hg)	Vapor Pressure (Pa)
0	0.000185	0.0247	28	0.002359	0.3145
10	0.000490	0.0653	30	0.002777	0.3702
20	0.001201	0.1601	40	0.006079	0.8105
22	0.001426	0.1901	50	0.01267	1.689
24	0.001691	0.2254	100	0.273	36.4
26	0.002000	0.2666			

FLAMMABILITY HAZARDS OF COMMON SOLVENTS

The following table lists relevant data regarding the flammability of common organic liquids [1–3].

REFERENCES

1. Turner, C.F. and McCreery, J.W., *The Chemistry of Fire and Hazardous Materials*, Allyn and Bacon, Boston, 1981.
2. NIOSH Pocket Guide to Chemical Hazards, Centers for Disease Control and Prevention 1600 Clifton Rd. Atlanta, GA 30333, USA (http://www.cdc.gov/niosh/npg/pgintrod.html, accessed January 2020).
3. Schnepp, R., *Hazardous Materials: Awareness and Operations*, Jones and Bartlett Learning, Burlington, MA, 2019.

Solvent	Formula	Specific Gravity	Boiling Temperature (°C)	Flash Point (°C)	Auto-ignition Point (°C)	How to Extinguish Fires*
Acetaldehyde	CH₃CHO	0.8	20.8	−38	185	a, b, c
Acetone	(CH₃)₂CO	0.8	56.08	−18	538	a, b
Acetonitrile	CH₃C≡N	0.79	81.6	6	−	a, c, d
Acetylacetone	CH₃COCH₂COCH₃	1.0	138.1	41	−	a, b, c
Acrolein	CH₂=CHCHO	0.8	52.3	−26	277	a, b, c
Acrylonitrile	CH₂=CH-CHC≡N	0.81	77.2	0	481	a, c, d
Allylamine	CH₂=CHCH₂NH₂	0.8	54	−29	374	a, b
1-Pentanethiol (amylmercaptan)	CH₃(CH₂)₄SH	0.8	127	18	−	a, b
Aniline	C₆H₅NH₂ H₂N—◯	1.0	184.1	70	768	a, b, c use masks
Anisole	C₆H₅OCH₃ O—◯	1.0	153.6	52	−	a, b, c
Benzaldehyde	C₆H₅CHO	1.1	178.7	65	192	a, b, c
Benzene	C₆H₆ ◯	0.88	80.2	−11	563	a, b, c
1-Butanol (butyl alcohol)	C₄H₉OH	0.8	117.6	29	366	a, b, c
t-Butylperacetate	CH₃CO(O₂)C(CH₃)₃	−	−	<27	−	b, c
t-Butylperbenzoate	C₆H₅CO(O₂)C(CH₃)₃	>1.0	112	75.6	8	a, b, c
Butyraldehyde	CH₃(CH₂)₂CHO	0.8	75.1	7	230	a, b, c
Carbon disulfide	CS₂	1.3	46.2	−30	100	b, d use masks
Crotonaldehyde	CH₃CH=CHCHO	0.9	102.35	7.2	232	a, b, c

(Continued)

Solvent	Formula	Specific Gravity	Boiling Temperature (°C)	Flash Point (°C)	Auto-ignition Point (°C)	How to Extinguish Fires*
Cumene hydroperoxide (hydroperoxide, 1-methyl-1-phenylethyl)	$C_6H_5C(CH_3)_2O_2H$	1.0	153 (100.6 at 1 kPa)	175	–	a, b, c
Cyclohexanone	$C_6H_{10}O$	0.9	155.4	43	420	a, b, c
Diacetyl(2,3-butanedione)	$(CH_3CO)_2$	1.0	88.1	27	–	a, b, c
Diethanolamine	$(HOCH_2CH_2)_2NH$	1.1	271.2	152	662	b, c
Diethylene glycol diethyl ether	$CH_3(CH_2OCH_2)_3CH_3$	0.9	189	83	–	a, halons
Diethylether	$(C_2H_5)_2O$	0.7	34.4	–45	180	a, b, halons
Diethylketone	$(C_2H_5)_2CO$	0.8	101.9	13	452	a, b, c
Dimethyl sulfate	$(CH_3)_2SO_4$	1.3	186 (decomposes)	83	188	a, b, c, d
Dimethyl sulfide	$((CH_3)_2S$	0.8	38	–18	206	b, c
1,4-Dioxane	$(CH_2CH_2O)_2$	1.0	101.3	12.7	180	a, b, c
Ethanol	C_2H_5OH	0.8	78.24	13	423	a, b, c
Ethylacetone (2-pentanone)	$CH_3COCH_2CH_2CH_3$	0.8	102	7	504	a, b, c
Ethylamine	$C_2H_5NH_2$	0.7	16.6	–18	384	a, b, c
Ethylenediamine	$H_2NCH_2CH_2NH_2$	0.9	118	38.9	385	a, b, c
Ethylene glycol	$HOCH_2CH_2OH$	1.1	197.4	111	413	a, b, c, d
Formaldehyde (aqueous solution)	$HCHO$	1.0	99	88	427	a, b, c
Furfural	C_5H_4O	1.2	161.5	60	316	a, b, c, d
Furfuryl alcohol	$C_5H_6O_2$	1.1	168	65	491	a, b, c
Gasoline	C_7H_{16} (isomers)	<1.0	38–218	–43	257	a, b, c
Hexylamine	$C_6H_{13}NH_2$	0.8	130	29	–	a, b
Isopropanol	$(CH_3)_2CHOH$	0.8	82.4	12	399	b, c
Isopropyl ether	$((CH_3)_2CH)_2O$	0.7	68.3	–28	443	a, b
Kerosene	–	<1.0	149–316	38–71	229	a, b, c
Methanol	CH_3OH	0.8	64.5	11	464	a, b
Methylamine (aq)	CH_3NH_2	0.7	31	–18	384	a, b, c
Methylaniline	$CH_3NHC_6H_5$	0.8	195.6	79.4	533	a, b
Methylethyl ketone	$CH_3COCH_2CH_3$	0.8	79.6	–6	516	a, b, c
Methylethyl ketone peroxide	$CH_3C(OOH)_2CH_2CH_3$	1.2	118 (decomposes)	51–93	–	a, b

(Continued)

Solvent	Formula	Specific Gravity	Boiling Temperature (°C)	Flash Point (°C)	Auto-ignition Point (°C)	How to Extinguish Fires*
Naphtha (mixture)	–	0.8–0.9	149–216	38–46	227–496	a, b, c
Paraldehyde	$(CH_3CHO)_3$, cyclic	1.0	124	36	238	a, b, c
2-Pentanone	See ethylacetone	–	–	–	–	–
3-Pentene nitrile	$CH_3CH=CHCH_2C\equiv N$	0.83	145.7	40	–	b, c, d
Peracetic acid	CH_3COOOH	1.2	105	40	–	a, b, c
Petroleum ether	–	<0.7	38–79	<0	288	a, b, c
Propionaldehyde (propanal)	CH_3CH_2CHO	0.8	48.0	8	207	a, b, c
Propylamine	$CH_3CH_2CH_2NH_2$	0.7	47.21	–37	318	a, b, c
Propylene glycol	$CH_3CHOHCH_2OH$	1.0	187	99	421	a, b
Sulfur chloride	S_2Cl_2	1.7	138	118	234	b, c
Sulfuryl chloride	$SOCl_2$	1.7	69.4	–	–	a, b, c
Tetrahydrofuran	C_4H_8O	0.9	66.0	–14	321	a, b, c
Thionylchloride	$SOCl_2$	1.6	75.6	–	–	–
Toluene	$C_6H_5CH_3$	0.87	110.60	4	510	a, b, c
Triethanolamine	$(HOCH_2CH_2)_3N$	1.1	350	179	–	b, c, d
Triethylamine	$(C_2H_5)_3N$	0.7	88.8	7	–	a, b, c
Xylene (o-)	$C_6H_4(CH_3)_2$	0.88	144.4	32	463	a, b, c
Xylene (m-)	$C_6H_4(CH_3)_2$	0.86	139.1	28	527	a, b, c
Xylene (p-)	$C_6H_4(CH_3)_2$	0.86	138.3	27	529	a, b, c

* a, foam; b carbon dioxide; c, dry chemical (ABC); and d, water.

ABBREVIATIONS USED IN THE ASSESSMENT AND
PRESENTATION OF LABORATORY HAZARDS

The following abbreviations are commonly encountered in presentations of laboratory and industrial hazards. The reader is urged to consult Ref. [1] for additional information.

REFERENCE

1. Furr, A.K., ed., *CRC Handbook of Laboratory Safety*, 5th ed., CRC Press, Boca Raton, 2000.

CC: Closed Cup; method for the measurement of the flash point. With this method, sample vapors are not allowed to escape as they can with the open cup method. Because of this, flash points measured with the CC method are usually a few degrees lower than those measured with the OC. The choice between CC and OC is dependent on the (usually ASTM) standard method chosen for the test.

COC: Cleveland Open Cup, see Open Cup.

IDLH: Immediately Dangerous to Life and Health; the maximum concentration of chemical contaminants, normally expressed as parts per million (ppm, mass/mass), from which one could escape within 30 minutes without a respirator and without experiencing any escape impairing (severe eye irritation) or irreversible health effects. Set by National Institute of Occupational Safety and Health (NIOSH). Note that this term is also used to describe electrical hazards.

LEL: Lower Explosion Limit; the minimum concentration of a chemical in air at which detonation can occur.

LFL: Lower Flammability Limit; the minimum concentration of a chemical in air at which flame propagation occurs.

MSDS: Material Safety Data Sheet; a (legal) document that must accompany any supplied chemical that provides information on chemical content, physical properties, hazards, and treatment of hazards. The MSDS should be considered only a minimal source of information and cannot replace additional information available in other, more comprehensive sources.

NOEL: No Observed Effect Level; the maximum dose of a chemical at which no signs of harm are observed. This term can also be used to describe electrical hazards.

OC: Open Cup; also called Cleveland Open Cup. This refers to the test method for determining the flash point of common compounds. It consists of a brass, aluminum or stainless steel cup, a heater base to heat the cup, a thermometer in a fixture, and a test flame applicator. The flash point is the lowest temperature at which a material will form a flammable mixture with air above its surface. The lower the flash point, the easier it is to ignite.

PEL: Permissible Exposure Level; an exposure limit that is published and enforced by OSHA as a legal standard. The PEL may be expressed as a time-weighted average (TWA) exposure limit (for an 8-hour workday), a 15-minute short-term exposure limit (STEL), or a ceiling (C, or CEIL, or TLV-C).

RTECS: Registry of Toxic Effects of Chemical Substances; a database maintained by the NIOSH. The goal of the database is to include data on all known toxic substances, along with the concentration at which toxicity is known to occur. There are approximately 140,000 compounds listed.

STEL: Short-Term Exposure Level; an exposure limit for a short term; 15-minute exposure that cannot be exceeded during the workday, enforced by OSHA as a legal standard. Short-term exposures below the STEL level generally will not cause irritation, chronic or reversible tissue damage, or narcosis.

REL: Recommended Exposure Level; average concentration limit recommended for up to a 10-hour workday during a 40-hour workweek, by NIOSH.

TLV: Threshold Limit Value; guidelines suggested by the American Conference of Governmental Industrial Hygienists to assist industrial hygienists with limiting hazards of chemical exposures in the workplace.

TLV-C: Threshold Limit Ceiling Value; an exposure limit which should not be exceeded under any circumstances.

TWA: Time-weighted average concentration for a conventional 8-hour workday and a 40-hour work-week. It is the concentration to which it is believed possible that nearly all workers can be exposed without adverse health effects.

UEL: Upper Explosion Limit; the maximum concentration of a chemical in air at which detonation can occur.

UFL: Upper Flammability Limit; the maximum concentration of a chemical in air at which flame propagation can occur.

WEEL: Workplace Environmental Exposure Limit; set by the American Industrial Hygiene Association (AIHA).

Some abbreviations that are sometimes used on MSDS, and in other sources, are ambiguous. The most common meanings of some of these vague abbreviations are provided below, but the reader is cautioned that these are only suggestions:

EST: Established; estimated
MST: Mist
N/A, NA: Not applicable
ND: None determined; not determined
NE: None established; not established
NEGL: Negligible
NF: None found; not found
N/K, NK: Not known
N/P, NP: Not provided
SKN: Skin
TS: Trade secret
UKN: Unknown.

CHEMICAL CARCINOGENS

The following table contains data on chemicals often used in the analytical laboratory that have come under scrutiny for their suspected or observed carcinogenicity [1–9]. Note that some of the references that these tables were compiled from are historical. In some cases, the chemical structure is provided to avoid ambiguity. The reader is advised to use these tables with care, as there is a great deal of variability in the classifications as new data become available. It is suggested that the reader maintain a current file of data from the appropriate regulatory agencies, along with a complete set of MSDS and to err on the side of caution.

Most of the chemicals listed are linked with only certain target organs and not all cancer types. Cancer risk is often dependent upon the kind of exposure (inhalation, ingestion, touch, etc.). Cancer risk is not necessarily universal but may only be present for people with a genetic predisposition. Thus, a carcinogen may not always cause cancer in every person exposed. The risk of exposure must be weighed against benefits; one should not avoid exposure to a medication, for example, if such medication is advised. For example, metronidazole, which is listed, is often used in the treatment of protozoan infection. Indeed, some cancer treatment drugs are listed.

Interpretation of the codes used in this table is based on the following key:

a. Compiled from monographs of the International Agency for Research on Cancer, IARC (part of the United Nations World Health Organization), as data becomes available. Classifications are as follows:
Group 1: carcinogenic to humans
Group 2A: probably carcinogenic to humans
Group 2B: possibly carcinogenic to humans
Group 3: not classifiable as to carcinogenicity to humans
Group 4: probably not carcinogenic to humans.
b. Compiled from data of the National Toxicology Program, NTP, whose reports are updated every 2 years (branch of the US Department of Health and Human Services). Classifications are as follows:
Y: reasonably anticipated to be human carcinogen
#: known human carcinogen.
c. Compiled from data of the Occupational Safety and Health Administration, OSHA, with standards set by the legislative process (part of the US Department of Labor). Classifications are as follows:
Y: possibly a human carcinogen
*: substance for which OSHA has promulgated expanded health standards that govern health concerns in addition to carcinogenesis.

REFERENCES

1. "Partial List of Selected Carcinogens," www.people.memphis.edu/~ehas/carcinogen.html, 2003.
2. Ruth, J.H., Odor thresholds and irritation levels of several chemical substances: A review, *Am. Ind. Hyg. Assoc. J.*, 47, A-142, 1986.
3. "Chemical Carcinogens," New Jersey Department of Health and Senior Services, Occupational Health Service, www.state.nj.health.eoh, 2003.
4. "Hazard Database: List of Carcinogens," www.ephb.nw.ru/~spirov/carcinogen_1st.html, 2003.
5. Fourteenth Report on Carcinogens, U.S. Department of Health and Human Services, Public Health Service, National Toxicology Program, National Institute of Environmental Health Sciences, Research triangle Park, Durham, NC, https://ntp.niehs.nih.gov/whatwestudy/assessments/cancer/roc/index.html, 2016.
6. EHP Online, www.ehp.niehs.nih.gov/roc/toc10.html, 2003.

7. NIOSH Pocket Guide to Chemical Hazards, National Institute of occupational Safety and Health, DHHS (NIOSH) Publication No. 2005-1492007, also available online: www.cdc.gov/niosh/docs/2005-149/default.html, accessed February 2020.

8. Moschel, R.C., "Carcinogens," in *Encyclopedia of Genetics*, S. Brenner and J. H. Miller, ed., Elsevier, 2001.

9. Oliveira, P.A., Colaco, A., Chaves, R., Guedes-Pinto, H., De-La-Cruz, P.L.F., and Lopes, C., Chemical carcinogenesis, *An. Acad. Bras. Cienc.*, 79(4), 593–616, 2007.

Chemical Name	CAS Number	IARC[a]	NTP[b]	OSHA[c]	Odor Low mg/m³	Irritating Conc. mg/m³
Acetaldehyde	75-07-0	2B	Y			
Acetamide	60-35-5	2B				
2-Acetylamino-fluorene	53-96-3		Y	Y		
Acrylamide	79-06-1	2A	Y			
Acrylonitrile	107-13-1	2B	Y	*	8.1	
Actinomycin D	50-76-0	C				
Adriamycin	23214-92-8	2A	Y			
Aflatoxins	1402-68-2	1	#			
2-Aminoanthra quinone	117-79-3	3	Y			

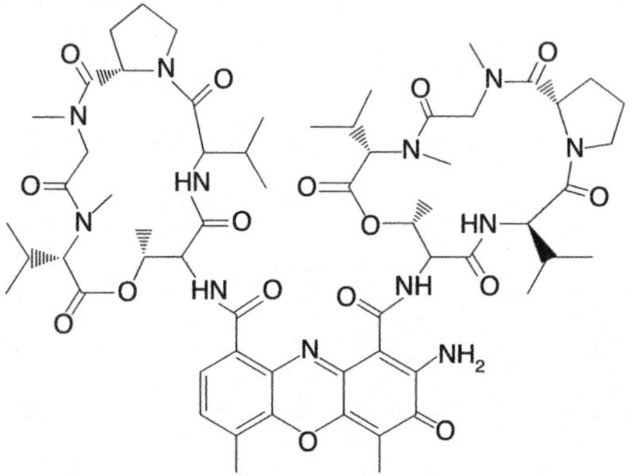

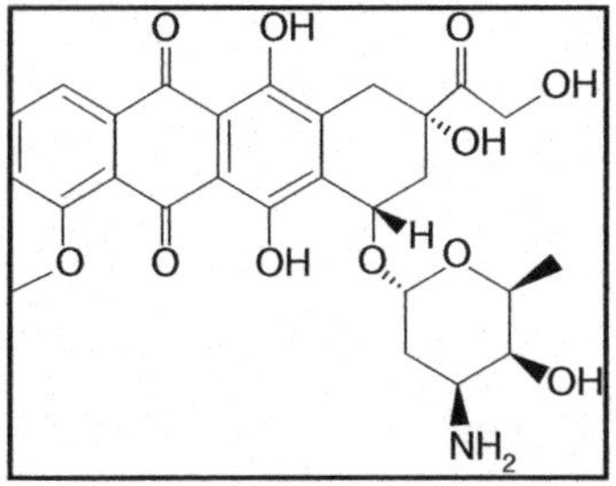

Actinomycin D

Adriamycin

(Continued)

Chemical Name	CAS Number	IARC[a]	NTP[b]	OSHA[c]	Odor Low mg/m³	Irritating Conc. mg/m³
p-Aminoazobenzene	60-09-3	2B				
4-Aminobiphenyl	92-67-1	1	#	Y		
1-Amino-2-methyl anthraquinone	82-28-0	3	Y			
o-Aminoazotoluene	97-56-3	2B	Y			
2-Aminonaphthalene	91-59-8	1	#	*		
Ammonium dichromate	7789-09-5	2B				
Amitrole	61-82-5	3	Y			
Androgenic steroids		2A				
o-Anisidine	90-04-0	2B	Y			
o-Anisidine hydrochloride	134-29-2	2B	Y			
Antimony trioxide	1309-64-4	2B				
Aramite	140-57-8	2B				

Chemical Name	CAS Number	IARC[a]	NTP[b]	OSHA[c]	Odor Low mg/m³	Irritating Conc. mg/m³
Arsenic compounds (certain)		1	#	*		
Arsenic, metal	7440-38-2	1	#	*		
Asbestos	1332-21-4	1	#	*		
Auramine, technical	2465-27-2	2B				
Azathioprine	446-86-6	1	#			

Chemical Name	CAS Number	IARC[a]	NTP[b]	OSHA[c]	Odor Low mg/m³	Irritating Conc. mg/m³
Aziridine		2B				
Benz[a]anthracene	56-55-3	2A	Y			

(Continued)

Chemical Name	CAS Number	IARC[a]	NTP[b]	OSHA[c]	Odor Low mg/m^3	Irritating Conc. mg/m^3
Benzene	71-43-2	1	#	*	4.5	9,000
Benzidine	92-87-5	1	#	Y		
Benzidine-based dyes		2A	Y			
Benzo[b]fluoranthene	205-99-2	2B	Y			
Benzo[j]fluoranthene	205-82-3	2B	Y			
Benzo[k]fluoranthene	207-08-9	2B	Y			
Benzofuran	271-89-6	2B				
Benzo[a]pyrene	50-32-8	1	Y			

(Continued)

Chemical Name	CAS Number	IARC[a]	NTP[b]	OSHA[c]	Odor Low mg/m^3	Irritating Conc. mg/m^3
Benzotrichloride	98-07-7		Y			
Benzyl violet 4B	1694-09-3	2B				
Beryllium	7440-41-7	1	#	Y		
Beryllium sulfate	13510-49-1	1	#			
Beryllium compounds (certain)		1	#			
N,N-bis(2-chloroethyl)-2-naphthylamine (chlornaphazine)	494-03-1	1				

Bis-chloroethyl nitrosourea (BCNU)	154-93-8	2A				
Bis-chloromethyl ether (BCME)	542-88-1	1	#	Y		
Bromodichloromethane	75-27-4	2B	Y			
1,3-Butadiene	106-99-0	1			0.352	
1,4-Butanediol di-methanesulfonate	55-98-1	1	#			
Butylated hydroxyanisole (BHA)	25013-16-5	2B	Y			
Butyrolactone, beta	3068-88-0	3				
Cadmium	7440-43-9	1	#	*		
Caffeic acid	331-39-5	2B				

Carbon black	1333-86-4	2B				
Carbon tetrachloride	56-23-5	2B	Y		60–300	
Catechols		2B				

(Continued)

Chemical Name	CAS Number	IARC[a]	NTP[b]	OSHA[c]	Odor Low mg/m^3	Irritating Conc. mg/m^3
Chlorambucil	305-03-3	1	#			
Chloramphenicol	56-75-7	2A	Y			
Chlordane	57-74-9	2B				

(Continued)

Chemical Name	CAS Number	IARC[a]	NTP[b]	OSHA[c]	Odor Low mg/m³	Irritating Conc. mg/m³
Chlordecone (Kepone)	143-50-0	2B				
Chlorinated toluenes, alpha		2A				
p-Chloroaniline		2B				
1-(2-Chloroethyl)-3-cyclohexyl-1-nitrosourea [CCNU]	13010-47-4	2A	#			
Chloroform	67-66-3	2B	Y		250	20,480
Chloromethyl methyl ether	107-30-2		#			
1-Chloro-2-methylpropene	513-37-1	2B				
3-Chloro-2-methylpropene	563-47-3	3	Y			
Chlorophenols		2B			0.0189	6,800
4-Chloro-o-phenylenediamine	95-83-0	2B	Y			
Chloroprene	126-99-8	2B	Y			
p-Chloro-o-toluidine	95-69-2	2A	Y			

(Continued)

Chemical Name	CAS Number	IARC[a]	NTP[b]	OSHA[c]	Odor Low mg/m³	Irritating Conc. mg/m³
Chlorozotocin	54749-90-5	2A	Y			
Chromium	7440-47-3	1	#			
Chromium compounds (certain)		1	#			
Chromium (VI) ions	18540-29-9	1	#			
Chrysene	218-01-9	2B	Y			
Cisplatin	15663-27-1	2A	Y			
Cobalt and compounds	7440-48-4	2B				
p-Cresidine	120-71-8	2B	Y			

(Continued)

Chemical Name	CAS Number	IARC[a]	NTP[b]	OSHA[c]	Odor Low mg/m³	Irritating Conc. mg/m³
Cupferron	135-20-6		Y			
Cycasin	14901-08-7	2B	B			
Cyclophosphamide	50-18-0	1	#			
Dacarbazine	4342-03-4	2B	Y			
DDT (dichlorodiphenyl-trichloroethane)	50-29-3	2B	Y		5.0725	
N,N′-Diacetyl-benzidine	613-35-4	2B				
2,4-Diaminoanisole	615-05-4	2B				
2,4-Diaminoanisole sulfate	39156-41-7		Y			
4,4′-Diaminodiphenyl ether	101-80-4	2B				
2,4-Diaminotoluene	95-80-7	2B	Y			

(Continued)

Chemical Name	CAS Number	IARC[a]	NTP[b]	OSHA[c]	Odor Low mg/m^3	Irritating Conc. mg/m^3
Dibenz[a,h]acridine	226-36-8	2B	Y			
Dibenz[a,j]acridine	224-42-0	2B	Y			
Dibenz[a,h]anthracene	53-70-3	2A	Y			
7H-Dibenzo[c,g]carbazole	194-59-2	2B	Y			

(Continued)

Chemical Name	CAS Number	IARC[a]	NTP[b]	OSHA[c]	Odor Low mg/m³	Irritating Conc. mg/m³
Dibenzo[a,e] pyrene	192-65-4	3	Y			
Dibenzo[a,h]pyrene	189-64-0	2B	Y			
Dibenzo[a,i]pyrene	189-55-9	2B	Y			
Dibenzo[a,l]pyrene	191-30-0	2A	Y			
1,2-Dibromo-3-chloropropane [DBCP]	96-12-8	2B	Y	*	0.965	1.93
p-Dichlorobenzene	106-46-7	2A	Y			
3,3'-Dichloro-benzidine	91-94-1	2B	Y	*		
3,3'-Dichloro-benzidine salts				*		
3,3'-Dichloro-4,4'-diaminodiphenyl ether	28434-86-8	2B				
1,2-Dichloroethane	107-06-2	2B	Y			

(Continued)

Chemical Name	CAS Number	IARC[a]	NTP[b]	OSHA[c]	Odor Low mg/m^3	Irritating Conc. mg/m^3
Dichloromethane	75-09-2	2B	Y	*		
1,3-Dichloropropene	542-75-6	2B	Y			
Di(2-ethylhexyl) phthalate	103-23-1	3				
1,2-Diethylhydrazine	1615-80-1	2B				
Dieldrin	60-57-1	3		Y		
Dienoestrol	84-17-3	1				
Diepoxybutane	1464-53-5	2B	Y			
Di(2,3-ethylhexyl) phthalate	117-81-7	2B	Y			
Diethylstilbestrol [DES]	56-53-1	1	#			
Diethyl sulfate	64-67-5	2A	Y			
Dihydrosafrole	94-58-6	2B				
1,8-Dihydroxyanthra-quinone	117-10-2		Y			
Diisopropyl sulfate	2973-10-6	2B				
3,3′-Dimethoxy-benzidine	119-90-4	2B	Y			
4-Dimethylamino-azobenzene	60-11-7	2B	Y	Y		
2,6-Dimethylaniline	87-62-7	2B	Y			
3,3′-Dimethyl-benzidine	119-93-7	2B	Y			

(Continued)

Chemical Name	CAS Number	IARC[a]	NTP[b]	OSHA[c]	Odor Low mg/m³	Irritating Conc. mg/m³
Dimethylcarbamoyl chloride	79-44-7	2A	Y			
1,1-Dimethyl hydrazine	57-14-7	2B	Y	d		
Dimethyl sulfate	77-78-1	2A	Y	d		
Dinitrofluoranthrene, isomers		2B				
1,8-Dinitropyrene	42397-65-9	2B				
2,4-Dinitrotoluene	121-14-2	2B				
2,6-Dinitrotoluene	606-20-2	2B				
1,4-Dioxane	123-91-1	2B	Y		0.0108	792

(Continued)

Chemical Name	CAS Number	IARC[a]	NTP[b]	OSHA[c]	Odor Low mg/m^3	Irritating Conc. mg/m^3
Direct Black 38, technical	1937-37-7					
Direct Blue 6, technical	2602-46-2					

(Continued)

Chemical Name	CAS Number	IARC[a]	NTP[b]	OSHA[c]	Odor Low mg/m^3	Irritating Conc. mg/m^3
Direct Brown 95, technical	16071-86-6	1				
Disperse Blue 1	2475-45-8	2B	Y			
Epichlorohydrin	106-89-8	2A	Y	d	50	335
1,2-Epoxybutane	106-88-7	2B				
Erionite (zeolite mineral)	12510-42-8	1	#			

(Continued)

Chemical Name	CAS Number	IARC[a]	NTP[b]	OSHA[c]	Odor Low mg/m^3	Irritating Conc. mg/m^3
Ethinylestradiol	57-63-6	1				
Ethyl acrylate	140-88-5	2B				
Ethylene dibromide [EDB]	106-93-4	2A			76.8	
Ethylene dichloride [EDC]	107-06-2	2B		Y	24	
Ethyleinimine	151-56-4			Y	4	200
Ethylene oxide	72-21-8	1	#	*	520	
Ethylene thiourea	96-45-7	3	Y			
Ethyl methanesulfonate	62-50-0	2B	Y			
N-Ethyl-N-nitroso-urea	759-73-9	2A				
Formaldehyde	50-00-0	1	Y	*	1.47	1.50
2-(2-Furyl)-3-(5-nitro-2-furyl) acrylamide	3688-53-7	2B				
2-(2-Formyl-hydrazino)-4-(5-nitro-2-furyl) thiazole	3570-75-0	2B				
Furan	110-00-9	2B	Y			
Gyromitrin	16568-02-8	1				
Hexachlorobenzene	118-74-1	2B	Y			
Hexachloroethane	67-72-1	2B	Y			
Hexamethyl phosphoramide	680-31-9	2B	Y			

(Continued)

Chemical Name	CAS Number	IARC[a]	NTP[b]	OSHA[c]	Odor Low mg/m³	Irritating Conc. mg/m³
Hydrazine	302-01-2	2B	Y	Y	3	
Hydrazine sulfate	10034-93-2		Y			
Hydrazobenzene	122-66-7		Y			
Indeno [1,2,3-cd] pyrene	193-39-5	2B	Y			
Isoprene	78-79-5	2B				
Lead (II) acetate	301-04-2	2B	Y	*		
Lead chromate	7758-97-6	2A	#			
Lead phosphate	7446-27-7	2B	Y	*		
Lindane (and other hexachloro-cyclohexane isomers)	58-89-9		Y	Y		
Melphalan	148-82-3	1	#			
Merphalan	531-76-0	2B				

(Continued)

Chemical Name	CAS Number	IARC[a]	NTP[b]	OSHA[c]	Odor Low mg/m³	Irritating Conc. mg/m³
Mestranol	72-33-3	1				
2-Methylaziridine	75-55-8	2B	Y			
Methylazoxy-methanol acetate	592-62-1	2B				
Methyl chloromethyl ether				Y		
5-Methylchrysene	3697-24-3	2B	Y			
4,4'-Methylenebis(2-chloroaniline [MOCA]	101-14-4	1	Y			
4,4'-Methylenebis(N, N-dimethyl-benzenamine	69522-43-6		Y			
4,4'-Methylene dianiline	101-77-9	2B	Y			
Methyl bromide	74-83-9	3		Y	80	
Methyl chloride	74-87-3	3			21	1,050
Methyl hydrazine	60-34-4	1			1.75	
Methyl iodide	74-88-4	3		Y		21,500
Methylmercury compounds		2B				
Methyl methanesulfonate	66-27-3	2A	Y			
N-Methyl-N-nitrosourea	684-93-5	2A				
N-Methyl-N-nitrosourethane		2B				

(Continued)

Chemical Name	CAS Number	IARC[a]	NTP[b]	OSHA[c]	Odor Low mg/m³	Irritating Conc. mg/m³
Methylthiouracil	56-04-2	2B				
Metronidazole	443-48-1	2B	Y			
Michler's ketone	90-94-8		Y			
Mirex	2385-85-5	2B	Y			

(Continued)

Chemical Name	CAS Number	IARC[a]	NTP[b]	OSHA[c]	Odor Low mg/m^3	Irritating Conc. mg/m^3
Mustard gas	505-60-2	1	#		0.015	
α-Naphthylamine	134-32-7	3		Y		
β-Naphthylamine	91-59-8	1	#	Y		
Nickel carbonyl	13463-39-3	1	Y		0.21	
Nickel	7440-02-0	2B	Y			
Nickel compounds (certain)		1	#	*		
Nickel, metallic and inorganic compounds	7440-02-0	2B	#			
Nitrilotriacetic acid	139-13-9	2B	Y			
5-Nitroacenaphthene	602-87-9	2B				
2-Nitroanisole	91-23-6	2B	Y			
5-Nitro-o-anisidine	99-59-2					
Nitrobenzene	98-95-3	2B				
4-Nitrobiphenyl	92-93-3			Y		
6-Nitrochrysene	7496-02-8	2B	Y			
Nitrofen	1836-75-5	2B	Y			
Nitrofluorene	607-57-8	2B				
Nitrogen mustards		2A	Y			
Nitrogen mustard N-oxide	126-85-2	2B				
2-Nitropropane	79-46-9	2B	Y		17.5	
1-Nitropyrene	5522-43-0	2B	Y			

(Continued)

Chemical Name	CAS Number	IARC[a]	NTP[b]	OSHA[c]	Odor Low mg/m³	Irritating Conc. mg/m³
4-Nitropyrene	57835-92-4	2B	Y			
N-nitroso-di-n-butylamine	924-16-3	2B	Y			
N-nitroso-n-propylamine	621-64-7	2B	Y			
N-nitroso-diethanolamine	1116-54-7	2B	Y			
N-nitroso-diethylamine	55-18-5	2A	Y			
N-nitroso-dimethylamine	62-75-9	2A	Y	Y		
p-Nitroso-di-phenylamine	156-10-5	3				
N-nitroso-di-n-propylamine	621-64-7		Y			
N-nitroso-N-ethyl urea	759-73-9		Y			
N-nitroso-N-methyl urea	684-93-5		Y			
N-nitrosomethylvinylamine	4549-40-0	2B	Y			
N-nitrosomorpholine	59-89-2	2B	Y			
N-nitrosonornicotine	16543-55-8	1	Y			
N-nitrosopiperidine	100-75-4	2B	Y			
N-nitrosopyrrolidine	930-55-2	2B	Y			
N-nitrososarcosine	13256-22-9	2B	Y			
Norethisterone	68-22-4		Y			
Estradiol-17B	50-28-2	1				

(Continued)

Chemical Name	CAS Number	IARC[a]	NTP[b]	OSHA[c]	Odor Low mg/m^3	Irritating Conc. mg/m^3
Estrone	53-16-7	1				
4,4'-Oxydianiline	101-80-4		Y			
Oxymetholone	434-07-1		Y			
Phenacetin	62-44-2	1	Y			
Phenazopyridine	94-78-0	2B	Y			
Phenobarbital	50-06-6	2B				
Phenazopyridine hydrochloride	136-40-3	2B	Y			
Phenoxyacetic acid derivatives		1		Y		

(Continued)

Chemical Name	CAS Number	IARC[a]	NTP[b]	OSHA[c]	Odor Low mg/m^3	Irritating Conc. mg/m^3
Phenoxybenzamine hydrochloride	63-92-3	2B	Y			
Phenyl glycidyl ether	122-60-1	2B				
N-phenyl-β-naphthylamine	135-88-6	1				
Phenylhydrazine	100-63-0			Y		
Phenytoin	57-41-0	2B	Y			
Phenytoin, sodium salt	630-93-3		Y			
Polybrominated biphenyls [PBBs]	36355-01-8	2B	Y			
Polychlorinated biphenyls [PCBs]		2A	Y	*		
Polychlorinated camphenes		2A				
Polychlorophenols		2B				
Polycyclic aromatic compounds, general		2A	Y			
Potassium bromate	7758-01-2	2B				
Potassium chromate	7789-00-6	1				
Potassium dichromate	7778-50-9	1				
Procarbazine	671-16-9	2A	Y			
Procarbazine hydrochloride	366-70-1	2A				

(Continued)

Chemical Name	CAS Number	IARC[a]	NTP[b]	OSHA[c]	Odor Low mg/m³	Irritating Conc. mg/m³
Propane sultone	1120-71-4	2B	Y			
β-Propiolactone	57-57-8	2B	Y	Y		
Propyleneimine	75-55-8	2B				
Propylene oxide	75-56-9	2B	Y			
Propylthiouracil 78>>	51-52-5	2B	Y			
Reserpine	50-55-5	3	Y			
Saccharin	81-07-2	3				
Safrole	94-59-7	2B	Y		1.4586	
Selenium sulfide	7446-34-6	1	Y	*		
Silica crystalline (respirable)			#			

(Continued)

Chemical Name	CAS Number	IARC[a]	NTP[b]	OSHA[c]	Odor Low mg/m³	Irritating Conc. mg/m³
Streptozotocin	18883-66-4	2B	Y			
Strontium chromate	7789-06-2		#			
Sulfallate	95-06-7	2B	Y			
2,3,7,8-Tetrachloro-dibenzo-p-dioxin [TCDD]	1746-01-6	1	#			
1,1,2,2-Tetrachloroethane	79-34-5	3		*		
Tetrachloroethylene	127-18-4	2	Y	*	21	1,302
Tetrafluoroethylene	116-14-3	2B	Y		31.3561	710.2
Tetranitromethane	509-14-8	2B	Y			
Thioacetamide	62-55-5	2B	Y			
4,4'Thiodianiline	139-65-1	2B	Y			
Thiourea	62-56-6	3	Y			
Thorium (232) dioxide	1314-20-1		#			
p-Toluidine	119-93-7		Y			
o-Toluidine	95-53-4	1	Y			
p-Toluidine	106-49-0	1	Y			

(Continued)

Chemical Name	CAS Number	IARC[a]	NTP[b]	OSHA[c]	Odor Low mg/m³	Irritating Conc. mg/m³
o-Toluidine hydrochloride	636-21-5		Y			
p-Toluidine	106-49-0		Y			
Toxaphene	8001-35-2	2B	Y	*	2.366	
Treosulfan	299-75-2	1				
1,1,2-Trichloroethane	79-00-5	3		*		
Trichloroethylene	79-01-6	2A	Y	*		
2,4,6-Trichlorophenol	88-06-2	2B	Y			
Tris(aziridinyl)-p-benzoquinone [triaziquinone]	68-76-8	1	#			
Tris(2,3-dibromo-propyl) phosphate	126-72-7	2A	Y			
Tryptophan P1	62450-06-0	2B				
Tryptophan P2	62450-07-1	2B				
Trypan blue	72-57-1	2B				

(Continued)

Chemical Name	CAS Number	IARC[a]	NTP[b]	OSHA[c]	Odor Low mg/m³	Irritating Conc. mg/m³
Uracil mustard	66-75-1	2B				
Urethane	51-79-6	2B	Y			
Vinyl acetate	108-05-4	2B				
Vinyl bromide	593-60-2	2A	Y			
Vinyl chloride	75-01-4	1	#	*		
Vinyl cyclohexene diepoxide	106-87-6	2B	Y			
Vinyl fluoride	75-02-5	2A	Y			
Vinylidene chloride	75-35-4	2A			2.000	
Vinylidene fluoride (monomer)	75-38-7	2A				

ORGANIC PEROXIDES

The following ethers have been tested for the potential to undergo conversion to peroxides [1,2].

REFERENCES

1. Ramsey, J.B. and Aldridge, F.T., Removal of peroxides from ethers with cerous hydroxide, *J. Am. Chem. Soc.*, 77, 2561, 1955.
2. Furr, A.K., ed., *CRC Handbook of Laboratory Safety*, 5th ed., CRC Press, Boca Raton, FL, 2000.

Ether	Quantities of Peroxides Found
Allyl ethyl ether	Moderate
Allyl phenyl ether	Moderate
Benzyl ether	Moderate
Benzyl n-butyl ether	Moderate
o-Bromophenetole	Very small
p-Bromophenetole	Very small
n-Butyl ether	Moderate
t-Butyl ether	Moderate
p-Chloroanisole	Very small
o-Chloroanisole	Very small
Bis(2-ethoxyethyl) ether (diethylene glycol diethyl ether)	Considerable
2-(2-Butoxyethoxy) ethanol (diethylene glycol mono-*n*-butyl ether)	Moderate
1,4-Dioxane	Moderate
Diphenyl ether	Moderate
Ethyl ether[a]	Very small
Ethyl ether[b]	Considerable
Ethyl ether[c]	Moderate
Isopropyl ether	Considerable
o-Methylanisole	Very small
m-Methylphenetole	Very small
Phenetole	Very small
Tetrahydrofuran	Moderate

[a] Obtained from sealed can of anhydrous ether, analytical reagent, immediately after opening.
[b] Obtained from a partially filled tin can (well stopped) containing the same grade of anhydrous ether as that described in note [a] but allowed to stand for an appreciable time.
[c] From a galvanized iron container used for dispensing ether.

TESTING REQUIREMENTS FOR PEROXIDIZABLE COMPOUNDS

Because some compounds form peroxides more easily or faster than others, prudent practices require testing the supply on hand in the laboratory on a periodic basis. The following list provides guidelines on test scheduling [1]. The peroxide hazard of the compounds listed in Group 1 is on the basis of time in storage. The compounds in Group 2 present a peroxide hazard primarily due to concentration, mainly by evaporation of the liquid. The compounds listed in Group 3 are hazardous because of the potential of peroxide-initiated polymerization. When stored as liquids, the peroxide formation may increase, and therefore, these compounds should be treated as Group 1 peroxidizable compounds.

REFERENCE

1. Ringen, S., *Environmental Health and Safety Manual: Chemical Safety*, Sec. 4–50, University of Wyoming, June 2000.

Group 1   Test every 3 months

   Divinyl acetylene
   Isopropyl ether
   Potassium
   Sodium amide
   Vinylidene chloride

Group 2   Test every 6 months

   Acetal
   Cumene
   Cyclohexene
   Diacetylene
   Dicyclopentadiene
   Diethyl ether
   Dimethyl ether
   1,4-Dioxane
   Ethylene glycol dimethyl ether (glyme)
   Methyl acetylene
   Methyl isobutyl ketone
   Methyl cyclopentane
   Tetrahydrofuran
   Tetrahydronaphthalene (tetralin)
   Vinyl ethers

Group 3   Test every 12 months

   Acrylic acid
   Acrylonitrile
   Butadiene
   Chloroprene
   Chlorotrifluoroethene

Methyl methacrylate
Styrene
Tetrafluoroethylene
Vinyl acetate
Vinyl acetylene
Vinyl chloride
Vinyl pyridine

TESTS FOR THE PRESENCE OF PEROXIDES

Peroxides may be detected qualitatively with one of the following test procedures [1].

REFERENCE

1. Gordon, A.J. and Ford, R.A., *The Chemist's Companion*, John Wiley and Sons, New York, 1972.

Ferrothiocyanate Test

Reagent preparation, in sequence:

Add 9 g $FeSO_4 \cdot 7H_2O$ to 50 mL 18 % (vol/vol) $HCl_{(aq)}$
Add 1–3 mg granular Zn
Add 5 g NaSCN.

After the red color fades, add an additional 12 g NaSCN, decant leaving unreacted Zn.

Upon mixing this reagent with a peroxide-containing liquid, the colorless solution will produce a red color, the result of the conversion of ferrothiocyanide to ferrithiocyanide. This test is very sensitive and can be used to detect peroxides at a concentration of 0.001 % (mass/mass).

Potassium Iodide Test

Reagent preparation:

Make a 10 % (mass/mass) solution of KI in water.

Upon mixing this reagent with a peroxide-containing liquid, a yellow color will appear within 1 minute.

Acidic Iodide Test

Reagent preparation:

To 1 mL of glacial acetic acid, add 100 mg KI or NaI.

Upon mixing this reagent with an equal volume of a peroxide-containing liquid, a yellow coloration will appear. The color will appear dark or even brown if the peroxide concentration is very high.

Perchromate Test

Reagent preparation:

Dissolve 1 mg of $Na_2Cr_2O_7$ in 1 mL of water, add a drop of dilute $H_2SO_{4(aq)}$.

Upon mixing this reagent with a peroxide-containing liquid, a blue color will develop in the organic layer indicating the formation of the perchromate ion.

CHARACTERISTICS OF CHEMICAL-RESISTANT MATERIALS

The following table provides guidance in the selection of materials that provide some degree of chemical resistance for common laboratory tasks [1].

REFERENCE

1. Furr, A.K., ed., *CRC Handbook of Laboratory Safety*, 5th ed., CRC Press, Boca Raton, 2000.

Physical Characteristics of Chemical-Resistant Materials

lity|Heat Resistance|Ozone Resistance|Puncture Resistance|Tear Resistance|Relative Cost|
|---|---|---|---|---|---|---|---|---|
|Butyl rubber|F|G|G|E|E|G|G|High|
|Chlorinated Polyethylene (CPE)|E|G|G|G|E|G|G|Low|
|Natural rubber|E|E|E|F|P|E|E|Medium|
|Nitrile-butadiene rubber (NBR)|E|E|E|G|F|E|G|Medium|
|Neoprene|E|E|G|G|E|G|G|Medium|
|Nitrile rubber (nitrile)|E|E|E|G|F|E|G|Medium|
|Nitrile rubber + polyvinylchloride (nitrile + PVC)|G|G|G|F|E|G|G|Medium|
|Polyethylene|F|F|G|F|F|P|F|Low|
|Polyurethane|E|G|E|G|G|G|G|High|
|Polyvinyl alcohol (PVA)|F|F|P|G|E|F|G|Very high|
|Polyvinyl chloride (PVC)|G|P|F|P|E|G|G|Low|
|Styrene-butadiene rubber (SBR)|E|G|G|G|F|F|F|Low|
|Viton|G|G|G|G|E|G|G|Very high|

Note: E=excellent, G=good, F=fair, P=poor.

SELECTION OF PROTECTIVE LABORATORY GARMENTS

The following table provides guidance in the selection of special protective garments that are used in the laboratory for specific tasks [1].

REFERENCE

1. Mount Sinai School of Medicine Personal Protective Equipment Guide, https://www.mountsinai. org/health-library/special-topic/personal-protective-equipment, accessed January 2020.

Material	Type of Garment	Common Use
Cotton/natural Fiber/ blends	Coveralls; lab coats; sleeve protectors; aprons	For dry dusts, particulates, and aerosols
Tyvek	Coveralls; lab coats; sleeve protectors; aprons; hoods	For dry dusts and aerosols
Saranax/Tyvek SL	Coveralls; lab coats; sleeve protectors; aprons; hoods; level B suits	Aerosols; liquids; solvents
Polyethylene	Barrier gowns; aprons	Body fluids
Polypropylene	Clean room suits; coveralls; lab coats	For dry dusts; non-toxic particulates
Polyethylene/Tyvek (QC)	Coveralls; aprons; lab coats; shoe covers	Moisture; solvents
Polypropylene	Coveralls; lab coats; shoe covers; caps; clean room suits	Non-toxic particulates; dry dusts
Tychem BR; Tychem TK	Full level A and level B suits	Highly toxic particulates; dry dusts
CPF	Full level A and level B suits; splash suits	Highly toxic chemicals; gases; aerosols
PVC	Full level A suits	Highly toxic chemicals; gases; aerosols

PROTECTIVE CLOTHING LEVELS

In the United States, OSHA defines various levels of protective clothing and sets parameters that govern their use with chemical spills and in environments where chemical exposure is a possibility. A summary of the definitions is provided below [1].

REFERENCE

1. OSHA Technical Manual, Section VIII, Chapter 1, Chemical Protective Clothing, www.osha.gov/dts/osta/otm/otm_viii/otm_viii_1.html, accessed January 2020.

Level A
- Vapor protective suit (meets NFPA 1991), pressure demand, full-face self-contained breathing apparatus (SCBA), inner chemical-resistant gloves, chemical-resistant safety boots, two-way radio communication.
- Protection Provided: Highest available level of respiratory, skin, and eye protection from solid, liquid, and gaseous chemicals.
- Used When: The chemical(s) have been identified and have high level of hazards to respiratory system, skin, and eyes; substances are present with known or suspected skin toxicity or carcinogenicity; operations must be conducted in confined or poorly ventilated areas.
- Limitations: Protective clothing must resist permeation by the chemical or mixtures present.

Level B
- Liquid splash-protective suit (meets NFPA 1992), pressure-demand, full-facepiece SCBA, inner chemical-resistant gloves, chemical-resistant safety boots, two-way radio communications.
- Protection Provided: Provides same level of respiratory protection as Level A but somewhat less skin protection. Liquid splash protection is provided but not protection against chemical vapors or gases.
- Used When: The chemical(s) have been identified but do not require a high level of skin protection; the primary hazards associated with site entry are from liquid and not vapor contact.
- Limitations: Protective clothing items must resist penetration by the chemicals or mixtures present.

Level C
- Support function protective garment (meets NFPA 1993), full-facepiece, air-purifying, canister-equipped respirator, chemical-resistant gloves and safety boots, two-way communications system.
- Protection Provided: The same level of skin protection as Level B but a lower level of respiratory protection; liquid splash protection but no protection to chemical vapors or gases.
- Used When: Contact with site chemical(s) will not affect the skin; air contaminants have been identified and concentrations measured; a canister is available which can remove the contaminant; the site and its hazards have been completely characterized.
- Limitations: Protective clothing items must resist penetration by the chemical or mixtures present; chemical airborne concentration must be less than IDLH levels; the atmosphere must contain at least 19.5 % oxygen.
- Not acceptable for chemical emergency response.

Level D
- Coveralls, safety boots/shoes, safety glasses, or chemical splash goggles.
- Protection Provided: No respiratory protection, minimal skin protection.
- Used When: The atmosphere contains no known hazard; work functions preclude splashes, immersion, potential for inhalation, or direct contact with hazard chemicals.
- Limitations: The atmosphere must contain at least 19.5 % oxygen.
- Not acceptable for chemical emergency response.

Optional items may be added to each level of protective clothing. Options include items from higher levels of protection, as well as hard hats, hearing protection, outer gloves, a cooling system, etc.

SELECTION OF LABORATORY GLOVES

The following tables provide guidance in the selection of protective gloves for laboratory use [1–5]. If protection from more than one class of chemical is required, double gloving should be considered. The first table covers general hand protection from scrapes, burns, ergonomic issues, cuts, and abrasions, while the second specifically deals with chemical hazards. In selecting the appropriate hand protection, one should identify the hazard, and where possible use engineering controls (drip controls on bottles, elimination of sharps, the use of ergonomically designed tools, etc.).

REFERENCES

1. Garrod, A.N., Martinez, M., and Pearson, J., Exposure to preservatives used in the industrial pre-treatment of timber, *Ann. Occup. Hyg.*, 43, 543–555, 1999.
2. Garrod, A.N., Phillips, A.M., and Pemberton, J.A., Potential exposure of hands inside protective gloves—a summary of data from non-agricultural pesticide surveys, *Ann Occup. Hyg*, 45, 55–60, 2001.
3. Mockelsen, R.L. and Hall, R.C., A breakthrough time comparison of nitrile and neoprene glove materials produced by different glove manufacturers, *Am. Ind. Hyg. Assoc. J.*, 48, 941–947, 1987.
4. OSHA, Federal Register, Vol 59, No. 66, 16334–16364, 29 CFR 1910, 1994.
5. Bruno, T.J., Svoronos, P.D.N., and Ringen, S.G., in Rumble, J., Ed., Health and safety information (Chapter 16), *CRC Handbook of Chemistry and Physics*, 100th Ed. CRC Press, Boca Raton, FL, 2019.

General Hand Protection Selection Criteria

Glove	Application Examples
Cotton	Weighing, glass handling (avoiding contamination with skin oils); note that these gloves can be used as a first layer when also using other gloves such as for chemical hazards.
Leather	Moderate hot or cold material handling; moving equipment.
Gel-filled (anti-vibration)	Operation of vibrating equipment.
Kevlar or fine mesh	Work with sheet metal, glass, or heavy cutting; note that these will not protect against punctures.
Chemical resistant	See table below for specifics.
Insulated for heat	Furnace work, handling hot glass or metal.
Insulated for cold	Cryogenic work, filling Dewars, replenishing nuclear magnetic resonance (NMR) magnets, etc.

Hand Protection for Chemical Hazards

Glove Material	Resistant to
Viton	PCBs, chlorinated solvents, aromatic solvents
Viton/butyl	Acetone, toluene, aromatics, aliphatic hydrocarbons, chlorinated solvents, ketones, amines, and aldehydes
SilverShield and 4H (PE/EVAL)	Morpholine, vinyl chloride, acetone, ethyl ether, many toxic solvents, and caustics
Barrier	Wide range of chlorinated solvents, aromatic acids
PVA	Ketones, aromatics, chlorinated solvents, , xylene, MIBK, trichloroethylene; do not use with water/aqueous solutions
Butyl	Aldehydes, ketones, esters, alcohols, most inorganic acids, caustics, dioxane
Neoprene	Oils, grease, petroleum-based solvents, detergents, acids, caustics, alcohols, solvents
PVC	Acids, caustics, solvents, grease, oil
Nitrile	Oils, fats, acids, caustics, alcohols
Latex	Body fluids, blood, acids, alcohols, alkalis
Vinyl	Body fluids, blood, acids, alcohols, alkalis
Rubber	Organic acids, some mineral acids, caustics, alcohols; not recommended for aromatic solvents, chlorinated solvents

SELECTION OF HEARING PROTECTION DEVICES

There are many sources of noise in the laboratory, and generally, employers are required to identify employees exposed to noise in excess of 85 decibels (dB) averaged over eight working hours. It is best to apply engineering and administrative controls to mitigate these exposures. If, after applying engineering and administrative controls, sound levels in the local environment are higher than regulatory or recommended levels, hearing protection devices (HPDs) should be used to reduce noise exposure [1,2].

REFERENCES

1. Schulz, T. and Madison, T., ed., *Hearing Conservation Manual,* 5th Ed., Council for Accreditation in Occupational Hearing Conservation, Milwaukee, WI, 53202, 2014.
2. Ringen, S.G. and Bruno, T.J., in Rumble, J. Ed., Health and safety information (Chapter 16), CRC Handbook of Chemistry and Physics, 100th Ed., Boca Raton, FL, 2019.

The following table provides a range of sound levels (in decibels, dB) for common laboratory equipment.

Equipment	Sound Level Range (dB)
Air compressor	75–95
Shaker table	75–85
Pressure-relief valves	75–120
Pressurized vortex tubes	75–85
Sonicator (probe type)	70–110
Fume hood	30–70
Vacuum pump	50–70
Room air conditioner	30–65

In addition to the noise generated by various devices in the laboratory, additional noise can be generated by devices not directly related to the function of the lab. Noise from radios, piped-in music, heating, ventilation and air conditioning (HVAC) systems, and telephones are also present in many laboratories.

It is not necessarily optimal to use hearing protection that simply maximizes sound attenuation; it remains important for workers to communicate with one another. Thus, devices that prevent intelligible speech can be problematic. Moreover, one must be able to recognize laboratory equipment alarms (such as low flow alarms on fume hoods, temperature alarms on ovens, and in some environments, back-up alarms on forklifts). In many cases, personnel will remove hearing protection to communicate with one another or to listen for an alarm, even though removal of an HPD in a high noise environment can substantially reduce hearing protection. By choosing the right hearing protection, removal of the HPD by personnel can be minimized. In the United States and other industrialized countries, sound levels must be determined so that the correct noise reduction rating (NRR) may be determined, and training must be provided by industrial hygienist or other appropriate personnel.

Clearly, hearing protection can only be effective if it is actually used, so user preference (between ear plugs and ear muffs, for example) must be considered. Since the anatomy of the ear varies with each individual, the effectiveness of plugs might not be universal, especially if inserted incorrectly. Moreover, in some environments, personnel inserting and removing ear plugs through the course of a workday can introduce dirt and bacteria and cause ear infections. When a higher NRR is needed, sometimes using two styles of HPDs simultaneously is effective, for example, ear muffs and disposable foam earplugs, although the NRR is not simply additive. The following table provides guidance in choosing the appropriate HPD. Note that in addition to the equipment shown below, there are combination devices available. One can obtain, for example, a passive earmuff combined with a protective hard hat and various types of face shields.

Type	Illustration	Notes
Passive earmuff	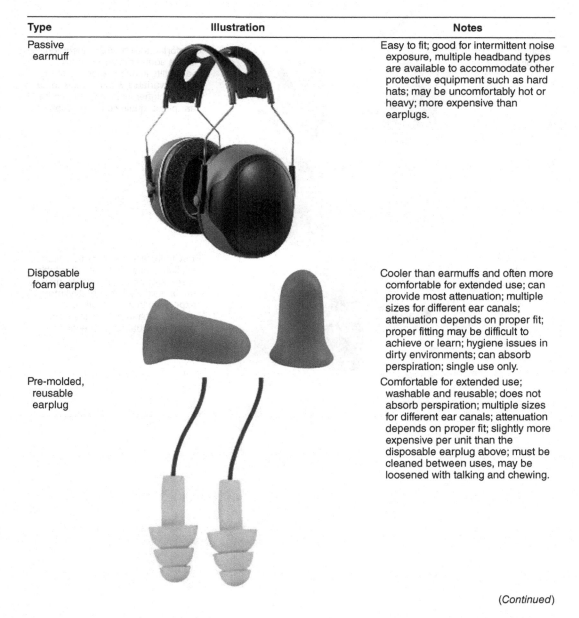	Easy to fit; good for intermittent noise exposure, multiple headband types are available to accommodate other protective equipment such as hard hats; may be uncomfortably hot or heavy; more expensive than earplugs.
Disposable foam earplug		Cooler than earmuffs and often more comfortable for extended use; can provide most attenuation; multiple sizes for different ear canals; attenuation depends on proper fit; proper fitting may be difficult to achieve or learn; hygiene issues in dirty environments; can absorb perspiration; single use only.
Pre-molded, reusable earplug		Comfortable for extended use; washable and reusable; does not absorb perspiration; multiple sizes for different ear canals; attenuation depends on proper fit; slightly more expensive per unit than the disposable earplug above; must be cleaned between uses, may be loosened with talking and chewing.

(Continued)

Type	Illustration	Notes
Canal caps/ semi-insert earplug	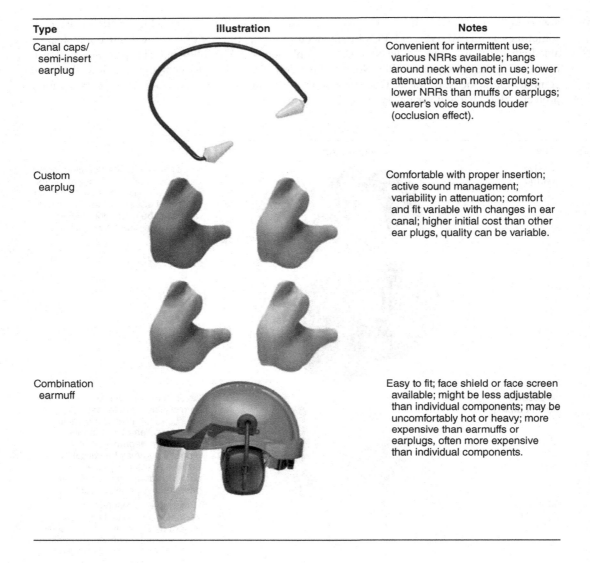	Convenient for intermittent use; various NRRs available; hangs around neck when not in use; lower attenuation than most earplugs; lower NRRs than muffs or earplugs; wearer's voice sounds louder (occlusion effect).
Custom earplug		Comfortable with proper insertion; active sound management; variability in attenuation; comfort and fit variable with changes in ear canal; higher initial cost than other ear plugs, quality can be variable.
Combination earmuff		Easy to fit; face shield or face screen available; might be less adjustable than individual components; may be uncomfortably hot or heavy; more expensive than earmuffs or earplugs, often more expensive than individual components.

LADDER SAFETY

In analytical laboratories with instrumentation, it is common for personnel to require access to above-bench heights and to areas behind instruments. Often, a ladder is employed to gain access to such places. Indeed, one often finds ladders in place semi-permanently in some laboratories, such as in front of NMR magnets. As common as ladders are in the laboratory, one of the most common type of lab injury is the result of falling from a ladder. Some guidelines for the safe use of ladders are provided below [1,2].

REFERENCES

1. Subpart X—Stairways and Ladders, Appendix A (American National Standards Institute (ANSI) 14.1, 14.2, 14.5 (1982)) of OSHA's Construction standards, Source for Type IAA: ANSI 14.1, 14.2, 14.5, 2009.
2. OSHA Quick Card-Portable Ladder Safety, https://www.osha.gov/Publications/portable_ladder_qc.html, accessed October, 2019.

STEP LADDERS

When possible, it is best to use a step ladder, which is a portable, self-supporting, A-frame ladder with two front side rails and two rear side rails (note that some older tripod type ladders are still in use). Generally, there are steps mounted between the front side rails and bracing between the rear side rails, but sometimes there are steps on both sides. While aluminum ladders are lightweight and are easy to carry, they pose a hazard when used near electrical equipment and electronics. It is generally better to use the heavier and more costly fiberglass ladders. Different ladder types are provided in Table 15.1:

When climbing or descending from a step ladder, it is important to maintain three points of contact: two hands and a foot, two feet, and one hand. Stand on the middle of the step, only use the ladder on a level surface, and be cautious of other personnel that may be present. The semi-permanent installations, around NMR magnets for example, should be set off with barricades. If you carried tools to the top of the ladder to work on an instrument, be sure to take them down with you before folding and removing the ladder. Never try to move the ladder while standing on it, by trying to "walk" the legs; never have more than one person on the ladder, and never stand on the top step of a ladder.

Straight, Combination, and Extension Ladders

While step ladders are to be preferred in the lab, sometimes a straight or combination ladder is needed. Combination ladders are a hybrid of the step ladder and straight ladder, and can fold or unfold, sometimes furnishing a small scaffold. An extension ladder is essentially two locking units of straight ladder sections. All of the guidelines for step ladders apply to straight and combination ladders, with some additional considerations. The angle of the straight ladder is of critical importance for safe operation. The proper angle for setting up a straight or combination ladder is to place its base a quarter of the working length of the ladder from the wall or other vertical surface. When accessing an elevated surface with a straight or combination ladder, the ladder must extend at least 3 feet above that surface. Never stand on the three top rungs of a straight, combination, or extension ladder. Finally, ladders must be free of slippery coatings such as oils and greases, and must be used on a level surface unless appropriate adjustment jacks are provided.

Table 15.1   Ladder Types and Ratings

Type	Duty Rating	Use	Load lbs (kg)
1AA	Special duty	Rugged	375 (170)
1A	Extra heavy duty	Industrial	300 (136)
I	Heavy duty	Industrial	250 (113)
II	Medium duty	Commercial	225 (102)
III	Light duty	Household	200 (91)

COMPRESSED AIR SAFETY

Compressed air is commonly used in laboratories and is typically piped in as a utility. One commonly sees utility pods with laboratory valves or cocks for air, hot and cold water, natural gas, lab vacuum, and occasionally steam. Compressed air may be hard-plumbed into instrumentation for valve actuation, vortex tube chilling, and as a source of air for flame ionization detectors. There are specific hazards associated with the use of compressed air sources and lines in laboratories, and safety precautions must be observed.

All pipes, hoses, and fittings must have at least the rating of the maximum pressure of the compressor. Compressed air pipelines should be identified as such with labeling affixed to the pipe (high pressure air or plant air are also acceptable labels).

The maximum working pressure should be known and not exceeded. Typically, the pressure of laboratory compressed air lines is no more than 125 psig (862 kPa), and reduced pressures are provided by diaphragm regulators.

Isolation valves (on/off ball-cock valves) should be provided for each instrument plumbed to compressed air.

Flexible hoses used for compressed air delivery must be appropriately rated, must be kept free from grease and corrosives, and must be secured to fixtures with hose clamps. The hoses themselves must be secured to prevent hose whip in the event of a rupture. Care must be taken to prevent kinking of flexible air hoses.

Compressed air must not be used to clean dirt and dust from clothing or personnel. Laboratory compressed air used for cleaning should be regulated to 15 psi (103 kPa) unless the blowguns used are equipped with diffuser nozzles to lower exit pressure and velocity.

Static electricity can be generated through the use of pneumatic devices and tools. This type of equipment must be grounded or bonded if it is used where fuel, flammable vapors, or explosive atmospheres are present.

All sources of laboratory or plant compressed air will contain water vapor, even if coalescence filters and particulate filters are provided in-line. This water vapor can condense into liquid or ice and can cause if the air contacts electrical or electronic devices. If the laboratory air supply is dried with a refrigerated conditioner, the dew point is still typically only dropped to 40°F (4 °C).

In some older facilities, compressed air systems can be severely contaminated with oil (usually from the compressor), rust (from steel piping), and liquid water. In such installations, it is important to provide coalescence filters and particulate filters in-line.

SAFETY IN THE USE OF CRYOGENS

Cryogens (liquified gases or cryogenic liquids) are used extensively in laboratories and in analytical instruments. A cryogen is typically defined as a liquefied gas with a normal boiling temperature of no higher than −90°C (−130°F, different sources cite differing upper limits). Dry ice (solid carbon dioxide, which sublimes at 194.65 K, −78.5°C) is also used to achieve low temperatures in the laboratory but is not considered to be a cryogen. The hazards associated with the use of cryogens are actually two-fold. There are hazards associated with the cryogenic fluids themselves, and there are hazards associated with the containers used to store and transport the cryogenic fluids. Here, unlike other sources, we will treat these separately [1–6].

REFERENCES

1. Lemmon, E.W., Properties of Cryogenic Fluids, in Rumble, J., Ed., *CRC Handbook of Chemistry and Physics*, 100th ed., 2019.
2. Safetygram 27, *Cryogenic Liquid Containers, Air Products and Chemicals*, Allentown PA, online media, accessed November 2019.
3. Safetygram 8, Liquid Argon, Air Products and Chemicals, Allentown PA, online media accessed, November 2019.
4. Safetygram 16, Safe Handling of Cryogenic Liquids, Air Products and Chemicals, Allentown PA, online media, accessed November 2019.
5. Safetygram 7, Liquid Nitrogen, Air Products and Chemicals, Allentown PA, online media, accessed November 2019.
6. Laboratory Safety and Chemical Hygiene Plan, Queensborough Community College of City University of New York, Bayside, NY, 2012.

CRYOGENS

The major hazards associated with the use of cryogens stem from their low temperatures, their high liquid-to-gas expansion ratios, toxicity, and air displacement. Some important properties germane to the handling of common cryogens are provided below. The gas-to-liquid expansion ratios are on the basis of 1:R and have a variability of up to ±5 depending on the local ambient temperature. The boiling temperatures are at atmospheric pressure (101.325 kPa).

Cryogen Fluid	Expansion Ratio R (Approximate)	Boiling Temperature (K)	Boiling Temperature (°C)
Argon	847	87.302	−185.84
Helium (^{4}He)	757	4.2238	−268.92
Hydrogen	850	20.271	−252.87
Nitrogen	696	77.355	−195.79
Oxygen	860	90.188	−182.96

Low Temperature

Clearly the main purpose of the use of a cryogen stems from the low temperatures at their boiling points. This leads to the potential of severe frostbite burn if a cryogen comes into contact with skin. While a thin layer of vapor formation will protect the skin initially, if a cryogen is allowed to pool (such as in clothing), the frostbite danger is severe. Protective clothing is essential and should include cryogen gloves and safety glasses (a full face shield is recommended). In addition, canvas shoes are discouraged since pooling can occur in the case of spillage. Pooling can also occur in cuffs on pant legs, which should also be avoided. Pants should not be tucked into shoes. A lab coat or shirt cuffs should be tucked under glove gauntlets. If skin should come in contact with a cryogen, it should be rinsed with cold water; do not apply dry heat to the affected area. If clothing has frozen to an individual due to cryogen exposure, cold water should be used to free the clothing, and emergency personnel must be summoned.

Another hazard due to the low temperature results from the ability of these fluids to embrittle materials including hoses, floor mats, and other surfaces in the laboratory. Of particular concern is the embrittlement of electrical insulation, which can lead to a fire hazard.

Since the boiling temperature of liquid nitrogen is below that of liquid oxygen, it is possible for oxygen to condense on any surface or vessel cooled by liquid nitrogen. Liquid oxygen is an oxidizer that can enhance the flammability characteristics of liquids and solids that it contacts. The liquid air that is seen dripping from lines transferring liquid nitrogen can be up to 50 % oxygen. If a blue tint is observed in a vessel being used with liquid nitrogen, the presence of liquid oxygen must be assumed. Additional hazards of liquid oxygen will be discussed below.

Asphyxiation

The high liquid-to-gas expansion ratios listed in the table above show that when a cryogen is vaporized, it has the potential to displace air in a laboratory. While most cryogens are not toxic *per se*, they can act as simple asphyxiants. Laboratories that use cryogens should be adequately ventilated. In labs with large containers of cryogen, it is important to have an oxygen monitor with an audible alarm. Personnel, including rescue workers, should not enter areas where the oxygen concentration is below 19.5 % (vol/vol), unless provided with a self-contained breathing apparatus or air-line respirator. The safe range as indicated on an oxygen monitor is between 19.5 % and 23 % (vol/vol). Personnel in an area of low oxygen concentration may be unaware of the condition, thus monitoring is critical.

If a person seems to become dizzy or loses consciousness while working with cryogens, they should be moved to a well-ventilated area immediately. If breathing has stopped, apply cardiopulmonary resuscitation (CPR), and emergency personnel must be summoned.

Overpressure

Related to the liquid-to-gas expansion ratio is the overpressure that can result if a cryogen is allowed to warm within an enclosure. Indeed, no cryogen can remain liquid within a container; some venting must be provided. If a vent becomes disabled or is not present, warming cryogen will vaporize and produce very high pressures based on the pressure–volume–temperature (PVT) surface of the fluid.

Reactivity and Toxicity

Some cryogens pose specific hazards or handling requirements based on their chemistry.

- *Liquid oxygen* (LOx) cannot be permitted to contact organic materials; some common cautions include solvents and vacuum pump oil. Organic materials can be readily ignited by spark or shock after exposure to LOx, including fingerprints on a surface. Clothing saturated with oxygen is readily ignitable and will vigorously burn. If LOx spills on an asphalt surface, do not walk over or roll equipment over that surface for 1 h. While not having specific toxicity issues, if LOx is exposed to high-energy electromagnetic radiation, it can produce ozone, which will solidify at LOx temperatures. Solid ozone is unstable, toxic (upon vaporization), and can explode if disturbed.
- *Liquid hydrogen* handling requires all of the precautions used for hydrogen gas. Liquid hydrogen should not be transferred in an atmosphere of air since it will readily condense in the liquid hydrogen, resulting in a potential explosive mixture. Liquid hydrogen must be transferred by helium pressurization in properly designed vacuum-insulated transfer lines pre-purged with helium or gaseous hydrogen. Liquid hydrogen, like liquid helium, can solidify air, which can block vents and safety relief devices. Dewars and other containers made of glass should not be used for liquid hydrogen service. Breakage makes the possibility of explosion too hazardous to risk.

CRYOGENIC LIQUID CONTAINERS

Several different types of cryogenic liquid containers are encountered in the laboratory during routine chemical analyses, and each of them have their own associated hazards and precautions. It is common parlance to refer to all of these containers as Dewars, but this is imprecise. The small portable containers used to assemble laboratory cold baths and the small transport containers (with loose-fitting lids and carry handles) are Dewars. These containers are used at ambient pressure. The larger supply containers (from which Dewars are filled) are called liquid cylinders. These containers are pressurized, with different pressure ratings available.

Liquid Cylinders

Liquid cylinders are large heavy containers with integrated casters or a dolly to facilitate movement. At least two of these casters should be equipped with a braking mechanism. Typical volumes and weights are provided below for liquid cylinders for nitrogen, oxygen, and argon.

Volume Capacity L (Nominal)	160	180	230
Tare weight, lbs (kg)	250 (114)	260 (118)	310 (141)
Filled weight, N_2 lbs (kg)	513 (233)	556 (253)	667 (303)
Filled weight, O_2 lbs (kg)	662 (301)	627 (285)	825 (375)
Filled weight, Ar lbs (kg)	695 (316)	753 (342)	936 (425)

Liquid helium cylinders, which often incorporate a liquid nitrogen jacket, are usually heavier than those for the common cryogens listed in the above table. Moreover, there is a larger variety of available sizes, ranging from 50 to 500 L. The above filled weights are typical as-filled weights. Some losses occur in transport, and the losses noticed for helium liquid cylinders can be considerable.

The weight of these cylinders can make them challenging to handle. The PPE discussed above (cryogens) must be used when handling liquid cylinders. Cylinders should be moved by pushing, not pulling, to reduce the potential of an upset. In locations of frequent transport of liquid cylinders, bottom door sills should be removed to eliminate the potential of bouncing or rough handling. If a cylinder must be transported by elevator, a freight elevator is preferred. Personnel should not ride in the elevator car with the cylinder. The cylinder should be transported in the elevator with no personnel and be met at the receiving floor. A placard reading "CRYOGEN TRANSFER—DO NOT ENTER ELEVATOR" should be posted facing the door if the elevator is to travel more than one story.

If liquid cylinders must be transported between buildings, it is critical to use ramps and ensure that there are no large cracks in paving that must be traversed. Note also that the casters commonly found on liquid cylinders are not rated for travel along long distances of pavement.

In the event loss of control of a liquid cylinder during transport, if the cylinder begins to fall it is usually best to simply let it go and summon qualified help as defined in the organization standard operating procedures.

Liquid cylinders are pressurized and can contain up to 350 psi (2,411 kPa), depending on the cylinder specifications. One commonly encounters pressure specifications on cylinders that may sometimes be confusing. The commonly encountered specifications are as follows:

- psia (pounds-force per square inch absolute) gauge pressure plus local atmospheric pressure.
- psid (psi difference) difference between two pressures, specified on the cylinder label.
- psig pounds (force) per square inch, gauge.

- psi-vg (psi vented gauge) difference between the measuring point and the local pressure.
- psi-sg (psi sealed gauge) difference between a chamber of air sealed at atmospheric pressure and the pressure at the measuring point.

Pressure relief devices are integral to all liquid cylinders and must remain un-obstructed with frost. If the outlet fitting on a pressure relief valve is facing the same direction as the liquid or gas dispensing valve, a fitting directing vented gas away from users should be added to protect personnel. Overpressurization of liquid cylinders is a serious hazard; cylinders can rupture if a pressure relief valve becomes impaired or inoperative.

Dewar Flasks

Dewar flasks or simply, Dewars, are small cryogen containers used at atmospheric pressure, with or without a loose-fitting cover or cap. Smaller Dewars are usually made from an evacuated silvered glass insert set into a metal jacket. Any exposed glass should be taped to prevent flying glass in the event of a catastrophic rupture.

When filling a small Dewar from a liquid cylinder, it is best to pre-cool the interior of the Dewar with a small amount of cryogen first, before completing the fill. Boiling and splashing generally occur when filling a warm container, so stand clear and wear appropriate PPE as discussed above. The flask should be clean and dry before filling.

A Dewar flask should not be filled to beyond 80 % of its capacity. Overfilling increases the risk of splashing and spillage. A beverage thermos bottle is not a substitute for a Dewar flask in the laboratory.

When carrying a small Dewar flask, make sure it is the only item you are carrying. Hold the Dewar flask as far away from the face as possible. Be aware of other personnel in the area.

Small Dewar flasks with liquid nitrogen are often used as cold traps in the laboratory. When instrument components are placed in the filled Dewar cold bath, it is important to avoid splashing and excessive boiling by inserting components slowly. If Pyrex wool insulation is placed around the flask, do not let the wool dip into the cryogen or become a vapor barrier. If the liquid nitrogen acquires a blue tint, it has become contaminated with liquid oxygen, and the discussion of LOx hazards above applies.

If a Dewar cold trap is used in association with a vacuum pump, it is important to periodically empty the trap carefully to avoid exposure to toxic chemicals and to prevent overpressurization should the trap run dry. Note also that venting liquid nitrogen near a vacuum pump v-belt can embrittle the belt and shorten its service life. Likewise, if the venting is near electrical cables, embrittlement of the electrical insulation can result in a fire hazard.

SELECTION OF RESPIRATOR CARTRIDGES AND FILTERS

Respirators are sometimes desirable or required when performing certain tasks in the chemical analysis laboratory. There is a standardized color code system used by all manufacturers for the specification and selection of the cartridges and filters that are used with respirators. The following table provides guidance in the selection of the proper cartridge using the color code.

Color Code	Application
Gray	Organic vapors, ammonia, methylamine, chlorine, hydrogen chloride, and sulfur dioxide or hydrogen sulfide (for escape only) or hydrogen fluoride or formaldehyde.
Black	Organic vapors, not to exceed regulatory standards.
Yellow	Organic vapors, chlorine, chlorine dioxide, hydrogen chloride, hydrogen fluoride, sulfur dioxide, or hydrogen sulfide (for escape only).
White	Chlorine, hydrogen chloride, hydrogen chloride, hydrogen fluoride, sulfur dioxide, or hydrogen sulfide (for escape only).
Green	Ammonia and methylamine.
Orange	Mercury and/or chlorine.
Purple	Solid and liquid aerosols and mists.
Purple + gray	Organic vapors, ammonia, methylamine, chlorine, hydrogen chloride, and sulfur dioxide or hydrogen sulfide (for escape only) or hydrogen fluoride or formaldehyde; solid and liquid aerosols and mists.
Purple + black	Organic vapors, and solid and liquid aerosols and mists.
Purple + yellow	Organic vapors, chlorine, chlorine dioxide, hydrogen chloride, hydrogen fluoride, sulfur dioxide, or hydrogen sulfide (for escape only); solid and liquid aerosols and mists.
Purple + white	Chlorine, hydrogen chloride, hydrogen chloride, hydrogen fluoride, sulfur dioxide, or hydrogen sulfide (for escape only); solid and liquid aerosols and mists.
Purple + green	Ammonia, methylamine, and solid and liquid aerosols and mists.

In addition to the cartridges specified in the table, particulate filters are available that can be used alone or in combination.

CHEMICAL FUME HOODS AND BIOLOGICAL SAFETY CABINETS

Engineered safety equipment is preferred over the reliance on PPE, and the fume hood is one such engineered safety device that is nearly ubiquitous in laboratories. Laboratories concerned with biological specimens (microbes, spores) also commonly are equipped with BSCs. The following section provides basic information on the function and application of these devices. The purpose here is not to provide design or installation instructions, since most users will find this equipment already installed in their work spaces. Rather, this information is to allow optimal use to be made of the installation that is preexisting [1].

REFERENCE

1. Bruno, T.J., *CRC Handbook of Chemistry and Physics*, 100th Ed., CRC Press, Boca Raton, FL, 2019.

Types of Chemical Fume Hoods

Most of the chemical fume hoods considered here consist of a cabinet or enclosure set at waist level (above a table or storage cabinet) that is connected to a blower located above the hood or external to the hood through a duct system. The cabinet has an open side (or sides) to allow a user to perform work within. A movable transparent sash separates the user from the work. Most chemical fume hoods have a sill that functions as an airfoil at the work surface below the sash. The connection to the blower might be by use of a v-belt, or it may be direct drive. This allows provision of a smooth flow of air with minimal turbulence. In some installations, axially mounted blowers are used, especially if multiple hoods are ducted into a common blower. Baffles located in the rear of the cabinet provide control of the air flow patterns and can usually be adjusted to provide the best air flow around the experiment or procedure being performed. Many chemical fume hoods are equipped with air flow indicators, low flow monitors and alarms, and differential pressure sensors to allow the user to operate safely. The major types of chemical fume hoods include the standard/conventional, walk-in, bypass, variable air volume, auxiliary air, or ductless types. Additional types include snorkels and canopies that are portable. Each type must be understood to be operated most efficiently within specifications (see the section below on safe operation).

Standard or Conventional

The standard chemical fume hood utilizes a constant speed motor, and for this reason, the volume of air drawn into the hood will change with movement of the sash position. As the sash is lowered, the velocity of the air drawn into the hood will increase.

Bypass

The bypass chemical fume hood is very similar to the standard/conventional hood except that as the sash is lowered, a vent is opened above the sash to allow additional air flow into the hood. This prevents a large increase in velocity in the working area inside the hood.

Variable Air Volume

The variable volume chemical fume hood controls the volume of air drawn into the hood as a function of sash position, while maintaining the face velocity of the air at a constant rate, within the specifications required. These types of chemical fume hoods are more energy efficient than the standard or bypass hoods because they minimize costs incurred by laboratory heating and cooling.

Auxiliary Air

The auxiliary air chemical fume hood includes an additional blower that injects air into or at the face of the hood, providing additional flow inside the enclosed cabinet. These types of hoods are rarely installed in renovations or new constructions but may be encountered in older laboratories. They are less desirable than the standard/conventional, bypass or variable volume types because they require a great deal of energy to operate (although the early designs featured the addition of an auxiliary air stream that was not air conditioned). These devices are mechanically more complex than other types and consequently more prone to maintenance problems.

Walk-In Hood

The walk-in hood is a chemical fume hood that is mounted directly on the laboratory floor or a slightly raised chemical resistance platform. It is used for the ventilation of larger pieces of equipment, with the advantage that these pieces of equipment can be wheeled in and out of the walk-in hood. The walk-in hood typically used two separate sashes.

Ductless Hoods

This type of chemical fume hood does not duct the air flow to outside the laboratory, but rather the air flow is returned to the room or interstitial space after passing through a means to remove contaminants. The contaminants may be removed by HEPA filters, activated carbon cartridges, adsorbents, or catalyst beds. The means of contaminant removal must be inspected and serviced at regular intervals.

Laminar Flow Hood (Clean Bench)

Related at least in principle to ductless hoods are laminar flow hoods, sometimes called clean hoods. These are devices intended to protect the work being performed from particulates in the air, which is accomplished by bathing the work area with HEPA-filtered air either blown at low velocity over the work area or blown from the bottom of the hood as an air curtain. Only approximately 10 % of the air flow is through the face of the hood. These are intended to protect the work or samples inside the hood, not primarily the user. These units should not be used in place of chemical fume hoods, rather they are used to protect the work from dust or pollen.

Snorkels and Canopies

Snorkels are flexible ducts routed from a blower duct that can be placed temporarily atop or near an experiment to provide some measure of protection. A canopy is similar, but it incorporates an additional bell-shaped collector that might be suspended above an experiment, but for the same purpose as that for the snorkel.

Chemical Fume Hood Operations

While the operation of chemical fume hoods is straightforward, safe operating practices must be observed. The face velocities should be optimized between 80 and 120 fpm. Face velocities in excess of 125 fpm can cause turbulent flow and allow outflow of contaminants from the hood and potentially expose the user to hazards. Face velocities are checked periodically by facilities managers. Ideally, chemical fume hoods are located in low traffic areas of the laboratory, away from entry doors. Safe operation of chemical fume hoods requires observance of the following practices, divided into primary (applicable to all installations) and secondary (applicable on a case-by-case basis):

Primary Guidelines

Before using the chemical fume hood, ensure that it is in working order. A simple air flow test can be done with a laboratory wipe or one can observe the flow monitor if one is present. Be aware of clattering sounds that might indicate a broken belt or screeching sounds that might indicate a failed or failing bearing.

The motor of the chemical fume hood should be running at all times, except for maintenance. If a switch is located inside the laboratory, it should be equipped with a lockout to protect maintenance staff.

Users should perform all work at least six inches inside the plane of the sash.

The user should avoid placing his/her head inside the chemical fume hood, beyond the plane of the sash.

Laboratory occupants should avoid traffic in front of the chemical fume hood to minimize turbulence.

Users should avoid rapid movements in front of the chemical fume hood; this includes rapidly raising and lowering the sash.

Keep the sash closed down as far as possible.

The baffles of the chemical fume hood should be free of obstruction.

If a heating device such as a hot plate is being used in the hood, the interior air flow can be dramatically changed by convection. In those instances, the lower baffles of the hood should be closed or minimized and the middle baffles fully opened to accommodate the convection.

Equipment located inside the cabinet that can potentially disrupt the air flow should be raised above the level of the lower baffles by a small shelf or blocks.

Hoses and power cords that must be run into the chemical fume hood from the outside should be run through the airfoil beneath the sash.

Secondary Guidelines

Some chemical fume hoods have a small cup sink located inside the cabinet. Water should be run into this sink periodically to maintain it free of obstruction and to keep the P-trap full and thereby prevent sewer gas back-up.

It is imperative to prevent the discharge of chemicals into the cup sink.

Some chemical fume hoods have compressed air lines that allow the use of air operated equipment such as vortex tubes (for heating and cooling). When using vortex tubes inside a fume hood, the sash should be fully closed because additional air flow inside the hood is present.

In a power outage, lower the sash to within an inch of the fully closed position.

Evaporations and digestions with perchloric acid must only be done in a specifically designed and designated perchloric acid chemical fume hood. The same considerations apply for the use of radionuclides and infectious agents.

Types of Biological Safety Cabinets (BSC)

BSCs, as distinct from chemical fume hoods, are enclosed ventilated cabinets intended to provide both a clean and a safe working environment for aerosols and biological hazards. All exhaust air is HEPA-filtered as it exits the biosafety cabinet, removing harmful bacteria and some viruses. The Centers for Disease Control (US) lists three classes of BSCs, namely, Classes I, II and III,.

Class I

Class I cabinets are open-front negative pressure cabinets that provide personnel and environmental protection but do not provide protection to the sample or media (the product) being used in the cabinet. There is air flow into the cabinet (at a face velocity of 75 fpm) that can potentially cause sample contamination. Class I cabinets are often used to enclose specific equipment (centrifuges, harvesting equipment. or fermenters) or ongoing procedures (cultures) that potentially generate aerosols. BSCs of this class are either ducted (connected to the building exhaust system) or unducted (recirculating HEPA-filtered exhaust back into the laboratory, provided there is an interlock with the building exhaust system). Some Class I cabinets are used for animal cage changing, and these typically require frequent HEPA filter changes due to odoriferous compounds saturating the filter.

Class II

A Class II BSC provides protection for both the worker and the sample or product, making it suitable as a sterile compartment for cell culture. This type of cabinet is the most versatile and most common, with face velocities similar to those of Class I. There are four types of Class II cabinets, the main features of which are discussed below. Note that there are additional differences among the types in Class II (types A1, A2, B1, and B2), primarily concerning the geometry of the air flows and placement of HEPA filters.

Type A1 (formerly Type A) does not have to be duct vented (although it is possible to connect to building ventilation systems by use of canopy exhaust connections), which makes it suitable for use in laboratories inaccessible to ductwork. This cabinet can be used for use of low to moderate hazard agents that do not include volatile toxic chemicals and volatile radionuclides. The supply air is HEPA filtered to present the sample or media being used with a particulate free air stream with a face velocity of at least 75 fps. This type of BSC cannot be used for volatile and toxic compounds and solvents because small quantities of these materials can quickly load the filter. Type A2 differs from Type A1 in that protection of the operator, and the environment is only afforded if the exhaust line is canopy vented to the building exhaust. The face velocities of these units are at least 100 fps.

Type B1 cabinets must be hard vented, with 50 % of the air exhausted from the cabinet while 50 % can be recirculated back into the room. This cabinet may be used with etiologic agents and traces of volatile and toxic chemicals and radionuclides required as an adjunct to microbiological studies (if the work is done in the directly exhausted portion of the cabinet). The air intake velocity of the B1 type is specified to be 100 fps. Type B2 cabinets must be 100 % exhausted through a dedicated duct. The air intake velocity is specified to be 100 fps. This cabinet may be used with etiologic agents treated with toxic chemicals and radionuclides required as an adjunct to microbiological studies.

Class III

The Class III cabinet is designed for highly infectious microbial agents. It is entirely gas tight, with a non-operating view window (cannot be opened). Access to the interior is through a dunk tank accessible through the floor of the cabinet or through a double door system. Both the supply and exhaust gas streams pass through HEPA filters. Heavy duty rubber gloves are used for manipulations in the interior of the cabinet.

LASER HAZARDS IN THE LABORATORY

Lasers are commonly used in the laboratory, although in many instruments most lasers are embedded in instrumentation and are therefore shielded or protected by optical barriers and interlocks that, when functioning properly, prevent accidental exposure. Care must be exercised when performing maintenance or when changing samples in such instruments. In this section, we provide basic information on laser safety and hazards [1–3]. This is by no means exhaustive nor is it meant to substitute for an understanding of the specific safety requirements of instrumentation or applicable law or regulations. The special case of common laser pointers has received considerable attention recently and is treated separately [4]. We note that as of 2007, the general practice in the United States is to use the ion-exchange chromatography (IEC) definitions.

REFERENCES

1. American National Standard for Safe Use of Lasers, American National Standards Institute, ANSI Z136.1, 2007.
2. Safety of Laser Products—Part 1: Equipment Classification and Requirements, International Electrotechnical Commission, IEC 60825-1, 2nd Ed. 2007.
3. Bruno, T.J. and Svoronos, P.D.N., *CRC Handbook of Chemistry and Physics*, 100th Ed. CRC Press, Boca Raton, FL, 2019.
4. Hadler, J. and Dowell, M., Accurate, inexpensive testing of laser pointer power for safe operation, *Meas. Sci. Tech.*, 24, 1–7, 2013.

Classes of Lasers

The following is a summary for the laser classes following the ANSI guidelines used in the United States:

Class I

Class I lasers are inherently safe with no possibility of eye damage under conditions of normal use. The safety can result from a low output power (in which case eye damage is impossible even after prolonged exposure) or due to an enclosure preventing user access to the laser beam during normal operation, such as in CD players, laser printers, surveying transits, or measurement instruments.

Class II

The blink reflex of the human eye will prevent eye damage, unless the person deliberately stares into the beam for an extended period. Thus, a Class II laser can cause some eye damage if this is done. Output power may be up to 1 mW. This class includes only lasers that emit visible light. Some laser pointers are in this category.

Class IIIa

Lasers in this class are mostly dangerous in combination with certain optical instruments that change the beam diameter or power density. Output power does not exceed 5 mW. Beam power density may not exceed 2.5 mW/cm². Many laser sights for firearms and some laser pointers are in this category.

Class IIIb

Lasers in this class may cause damage if the beam enters the eye directly. This generally applies to lasers powered from 5 to 500 mW. Lasers in this category can cause permanent eye damage with exposures of 1/100th of a second or less depending on the strength of the laser. A diffuse reflection (on paper or from a matte surface) is generally not hazardous, but a specular reflection from a highly reflective surface can be just as dangerous as direct exposures. Protective eyewear is recommended when direct beam viewing of Class IIIb lasers may occur. Lasers at the high-power end of this class may also present a fire hazard and can lightly burn skin.

Class IV

Lasers in this class have output powers of more than 500 mW in the beam and may cause severe, permanent damage to eye or skin without being magnified by optics of eye or instrumentation. Diffuse reflections of the laser beam can be hazardous to skin or eye within the Nominal Hazard Zone. Many industrial, scientific, military, and medical lasers are in this category.

The following is a summary of the laser classes following the IEC guidelines.

Class 1

A Class 1 laser is safe under all conditions of normal use, with no known biological hazard present. This class includes high-power lasers within an enclosure that prevents exposure to the radiation and that cannot be opened without shutting down the laser. This typically requires an interlocking.

Class 1M

A Class 1M laser is safe for all conditions of normal use except when passed through magnifying optics such as microscopes, telescopes, or on optical benches. Class 1M lasers typically produce large-diameter beams or beams that are divergent. The classification of a Class 1M laser must be changed if the emergent light is refocused.

Class 2

A Class 2 laser is safe for all conditions of normal use because the blink reflex will limit the exposure to no more than 0.25 seconds. It only applies to visible light lasers (400–700 nm) limited to 1 mW continuous wave or more if the emission time is less than 0.25 seconds or if the light is not spatially coherent. Intentional suppression of the blink reflex could lead to eye injury. Many laser pointers are Class 2.

Class 2M

A Class 2M laser is similar to a Class 2, but it is used in an instrument that may focus the beam. This laser is safe because of the human blink reflex, provided the beam is not viewed through optical instruments as described above for Class 1M.

Class 3R

A Class 3R laser is considered safe if handled carefully, with restricted beam viewing. These lasers can be hazardous to the human eye if the beam is viewed for extended periods of time or under fixated conditions. Continuous beam Class 3R lasers operating in the visible region are limited in power output to 5 mW. For other wavelengths and for pulsed lasers, other limits will apply.

Class 3B

A Class 3B laser is hazardous if the eye is exposed directly, but diffuse reflections such as from paper surfaces are not harmful. Continuous lasers in the wavelength range from 315 nm to far infrared are limited in power output to 0.5 W. For pulsed lasers between 400 and 700 nm, the limit is 30 mJ. Other limits apply to other wavelengths and to short pulse lasers. Protective eyewear is typically required where direct viewing of a class 3B laser beam may occur. Class 3B lasers must be equipped with a key switch and a safety interlock.

Class 4

Class 4 lasers include all lasers with beam power greater than those covered in class 3B. By definition, a Class 4 laser can burn the skin, in addition to causing severe and permanent eye damage. This eye damage can result from direct or diffuse beam viewing. These lasers may ignite combustible materials and thus may represent a fire risk. Class 4 lasers must be equipped with a key switch and a safety interlock. Many industrial, scientific, military, and medical lasers are in this category.

Laser Pointers

Laser pointers, ubiquitous at meetings, shows, and in the classroom, deserve separate consideration because of recent work on actual observed power output. For purposes of classification into the levels discussed above, the typical laser pointer is classified as 3R. The human light aversion response can generally protect against 3R lasers; however, this response is less sensitive in the near-infrared range (700–1,400 nm). Thus, to prevent retinal burns, laser pointers must not emit hazardous levels of infrared, and must be Class 1 compliant in terms of accessible emission level (AEL) in that wavelength range. In a testing program undertaken at NIST, cited in Reference 4, it was found that of the laser pointers randomly chosen and tested, all but two pointers failed to comply by more than 15 % of the specified AEL, and 48 % emitted more than twice the specified AEL at one or more specified wavelength. This indicates a risk of 3B exposure from these devices that are nominally classified as 3R.

EFFECTS OF ELECTRICAL CURRENT ON THE HUMAN BODY

The following table provides information on the effects of electrical shock on the human body [1]. The table lists current values in milliamperes. The voltage is an important consideration as well because of the relationship with resistance:

$$I = V/R,$$

where I is the current, V is the voltage, and R is the resistance. The presence of moisture can significantly decrease the resistance of the human skin and thereby increase the hazard of an electrical shock. The current difference between a barely noticeable shock and a lethal shock is only a factor of 100. In individuals with cardiac problems, the difference may be lower.

REFERENCE

1. Furr, A.K., ed., *CRC Handbook of Laboratory Safety*, 5th ed., CRC Press, Boca Raton, FL, 2000.

Current (Milliamperes)	Reaction
1	Perception level, a faint tingle.
5	Slight shock felt; disturbing but not painful. Average person can let go. However, vigorous involuntary reactions to shocks in this range can cause accidents.
6–25 (women) 9–30 (men)	Painful shock, muscular control is lost. Called freezing or "let-go" Range.[a]
50–150	Extreme pain, respiratory arrest, severe muscular contractions, individual normally cannot let go unless knocked away by muscle action. Death is possible.
1,000–4,300	Ventricular fibrillation (the rhythmic pumping action of the heart ceases). Muscular contraction and nerve damage occur. Death is most likely.
10,000+	Cardiac arrest, severe burns, and probable death.

[a] The person may be forcibly thrown away from the contact if the extensor muscles are excited by the shock.

ELECTRICAL REQUIREMENTS OF COMMON LABORATORY DEVICES

The following table lists some common laboratory devices along with the current and power requirements for the operation of the device [1]. This information is important to consider when instrumentation is being installed, relocated, or used on the same circuit. Common 120 V circuits in laboratories are typically rated at 10 or 15 amperes. The reader should note that the current draw often spikes to a high level in first few microseconds after a device is energized. This is especially true for devices that have electric motors.

REFERENCE

1. Furr, A.K., ed., *CRC Handbook of Laboratory Safety*, 5th ed., CRC Press, Boca Raton, FL, 2000.

Instrument	Current (Amperes)	Power (Watts)
Balance (electronic)	0.1–0.5	12–60
Biological safety cabinet	15	1,800
Blender	3–15	400–1,800
Centrifuge	3–30	400–6,000
Chromatograph	15	1,800
Computer (PC)	2–4	400–6,000
Freeze dryer	20	4,500
Fume hood blower	5–15	600–1,800
Furnace/oven	3–15	500–3,000
Heat gun	8–16	1,000–2,000
Heat mantle	0.4–5	50–600
Hot plate	4–12	450–1,400
Kjeldahl digester	15–35	1,800–4,500
Refrigerator/freezer	2–10	250–1,200
STILLS	8–30	1,000–5,000
Sterilizer	12–50	1,400–12,000
Vacuum pump (mechanical)	4–20	500–2,500
Vacuum pump (diffusion)	4	500

ELECTRICAL EXTENSION CABLES AND OUTLET STRIPS

It is common in the laboratory to require electrical power at a location that is too far from a receptacle. In such cases, it is common practice to temporarily use an extension cable. The improper and unwise application of electrical cables is a major cause of electrical fires. A heavy reliance on extension cables in a laboratory is an indication that too few receptacles are available. The only remedy to this is the installation of additional circuits in the laboratory. Most organizations have policies and standard operating procedures that cover the use of electrical extension cables, so this section serves as a supplement to those guidelines [1,2].

REFERENCES

1. Use Extension Cords Properly, The Electrical Safety Foundation International (ESFI), online resource, www.esfi.org/about-us, accessed November, 2019.
2. Focus 4 Construction Safety and Health, Occupational Safety and Health Administration, U.S. Department of Labor, online resource, www.osha.gov/sites/default/files/2018-12/fy07_sh-16586-07_4_electrical_safety_participant_guide.pdf, accessed November, 2019.

Most of the guidelines below follow from good industrial practice as well as common sense, and they serve as a starting point:

1. Use only electrical cables that are properly rated for the load. Earlier, we discussed the approximate load of common laboratory devices, and the equipment in the laboratory may also be labeled with the electrical load. The table below can be used as a guide to the choice of the appropriate cable.
2. Do not run electrical cables through walls, doorways, ceilings, or floors. If an electrical cable is obscured, it cannot be observed for problems.
3. Do not cover electrical cables with floor mats, since this can hamper heat dissipation and can obscure signs of wear.
4. If an electrical cable must be run along a floor temporarily, it must be taped securely to the floor to prevent movement and to minimize the tripping hazard. It must be clearly visible, to prevent a tripping hazard. Never staple or nail an electrical cable to any surface.
5. Never use an electrical cable that is damaged in any way; check it beforehand for any signs of deterioration. This includes broken or cracked insulation. Do not use electrical cables that have been "repaired" with duct tape or even electrical tape. Instead obtain a serviceable replacement.
6. Always use a grounded three-pronged extension cable, and always use a grounded outlet.
7. Avoid the temptation of grinding polarized plugs to make them more convenient.
8. Fully insert the plug of an extension cable into an outlet.
9. Keep extension cables away from water. In general, if you can touch an electrical appliance or outlet with one hand, and a water source (such as a faucet) with the other, a ground fault circuit interrupter (GFCI) should be used. A GFCI should be used in all damp environments and when using a circuit outdoors.
10. Electrical extension cords must be connected to prevent tension at joints and terminal screws; straining a cord can cause the strands of one conductor to loosen.
11. Do not use an extension cable with such power-intensive devices as heat guns or space heaters.
12. Unused extension cables should be unplugged and stored indoors, preferably wrapped or tied with string or rubber band to prevent them from lying on the floor where they can be damaged.
13. Do not "daisy chain" multiple electrical extension cables.
14. Power strips should not be permanently secured to a bench or instrument. These are temporary connection devices, and any fastening implies a more permanent installation.
15. An extension cable should not be used as a disconnect means for an apparatus. To deenergize a piece of equipment, first turn off the on/off switch before unplugging the power cable or extension cable.

In the table below, we provide some basic information on electrical extension cable size and nominal capacity. The AWG is the American wire gauge scale. To convert to the metric gauge scale, multiply the wire diameter in mm by a factor of 10 to obtain the nominal metric gauge. Note that in AWG, the diameter increases as the gauge decreases, but for metric gauges, it is the opposite.

Additional details and guidance should be sought in your institutional safety plan or local codes. Since this information is geared for laboratory use, we present information only for 25 foot (7.6 m) lengths. Note that some laboratory apparatus will draw large current upon start-up, such as vacuum pumps. This must be taken into account when choosing an electrical extension cable. If an electrical extension cable is used for more than 3 hours, the user should de-rate the amperage by 1.25 %. This might require choosing a cable of heavier gauge (larger diameter conductors).

AWG	Diameter, in (mm)	Nominal Current Rating (Amperes)	Comments
12	0.0808 (2.052)	20	Good for most medium and heavier duty applications.
14	0.0641 (1.628)	15	Good for light and medium duty laboratory apparatus.
16	0.0508 (1.290)	10	Used in computers and monitors as well as other light duty applications.
18	0.0453 (1.024)	7	Used in laptop computer power supplies and other very light duty applications.

Ubiquitous in most laboratories is a drawer or cabinet with unused computer cables that staff members have refused to part with. It is often tempting and expedient to cut off the connectors and use them to hard-wire or re-wire laboratory equipment. This may be prohibited by your organization's standard operating procedures. If not prohibited, pay close attention to the loads being served, noting that these electrical cables are almost always intended for very low loads.

GENERAL CHARACTERISTICS OF IONIZING RADIATION FOR THE PURPOSE OF PRACTICAL APPLICATION OF RADIATION PROTECTION

The following table provides practical information to allow the design and implementation of radiation protection in laboratory and industrial environments. Additional information and details can be found in the Refs. [1–4].

REFERENCES

1. Johnson, T.E. and Birky, B. K., *Health Physics Radiological Health*, 4th Ed., Lippincott Williams and Wilkins, Baltimore, MD, 2012
2. Cember, H. and Johnson, T. E., *Introduction to Health Physics*, 4th Ed., McGraw Hill, New York, 2011.
3. 1990 Recommendations of the International Commission on Radiological Protection. ICRP Publication 60. Ann. ICRP 21 (1–3).
4. Grove, T.W. and Bruno, T.J., General characteristics of ionizing radiation for the purposes of radiation protection, in Rumble, J., Ed., *CRC Handbook of Chemistry and Physics*, 100th Ed., CRC Press, Boca Raton FL, 2019.

Type (Symbol)	Physical Properties	Range	Shielding	Biological Hazards	Comments
Alpha particle (α)	• Very large mass (2 protons, 2 neutrons, 0 electrons) • +2 charge	• Very short 3–6 cm (~1–2 inches) in air	• Few centimeters of air • Sheet of paper • Dead (outer) layer of skin	• Internally, the source of alpha radiation is in close contact with live body tissue. It can deposit large amounts of energy in a small amount of body tissue. • Rarely an external hazard.	• Alpha particles of at least 7.5 MeV are required to penetrate the epidermis, the protective layer of skin, which is about 0.07 mm (70 μm) thick. • The range, R, of most particles of common emitters (4.5–5.5 MeV) is 3–4 cm in air. • Range in air = $0.322E^{3/2}$ cm, when E is expressed in in MeV.
Beta particle (β)	• Small mass • 1 or +1 charge • –1 charge particle is an electron, +1 charge particle is a positron	• Short 6–600 cm (1 inch to 20 feet) in air	• Low atomic number materials • Plastic • Glass • Aluminum	• Internal hazard • Externally, may be hazardous to the skin and eyes	• Beta particles of at least 70 keV are required to penetrate the epidermis. • $R_{air} \approx 3.65$ meters per MeV. • The range of beta particles in material in g/cm^2 (thickness in cm multiplied by the density in g/cm^3) is approximately half the maximum energy in MeV. ($R \approx E_{max}/2$). • Dose rate (in rad/hr) at 1 cm from a beta point source; ≈ 300 rad/hr per Curie. • Dose rate (in rad/hr) in a solution ≈ 2.12 $\bar{E}C/\rho$; where $\bar{E}$ = average energy in MeV, C = concentration in μCi/cm^3, and ρ = density of the solution in g/cm^3. The dose rate is about one-half this value at the surface. • An aqueous solution of 1 Curie ^{32}P in a glass vial typically produces 3 mrad/hr at 1 meter from Bremsstrahlung. • Shielding causes Bremsstrahlung radiation similar to X-rays and gamma rays; generally the higher the atomic number of a material, the more intense the Bremsstrahlung radiation.

(*Continued*)

Type (Symbol)	Physical Properties	Range	Shielding	Biological Hazards	Comments
Neutron (n or n⁰)	• Large mass • No charge	• Very far in air • Easily travel several hundred meters • High penetrating power due to lack of charge	• High hydrogen content material • Water • Concrete • Plastic	• External whole-body exposure hazard • May be external and/or internal hazard • Depends on whether source is inside or outside the body • Energy dependent	• Shielding can be provided by hydrogen-rich materials such as hydrocarbons, water, waxes, high water concrete. • Can cause neutron activation, in which radionuclides are formed. • Particularly damaging to soft tissue such as the cornea.
X-ray, gamma ray (γ)	• No mass • No charge • Electromagnetic wave • X-rays and gamma rays are similar, but place of origin and energy levels may differ	• Very far in air • Easily travel several hundred meters • Very high penetrating power since it has no mass and no charge	• High atomic number materials • Depleted uranium • Lead • Steel • Concrete • Water	• Whole body exposure hazard • May be external and/or internal hazard • Can be a skin and/or eye hazard • Depends on whether source is inside or outside the body • Energy dependent	• Shielding requires large mass and density materials; lead or depleted uranium is commonly used. • Doubling the distance from a point source will result in a reduction of exposure (or dose) by a factor of four. • Protective clothing and other PPE can effectively guard against ingestion or absorption of radioactive material but is not usually practical for protecting against X-rays or gamma rays.

RADIATION SAFETY UNITS

Ionizing radiation, consisting of X-rays, gamma rays, alpha particles, beta particles, and neutron particles, is measured and quantified in units of radioactivity source and dose [1–4]. The radioactivity measures the strength of a source in terms of events of emission per second. Dose is a measure of the energy that is actually absorbed into matter.

REFERENCES

1. Radiation—Quantities and Units of Ionizing Radiation, Canadian Centre for Occupational Health and Safety OSH Answer List Series, 2008.
2. Radiation Fact Sheets, Health Physics Society, http://hps.org/hpspublications/radiationfactsheets. html, accessed January 2020.
3. Furr, A.K., *CRC Handbook of Laboratory Safety*, 5th ed. CRC Press, Boca Raton, FL, 2000.
4. Grove, T.W. and Bruno, T.J., *CRC Handbook of Chemistry and Physics*, 100th Ed., CRC Press, Boca Raton, FL, 2019.

RADIOACTIVITY

In the SI system, the becquerel (Bq) has replaced the Curie (Ci) as the accepted unit of radio-activity (or simply activity). One Bq is one event of radiation emission (such as a disintegration) per second. It is related to the older unit by:

$$1 \text{ Ci} = 3.7 \times 10^{10} \text{ Bq}$$

$$1 \text{ Ci} = 37 \text{ GBq} = 37,000 \text{ MBq}$$

The following chart provides a practical guide to convert between the two units:

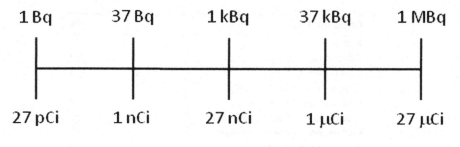

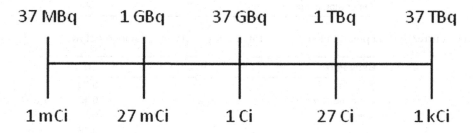

Class A radionuclides: 0.3 Bq/cm^2=8.1 pCi/ cm^2
Class B radionuclides: 3 Bq/ cm^2=81 pCi/ cm^2
Class C radionuclides: 30 Bq/ cm^2=810 pCi/ cm^2.

Energy

For ionizing radiation, the energy is often measured in electron volts (eV), which is related to other energy quantities by:

$$1 \text{ eV} = 1.6021766208(98) \times 10^{-19} \text{ J.}$$

Dose

The older unit of dose, which is defined as the energy that is actually absorbed, is the radiation absorbed dose (RAD). The RAD was defined as the dose that would cause 0.01 J to be absorbed in 1 kg of matter (or 100 ergs/g). The modern SI unit is the Gray (Gy):

$$100 \text{ RAD} = 1 \text{ Gy}$$

Equivalent Dose

The committed dose (or more properly, the committed dose equivalent, $H_{T, 50}$) is the total dose accumulated over a 50-year period after the ingestion or inhalation. The equivalent dose (also called the dose equivalent or biological dose) describes the effect of radiation on human tissue, rather than the physical effects of the radiation alone. This quantity is expressed in Sieverts (Sv) and is found by multiplying the absorbed dose, in grays, by a dimensionless quality factor Q (which depends on the radiation type) and by another dimensionless factor N (the tissue weighting factor). Q is also called the relative biological effectiveness (RBE). The factor N depends upon the part of the body irradiated, the time and volume over which the dose was spread, and the species of the subject.

The currently accepted, approximate Q factors are provided below:

Radiation Type	Q
X-rays	1
Gamma rays	1
Beta particles	1
Thermal neutrons (<10 keV)	5
Fast neutrons (10–100 keV)	10
Fast neutrons (100 keV–2 MeV)	20
Fast neutrons (2 MeV–20 MeV)	10
Fast neutrons (>20 MeV)	5
Protons (>2 MeV)	5
Alpha particles	20
Other atomic nuclei	20

The currently accepted N factors for human body parts are provided below:

Body Part	N
Gonads	0.20
Bone marrow	0.12
Colon	0.12
Lung	0.12
Stomach	0.12
Bladder	0.05
Brain	0.05
Breast	0.05
Kidney	0.05
Liver	0.05
Muscle	0.05
Esophagus	0.05
Pancreas	0.05
Small intestine	0.05
Spleen	0.05
Thyroid	0.05
Uterus	0.05
Bone surface	0.01
Skin[a]	0.01
Organs not listed above, collectively	0.05
Whole body (scale definition)	≡1

[a] The weighting factor for skin implies a whole-body exposure.

Relative to the effect on humans, the following N factors have been suggested for other organisms:

Organism	N
Viruses	0.03–0.0003
Bacteria	0.03–0.0003
Single-cell organisms	0.03–0.0003
Insects	0.1–0.002
Mollusks	0.06–0.006
Plants	2–0.02
Fish	0.75–0.03
Amphibians	0.4–0.14
Reptiles	1–0.075
Birds	0.6–0.15
Humans (scale definition)	≡1

In terms of the older unit, rem (roentgen equivalent in man):

$$1 \text{ rem} = 0.01 \text{ Sv, assuming } Q = 1.$$

The following chart provides a practical guide for converting between the two units:

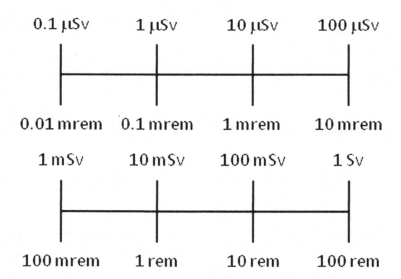

The approximate effects of full-body dosages are summarized below:

Dose (Sv)	Effect
1	Nausea
2–5	Hair loss, hemorrhage, death is possible
>3	Death is likely in 50 % of cases within 30 days
6	Death is likely in all cases

The relationship between nuclide half-lives elapsed and the remaining radioactivity is provided below:

Half-Lives Elapsed	Percent Remaining
0	100
1	50
2	25
3	12.5
4	6.25
5	3.125
n	$(1/2)^n$ (100 %)

RELATIVE DOSE RANGES FROM IONIZING RADIATION

It is important to place in perspective the relative ionizing radiation dose acquired in common laboratory settings. The most commonly encountered source is a ^{63}Ni source used in gas chromatographic electron capture detectors (ECDs) and in ion mobility spectrometers (IMSs) [1]. In both instruments, the source is sealed and has a radioactivity of 15 mCi. The exposures cited refer only to normal operation; it does not consider exposures if the device is dismantled or allowed to overheat.

Background

Natural background consists of the highly variable sum of all ubiquitous sources of ionizing radiation encountered on the planet [2]. Background in general can be divided into the following four major contributions:

Contribution	Average Dose, mrem/year, United States
Terrestrial contribution	21
Cosmic contribution	33
Airborne radon (and daughter) contribution	228
Internal consumption contribution	29
Total natural background	**311**

In the United States, the average natural background ionizing radiation level is 311 mrem. This is variable due primarily to differences in altitude and primordial radionuclides and their daughters. For example, the averages in the United Kingdom and Finland are 200 and 700 mrem, respectively. Higher levels are found at higher altitudes and regions with a larger radon budget. Within the United States, for example, the background in Denver, Colorado, is approximately 450 mrem, while in most of Florida, it is closer to 230 mrem. The terrestrial contribution primarily arises from radionuclides of potassium, uranium, and thorium, and their daughters. The cosmic contribution arises primarily from muons, neutrons, and electrons and varies with terrestrial magnetic field and altitude. The internal contribution results from consuming radionuclides of potassium and carbon in food and water. By far, the largest contribution is from radon and radon daughters. The radon budget results from terrestrial sources of uranium [3]. Within the United States, the action level requiring indoor radon mitigation is reached when a measurement results in 4 pCi/L (150 Bq/m^3) or higher. This level of radioactivity results in a dose of between 300 mrem and 700 mrem/year, assuming 80 % indoor occupancy. The range cited results from different dose conversion coefficients and dosimetric models used by different agencies [4].

Typical Incremental Increases above Background

Exposure to ionizing radiation in the laboratory results in a dose level in excess of the background levels discussed above [5]. The following charts place in perspective the additional dose received above background for some common exposures. Since in Figure 15.1 the ranges are dwarfed by tobacco use, Figure 15.2 is presented with this contribution removed. This is significant because the tobacco dose is specific to the lungs and not the whole body. A more relevant comparison for Figure 15.1 would be obtained by multiplying the listed 8,000 mrem by the tissue weighting factor for the lungs, 0.12.

The high incremental level associated with tobacco use (1 mrem/hr while smoking) results from the accumulation of ^{210}Po and ^{210}Pb (radon daughters that are alpha and gamma emitters) on tobacco leaves. The incremental dose accrued by air travel is dependent on altitude, with higher levels

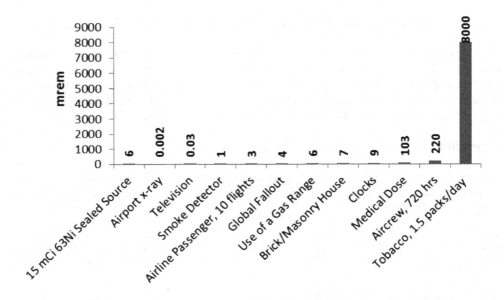

Figure 15.1   A comparison of increments to natural background levels, explicit for the 15 mCi ⁶³Ni sources used in ECDs and IMSs.

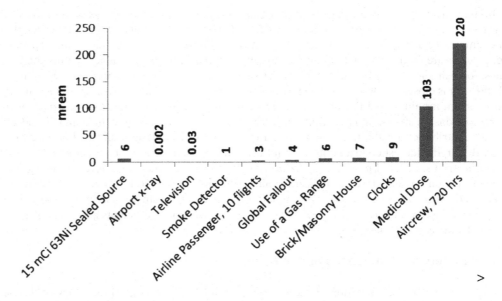

Figure 15.2   The same comparison as in Figure 15.1 but with tobacco use removed. Thus, the data shown above are all whole-body dosages.

associated with higher altitudes, and will range between 0.3 and 0.5 mrem/hr. The incremental dose due to medical imaging or radiation treatment can be misleading. For many individuals, the dose can be close to zero, but in the case of radiation treatment, it can be much higher. Indeed, patients given certain radiation treatments become incremental sources themselves, resulting in incremental dosage above background to attending medical personnel, caregivers, and the general public.

The listing for clocks in both Figures 15.1 and 15.2 is for older (even antique) clocks that have dials coated with Ra/ZnS paint to provide illumination.

In instrumentation such as ECDs and IMSs, the devices are sealed sources in shielded enclosures and are covered by general licenses in the United States. They are designed with inherent radiation safety features so that they can be used by persons with no radiation training or experience [6].

REFERENCES

1. Bruno, T.J. and Svoronos, P.D.N, *CRC Handbook of Chemistry and Physics*, 100th Ed., CRC Press/ Taylor & Francis Group, Boca Raton, 2018.
2. Metting, N.F., *Ionizing Radiation Dose Ranges*, Office of Science, United States Department of Energy, www.nrc.gov/docs/ML1209/ML120970113.pdf, 2010.
3. NCRP Report 160, Ionizing radiation exposure of the population of the United States: Recommendations of the National Council on Radiation Protection and Measurements, Bethesda, MD, 2009.
4. Background Information on "Update on Perspectives and Recommendations on Indoor Radiation," Position statement of the Health Physics Society, 2009.
5. Johnson, T.E. and A. Fellman, Estimated dose and risk from 15 mCi 63Ni sealed source Type NR-348-D-111-B, Report Prepared for Hewlett Packard Co., CSI Radiation Safety, Gaithersburg, MD, 1999.
6. https://www.nrc.gov/materials/miau/general-use.html, accessed July 2018.

CHAPTER 16

Miscellaneous Tables

UNIT CONVERSIONS

The international system of units is described in detail in NIST Special Publication 811 [1] and lists of physical constants and conversions factors. Selected unit conversions [1–5] are presented in the following tables. The conversions are presented in matrix format when all of the units are of a convenient order of magnitude. When some of the unit conversions are of little value (such as the conversion between metric tons and grains), tabular form is followed, with the less useful units omitted.

REFERENCES

1. Thompson, A. and Taylor, B.N., *Guide for the Use of the International System of Units*, National Institute of Standards and Technology (U.S.), Gaithersburg, MD, Special Publication SP-811, 2008.
2. Chiu, Y., *A Dictionary for Unit Conversion*, School of Engineering and Applied Science, The George Washington University, Washington, DC, 20052, 1975.
3. Rumble, J., ed., *CRC Handbook for Chemistry and Physics*, 100th ed., CRC Press, Boca Raton, FL, 2019.
4. Bruno, T.J. and Svoronos, P.D.N., *CRC Handbook of Basic Tables for Chemical Analysis*, 3rd ed., CRC Press, Boca Raton, FL, 2011.
5. Kimball's Biology Page, www.biology_pages.info, accessed December 2019.

Area

Multiply	By	To Obtain
Square millimeters	0.00155	square inches (US)
	1×10^{-6}	square meters
	0.01	square centimeters
	1.2732	circular millimeters
Square centimeters	1.196×10^{-4}	square yards
	0.00108	square feet
	0.15500	square inches
	1×10^{-4}	square meters
	100	square millimeters
Square kilometers	0.38610	square miles (US)
	1.1960×10^{6}	square yards
	1.0764×10^{7}	square feet
	1×10^{6}	square meters
	247.10	acres (US)
Square inches (US)	0.00694	square feet
	0.00077	square yards
	6.4516×10^{-4}	square meters
	6.4516	square centimeters
	645.15	square millimeters
Square feet (US)	3.5870×10^{-8}	square miles
	0.11111	square yards
	144	square inches
	0.09290	square meters
	929.03	square centimeters
	2.2957×10^{-5}	acres
Square miles	640	acres
	3.0967×10^{6}	square yards
	2.7878×10^{7}	square feet
	2.5900	square kilometers

Density

kg/m^3	g/cm^3	lb/ft^3
16.018	0.016018	1
1	0.001	0.062428
1 000	1	62.428
2,015.9	2.0159	125.85

Enthalpy, Heat of Vaporization, Heat of Conversion, Specific Energies

kJ/kg (J/g)	cal/g	Btu/lb
2.3244	0.55556	1
1	0.23901	0.43022
4.1840	1	1.8

Length

Multiply	By	To Obtain
Angstroms	1×10^{-10}	meters
	3.9370×10^{-9}	inches (US)
	1×10^{-4}	micrometers
	1×10^{-8}	centimeters
	0.1	nanometers
Nanometers	1×10^{-9}	meters
	1×10^{-7}	centimeters
	10	angstroms
Micrometers (μm)	3.9370×10^{-5}	inches (US)
	1×10^{-6}	meters
	1×10^{-4}	centimeters
	1×10^{4}	angstroms
Millimeters	0.03937	inches (US)
	1,000	micrometers
Centimeters	0.39370	inches (US)
	1×10^{4}	micrometers (μm)
	1×10^{7}	nanometers
	1×10^{8}	angstroms
Meters	6.2137×10^{-4}	miles (statute)
	1.0936	yards (US)
	39.370	inches (US)
	1×10^{9}	millimicrons
	1×10^{10}	angstroms
Kilometers	0.53961	miles (nautical)
	0.62137	miles (statute)
	1,093.6	yards
	3,280.8	feet
Inches (US)	0.02778	yards
	2.5400	centimeters
	2.5400×10^{8}	angstroms
Feet (US)	0.30480	meters
	30.480	centimeters
Yards (US)	5.6818×10^{-4}	miles
	0.91440	meters
	91.440	centimeters
Miles (nautical)	1.1516	statute miles
	2026.8	yards
	1.8533	kilometers
Miles (US statute)	320	rods
	0.86836	nautical miles
	1.6094	kilometers
	1,609.4	meters

Pressure

MPa	atm	Torr (mm Hg)	bar	lbs/in² (psi)
6.8948×10^{-3}	0.068046	51.715	6.8948×10^{-2}	1
1	9.8692	7,500.6	10.0	145.04
0.101325	1	760.0	1.01325	14.696
1.3332×10^{-4}	1.3158×10^{-3}	1	1.332×10^{-3}	0.019337
0.1	0.98692	750.06	1	14.504

Specific Heat, Entropy

kJ/(kgK) J/(g-K)	Btu/(°R-lb)
4.184	1
1	0.23901

Specific Volume

m³/kg (L/g)	cm³/g	ft³/lb
0.062428	62.428	1
1	1,000	16.018
0.001	1	0.016018

Surface Tension

N/m	dyne/cm	lb/in
175.13	175.13×10^3	1
1	1,000	5.7102×10^{-6}
0.001	1	5.7102×10^{-3}

Temperature

$$T(\text{Rankine}) = 1.8\,T(\text{Kelvin})$$
$$T(\text{Celsius}) = T(\text{Kelvin}) - 273.15$$
$$T(\text{Fahrenheit}) = T(\text{Rankine}) - 459.67$$
$$T(\text{Fahrenheit}) = 1.8\,T(\text{Celsius}) + 32$$

Thermal Conductivity

mW/(cm-K)	J/(s-cm-K)	cal/(s-cm-K)	Btu/(ft-h-R)
17.296	0.017296	0.0041338	1
1	0.001	2.3901×10^{-4}	0.057816
1,000	1	0.23901	57.816
4,184	4.184	1	241.90

Velocity

Multiply	By	To Obtain
Feet per minute	0.01136	miles per hour
	0.01829	kilometers per hour
	0.5080	centimeters per second
	0.01667	feet per second
Feet per second	0.6818	miles per hour
	1.097	kilometers per hour
	30.48	centimeters per second
	0.3048	meters per second
	0.5921	knots
Knots (Br)	1.0	nautical miles per hour
	1.6889	feet per second
	1.1515	miles per hour
	1.8532	kilometers per hour
	0.5148	meters per second
Meters per second	3.281	feet per second
	2.237	miles per hour
	3.600	kilometers per hour
Miles per hour	1.467	feet per second
	0.4470	meters per second
	1.609	kilometers per hour
	0.8684	knots

Speed of Sound

m/s	ft/s
0.3048	1
1	3.2808

Viscosity

kg/(m-s) (N-s/m², Pa·s)	cP (10^{-2}g/(cm-s))	lb-s/ft² (slug/(ft-s))	lb/(ft-s)
1.48816	1,488.16	0.31081	1
1	1,000	0.020885	0.67197
0.001	1	2.0885×10^{-5}	6.7197×10^{-4}
47.881	4.7881×10^{-4}	1	32.175

Volume

Multiply	By	To Obtain
Barrels (pet)	42	gallons (US)
	34.97	gallons (Br.)
Cubic centimeters	10^{-3}	liters
	0.0610	cubic inches
Cubic feet	28,317	cubic centimeters
	1,728	cubic inches
	0.03704	cubic yards
	7.481	gallons (US, liq.)
	28.317	liters
Cubic inches	16.387	cubic centimeters
	0.016387	liters
	4.329×10^{-3}	gallons (US, liq.)
	0.01732	quarts (US, liq.)
Gallons, imperial	277.4	cubic inches
	1.201	US gallons
	4.546	liters
Gallons, (US, liquid)	231	cubic inches
	0.1337	cubic feet
	3.785	liters
	0.8327	imperial gallons
	128	fluid ounces (US)
Ounces, fluid	29.57	cubic centimeters
	1.805	cubic inches
Liters	0.2642	gallons
	0.0353	cubic feet
	1.0567	quarts (US, liq.)
	61.025	cubic inches
Quarts, (US, liquid)	0.0334	cubic feet
	57.749	cubic inches
	0.9463	liters

Mass (Weight)

Multiply	By	To Obtain
Milligrams	2.2046×10^{-6}	pounds (avoirdupois)
	3.5274×10^{-5}	ounces (avoirdupois)
	0.01543	grains
	1×10^{-6}	kilograms
Micrograms	1×10^{-6}	grams
Grams	0.00220	pounds (avoirdupois)
	0.03527	ounces (avoirdupois)
	15.432	grains
	1×10^{6}	micrograms
Kilograms	0.00110	tons (short)
	2.2046	pounds (avoirdupois)
	35.274	ounces (avoirdupois)
	1.5432×10^{4}	grains
Grains	1.4286×10^{-4}	pounds (avoirdupois)
	0.00229	ounces (avoirdupois)
	0.06480	grams
	64.799	milligrams
Ounces (avoirdupois)	3.1250×10^{-5}	tons (short)
	0.06250	pounds (avoirdupois)
	437.50	grains
	28.350	grams
Pounds (avoirdupois)	5×10^{-4}	tons (short)
	16	ounces (avoirdupois)
	7,000	grains
	0.45359	kilograms
	453.59	grams
Tons (short, US)	2,000	pounds (avoirdupois)
	3.200×10^{4}	ounces (avoirdupois)
	907.19	kilograms
Tons (long)	2,240	pounds (avoirdupois)
	1,016	kilograms
Tons (metric)	1,000	kilograms
	2,205	pounds (avoirdupois)
	1.102	tons (short)

MASS AND VOLUME-BASED CONCENTRATION UNITS

Because the mass of 1 L of water is approximately 1 kg, mg/liter units of aqueous solution are nearly equal to ppm units. The precise equivalence is obtained by dividing by the density ρ:

$$ppm = (mg/liter)/\rho$$

where the solution density, ρ, is in g/cm^3. Some sources will substitute specific gravity for density in the above equation. The specific gravity is the ratio of the solution density to that of the density of pure water at 4 C. Since the density of pure water at 4 C is 1 g/cm^3, the specific gravity is equal to the solution density when expressed in metric units of g/cm^3.

Parts per Million

Parts per Million	vs.	Percent
1 ppm	=	0.0001 %
10 ppm	=	0.001 %
100 ppm	=	0.01 %
1,000 ppm	=	0.1 %
10,000 ppm	=	1.0 %
100,000 ppm	=	10.0 %
1,000,000 ppm	=	100.0 %

Parts per Billion

Parts per Billion	vs.	Percent
10	=	0.000001 %
100	=	0.00001 %
1,000	=	0.0001 %
10,000	=	0.001 %
100,000	=	0.01 %
1,000,000	=	0.1 %

Parts per Trillion

Parts per Trillion	vs.	Percent
100	=	1×10^{-8} %
10,000	=	0.000001 %
1,000,000	=	0.0001 %
100,000,000	=	0.01 %

CONCENTRATION UNITS NOMENCLATURE

The following table provides guidance in the use of base-ten concentration units (presented in the three preceding tables), since there are differences in usage worldwide.

Number	Number of Zeros	Name (Scientific Community)	Name (United Kingdom, France, Germany)
1,000.	3	Thousand	Thousand
1,000,000.	6	Million	Million
1,000,000,000.	9	Billion	Milliard, or thousand million
1,000,000,000,000.	12	Trillion	Billion
1,000,000,000,000,000.	15	Quadrillion	Thousand billion

MOLAR-BASED CONCENTRATION UNITS

Molarity, M: (moles of solute)/(liters of solution)
Molality, m: (moles of solute)/(kilograms of solvent)
Normality, N: (equivalents* of solute)/(liters of solution)
Formality, F: (moles of solute)/(kilograms of solution)
*Reaction dependent; based on the number of protons exchanged in a given reaction.

To Convert from ppm to Formality Units

$$F = ppm/(1,000 \ RMM),$$

where RMM is the relative molecular mass of the solute i.

To Convert from ppm to Molality Units

$$m = [ppm/(1,000 \ RMM)][1/(1 - tds/1,000,000)]$$

where tds is the mass of total dissolved solids in ppm in the solution.

To Convert from ppm to Molarity Units

$$M = [ppm/(1,000 \ RMM)]\rho$$

where ρ is the solution density.

Prefixes for SI Units

Fraction	Prefix	Symbol
10^{-1}	deci	d
10^{-2}	centi	c
10^{-3}	milli	m
10^{-6}	micro	μ
10^{-9}	nano	n
10^{-12}	pico	p
10^{-15}	femto	f
10^{-18}	atto	a

Multiple	Prefix	Symbol
10	deka	da
10^{2}	hecto	h
10^{3}	kilo	k
10^{6}	mega	M
10^{9}	giga	G
10^{12}	tera	T
10^{15}	peta	P
10^{18}	exa	E

RECOMMENDED VALUES OF SELECTED PHYSICAL CONSTANTS

The following table provides some commonly used physical constants that are of value in thermodynamic and spectroscopic calculations [1,2].

REFERENCES

1. Rumble, J., ed., *CRC Handbook for Chemistry and Physics*, 100th ed., CRC Press, Boca Raton, FL, 2019.
2. Fundamental Physical Constants, https://www.nist.gov/pml/fundamental-physical-constants, accessed December 2019.

Physical Constant	Symbol	Value
Avogadro constant	N_A	$6.02214199 \times 10^{23}$ mol^{-1}
Boltzmann constant	k	$1.3806503 \times 10^{-23}$ J K^{-1}
Charge to mass ratio	e/m	$-1.758820174 \times 10^{11}$ C kg^{-1}
Elementary charge	e	1.60218×10^{-19} C
Faraday constant	F	96485.3415 C mol^{-1}
Molar gas constant	R	8.314472 J mol^{-1}·K^{-1}
"Ice point" temperature	T_{ice}	273.150 K (exactly)
Molar volume of ideal gas (STP)	V_m	2.24138×10^{-2} m^3·mol^{-1}
Permittivity of vacuum	ε_o	8.854188×10^{-12} kg^{-1} m^{-3}·s^4·A^2 (F·m^{-1})
Planck constant	h	$6.62606876 \times 10^{-34}$ J·s
Standard atmosphere pressure	p	$101,325$ N·m^{-2} (exactly)
Atomic mass constant	m_u	$1.66053873 \times 10^{-27}$ kg
Speed of light in vacuum	c	$299,792,458$ m s^{-1} (exactly)

STANDARDS FOR LABORATORY WEIGHTS

The following table provides a summary of the requirements for metric weights and mass standards commonly used in chemical analysis [1,2]. The actual specifications are under the jurisdiction of ASTM Committee E-41 on General Laboratory Apparatus and are the direct responsibility of subcommittee E-41.06 that deals with weighing devices. These standards do not generally refer to instruments used in commerce. Weights are classified according to Type (either Type I or Type II), Grade (S, O, P, or Q), and Class (1, 2, 3, 4, 5 or 6). Information on these mass standards is presented to allow the user to make appropriate choices when using analytical weights for the calibration of electronic analytical balances, for making large-scale mass measurements (such as those involving gas cylinders) and in the use of dead weight pressure balances.

REFERENCES

1. Annual Book of ASTM Standards, ANSI/ASTM E617-18, *Standard Specification for Laboratory Weights and Precision Mass Standards*, Book of Standards Volume: 14.04, 2018.
2. Battino, R. and Williamson, A.G., Single pan balances, buoyancy and gravity, or "a mass of confusion", *J. Chem. Educ.*, 61(1), 51, 1984.

TYPE—CLASSIFICATION BY DESIGN

Type I

One piece construction; contains no added adjusting material; used for highest accuracy work.

Type II

Can be of any appropriate and convenient design, incorporating plugs, knobs, rings, etc.; adjusting material can be added if it is contained so that it cannot become separated from the weight.

Grade—Classification by Physical Property

Grade S	Density	7.7–8.1 g/cm³ (for 50 mg and larger)
	Surface area	Not to exceed that of a cylinder of equal height and diameter
	Surface finish	Highly polished
	Surface protection	None permitted
	Magnetic properties	No more magnetic than 300 series stainless steels
	Corrosion resistance	Same as 303 stainless steel
	Hardness	At least as hard as brass
Grade O	Density	7.7–9.1 g/cm³ (for 1 g and larger)
	Surface area	Same as Grade S
	Surface finish	Same as Grade S
	Surface protection	May be plated with suitable material such as platinum or rhodium
	Magnetic properties	Same as Grade S
	Corrosion resistance	Same as Grade S
	Hardness	At least as hard as brass when coated; smaller weights at least as hard as aluminum
Grade P	Density	7.2–10 g/cm³ (for 1 g or larger)
	Surface area	No restriction
	Surface finish	Smooth, no irregularities
	Surface protection	May be plated or lacquered
	Magnetic properties	Same as Grades S and O
	Corrosion resistance	Surface must resist corrosion and oxidation
	Hardness	Same as Grade O
Grade Q	Density	7.2–10 g/cm³ (for 1 g or larger)
	Surface area	Same as Grade P
	Surface finish	Same as Grade P
	Surface protection	May be plated, lacquered, or painted
	Magnetic properties	No more magnetic than unhardened unmagnetized steel
	Corrosion resistance	Same as Grade P
	Hardness	Same as Grades O and P

Tolerance—Classification by Deviation[a]

	Class 1			Class 2	
Grams	Individual Tolerance (mg)	Group Tolerance (mg)	Grams	Individual Tolerance (mg)	Group Tolerance (mg)
500	1.2		500	2.5	
300	0.75		300	1.5	
200	0.50		200	1.0	
100	0.25	1.35	100	0.5	2.7
50	0.12		50	0.25	
30	0.074		30	0.15	
20	0.074		20	0.10	
10	0.050	0.16	10	0.074	0.29
5	0.034		5	0.054	
3	0.034		3	0.054	
2	0.034		2	0.054	
1	0.034	0.065	1	0.054	0.105

Class 3		Class 4		Class 5		Class 6	
Grams	Tolerance (mg)	Grams	Tolerance (mg)	Grams	Tolerance (mg)	Grams	Tolerance (mg)
500	5.0	500	10	500	30	500	50
300	3.0	300	6.0	300	20	300	30
200	2.0	200	4.0	200	15	200	20
100	1.0	100	2.0	100	9	100	10
50	0.6	50	1.2	50	5.6	50	7
30	0.45	30	0.9	30	4.0	30	5
20	0.35	20	0.7	20	3.0	20	3
10	0.25	10	0.5	10	2.0	10	2
				5	1.3	5	2
				3	0.95	3	2
				2	0.75	2	2
				1	0.50	1	2

[a] In simple terms, the permitted deviation between the assigned nominal mass value of the weight and the actual mass of the weight. Verification of tolerance should be possible on reasonably precise equipment, without using a buoyancy correction, within the political jurisdiction or organizational bounds of a given weight specification.

Applications for Weights and Mass Standards[a]

Application	Type	Grade	Class
Reference standards used for calibrating other weights	I	S	1,2,3, or 4[a]
High-precision standards for calibration of weights and precision balances	I or II[b]	S or O[b]	1 or 2[c]
Working standards for calibration and precision analytical work, dead weight pressure balances	I or II[b]	S or O	2
Laboratory weights for routine analytical work	II	O	2 or 3
Built-in weights, high-quality analytical balances	I or II	S	2
Moderate precision laboratory balances	II	P	3 or 4
Dial scales and trip balances	II	Q	4 or 5
Platform scales	II	Q	5 or 6

[a] Primary standards are for reference use only and should be calibrated. Since the actual values for each weight are stated, close tolerances are neither required nor desirable.

[b] Type I and Grade S will have a higher constancy but will probably be higher priced.

[c] Since working standards are used for the calibration of measuring instruments, the choice of tolerance depends upon the requirements of the instrument. The weights are usually used at the assumed nominal values, and appropriate tolerances should be chosen.

Reprinted (with modification) with permission of the ASTM International (formerly American Society for Testing and Materials), 100 Barr Harbor Drive, West Conshohocken, Pennsylvania, USA.

GENERAL THERMOCOUPLE DATA

The following tables provide some basic information about common thermocouples used in laboratory instruments. It is critical when replacing thermocouples in instrumentation that the appropriate junction is chosen, and that the installation is done properly. These tables are to aid in those decisions.

REFERENCE

1. Benedict, R.P., *Fundamentals of Temperature Pressure and Flow Measurements*, 3rd ed., John Wiley & Sons, New York, 1984.

Types and Applications of Thermocouples

Thermocouple Type	+ Wire	− Wire	Application Range (°C)
B	Platinum (30 %), rhodium (70 %)	Platinum (6 %) rhodium (94 %)	1,370–1,700
C	W5Re tungsten (5 %) rhenium, (95 %)	W26Re tungsten (26 %) rhenium (74 %)	1,650–2,315
E	Chromel	Constantan	95–900
J	Iron	Constantan	95–760
K	Chromel	Alumel	95–1,260
N	Nicrosil	Nisil	650–1,260
R	Platinum (13 %) rhodium (87 %)	Platinum	870–1,450
S	Platinum (10 %) rhodium (90 %)	Platinum	980–1,450
T	Copper	Constantan	−200 to 350

Notes: Chromel is an alloy consisting of approximately 90 % nickel and 10 % chromium.
Alumel is a magnetic alloy consisting of approximately 95 % nickel, 2 % manganese, 2 % aluminum, and 1 % silicon.
Constantan is a copper–nickel alloy usually consisting of approximately 55 % copper and 45 % nickel.
Nicrosil is a nickel alloy containing 14.4 % chromium, 1.4 % silicon, and 0.1 % magnesium.
Nisil is a nickel alloy containing approximately 4.4 % silicon.

ANSI color codes for thermocouple wires in the United States. Other countries may have different conventions. Some manufacturers code the wires with a colored stripe instead of a solid color.

Thermocouple Type	+ Wire Color	− Wire Color
B	Gray	Red
J	White	Red
K	Yellow	Red
R	Blue	Red
S	Blue	Red
T	Blue	Red

THERMOCOUPLE REFERENCE VOLTAGES

The following table provides power series expansions for the most common types of thermocouples used in the laboratory for temperature measurement [1,2]. It is best to use the thermocouple voltages in gradient mode, with the temperature of interest referenced to an additional thermocouple junction at some known temperature. Note that the temperature ranges differ with the previous tables; here, the temperature range is provided for the correlation, not the applicability of the couple.

REFERENCES

1. Powell, R.L., Hall, W.J., Hyink, C.H., Sparks, L.L., Burns, G.W., Scroger, M.G., and Plumb, H.H., Thermocouple Reference Tables based on the IPTS-68, NBS Monograph 125, March 1974.
2. Benedict, R.P., *Fundamentals of Temperature Pressure and Flow Measurements*, 3rd ed., John Wiley & Sons, New York, 1984.

Type T Thermocouples, Copper/Constantan

Temperature Range (C°)	Exact Reference Voltage (mV) E
0–400	$+3.8740773840 \times 10 \times T$
	$+3.3190198092 \times 10^{-2} \times T^2$
	$+2.0714183645 \times 10^{-4} \times T^3$
	$-2.1945834823 \times 10^{-6} \times T^4$
	$+1.1031900550 \times 10^{-8} \times T^5$
	$-3.0927581898 \times 10^{-11} \times T^6$
	$+4.5653337165 \times 10^{-14} \times T^7$
	$-2.7616878040 \times 10^{-17} \times T^8 \times 10^{-3}$

Type J Thermocouples, Iron/Constantan

Temperature Range (C°)	Exact Reference Voltage (mV) E
0–760	$+5.0372753027 \times 10 \times T$
	$+3.0425491284 \times 10^{-2} \times T^2$
	$-8.5669750464 \times 10^{-5} \times T^3$
	$+1.3348825725 \times 10^{-7} \times T^4$
	$-1.7022405966 \times 10^{-10} \times T^5$
	$+1.9416091001 \times 10^{-13} \times T^6$
	$-9.6391844859 \times 10^{-17} \times T^7 \times 10^{-3}$

Type E Thermocouples, Chromel/Constantan

Temperature Range (C°)	Exact Reference Voltage (mV) E
0–1,000	$+5.8695857799 \times 10 \times T$
	$+4.3110945462 \times 10^{-2} \times T^2$
	$+5.7220358202 \times 10^{-5} \times T^3$
	$-5.4020668085 \times 10^{-7} \times T^4$
	$+1.5425922111 \times 10^{-9} \times T^5$
	$-2.4850089136 \times 10^{-12} \times T^6$
	$+2.3389721459 \times 10^{-15} \times T^7$
	$-1.1946296815 \times 10^{-18} \times T^8$
	$+2.5561127497 \times 10^{-22} \times T^9 \times 10^{-3}$

Type K Thermocouples, Chromel/Alumel

Temperature Range (C°)	Exact Reference Voltage (mV) E
0–1,100	-1.8533063273×10
	$+3.8918344612 \times 10 \times T$
	$+1.6645154356 \times 10^{-2} \times T^2$
	$-7.8702374448 \times 10^{-5} \times T^3$
	$+2.2835785557 \times 10^{-7} \times T^4$
	$-3.5700231258 \times 10^{-10} \times T^5$
	$+2.9932909136 \times 10^{-13} \times T^6$
	$-1.2849848789 \times 10^{-16} \times T^7$
	$+2.2239974336 \times 10^{-20} \times T^8$
	$+125 \exp\left(-\frac{1}{2}\left\{ \frac{T-127}{65} \right\}^2 \right) \times 10^{-3}$

STANDARD CGA FITTINGS FOR COMPRESSED GAS CYLINDERS

The following table presents a partial list of gases and the Compressed Gas Association (CGA) fittings that are required to use those gases when they are stored in, and dispensed from, compressed gas cylinders [1].

REFERENCE

1. CGA Pamphlet V-1-87, American Canadian and Compressed Gas Association Standard for Compressed Gas Cylinder Valve Outlet and Inlet Connections, ANSI,B57.1; CSA B96, 1987.

Gas	Fitting
Acetylene	510
Air	346
Carbon dioxide	320
Carbon monoxide	350
Chlorine	660
Ethane	350
Ethylene	350
Ethylene oxide	510
Helium	580
Hydrogen	350
Hydrogen chloride	330
Methane	350
Neon	580
Nitrogen	580
Nitrous oxide	326
Oxygen	540
Sulfur dioxide	660
Sulfur hexafluoride	590
Xenon	580

The following graphic shows the geometry and dimensions of common CGA fittings for compressed gas cylinders[a].

[a] Reproduced from the CGA Pamphlet V-1-87, American Canadian and Compressed Gas Association Standard for Compressed Gas Cylinder Valve Outlet and Inlet Connections, ANSI,B57.1; CSA B96, by permission of the Compressed Gas Association.

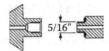

CONNECTION 110 - Lecture Bottle Outlet for Corrosive Gases - 5/16" - 32 RH INT., with Gasket

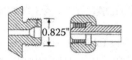

CONNECTION 326 - 0.825" - 14 RH EXT.

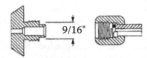

CONNECTION 170 - Lecture Bottle Outlet for Non-Corrosive Gases 9/16" - 18 RH EXT. and 5/16" - 32 RH INT., with Gasket

CONNECTION 350 - 0.825" - 14 LH EXT.

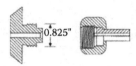

CONNECTION 320 - 0.825" - 14 RH EXT., with Gasket

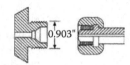

CONNECTION 540 - 0.903" - 14 RH EXT.

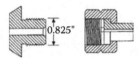

CONNECTION 330 - 0.825" - 14 LH EXT., with Gasket

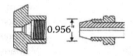

CONNECTION 590 - 0.956" - 14 LH INT.

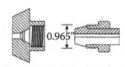

CONNECTION 510 - 0.885" - 14 LH INT.

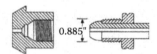

CONNECTION 580 - 0.965" - 14 RH INT.

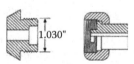

CONNECTION 660 - 1.030" - 14 RH EXT., with Gasket

GAS CYLINDER STAMPED MARKINGS

The graphic below describes the permanent, stamped markings that are used on high-pressure gas cylinders commonly found in analytical laboratories. Note that individual jurisdictions and institutions have requirements for marking the cylinder contents as well. These requirements are in addition to the stamped markings, which pertain to the cylinder itself rather than to the fill contents.

There are four fields of markings on cylinders that are used in the United States, labeled 1–4 on the figure.

Field 1—Cylinder Specifications:

    DOT stands for the US Department of Transportation, the agency that regulates the transport and specification of gas cylinders in the United States. The next entry, for example, 3AA, is the specification for the type and material of the cylinder. The most common cylinders are 3A, 3AA, 3AX, 3AAX, 3T, and 3AL. All but the last refer to steel cylinders, while 3AL refers to aluminum. The individual specifications differ mainly in chemical composition of the steel and the gases that are approved for containment and transport. The 3T deals with large bundles of tube trailer cylinders.

    The next entry in this field is the service pressure, in psig.

Field 2—Serial Number:

    This is a unique number assigned by the manufacturer.

Field 3—Identifying Symbol:

    The manufacturer identifying symbol historically can be a series of letters or a unique graphical symbol. In recent years, the DOT has standardized this identification with the "M" number, for example, M1004. This is a number issued by DOT that identifies the cylinder manufacturer.

Field 4—Manufacturing Data:

    The data of manufacture is provided as a month and year. With this date is the inspector's official mark, for example, H. In recent years, this letter has been replaced with an IA number, for example, IA02, pertaining to an independent agency that is approved by DOT as an inspector.

    If "+" is present, the cylinder qualifies for an overfill of 10 % in service pressure.

    If a star is present, the cylinder qualifies for a 10-year rather than a 5-year retest interval.

    Also stamped on the cylinder will be the retest dates. A cylinder must have a current (that is, within 5 or 10 years) test stamp. On the collar of the cylinder, the owner of the cylinder may be stamped.

REFERENCE

1. Hazardous Materials: Requirements for Maintenance, Requalification, Repair and Use of DOT Specification Cylinders, 49 CFR Parts 107, 171, 172, 173, 177, 178, 179, and 180; [Docket No. RSPA-01-10373 (HM-220D)]RIN 2137-AD58, August 8, 2002.

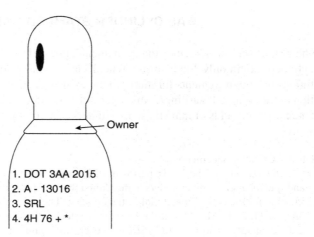

1. DOT 3AA 2015
2. A - 13016
3. SRL
4. 4H 76 + *

PLUG AND OUTLET CONFIGURATIONS FOR COMMON LABORATORY DEVICES

The following schematic diagrams show typical plug and outlet configurations used on common laboratory instruments and devices [1]. These figures will assist in identifying which circuits and capacities will be needed to operate different pieces of equipment.

REFERENCE

1. Plugs, Receptacles, and Connectors of the Pin and Sleeve Type for Hazardous Locations, National Electrical Manufacturer Association, Standard FB 11, 2000.

2 pole, 2 wire

Current / Voltage	15 Amp R　　P	20 Amp R　　P	30 Amp R　　P	50 Amp R　　P
125	⊙⊙ , ⊙⊙	⊙⊙ , ⊙⊙		
250	⊙⊙ , ⊙⊙	⊙⊙ , ⊙⊙		

2 pole, 3 wire (grounding)

Current / Voltage	15 Amp R　　P	20 Amp R　　P	30 Amp R　　P	50 Amp R　　P
125	⊙⊙ , ⊙⊙	⊙⊙ , ⊙⊙		
250	⊙⊙ , ⊙⊙	⊙⊙ , ⊙⊙	⊙⊙ , ⊙⊙	⊙⊙ , ⊙⊙

In addition to the receptacles and plugs, one will often find symbols in instrument operating and service manuals that are indicative of the type of power that is compatible with the device. The following symbols are commonly encountered:

Alternating current

Alternating/direct current (selectable)

Direct current.

Subject Index

Chemical Index

A

Abietic acid, surfactant properties, 695
Acetaldehyde, 308
    as chemical carcinogen, 720
    flammability hazards of, 713
Acetal, peroxide testing requirements, 747
Acetamide, as chemical carcinogen, 720
Acetate, 308
Acetate buffer, 674–677
Acetic acid, 299
    eluotropic values on octadecylsilane, 188
    HPLC column regeneration, 192
    incompatible chemicals with, 702
    mass spectrum peaks, 578
    recrystallization solvent, 692
    as solvent for HPLC, 172
    as solvent for UV spectrophotometry, 321
    thin layer chromatography solvent, 242
    UV absorbance of reverse phase mobile phases, 176
Acetic acid-d_4, residual ^{1}H NMR spectrum of, 474
Acetic acid, pK_a values in buffer systems, 671
Acetic anhydride, derivatizing reagent for GC, 135
Acetone
    eluotropic values on octadecylsilane (ODS), 188
    flammability hazards of, 713
    ^{1}H NMR chemical shifts, 476
    mass spectrum peaks, 578
    mass spectrometer leak detection, 584
    NMR water signal, 476
    partition coefficient, air/water system, 128
    recrystallization solvent, 692
    solubility parameters for, 284, 285
    as solvent for GC, 131
    as solvent for HPLC, 172
        instability of, 174, 285
    as solvent for UV spectrophotometry, 321
    spurious mass spectra, 585
    stabilizers, 174
    as a supercritical fluid solvent, 280
    thin layer chromatography solvent, 242
    UV spectrum of, 323
Acetone-d_6
    residual ^{1}H NMR spectrum of, 474
    solvent chemical shift, 472
Acetonitrile
    eluotropic values on octadecylsilane, 188
    flammability hazards of, 713
    ^{1}H NMR chemical shifts, 476
    HPLC solvent, 172
    mass spectrum peaks, 578
    NMR water signal, 476
    physical properties, 373
    recrystallization solvent, 692
    as solvent for GC, 131
    as solvent for infrared spectrophotometry, 372
    thin layer chromatography solvent, 242
    UV absorbance of reverse phase mobile phases, 176
    UV solvent, 321
    UV spectrum of, 323
Acetonitrile-d_3
    residual ^{1}H NMR spectrum of, 474
    solvent chemical shift, 472
Acetonyl acetone, as packed column stationary phase, 52
Acetonyl acetone (2,5-hexanedione), GC liquid phase, 52
3-(p-Acetophenoxy) propyl bonded phase, as HPLC
    stationary phase, 197
Acetylacetonate, chelating agent for inorganic species, 120
Acetylacetone, flammability hazards of, 713
2-Acetylaminofluorene, as chemical carcinogen, 720
Acetylcellulose, HPLC stationary phase, 210
Acetylene
    CGA fittings, 809
    HaySep polymer relative retention, 75
    incompatible chemicals with, 702
Acetylide
    UV active functionalities, 332
    UV detection of chromophoric groups, 229
Acetyl peroxide, hazards of, 704
Acetyl tributyl citrate, GC liquid phase, 52
α-Acid glycoprotein, as HPLC stationary phase, 208
Aconitate buffer, 673–677
Acridine orange, electrophoresis stain, 291, 292
N-(9-Acridinyl) malemide, derivatizing reagent
    for HPLC, 233
Acrolein, flammability hazards of, 713
Acrylamide
    as chemical carcinogen, 720
    in electrophoresis, 294
    polyacrylamide gel, 288
Acrylic acid, peroxide testing requirements, 747
Acrylonitrile
    as chemical carcinogen, 720
    flammability hazards of, 713
    peroxide testing requirements, 747
Actinomycin D, as chemical carcinogen, 720
Activated alumina, trapping sorbent for GC, 114
Activated carbon
    adsorbents for gas-solid chromatography, 66
    as GC stationary phase, 66
    incompatible chemicals in, 702
    trapping sorbent, 113
    as trapping sorbent
        for AAS, 632
        for trace metals, 117
Acyl fluoride, ^{13}C-^{19}F coupling constants, 549
Adipic acid, hazards of, 704
Adiponitrile, GC liquid phase, 52
Adrenaline
    spray reagents for, 261
    stationary and mobile phases, 249
Adriamycin, as chemical carcinogen, 720
Aflatoxins, as chemical carcinogen, 720
Agarose, as HPLC stationary phases, 184

Printed in the United States
by Baker & Taylor Publisher Services